ESTATES GAZETTE LAW REPORTS

Estates Gazette Law Reports

1991

Volume 1

Edited by
J Muir Watt OBE MA
of the Inner Temple, barrister

Audrey Boyle BA
Publisher's editor

Estates Gazette

A member of Reed Business Publishing Group

THE ESTATES GAZETTE LIMITED
151 WARDOUR STREET, LONDON W1V 4BN

First published 1991
ISBN for complete set of 2 volumes: 0 7282 0169 0
ISBN for this volume: 0 7282 0170 4

ISSN 0951-9289

Typesetting by Digital Graphics, 147-149 Wardour Street, London
Printed and bound at the Bath Press, Avon

CONTENTS

SUPPLEMENT

Cases not reported in "Estates Gazette"

TABLE OF CASES

Titles of cases shown in bold type
Names of parties reversed shown in ordinary type

INDEX OF SUBJECT-MATTER

LANDLORD AND TENANT — *continued*
Forfeiture — *continued*
Relief — continued

equitable jurisdiction also excluded — Held, by majority, tenants not entitled to relief under either statutory provisions or equitable jurisdiction — Nicholls LJ, dissenting, would have held relief could be claimed in principle under equitable jurisdiction.
***Billson* v *Residential Apartments Ltd (CA),* 70**

Physical re-entry by lessors — 999-years' lease — Alleged breaches by lessees of insurance covenant — Block of residential flats in squalid condition after sublease to London borough for temporary housing — Notice by lessors under s146 of Law of Property Act 1925 complaining of breach of covenant — Lessees claimed they had not been given reasonable time to comply with notice, that re-entry was a trespass and lease had not been forfeit — Held, reasonable time not allowed and re-entry accordingly unlawful.
***Cardigan Properties Ltd* v *Consolidated Property Investments Ltd (QB),* 64**

Implied covenant — See **Landlord and tenant** (Repairs — Implied covenant)
Landlord and Tenant Act 1927 — See **Landlord and tenant** (Assignment)
Landlord and Tenant Act 1954 Part II
"Business"

Continuity of business tenancy — Link with landlords' repairing obligations — Premises virtually uninhabitable from disrepair — Tenant's business was subletting rooms — Long period of no occupation owing to disrepair — Held, no repairing obligations, express or implied, on landlords — No protection against lack of continuity in occupation for business purposes — Tenant failed to make good claim to business tenancy.
***Demetriou* v *Poolaction Ltd (CA),* 100**

Whether tenancy business or residential — Part of premises used as kennels for period of years — *Cheryl Investments Ltd* v *Saldanha* followed — Held, running of business on subject property was merely incidental, akin to hobby — Tenancy residential and protected by Rent Act 1977.
***Gurton* v *Parrott (CA),* 98**

Landlord's opposition to new tenancy

S30(1)(*f*) and s31A (1) (*a*) and (*b*) — Whether tenant had to make advance election to claim tenancy of whole property demised or part only — Formulation of preliminary issue ordered by county court registrar — Held, s31A provides escape route or fall-back position — If para (*a*) of section 31A(1) is satisfied tenant will obtain tenancy of entire holding — If para (*a*) is not satisfied tenant may still, under para (*b*), obtain tenancy of part of holding.
***Romulus Trading Co Ltd* v *Henry Smith's Charity Trustees (No 2) (CA),* 95**

Landlord and Tenant Act 1987
Interim management order

Application by interlocutory motion for interim management order under s24 pending main proceedings for acquisition order — Management by landlords "simply non-existent" — Damp penetration from leaking roof causing further deterioration — Judge satisfied that landlords unlikely to take steps to improve management — Interim order made despite number of technical objections by defendant landlords.
***Howard* v *Midrome Ltd (Ch),* 58**

Tenants' rights of first refusal

Regulated tenancies — Original disposal a private treaty sale by vendors who had purchased at auction 18 months before — Five situations identified as possible instances where tribunal might investigate surrounding circumstances of original disposal — Construction of s12(3)(*b*) — *Cousins* v *Metropolitan Guarantee* followed — Held, original disposal price not out of line with open market value.
***Gregory* v *Saddiq (LVT),* 237**

Reversion assigned by landlords without giving tenants first refusal — Validity of tenants' purchase notice served on new landlords — Whether tenants "qualifying tenants" — Questions as to what interests tenants entitled to acquire — Jurisdiction of courts and rent assessment committees — Held, tenants were "qualifying tenants" — Notice valid — Rent assessment committees have no jurisdiction to determine validity of notice — Landlord and Tenant Act 1987, ss 12 and 13.
***Denetower Ltd* v *Toop (CA),* 84**

Leasehold enfranchisement
"Other residence"

Whether leaseholder had occupied premises as residence for requisite period — Extent of occupation of "other residence" within meaning of para 6(1)(*d*)(ii) of Sched 3 to Leasehold Reform Act 1967 — Allegations of misrepresentation or concealment of material facts — County court judge's finding of deliberate omission of reference to occupation of cottage owned by applicant upheld — Order made under s20(6) of 1967 Act making void any further application for enfranchisement within five years.
***Dymond* v *Arundel-Timms (CA),* 109**

Rateable value limits

Surrender of existing long lease and grant of new one — Effort to secure lower rateable value because of improvements — Scheme to take advantage of Leasehold Reform Act 1967, s23(3) — Distinction between agreement to surrender and completed surrender — Held, relief under s23(3) not available where agreement to surrender has been executed — Leaseholder had no *locus standi* to apply for reduction of rateable value.
***Woodruff* v *Hambro (CA),* 107**

Notice to quit — See also **Agricultural holdings** (Notice to quit)
Validity

Clause in agreement providing for tenancy to continue until landlord, being LCC as highway authority, required land for road widening — Reversion eventually passed to London Residuary Body, not a highway authority, and then to purchasers — Notice to quit given to tenants before sale — Whether clause providing for tenancy to continue until land required for road widening repugnant to tenancy — Held, once landlord and highway authority were not same person, clause ceased to be effective — Notice to quit valid.
***Prudential Assurance Co Ltd* v *London Residuary Body (Ch),* 90**

Periodic tenancies — Notice given by one only of two joint tenants without authority of other — Two appeals raising same point — Whether notices valid under general law and provisions of respective tenancy agreements — Held, *Greenwich London Borough Council* v *McGrady* binding authority in favour of view that notice by one only of joint tenants was valid and effective — *Howson* v *Buxton* not authority for contrary view.
***Hammersmith and Fulham London Borough Council* v *Monk; Barnet London Borough Council* v *Smith (CA),* 263**

Possession
Procedure

Form of notice of possession proceedings prescribed by regulations — Grounds for possession order — Whether notice given by council landlord in accordance with prescribed form — Whether proper particulars given — Notice incorrectly stated rent lawfully due not paid — Figure included amounts for rates and water rates — County court judge held notice defective and he was deprived of jurisdiction — Held, error did not deprive judge of jurisdiction, although it might affect action on merits of case — Case remitted — Housing Act 1985, Part IV, s83.
***Dudley Metropolitan Borough Council* v *Bailey (CA),* 53**

Unlawful occupation

Claim to council tenancy — Conflicting indications by council — Assessments of rent rebate — Issue of rent book — Suggestion that three indicea recognised by *Street* v *Mountford* present — Council in letter referred to appellant's "unlawful occupation" and stated legal proceedings would be instituted — Held, despite conflicting signals by council, appellant not a tenant — Possession order correctly made.
***Westminster City Council* v *Basson (CA),* 277**

Rent Acts
Possession

Case 1 in Sched 15 to Rent Act 1977 — Rent in arrear — Tenant's counterclaim for damages for breach of covenant to repair implied by s11 of Landlord and Tenant Act 1985 — County court judge's award on counterclaim greater than rent arrears — Judge nevertheless made possession order — Whether order justified — Equitable set-off — Whether tenant came to court with clean hands — Held, matter alleged against equitable set-off nowhere near enough to attract equity principle — Tenant's appeal allowed.
***Televantos* v *McCulloch (CA),* 123**

Case 16 in Sched 15 to Rent Act 1977 — Cottage required for agricultural worker — Whether condition in Case 16 (*b*) satisfied — Whether copy given to tenant of certificate of fair rent obtained by landlords constituted "notice" for purpose of condition — Held, giving of certificate to tenant did constitute notice — Certificate drew attention to Case 16 and possibility of recovery of possession.
***Springfield Investments Ltd* v *Bell (CA),* 115**

Resident landlord

Room retained by landlord did not contain cooker and landlord did not sleep there — He did not want cooker and bought ready-cooked food or food which did not require cooking — Nor did he sleep in flat, although spending day there, because of physical disability which necessitated assistance in dressing and undressing — Held, conditions for occupying "as his residence" satisfied — Rent Act 1977, s12.
***Palmer* v *McNamara (CA),* 121**

Succession to statutory tenancy

Meaning of "residing with" — Period of physical separation during six months immediately before original tenant's death — Original tenant absent temporarily through ill health during part of six months — Held, following reasoning in *Foreman* v *Beagley,* that appellant qualified as statutory tenant by succession despite temporary absence of original tenant — Rent Act 1977, Sched 1, para 3.
***Hedgedale Ltd* v *Hards (CA),* 118**

Rent review clause — See **also Arbitration** (Appeal); **Compensation for acquisition of land** (Valuation — Rent review clause); **Negligence** (Duty of care and liability — Receiver)
Construction: general

Underlease — Whether reference to "said term" to be interpreted as term of underlease or head lease — If latter, trigger notice served by underlessees on underlessors was out of time, time being of the essence of the provisions — Held, "said term" referred to underlease itself, not head lease — Review notice had been served in time and was valid.
***FLC Management Ltd* v *Ganton House Investments Ltd (CA),* 132**

LANDLORD AND TENANT — *continued*
Rent review clause — *continued*
Construction: hypothetical lease
Hypothetical tenants — Assumptions to be made by arbitrator — Interlinked interests in adjoining properties, the Greyhound Hotel, Croydon, and extension — Reversions in separate ownership — Rent of extension under review — *F R Evans (Leeds) Ltd* v *English Electric Ltd* analysed — Held, strictly arbitrator not entitled to assume actual hotel lessee either was or was not possible hypothetical tenant — He was, however, entitled to find as a fact that hotel lessee, whoever he might be, would be potential tenant.
***First Leisure Trading Ltd* v *Dorita Properties Ltd (Ch),* 133**
Reservation of "additional rent" — Whether hypothetical letting should take account of existence of additional obligation in calculation of reviewed rent — Point similar to *Guys "n" Dolls Ltd* v *Sade Brothers Catering Ltd* — Tenants' proposed construction would result in eliminating additional rent at first review — Held, additional rent to be disregarded in reviews of basic rent.
***Buffalo Enterprises Ltd* v *Golden Wonder Ltd (Ch),* 141**
Construction: time and date
Negligence — Landlords' trigger notice served out of time by surveyors, but principle of *United Scientific Holdings* case applied and notice effective to trigger rent increase — Questions as to date from which increase operated and date at which calculation of rent should be made — Held, on preliminary construction issue, that increase should operate as from date from which it would have operated if time-table observed — Rent should be calculated as on last day on which trigger notice could have been served according to strict time-limit.
***Henniker-Major* v *Daniel Smith (a firm) (CA),* 128**
Forfeiture, relief from — See **Landlord and tenant** (Forfeiture — Relief)
Repairs
Implied covenant — See also **Landlord and tenant** (Rent Acts — Possession (Case 1))
Construction and application of s32 of Housing Act 1961 (re-enacted in s11 of Landlord and Tenant Act 1985) — Preliminary issues of law in action for damages — Meaning of "structure" of dwelling-house — Impact of landlords' implied obligations on tenant's express obligations — Analysis of s32 and rulings on number of questions of detail.
***Irvine* v *Moran (QB),* 261**
Secure tenancy — See **Housing** (Secure tenancy)
Service occupancy — See **Landlord and tenant** (Forfeiture)
User covenant
Breach
Use of shop as "public house beer shop" prohibited — Use without consent for purposes other than sale or hire of videos restricted — Whether consent to change of use unreasonably withheld — Whether relief against forfeiture should have been refused — Held, there was ample evidence to justify county court's decision rejecting tenant's submissions and granting landlord possession.
***Sood* v *Barker (CA),* 87**
Wrongful eviction
Damages
Tenant locked out of bed-sitting-room and locks changed — Offer of key not genuine restitution — Room wrecked in tenant's absence — History of harassment — Landlord's appeal from award to tenant of £31,000 damages — Tenant's valuation evidence undisputed in county court, where landlord not represented — Held, Court of Appeal had no grounds for interfering with award — Housing Act 1988, ss27 and 28.
***Tagro* v *Cafane (CA),* 279**
LANDLORD AND TENANT ACT 1927 — See **Landlord and tenant** (Assignment)
LANDLORD AND TENANT ACT 1954 — See **Landlord and tenant** (Landlord and Tenant Act 1954 Part II)
LANDLORD AND TENANT ACT 1985 — See also **Landlord and tenant** (Rent Acts — Possession (Case 1)) (Repairs)
LANDLORD AND TENANT ACT 1987 — See **Landlord and tenant** (Landlord and Tenant Act 1987)
LANDS TRIBUNAL
Interim award
Power to amend
Errors and oversights by parties in claims and evidence — Application of slip rule — Held, interim award decisive of issues dealt with and final and binding — Distinction between purely clerical error or miscalculation and error or oversight in production of evidence or argument — Tribunal entitled to correct former but not remedy latter — Slip rule applied to measurement error and consequent valuation of goodwill — Failure to claim for rent loss and termination payments not reopenable.
***Craske* v *Norfolk County Council (LT),* 221**
LAW OF PROPERTY (MISCELLANEOUS PROVISIONS) ACT 1989 — See **Vendor and purchaser** (Option)
LEASEHOLD ENFRANCHISEMENT — See **Landlord and tenant** (Leasehold enfranchisement)
MISDESCRIPTION — See **Auctions**
MISREPRESENTATION — See **Vendor and purchaser**

NEGLIGENCE
Damages: measure and quantification
Solicitors
Leasehold enfranchisement — Failure to claim extended tenancy within statutory time-limit — Assessment of damages — Plaintiff lessee mitigated damages by compromising proceedings in county court — Defendents argued damages should be assessed at date of breach but amount reduced by reference to subsequent inflationary increases in house values — Held, if plaintiff's compromise in pursuance of duty to mitigate to be taken into account, her loss should be assessed as at date of compromise.
***Snipper* v *Enever Freeman & Co (QB),* 0**
Surveyors
Failure by building surveyor to report serious defect calling for substantial expenditure — Purchasers wanted trouble-free country property — Complaint that report unduly optimistic and reassuring — Report stated house was sound and such defects as were disclosed could be dealt with on on-going maintenance basis — Held, negligence proved — Alternative bases of calculating damages, cost of repairs or difference in value, set out — Judge decided in favour of cost of repairs basis and added damages for distress and inconvenience, plus interest.
***Watts* v *Morrow (Off Ref),* 150**
Valuers
Mortgage valuation — Accrued mortgage interest — Held, lenders entitled to be put in same position as if negligence had not occurred but not entitled to receive compensation for lost interest at rate borrowers contracted to pay — Lenders wrongly trying to obtain compensation from valuer for borrowers failure and not proper damage for valuer's negligence — Decision of Court of Appeal reversed — *Baxter* v *FW Gapp & Co* in main overruled — Judgment for valuer without fresh trial.
***Swingcastle Ltd* v *Alastair Gibson (a firm) (HL),* 157**
Duty of care and liability
Heating engineers
Complaint about inspection of aged heating system — Defendants, a specialist firm, instructed by surveyors to report on system before house purchase — Report indicated nothing wrong, but soon unsatisfactory indications arose — Another specialist firm reported cracks in heat exchanger, which led to unit being condemned — Held, in case of system reaching suspect age it was not sufficient to rely on standard combustion test — Damages awarded.
***Pfeiffer* v *E & E Installations (CA),* 162**
Receiver
Receiver appointed by mortgagees — Borrowers mortgaged tenanted properties, in many cases subject to rent review provisions — Failure by receiver to serve trigger notices in time — Sale by mortgagees under power of sale — Sale prices at auction depressed because rental income lower owing to absence of recent reviews of rent — Held, receiver negligence in not serving trigger notices in time — Allegation of contributory negligence against plaintiff borrowers dismissed — Inquiry into damages directed, with interim award of minimum sum.
***Knight* v *Lawrence (Ch),* 143**
Tree roots damage — See **Nuisance** (Tree)
Estate agents — See **Estate agents** (Negotiations for purchase)
Rent review clause
Surveyors — See **Landlord and tenant** (Rent review clause — Construction: time and date)
Surveyors — See **Negligence** (Damages: measure and quantification — Surveyors)
Valuers — See **Negligence** (Damages: measure and quantification — Valuers)
NUISANCE
Tree
Invasion of adjoining property
Tree belonging to local authority, not subject to tree preservation order — Authority "passionately" desirous of keeping tree — Tree pushing adjoining owner's wall further into property — Adjoining owner secured mandatory injunction requiring authority to remove tree — *Shelfer* v *City of London Electric Lighting Co* referred to — Authority on appeal argued injunction oppressive and damages sufficient remedy — Held, not function of courts to licence breaches of normal rights of citizens.
***Elliott* v *Islington London Borough Council (CA),* 167**
OPTION TO PURCHASE — See **Vendor and purchaser** (Option)
POSSESSION CLAIMS — See **Housing** (Secure tenancy); **Landlord and tenant** (Possession)(Rent Acts — Possession)
RANSOM STRIP — See **Compensation for acquisition of land** (Valuation)
RATES AND RATING
Proposal
Validity
Error in completion of proposal form by water authority — Proposal to expunge sewage treatment works from valuation list with view to securing assessment on cumulo-value basis — Proposal named works closed down years previously— Valuation officer understood works which authority intended — Local valuation court held disputed proposal invalid — Divisional Court quashed decision on judicial review — Court of Appeal held proposal invalid — General Rate Act 1967, s31 and Sched 4.
***R* v *Northamptonshire Local Valuation Court, ex parte Anglian Water Authority (CA),* 171**

A Court of Appeal

July 31 1990

(Before Lord Justice PURCHAS, Lord Justice NOURSE and Lord Justice GLIDEWELL)

STRADBROKE (EARL OF) v MITCHELL

Estates Gazette January 26 1991 and February 2 1991

[1991] 03 EG 128 and [1991] 04 EG 132

Agricultural holding — Notice to quit purporting to be served B in accordance with Case E in Schedule 3 to the Agricultural Holdings Act 1986 — Notice to quit found to have been served by the landlord with no honest belief in its grounds — Appeal by landlord from decision of Aldous J holding that the notice was fraudulent and therefore invalid — Judgment confined to this issue — Argument raised in respondent's notice that tenant had served a valid counternotice not considered in view of court's decision dismissing landlord's appeal on issue of fraud

The appellant landlord (plaintiff below) was the freehold owner of the Henham Estate in Suffolk, which included two farms, Valley and Green Farms, of which the respondent was C the tenant — The combined farms included a number of cottages — There was a covenant in the respondent's tenancy agreement prohibiting the tenant from assigning, subletting or parting with possession of the farms or any part thereof "except cottages and gardens to workmen on the farm" — The appellant served a notice to quit on the respondent, referring to Case E in Schedule 3 to the 1986 Act, and stating as the grounds thereof alleged breaches of the above covenant by letting farm cottages to persons who were not workmen on the farms — The respondent replied, not, as would have been expected, by serving a notice requiring arbitration but by a D counternotice stating that he required section 26(1) of the 1986 Act to apply to the notice to quit — Before Aldous J there was argument as to whether this counternotice could be treated as a notice demanding arbitration — The argument was unsuccessful and, as mentioned above, it was not raised in the Court of Appeal

The facts in regard to the cottages, examined in great detail both by Aldous J and by the Court of Appeal, were that at the date of the appellant's notice to quit the respondent had not been guilty of the breaches of covenant alleged — If the respondent's solicitors had served a notice requiring arbitration, in accordance with article 9 of the Agricultural E Holdings (Arbitration on Notice) Order 1987, the arbitrator would inevitably have found that the tenant was not in breach and the notice to quit would have failed — The respondent, however, had claimed, and Aldous J had decided, that the notice to quit failed for a different reason, namely, that it contained misrepresentations which were not only false but fraudulent, being made without any honest belief in their truth

It was, of course, for the tenant to make good an allegation of fraud, a serious allegation — As the defendant below he had established his case to the satisfaction of Aldous J — The F questions now formulated by the Court of Appeal were the following:

1 Did the court have jurisdiction to entertain a challenge to the validity of the notice to quit on the ground of fraud?
2 If so, what constituted fraud for this purpose?
3 Is a notice to quit under section 26(2) of and Schedule 3 to the 1986 Act a document containing representations which, if false, could amount to fraudulent misrepresentations?
4 If so, did the evidence before Aldous J entitle him to find that fraud by recklessness on the part of the appellant was established? G

In answer to the first question the court decided that they did have jurisdiction, rejecting an argument by the appellant that the allegation of fraud raised "a question arising under the provisions of section 26(2) of, and Schedule 3 to, the 1986 Act," and thus, because of article 9 of the 1987 order, determinable only by arbitration — If, as the respondent alleged, the notice was a nullity because of fraud, it was not a notice to which the 1987 order applied — As regards the second question, fraud was present when a false representation was made (1) knowingly or (2) without belief in its truth or (3) recklessly, careless whether it be true or false — In this connection the court rejected a submission that the H false statement must have actually deceived, or at least have been capable of deceiving, the person to whom it was addressed — In the present case the respondent was not deceived; he knew that the allegations in the notice to quit were unfounded — However, although proof of actual deceit may be essential in the context of some crimes and torts, it is not an essential element in the concept of fraud: see eg *per* Viscount Dilhorne in *R* v *Scott* — The third question was answered by analogy with authorities on the Landlord and Tenant Act 1954, *Betty's Cafés Ltd* v *Phillips Furnishing Stores Ltd* and *Marks* v *British Waterways Board* — These authorities J established that a notice is invalid and of no effect if a material statement contained in it is false and made fraudulently by the giver, knowing it to be untrue or reckless as to its truth or falsity — Finally, in answer to the fourth question, the evidence fully justified Aldous J in concluding, after a painstaking review, that the false statements made in the appellant's notice to quit were made recklessly, without any belief in their truth

For these reasons the appeal had to be dismissed

The following cases are referred to in this report.

Angus v *Clifford* [1891] 2 Ch 449 K
Betty's Cafés Ltd v *Phillips Furnishing Stores Ltd* [1959] AC 20; [1958] 2 WLR 513; [1958] 1 All ER 607; [1958] EGD 92; (1958) 171 EG 319, HL
Chapman v *Honig* [1963] 2 QB 502; [1963] 3 WLR 19; [1963] 2 All ER 513, CA
Derry v *Peek* (1889) 14 App Cas 337
Lazarus Estates Ltd v *Beasley* [1956] 1 QB 702; [1956] 2 WLR 502; [1956] 1 All ER 341, CA
London & Globe Finance Corporation Ltd, Re [1903] 1 Ch 728
Magdalen College, Oxford v *Heritage* [1974] 1 WLR 441; [1974] 1 All ER 1065; (1974) 27 P&CR 169; [1974] EGD 1; 230 EG 219, CA
Marks v *British Waterways Board* [1963] 1 WLR 1008; [1963] 3 All ER 28, CA
Pyx Granite Co Ltd v *Ministry of Housing and Local Government* [1960] AC 260; [1959] 3 WLR 346; [1959] 3 All ER 1; (1959) 58 LGR 1; 10 P&CR 319, HL

R v *Scott* [1975] AC 819; [1974] 3 WLR 741; [1974] 3 All ER 1032, HL
Smith v *East Elloe Rural District Council* [1956] AC 736; [1956] 2 WLR 888; [1956] 1 All ER 855; (1956) 54 LGR 233; 6 P&CR 102, HL

This was an appeal by the plaintiff landlord, The Rt Hon Robert Keith Rous, sixth Earl of Stradbroke, from the decision of Aldous J, holding that a notice to quit given by the plaintiff purporting to terminate the tenancy of two farms held by the defendant tenant (the present respondent), Matthew Miller Mitchell, was a nullity because of fraudulent misrepresentations contained in the notice. The decision of Aldous J was reported at [1989] 2 EGLR 5; [1989] 49 EG 59 and 50 EG 45.

Robert Pryor QC and Martin Rodger (instructed by Wedlake Saint, agents for Leathes Prior, of Norwich) appeared on behalf of the appellant; Derek Wood QC and John Male (instructed by Robbins Olivey & Blake Lapthorn, agents for Burges Salmon, of Bristol) represented the respondent.

Giving judgment, **GLIDEWELL LJ** said: The plaintiff is the freeholder of the Henham Estate, which includes farms known as Valley and Green Farms, Henham in Suffolk, The defendant is, and has since 1953 been, the tenant of those combined farms, which he has farmed since that time. The farms include a farmhouse, which was erected in about 1956, and a number of cottages. By a notice dated June 3 1988 the plaintiff required the defendant to quit and deliver up possession of the farms on October 11 1989. In this action, begun by writ issued on November 10 1988, the plaintiff claimed that no valid counternotice requiring arbitration was served by the defendant. The plaintiff, therefore, claimed a declaration that the defendant's tenancy would determine on October 11 1989.

By his defence the defendant contended that the purported notice to quit was not a valid notice on three grounds, namely that it was uncertain, that the plaintiff acted fraudulently in serving the notice and that there was an agreement between the parties that the plaintiff would not serve a notice to quit for the reasons relied on in the notice. The defendant also contended that he had served a valid counternotice, and thus counterclaimed an injunction restraining the plaintiff from taking any steps to enforce the purported notice to quit and a declaration that a letter and counternotice dated June 16 1988 served on his behalf constituted a valid counternotice.

The action came for trial before Aldous J who, on July 14 1989, gave judgment in favour of the defendant. The judge found in favour of the plaintiff on three of the arguments advanced on behalf of the defendant, namely, he found that the notice was not uncertain, that there was no agreement not to serve a notice to quit, and that no valid counternotice had been served. However, he concluded that the plaintiff acted fraudulently in instructing his solicitors to serve the purported notice to quit, which therefore was invalid and of no effect.

The plaintiff now appeals against this decision. By a respondent's notice, the defendant indicated that he wished to re-argue all the matters on which the judge had concluded against him. However, at the hearing before us Mr Derek Wood QC, for the defendant, expressly did not advance any argument against the judge's conclusions on uncertainty and on the issue of an agreement not to serve a notice to quit. He does, however, maintain his argument that a valid counternotice was served. We have not so far heard submissions on this issue, which will arise only if we decide to allow the appeal in relation to the issue of fraud. This judgment is, therefore, confined to that last issue.

The background facts

These were carefully set out by the learned judge and for the most part I gratefully adopt and repeat what he said.

In 1953 the defendant took a tenancy of Valley Farm, Henham, from the fourth Earl of Stradbroke, who then owned the Henham Estate. A farmhouse was built for the defendant in or about 1956, and Green Farm was added to his tenancy. The holding now comprises some 647 acres. On August 9 1958, the defendant entered into a tenancy agreement granting him a tenancy of the combined farms. The agreement includes a covenant not to assign, let or part with possession of the farms or any part thereof, except cottages and gardens let to workmen on the farms. As I have said, the tenancy includes a number of cottages. From time to time cottages have been added to and deleted from the holding.

In about 1980 the fourth Earl granted a tenancy of that part of the Henham Estate which was not already tenanted to a company. The fourth Earl's brother, who was the plaintiff's father, objected to the grant of the lease. In 1981 the plaintiff started proceedings in his father's name and on his instructions seeking to set aside that tenancy. That litigation was settled in 1984, with the result that the tenancy was set aside; and from October 1984 the plaintiff became responsible for running the estate and for the farming of the non-tenanted part.

On July 14 1983 the fourth Earl of Stradbroke died, leaving no son, and four days later his brother died. The plaintiff, as his eldest son, inherited the title and the estate and therefore became the freehold owner of Valley and Green Farms.

The plaintiff has lived in Australia for 32 years and is resident there. He has a number of business interests in Australia. Prior to July 1983 he visited England on a number of occasions, but most of his time was spent in Australia with his family.

In December 1983 the plaintiff came to England and it was made clear to him that he would have to pay a considerable amount of capital transfer tax. He paid about £300,000 and was advised that he would have to make further payments of about £66,000 per annum. Towards the end of 1984 it became clear that he would obtain possession of the land which had been in-hand prior to the grant of the tenancy to the company, and therefore had to make arrangements for it to be managed and farmed. He met the defendant, liked him and his family and was impressed by the way that he farmed. The plaintiff and the defendant discussed the idea that they would farm the in-hand land in partnership, with the plaintiff having first call on the profits to enable him to make the CTT payments. While this was being considered, arrangements were made for the defendant to manage the farming of the in-hand land; this agreement was recorded in a letter of October 2 1984 written by the plaintiff to the defendant. Under that agreement the defendant was to act as farm manager for £2,000 a month; he would provide all the necessary machinery and labour for which he would be paid normal contract rates; he would supply all the materials which the plaintiff would pay for at cost, and as it was an interim arrangement either party could give seven days' notice to determine the agreement. This appeared to be a satisfactory arrangement for both parties, as the plaintiff contemplated that the profit would enable him to meet the CTT payments, and they appeared to be in complete agreement as to what they hoped could be achieved.

About this time the plaintiff started to exploit certain ideas by which further revenue would be forthcoming from the estate, and was also advised not to enter into a partnership agreement with the defendant. The matter was discussed with the defendant and in a letter dated June 26 1985 the terms they agreed were recorded. The defendant was to manage the estate and would receive £2,000 a month and 10% of all income from new projects, such as caravans, steam rallies and a skeet-shooting club. The defendant was to make any decision he saw fit as regards the farm, lands, wood, shoot and all items on the estate, except for tenanted cottages and houses, which the plaintiff would control from Sydney. To enable this to be done the plaintiff made arrangements for banking facilities at Barclays Bank with an overdraft limited to £500,000.

In accordance with those arrangements the defendant managed the estate with the support of the plaintiff from June 1985. However, the profit anticipated by the plaintiff did not materialise and by December 1985 the full overdraft facility had been taken up. The plaintiff, in January 1986, considered a number of ways to prevent more money being spent on the estate and to achieve a satisfactory return. In March he terminated his arrangement with the defendant for managing the estate, but continued to use him as the contractor. Soon after, upon advice which he seems to have obtained from a Mr Rod Jones [an independent agricultural contractor], he came to the conclusion that the defendant's charges for contracting were too high. By harvest time of 1986 the relationship between the plaintiff and the defendant had broken down. The defendant's services were dispensed with and he ceased to have any dealings with the in-hand land. The plaintiff refused to pay the defendant his contracting charges and issued proceedings, claiming over £100,000 because of alleged overcharging. The defendant counterclaimed for the amount he alleged that he was owed for his contracting charges. This action has now been settled by agreement between the parties.

In August 1986 the plaintiff instructed William H Brown, a firm of chartered surveyors and land agents, to seek from the defendant an increase of rent in respect of his holding. William H Brown served on

A the defendant the appropriate notice requiring that the rent payable with effect from October 11 1987 be referred to arbitration. From July 1987 onwards Mr Nigel Wild [FRICS] of William H Brown was in charge of these negotiations on behalf of the plaintiff. An arbitrator was appointed, but the arbitration was deferred while negotiations took place. By May 1988 the negotiations had resulted in the basis of an agreement being reached, but at that stage the plaintiff's solicitors on his instructions served the notice to quit upon which this action is based and also, as an alternative, a notice to remedy. Both notices relate to alleged breaches of his tenancy agreement by the defendant consisting of the subletting of five cottages allegedly in breach of the covenant against subletting to which I have referred. I must, therefore, consider the facts relating to these alleged breaches and the B notices in more detail, but before I do it is convenient to set out the relevant legal provisions.

The law applicable

Lettings of agricultural holdings have for many years been controlled by statute. The Act in force in relation to these proceedings is the Agricultural Holdings Act 1986. Section 2 of that Act provides, in general, that a letting of an agricultural holding may not be for a lesser interest than a tenancy from year to year. By section 25(1) of the Act, the normal rule that a tenancy from year to year may be terminated by at least six months' notice is amended so that an agricultural tenancy requires at least 12 months' notice terminating C at the end of a year of the tenancy.

Section 26 of the 1986 Act places restrictions on the operation of notices to quit, which in effect divides such notices into two classes. The difference between the two classes is that notices to quit in the first class need not state or be based upon the reason why the landlord seeks to recover possession of the holding, whereas a notice to quit in the second class must be based upon such a reason and must state the reason in the notice itself. If a notice of the first class is served, then section 26(1) of the Act provides that if, within one month after the service of the notice to quit,

> the tenant serves on the landlord a counter-notice in writing requiring that this subsection shall apply to the notice to quit, then . . . the notice to quit shall not have effect unless, on an application by the landlord, the [Agricultural Land] Tribunal consent to its operation.

D By section 27(1) the tribunal may consent only if:

(i) they are satisfied that one of the matters set out in section 27(3) is proved and [even if they are so satisfied]

(ii) the tribunal shall withhold consent under section 26 above to the operation of the notice to quit if in all the circumstances it appears to them that a fair and reasonable landlord would not insist on possession.

Obtaining the consent of the tribunal to the operation of a notice to quit is not often an easy task.

The second class of notice to quit is governed by section 26(2) of the 1986 Act, which provides in effect that the consent of the agricultural land tribunal to the operation of the notice is not necessary "in any of E the cases set out in Part I of Schedule 3 to this Act". Those cases relate to a number of disparate situations. For instance, Case B applies where the land is required for a use other than agriculture for which planning permission has been granted. In each case the facts which satisfy the provisions of the particular case, ie the landlord's reason for seeking possession, must be stated in the notice to quit itself.

In this action Cases D and E are relevant and I therefore set them out.

Case D

> At the date of the giving of the notice to quit the tenant had failed to comply with a notice in writing served on him by the landlord, being either —
>
> (*a*) a notice requiring him within two months from the service of the notice to pay any rent due in respect of the agricultural holding to which the F notice to quit relates, or
>
> (*b*) a notice requiring him within a reasonable period specified in the notice to remedy any breach by the tenant that was capable of being remedied of any term or condition of his tenancy which was not inconsistent with his responsibilities to farm in accordance with the rules of good husbandry
>
> and it is stated in the notice to quit that it is given by reason of the said matter.

Case E

> At the date of the giving of the notice to quit the interest of the landlord in the agricultural holding had been materially prejudiced by the commission by the tenant of a breach, which was not capable of being remedied, of any term or condition of the tenancy that was not inconsistent with the tenant's responsibilities to farm in accordance with the rules of good husbandry, and it is stated in the notice that it is given by reason of the said matter.

It will be seen that if a breach is capable of being remedied, the G landlord may claim possession under Case D only if he has first served a notice requiring the tenant to remedy the breach and the tenant has failed to do so within a reasonable period. If, however, the breach is not capable of remedy, and the landlord's interest has been materially prejudiced, the landlord may serve notice to quit under Case E.

The combined effect of section 29 of and Schedule 4, para 1, to the 1986 Act is that the Lord Chancellor may make an order "requiring any question arising under the provisions of section 26(2) of, and Schedule 3(2) to, this Act to be determined by arbitration under this Act".

In exercise of that power the Lord Chancellor made the Agricultural Holdings (Arbitration on Notice) Order 1987. Article 9 H of that order provides:

> Where it is stated in a notice to quit an agricultural holding or part thereof that the notice is given for one or more of the reasons specified in Case A, B, D or E and the tenant wishes to contest any question arising under the provisions of section 26(2) of, and Schedule 3 to, the 1986 Act relating to any of the reasons so stated, he shall within one month after the service of the notice serve on the landlord notice in writing requiring the question to be determined by arbitration under the 1986 Act.

It will thus be seen that if a tenant wishes to contest any part of a reason stated in a notice to quit served on him under Case E of Schedule 3, he must, within one month of the service of that notice, serve a counternotice requiring the question to be determined, not by J the agricultural land tribunal but by arbitration.

The notices

Solicitors acting for the plaintiff served on the defendant a notice dated June 2 1988 which required him

> to remedy within two months from the date of service of this notice the breaches set out below of the terms and conditions of your tenancy, being breaches which are capable of being remedied . . .

The breach alleged was a breach of clause 4(13) of the tenancy agreement, the covenant to which I have referred, against assigning or underletting any part of the farm except cottages let to workmen on the farm. The particulars of breach were:

> That you let or entered into agreement to let to tenants who are not workmen K on the farm (1) the pair of cottages Nos 20 and 21 King's Lane, Sotherton (2) the pair of cottages which make up what was once the Old Valley Farmhouse known as Nos 2 and 3 Old Valley Farm Cottages, Henham and (3) 24 Green Farm, Sotherton.

The plaintiff's solicitors also served on the defendant a notice to quit dated June 3 1988. This required him to quit

> and deliver up possession of the holding of which you are my tenant at the expiration of the year of your tenancy which shall expire next after the end of 12 months from the service of this notice.

That, if it were effective, would bring the tenancy to an end, as the statement of claim alleges, on October 11 1989.

The notice stated that it was given in accordance with section 26(2) L of, and Case E in Schedule 3 to, the Agricultural Holdings Act 1986. The reason stated in the notice was a breach of clause 4(13) of the tenancy agreement, and the details of the alleged breach were the same as those set out in the notice to remedy of June 2 1988, ie the letting in alleged breach of agreement of the five cottages specified in that notice and again described in the notice to quit.

On June 16 1988, solicitors acting for the defendant served on the plaintiff a counternotice in the following terms:

> I hereby give you a counternotice that I require subsection (1) of section 26 of the Agricultural Holdings, Act, 1986, to apply to the Notice to Quit the farm, land and premises situate at Henham in the County of Suffolk, known as Valley and Green Farms, which I hold as Tenant thereof, and which Notice to Quit is dated the 3rd June, 1988. M

It will be seen that the counternotice was expressed to apply section 26(1) and was not in terms a notice under the 1987 order requiring reference to arbitration of the matters challenged in the landlord's notice to quit.

The subletting of the cottages

At one time there were 13 cottages on the defendant's holding, but by June 1988 four, at Sotherton Corner, had been surrendered to the landlord. There were thus nine cottages remaining in the tenancy. Confusingly, some of these are referred to by two different names. Despite this, the judge held that the notice was sufficiently clear to enable the defendant to know to which cottages the notice to quit was intended to relate.

On October 1 1981 agents acting for the fourth Earl wrote to the defendant telling him that it was "in order for you to let the Old Valley Farmhouse as a holiday cottage". However, it is accepted that the plaintiff did not know of this letter and nothing therefore turns upon it.

However, on March 16 1984 the plaintiff himself wrote to the defendant a brief letter which said:

Re SUBLETTING
We hereby agree to you subletting the below mentioned properties:
20 Valley Farm Cottages, King's Lane
24 Henham Green, Henham

Moreover, on November 1 1985, the plaintiff wrote to the defendant a letter confirming the defendant's readiness to surrender the cottages at Sotherton Corner and to waive claims in respect of improvements he had made to those cottages. In return the letter allowed him the use of one alternative cottage without increase of rent. The letter continued:

I note it was further agreed that you may sublet all or any of your cottages for holiday let situations as and when required and no extra rent from you is required.

The facts of the occupation of the cottages referred to in the notice to quit at the date of the service of that notice, and at material times beforehand, were not in dispute and were found by the judge as follows.

(i) 20 King's Lane, Sotherton. This is the cottage otherwise called 20 Valley Farm Cottages, King's Lane, Henham, for which the plaintiff had given consent to sublet in the letter of March 16 1984. Following the giving of that consent the cottage had been sublet to three tenants in turn, each on protected shorthold tenancies. At the date of the notice to quit it was occupied by a Mr Warren under a two-year shorthold tenancy dated March 18 1988.

(ii) 21 King's Lane, Sotherton, otherwise known as 21 Valley Farm Cottages, King's Lane, Henham. At all material times this was occupied by a Mr Briggs, who was a semi-retired farm worker working for the defendant. It is agreed that his tenancy was not a breach of the covenant in the lease.

(iii) 1 Old Valley Farmhouse. Permission had been granted to let this as a holiday cottage in the letter of November 1 1985. However, on September 14 1986 it was let by the defendant on a two-year protected shorthold tenancy to a Mr Marsham. Mr Marsham gave notice to terminate that tenancy and it duly terminated at the end of May 1986. The cottage was vacant at the date of service of the notice to quit.

(iv) 2 Old Valley Farmhouse. This also was the subject of the permission of November 1 1985 to let as a holiday cottage. It had been so sublet, but not let in any other way. It was also vacant at the date of the service of the notice to quit.

(v) The cottage described in the notice to quit as 24 Green Farm should properly have been described as 24 Henham Green, Henham. This is the second cottage in respect of which the plaintiff had given to the defendant consent to sublet without restriction in his letter of March 16 1984. It also had been sublet by the defendant on a series of shorthold tenancies, the latest of which was a two-year protected shorthold tenancy granted on October 1 1985 to a Mr Willard.

It is accepted by Mr Pryor, for the plaintiff, that the subletting to Mr Briggs of 21 Valley Farm Cottages was not a breach of the covenant in the tenancy agreement; that the sublettings of 20 Valley Farm Cottages and 24 Henham Green were not breaches, since they were permitted by the letter of March 16 1984; and that the holiday lettings of 2 Old Valley Farmhouse were also not breaches, since they were permitted by the letter of November 1 1985. The only apparent breach was the shorthold subletting to Mr Marsham of 1 Old Valley Farmhouse but, as I have said, this had come to an end at the date of the service of the notice to quit. It follows, in my view, that this breach was clearly remediable (though Mr Pryor argues somewhat faintly to the contrary) and thus could not properly be the subject of a Case E notice to quit.

It thus also follows on the facts that if the defendant's solicitors had served a counternotice requiring arbitration in accordance with the 1987 order and section 26(2), the arbitrator would inevitably have found that the defendant was not in breach as alleged in the notice to quit, and thus would not have upheld the notice.

The issues

The argument on behalf of the defendant is that the notice to quit dated June 3 1988 was invalid and of no effect because it contained misrepresentations which were not only untrue but fraudulent. It was, of course, for the defendant to establish this case before Aldous J. In order to decide whether the defendant had made out his case of fraud the judge was required, and this court is now required, to answer four questions. These are:

1 Does the court have jurisdiction to entertain a challenge by the defendant to the validity of the notice to quit on the ground of fraud?

2 If so, what constitutes fraud of this purpose?

3 Is a notice to quit under section 26(2) and Schedule 3 to the 1986 Act a document which contains representations which, if false, may amount to a fraudulent misrepresentation?

4 If so, did the evidence before the judge entitle him to find that fraud by recklessness on the part of the plaintiff was established?

I will consider each of these matters in turn.

1 *The court's jurisdiction*

Mr Pryor submits that an allegation that a notice to quit given under section 26(2) of and Schedule 3 to the 1986 Act contains fraudulent misrepresentations raises a "question arising under the provisions of section 26(2) of, and Schedule 3(2) to, the 1986 Act relating to any of the reasons so stated". Thus, in accordance with article 9 of the 1987 order, this question can be determined only by arbitration. The court has no jurisdiction to determine the question.

In support of this submission, Mr Pryor relies principally on the decision of this court in *Magdalen College, Oxford* v *Heritage* [1974] 1 WLR 441. In that case the defendant, who was tenant of a farm, fell into arrears with her rent. The plaintiff, her landlord, served on her a notice requiring her to pay the rent, which did not specify any time within which the rent must be paid. On her failure to pay the arrears, after five months the landlord served a notice to quit under Case D. The defendant did not serve a counternotice requiring arbitration. In an action for possession of the farm the defendant argued that, as the notice to pay the rent was defective, the subsequent notice to quit was invalid. This court held that the issue was one which, under article 9 of the 1987 order, could be determined only by arbitration and that the court had no jurisdiction to decide it. Megaw LJ, in a judgment with which Buckley and Roskill LJJ agreed, summarised his conclusion in the following words at p 448:

To my mind, the construction of article 9 is clear, and is conclusive against the defendant on the main point put forward on her behalf. Where the tenant wishes to object to the notice to quit by reason of an objection in relation to the form of the notice to pay, such as the objection here, the tenant is given his opportunity of raising it to be dealt with by arbitration under the article 9 procedure, within the time limit therein provided. That remedy is exclusive of any other remedy, such as the remedy upon which the defendant now seeks to rely; the raising of the issue as a defence in an action for possession.

However, early in his judgment, at p 446, Megaw LJ said:

There is no suggestion in this case of any fraud or dishonesty or misrepresentation on the part of the plaintiffs. Nothing that I say is to be taken as having any necessary application in any case where any such factor exists.

Mr Pryor also reminds us that the decision in *Smith* v *East Elloe Rural District Council* [1956] AC 736 shows that Parliament can by appropriate language deprive the court of jurisdiction even when fraud is alleged. In that case the council made a compulsory purchase order relating to land owned by Mrs Smith. After a public inquiry the order was confirmed by the minister. By para 15 of Schedule 1 to the Acquisition of Land (Authorisation Procedure) Act 1946, Mrs Smith was entitled to make an application to the High Court within six weeks of the confirmation of the order challenging its validity. She did not do so. By para 16 of Schedule 1:

Subject to the provisions of the last foregoing paragraph, a compulsory purchase order . . . shall not . . . be questioned in any legal proceedings whatsoever . . .

Mrs Smith sought to bring an action against the council for trespass, alleging that the CPO was made in bad faith and claiming a declaration to that effect. The House of Lords held that the clear words of article 16 precluded her from challenging the validity of the CPO in this way, even though she was alleging bad faith.

It follows, in my view, that *Smith* v *East Elloe Rural District Council* does establish Mr Pryor's proposition, but subject to the qualification that the words of the Act which are said to exclude the jurisdiction of the court to determine an allegation of fraud will have this effect only if that is clearly their meaning. In *Pyx Granite Co Ltd* v *Ministry of Housing and Local Government* [1960] AC 260, Viscount Simonds said at p 286:

> A It is a principle not by any means to be whittled down that the subject's recourse to Her Majesty's courts for the determination of his rights is not to be excluded except by clear words. That is, as McNair J called it in *Francis* v *Yiewsley and West Drayton UDC*, a "fundamental rule" from which I would not for my part sanction any departure.

This court in *Magdalen College, Oxford* v *Heritage* held that the words of article 9 were clear enough to exclude the power of the court to decide the issue in that case, but expressly reserved the determination of an allegation of fraud.

In *Lazarus Estates Ltd* v *Beasley* [1956] 1 QB 702 a landlord had served on the tenant of a rent-controlled dwelling-house a notice of increase of rent allegedly justified by the carrying out of work of repair to the house. Schedule 2, para 4 to the Housing Repairs and B Rents Act 1954 entitled the tenant to apply to the county court within 28 days to determine whether the work of repair had been carried out and, if so, its value, but she made no application. She did not pay the increase in rent. In an action claiming the unpaid increase as arrears the tenant in her defence alleged that the landlord's declaration that the work had been carried out was fraudulent as only about half of the alleged work had been done. Para 5 of the Schedule provided that, subject to the right to apply to the court under para 4, the validity of the landlord's declaration "shall not be questioned". This court nevertheless held that the court had jurisdiction to determine the allegation of fraud. Denning LJ said at p 712:

> We are in this case concerned only with this point: can the declaration be C challenged on the ground that it was false and fraudulent? It can clearly be challenged in the criminal courts. The landlord can be taken before the magistrate and fined £30: see Second Schedule (paragraph 6); or he can be prosecuted on indictment, and (if he is an individual) sent to prison: see section 5 of the Perjury Act, 1911. But the landlords argue before us that the declaration cannot be challenged in the civil courts at all even though it was false and fraudulent; and that the landlords can recover and keep the increased rent even though it was obtained by fraud. If this argument is correct, the landlords would profit greatly from their fraud. The increase in rent would pay the fine many times over. I cannot accede to this argument for a moment. No court in this land will allow a person to keep an advantage which he has obtained by fraud. No judgment of a court, no order of a minister, can be allowed to stand if it has been obtained by fraud. Fraud unravels everything. The court is careful not to find fraud unless it is distinctly D pleaded and proved; but once it is proved, it vitiates judgments, contracts and all transactions whatsoever: see as to deeds, *Collins* v *Blantern*, as to judgments, *Duchess of Kingston*'s case, and as to contracts, *Mater* v *Miller*. So here I am of opinion that if this declaration is proved to have been false and fraudulent, it is a nullity and void and the landlords cannot recover any increase of rent by virtue of it.

Parker LJ said at p 721:

> Accordingly, the question here is whether the tenant is seeking to challenge the validity on some ground other than that repairs had not been carried out during the period specified to a value not less than that specified. That the tenant is seeking to challenge the validity on that ground is clear, but is she also seeking to challenge it on another ground? The contention on her behalf is that should she succeed in proving fraud on the part of the landlords the declaration would be a nullity, whereas mere proof that repairs had not been E done to the value specified would not make the declaration a nullity but would merely make it cease to have effect. Therefore, it is said, the tenant is seeking to do something more than challenge the validity of the declaration on the ground that repairs had not been done to the value specified. I think that this contention is correct. No doubt it can be said that the real question in any case is whether repairs to the value specified have in fact been done, and that proof of fraud in the making of the declaration is merely proof of the quality of the act or its motive. Nevertheless, that quality, if proved, vitiates all transactions known to the law of however high a degree of solemnity. Suppose that on an application under paragraph 4 of the Second Schedule the landlord by fraud persuades the county court to uphold a declaration and that months later the tenant discovers this and is in a position to prove that fraud. Surely the tenant could refuse to pay the increase in rent, and when sued could allege that the decision of the county court was obtained by fraud. If that be the true F position, why cannot a tenant who has not adopted the procedure of paragraph 4 equally claim that on proof of fraud the declaration is not satisfactory evidence for the purpose of section 23?

Mr Wood, for the defendant, submits:

(i) that the question whether the landlord's notice to quit contained fraudulent misrepresentations is not a "question arising under the provisions of section 26(2) of, and Schedule 3(2) to, the 1986 Act relating to any of the reasons . . .", and thus is not a question to be referred to arbitration under article 9 of the 1987 order;

(ii) that the wording of article 9 is certainly not sufficiently clear and explicit to exclude the jurisdiction of the court to decide the allegation of fraud; and

(iii) that the alleged misrepresentations in the notice to quit were statements of the same nature as those in the notice of increase of rent G in *Lazarus Estates* v *Beasley*, and therefore we should follow that decision and the passages from the judgments in this court which I have set out.

In my view, all these submissions by Mr Wood are correct. I therefore conclude that Aldous J was right to hold that he had jurisdiction to decide the issue. I agree with the words in which he expressed this conclusion*:

> There does not appear to me to be any provision in the Agricultural Holdings Act 1986 or the order of 1987 which would prevent a tenant asserting in the High Court that a notice to quit was a nullity. It is only if he accepts that the notice is a notice that his right to challenge it is confined to arbitration. I therefore hold that the defendant is not precluded from challenging the notice on the ground of fraud and in particular to contend that the notice is a nullity H and should not be given effect to.

2 *What constitutes fraud?*

Aldous J in his judgment† cited two passages from the speech of Lord Herschell in *Derry* v *Peek* (1889) 14 App Cas 337, which he rightly described as the "classic statement" of what constitutes fraud. Lord Herschell said at p 374:

> Secondly, fraud is proved when it is shown that a false representation has been made (1) knowingly, or (2) without belief in its truth, or (3) recklessly, careless whether it be true or false. Although I have treated the second and third as distinct cases, I think the third is but an instance of the second, for one who makes the statement under such circumstances can have no real belief in the truth of what he states. J To prevent a false statement being fraudulent, there must, I think, always be an honest belief in its truth. And this probably covers the whole ground, for one who knowingly alleges that which is false, has obviously no such honest belief.

On p 375 Lord Herschell went on to state:

> At the same time I desire to say distinctly that when a false statement has been made the questions whether there were reasonable grounds for believing it, and what were the means of knowledge in the possession of the person making it, are most weighty matters for consideration. The ground upon which an alleged belief was founded is a most important test of its reality. I can conceive many cases where the fact that an alleged belief was destitute of all reasonable foundation would suffice of itself to convince the court that it was not really entertained, and that the representation was a fraudulent one. So, too, although means of knowledge are, as was pointed out by Lord Blackburn in K *Brownlie* v *Campbell*, a very different thing from knowledge, if I thought that a person making a false statement had shut his eyes to the facts, or purposely abstained from inquiring into them, I should hold that honest belief was absent, and that he was just as fraudulent as if he had knowingly stated that which was false.

Aldous J also quoted a passage from the judgment of Bowen LJ in *Angus* v *Clifford* [1891] 2 Ch 449 at pp 471-2, from which I think it necessary to quote only a short extract:

> Now whether you take the enquiry in the one order or in the other, whether you regard it from the point of view which is indicated in the expressions which are well known to the common law, and which have been used by Lord Herschell and Lord Bramwell, that a man is bound to have some honest belief in a statement if he makes it, or whether you treat the matter in the inverse L order, with regard to the necessity of finding at least some recklessness to truth, that is to say some indifference to truth which amounts to dishonesty, in either view it seems to me the result is the same. A man ought to have a belief that what he is saying is true.

Mr Pryor accepts that a statement which is made recklessly, without an honest belief in its truth, is made fraudulently, but he submits that, in order to render the document in which the statement is made invalid and of no effect, the statement must deceive, or at least be capable of deceiving, the person to whom it is addressed. Here, he submits, Mr Mitchell was not deceived. He knew that he had been granted consent to sublet the cottages, and thus that the allegations in the notice to quit were unfounded. All he had to do was to serve an appropriate counternotice under article 9 requiring the M issue to be referred to arbitration, and he would have suffered no loss. The risk that he, or his solicitors, might serve the wrong counternotice can be disregarded. Thus the essential requirement of deceit is not made out.

In support of this proposition Mr Pryor relies, I believe, on the well-known passage from the judgment of Buckley J in *London & Globe Finance Corporation Ltd, Re* [1903] 1 Ch 728 at p 732:

> To deceive is, I apprehend, to induce a man to believe that a thing is true which is false, and which the person practising the deceit knows or believes to be

*Editor's note: See [1989] 2 EGLR 5 at p 10M.
†Editor's note: See [1989] 2 EGLR 5 at p 11B.

A false. To defraud is to deprive by deceit: it is by deceit to induce a man to act to his injury. More tersely it may be put, that to deceive is by falsehood to induce a state of mind; to defraud is by deceit to induce a course of action.

R v *Scott* [1975] AC 819 was an appeal against a conviction of conspiracy to defraud. One of the arguments for the appellant was that there could not be a conspiracy to defraud without deceit. Viscount Dilhorne, in a speech with which all their lordships agreed, quoted the passage from the judgment of Buckley J set out above, and said of it at p 836B:

> In a great many and it may be the vast majority of fraud cases the fraud has been perpetrated by deceit and in many cases Buckley J's dicta have been quoted in charges to juries. It does not, however, follow that it is an exhaustive definition of what is meant by "defraud". Buckley J had to decide when a
> B *prima facie* case had been shown "of doing some or one of the acts" mentioned in sections 83 and 84 of the Larceny Act 1861 "with intent to deceive or defraud". He did not have to make or have to attempt to make an exhaustive definition of what was meant by "defraud".

Later, Viscount Dilhorne said at p 838:

> In *Welham* v *Director of Public Prosecutions* [1961] AC 103 this House had to consider the meaning of "intent to defraud" in relation to forgery. In the course of his speech Lord Radcliffe said, at pp 123, 124:
> "Now, I think that there are one or two things that can be said with confidence about the meaning of this word 'defraud'. It requires a person as its object: that is, defrauding involves doing something to someone. Although in the nature of things it is almost invariably associated with the obtaining of an advantage for the person who commits the fraud, it is the effect upon the
> C person who is the object of the fraud that ultimately determines its meaning. . . .
> "Secondly, popular speech does not give, and I do not think ever has given, any sure guide as to the limits of what is meant by 'to defraud'. It may mean to cheat someone. It may mean to practise a fraud upon someone. It may mean to deprive someone by deceit of something which is regarded as belonging to him or, though not belonging to him, is due to him or his right".
> Later, Lord Radcliffe said, at p 126, that he was unable to accept Buckley J's observations in *In re London & Globe Finance Corporation Ltd* [1903] 1 Ch 728, which he said were *obiter*, as an authoritative exposition of words employed in subsequent statutes.
> While the meaning to be given to words may be affected by their context and Lord Radcliffe was only considering the meaning of intent to defraud in section 4 of the Forgery Act 1913, the passages which I have cited from his
> D speech are, I think, of general application; and certainly those passages and his speech lend no support to the contention that there cannot be a conspiracy to defraud which does not involve deceit.
> In the course of delivering the judgment of the Court of Appeal in *Reg* v *Sinclair* [1968] WLR 1246, where the defendants had been convicted of conspiracy to cheat and defraud a company, its shareholders and creditors by fraudulently using its assets for purposes other than those of the company and by fraudulently concealing such use, James J said, at p 1250: "To cheat and defraud is to act with deliberate dishonesty to the prejudice of another person's proprietary right". Again, one finds in this case no support for the view that in order to defraud a person that person must be deceived.

The learned judge's conclusion on this aspect of the case was expressed in the following words*:

> E Mr Pryor, although accepting the principle in the speech of Lord Herschell which I have just cited, submitted that before fraud could be found the court must also hold that the dishonest and false representation was calculated to deceive and did deceive the recipient. That may be necessary in actions in which a party seeks to recover damages based on fraudulent statements, but in a case where the party committing the fraud seeks to rely upon his fraudulent conduct a court will not give effect to that conduct whether or not it deceives the recipient. In practice all fraudulent statements and notices to quit under Schedule E are calculated to deceive in that they are assertions that the landlord honestly believes that he has a good ground of complaint, honestly believes the facts stated as reasons for the notice and honestly believes that he has a reasonable case to terminate the tenancy because his interest has been materially prejudiced. But the fact that the tenant is not deceived is irrelevant. The notice if fraudulent is a nullity and the court will refuse to give it effect.

F Subject to the answer to the third question, to which I now turn, I entirely agree with Aldous J's conclusion on the second issue.

3 *Can the statement of reasons in a Schedule 3 notice to quit constitute a fraudulent misrepresentation?*

Mr Pryor submits that the answer to this question is "no". He argues that the effective part of the document is the notice of termination of the tenancy, which is simply a notice to quit at common law; the statement of reasons included in the notice in order to satisfy Case E of Schedule 3 is no more than an indication of the facts which the landlord will seek to prove if the tenant serves a counternotice requiring arbitration; the truth or falsity of that statement is irrelevant to the validity of the notice. G

In support of this submission Mr Pryor relies on the decision of this court in *Chapman* v *Honig* [1963] 2 QB 502. I mean no discourtesy to him when I say that in my view this authority does not support his argument. Indeed, there is a short passage in the judgment of Lord Denning MR at p 513 which runs counter to Mr Pryor's argument.

Mr Wood has referred us to a number of cases in which the problem was similar to that which I am now discussing. These include *Betty's Cafés Ltd* v *Phillips Furnishing Stores Ltd* [1959] AC 20 and *Marks* v *British Waterways Board* [1963] 1 WLR 1008. Both decisions related to proceedings under Part II of the Landlord and Tenant Act 1954. In *Betty's Cafés* a landlord opposed the tenant's application for a new lease of a café in Bradford on the ground provided by section 30(1)(*f*) of the 1954 Act, namely that on H termination of the current tenancy the landlord intended to reconstruct the premises, and could not reasonably do so without obtaining possession. That ground was, as required by section 26(6) of the Act, stated in the landlord's notice of opposition to the grant of the new tenancy. The question at issue was whether the landord was required to have formed that intention at the date of giving notice of opposition or whether, as this court and the House of Lords held, the landlord could succeed in his opposition if he proved that he held the intention at the date of the hearing. However, though agreeing with this conclusion, Lord Denning expressed the view that the landlord's statement of his grounds of opposition in his notice must be honest and truthful. He said at pp 50-51: J

> Such being the true interpretation of these notices, I am of opinion that they must be given honestly and truthfully. They are not to be regarded merely as pleadings preparatory to a trial — in which parties, I regret to say, sometimes deny the truth, or refuse to admit it, if it suits their plan of campaign. These notices are intended to be acted upon before there is a trial at all. On the receipt of such a notice, the tenant has to decide his course of action — for instance, whether to accept the alternative accommodation that is offered, or whether to accept the landlord's word that he intends to occupy the premises himself, or as the case may be. In every case he has to decide whether to apply for a new lease or not. It would be deplorable if a landlord could be allowed to get an advantage by misrepresenting his state of mind or any other fact. Suppose he said in his notice: "I intend to reconstruct the premises", or "I intend to occupy for the purpose of my own business", when he, in fact, had no such intention at all. On the faith of such a statement, the tenant might be induced K to abstain from applying to the court for a new tenancy, because he would think it no use to do so. He would know that he would have to pay the costs if he lost. Just imagine the tenant's consternation if at the end of the tenancy, after he had left, the landlord did not reconstruct the premises or occupy them himself, but straightaway let in someone else. Would the tenant have no redress? I should have thought it clear that the notice would be bad — voidable — liable to be set aside for fraudulent misrepresentation: see *Lazarus Estates Ltd* v *Beasley*. If it was avoided, the original tenancy would continue. The landlord would get no advantage from his misrepresentation — which is as it ought to be. If it was too late to avoid the notice, the landlord would be liable at common law in damages for fraud: just as he would be under section 55 if the misrepresentation was made to the court.
> Provided, however, that the notice is a good and honest notice when it is given, then it is clear to my mind that the ground stated therein must be L established to exist at the time of the hearing.

At p 52 he repeated the point shortly:

> What is the result of this? If the notice had been a dishonest notice in which the landlords had fraudulently misrepresented their intention — or, I would add, if there had been a material misrepresentation in it — I should have thought it would be a bad notice.

In *Marks* v *British Waterways Board* Mr Marks, the tenant, held a sublease of business premises, of which the British Waterways Board was the freeholder. Mr Marks served notice on his lessor requesting a new tenancy under the 1954 Act. At that time, the sublessor had agreed to surrender his lease to British Waterways Board. The board wished to obtain possession of the premises in order to demolish and reconstruct. The sublessor served a counternotice stating that this M was the intention of "the landlords".

By the time of the hearing, the surrender had been effected, and British Waterways Board was Mr Mark's landlord. This court held that the counternotice was valid and effective.

In his judgment Lord Denning (by then MR) said:

> . . . the notice opposing the new tenancy, a landlord's notice, is to be regarded as in the nature of a pleading . . . . It is sufficient as long as it gives notice to the tenant of the case he has to meet. So long as it is not deceptive or misleading, it avails the subsequent owner of the property who is the landlord at the date of the hearing.

Harman LJ said at p 1018:

*Editor's note: See [1989] 2 EGLR 5 at p 10M.

A The only object, as Lord Simons said in the *Betty's Cafés Ltd* case, of a counternotice is to inform the tenant of the case which will be made against him when the hearing comes on, and providing it does that, it is, I think, a good counternotice. I think that is the effect of the *Betty's Cafés* case. If I may say so with the greatest respect, you can look on my Lord's, the present Master of the Rolls, observations in that case as going rather further; but I do not think they were intended to and, with respect to him, I do not think they should do. You must not mislead the tenant: you must not say anything which is fraudulent, but if your notice is given in good faith and the facts about reconstruction can be substantiated by the person who is the landlord when the hearing comes on, I think the counternotice really has served the purpose which the legislature can be said to have required of it.

Pearson LJ made the same point at p 1020:

> B What one has to inquire into in regard to the notice given under section 26 is whether it was given bona fide; whether it was an honest notice. It may be that there are some further requirements with regard to it. It may be that it would be void if it were deceptive or misleading or if it contained some material misrepresentation.

In my view, these authorities establish the proposition that a landlord's counternotice under section 26(6) of the 1954 Act is invalid and of no effect if the statement contained in it of the landlord's intention is fraudulent, not honestly made. The notice with which we are here concerned is a notice of the same sort as a notice under section 26(6). It is not a mere notice to quit. In my view, a notice to quit which states that it is given for one of the reasons set out in the various cases in Schedule 3 to the 1986 Act is invalid and of C no effect if it contains a statement which is false and made fraudulently by the giver of the notice, ie knowing the statement to be untrue, or reckless whether it is true or false.

4 *Did the evidence establish that the plaintiff had been recklessly fraudulent?*

In order to answer this question, it is necessary to recapitulate the events which led up to the service of the notice to quit.

In his judgment Aldous J summarised the steps the plaintiff had taken after he inherited the Henman Estate to familiarise himself with the estate, including visiting the defendant's farms and the cottages in the summer and early autumn of 1984.

> . . . Clearly the plaintiff was correct when he stated in his evidence that during D 1984 he was becoming more and more familiar with details of the estate and who was living in the cottages. He certainly knew that Mr Briggs was living at 21 Valley Farm Cottage and must have known that 24 Henham Green had been sublet to a person who was not a farmworker . . . .
>
> . . . when he agreed in November 1985 to allow the defendant to sublet all the cottages as holiday lets he was the person in charge and was fully aware of the names of the cottages, and, as he said in cross-examination, was taking a day-to-day interest in them.

Mr Prior did not challenge this conclusion of the learned judge.

I have referred earlier to the notice served on the defendant by William H Brown in September 1986 requiring that the rent for the farms from October 1987 should be referred to arbitration. That notice was served following instructions given by the plaintiff to Mr E Derek Turnbull [FRICS] of William H Brown in a letter dated September 11 1986. In that letter the plaintiff referred to the fact that some of the cottages were let and said:

> I may have written him a letter allowing him to lease his cottages out as holiday lets — this was when he told me what a super job he was going to do on the land in hand.

The judge commented:

> Thus at that time he did recollect the letter he had written in 1985.

The judge also referred to a letter written by the plaintiff to Mr Turnbull on January 21 1987, in which he said:

> He is said to have three or four of his cottages rented out at around £40 a week to non-ag workers — I may have written him allowing this three or four years F ago — please revoke it immediately or bring it to his notice he appears to be in breach.

The judge commented:

> Again this letter shows that the plaintiff had a recollection of a written permission he had given to the defendant to sublet.

It seems that in the summer of 1987 Mr N R Wild [FRICS], a partner in William H Brown, took over from Mr Turnbull the conduct of the negotiations relating to the rent of the defendant's farms. On Monday July 20 1987, Mr Wild visited Valley and Green Farms and was shown round by the defendant. He inspected the farms and the cottages and produced a report for his own benefit in which each of the cottages was accurately recorded, together with the nature of the tenancy where the cottage was sublet. The defendant showed Mr G Wild a copy of the 1984 letter from the plaintiff giving him permission to sublet two of the cottages.

In a letter to the defendant dated September 9 1987, written on the plaintiff's instructions, Mr Wild put forward proposals for a new agreement between the plaintiff and the defendant. These included the defendant giving up all the residential properties with the exception of his own farmhouse and the two cottages at Henham Green, surrendering some of the buildings at the farms in return for the provision by the plaintiff of a new modern building, and entering into a full repairing lease, with a reduction in his rent to around £20,000 pa. To this the defendant replied in a letter dated September 16 1987 that:

> The cottages need not be too much of a problem, but the ideas you've put H forward regarding the buildings need to be gone into in much more detail.

Early in October 1987 Mr Wild met the defendant and the agents acting on his behalf, and an agreement was reached in principle that the defendant would surrender the cottages with the exception of the two at Henham Green and some of the farm buildings, and that the plaintiff would provide the defendant with a new farm building. A tentative agreement as to a reduction of rent to around £20,000 pa was also reached. However, when Mr Wild obtained a quotation for the cost of erecting the new building, this proved unacceptable to the plaintiff, who instructed Mr Wild to proceed with the arbitration.

On October 20 1987 Mr Wild wrote to the plaintiff a letter in which he said:

> J We cannot force MM to give up the cottages. He is allowed to occupy them with his own employees and I have seen a letter from you giving MM permission to let out two of the properties. He maintains this permission is not rescinded by the termination of your joint farming arrangements and certainly your letter does not contain any conditions. The whole question of what letting arrangements are permitted is fundamental to the rent arbitration and I do need more information from you.

On November 17 1987 the plaintiff, while in Australia, wrote a letter to Mr Wild saying that he had found the original file and would be sending it to Mr Wild as soon as he could. The letter contained a postscript in which the plaintiff said:

> See my letter dated 3/84 — 20 Valley Farm cottages, King's Lane, and 24 Henham Green. Two cottages only agreed to be rented out.

K The judge commented here:

> It is quite clear that the plaintiff looked into the file and read the letter of March 1984 in which he agreed that the defendant could sublet 20 Valley Farm and 24 Henham Green.

On January 25 1988 Mr Wild wrote to the plaintiff saying that he was able to reach agreement with the defendant on the terms previously discussed. He estimated that the new building would cost approximately £65,000. He said that he had not reached final agreement on the rent but believed that the defendant would not agree to pay more than £22,000 pa. He asked for further instructions.

Meanwhile the arbitration was stayed pending the outcome of the negotiations, and the plaintiff instructed Mr Wild to defer it further L until he came to England in May 1988, which was done.

In May 1988 the plaintiff consulted Mr Chadd, a partner in Leathes Prior, solicitors of Norwich. In a memorandum to Mr Chadd dated May 4 1988 the plaintiff said:

> Please check the lease and see if there is any reason why Mitchell should let out six of his eight cottages/houses to people outside farm labourers. Each cottage needs about £20,000 minimum spending on it and I want them back. Please note I did give him permission to do what he liked re letting them out whilst he was manager for THKPL, but not afterwards ie August 1986. Even if we made him keep them empty it would save repairs etc. And cost him rates.

On this Aldous J commented that the plaintiff

> . . . did not record the fact that permission had been given in writing in 1984 and 1985 and this is particularly surprising as on 17th November 1987 the M plaintiff had sent to Mr Wild the letter containing the 1984 permission to sublet and draw his attention to it.

The plaintiff in his evidence said that at this stage he had forgotten the two letters of 1984 and 1985.

Following his instructions from the plaintiff, Mr Chadd sent instructions to counsel to advise. In those instructions he said:

> The tenant is at present in breach of the terms of the lease in that he has sublet a number of cottages on the farms to tenants who are not agricultural workmen. From them he is receiving a market rent but the landlord is still responsible for the repairs. From October 1984 to May 1986 the tenant acted as the landlord's manager in respect of the land in hand on the estate. During that time the landlord orally indicated to him that he could do as he wished so far as the cottages were concerned, so long as he remained manager of the

estate. Counsel is asked to consider whether this oral waiver of the covenant contained in clause 4.13 of the lease would provide the tenant with grounds for resisting successfully a notice to remedy the breach.

The assertion in these instructions that the tenant was in breach of the terms of his lease was, of course, incorrect, but it was based upon the instructions which the plaintiff had given to Mr Chadd in his memorandum of May 4 1988. Mr Chadd appears not to have obtained any more detailed instructions from the plaintiff before sending his instructions to counsel. It is noticeable, however, that counsel was asked to advise about the possibility of serving a notice to remedy the breach, not a notice to quit.

Counsel duly advised Mr Chadd in a telephone conference on May 20 1988 and the advice was conveyed to the plaintiff by Mr Chadd in a letter dated May 23 1988. I think it necessary to set out the relevant part of this letter in full, as did Aldous J in his judgment. This read as follows:

The Cottages

6.1 Under the terms of the lease Mr Mitchell is not allowed to assign or part with possession of any part of the farm (except cottages and gardens to workmen on the farm). If he is renting the cottages out to tenants who are not workmen on the farm then he is in breach of the terms of the lease. The prohibition in the lease is absolute and is not subject to his obtaining your approval.

6.2 A difficulty may arise, however, in the event of your having given him permission to sub-let them. I understand that such permission was given orally to him and was limited to the period during which he was managing the estate on your behalf under the contractorship arrangement. If he acted upon your consent to non-workmen occupying the cottages and allowed them into occupation he may now find it difficult to remove them. He may, therefore, plead estoppel ie that you cannot now say that the tenants should not be there. In addition, we may have some evidential difficulties in establishing exactly what the ambit of the permission given to him was in the absence of anything in writing. He, of course, has the same difficulty but the point is that it gives scope for argument which means delay and, inevitably, expense.

6.3 The first step would be to establish exactly what the position is in respect of the tenancies of the various cottages which he has let to non-agricultural workmen. If he is offering to hand them back to you as part of the settlement in respect of the rent review currently under arbitration then it suggests that he is able to recover possession of them at any time he wants. If this is so, then his breach is capable of remedy. This is important because the procedure which would need to be adopted if you wish action to be taken on this item would be for notices to be served on Mr Mitchell and those notices must either specify (under Case D of Schedule 3 to the 1986 Act) that the breach be remedied or under Case E where the breach cannot be remedied. If we are in any doubt about the exact nature of the tenancies of the cottages we should serve notices under both cases in respect of all of them.

6.4 The tenant in any event has a right to serve counter-notice requiring arbitration and it would be necessary for you to establish damage and prejudice from the breaches which have occurred. It would be possible to establish damage to your reversionary interest in those cottages if the tenants cannot be removed and have obtained statutory or protected tenancies.

Aldous J said:

I draw attention to the fact that the plaintiff was advised that the first step was to find out what the position was in respect of the tenancies and if the defendant was able to recover possession his breach was capable of remedy and only if there was doubt should a Case E notice be served. He was also advised that it would be necessary to establish damage and prejudice from the breaches.

In a note written by the plaintiff on May 25 1988 on Mr Chadd's letter of May 23 he instructed Mr Chadd to get the cottages back. On May 31 1988 the plaintiff spoke to Mr Chadd on the telephone and gave him details of who was living in the cottages. Mr Chadd recorded this information in a note, which he used in order to prepare the notices to remedy and to quit. The note reads:

Mitchell's tenancies: No 2 Uggeshall. Mr Hartshorn genuine agric. 23 Green Farm Cottage, empty. 24 Green Farm Green, Sotherton Mr Willard non-agric. 20 King's Lane, Sotherton Mr Briggs ret. agric. OK. 21 King's Lane, Mr Warren *out*. Old Valley Farm Cottages, Nos 2 and 3 (*both empty*, let for holidays).

The notices to remedy the alleged breach and to quit were then served on the defendant by Leathes Prior on June 2 and 3 1988 respectively.

The judge found that the plaintiff did not see the notice to quit before it was served but that it did go to him shortly thereafter. He said:

On the plaintiff's evidence it seems clear that he read it soon after it was served and must be taken to have affirmed its contents, as it was signed on his behalf. He took no action to draw any mistake in it to anybody's attention.

Aldous J in his judgment then said:

One of the issues I have to decide is whether or not the plaintiff had by May 1988 forgotten that he had given written permission for the defendant to sublet cottages. If he had not, then the information given by him to Mr Chadd upon which counsel's advice was sought was so incomplete as to be false and known to be false and the whole basis of the notice to quit was also false in that 20 Valley Farm Cottage was sublet pursuant to the 1984 permission, as was 24 Henham Green, and all the cottages were covered by the 1985 permission to let for holiday situations.

The written permissions to sublet were made in 1984 and 1985. The plaintiff remembered that something had been put into writing on September 11 1986 and again on January 21 1987, and saw the 1984 letter of permission on November 17 1987 when he sent it to Mr Wild. It is therefore very surprising that by May 1988 he had forgotten. Despite this the plaintiff was clear in his evidence that he had forgotten. His excuse was that he had a bad memory. Mr Wood submitted that I should not accept this evidence, on the basis that the plaintiff should not be believed.

I do not believe that the plaintiff deliberately sought to mislead me by giving false evidence. He said that he had a bad memory and this I accept. He also struck me as a person who did not bother himself with details which did not coincide with that which he considered important. At the time when the question of the notices was being considered the defendant was irritating the plaintiff by not paying his rent. Also the plaintiff was irritated by the fact that Valley and Green Farms were (as he put it) a weak link in the chain for Heritage relief. At that time the negotiations had reached the stage when the defendant was prepared to return certain buildings, and all but two of the cottages, but the plaintiff would have to pay for a new building and for repairs to the two cottages left in the tenancy. He saw the opportunity based on the advice of Mr Chadd to exert pressure upon the defendant by way of the notices to remedy and quit. His motive was to use any means that occurred to him to get what he wanted, namely the cottages and the buildings, without paying anything.

The judge expressed his overall conclusion on this part of the case in the following words:

As I have said, I do not believe that the plaintiff deliberately set out to mislead me by giving false evidence. However, I cannot accept his evidence as to what happened and what was in his mind as his recollection cannot be relied on, particularly where it did not accord with the documents and was contrary to reasonable explanation. Clearly Mr Chadd was under the impression that the permission that the plaintiff had given to sublet was oral and I see no reason why the plaintiff would have deliberately given him false information. The conclusion must be that either the plaintiff did not remember that written permission had been granted or that the information he gave Mr Chadd was imprecise, which led Mr Chadd to conclude that the permission had been given orally. I do not believe that Mr Chadd's recollection can be relied on as to what was said unless recorded in a document, as I came to the firm conclusion from his evidence that his recollection of what happened was a reconstruction of what he believed happened on the basis of documents and what he did.

I suspect that what did happen was that the plaintiff gave to Mr Chadd information similar to that recorded in his note of May 4 1988, namely that he had given the defendant permission to do what he liked about letting the cottages out while he was manager but not afterwards. From that Mr Chadd assumed it was an oral permission. However, I believe that the plaintiff's mind was fixed on his objective of getting the cottages back by pressure and that if he had paused to think about it he would have remembered that something had been put into writing. He never bothered to think; he impulsively went ahead with an intention to achieve his aims. He was indifferent as to whether his instructions on this point were properly understood and whether written permission had been given. He was sent the instructions to counsel and did not bother to read them carefully enough or think about what was said to form a view that they were correct.

Mr Pryor submits that the judge's conclusion in the last paragraph I have quoted was inconsistent with his finding that the plaintiff did not deliberately give false evidence to him. Moreover, submits Mr Pryor, even if the plaintiff had given thought to the matter and had remembered putting some permission in writing, it does not follow that he would have remembered the exact extent of the permissions he had given to the defendant.

I am not persuaded by these arguments. The conclusion to which the judge came that in May 1988 the plaintiff's mind was so fixed on his objective of getting the cottages back by pressure that he was oblivious to any other consideration, and thus did not pause to consider whether he had given consent in writing to the subletting, is not inconsistent with a finding that at the time of the trial in July 1989 the plaintiff was not lying when he said in evidence that in May 1988 he had not remembered the two letters of 1984 and 1985. Moreover, if he had remembered the letters, it would have been inevitable that he would have told Mr Chadd of their existence. Mr Chadd in turn would then have required to have copies of the letters, and once he saw them he would have known the true position. It is inconceivable that Mr Chadd would have served the notice to quit in the form in which he did serve it if he had seen the two letters. Accordingly, I am of the view that the judge was perfectly entitled on the evidence before him, which I have sought to summarise, to reach the conclusion to which he came in the passage I have set out.

The judge then turned to consider the statements made in the notices to quit. The first two properties referred to in the notice are the two cottages which are properly known as 20 and 21 King's Lane, Henham. No 21 was the cottage occupied by Mr Briggs, the agricultural worker. It is accepted that there was no breach of the tenancy agreement in the defendant's subletting this cottage to Mr Briggs. The judge recorded the plaintiff as giving evidence that:

> . . . he never authorised Mr Chadd to serve a notice in relation to the cottage occupied by Mr Briggs . . . . The only conclusion that I can draw is that the plaintiff had no honest belief that the defendant was in breach of the tenancy in letting the cottage to Mr Briggs. In fact he believed to the contrary. Despite that, he took no step to correct the position when he saw the notice. He was not interested in seeing that the notice was correct but only in achieving his aims. I hold that the statement that the defendant was in breach of his lease by subletting 21 Valley Farm Cottages was false, was known to the plaintiff to be false and was approved by him. It therefore was a fraudulent statement.

This is the sole point on which I disagree with the learned judge. It is clear from Mr Chadd's note of his conversation on May 31 1988 with the plaintiff that he was instructed that one of the King's Lane cottages was occupied by Mr Briggs who was a retired agricultural worker. I apprehend that the expression "OK" was intended to mean that the plaintiff had in effect said to Mr Chadd that there was no breach in this case. Mr Chadd seems, when he came to have read his own note, to have misunderstood the position. It is correct, as the judge said, that the plaintiff never made any attempt to correct this misstatement when he saw the notice to quit, but that cannot make him guilty of fraud in my view. It is the plaintiff's state of mind at the time he gave the instructions to Mr Chadd, ie on and before May 31 1988, which is vital. If he gave correct instructions then, his failure to correct a mistake in the notice thereafter does not amount to a deliberate intention to mislead. On the other hand, it is a very relevant piece of evidence as to the plaintiff's general state of mind and amply justifies the judge's conclusion in relation both to this and to the other properties that the plaintiff had shut his mind to the true position and was indifferent, ie reckless, as to whether he had given consent to the defendant to sublet.

As to 20 Valley Farm Cottage, this was let pursuant to the 1984 written permission to a Mr Warren under a protected shorthold tenancy. The judge said in relation to this:

> I accept that the plaintiff, if he had bothered to remember that written permission had been given, would not have remembered the details of the permission and would have needed to consult the document to find out that 20 Valley Farm Cottage was within the permission. However, I believe that the statement in the notice relating to 20 Valley Farm Cottage was based on the understanding of Mr Chadd that the permission was oral and that the plaintiff knew as he read the instructions. The plaintiff did not have any honest belief that it was true in the sense that he did not care whether it was true or false. The statement in the notice was based on a fraudulent statement. It would probably not have been made except for that statement. In such circumstances it seems to me that the whole basis for the statement was fraudulent.

No 24 Henham Green was the other cottage covered by the 1984 permission. In May 1985 this was let on a protected shorthold tenancy to a Mr Willard. The judge reached exactly the same conclusion in relation to this cottage as he had in relation to 20 Valley Farm Cottage. In my view, on the evidence before him, Aldous J was perfectly entitled to reach the decision that the statements in the notice to quit about these two cottages were false and that their falsity was the result of the plaintiff's recklessness which amounted to fraud.

The other two cottages referred to in the notice to quit were nos 1 and 2 Old Valley Farmhouse. They were both covered by the 1985 permission to let for holiday purposes. The evidence establishes that no 1 had been let on a protected shorthold tenancy to a Mr Marsham, which was on the face of it a breach of the tenancy agreement. However, this tenancy had terminated before the end of May 1988, at which date the cottage was vacant. Thus there was no breach that could properly be the subject of a notice to quit. No 2 had been let for holiday purposes only, in accordance with the 1985 permission, and thus in this case there was no breach. In my view, the judge might well have reached the same conclusion with regard to these two cottages as he did in relation to the cottages covered by the 1984 permission.

However, he dealt with the Old Valley Farmhouse cottages in a different way. He quoted the plaintiff as saying in his evidence that he did not believe that he could suffer material prejudice provided the cottages were vacant. The judge held that it was clear that the plaintiff had no belief that these cottages were let, nor that it was right to serve a notice to quit in respect of cottages that were vacant. He found that the plaintiff was fraudulent in relation to these two cottages because

> He just did not bother to consider whether the statements made in the notice to quit in relation to those cottages were true or false.

He therefore held that the notice to quit in respect of these two cottages was fraudulent in this respect. In my view, on the evidence, this was a perfectly proper conclusion and one that cannot in any way be faulted.

Conclusion

The statement in the notice to quit that the defendant was in breach of his tenancy agreement in subletting the five cottages referred to was false in relation to each of the cottages. In relation to four of the cottages, the inclusion in the notice to quit of the false statements was the result of the instructions given by the plaintiff to Mr Chadd being inaccurate. The judge's conclusion that the inaccuracy in the instructions was the result of the plaintiff's reckless fraud was one which was amply justified by the evidence before him. For these reasons I would dismiss the appeal.

Court of Appeal

June 14 1990

(Before Lord Justice PURCHAS, Lord Justice NOURSE and Lord Justice GLIDEWELL)

MANN v GARDNER AND ANOTHER

Estates Gazette March 23 1991

[1991] 11 EG 107

Agricultural Holdings Act 1986, section 12 and Sched 2, para 4 — Frequency of rent arbitrations — Three-year interval and disregards thereof — Disregard of any reduction of rent agreed in consequence of a change in the fixed equipment — Meaning of fixed equipment — Whether a farmhouse constituted fixed equipment for this purpose — Appeal from decision of county court judge who held that it did — Decision reversed — Surrender of farmhouse not a change in the fixed equipment but a change in the holding itself — Para 4(2)(*b*) of Sched 2 not applicable

The question in this case was whether a demand for arbitration by the tenants in respect of the rent of the holding was effective or not — The rent of the holding had been fixed by agreement in 1983 at £21,650 pa from October 11 1983 — By an agreement in 1985 it was agreed that the tenants would surrender the cottage known as Brick Kiln Farmhouse (for which they had no further use) and the rent of the holding would be decreased by £100 from £21,650 to £21,550 pa — Notices demanding arbitration were served by both landlord and tenants, on October 2 and October 8 1986 respectively, but it was the tenants' notice which was material for the purpose of the present proceedings — The landlord in fact, after an arbitrator had been appointed, objected to the arbitration proceeding, on the ground that the arbitrator had no jurisdiction (an objection inconsistent with the landlord's own notice and statement of case) — The arbitrator, not surprisingly, stated a special case — The county court judge decided in favour of the tenants and remitted the matter to the arbitrator for him to proceed — The landlord appealed

It was common ground that for the purpose of section 12 of the 1986 Act "the next termination date following the date of the demand for arbitration" was October 11 1987 — The dispute was as to what, with reference to that date and the three-year bar, was the date from which there took effect a previous increase or reduction of rent — The landlord contended that the date from which there took effect a

previous reduction was the date of the 1985 agreement to surrender the farmhouse, April 29 1985, which resulted in a reduction of £100 pa in the rent — This date was less than three years before October 11 1987, so that, as a result of para 4(1)(*b*) of Sched 2, the demand for arbitration was ineffective — The tenants, on the other hand, contended that the date from which there took effect a previous increase of rent was October 11 1983, which was more than three years before October 11 1987, so that the demand for arbitration was effective — The tenants' reply to the landlord's contention was that April 29 1985 could not be used as a point of reference for the application of the three-year bar, as para 4 (2)(*b*) provided that a reduction of rent agreed in consequence of a change in the fixed equipment should be disregarded for that purpose — The landlord's answer, which was accepted by the Court of Appeal as correct and concluded the matter, was that the disregard provision in para 4(2)(*b*) did not apply because a surrender of part of a holding, whether or not it includes, or even consists of, a building or other fixed equipment, is not a "change in the fixed equipment" but a change in the holding itself — It is no different from a surrender of bare land forming part of the holding — The result therefore was, as the landlord had submitted, that the date from which there took effect a previous reduction of rent was April 29 1985, less than three years before October 11 1987 — Consequently the tenants' demand for arbitration was ineffective — Appeal allowed

No cases are referred to in this report.

This was an appeal by Sir Rupert Mann Bt, the landlord of an agricultural holding known as Common Farm, Billingford, Norfolk, from a decision of Judge Hyam, in Norwich County Court, in favour of the tenants, Edgar Stanley Gardner and John Edgar Gardner, on a case stated by an arbitrator under the Agricultural Holdings Act 1986. The question concerned the effectiveness of a demand by the tenants for arbitration in regard to the rent of the holding.

Paul Morgan (instructed by Burges Salmon, of Bristol) appeared on behalf of the landlord; Robert Bailey-King (instructed by Sharpe Pritchard, agents for Howes Percival, of Norwich) represented the respondent tenants.

Giving the first judgment at the invitation of Purchas LJ, **Nourse LJ** said: This is an appeal against a decision of His Honour Judge Hyam given in the Norwich County Court on a special case stated by an arbitrator under the Agricultural Holdings Act 1986. Shortly stated, the question for decision is whether a demand that the amount of an agricultural rent should be referred to arbitration is effective or not.

The landlord is Sir Rupert Mann. The subject-matter of the tenancy is an agricultural holding known as Common Farm, Billingford in Norfolk. The tenancy derives from a tenancy agreement dated September 28 1970 and made between the landlord's predecessor in title, Sir Edward John Mann, of the one part and Edgar Stanley Gardner of the other part. The term of the tenancy was limited to begin on October 11 1970 and to continue from year to year until determined by either party giving to the other not less than 12 months' previous notice in writing expiring at the end of the second or any subsequent year. By a memorandum dated October 7 1983 and made between the landlord of the one part and Edgar Stanley Gardner and John Edgar Gardner ("the tenants") of the other part it was agreed, first, that the tenants would thereafter be the joint tenants of the holding and, second, that the rent would be £21,650 pa with effect from October 11 1983.

By a memorandum dated April 29 1985 and made between the landlord of the one part and the tenants of the other part the tenants agreed to surrender and deliver up possession as from May 1 1985 of "the cottage known as Brick Kiln Farmhouse, Billingford" and the landlord agreed to accept such surrender. It was also agreed that the landlord should not claim dilapidations against the tenants on quitting the cottage and that the tenants should not seek compensation from the landlord for any matters arising on their quitting the same. Further (and this is important), it was agreed that the rent of the holding should be decreased by the sum of £100 from £21,650 pa to £21,550. Brick Kiln Farmhouse had previously been occupied by an employee of the tenants who had moved elsewhere, so that the tenants no longer had a use for it.

Pursuant to section 12(1) of the 1986 Act, notices demanding arbitration dated October 2 and 8 1986 on behalf of the tenants and the landlord respectively were duly served. The latter notice did not add anything to the former and, save for one purpose which I will mention, it can be ignored. At that time the total area of the holding was 434.875 acres. The narrative can be taken up at para 1(5) of the special case, which is part of the agreed statement of facts and reads as follows:

> With effect from 11th October 1987, two additional parcels of unreclaimed land were included in the holding and the rent was varied. The parcels were: —
> (a) 3.738 Acres approximately, outlined red on the attached plan (No. 2).
> (b) 1.081 Acres approximately, outlined blue on the attached plan (No. 3).
> The terms of the tenancy as revised were reduced into writing in a memorandum undated and not signed by the Tenants. This, however, contained a mathematical error. The existing rent prior to the variation was incorrectly stated to be £21,650 and not £21,550.
> The rent increase wef 11 10 87 was from £21,550 to £21,640. The payments actually made were adjusted and revised accordingly. Subject to that amendment the memorandum in this respect accurately reflects the matters agreed between the parties.

That undated memorandum of October 1987 was signed by the landlord but not by the tenants. It was expressed as an agreement to take effect from October 11 1987. By clauses 1 to 3 it was agreed that "the landlord shall add to the Holding" the two parcels of land and that "the rent for the Holding" should be increased in the first case by an initial sum of £37 subject to annual increases thereafter until October 10 1992 and in the second case by the sum of £53. Thus the initial annual increase was £90. Clause 4 of the memorandum, amended as mentioned in the special case, said:

> The rent for the Holding shall therefore be increased by the sum of £90.00 from £21,550 to £21,640.

By clause 6 it was provided that except as varied the terms and conditions of the original tenancy agreement and supplementary memoranda should remain in full force and effect.

On November 26 1987, pursuant to the tenants' notice demanding arbitration, Mr J H Harris FRICS FAAV was appointed as arbitrator and statements of case were subsequently put in on both sides. The tenants' contention was that the rent properly payable from October 11 1987 ought to be less than the existing rent. The landlord contended that it ought to be more than the existing rent. In due course a hearing was fixed for November 14 1988, but about two weeks beforehand solicitors acting on behalf of the landlord, relying on the 1986 Act, Schedule 2, para 4(1)(*b*), objected to the arbitration's proceeding. Although that objection was inconsistent both with the landlord's own notice demanding arbitration and with his statement of case, it has not been suggested, at any rate in this court, that it is not open to him to take it. This change of tack suggests an apprehension on the landlord's part that the arbitrator might reduce the rent below its existing level.

The practical result of the objection was the stating of the special case, which came before Judge Hyam on November 22 1989, when the questions raised were answered in favour of the tenants and the matter remitted to the arbitrator for him to proceed accordingly. The landlord has now appealed to this court.

The provisions for arbitration of rent supplementary to section 12 of the 1986 Act are contained in Schedule 2. Paras 4, 5 and 6 of that Schedule are headed "Frequency of arbitrations under section 12". So far as material, para 4 is in these terms:

> (1) Subject to the following provisions of this Schedule, a demand for arbitration shall not be effective for the purposes of section 12 of this Act if the next termination date following the date of the demand falls earlier than the end of three years from any of the following dates, that is to say — . . .
> (*b*) the date as from which there took effect a previous increase or reduction of rent (whether made under that section or otherwise), . . .
> (2) The following shall be disregarded for the purposes of subpara (1)(*b*) above — . . .
> (*b*) an increase of rent under subsection (1) of section 13 of this Act or such an increase as is referred to in subsection (3) of that section, or any reduction of rent agreed between the landlord and the tenant of the holding in consequence of any change in the fixed equipment provided on the holding by the landlord; . . .

In regard to that paragraph, it is agreed that the next termination date following the date of the demand — see section 12(4) — was October 10 or perhaps 11 1987; it matters not which. If, for the

A purposes of subpara (1)(*b*) of para 4, the correct view is that the date as from which there took effect a previous increase of rent was October 11 1983, then that was more than three years before the relevant termination date and para 4(1) does not render the demand for arbitration ineffective. That indeed is what the tenants, basing themselves on the memorandum of October 7 1983, have at all times contended to be the case. But the landlord, basing himself on the memorandum of April 29 1985, contends that the date as from which there took effect a previous reduction of rent was May 1 1985, so that the demand for arbitration is rendered ineffective by para 4(1)(*b*). Alternatively, he contends that the demand was, in one way or another, rendered ineffective by the oral agreement which is recorded B in the undated memorandum of October 1987.

As to the first of these contentions, Mr Bailey-King for the tenants relies on that part of para 4(2)(*b*) which requires that, for the purposes of subpara (1)(*b*), there shall be disregarded any reduction of rent agreed "in consequence of any change in the fixed equipment provided on the holding by the landlord". The expression "fixed equipment" is defined by section 96(1) of the Act to include:

> . . . any building or structure affixed to land and any works on, in, over or under land, and also includes anything grown on land for a purpose other than use after severance from the land, consumption of the thing grown or of its produce, or amenity, and any reference to fixed equipment on land shall be construed accordingly;

In the light of that definition it is clear to me that Brick Kiln C Farmhouse, being a building or structure affixed to land, was part of the fixed equipment provided on the holding by the landlord. On that footing Mr Bailey-King submits that the agreement of April 29 1985, by surrendering the farmhouse back to the landlord, effected a change in that equipment within para 4(2)(*b*). Immediately before the agreement it was part of that equipment. Immediately afterwards it was not. There was therefore a change in the fixed equipment. He also submits that the £100 reduction of rent was agreed "in consequence" of that change. And he adds that the application of para 4(2)(*b*) is not disturbed by the fact that the effect of the agreement was, as the learned judge correctly held, to surrender not only the farmhouse but also the land within its curtilage, which is thought to be of an area of about 0.603 acres. It was these D submissions which appealed to Judge Hyam when he decided the first question in favour of the tenants.

The contrary argument of Mr Morgan in regard to para 4(2)(*b*) is to this effect. He submits, first, that the farmhouse and the land within its curtilage did not fall within the definition of fixed equipment and, second, that even if it did and there was a change in the fixed equipment within para 4(2)(*b*), the reduction of rent was not made in consequence of it. I have already made it clear enough that I would reject the first of these submissions and, on a view of the agreement of April 29 1985 as a whole, I would also reject the second.

Mr Morgan's third submission is more impressive. He says that a surrender of part of the holding, albeit of part which includes or even E consists only of a building or structure affixed to the land, is not a change in the fixed equipment provided on the holding by the landlord. It is a change in the holding itself and, as such, is no more within para 4(2)(*b*) than is a surrender of land to which no building or structure is affixed.

Although my mind has wavered on this question, I have come to a clear, if somewhat reluctant, conclusion that Mr Morgan's third submission is correct. Two further considerations which he advances support it. First, the definition of fixed equipment distinguishes between the building or structure on the one hand and the land to which it is affixed on the other, a distinction which appears also in sections 7 to 11 of the Act (which deal specifically with fixed equipment), thus suggesting that the changes contemplated by para F 4(2)(*b*) are changes to the physique of the building or structure itself. Second, the first part of subpara (2)(*b*) requires there to be disregarded any increase of rent under section 13(1) of the Act or such an increase as is referred to in section 13(3). Broadly stated, those are increases which result from improvements carried out on the holding by the landlord, an exercise which manifestly does not involve the addition of any land to the holding. That supports the view that the changes in fixed equipment contemplated by para 4(2)(*b*) do not include those which involve the subtraction of any land from the holding.

My reluctance to arrive at this conclusion arises from a belief that it is unsatisfactory that an agreement for a rent of £21,650 to be reduced by no more than £100 should have the result of putting back for up to three years the right of either party to demand an arbitration of rent G for the remainder of the holding. In reality, there was no agreement on April 29 1985 that a rent of £21,550 was the rent properly payable for the remainder of the holding as at May 1. That figure was simply the result of an agreement that the amount deductible in consequence of the surrender should be £100. Moreover, if the surrender had been made without any reduction of rent, the date as from which there took effect a previous increase of rent would have remained October 11 1983 and the demand for arbitration would have remained effective. Be that as it may, Mr Bailey-King has not argued that the £100 was *de minimis* and I have no reason to think that such an argument, had it been put, would have succeeded. In the end, therefore, my reluctance cannot be translated into any decision contrary to that which I have indicated. H

That makes it necessary to consider an alternative argument on the first question which, although rejected by the judge, has been revived by Mr Bailey-King under a respondent's notice put in out of time with the leave of the court. This argument is founded on para 6 of Schedule 2 to the Act, which is in these terms:

> Where under an agreement between the landlord and the tenant of the holding (not being an agreement expressed to take effect as a new contract of tenancy between the parties) provision is made for adjustment of the boundaries of the holding or for any other variation of the terms of the tenancy, exclusive of those relating to rent, then, unless the agreement otherwise provides —
>
> (*a*) that provision shall for the purposes of sub-paragraph (1) of paragraph 4 above be treated as not operating to terminate the tenancy, and accordingly as not resulting in the commencement of a new contract of J tenancy between the parties, and
>
> (*b*) any increase or reduction of rent solely attributable to any such adjustment or variation as aforesaid shall be disregarded for the purposes of paragraph (*b*) of that sub-paragraph.

Mr Bailey-King submits that the agreement of April 29 1985 was one by which provision was made for adjustment of the boundaries of the holding within that paragraph. I am entirely satisfied that it was not. Although there is neither definition nor authority to guide us as to what is or is not an adjustment of the boundaries of the holding, some assistance is gained from section 31(1)(*a*) of the Act, which enables a landlord to give a notice to quit part of the holding if it is given "for the purpose of adjusting the boundaries between agricultural units or amalgamating agricultural units or parts of such K units". Mr Morgan correctly submits that the landlord would not have been able to serve the tenant with a notice to quit Brick Kiln Farmhouse under that provision. That does not quite answer the question under para 6, because the conclusive objection under section 31(1)(*a*) would have been that the notice was not given "for the purpose" of adjusting boundaries. It would have been given for the purpose of taking back a cottage for which the tenants no longer had a use. However, section 31(1)(*a*) does confirm the expectation which one might anyway have entertained, namely that a surrender of part of the holding to the landlord which is not an arrangement, putting in order, harmonisation or adaptation (those are the dictionary meanings of "adjustment") of the boundaries between L the holding and some other property cannot fall within para 6. It is impossible to describe the surrender which took place here as having been any of those things.

On the facts of this case the point is put beyond argument by the location of Brick Kiln Farmhouse not on the boundary of the holding but at its centre. Even if it had been on the boundary, my conclusion would have been the same. If Mr Bailey-King's argument were correct, it would defy not only the natural and ordinary meaning of the expression which has to be applied; it would enable para 6 to be used as a wholesale means of escaping from the effect of para 4 as I have construed it.

For these reasons, I think that Judge Hyam was right to reject the tenants' alternative argument on the first question. But since I M respectfully disagree with his view that the tenants can rely on para 4(2)(*b*), I would allow the landlord's appeal on that question. I should add that it seems very doubtful whether Mr Morgan's third and decisive submission was put to the learned judge or, if it was, whether it was developed to the extent in which it has been developed by Mr Morgan in his excellent argument in this court.

I would therefore answer the three questions raised in para 4(a) of the special case as follows:

The reduction of rent of £100 per annum as from 1st May 1985 arising from the tenants' surrender of the Brick Kiln Farmhouse —

(i) is a reduction within the terms of para 4(1)(*b*) of Schedule 2 to the Act;

A (ii) is not a reduction of rent within the terms of para 4(2)(*b*) of Schedule 2 to the Act;

(iii) is not a reduction of rent attributable to an adjustment of the boundaries of the holding under the terms of para 6 of Schedule 2 to the Act.

The demands for arbitration having been ineffective, we come to the questions in para 4(b), the first of which is whether the arbitrator has jurisdiction to determine the rent properly payable in accordance with his appointment. That question must be answered in the negative and the second and third questions are now dead in any event.

The questions raised in para 4(c), which go to the second question
B argued on this appeal, are expressed to arise only if the demands for arbitration were initially effective, which, as I have now held, they were not. Having heard full argument on the second question, we could certainly express a view on it, if we thought it appropriate to do so. For my part, I do not think that it is. Although the question has given rise to some interesting arguments on both sides, its resolution depends mainly on the effect in law of the particular oral agreement recorded in the undated memorandum of October 1987 and only very little, if at all, on the construction of the 1986 Act. I have not been able to see that an answer to it would be of any significant additional value in the present case or of any general value outside it.

I would allow this appeal and answer the material questions in the special case in the senses which I have indicated.

C

Agreeing, **Glidewell LJ** said: In relation to the facts found by the arbitrator in this case stated, the effect of para 4(1)(*b*) of Schedule 2, to the Agricultural Holdings Act 1986, is that unless the reduction of rent agreed on April 29 1985 between the landlord and the tenants in regard to the surrender by the tenants on May 1 1985 of Brick Kiln Farmhouse fell within one of the exceptional categories in para 4(2) of that Schedule, then the demands for arbitration made in October 1986 were ineffective, being made less than three years before the next termination date following that reduction, that is to say October 10 or 11 1987.

The argument of Mr Morgan on behalf of the landlord is to the effect that that decrease in rent did not come within para 4(2)(*b*) of
D Schedule 2. As Nourse LJ has said, the question is thus: was it a reduction of rent agreed between the landlord and the tenants "in consequence of any change in the fixed equipment provided on the holding by the landlord"? "Fixed equipment" as defined in section 96(1) of the Act can, I agree with Nourse LJ, clearly include a house on the holding. But, to my mind, putting it quite shortly, the effect of the agreement of April 29 1985 was that there was no change in that piece of fixed equipment, that is to say the house, on May 1 1985. The house was unaltered. It ceased to be part of the holding because of the surrender. In my view, the phrase "change in the fixed equipment" means some alteration of the equipment itself, which may include its total removal or demolition, but a change which leaves the holding
E otherwise unaltered.

I agree, too, with the point, to which Nourse LJ has referred, that the first part of subpara (2)(*b*) relates to an increase of rent under subsections (1) or (3) of section 13 of the Act, that is to say an increase resulting from an improvement carried out by the landlord. Such an improvement will include the addition of fixed equipment but not an addition to the holding. It seems appropriate, therefore, to construe this subpara as dealing with additions to and subtractions from the fixed equipment, or alterations to the fixed equipment which have the effect of subtracting from it, with consequent increases and decreases in the rent which leave the holding itself otherwise unaltered.

F For those reasons, which really do no more than express in less felicitous language what Nourse LJ has already said, I agree with him that this appeal should be allowed.

On the other points argued before us, namely the point argued by Mr Bailey-King that the reduction of rent stemmed from an adjustment of the boundaries of the holding under para 6 of Schedule 2, I agree with and add nothing to what Nourse LJ has said. Like him, I do not think it necessary to answer the question arising out of the oral agreement of October 1987. I, too, would therefore allow the appeal and give the same answers to the questions as those proposed by Nourse LJ.

Also agreeing, **Purchas LJ** said: The only questions which it is necessary to answer for the disposal of this appeal are questions 4(a)(i) and (ii) and 4(b)(i). The judge came to the conclusion that the G
reduction in rent of £100 fell within the provisions of para 4(2)(*b*) of Schedule 2 to the Agricultural Holdings Act 1986 and thus that para 4(1)(*b*) did not apply to render the demand for arbitration effective.

The reduction in rent involved arose out of an agreement of April 29 1985 under which the tenant "agreed to surrender and deliver up possession of the cottage known as Brick Kiln Farmhouse in all respects as though he had served on the landlord full and sufficient notice to quit expiring on that date". Para 4(2)(*b*) of the Schedule provides that:

> an increase of rent under subsection (1) of section 13 of this Act or such an increase as is referred to in subsection (3) of that section, or any reduction of rent agreed between the landlord and the tenant of the holding in consequence of any change in the fixed equipment provided on the holding by the landlord; H
> . . .

shall be disregarded for the purposes of subpara (1)(*b*). The issue is, therefore: did the surrender effected by the agreement amount to a change in the fixed equipment within the meaning of that phrase in subpara (2)(*b*) of the Schedule? It is necessary to look at two other provisions of the Act.

Section 96(1) provides a definition of "fixed equipment":

> "fixed equipment" includes any building or structure affixed to land and any works on, in, over or under land, and also includes anything grown on land for a purpose other than use after severance from the land, consumption of the thing grown or of its produce, or amenity, and any reference to fixed equipment on land shall be construed accordingly; J

This definition clearly draws a distinction, Mr Morgan submitted, between the land "as such and the things fixed to it or growing on it or in its upper surfaces". The specific reference to building or structure in that definition, in my judgment, is included to counteract the effect of the law which establishes that anything fixed to the land becomes for the purposes of real property law part of the land. The distinction, therefore, is drawn by this definition for the purposes of the provisions of this Act.

Section 13 is the other provision to which I will refer. That provides:

> (1) Where the landlord of an agricultural holding has carried out on the holding any improvement to which this section applies he may by notice in writing served on the tenant within six months from the completion of the K
> improvement increase the rent of the holding as from the completion of the improvement . . .

The section then proceeds to deal in detail with the effect of that approach. It has not been argued to the contrary that that section does not apply to a change in the extent of the land comprised in the holding.

Para 4(2)(*b*) in its first part deals with increases in rent and in its second part deals with reductions in rent. Quite apart from the ordinary construction of the definition of fixed equipment in section 96(1), it would be surprising, to say the least, if in para 4(2)(*b*) the first part did not apply to changes in the extent of the holding while the second part did. In my judgment, there is no need for such a strained L
interpretation of para 4(2)(*b*). Its plain intention is directed to exclude alterations to buildings etc on the land but not to exclude changes in the extent of the land comprised in the holding. This is precisely what is effected by the surrender by the tenants in accordance with the agreement of April 29 1985 of Brick Kiln Cottage and its immediate curtilage. I do not consider that there is anything to be gained by drawing distinctions between the land immediately under the building and the land surrounding it. This was a perfectly ordinary surrender of a tenancy and, as such, did not come within the particular provisions of the Schedule with which we have been dealing in this appeal.

I have also come to the same conclusion as Nourse LJ on the issue raised in the respondent's notice, namely that para 6(*b*) of the M
Schedule does not apply to the surrender of the farm. That provision is clearly directed towards minor adjustments in boundaries without any substantial alteration in the extent of the holding involved.

For those reasons I also would answer the questions, which I mentioned earlier in this judgment, to the effect that whereas para 4(1)(*b*) of Schedule 2 does apply, para 4(2)(*b*) of Schedule 2 does not apply, and that para 6(*b*) of Schedule 2 does not apply, and I would answer the questions 4(a)(i), (ii) and (iii) in those respects.

As to the final question, 4(b)(i), I would answer that in the negative.

For the reasons given already by Nourse LJ, I do not consider it either useful or appropriate to proceed to consider the issues which were argued on the other matters raised in this appeal. I also would allow this appeal.

The appeal was allowed, with costs of appeal; leave to appeal to House of Lords was refused.

Court of Appeal

November 8 1990

(Before Lord Justice MANN and Sir David CROOM-JOHNSON)

CRAWFORD v ELLIOTT

Estates Gazette April 6 1991

[1991] 13 EG 163

Agricultural Holdings Acts 1948 and 1986 — Grant of limited right of occupation — Application of section 2 — Document purporting to be a notice to quit held to be invalid by county court judge — No counternotice served in reply — Held by Court of Appeal, reversing the decision of the judge, that the position was as follows — As a result of section 2, whether of the 1948 or 1986 Act, the occupier had acquired a protected agricultural tenancy from year to year; the disputed document was a perfectly good notice to quit; and, as no counternotice had been served by the tenant in response to it in accordance with section 26(1) of the 1986 Act, the notice became valid and effective — Appeal by landlord was allowed

The facts were that the trustees of a deceased landlord, having recovered possession of an agricultural holding following the death of a tenant, permitted the present respondent to go into occupation of the farm — The terms of the occupation were set out in an "acknowledgement" by the respondent, who stated that his occupation was limited to the lifetime of one of the trustees, that it was personal to the respondent, that it could not be assigned and that it would cease on the respondent's death — It was also conditional on the payment of £850 pa and on compliance with certain stated terms — It was not disputed at the appeal that the effect of this arrangement was, as a result of section 2 of the 1948 Act (in substance subsequently reproduced by section 2 of the 1986 Act), to create a protected agricultural tenancy from year to year — The particular trustee named in this document died in 1987 and the present appellant (the last surviving trustee) in 1988 sent the respondent a document purporting to terminate his tenancy at Candlemas (February 2) 1989 — The nature of this document was critical — Was it a valid and effective notice to quit? — No counternotice in response to it under section 26(1) of the 1986 Act had been served by the respondent — If the document was valid, then *cadit quaestio*: the respondent's tenancy had ended

The judge below had held that the document was not a valid notice to quit — He considered that it was defective because it failed to set out what really happened, namely that the tenant was a protected agricultural tenant and had a right to serve a counternotice in reply to the landlord's notice to quit but had failed to exercise that right — The Court of Appeal, however, agreed with the appellant's submission that there was no statutory requirement for the landlord to set out the tenant's right to serve a counternotice and the consequences of not doing so — The court could not imply such a requirement — As far as the common law was concerned, the landlord's notice to quit satisfied its requirements — The court's correct approach was that explained in *Carradine Properties Ltd* v *Aslam* and *Germax Securities Ltd* v *Spiegal* — The respondent's submission that the document was misleading as mentioning a licensee and failing to refer to the agricultural holdings legislation was rejected — The notice to quit was valid and effective — Appeal allowed

The following cases are referred to in this report.

Carradine Properties Ltd v *Aslam* [1976] 1 WLR 442; [1976] 1 All ER 573; (1975) 32 P&CR 12

Germax Securities Ltd v *Spiegal* (1978) 37 P&CR 204; 250 EG 449, CA

This was an appeal by Rosalie E Crawford, the surviving trustee of William Graham deceased, from the decision of Judge Forster, at Carlisle County Court, dismissing the appellant's claim for possession of a farm known as The Green, Irthington, Carlisle, Cumbria, occupied by the respondent (defendant in the action), Ronald Elliott.

Richard Gordon (instructed by Cartmell Shepherd, of Penrith) appeared on behalf of the appellant; John R Gregory (instructed by Hewitson & Harker, of Kirkby Stephen) represented the respondent.

Giving judgment, **MANN LJ** said: There is before the court an appeal against the judgment of His Honour Judge Forster given in the Carlisle County Court on December 12 1989, whereby the learned judge dismissed the plaintiff/appellant's claim for possession of a farm known as The Green, Irthington, Carlisle, Cumbria. There was also an order for the payment by the plaintiff of the defendant's costs on scale 3. The issue in the appeal can be shortly stated. It is whether a notice given by the plaintiff was effective to terminate the defendant's agricultural tenancy of the farm. I shall come to the notice later.

The background facts can be shortly stated. The farm was a freehold belonging to one William Graham. He, on December 31 1964, let the farm to one John Prudham on a tenancy from year to year which was subject to the protection of the Agricultural Holdings Act 1948 (as it then was). In 1967 William Graham died and his property was held by the three trustees of his will, one Margaret Graham, one Mary Graham and the plaintiff in this action, who is a solicitor. She brings the action as the sole survivor of those three trustees. John Prudham, the agricultural tenant, died in March 1973. Possession of the farm was recovered by the trustees under and in conformity with the Agricultural Holdings Act.

In January 1975 the trustees of William Graham permitted an "occupancy" of the farm to Ronald Elliott, who was in fact, though it matters not, married to the great niece of Mary Graham, one of the trustees. There was a permission and an acknowledgement. The permission we have not seen. The acknowledgement we have seen and it is in these terms:

> I RONALD ELLIOTT hereby acknowledge
>
> (1) that my occupancy of the farm known as The Green situate at Irthington, Cumbria, is based upon an equity arising from the events which have occurred
>
> (2) And that my right to such occupancy is limited to the lifetime of Miss Mary Eleanor Graham and is conditional upon
>
> (a) payment of £850 per annum by equal half-yearly payments on 2nd February and 2nd August in each year and
>
> (b) compliance with the terms (except as to rent and duration) of an Agreement dated 31st December 1964 made between William Graham and John James Prudham
>
> AND I FURTHER ACKNOWLEDGE that the aforesaid right is personal to me and will cease on my death and that it cannot be assigned by me.

That acknowledgement is dated January 27 1975.

Ronald Elliott entered upon the farm and doubtless farmed it and doubtless paid "the rent". He became — although perhaps he did not know it and the trustees certainly did not appreciate it until shortly before the present action was commenced — an agricultural tenant by virtue of the Act of 1948, subsequently the Agricultural Holdings Act 1986. Mary Graham, who was referred to in the acknowledgement, died on December 24 1987.

On January 15 1988 the plaintiff in the action, as surviving trustee, wrote to Mr Elliott in these terms:

> You will be aware that the death of Miss M E Graham brings to an end your occupancy of The Green. However, it was agreed that you could remain in occupation until the Candlemas following the expiration of one year from Miss Graham's death. As Miss Graham died on December 24 last, the date of vacation will be February 2 1989. I am enclosing formal Notice to Quit on that date and should be obliged if you would sign the attached copy Notice and return this to me by way of acknowledgement.

There is then a paragraph which is immaterial.

What was described as "formal Notice to Quit" is in these terms:

> I the undersigned being the surviving trustee of the Will of William Graham deceased and as such your Landlord hereby give you Notice to Quit and deliver up to me on the 2nd day of February 1989 possession of the farm

A known as The Green, Irthington, Carlisle, held by you as occupier in accordance with an acknowledgement under your hand dated 27th January 1975 and a permission of the same date under the hands of myself and the late Mary Ellinor Graham.
Dated the 15th day of January 1988.

It was signed by the plaintiff.

It is that last document which is critical to these proceedings. It is critical not only because of the position at common law, but, more important, by reason of the provisions of the Act of 1986. Section 2 of that Act, echoing, albeit in slightly different language, section 2 of the Act of 1948, provides:

(1) An agreement to which this section applies shall take effect, with the necessary modifications, as if it were an agreement for the letting of land for a
B tenancy from year to year

— and I omit certain words —

(2) Subject to subsection (3) below, this section applies to an agreement under which

(a) any land is let to a person for use as agricultural land for an interest less than a tenancy from year to year, or
(b) a person is granted a licence to occupy land for use as agricultural land,

if the circumstances are such that if his interest were a tenancy from year to year he would in respect of that land be the tenant of an agricultural holding.

Subsection (3) is immaterial.

It is no longer in dispute, whatever the date of appreciation may
C have been, that the instrument of 1975 was, by reason of section 2 and its predecessor, the subject of section 2.

Under section 26 of the Act it is provided:

(1) Where —
(a) notice to quit an agricultural holding or part of an agricultural holding is given to the tenant, and
(b) not later than one month from the giving of the notice to quit the tenant serves on the landlord a counter-notice in writing requiring that this subsection shall apply to the notice to quit

then, subject to subsection (2) below, the notice to quit shall not have effect unless, on an application by the landlord, the Tribunal consent to its operation.

Reference to "the Tribunal" is a reference to the Agricultural Land
D Tribunal.

Even a cursory glance at the section shows that, if there be no counternotice responding to the notice to quit, that notice to quit can take effect according to its terms. The notice of January 15 1988 did not provoke a counternotice. Thus if it was a notice to quit, it can take effect according to its terms.

The learned judge in the court below held that it was not a notice to quit. He said this — according to the agreed notes of his judgment:

My final view, therefore, is as follows:
1. I agree that normally the information which is required to be given by the landlord in the Notice to Quit is very bare and minimal.
2. I agree there is no authority to say that the landlord must specify the statutory rights.
E 3. I do find, however (i) that the Notice to Quit completely fails to set out what really happened here. That is that the Tenants become protected tenants and the Defendant had certain rights which subsequently he, of course, failed to exercise, and (ii) because of the ambiguity of language, I think that there was a duty on the Landlord in the peculiar circumstances of this case to make it clear what rights he was purporting to ask the Defendant to give up. This he failed to do.
Therefore, in those circumstances, I am not satisfied that the Notice to Quit was clear and valid.

Mr Gordon, on behalf of the appellant, says that a notice to quit is assailable by reference either to a rule of common law or to a statutory provision. It may be said at once that there is no statutory requirement in regard to the giving of a notice to quit concerning an
F agricultural holding. There is no statutory requirement that a tenant should be alerted as to the legislation or, in particular, as to his right to serve a notice under that legislation. It is perhaps a matter for remark that there is no legislative provision to that effect, but there is not and this court cannot imply one.

So far as the common law is concerned, Mr Gordon says that all that is requisite is that what should be specified is the land to which the notice relates, that the land is required back by the server on a specified date, and no more. If those critical components are to be found, then what one has to do is to stand back and look at the notice, if there be complaint about its language, and ask oneself the question posed by Goulding J in *Carradine Properties Ltd* v *Aslam* [1976] 1 WLR 442 at p 444G. The learned judge said:

I would put the test generally applicable as being this: "Is the notice quite clear G
to a reasonable tenant reading it? Is it plain that he cannot be misled by it?"

That is a method of approach which was adopted by this court in *Germax Securities Ltd* v *Spiegal* (1978) 37 P&CR 204 at p 206. It is an approach which I would propose to adopt in this case.

Mr Gregory, for the respondent, contends that the document is misleading. He says take the letter and the notice. They either in terms or by implication describe the recipient as a licensee, and that express or implicit description could have created in the mind of the reader that what the landlord intended was not a notice to quit. Mr Gregory points out that there was no reference in the notice or the accompanying letter to the Agricultural Holdings Act. Our attention was drawn to the precedent books, all of which prudently do refer to H
the agricultural holdings legislation. But, as I have indicated, there is no obligation imposed under that legislation that it should be referred to.

One is left, therefore, with what is required at common law. The components of that requirement I have indicated. For my part I do not think, in regard to those critical components, that there is anything in the notice or the letter which can be described as misleading, I emphasise that there is in this case no suggestion of bad faith or fraud, or anything of that sort.

For my part I would regard the notice which was served as a valid notice to quit to take effect in accordance with its terms. It took effect, in my judgment, in accordance with its terms because the opportunity under the legislation to serve a notice has now passed. I J
would allow this appeal.

Sir David Croom-Johnson agreed and did not add anything.

The appeal was allowed with costs in Court of Appeal and below. The order of court below was set aside and possession ordered three months from November 8 1990. Leave to appeal to the House of Lords was refused.

Court of Appeal K

November 20 1990

(Before Lord Justice NEILL, Lord Justice RUSSELL and Lord Justice FARQUHARSON)

BURGOYNE v GRIFFITHS AND ANOTHER

Estates Gazette April 6 1991
[1991] 13 EG 164

**Rent (Agriculture) Act 1976 — Possession of cottage — Whether licence to occupy cottage was a term of occupier's contract of employment or was an arrangement quite L
independent of the contract of employment — Whether licence to occupy came to an end with ending of employment and protected occupancy was converted into a statutory tenancy within the Rent (Agriculture) Act — Fair rent registered under that Act — County court judge's grant of possession to owner of cottage in consequence of non-payment of rent upheld by Court of Appeal**

**The appellants in this case (defendants in the county court) had been living since 1960 in a cottage on a farm where the husband had worked as a foreman or bailiff — They had occupied the cottage rent-free — The husband unfortunately began to suffer severely from rheumatoid arthritis and from M
1977 was unable to work full time — Eventually he ceased to be able to work and in 1985 his employment was terminated — The couple continued to live in the cottage rent-free until, in 1987, the owner of the farm and cottage applied to the rent officer to have a fair rent registered for the cottage under the provisions in sections 13 and 14 of the Rent (Agriculture) Act 1976 — The rent officer determined a fair rent of £27 a week, but the defendants declined to pay it, asserting that the original licence permitted them, and the survivor of them, to live in the cottage rent-free for the rest of their lives — The owner, the present respondent, then instituted proceedings**

A **for possession and other relief — The county court judge granted the owner a possession order, suspended on terms that the owner should be paid £1,500 within two months and should also be paid the current rent, together with mesne profits and arrears of rent by instalments of £10 per week — The defendants appealed**

On appeal the appellants submitted that the original arrangement relating to the cottage was in effect a licence wholly independent of the husband's contract of employment, not a term of that contract — The respondent argued that the licence to occupy was an integral part of the
B **husband's contract of employment and ceased to exist as such when that employment terminated — The case of *Ivory* v *Palmer* was cited in this connection — Although the licence ceased to exist, section 4 of the Rent (Agriculture) Act converted what had been a protected occupancy into a statutory tenancy — As a statutory tenant the husband was liable to have the rent of the cottage registered under section 13 of that Act — The Court of Appeal agreed with the respondent's submission — Under the 1976 Act, section 7, and Schedule 4, Case III, the judge was entitled to make an order for possession provided that he considered it reasonable to do so — The court rejected a suggestion that the judge had**
C **not given consideration to the question of reasonableness — It was true that he had not expressly referred to it, but the court was satisfied that he had applied his mind to it — Appeal dismissed**

The following case is referred to in this report.

Ivory v *Palmer* [1975] ICR 340, CA

This was an appeal by the defendants, Arthur John Griffiths and his wife, Margaret Jane Griffiths, from the decision of Judge Smyth, at Ludlow County Court, granting the plaintiff, Marjorie Anne Burgoyne, possession of Middleton House Cottage, in Middleton
D Farm, near Ludlow, Shropshire.

John West (instructed by Lloyds & Cooper, of Leominster) appeared on behalf of the appellants; Edward B Jones (instructed by Emrys Jones & Co, of Welshpool) represented the respondent.

Giving the first judgment at the invitation of Neill LJ, **Russell LJ** said: Middleton House Cottage is part of Middleton Farm, Middleton on the Hill near Ludlow, Shropshire. The farm is now owned by the plaintiff, Marjorie Anne Burgoyne. This is an appeal against a judgment in her favour, whereby upon terms she was granted possession of Middleton House Cottage by virtue of a judgment of His Honour Judge Smyth delivered on September 11
E 1989.

The plaintiff acquired the farm by purchase, her predecessor in title being the partnership carrying on business as W Burgoyne & Son. The partnership was between Lionel and Michael Burgoyne. In about 1960 the first defendant, Mr Arthur John Griffiths, commenced employment with the partnership as a working foreman or bailiff and together with his wife, who ultimately became the second defendant in these proceedings, he moved into the cottage. The terms of his occupancy were that he and his wife should live there rent-free during the term of his employment, the employers paying the rates.

Unhappily, Mr Griffiths, with the passage of years, became
F disabled as a result of his suffering from rheumatoid arthritis. In about 1977 he found himself unable to continue his duties fulltime and eventually on or about May 31 1985 his employment came to an end because of his ill-health. By that time there had come into operation on January 1 1977 the Rent (Agriculture) Act 1976. By virtue of the provisions of that Act there is no doubt but that the defendant had become what in the Act is described as "a protected occupier". Section 2 of the Act, so far as material to this appeal, reads:

> Where a person has, in relation to a dwelling-house, a relevant licence . . . and the dwelling-house is in qualifying ownership, or has been in qualifying ownership at any time during the subsistence of the licence . . . he shall be a protected occupier of the dwelling-house if —
> (*a*) he is a qualifying worker, or
> (*b*) he has been a qualifying worker at any time during the subsistence of the G
> licence or tenancy . . .

Section 4 of the Act, so far as is material, reads:

> where a person ceases to be a protected occupier of a dwelling-house on the termination, whether by notice to quit or by virtue of section 16(3) of this Act or otherwise, of his licence . . . he shall, if and so long as he occupies the dwelling-house as his residence, be the statutory tenant of it.

After the termination of his employment, Mr Griffiths and his wife continued to occupy the cottage free of all payment until in 1987 the plaintiff, who had acquired the freehold of the cottage, applied to have a rent registered in respect of it. On July 3 1987 the rent officer determined a fair rent at £27 a week. The defendant declined to make any payment in respect of rent, it being his contention that after the H
termination of his employment or perhaps before the termination of his employment — it is not entirely clear — he had been granted by Lionel Burgoyne a licence permitting him and/or his wife to continue to live in the cottage rent-free for the rest of his life or alternatively the life surviving as between the first and second defendants. It was as a result of the defendant's refusal to make any payments to the plaintiff that she instituted proceedings seeking possession, arrears of rent and mesne profits, in consequence of which the judge, after a hearing which occupied three days, made on September 11 last, the order, the effect of which I have earlier indicated.

Para 2 of the particulars of claim reads as follows:

> By an agreement made orally in or around March 1960 the Defendant in consequence of his employment as an agricultural worker of the Plaintiff's J
> predecessor in title was permitted to enter and thereafter live in Middleton House Cottage aforesaid free of rent and rates if and so long as he remained employed in agriculture as aforesaid.

A defence was filed. In para 3 the employment of the first defendant was set out and no admission was made as to para 2 of the particulars of claim. Para 5 of the defence alleged that Mr Burgoyne had given the defendants leave to live in the cottage rent-free so long as both or either remained alive and wished to remain.

In a sentence, Mr West, on behalf of the defendants, submits that the original arrangement made between the partnership and the first defendant was an arrangement quite independent of the contract of employment to the extent that the licence granted by the partnership K
was determinable by the partnership at any time after 1960 upon reasonable notice served upon the defendant by his employers. The contrary argument advanced on behalf of the plaintiff was and is that the licence to occupy this cottage formed part and parcel of and was a term of the defendant's contract of employment with the partnership and that it was in point of time coterminous with the contract of employment. Accordingly, so submits the plaintiff, once the employment came to an end so did the licence to occupy the cottage. There was no need to serve any notice upon the licensees; the cottage was part of Middleton Farm upon which the defendant was employed; that fact, and the fact that no payment of any kind was required, created the plain inference that the employers were intending to create and did create an occupancy at the cottage which L
was an integral part of the contract of employment. If that be the case, then upon the termination of the employment the licence to occupy the cottage came to an end and section 4 of the 1976 Act converted what had been a protected occupancy into a statutory tenancy. When that happened, the plaintiff, as the owner of the cottage, was entitled to register a fair rent and, in the event of the defendants not paying that rent, the plaintiff was entitled to recover possession subject to the limitations imposed upon the court by other provisions of the 1976 Act, to which I must later refer.

The court has available to it not only a very lengthy judgment of the learned judge, which he reserved, but extensive notes of evidence. During the course of the hearing the defendant gave evidence on oath M
to the effect that he believed he had a licence to occupy the property for the rest of his life. Upon that aspect of the case he was expressly disbelieved by the learned judge. Furthermore, during the course of the notes prepared by the judge there is an indication that a concession was made during the course of the hearing that the original occupation of the first defendant amounted to a service occupancy. In the course of the judge's findings, that concession is referred to again when, at p 26 of the transcript, para 4 of the judgment, the judge said:

> In 1960 still in the same employment he

— that is the defendant —

> and his family moved to Middleton House Cottage under the same terms. In

the initial "holding" defence, it is pleaded that in 1960 it was agreed between Lionel Burgoyne on behalf of W Burgoyne & Sons and Mr Griffiths and his wife, that they should occupy the cottage for the remainder of their lives free of rent and rates. This allegation so far as 1960 is concerned is not repeated in the defence signed by counsel, but it was not made clear to me that it was abandoned, until halfway through Mr Lionel Burgoyne's evidence in chief, when the concession was made that at that time it was a service agreement.

Finally, it is to be observed that in the course of Mr Burgoyne's evidence, according to the judge's note of it, it was never put to him that when the agreement was made in 1960 he contemplated that he could bring the licence to occupy the cottage to an end irrespective of the contract of employment and at any time upon reasonable notice after the licence was created.

Mr West, on behalf of the defendants, submits that there was no real evidence enabling the judge to make a finding that the occupation of the cottage created in 1960 was a term of the contract of employment. In my judgment, that submission is misconceived. I take the view that the learned judge was clearly at liberty, in the light of all the circumstances that I have outlined, to come to the conclusion, as he did, that in 1960 there was what he loosely described as a "service occupancy" by the first defendant, in other words an occupancy of this cottage which formed a term of the contract of employment. Once the finding was made that the occupancy was truly a term of the contract of employment, then plainly, in my view, it follows that it came to an end so soon as the employment came to an end in May 1985.

We were referred to a number of authorities, but it is necessary to cite from only one of them, the case of *Ivory* v *Palmer* [1975] ICR 340. The first part of the headnote reads:

The plaintiffs employed the defendant, who was aged about 53 years, as a foreman carpenter. While so employed, he was to occupy a house belonging to the plaintiffs rent free. At an interview preceding his employment, the plaintiffs' agent told the defendant that he would have a job for life, security for life and a home for his widow. Nothing else was said as to the duration of the employment or circumstances in which it would be terminated. After six months' service the plaintiffs dismissed the defendant summarily and he took up some other work but continued to occupy the house. The plaintiffs brought an action for possession of the house, and Judge Chope, granting the order, held that the defendant was employed for life subject to his working satisfactorily; that the defendant's dismissal was wrongful but, since the licence to occupy the house was ancillary to the contract of employment, the plaintiffs' repudiation of the contract revoked that licence.

At p 352 of the report, Roskill LJ, who dissented upon one point not material in the current appeal, said:

Once the defendant was dismissed, under what was an ordinary service agreement between employer and employee, he had, as a matter of the ordinary contract law of master and servant, no alternative but to accept that dismissal as an accomplished fact, whether it was right or wrong. If it was right, he had no remedy; if it was wrong, his remedy, as was pointed out in *Vine's* case, lay in damages. But he cannot, in one and the same breath, as I ventured to point out to Mr Godfrey yesterday, accept that his service agreement is at an end, and yet assert that his right to the continued occupation of the cottage remains unaffected. This is not a severable contract. It is a single contract of employment between employer and employee. Once that contract goes, it goes as a whole, and that, with the greatest respect to Mr Godfrey's argument, is the short answer to this appeal.

Browne LJ, at p 354 of the report, said:

Mr Godfrey accepts that if the defendant had himself put an end to his employment, he would have had no right to go on living in the house, and indeed the defendant himself agreed that this was so in the course of his evidence. Mr Godfrey also agreed that the defendant's right to live in the house would come to an end if he had been dismissed for good cause. The judge held that the licence came to an end when the contract of employment was terminated, even though such termination was wrongful. In my view, this was clearly right. The contract between the plaintiffs and the defendant was one contract, a single contract to employ him for life and allow him to live in this cottage as a licensee so long as the employment continued. The plaintiffs repudiated the contract. In the case of an ordinary contract, the defendant would, of course, have had an election whether to accept the repudiation or to hold the plaintiffs to the contract; but contracts of employment are different. They are the exception to this general rule. When the plaintiffs repudiated the contract, the defendant had no choice but to accept the repudiation of the contract, and he himself accepted that the employment had come to an end. The result was that the contract between the plaintiffs and the defendant was terminated. This must mean that the whole contract was terminated, not only the contract of employment but also the licence to occupy the house, because these were both one and the same contract. Even though this termination was wrongful, the defendant's only remedy is in damages. It would be most extraordinary if the defendant were entitled to continue to live in this house rent free for the rest of his life even though the employment had been terminated.

As it seems to me, it is plain in the instant case that when this defendant's employment came to an end, so at that moment did his licence to occupy the cottage come to an end, and it was not necessary that there should have been at the same time or thereafter any notice to him, formal or informal, seeking possession of the cottage. Accordingly, as it appears to me, the plaintiff thereafter was entitled to regard the defendant as a statutory tenant and there is no dispute but that thereafter she was entitled to have a fair rent assessed and registered and to seek payment of it.

The right to possession of the property in the event of non-payment of rent is set out in the 1976 Act. Section 7 provides:

(1) This section applies in the Cases in Part 1 of Schedule 4 to this Act

Among those cases in Part 1 of Schedule 4 to the Act are cases where rent remains unpaid.

Subsection (2) provides:

In those Cases the court shall not make an order unless it considers it reasonable to do so.

The submission made here is that the learned judge, while he was entitled on his findings to make the order involving monthly payment, was not entitled, unless he found it reasonable to do so, to go on and make an order for possession, albeit a suspended one. That submission is plainly right, for the terms of the Act are mandatory and require the judge to test the case by standards of reasonableness before he makes an order for possession. It is pertinent to observe that nowhere in the judgment does the learned judge expressly apply himself to the provisions of section 7 and the requirement that an order for possession should be made only if it is reasonable to do so. But for my part I remind myself that the judge here had listened to this case over a period of three days, albeit not consecutive days, had satisfied himself that the defendant had not been as frank as he ought to have been in relating the factual background and had satisfied himself that there had been no effort to make any sort of payment to the plaintiff once the fair rent had been assessed. Accordingly, while it is justifiable to express the criticism that Mr West has done, that the judgment is silent as to the question of reasonableness, for my part I am satisfied that, while unexpressed, the learned judge must have applied his mind in the circumstances of this case to the test of reasonableness. Accordingly, I would not interfere with the order that he made that possession of this dwelling-house should be granted to the plaintiff, suspended on the terms appearing in the order. Those terms were to the following effect: that the defendant should pay the plaintiff the sum of £1,500 within two months of the order and that he should pay the current rent, together with mesne profits and arrears of rent by instalments of £10 per week in addition to the current rent. The overall figure found by the judge to be due was £2,808. There is no dispute as to the arithmetic.

For the reasons that I have endeavoured to outline in this judgment, I take the view that the learned judge was entitled to make the findings that he did and that once he made those findings of fact upon material that was available to him he was entitled to make the order to which I have referred.

I would, for my part, therefore dismiss this appeal.

Neill and **Farquharson LJJ** agreed and did not add anything.

The appeal was dismissed with costs; the order for costs not to be enforced without leave of the court; liberty for the plaintiff to apply in respect of those costs; legal aid taxation for the defendants, if necessary; further terms of order to be agreed between the parties.

A
Chancery Division

December 17 1990

(Before Mr John MOWBRAY QC, sitting as a deputy judge of the division)

DIVALL v HARRISON AND ANOTHER

Estates Gazette April 13 1991

[1991] 14 EG 108

Agricultural holding — Notice to quit — Validity of notice —
B **Complications in regard to landlord's title — Notice given by solicitors two of whom were executors holding the legal title to the reversion — Notice was, however, expressed to be given on behalf of the plaintiff, who was entitled in equity to the land as residuary beneficiary subject to the completion of the administration of the estate — Whether the notice was valid, particularly as it described the plaintiff as "your landlord" — Notice held to be valid in spite of criticisms of its form — Tenants not likely to have been misled**

The problem in regard to the notice to quit in this case arose from the devolution of the title to the legal estate in the
C **reversion — The testatrix, having appointed the partners in a firm of solicitors as her executors, left the whole of her estate, subject to the payment of debts and testamentary expenses etc, to the present plaintiff — Two of the partners proved the will, power being reserved to the other partners in the firm — The notice to quit which was at the centre of the litigation was served the day after the probate was granted but some months before an assent was executed in favour of the plaintiff as devisee in respect of the land**

The position when the critical notice to quit was given was that the legal estate in the land in question was vested in the two partners who had obtained probate, while the plaintiff was
D **entitled to the land in equity subject to the administration of the testatrix's estate — The notice to quit was on the solicitors' writing paper and was headed with the firm's name and address and the list of six partners (including the two executors) and associates — The notice went on to state that it was written by the firm as solicitors dealing with the estate of the testatrix "on behalf of the residuary beneficiary thereof, Mr John Jack Divall" — Was a notice in this form invalid? The question had an importance from the agricultural holdings aspect because the tenancy was a protected agricultural tenancy subject to the Agricultural Holdings Act**
E **1986 and no counternotice invoking section 26(1) of the Act had been served by the tenants in response to the notice to quit — In fact a later notice to quit was served without prejudice to the validity of the earlier notice and a counternotice had been served in response to that later notice; but no issue arises on this point**

The validity of the first notice was challenged on the ground that it was given on behalf of someone who was not the landlord — In law the landlords were the two partners who had obtained probate, but the objection was that they had given the notice expressly on behalf of the plaintiff, who was
F **entitled only to the land in equity and indeed only so entitled subject to the clearing of the testatrix's estate — It was accepted that at common law an equitable owner was not entitled to give a notice to quit determinative of a legal tenancy**

The judge nevertheless concluded that the notice to quit was valid — It was given by the solicitors dealing with the deceased's estate and with the authority of the two partners who were the executors and who held the legal title to the reversion — It was a notice which the tenants could act upon with safety and by which they were unlikely to have been seriously misled, although the plaintiff had been erroneously described as the landlord — The case of *Lemon* v *Lardeur*, cited on behalf of the tenants, was distinguishable — Notice to G
quit declared to be valid

The following cases are referred to in this report.

Harmond Properties Ltd v *Gajdzis* [1968] 1 WLR 1858; [1968] 3 All ER 263, CA
Jones v *Phipps* (1868) LR 3 QB 567
Lemon v *Lardeur* [1946] KB 613

This was an originating summons to determine the validity of a notice to quit. The notice to quit was in respect of the tenancy of 57.89 acres of land at Brakey Hills Farm, Lakenheath, Suffolk, the land being an agricultural holding subject to the Agricultural Holdings
Act 1986. The plaintiff was John Divall, who was the residuary H
beneficiary under the will of Mrs Gladys Victoria Bruce, and the defendants were the tenants, John and Gordon Harrison.

Robert Wakefield (instructed by Metcalfe Copeman & Pettefor of Thetford) appeared on behalf of the plaintiff; John Holt (instructed by Rudlings & Wakelam of Bury St Edmunds) represented the defendants.

Giving judgment, **MR JOHN MOWBRAY QC** said: Before me is an originating summons asking whether a notice to quit was valid. One of the documents relied on by the landlord as a valid notice to quit is dated June 2 1988. To see how the question about its validity arises, I must give an account of the position as regards the landlord's
interest. J

I can start with the relevant tenancy agreement. It was dated January 9 1985 and it was a tenancy agreement granting a tenancy over 57.89 acres of land at Brakey Hills Farm, Lakenheath, Suffolk. The landlord who granted the agreement was Thomas Donaldson Bruce. The tenants under the agreement were John Harrison and Gordon Harrison. The first clause of the tenancy agreement reads:

> 1. The tenancy shall be from the eleventh day of October One thousand nine hundred and eighty-four to the eleventh day of October One thousand nine hundred and eighty-five and thereafter from year to year until it is determined at the end of the second or any subsequent year by either party giving to the other of them not less than twelve calendar months' notice in writing.

Mr Bruce died on August 22 1986. By his will he had appointed his
daughter, Mrs Helen Drake, and his stepson, the plaintiff, Mr Divall, K
to be his executors. They proved his will on November 5 1986. It contained a specific bequest of the land in question to his wife, Gladys Victoria Bruce. The executors executed an assent in her favour absolutely on February 10 1987. Mrs Bruce died on September 13 1987, having by her will appointed to be her executors the partners in a firm of solicitors, Cunningham John & Co. Two of the partners in that firm, Mr Cunningham and Mr Sheehan, proved that will on June 1 1988, power being reserved to the other partners in the firm. Mr Divall took the whole estate under a general devise and bequest, subject only to the payment of debts and the general clearing of the estate. It will be recalled that the notice to quit that I am
considering was given on June 2, the day after probate. It was given, L
of course, before any assent was executed in favour of Mr Divall. That came on October 3 1988. It was only then, as far as the evidence before me is concerned, that the estate was cleared.

The position when the notice to quit was given on June 2 1988 was this: that the legal estate in the land in question was vested in the two partners as executors of Mrs Bruce and Mr Divall was entitled to the land in equity as residuary legatee, subject only to the clearing of the estate.

The notice to quit whose validity I am considering is on the firm's headed writing paper. It is headed "Cunningham John & Co, Solicitors". It is addressed to the tenants. The firm's address is given, and then a list of partners, including Mr Cunningham and Mr
Sheehan, four other partners and six associates. The letter reads: M

> Re: The estate of Thomas Donaldson Bruce deceased and Gladys Victoria Bruce deceased. Residuary beneficiary Mr John Jack Divall. Tenancy agreement dated 9.1.1985 in respect of 57.89 acres of land at Brakey Hills Farm, Lakenheath, Suffolk.
> We now write as solicitors dealing with the estates of Mr T D Bruce deceased and Mrs Gladys Victoria Bruce deceased on behalf of the residuary beneficiary thereof, Mr John Jack Divall, your landlord, to give you notice to quit and to deliver up to him on the 11th day of October 1989 the possession of the land comprising 57.89 acres or thereabouts situate at Brakey Hills Farm, Lakenheath, Suffolk held by you as tenant under the above tenancy agreement.
> Dated: 2nd June 1988
> Signed: Cunningham John & Co, Solicitors.

A Then at the bottom it reads:

> All correspondence to be addressed to Ref: A. J. Cowlishaw, Cunningham John & Company . . .

And their address is given again.

Whether that notice is valid is important to the parties because the tenancy was an agricultural tenancy and no counternotice was given under the Agricultural Holdings Act 1986 by reference to that notice. A later notice to quit was given without prejudice to the validity of the notice I have just read and a counternotice was served in regard to that later notice. It will be seen, therefore, that whether the machinery under the Act came into operation depends on whether this notice of June 2 1988 was valid.

B The main criticisms of the notice to quit of June 2 1988, which have been directed at it in a very interesting and, if I may say so, able argument, which took every point that could have been taken, were these. The notice was not given, it is said, by the landlords but was given on behalf of someone who was not the landlord. The landlords were the two partners, Mr Cunningham and Mr Sheehan. In this notice to quit they said that they were giving the notice on behalf of Mr Divall. He was not the landlord. He was entitled only in equity and he was not, so far as the evidence before me goes, absolutely entitled in equity but only subject to the clearing of the estate. It was argued for the tenants that a mere equitable owner is not able to give a notice to quit under a common law tenancy. That is accepted on the other side and, in my judgment, is correct, notwithstanding an old
C dictum to the contrary.

None the less, I have concluded that this notice to quit was valid. It seems to me that it was given by, or at any rate with the authority of, the freeholders, Mr Cunningham and Mr Sheehan. It was given by their firm and thereby on behalf of all the partners in the firm, of whom they were two. It must also have been given with the authority of the freeholders, the two partners, because Cunningham John & Co wrote, "as solicitors dealing with the estates of Mr T D Bruce deceased and Mrs Gladys Victoria Bruce deceased". They were writing as solicitors dealing with Mrs Bruce's estate, and thereby with the authority of her executors, two of their own number.

The letter continues, ". . . on behalf of . . . Mr John Jack Divall,
D your landlord". It is on those words that the real question arises about the validity of this notice. First of all, those words state that the plaintiff, Mr Divall, is the landlord. That was erroneous. Mr Divall was not the landlord. He had only an equitable interest and the two partners were the landlords. Moreover, second, the notice purported to be given on behalf of Mr Divall with a possible implication that it was not given by or on behalf of the true landlords, Mr Cunningham and Mr Sheehan. I do not consider that that invalidates the notice. In *Jones* v *Phipps* (1868) LR 3 QB 567, at p 573 Lush J, giving the judgment of the Divisional Court of Queen's Bench, said:

> It is clear that the notice must be such as the tenant may act upon with safety, that is, one which is in fact, and which the tenant has reason to believe to be,
> E binding on the landlord.

In a very similar passage, Russell LJ described that as "the crucial matter" in *Harmond Properties Ltd* v *Gajdzis* [1968] 1 WLR 1858 at p 1864. Russell LJ disapproved too technical an approach to a notice to quit and said that the crucial point in that case was that, at the moment it was given, it bound the true landlords.

It is clear that the notice of June 2 1988 bound the true landlords, the two partners in the firm, because it was given by the firm as solicitors dealing with Mrs Bruce's estate. I think it could possibly have misled the tenants into thinking that Mr Divall was properly to be described as their landlord, but as it described him as the "residuary beneficiary" of Mrs Bruce's estate, I do not think there was a possibility of the tenants being seriously misled about his true
F position.

There is a principle under which someone who is not the true landlord can give notice to quit in his own name without revealing the true position, if the person giving notice is the general agent of the landlord. It was persuasively argued for the tenants that that principle could not apply in the present case because there is no evidence at all that Mr Divall was any kind of agent for the two partners who held the legal estate. That is perfectly correct, but the landlord has not based any argument on any general agency and, for the reasons I have already given, the landlord is able to get home by a different route.

For the tenants the case of *Lemon* v *Lardeur* [1946] KB 613 was cited. That case certainly bears out the submission I have mentioned, that it is only a general agent who can rely on the principle I have G mentioned. That was a case where a husband gave the notice, whereas his wife was the landlord. That notice was held by the Court of Appeal to be invalid because it was not given by the landlord and because there was insufficient evidence that the husband was the general agent of his wife, the landlord. It seems to me that if in *Lemon* v *Lardeur* the notice had been countersigned by the wife there could have been no doubt about its validity. In the present case we have the equivalent, because the landlords joined in the notice in the ways that I have described. I conclude that the authority of *Lemon* v *Lardeur* does not assist the tenants in the present case.

I will, therefore, make a declaration that the notice to quit which I have described was valid and I will ask counsel about the form of
order. H

Chancery Division

October 10 1990

(Before Mr Timothy LLOYD QC, sitting as a deputy judge of the division)

LAND v SYKES AND ANOTHER

Estates Gazette April 27 1991
[1991] 16 EG 125 J

Agricultural Holdings Act 1986 — Landlord's action for possession — Notices to quit served in reliance on Case E and Case G in Schedule 3 to the Act — Effect of severance of the reversion — At first sight, odd position of tenants arguing in favour of notice to quit served on them — Whether notice to quit, which failed to take account of severance of the reversion, invalid — Cases discussed on implication of new tenancy after termination of previous tenancy — Whether only "objective" evidence of intention admissible — Held Case E notice to quit invalid and Case G notice effective to K terminate tenancy — Judgment for landlord

In a case with curious and interesting features the landlord sought an order for possession of four farms and adjoining land against the executors of a deceased tenant — There had been some previous litigation between the parties concerning agricultural holdings and partnership problems, but the events leading to the present contest began with the transfer by the landlord of a narrow strip of land to a relative, thus bringing about a severance of the reversion — Later in the same year the landlord complained that the tenant, Mr Sykes, had cut down two substantial trees on the holding — A Case E L notice was served on Mr Sykes, expiring on April 1 1988 — This notice took no account of the severance of the reversion, being expressed in a way which included all the lands which were subject to the tenancy before the severance — It should be mentioned that although the tenants had served a notice requiring arbitration in regard to the Case E notice within the time-limit laid down by article 9 of the Agricultural Holdings (Arbitration on Notices) Order 1987, they failed to follow this up by proceeding to the appointment of an arbitrator within the time prescribed by article 10 — The result was to deprive the tenants of the right to contest the reasons stated in the notice to quit — It did not, however, affect the question of the M possible invalidity of the notice to quit because of the severance of the reversion

Mr Sykes died on March 5 1988 and the landlord's solicitors duly served a Case G notice to quit on all concerned — This notice, unlike the Case E notice previously served, took account of the severance of the reversion and was expressed so as to exclude the strip of land which had been transferred — No application for succession to the tenancy of Mr Sykes had been made by any family claimant within three months of his death as provided by section 39 of the 1986 Act — The Case G notice would thus take effect on April 1 1989, assuming that

A **the Case E notice had not already terminated the tenancy on April 1 1988 — Proceedings for possession of the holding, based on the Case G notice, were issued by the landlord on June 1 1989 — Although the statement of claim initially relied entirely on the Case G notice, it was amended to rely in the alternative on the Case E notice when it was found that the defendant tenants were relying on the Case E notice in their defence and counterclaim — The stage was then set for a debate on both notices and for the revelation of the strategy of the tenants in their, at first surprising, argument in favour of the validity of the Case E notice**

B **The landlord's case was short and simple — The Case E notice being in their submission invalid, the tenancy vested in the executors of Mr Sykes and the Case G notice, served properly in time, terminated that tenancy on April 1 1989 — The executor tenants' case was less simple — They submitted that the Case E notice was valid and terminated the tenancy on April 1 1988 — The tenants, however, remained in possession and the inference to be drawn was that they did so under an implied agreement with the landlord for a new tenancy — Even if they were in possession only under a licence (unless purely gratuitous), the effect of section 2 of the 1986 Act was to convert it into a fully protected agricultural tenancy from**
C **year to year — Unfortunately for this line of argument, the judge decided that the Case E notice had been invalid — The landlord had been entitled to serve such a notice only in respect of the land which remained in her ownership after the severance of the reversion — The notice actually served did not make it clear that the strip of land transferred out of her ownership had been excluded — The notice was at least ambiguous and uncertain — After considering the wording used, the submissions of counsel and a number of authorities, the judge held that the notice did not comply with the vital requirement of certainty — A reasonable tenant would be**
D **confused or left in doubt about it**

The result was that, the Case E notice having been invalid, the landlord's Case G notice had been effective to terminate the tenancy — This was enough to decide the case but, in deference to the argument on behalf of the tenants, the judge expressed his view on what the position would have been if the tenancy had been terminated by the Case E notice — After considering a number of authorities and referring to the question of subjective and objective indications of intention, he concluded that he would have found the evidence in support of the creation of a new tenancy after April 1 1988
E **unconvincing**

The defendants were ordered to give up possession and their counterclaim for a declaration that they held a protected agricultural tenancy was dismissed

The following cases are referred to in this report.

Cardiothoracic Institute v *Shrewdcrest Ltd* [1986] 1 WLR 368; [1986] 2 EGLR 57; (1986) 279 EG 69
Carradine Properties Ltd v *Aslam* [1976] 1 WLR 442; [1976] 1 All ER 573; (1975) 32 P&CR 12
Cawley v *Pratt* [1988] 2 EGLR 6; [1988] 33 EG 54, CA
Clarke v *Grant* [1950] 1 KB 104; [1949] 1 All ER 768, CA
Doe d Cheny v *Batten* (1775) 1 Cowp 243
Doe d Lord v *Crago* (1848) 6 CB 90; 17 LJCP 263; 12 Jur 705
F *Doe d Williams* v *Smith* (1836) 5 Ad & El 350
Frankland v *Capstick* [1959] 1 All ER 209; [1959] 1 WLR 204, CA
Hankey v *Clavering* [1942] 2 KB 326; [1942] 2 All ER 311, CA
Herongrove Ltd v *Wates City of London Properties plc* [1988] 1 EGLR 82; [1988] 24 EG 108
Javad v *Aqil* [1990] 2 EGLR 82; [1990] 41 EG 61
Legal & General Assurance Society v *General Metal Agencies* (1969) 20 P&CR 953
Longrigg, Burrough & Trounson v *Smith* [1979] EGD 472; (1979) 251 EG 847, CA
Maconochie Brothers v *Brand* [1946] 2 All ER 778
Marcroft Wagons Ltd v *Smith* [1951] 2 KB 496; [1951] 2 All ER 271
Mountford v *Hodkinson* [1956] 1 WLR 422; [1956] 2 All ER 17, CA
Sector Properties Ltd v *Meah* [1974] EGD 329; (1973) 229 EG 1097
Wride v *Dyer* [1900] 1 QB 23

This was an action by Mrs Edna Lydia Land, the landlord, G
claiming possession of four farms in Wyverstone, Stowmarket, Suffolk, from the defendants, Jasminia Felicia Sykes and Wallace John Sykes, the executors of Kenneth William Sykes, who had been the tenant of these farms. There was a counterclaim by the defendants for a declaration that they held a protected agricultural tenancy.

Jonathan Brock (instructed by Burges Salmon, of Bristol) appeared on behalf of the plaintiff; Paul Morgan (instructed by Reynolds Porter Chamberlain) represented the defendants.

Giving judgment, **MR TIMOTHY LLOYD QC** said: I have before me an action in which Mrs Edna Lydia Land claims possession of certain agricultural land at Wyverstone, Stowmarket, Suffolk, from H
Jasminia Felicia Sykes and Wallace John Sykes, who are the executors of, and respectively the widow and son of, the late Kenneth William Sykes. To that action there is a counterclaim by the defendants in which they claim a declaration that they are tenants of the land in question with the protection of the Agricultural Holdings Act 1986.

On December 17 1985 Mrs Land granted to the late Mr Sykes an agricultural tenancy of "all those farms, farmhouse buildings and land containing 291.24 acres or thereabouts situate in the parish of Wyverstone in the County of Suffolk and known as Grange Farm aforesaid Sudbourne Farm, Plough Farm, Crooksells Hall Farm and adjoining land". The tenancy ran for one year from April 1 1985 and J
thereafter from year to year until determined at the end of the second or some subsequent year by notice in writing. That lease, however, was by no means the beginning of the relationship between Mrs Land and the late Mr Sykes. Previously Mrs Land and the late Mr Sykes had been in partnership, farming the land in question and indeed other land. The partnership, however, was brought to an end by Mrs Land, and she sought at that stage to obtain possession of the agricultural land from Mr Sykes. He did not accept her entitlement to do that and proceedings in the High Court followed. These came to trial before the late Judge Finlay QC, whose decision in favour of Mr Sykes was affirmed by the Court of Appeal: see (1984) 271 EG 1264.

The grant of the lease dated December 17 1985 was, I was told by counsel, part of the working out of the consequences of the order K
made in those proceedings. Under that lease the late Mr Sykes was tenant of the new holding, of a somewhat reduced area compared to the previous holding, under a tenancy which could not be determined before April 1 1987 and which, in relation to any attempt by Mrs Land to determine it, enjoyed the significant protection afforded by the agricultural holdings legislation. The year of the term was to run from April 1 in every year and the initial rent was the yearly amount of £12,900, subject to increase in accordance with the provisions of the relevant legislation, and payable in arrear by equal half-yearly payments on October 11 and April 6 in each year. The lack of coincidence between the rent payments days and the periods to which each year of the term ran gives rise to a minor point in the case. I L
should say at once that, in my view, the position under that tenancy was that on each payment day, that is to say October 11 and April 6, there was payable by the tenant the sum of £6,450 but that those instalments of rent were payable in respect of the period, not up to the payment day, but up to October 1 and April 1 respectively. Thus the effect of using those particular payment days was to give the tenant a few days' grace after the end of the relevant half year in which to pay the relevant instalment of rent. No other point arises on the terms of the tenancy, which are in fairly standard modern form of an agricultural tenancy.

On April 4 1986 Mrs Land conveyed by way of gift to her grandson, a Mr G H Warner, in contemplation and consideration of his marriage, a small plot of land. I am told this was envisaged as a M
building plot and that it had formerly formed part of the holding farmed by the partnership, but it was not included in the 1985 tenancy. Six months later, on October 24 1986, Mrs Land executed a further deed of gift in favour of Mr Warner in relation to a narrow strip of land adjoining the plot which she had conveyed by the previous deed. This strip was conveyed expressly subject to the lease in favour of the late Mr Sykes but without any part of the rent due under the lease being apportioned to the land conveyed to Mr Warner. The effect of this second deed was that the reversion to the late Mr Sykes' lease was severed. By virtue of section 140 of the Law of Property Act 1925 each of the reversioners was entitled thereafter to exercise the rights of a landlord in respect of that part of the land

A owned by him or her respectively. It appears that Mr Warner did not himself take advantage of such planning or development potential as the plot may have had but on December 15 1986 conveyed the plot together with the narrow strip to a Mr and Mrs Chittenden. I was told that, in connection with the transfer of the land to them, the late Mr Sykes and his family became aware of the fact not only that the building plot had been conveyed but that there had also been conveyed the small strip of land subject to the tenancy.

In September 1986, solicitors then acting for Mrs Land, Rustons & Lloyds, told the solicitors then acting for the late Mr Sykes, Bankes Ashton & Co, that their client Mr Sykes had cut down two substantial trees within the holding and complained of this conduct as contrary to the terms of the lease. Messrs Bankes Ashton denied that there was

B any breach of the terms of the lease.

In March 1987 Mrs Land instructed new solicitors, Farrer & Co, both in relation to the issues which remained outstanding from the termination of the former partnership, and in relation to the tenancy. Mr Jessel, a partner in that firm, who had conduct of the matter on her behalf and who gave evidence at the trial, told me frankly that his instructions were to secure vacant possession of the holding if it could lawfully be done. He is a very experienced solicitor, particularly with reference to agricultural property matters.

The first step that he took in relation to the tenancy was to serve a notice to quit on the late Mr Sykes to expire on April 1 1988, relying, by reference on the cutting down of the two trees, on the ground set

C out in Case E of Part I of Schedule 3 to the Agricultural Holdings Act 1986. He served the notice, which I shall refer to as the Case E notice, under cover of a letter of March 17 1987. The first of the issues in the case turns on the Case E notice and arises from the fact that it took no account of the severance of the reversion that had by then taken place. This was known to both landlord and tenant, but not to Mr Jessel.

The notice is headed "Re The Holding known as Grange Farm, Sudbourne Farm, Plough Farm, Crooksells Hall Farm and all land held therewith". It is addressed to Mr Sykes and the crucial words in the operative part of the notice are:

> All that holding and premises known as Grange Farm, Sudbourne Farm, Plough Farm, Crooksells Hall Farm and all land held therewith and situate at
> D Wyverstone, Stowmarket aforesaid which you hold of [Mrs Land] as tenant thereof

No point is now taken on the validity or otherwise of the landlord's reliance on the ground in Case E. The point is taken, however, that the notice was invalid because it was ambiguous, in that did not make it clear to the tenant whether he was being required to give up possession of the whole holding including the narrow strip or of only that part of the holding which he held from Mrs Land, that is to say minus the narrow strip. One might expect that a point of that kind would be taken on behalf of the tenant, and indeed in correspondence so it was. However, a curious feature of this case, for reasons which will appear, is that it is now the tenant who asserts that that notice was

E valid and the landlord that the notice was invalid. I shall return to the question of its validity or otherwise when I have dealt further with the facts.

The Case E notice was sent both to the late Mr Sykes and to his solicitors Bankes Ashton. On April 8 1987 the latter replied to Farrer & Co confirming that the notices had been received and saying: "We understand that your client has disposed of part of the freehold reversion and if that be the case we believe the notices as served are invalid. We invite your comments on this point." That letter was sent by fax to Farrers on April 8 and on April 10 Mr Jessel replied confirming receipt of the letter and saying that he was investigating the points raised and hoped to come back shortly on them. At about the same time the half-year's rent due on April 6 1987 was paid.

F The next thing that happened was that on April 15 1987 Bankes Ashton sent both to Mrs Land and to Farrers a notice under the 1986 Act stating that Mr Sykes wished to contest certain matters arising out of the reliance on Case E and that he required all questions so arising to be determined by arbitration. Receipt of the notice was acknowledged by Farrers. On June 4 Mr Jessel wrote to Bankes Ashton saying: "We have so far been corresponding on certain matters relating to the existing tenancy and we appreciate that you are waiting to hear from us about the possible severance of the reversion and we are still investigating that." He went on, in a longish letter, to deal with a large number of points arising from the unresolved affairs of the former partnership between the parties. On July 6 1987 Mr Jessel wrote again to Bankes Ashton, this time replying substantively on the question of the severance. He debated whether the little strip G was actually included in the tenancy but said that even if it was he did not think that the disposal invalidated the notice and that he would be glad to know their reasons in law for taking the view that the notices were invalid. He stated that in his view the notice was valid. That challenge was never effectively taken up by Bankes Ashton.

Furthermore, although in April they had served a notice requiring disputes out of the Case E notice to be referred to arbitration, they failed to refer the matter to arbitration within the mandatory three-month time-limit imposed by the Agricultural Holdings Act, and it is not in dispute that the effect of that omission is that the notice of April 15 is itself ineffective and that the reliance on Case E became incontestable. On the other hand it is also not in dispute that that step H and that omission had no relevance to any invalidity of the notice depending on ambiguity or inconsistency resulting from the severance. In October 1987 Mr Jessel served a notice to pay rent due and in November the rent was paid within the two-month time-limit.

On March 5 1988 Mr Sykes died. Mrs Land came to hear of that shortly afterwards and thereby the news came to Mr Jessel. Mr Jessel realised that this gave a further opportunity to seek to obtain possession of the holding, by serving a notice to quit relying on Case G in Part I of Schedule 3 to the 1986 Act. Accordingly, on March 21 1988 he served such a notice. The notice was sent to Mrs Sykes, the widow, to Mr Sykes, the son, by name to the personal representatives, to the solicitors, Bankes Ashton, and to the treasury solicitor acting for the president of the Family Division, since Mr J Jessel did not know what the position was or would be as regards representation to the late Mr Sykes' estate. No point is taken on the service of the notice.

Mr Jessel was of course by then fully aware of the severance of the reversion and the notice is drafted accordingly. It is headed "Re so much of the holding known as Grange Farm, Sudbourne Farm, Plough Farm, Crooksells Hall Farm and land as is held from Edna Lydia Land". It is addressed to the personal representatives and to the President and says in its substantive part: "Notice to quit . . . all the holding and premises described above which was held of [Mrs Land] by the late Kenneth William Sykes as tenant thereof until his death on or about March 5 1988 and now held by you as his personal representatives"; the notice was to expire on April 1 1989. K

In his letters to Mrs Sykes, Mr Sykes and the personal representatives, Mr Jessel stated that the notice was served without prejudice to the contention that the tenancy had already been determined as a result of the Case E notice. The letter to the solicitors enclosed copies of the letters to their clients, so they saw the position in this respect as well. The statement that the notice was sent without prejudice was not mentioned in the letter to the treasury solicitor. It is not suggested that anything turns on that.

On April 1 1988 the Case E notice, if it was valid, expired. It would have been possible, as Mr Morgan for the defendants has submitted to me, for Mr Jessel to start proceedings there and then on behalf of Mrs Land in reliance on the Case E notice, either using the procedure L afforded by Ord 15 r 6A of the Rules of the Supreme Court or simply joining Mrs Sykes and Mr Sykes and anyone else who appeared to be in possession as defendants as being the occupiers. That, however, was not the course that was taken, partly, as Mr Jessel told me in evidence, for procedural reasons, namely that he preferred to wait until a grant of probate had been taken out, so that he knew who the correct defendants were, to be sued on behalf of the estate as the person in whom any tenancy had been vested immediately before April 1 1988. Mr Jessel did not in fact start proceedings until June 1989 following the grant of probate on April 27 1989.

Mr Jessel was, however, faced with the fact that the rent which was unquestionably due on April 6 1988 in respect of the previous half year was not paid. On April 13 he wrote a letter to Bankes Ashton M saying:

> We are concerned to note the rent due at the end of the half year of the tenancy which has just expired has not been paid. Our client does not wish unnecessarily to serve a further notice to pay but we should be glad if you could arrange for the executors to pay the rent to your firm as usual as soon as possible.

That produced no response and so, on April 26 he sent, addressed to the personal representatives of the late Mr Sykes and with a copy to Bankes Ashton and to the Treasury Solicitor, a notice to pay rent in the statutory form which is required if it is to be relied on later in support of a notice to quit relying on Case D. That notice is in the familiar statutory form headed "Agricultural Holdings Act 1986

A Schedule 3 Part I Case D Notice to Tenant to pay rent due''. It is headed "Re the Holding" by name, addressed to the personal representatives and to the president, and it contains in a prominent box warnings such as "If you want your tenancy to continue you must act quickly". Just within the two-month period, on June 27, the rent was paid. The rent was acknowledged by Mr Jessel by a letter of July 4 1988 in which he said: "Thank you for your letter of 27th June enclosing cheque for £6,450 representing the rent for the half year ending on 31st March 1988." As is clear from what I have already said, his view as to the period in respect of which the rent was due coincides with my view of the true interpretation of the lease.

In the meantime Mr Jessel set about trying to discover from Bankes B Ashton, who were the personal representatives of the late Mr Sykes, so that proceedings could be commenced. Thus on June 1 1988 he spoke to Mr Gair of that firm and said that he was now about to issue proceedings. Mr Gair told him that he had authority to accept service of any proceedings, that he did not have the names of the personal representatives, but that he would find them out from his probate colleague. Mr Jessel frankly said that he could not recall whether he said expressly that the proceedings which he was about to issue were proceedings for possession of the holding, and his attendance note simply refers to "the proceedings". Mr Morgan pointed out that there might also have been proceedings between the parties to do with the partnership. Mr Jessel was certain that what he, in fact, had in mind was proceedings for possession, particularly because he had C then just received a draft of such proceedings settled by counsel, but he could not recall whether in talking to Mr Gair he had actually said that the proceedings would be for possession. I find that Mr Gair knew that what was contemplated were proceedings for possession. I note that on September 9 1987 Mr Jessel had written to Mr Gair on the unresolved matters arising from the termination of the partnership saying among other things: "We have therefore been instructed that if there is further delay our client will regrettably have to return to the Courts for an account as to the sums due to her as a result of the termination of the [partnership]." While the partnership matters needed to be resolved, there was not the same reason for urgency on that account as there was in relation to possession proceedings. Moreover, so far as I am aware, nothing had changed on the question D of the partnership to render proceedings more necessary or appropriate by the summer of 1988, whereas in relation to the tenancy the Case E notice had expired, giving the landlord for the first time the basis for a claim to possession. In the context of the correspondence and communications that I have seen it seems to me clear that from June 1988 onwards both sides knew that the proceedings that were being talked about were proceedings for possession.

Some consideration was given within Mr Jessel's firm and indeed in discussion between a Miss Brookes of Farrers (in the litigation department) and Mr Gair on the telephone, for example on June 21 1988, as to the possibility of taking proceedings against the estate in E reliance on Ord 15, r 6A. For whatever reason, that possibility was not pursued. It took a significant time before the executors were even identified — it seems to have taken some time to find the late Mr Sykes' last will. Mr Jessel was not in fact told who the executors were until a letter of August 25 1988 written by new solicitors, Cunningham John, instructed to act for the Sykes family.

In the meantime, however, a further significant time-limit had elapsed. Under the Agricultural Holdings Act 1986, where an agricultural tenant has died, a member of his family may, in certain circumstances, apply to the agricultural lands tribunal for the right to succeed to his tenancy. Such an application must be made within three months of the death. If such an application is made it may F provide a bar to a notice relying on Case G. No such application was made in the present case on behalf of any member of the Sykes family within the three months following the death of Mr Sykes. Accordingly, by the middle of June 1988 Mr Jessel knew that his Case G notice would take effect on April 1 1989 (assuming that the Case E notice had not already determined the tenancy) and that there would be no defence to a claim for possession thereafter.

It was no secret at the trial that following the involvement of Cunningham John there were negotiations between the parties on a without prejudice basis for a resolution of issues between the parties. It was also no secret that there were or might be claims on behalf of the Sykes family against Bankes Ashton and that, for that reason, other solicitors, Reynolds Porter Chamberlain, became involved in the matter, on the instructions of Bankes Ashton's professional G indemnity insurers.

The next event in the history is that on January 23 1989 Cunningham John wrote to Mr Jessel the following letter:

Dear Sirs,
Kenneth William Sykes Deceased,
The personal representatives of the above-named deceased have asked us to forward to you the enclosed cheque for £6,450 being the rent due from 6th April to 11th October 1988.

That appears to have come out of the blue and was one of relatively few occasions on which rent was paid without prompting. Of course it was significantly late, but it was nevertheless tendered. Mr Jessel told me that, while as he understood it there was no love lost at all H between the Land family and the Sykes family, he gathered that Mr Sykes, the second defendant in the proceedings, felt less antagonism for the Land family than his late father had done and that, perhaps for that reason and also because as a matter of the effective working relationship between his firm and Cunningham John there was an express attempt on the part of both of them to reduce the antagonism, this payment might have been tendered as a relatively friendly gesture. At all events the cheque was tendered, but it put Mr Jessel in a quandary which was reflected in his letter in reply of January 25. He thanked Cunningham John for the letter enclosing the cheque and said: "We regret that so long as the notice to quit which was served in 1987 and expired in 1988 remains outstanding our clients cannot accept any rent for this holding. We shall therefore hold your cheque J unpresented." Now, of course, if the Case E notice which purportedly expired in 1988 was invalid, as Bankes Ashton had claimed, then rent was undoubtedly due and there need have been no inhibition on Mr Jessel's accepting the cheque. If, however, reliance was to be placed on the Case E notice (and he was not minded to abandon that possibility) it would be inconsistent with that to accept the cheque tendered as rent. Cunningham John on February 2 acknowledged and noted his letter of January 25.

The next significant event occurred on April 10 1989. What had happened was that there was a tragedy in the Land family which gave rise to an acute need for further cash. Mrs Warner, Mrs Land's daughter who acted for her throughout pursuant to powers of K attorney, told Mr Jessel of this and, I assume, asked him what he could do to help and, in particular, if there was any way in which the family could use the money represented by the outstanding cheque. He would clearly have preferred not to present the cheque and to have continued to take the cautious line that he had taken in January. However, in the circumstances what he did was to speak both to Mr Connolly of Cunningham John and to a Mr Hadfield of Reynolds Porter Chamberlain in an endeavour to obtain their agreement that the cheque could be presented on a neutral basis. Although clearly credit would be given for the payment, what he hoped to secure, on compassionate grounds, was that the payment itself would be entirely neutral and without legal consequences. As it happens he spoke first to Mr Hadfield, having been unable to get hold of Mr L Connolly when he first tried. He asked Mr Hadfield to take instructions on the point, but Mr Hadfield said that his instructions were such that he would deal with the point without reference to his clients. Mr Hadfield told him, first, that he was unsure whether he was in a position to agree, since he was acting not for the tenants but only for insurers, but he also said that even if he had relevant authority it was not an undertaking that he felt he could give. Effectively he said that Mr Jessel must take his own course. Mr Jessel then spoke to Mr Connolly, explained the upshot of his discussion with Mr Hadfield and made the same request. Not surprisingly, Mr Connolly said that in the circumstances Mr Jessel could hardly expect that he would agree to the proposition given that Reynolds Porter M had not. Nevertheless Mr Jessel, despite having failed to obtain the agreement that he had hoped for, proceeded to present the cheque and on April 11 wrote to Mr Connolly to say that he had done so.

Now, between the tender of the cheque in January and its presentation in April, certain correspondence had passed between Mr Jessel and Mr Connolly on which some reliance is placed. Apart from minor matters such as the unconditional surrender of an empty barn in Grange Farm, that correspondence is largely concerned on Mr Connolly's part with making Mr Jessel aware of the attempts that he was making to ensure that Reynolds Porter Chamberlain took up a definitive position in the matter. In the first of the relevant letters on February 10 1989, after referring to Reynolds Porter Chamberlain

A who had counsel's opinion but were waiting for instructions, Mr Connolly said:

> Mr Hadfield does not rule out the possibility that the insurers will instruct us to challenge any possession proceedings commenced by your client on the basis that the Case D Notice of 6th April 1988

I interpose that the reference must be to the notice to pay of April 26 1988

> has the effect of creating a new tenancy from year to year notwithstanding the admitted validity of the notice to quit dated 17th March 1987 —

The latter is, of course, the Case E notice.

On March 10 Mr Connolly wrote saying among other things that he was instructed to accept service of any process but asking that Mr
B Jessel would not commence possession proceedings without prior notice to him. Mr Jessel was content to agree to that. On April 7 1989 there was a telephone conversation between Mr Connolly and Mr Jessel in the course of which Mr Jessel records in his attendance note that he said: "Now that the anniversary had passed", by which he was referring to the expiry of the Case G notice on April 1 1989, "we needed to know the position". It is I think clear, as Mr Brock submitted, that throughout that period and certainly on April 10 1989 those acting for the tenants, and no doubt the tenants themselves, knew that the landlords had possession proceedings either prepared or contemplated as imminent, that they were pressing for the information which they saw as necessary for the institution of
C those proceedings and, after April 1 1989, they were all the more likely to take those proceedings, having by then an additional string to their bow in the form of the Case G notice. Indeed, in due course on May 2 1989, Mr Connolly wrote to Mr Jessel enclosing a copy of the probate which had been issued on April 27 and saying: "May we take it that your clients will now be commencing possession proceedings."

The proceedings followed in due course being issued on June 1 1989. Those proceedings, it is right to record, rely in the statement of claim solely on the Case G notice. They do not, as they could no doubt have done, rely in the alternative on the Case E notice and, if that be found invalid, then on the Case G notice. Reliance is, however, put in the alternative on the Case E notice in the reply and defence to counterclaim, the defendants having by their defence and
D counterclaim asserted the validity of the Case E notice and thereby for the first time, as Mr Jessel said, enabled him to realise the significance of the observation in the letter of February 10 1989 about the "admitted validity" of the Case E notice. There had been no admission of the validity of that notice prior to the service of the defence and counterclaim. Against that background the landlord's case is short and simple. It is said that the tenancy vested in the executors on the death of Mr Sykes, that the Case G notice was duly served, that it expired on April 1 1989 and thereby the executors' right to remain in possession was brought to an end.

Mr Morgan, for the tenants, says that the position is far from being so simple. First, he says, the Case E notice was valid and terminated
E the tenancy created by the agreement of December 17 1985 on April 1 1988. He goes on to say that the executors remained in possession of the farm after April 1 1988, as undoubtedly they did, and that, unless their status at that time was that of trespassers, they must have been in possession on one basis or another which would result in their now having the benefit of a new tenancy quite distinct from that vested in the late Mr Sykes. Such a tenancy would have the protection of the Agricultural Holdings Act 1986, so that it could not be terminated except in accordance with the provisions of that Act. He points out that if they were tenants of any kind for an interest less than from year to year, or licensees under anything other than a purely gratuitous licence, the effect of section 2 of the 1986 Act is to convert whatever interest they would have had contractually and at common law into a
F tenancy from year to year with the protection of the 1986 Act. Accordingly, he says the crucial question is to examine the status of the executors as between themselves and Mrs Land in relation to the holding following April 1 1988.

Mr Brock, for Mrs Land, says that the Case E notice is invalid, but even if it is valid, he says the executors can only be entitled to possession if there was created, by agreement between Mrs Land and them, after April 1 1988, a new separate consensual interest in the land by way of a tenancy (on the footing that there is really no case for anything other than a tenancy) and he says that the necessary basis for a creation of a tenancy is simply not made out on the facts of the present case.

I must consider first the issue as to the validity or otherwise of the Case E notice. I have already quoted the material parts of the Case E G
notice and referred to the circumstances of the severance of the reversion which is relied on in relation to the contention that the notice was invalid. The effect of that severance was that Mrs Land was entitled to serve such a notice only in relation to that part of the land comprised in the 1985 tenancy agreement the reversion to which was vested in her. She was not entitled to serve such a notice in relation to the narrow strip that had been conveyed away, at all events unless she did so on behalf of and with the authority of the then owners of that strip, Mr and Mrs Chittenden. It is not suggested that she had that authority or that she held herself out, or was held out, as having any such authority.

It is well known that the severance of the reversion to a single
tenancy often gives rise to problems. It is also well known that H
problems have often arisen, either at common law or in the context of one or other of the various legislative regimes affecting the relationship of landlord and tenant, from unclear, ambiguous or mistaken notices served by one party on the other. So far as I am aware, there is no reported case in which these two types of problems have coincided. There were cited to me various cases concerning problems arising on notices, some of which were concerned with errors as regards the date for which notice was given and others with problems as regards the extent of the premises in relation to which the notice was given. The latter, however, have always given rise to a contention that the notice related to less land than was comprised in
the tenancy. In this case, by contrast, the suggestion is that the notice J
relates to more land than that in relation to which the landlord was entitled to give a notice.

It is a basic principle that a notice to quit must be clear and certain, so as to bind the party who gives it and to enable the party to whom it is given to act upon it at the time when it is given. It may not be ambiguous. On the other hand the court will not strain to find doubts, difficulties or ambiguities but will construe the notice as a whole in the light of the relevant surrounding circumstances and by reference to its effect on a reasonable tenant reading it. There was cited to me Goulding J's decision in *Carradine Properties Ltd* v *Aslam* [1976] 1 WLR 442 where at p 444 the judge referred to a number of early authorities which he said showed a general principle
that the court allows common sense to enter into the interpretation of K
an inaccurate notice to quit. He relied on passages such as the following:

> It is not required that a notice should be worded with the accuracy of a plea. This is not drawn with strict precision; but I think it is sufficiently clear.

See *Doe d Williams* v *Smith* (1836) 5 Ad & El 350.

> I agree, however, with the rule laid down . . . that we must look at the intention of the landlord, and that when language is used which leaves the effect of the notice open to doubt, the rule of construction is to make it sensible and not insensible.

See *Wride* v *Dyer* [1900] 1 QB 23, at p 25.

> But what is the rule of construction of such documents as this? . . . The
> court . . . will give an interpretation to the notice consistent with the L
> intention of the party serving it, if clear.

Ibid at p 26.

The learned judge summarised the position in the following words:

> I would put the test generally applicable as being this: Is the notice quite clear to a reasonable tenant reading it? Is it plain that he cannot be misled by it?

That was a case in which it was possible for the court to say not only that it was obvious that there was a mistake but also that it would be obvious to a reasonable tenant what the correct text should have been. That is because the notice was served in 1974 and it was expressed to be served in relation to a date in 1973 and the error in the circumstances was clearly a slip for 1975. Thus at p 446D the learned
judge said: M

> If that is right, I think a benevolent approach could be applied in this case, as in the *Duke of Bedford's* case, (1796) 7 Term Rep 63, because reasonably read by a reasonable tenant the mistake is obvious on the face of it, and there is no doubt what the mistake was.

As examples of the application of this principle there were cited to me two more recent cases. The first is *Cawley* v *Pratt* [1988] 2 EGLR 6*. This was an agricultural case in which a notice to quit was headed "Re the ho‚ding known as The Old Cow Shed and Buildings at the rear of the Wentloog Castle, Castleton, Newport, Gwent" and it was suggested that that was invalid because it referred to only part of the

---

*Editor's note: Also reported at [1988] 33 EG 54.

A holding in circumstances in which it was not open to the landlord to serve notice in relation to part only. That argument was rejected by the county court judge and by the Court of Appeal. Stocker LJ, at p 8M, said:

It seems to me that any reasonable tenant receiving the notice would see that he had to give up possession of "all that holding" and "holding", though the learned judge did not cite the section, is that which I have already cited as defined in section 1(1) of the Act of 1986.

Accordingly the notice was held to be valid.

The second case was *Herongrove Ltd* v *Wates City of London Properties plc* [1988] 1 EGLR 82* which was concerned with a notice under section 25 of the Landlord and Tenant Act 1954. In that case B the premises demised consisted of the ninth floor of a building in the City of London and two sets of storage premises and car-parking spaces physically substantially removed in distance from the office premises on the ninth floor. The notice, however, stated that it related to ninth-floor premises at the relevant address. Again this was a case in which it was not open to the landlord as a matter of law to serve notice in relation to only part of the premises contained in the demise. Harman J, applying the test enunciated by Goulding J, asked himself whether a reasonable tenant in the circumstances of the actual tenant, in his own premises, knowing what he has and with access to his own lease, being a businessman or business company, and receiving a notice which says that it relates to ninth-floor premises at the relevant address, would know what was meant by it. C Would it be quite clear to such a person that the notice includes not only the ninth floor but also the other parts of the demised premises? He answered that question as follows, at p 83:

To my mind, it is not [clear]. To my mind, any reasonable tenant knowing his position and knowing that he had other premises [elsewhere] in the same building, might well say, "I wonder if they are meaning to take back these offices but not to take back the car-parking and the storage spaces. I know that there is a great deal of talk about redevelopment of office premises. Perhaps all they want is to redevelop this floor as a whole of which I have only a part, and that they are going perhaps to offer me some other premises in the building and allow me to keep my car-parking spaces.

Of course there is no evidence, and there could not be any evidence, that the tenant actually thought that. There is in fact a simple, straightforward D counternotice by the tenant . . . Such being the facts, it is pure speculation as to what result this notice had upon the tenant, but I have never had any doubt since I first looked at it that in fact this notice does not tell a reasonable tenant, knowing the position on the ground, that he is expected to give up the whole of his premises. It might leave him in genuine doubt as to whether he is to give up part, that is the ninth floor, or not, and a notice that has an ambiguity of that sort about it seems to me to be fatally flawed.

It seems to me that that case is of some assistance, bearing in mind that if the tenant had had access to legal advice he would have been told that the landlord was not in fact entitled to serve the notice except in relation to the whole of the demised premises, just as here the tenant would have been told that Mrs Land could serve a notice only in relation to the holding less the strip. There was also cited to me E *Frankland* v *Capstick* [1959] 1 WLR 204, which contains useful citations of two earlier cases, *Mountford* v *Hodkinson* [1956] 1 WLR 422 and *Hankey* v *Clavering* [1942] 2 All ER 311. Mr Morgan points out that the judgments of the Court of Appeal in *Frankland* v *Capstick* support the proposition that the court does not allow a notice to be invalidated on merely technical points if the meaning is so plain that it cannot be misunderstood by the reasonable recipient. There was some debate as to whether the case supports the proposition that one may have regard to what the parties in fact knew about the relevant facts and circumstances. In my judgment, nothing turns on that question in this case. It is clear, for example from *Mountford* v *Hodkinson*, that what matters is not the effect of the notice on the actual recipient but its effect on a reasonable recipient: F hence Harman J's observation "there could not be evidence" as to what the tenant actually thought. It is, however, also clear that one must have regard to the actual facts and circumstances, including, here, the severance of the reversion. Both landlord and tenant knew of the severance, and it seems to me, therefore, that it matters not whether one takes an objective test, by reference to the actual facts, whether known, or takes the reasonable recipient but endows him with the actual recipient's actual knowledge.

Mr Brock, arguing for ambiguity such as to mislead, confuse and leave in doubt a reasonable recipient tenant, says that there is a contradiction between that part of the notice which says that it relates to "the holding known as Grange Farm etc and all land held therewith" on the one hand and that part of it which says that it G relates to the land "which the tenant holds of Mrs Land as tenant". He says that "the holding" is a well-known term in agricultural law and means, as it is defined by the statute, the whole of the land comprised in the contract of tenancy. Therefore, on the one hand, the heading to the notice and the first part of the words defining the subject-matter of the notice in its operative part refer to the land including the narrow strip. On the other hand, he says that part of the notice which says "the land which you hold of her as tenant" refers to the severed reversion retained by Mrs Land and therefore excludes the narrow strip.

Mr Morgan by contrast, arguing for validity, says that that approach is not consistent with the attitude taken by the courts. He H says that the notice must be taken as a whole, and construed together and in favour of consistency rather than of inconsistency. He points out particularly, in relation to the operative part of the notice, that the phrase "All that holding and premises known as Grange Farm etc and all land held therewith and situate at Wyverstone etc which you hold of her as tenant thereof" must be taken as a whole, and that the words "all that holding and premises" belong as a matter of grammar with the words "which you hold of her as tenant". He therefore says that on its true construction the notice relates only to Mrs Land's reversion and therefore relates only to that part of the land in relation to which she was entitled to give a notice. Alternatively, he says that even if it is expressed to relate to the whole there is no reason why it should not be given effect in relation to the J part.

In my judgment, while I can see the force of Mr Morgan's contentions as a matter of the true construction of the notice, it is Mr Morgan who is using the wrong test. The question is not what construction is given to the notice by a judge after argument at a trial. The question is, as enunciated by Goulding J: is the notice quite clear to a reasonable tenant reading it? Is it plain that he cannot be misled by it? In my view, a reasonable tenant, following the severance, would be confused or misled or left in doubt by this notice. He would see an apparent contradiction between the heading and the first part of the operative phrase, which on its face relates to the whole of the land in his contract of tenancy, and the words at the end which on K their face relate to the whole minus the narrow strip. It seems to me that it is nothing to the point that the reasonable tenant might be aware that it was only in relation to the part that Mrs Land could give a valid notice unless, which he had no reason to suppose, she was doing so with the authority of the owners of the narrow strip, and would, therefore, conclude that that was all that was comprised in the notice. That could have been an answer to the argument for the invalidity of the notice in *Herongrove* v *Wates* (above). Mr Morgan's alternative approach, that though it is expressed to relate to the whole it could be given effect as regards part only, seems to me to put the recipient of the notice in just as much of a state of doubt and confusion.

Accordingly, I hold that the Case E notice was invalid, as not L complying with the requirements of certainty which the law imposes as a condition of the validity of a notice to quit. It follows that the tenancy created by the 1985 tenancy agreement did not expire on April 1 1988 and that it was eventually brought to an end by the Case G notice on April 1 1989. It is common ground that on that footing the plaintiff is entitled to possession.

That being so, it is not strictly necessary to consider the other matters, which were argued at length before me, as to the effect of what happened after April 1 1988 if the tenancy had been brought to an end by virtue of the Case E notice. In deference to the arguments of counsel, which on this as on all aspects of the case were admirably clear, cogent and economical, and having regard to the importance of the case to the parties I propose to address those issues briefly. M

It is clear, as Mr Morgan submits, that if the defendants remained in possession after April 1 1988, the tenancy having expired under the Case E notice, as anything other than trespassers or on a purely gratuitous basis, they must have acquired an interest, whether licence or tenancy, which, if it was not itself a tenancy from year to year, would be converted into such a tenancy by section 2 of the Agricultural Holdings Act 1986. It is, however, also clear that they can have acquired only such an interest by agreement with Mrs Land, albeit possibly by an agreement implied from the acts of the parties and the circumstances. It is therefore necessary to consider what was said and done after April 1 1988 in order to assess the defendants' contention that a new agreement was reached.

*Editor's note: Also reported at [1988] 24 EG 108.

A As regards the legal principles relevant on this point, there was little at issue between counsel on this aspect of the case save on one point. It is common ground that a new tenancy or other interest can arise only by virtue of an agreement. Mr Morgan contended that, in line with the general principles as regards formation of contracts, the acts and words of the parties have to be considered objectively, and that evidence of their subjective intention is irrelevant and inadmissible. For this reason, he said, he would not call any of those concerned with the relevant dealings on the defendants' part, and, though not objecting to it being heard *de bene esse*, he argued that evidence from Mr Jessel of his subjective intention, not communicated to the other side, was inadmissible. He relied in particular on a recent, and not fully reported, decision of the Court of
B Appeal in *Javad* v *Aqil* on May 15 1990, reported in the *New Law Journal* on September 7 1990 at p 1232*.

Mr Brock, however, pointed to a line of cases concerned with the allegations of the creation of a new tenancy. These have usually been concerned with the position after the expiry of a previous tenancy, although it is agreed between counsel that there should be no difference in principle between this situation and a case, such as *Javad* v *Aqil*, where the court is concerned to determine the legal position resulting from informal dealing in advance of a contemplated, but never concluded, formal tenancy. In these cases the court has repeatedly said that what matters is the real intention of the parties: the earliest of these is *Doe d Cheny* v *Batten* (1775) 1
C Cowp 243, in which Lord Mansfield made this much quoted observation, at p 245:

> The question therefore is, *quo animo* the rent was received, and what the real intention of both parties was?

In support of this argument Mr Brock also relied on *Maconochie Brothers Ltd* v *Brand* [1946] 2 All ER 778, *Legal & General Assurance Society* v *General Metal Agencies* (1969) 20 P&CR 953, *Sector Properties Ltd* v *Meah* (1973) 229 EG 1097, *Longrigg, Burrough & Trounson* v *Smith* (1979) 251 EG 847 and *Cardiothoracic Institute* v *Shrewdcrest* [1986] 1 WLR 368†. Mr Morgan filled out the citation of the authorities relevant to this point with *Doe d Lord* v *Crago* (1848) 6 CB 90, *Clarke* v *Grant* [1950] 1 KB 104, and *Marcroft Wagons Ltd*
D v *Smith* [1951] 2 KB 496 as well as the very recent Court of Appeal case *Javad* v *Aqil* mentioned above. It is fair to say that most of the cases can be explained consistently with ordinary principles of the law of contract: either the references to the intentions of the parties could be to those intentions ascertained objectively, or, where a mistake on the part of one side is involved and has been held relevant, its admission in evidence can be explained as one which the other party caused or contributed to (as in *Doe d Lord* v *Crago),* or one which the other party must have been aware of (as in *Legal & General* v *General Metal Agencies*). But this does not account for all the cases. In *Clarke* v *Grant* the landlord was allowed to adduce evidence that rent had been accepted in a mistaken belief that it was due in arrear
E for a period of the expired tenancy, although there was no indication that the tenant knew of or had caused or contributed to that mistake. Because of the mistake it was held that there was no new tenancy, because the landlord had not, in fact, had the intention to create one. That was a judgment of the Court of Appeal, as was *Sector Properties Ltd* v *Meah* in which it was specifically argued for the tenant that only an objective test could be used, but the Court of Appeal rejected this argument: see p 1103. Consistent with those cases is an observation of Lord Scarman in the Court of Appeal in *Longrigg, Burrough & Trounson* v *Smith* (1979) 251 EG 847 at p 849. In that case, after the expiry of a tenancy of business premises, money described as rent was paid and accepted over a considerable period, but a partner in the landlord firm said in evidence that he thought it
F was being accepted as mesne profits and Lord Scarman said that that was evidence to which the court ought to have regard. Mr Morgan suggested that, at most, there is a special exception from the ordinary rules of contract law allowing the admission of evidence of a unilateral mistake, but that evidence of subjective intention may not be admitted more widely. That, however, would not be consistent with Lord Scarman's observation in *Longrigg* and, while this area may require re-examination and rationalisation, it is not appropriate, especially as it is not strictly necessary, for me to undertake that task.

---

*Editor's note: Now reported at [1990] 2 EGLR 82.
†Editor's note: Also reported at [1986] 2 EGLR 57.

The defendants' case for saying that a new interest was created is G pleaded as relying on two different matters: the service of the notice to pay rent on April 26 1988 and the tender, and ultimately acceptance, of rent in 1989. I do not see how the first of those can afford the basis for such a contention. The rent demanded was due entirely in respect of a period during the already expired tenancy, up to April 1 1988. Nothing therefore turns on the demand and payment of the rent as such. The statutory notice is, of course, couched in terms appropriate to a continuing tenancy, since the point of it is to lead, if the notice is not complied with, to a further ground for obtaining possession. Mr Morgan does not say that he can use this notice as part of a sequence of offer and acceptance necessary to the creation of the contract, and I accept Mr Brock's contention that that is fatal to Mr Morgan's submission. It was an appropriate notice to H serve, both because, in the past, service of such a notice had often had to be resorted to in order to obtain payment, and because the tenant had contended that the tenancy had not been brought to an end by the Case E notice. Mr Morgan, in his submissions, relied on the landlord's failure to serve a notice under section 83 of the 1986 Act within two months after April 1 1988 as an indication that the landlord was treating the defendant as lawfully entitled to possession. Such an omission, however, is entirely equivocal.

The substantial part of Mr Morgan's case in this respect turns on the tender, by the letter of January 23 1989, of £6,450, expressed to be rent for the (in my view inappropriately calculated) period April 6 to October 11 1988 — after the expiry of the Case E notice — and the J presentation of the cheque on April 10 or 11 1989 in the circumstances which I have described. Mr Morgan says that that sum was tendered as rent for that period, that its character as such could not be altered without an agreement which was sought but not forthcoming, and that it was accordingly accepted as rent, with the result that the defendants cannot be treated as having been anything other than tenants for the period in respect of which rent was paid. The force of that argument is clear and if there were an irrebuttable presumption that payment and acceptance of a sum labelled as rent created a tenancy, the result would follow. But one has to look further, undoubtedly, at the surrounding circumstances, and also, according to the line of authority discussed above, at the real intentions of the parties — in particular here of Mr Jessel — even if not manifested to K the other party.

Highly relevant among the surrounding circumstances is the fact that the defendants undoubtedly had had the benefit of a tenancy with the protection of the Agricultural Holdings Acts, that the tenancy had (on this assumption) expired but that the tenants denied its expiry, that on any footing they were liable to pay for their use and occupation of the land, at not less than the rate tendered, and that the tender was consistent with the contention they made, namely that the original tenancy still continued. It was not, indeed, or might not be, consistent with a new and different tenancy, since rent should on that footing have been tendered as from April 1 or 2 1988 — but that is a minor point. I bear in mind in this context, particularly, the facts, the L decisions, and certain observations of the judges in both *Marcroft Wagons* v *Smith* and *Longrigg, Burrough & Trounson* v *Smith.* Among many relevant passages in those judgments, I refer particularly to Denning LJ in *Marcroft Wagons Ltd* v *Smith* [1951] 2 KB 496, at p 506:

> In these circumstances, it is no longer proper for the courts to infer a tenancy at will, or a weekly tenancy, as they would previously have done from the mere acceptance of rent. They should only infer a new tenancy when the facts truly warrant it. The test to be applied in Rent Restriction Acts cases is the same test as that laid down by Lord Mansfield in cases of holding over

— and he quoted the observation already quoted above.

> If the acceptance of rent can be explained on some other footing than that a contractual tenancy existed, as, for example by reason of an existing or M possible statutory right to remain, then a new tenancy should not be inferred.

From the judgments in *Longrigg, Burrough & Trounson* v *Smith* again there are a number of relevant passages and I refer in particular to the following passage in Lord Scarman's judgment at p 849:

> Indeed one would have thought that today, where tenants have in one respect or another the protection of the law for possession of premises to which they would have at common law no contractual entitlement, the courts would not be as quick to infer a new tenancy as in the old days they would have been where there was nothing to explain the presence of a defendant upon the premises or upon the land other than a trespass or a contract.

Mr Morgan makes the point that that approach may be all very well, but the question is, given the payment of money in respect of the

A period of occupation: what is the basis of the occupation by the former tenants? And he points out that in *Marcroft Wagons* v *Smith* Denning LJ said that the occupier was there as licensee, which was sufficient to exclude the application of the Rent Restrictions Acts, but would not have been sufficient to exclude the operation of the Agricultural Holdings Acts. I bear in mind that point but it seems to me that, where one is concerned with one single payment made substantially after the period in respect of which it was tendered, it is not necessary to conclude that the occupier is doing anything more than making a payment for or on account of the liability which undoubtedly existed to pay for the benefit that he has of the occupation of the premises, whether it be by way of rent or by way of mesne profits or damages for the use and occupation of the land, a B liability which, if not discharged, would or might well also attract a further liability to interest.

What is also relevant among the circumstances of this case is that the landlord was, to the knowledge of the tenant, wishing to commence possession proceedings, and intending to do so as soon as his solicitors had what they regarded as necessary information for the preparation and institution of the proceedings. That is clear even without regard to the correspondence between February 10 1989 and April 7 1989, but it is even more clear from those documents. To suggest that the landlord could be supposed, by the tenant, to have intended to create a new tenancy in those circumstances by the acceptance of rent, on April 10 1989, just as the issue of proceedings C was becoming, at last, very close, given the progress on the probate side, seems to me fanciful. I am reminded of the trenchant words of Henn Collins J in *Maconochie Brothers* v *Brand* [1946] 2 All ER 778 as follows at p 779:

> That they did receive and pay into their bank a cheque paid in respect of a quarter's rent in advance, that quarter being after the expiration of the notice to quit, is true, but it is equally true that the question is: Was a new tenancy thereby created? That is, not a tenancy to be discovered by some act of law, but a tenancy created by two assenting minds. It is obvious that the one thing which the landlords wanted was possession of these premises, and to tell me that they assented to the creation of a new tenancy is to tell me what I do not believe. Therefore, if I have to find that there was a new tenancy, it would have to be because of some legal ingenuity and in face of what my mind informs me D is not the fact.

If reference is made to Mr Jessel's evidence as to what was in his mind, the position is made that much clearer. However, in fact, it seems to me that it is already abundantly clear, even without regard to that evidence, from what passed between him and the various E solicitors acting for the defendants, especially Mr Connolly. I record that, if I had to decide the point, I would not reach a different conclusion if I had not heard that part of Mr Jessel's evidence which was concerned with his uncommunicated state of mind.

Thus it seems to me that Mr Morgan's submissions rely on too formalistic an approach and disregard matters to which, even on his contentions, the court is required to have regard. The observations of Ormrod LJ in *Longrigg, Burrough & Trounson* v *Smith* seem apposite in this context:

> The old common law presumption of a tenancy from the payment and acceptance of a sum in the nature of rent dies very hard. But I think the authorities make it quite clear that in these days of statutory controls over the landlord's rights of possession, this presumption is unsound and no longer F holds. The question now is a purely open question; it is simply: Is it right and proper to infer from all the circumstances of the case, including the payments, that the parties had reached an agreement for a tenancy? I think it does not now go any further than that.

In the light of what was said and done between the parties and what was known to each party, if the question had arisen as necessary for decision, I would have held that the tender and acceptance of £6,450 in 1989 did not in the circumstances demonstrate or constitute an agreement to treat the defendants as having been tenants, either for the period mentioned in the letter of January 23 1989 or, more straightforwardly, for six months from April 1 1988, but would have treated it as paid, in the circumstances then prevailing, on account of an undoubted liability on the defendants' part to pay for their use and G occupation of the land either, as they then contended but now deny, as rent under the 1985 tenancy or by way of mesne profits or damages for their use and occupation of the land, if that tenancy had already been brought to an end. I would not, therefore, have held that the defendants were entitled to a new tenancy by virtue of anything done after April 1 1988.

As it is, however, for the reason that the Case E notice was not valid, I hold that the tenancy of December 17 1985 was terminated (in respect of the plaintiff's part of the severed reversion) by the Case G notice with effect from April 1 1989 and I will order the defendants to give possession accordingly, and dismiss the counterclaim.

The defendants are liable to pay mesne profits since then at a rate H that has been agreed between the parties of £13,500 pa but taking credit for any sums that may have been paid on account as interim payments.

A

Queen's Bench Divisional Court

June 11 1990

(Before Lord Justice FARQUHARSON and Mr Justice ALLIOTT)

MAY v VINCENT

Estates Gazette February 2 1991

[1991] 04 EG 144

Trade Descriptions Act 1968 — Application to auctioneers — Attribution in auction catalogue of a painting to "J M W Turner RA" — (B) Painting not by Turner — Information preferred against auctioneer by county trading standards officer — Distinction between section 1(1)(*a*) and 1(1)(*b*) of 1968 Act — Disclaimer — Appeal to divisional court — 1968 Act applies to art world and to auctioneers — Disclaimer not applicable to an offence under section 1(1)(*a*) — Magistrates wrong to acquit auctioneer but court informed that county council intended to clarify the position and not to prosecute him to conviction

A husband and wife read an auction catalogue which described a particular lot as "J M W Turner RA a watercolour of moorland, stream, bridge and people" — (C) The wife authorised a bid of up to £700 and in fact secured the painting at the sale for £400 — The painting was not by J M W Turner RA and appears to have been virtually worthless — The catalogue contained a comprehensive disclaimer of liability on the part of the vendor and the auctioneers — It was made clear in the resulting proceedings that there was no reflection on the honesty and integrity of the respondent auctioneer — The information charged the respondent with an offence under section 1(1)(*a*) of the 1968 Act by applying a false trade description in the catalogue

The magistrates concluded, rejecting an argument to the (D) contrary, that the 1968 Act did apply to the art world and artistic products; that the application of the doctrine of disclaimer was not clear; and that, although the respondent auctioneer did apply the false description in the catalogue, he was in so doing acting as agent for the true seller — They also considered that prosecution after such a long delay in bringing the matter before the court was oppressive — The purchasers had seen the catalogue in the autumn of 1986, the picture was collected in November 1986, the first complaint was made in April 1988, information laid in April 1989, and the hearing before the magistrates took place in August 1989 — An appeal from the magistrates' decision was made by case stated for the (E) opinion of the divisional court

The appellant, a trade protection officer acting for Derbyshire County Council, referred to the cases of *Newman* v *Hackney London Borough Council* and *R* v *Southwood* and submitted that an offence under section 1(1)(*a*) of the Trade Descriptions Act 1968 was committed when a false trade description was applied to goods; that the doctrine of disclaimer had no application when the offence alleged was under section 1(1)(*a*); and that the courts had made it clear that in respect of disclaimer there was a distinction between section 1(1)(*a*) (application of a false description to goods) and section (F) 1(1)(*b*) (sale of goods to which a false description was applied) — The court accepted these submissions — The point about agency mentioned by the magistrates failed, as auctioneers were covered by section 1(1)(*a*) in the same way as any other trade or business — As regards delay, the court agreed that it was very considerable in this case, but it was not a ground for acquittal

The appeal was allowed but, as the appellant had come for purposes of clarification and not with the intention of going back to the magistrates to prosecute to conviction, and as there had been considerable delay, no order was made as to (G) costs

The following cases are referred to in this report.

Aitchinson v *Reith & Anderson (Dingwall & Tain) Ltd* 1974 SLT 282; 1974 JC 12

Newman v *Hackney London Borough Council* (1982) 80 LGR 611; [1982] RTR 296

R v *Southwood* [1987] 1 WLR 1361; [1987] 3 All ER 556; [1987] RTR 273; [1988] TLR 39

This was an appeal by Philip May, an assistant county public protection officer of the trading standards division, for Derbyshire County Council, by way of case stated from the decision of the (H) Ashbourne justices acquitting the respondent, William Thomas Duncan Vincent [FRICS], auctioneer, of an offence under section 1(1)(*a*) of the Trade Descriptions Act 1968.

Patrick Gallagher (instructed by Kingsford Stacey, agents for Head of Legal Services, County Director's Department, Derbyshire County Offices, Matlock) appeared on behalf of the appellant; the respondent appeared in person.

Giving judgment, **Alliott J** said: On April 21 1989 an information was preferred by the appellant against the respondent that he, on or about October 27 1986 at Ashbourne, in the course of a trade or business did apply to certain goods, namely a watercolour painting, a false trade description by means of an entry in an auction (J) catalogue, namely:

J M W Turner RA a watercolour of moorland, stream, bridge and people, unframed $6\frac{1}{2}'' \times 9\frac{1}{2}''$

to the effect that it was by the said J M W Turner RA contrary to section 1(1)(*a*) of the Trade Descriptions Act 1968. The information was heard by magistrates sitting for the Petty Sessional Division of West Derbyshire at Ashbourne on August 17 1989. They found the following facts:

(a) In the autumn of 1986 Mr D J Smith and Mrs S A Smith read an auction catalogue produced by the respondent in which lot 345 was described in the way I have just read and that at that time Mrs Smith contacted the auctioneer and authorised a bid of up to £700.

(b) That the sale took place on October 29 1986 and that Mrs Smith (K) had been successful at £400.

(c) That Mr Smith collected the picture on November 1 1986 and paid £400.

(d) That the picture attributed to J M W Turner R A was not in fact by J M W Turner RA.

(e) That the respondent was of the opinion that the painting was not by J M W Turner RA.

(f) That the picture was virtually worthless.

(g) That there was a comprehensive disclaimer at the front of the catalogue.

A I myself, when reading the papers, have read that as relating to a block which appears in the catalogue under the heading "Misrepresentation Act 1967". That reads:

> The auctioneers act as agents only. Each lot set out in the catalogue, or as divided or joined with any lots at the sale at the sole discretion of the auctioneers, is sold with all faults, imperfections and errors of description, and neither the vendor nor the auctioneers are responsible for the authenticity, attribution, genuineness, origin, authorship, date, age, period, condition or quality of any lot. All statements in the catalogue as to any such matters are statements of opinion, and are not to be taken as, or as implying statements or representations of fact. Purchasers are deemed to have satisfied themselves before bidding by inspection or otherwise as to all such matters, and as to the physical description of any lot. Neither the vendor nor the auctioneers make or give, nor has any person in the employment of the
> B auctioneers any authority to make or give, any representation or warranty, and in any event neither the vendor, nor the auctioneers are responsible for any representation or warranty or for any statement in the catalogue.

It may well be that the magistrates had in mind paras 5 and 8 of the conditions of sale, also printed in the catalogue, which echo those sentences I have just read and which it would be otiose to read out as well. Certainly there was, on any showing, a comprehensive disclaimer in the catalogue.

I go back to the findings of fact of the magistrates:

(h) That the respondent was selling the picture for a client.

(i) That the respondent was responsible for the description in the catalogue.

C At paras 4 and 5 of the case stated the rival contentions were summarised:

> 4 It was contended by the appellant that the disclaimer in the catalogue did not constitute a defence when the respondent was of the opinion that the description of the picture in the catalogue was untrue or he suspected it to be untrue even though the words J M W Turner RA were printed on the surround to the picture.
>
> 5 It was contended by the respondent that the Trade Descriptions Act 1968 did not and could not apply to the Art World which deals in opinions not facts and that even if it did the disclaimer in the catalogue protected him. [The respondent] also claimed it was unjust to be prosecuted so long after the event — particularly as he had long since accounted for the receipts of the sale. He further contended that he had only attached to the picture the description which appeared on the "frame".

D It is right to interpose at this stage that Mr Gallagher has been at pains to emphasise that not in the conduct of the original prosecution, nor in the prosecution of this appeal by way of case stated, is it intended to reflect in any way upon the honesty and integrity of the respondent.

The magistrates went on to say at para 7:

> We were of the opinion that:
>
> (a) The Trade Descriptions Act did apply to the Art World.

However dismayed the respondent may be, we have no doubt that that is a correct finding by the magistrates.

> (b) The part regarding the use of a disclaimer is not clear and might benefit if it is broken into two sentences.
>
> E Although there is authority for saying a disclaimer can protect a person making a false statement it cannot protect a person making a statement knowing it to be false but it can protect a person unaware of the true facts or uncertain of the true facts if the disclaimer is sufficiently bold and makes the buyer aware of the fact that the article may not be genuine and is not warranted to be genuine — in such a case the doctrine of Caviat [*sic*] Emptor could apply.
>
> (c) There had been long delay in this case being brought before the court and that it was oppressive to prosecute in a case such as this two years after the event and for the matter to be disposed of in very little under three years after the event. The picture was collected on the 1st November 1986 — complaint was made to the Trading Standards Officer on the 27th April 1988 and an information was laid on the 21st April 1989.
>
> (d) That although [the respondent] applied the description in the catalogue
> F to the painting he was in so doing acting as agent for the true seller.

Thereafter they posed for the opinion of this court:

> Were we correct in reaching the decision that we did when this matter was heard?

and then they get the date [of the notice of appeal] wrong. They say "1st June 1989" and it should be October 17 1989.

I am bound to say, in parenthesis, that it is rather difficult for this court to answer that question in the form that they have posed it because they do not expressly make their decision known.

In support of this appeal the appellant puts forward a number of propositions of law:

1 That the provisions of section 1(1)(*a*) of the Trade Descriptions Act 1968 apply to auctioneers. If authority be needed for this proposition he points to *Aitchinson* v *Reith & Anderson (Dingwall & Tain) Ltd* 1974 SLT 282 as persuasive authority. In fact nothing in the G legislation suggests that auctioneers should not be governed by the legislation like every other trade or business and I have no doubt they are so bound.

2 That an offence under section 1(1)(*a*) of the Trade Descriptions Act 1968 is committed when the false trade description is applied to the goods.

3 That the doctrine of disclaimer has no application to an offence of applying a false trade description, contrary to section 1(1)(*a*) of the Trade Descriptions Act 1968. In support of the second and third propositions the appellant relies upon two authorities. First, *Newman* v *Hackney London Borough Council* [1982] RTR 296 in which the Divisional Court held: H

> (1) that an offence under section 1(1)(*a*) of the Trade Descriptions Act 1968 was committed when the false trade description was applied to the goods . . .; so that, in the present case, the odometer reading having been altered to record the meaningful figure of 21,000 miles with the obvious object of deceiving somebody, the false trade description was applied when the odometer reading was so altered.
>
> (2) That the common law doctrine of disclaimer had no application to an offence of applying a false trade description contrary to section 1(1)(*a*) . . .

In giving the judgment of the court, Ormrod LJ said at p 302G:

> In the course of a long and very careful judgment, the circuit judge, sitting at the Crown Court, held as a matter of law that the disclaimer doctrine, which had been developed by the courts in connection with the second part of section 1(1) of the Trade Descriptions Act 1968, did not apply to cases of applying a J false trade description in contrast to those cases where the charge was one of supplying goods or articles to which a false trade description had been applied. This is the point with which the court is dealing at present.
>
> Section 1(1) provides:
>
> "Any person who, in the course of a trade or business,
> (*a*) applies a false trade description to any goods; or
> (*b*) supplies or offers to supply any goods to which a false trade description is applied;
> shall, subject to the provisions of this Act, be guilty of an offence."

The distinction which the Crown drew was the distinction between applying and supplying. Mr Buckhaven submits that there is no valid distinction between those two offences for this present purpose and K that the disclaimer doctrine should apply equally to those who apply a false trade description.

In my judgment, there is a world of difference between the two offences. It is perfectly true that the application of a false trade description must, in some way, be related to a sale or prospective sale but, looking at the Act itself, I am disposed to take the view that the offence is committed when the false trade description is applied to the vehicle or goods and that is at the time when the odometer reading is altered to read a meaningful figure like 21,000 miles. In that light, a disclaimer has no application at all.

The second case relied on by Mr Gallagher is that of *R* v *Southwood* [1987] RTR 273 in which the Lord Chief Justice, sitting in the Court of Appeal Criminal Division, expressly approved the L *Newman* case and specifically approved the distinction between section 1(1)(*a*) and section 1(1)(*b*) of the 1968 Act, and went on to hold that it is "not open to a person charged under section 1(1)(*a*) to rely on any disclaimer".

Those propositions, urged upon us by Mr Gallagher, are, in my view, well founded. Subject to the issue of delay, they would dispose of this appeal because the other point raised by the justices, that of agency, falls, in my view, as soon as we hold, as I have done, that the auctioneers are covered by section 1(1)(*a*) as in any other trade or business.

Newman's case was cited to the magistrates and had they applied it correctly they would have been bound to convict. The delay in bringing this matter to court, however, was indeed considerable. The M picture was collected in November 1986, the first complaint was made in April 1988, the information was laid on April 21 1989 and the hearing eventually in the magistrates' court was on August 17 1989. It is apparent that Mr Vincent, the respondent, acting, then as now, in person, did not specifically raise the preliminary issue of the delay amounting to abuse of the process of the court, but merely commented, one suspects in strong terms, when he came to give his evidence to the magistrates. In those circumstances, that is not a sufficient basis for the court holding that the delay is in any way a basis for the acquittal and equally that is a matter that the justices should not have allowed to sway them in coming to the decision that they did.

A However, Mr Gallagher, sensibly in my view, has indicated to the court quite openly that the purpose of coming to this court is not with the intention of going back to the magistrates to prosecute the respondent to a conviction but merely to clarify the position in law, which I hope this judgment will go some way to doing. So far as the delay is concerned, it may well be that Farquharson LJ will be able to reflect that in any ancillary order that may follow the judgment of B this court.

Farquharson LJ agreed and did not add anything.

The appeal was allowed. The court made no order as to costs.

For a further case on this subject see p 245

COMPULSORY PURCHASE AND COMPENSATION

A
House of Lords

December 13 1990

(Before Lord BRIDGE OF HARWICH, Lord TEMPLEMAN, Lord GRIFFITHS, Lord ACKNER and Lord LOWRY)

HUGHES AND ANOTHER v DONCASTER METROPOLITAN BOROUGH COUNCIL

Estates Gazette February 9 1991

[1991] 05 EG 133

B **Compensation for acquisition of land — Land Compensation Act 1961, rule (4) in section 5 — Appeal from decision of Court of Appeal — Whether rule (4) in section 5 (account not to be taken of increase in value of land due to use contrary to law) applies to compensation for disturbance — Whether use of land which was immune from enforcement action (having been carried out before 1963) could nevertheless be "contrary to law" within rule (4) — Minority view of Dillon LJ on both questions upheld by the House of Lords**

In the present case the compulsory acquisition resulted in the closing down of the claimants' business, that of scrap metal and rag merchants, which had been developing
C **successfully until the acquisition — The crucial factor, which gave rise to the questions raised in these proceedings, was that at no time had planning permission been sought or obtained for the uses of the business — An important distinction, however, had to be made between the "blue land" and the "green land" (areas identified by their colours on a plan) — The point of the distinction was that the blue land had, because of the date of its unauthorised development, become immune from enforcement proceedings, whereas the green land did not enjoy this immunity — The claimants claimed compensation on the basis of the extinguishment of their**
D **business — The case raised two questions of importance on which surprising differences of opinion arose**

The first question was whether rule (4) applied, as the acquiring authority contended, to the assessment of the compensation generally, including any element referable to disturbance, or whether, as the claimants contended, it applied only to the market value of the land acquired — The second question was whether a use of land begun between July 1 1948 and December 31 1963, without planning permission, was, as the authority contended but the claimants disputed, a use which was "contrary to law" under rule (4) notwithstanding that it was an established use immune by
E **statute from enforcement proceedings — The Lands Tribunal (Mr Victor Wellings QC) answered both questions in the claimants' favour — On appeal by case stated to the Court of Appeal the majority of the court (Staughton and Mann LJJ) affirmed the tribunal's answer to the first question in favour of the claimants, but answered the second question in favour of the authority, reducing the tribunal's valuation of the blue land accordingly — Dillon LJ, dissenting on both points, answered the first question in favour of the acquiring authority and the second in favour of the claimants**

In the House of Lords, Lord Bridge of Harwich, after F
referring to well-known authorities, such as *Horn* v *Sunderland Corporation*, pointed out that, although compensation in respect of the market value of land acquired and compensation for disturbance must in practice be separately assessed, these two elements have been consistently treated as parts of a single whole, together constituting "the value of the land" to the owner — This basic principle was perfectly well known to the draftsman of section 2 of the Acquisition of Land (Assessment of Compensation) Act 1919 (re-enacted in section 5 of the Land Compensation Act 1961) — When he used the phrase "the value of the land" in rule (4) G
he must have used it in the sense which it would have conveyed to any lawyer familiar with the 1845 code as judicially interpreted, ie as embracing both elements — Dillon LJ had correctly answered the first question in favour of the acquiring authority

On the second question, as to whether uses which had been rendered by statute immune from enforcement action could nevertheless be branded as "contrary to law", the majority of the Court of Appeal had felt constrained by the decision in *LTSS Print & Supply Services Ltd* v *Hackney London Borough Council* to hold that they could — Here again Dillon LJ's minority view was vindicated by the House of Lords — H
The House did not overrule this decision, which may have been correct in its narrow context, but the majority of the Court of Appeal erred in treating it as an authority for holding, in an entirely different context, that an established use could be "contrary to law" for the purpose of rule (4)

The result was that the acquiring authority's appeal and the claimants' cross-appeal were both allowed — The effect of allowing the cross-appeal was to restore the tribunal's valuation of the uncovered part of the blue land — The effect of allowing the appeal was that the compensation for disturbance would have to be reassessed on the basis that the J
claimants were entitled to be compensated for the effect of the acquisition on their business in relation only to the blue land

The following cases are referred to in this report.

Attorney-General v *Smith* [1958] 2 QB 173; [1958] 3 WLR 81; [1958] 2 All ER 557; (1958) 56 LGR 295; 9 P&CR 439
Colonial Sugar Refining Co Ltd v *Melbourne Harbour Trust Commissioners* [1927] AC 343
Commissioners of Inland Revenue v *Glasgow & South-Western Railway Co* (1887) 12 App Cas 315
Hartnell v *Minister of Housing and Local Government* [1965] AC 1134; [1965] 2 WLR 474; [1965] 1 All ER 490; (1965) 63 LGR 103; 17 P&CR 57, HL K
Horn v *Sunderland Corporation* [1941] 2 KB 26; [1941] 1 All ER 480, CA
LTSS Print & Supply Services Ltd v *Hackney London Borough Council* [1976] QB 663; [1976] 2 WLR 253; [1976] 1 All ER 311; (1975) 74 LGR 210; 31 P&CR 133; 240 EG 711, CA
M'Ardle v *Glasgow Corporation* 1972 SC 41
Ricket v *Metropolitan Board of Works* (1865) 31 LJ (QB) 257
Stebbing v *Metropolitan Board of Works* (1870) LR 6 QB 37

This was an appeal by Doncaster Metropolitan Borough Council and a cross-appeal by the claimants, Charles Arthur Hughes and Nora Louisa Hughes, from the decision of the Court of Appeal (reported at [1990] 1 EGLR 40; [1990] 26 EG 168) in relation to

A compensation for the compulsory acquisition of the claimants' freehold land and premises at Black Bank, Doncaster Carr, Doncaster. The Lands Tribunal (Mr Victor Wellings QC) had awarded the claimants compensation of £551,160, together with surveyors' fees.

Lionel Read QC and Simon Pickles (instructed by Sharpe Pritchard for Mr W R Bugler, director of legal services, Doncaster Metropolitan Borough Council) appeared on behalf of the council; Robert M K Gray QC, Robin Campbell and Neil Calver (instructed by Gregory Rowcliffe & Milners, for Taylor Bracewell, of Doncaster) represented the claimants.

In his speech, **Lord Bridge of Harwich** said: This appeal raises
B two questions of law in respect of the compensation payable pursuant to section 5 of the Land Compensation Act 1961 for land acquired by compulsory acquisition. The land in question was acquired by the Doncaster Metropolitan Borough Council ("the acquiring authority") from Mr and Mrs Hughes ("the claimants") pursuant to a compulsory purchase order made in 1973. Notice to treat was served in January 1976 and the acquiring authority took possession on November 1 1981. The total area of the site acquired was 3.872 acres. The whole site was used by the claimants until the date of acquisition for the purpose of their business as merchants dealing in scrap metal and rags. Part of the site ("the blue land") had been used for the same purpose since 1959. The claimants purchased the blue land in 1969 together with the goodwill of the scrap metal and
C rag merchant's business carried on there by their predecessors in title. The claimants purchased the remainder of the site ("the green land") in 1972 and the use of the green land for the purpose of the business began then. No permission had ever been granted under the Town and Country Planning Acts for the development involved in the material change of use of the blue land and the green land when they were first used for the purpose of the scrap metal and rag merchant's business in 1959 and 1972 respectively. Thus, when the site was acquired the position under the Town and Country Planning Act 1971 was that the use of the blue land, being a use begun before the end of 1963, was an "established use" as defined by section 94 and was immune from enforcement proceedings under Part V, but the use of the green land remained liable to enforcement proceedings
D requiring its discontinuance.

The two questions of law raised by the appeal depend upon the true construction of the rules for the assessment of compensation in respect of any compulsory acquisition originally enacted by section 2 of the Acquisition of Land (Assessment of Compensation) Act 1919 and now re-enacted, with no material amendment, by section 5 of the consolidating Act of 1961. They provide as follows:

(1) No allowance shall be made on account of the acquisition being compulsory;

(2) The value of land shall, subject as hereinafter provided, be taken to be the amount which the land if sold in the open market by a willing seller might be expected to realise;

E (3) The special suitability or adaptability of the land for any purpose shall not be taken into account if that purpose is a purpose to which it could be applied only in pursuance of statutory powers, or for which there is no market apart from the special needs of a particular purchaser or the requirements of any authority possessing compulsory purchase powers;

(4) Where the value of the land is increased by reason of the use thereof or of any premises thereon in a manner which could be restrained by any court, or is contrary to law, or is detrimental to the health of the occupants of the premises or to the public health, the amount of that increase shall not be taken into account;

(5) Where the land is, and but for the compulsory acquisition would continue to be, devoted to a purpose of such a nature that there is no general demand or market for land for that purpose, the compensation may, if the Lands Tribunal is satisfied that reinstatement in some other place is bona
F fide intended, be assessed on the basis of the reasonable cost of equivalent reinstatement:

(6) The provisions of rule (2) shall not affect the assessment of compensation for disturbance or any other matter not directly based on the value of land . . . ;

The first question is whether rule (4) applies, as the acquiring authority contend, to the assessment of compensation generally, including any element referable to disturbance, or whether, as the claimants contend, it applies only to the assessment of the market value of the land acquired and not to the assessment of compensation for disturbance.

The second question is whether a use of land begun between July 1 1948 and December 31 1963 without the grant of planning permission under the Town and Country Planning Acts is, as the acquiring authority contend but the claimants dispute, a use which is "contrary G
to law" within the meaning of rule (4) notwithstanding that it is an established use which is immune from enforcement proceedings.

The question of the amount of compensation to which the claimants were entitled was referred to the Lands Tribunal pursuant to the Act of 1961 and determined by Mr Victor Wellings QC. He answered both questions of law in the claimants' favour. Holding that rule (4) did not apply to compensation for disturbance, he included in his assessment a sum of £300,000 as compensation for loss of the goodwill of the scrap metal and rag merchant's business as carried on over the whole site. Holding that the business use of the blue land was not contrary to law, he valued the uncovered part of the blue land, one acre in extent, at £17,500 as reflecting the business use. H
It was, however, common ground that the business use of the green land, being liable to enforcement proceedings, fell within rule (4). Mr Wellings accordingly disregarded the enhancement in value which the business use attracted to the uncovered part of the green land. On this basis he valued the green land at £10,000 per acre. The total compensation, including the value of buildings and certain other items not relevant to the appeal, he determined in the sum of £551,160.

The acquiring authority appealed to the Court of Appeal by case stated. The majority of the court (Staughton and Mann LJJ) [1990] 1 WLR 845* affirmed Mr Wellings' answer to the first question of law in favour of the claimants and accordingly dismissed the appeal in relation to the award of compensation for disturbance, but answered J
the second question of law in favour of the acquiring authority and accordingly allowed the appeal to the extent of reducing the valuation of the blue land from £17,500 to £10,000 to exclude the enhancement attributable to the business use, which they held to be contrary to law. Dillon LJ dissented on both points. He answered the first question of law in favour of the acquiring authority and the second in favour of the claimants. By leave of the Court of Appeal the acquiring authority now appeal and the claimants cross-appeal to your lordships' House.

It is well-settled law that whatever compensation is payable to an owner on compulsory acquisition of his land in respect of disturbance is an element in assessing the value of the land to him, not a distinct and independent head of compensation. This is because, K
under section 63 of the Lands Clauses Consolidation Act 1845 (8 Vict c 18), the substance of which is now re-enacted by section 7 of the consolidating Compulsory Purchase Act 1965, "the value of the land to be purchased by the acquiring authority" is the only head of compensation under which compensation for disturbance is capable of being accommodated. The other heads of compensation for which the section provides, severance and injurious affection, relate only to the depreciatory effect of the acquisition on other land retained by the owner. Judicial interpretation of the Act of 1845 held that the value of the land meant its value to the owner, not its value to the acquiring authority: *Stebbing* v *Metropolitan Board of Works* (1870) LR 6 QB 37. This value was to be assessed as including all the loss L
which the owner suffered in consequence of being dispossessed: *Ricket* v *Metropolitan Railway Co* (1865) 34 LJQB 257.

In *Commissioners of Inland Revenue* v *Glasgow & South-Western Railway Co* (1887) 12 App Cas 315 a jury had assessed compensation payable to a landowner under the parallel provisions of the Lands Clauses Consolidation (Scotland) Act 1845 in separate sums representing the value of the land and buildings on the one hand and compensation for loss of business on the other. The conveyance on sale attracted *ad valorem* stamp duty under the Stamp Act 1870 and it was held that the sum allowed as compensation for loss of business was part of the consideration for the sale liable to the duty. Lord Halsbury LC said, at pp 320-321:

The two things, and the only two things, which are within the ambit and M
contemplation of the statute, are the value of the lands and such damages as may arise to other lands held therewith by reason of the particular land which is taken being taken from them.

. . . That was alone what the jury in this case had power to assess, because it is admitted that no question arises here upon the other part of the section — no question arises here about any damage from severance. It is admitted, therefore, impliedly, that the only thing which the jury had here to assess was the value of the land. . . . In treating of that value, the value under the circumstances to the person who is compelled to sell (because the statute compels him to do so) may be naturally and properly and justly taken into account; and when such phrases as "damages for loss of business" or "compensation for the goodwill" taken from the person are used in a loose

*Editor's note: Also reported at [1990] 1 EGLR 40.

A and general sense, they are not inaccurate for the purpose of giving verbal expression to what everybody understands as a matter of business; but in strictness the thing which is to be ascertained is the price to be paid for the land — that land with all the potentialities of it, with all the actual use of it by the person who holds it, is to be considered by those who have to assess the compensation.

In *Horn* v *Sunderland Corporation* [1941] 2 KB 26 the question posed was whether, in the application of rules (2) and (6) under section 2 of the Act of 1919, the owner of agricultural land acquired compulsorily which had a market value as land ripe for building development far in excess of its value as agricultural land was entitled, in addition to that value, to compensation for the loss of his farming business. By a majority (Sir Wilfrid Greene MR and Scott
B LJ, Goddard LJ dissenting) the court held that he was not. Sir Wilfrid Greene MR said, at pp 34-35:

> Now, rule 6 does not confer a right to claim compensation for disturbance. It merely leaves unaffected the right which the owner would before the Act of 1919 have had in a proper case to claim that the compensation to be paid for the land should be increased on the ground that he had been disturbed. It is true that the arbitrator now cannot well avoid doing what a jury under the Lands Clauses Acts was not bound to do, namely, arrive at one figure for the value of the land and (in a proper case) another figure for disturbance, but these two figures are still, in my opinion, merely the elements which go to build up the global figure of price or compensation payable. . . . The truth of the matter is that, as in cases under the Lands Clauses Acts alone, so in cases where the Act of 1919 applies, the sum to be ascertained is in essence one sum,
> C namely, the proper price or compensation payable in all the circumstances of the case. If those circumstances are such as to make it impossible for the owner to claim that he has suffered damage through disturbance for which he ought to be compensated, then he is not entitled to have the price or compensation for his land increased by an addition for disturbance even if he has in fact been disturbed. It is a mistake to construe rules 2 and 6 as though they conferred two separate and independent rights, one to receive the market value of the land and the other to receive compensation for disturbance, each of which must be ascertained in isolation.

Horn's case has been accepted ever since as settling an important principle in relation to the operation of the compensation rules.

In *M'Ardle* v *Glasgow Corporation* 1972 SC 41 the claimant landowner carried on business as a publican on the land acquired. In part of the public house premises the business was carried on under a
D temporary planning permission which, at the date of acquisition, had less than two years to run. Under the Land Compensation (Scotland) Act 1963, which embodies the same rules for assessing compensation as section 5 of the English Act of 1961, sections 22(1) and 24(4), also corresponding to parallel provisions in the Act of 1961, required that "in ascertaining the value of the relevant interests" it should be assumed that planning permission would be granted for the public house use. It was contended for the authority that, while the statutory planning assumption permitted the market value of the premises to be assessed under rule (2) as having full permission for use as a public house, compensation for disturbance of the publican's business should nevertheless be assessed on the basis that he would in any
E event have had to cease using part of the premises for the purpose of that business on expiry of the temporary planning permission. The Court of Session rejected this contention. The Lord President, Lord Clyde, in delivering the opinion of the court, said, at pp 47-48:

> But the fact is that nowhere in this series of statutory provisions since 1845 is a disturbance claim treated as something separate and distinct from the rest of the claim. It is an element in the total computation of the compensation. The distinction which the acquiring authority seek to make between a disturbance claim and a claim for the value of the land is thus a false distinction. They are both elements in the value of the relevant interest within the meaning of that phrase in this series of Acts of Parliament.

Thus, although compensation in respect of the market value of land acquired and compensation for disturbance must in practice be
F separately assessed, the courts have consistently adhered to the principle, both before and after the present rules were first introduced by the Act of 1919, that the two elements are inseparable parts of a single whole in that together they make up "the value of the land" to the owner, which, unless he retains other land depreciated by severance or injurious affection, was the only compensation which the 1845 code awarded to him.

This principle must have been perfectly well known to the draftsman of the Act of 1919 and when he used in rule (4) the phrase "the value of the land," meaning the value of the land acquired, it seems to me that he must have used it in the sense it would have conveyed to any lawyer familiar with the 1845 code as judicially interpreted, ie as embracing both elements of compensation, unless the context clearly calls for a narrower meaning. For my part I can see G nothing in the context which does call for a narrower meaning. Rule (2) introduces the new principle that "the value of land" (not, be it noted, "the value of *the* land") is to be the open market value obtainable by a willing seller. Rule (6) is a necessary qualification of that principle in order to preserve the disturbance element which is to be unaffected by rule (2); but rule (6) makes no reference to rule (4).

Over and above these considerations the most formidable obstacle which, to my mind, confronts the claimants' construction of rule (4) is that it imputes to the legislature a quite paradoxical intention. If, as I take to be obvious, it is just and right that the owner of land acquired compulsorily should not receive in compensation any enhancement of the market value of the land attributable to a use of the land in any of the categories to which rule (4) applies, I can conceive of no reason H why the legislature should nevertheless have thought it just and right that he should be compensated for being prevented from continuing any such use. Accordingly, in agreement with Dillon LJ, I would answer the first question in favour of the acquiring authority and allow the appeal.

I turn now to the cross-appeal. Ever since July 1 1948, when the Town and Country Planning Act 1947 came into force, planning permission has been required for the carrying out of any development of land and the definition of development has included the making of any material change in the use of land. Hence, a use of land begun without planning permission, although it could be stopped under the Act only by the somewhat cumbrous machinery J for the enforcement of planning control initiated by the service of an enforcement notice, was unlawful *ab initio: Attorney-General* v *Smith* [1958] 2 QB 173. But the Act of 1947 imposed a time-limit for the service of an enforcement notice of four years from the date when the development to which the notice related was carried out. This applied to any form of development, including making a material change of use. Under the 1947 Act regime it was customary to use the phrase "existing use right" to describe the right enjoyed by an occupier of land in relation either to a use of land begun before July 1 1948 or to a use begun after that date without planning permission which had continued for more than four years without service of an enforcement notice. So long as a use in either category continued, the nature of the right which the user enjoyed was for all practical K purposes the same.

Under the Caravan Sites and Control of Development Act 1960 express planning permission to use land as a caravan site was a prerequisite to the grant of a caravan site licence. In *Hartnell* v *Minister of Housing and Local Government* [1965] AC 1134 this House held that in granting such a permission in relation to land which had been used as a caravan site without planning permission for more than four years the local planning authority were not entitled to impose conditions which had the effect of cutting down the scope of the owner's existing use right. Lord Evershed said, at p 1163:

> . . . it will be observed that what has in the case before your Lordships been L referred to as the respondent's "prescriptive right" as regards the northern part of his field OSP 145 is derived from the effect of section 23 of the Act of 1947; for at the time of the coming into operation of the Act of 1960 the respondent, though he had never obtained permission for its use under the Act of 1947, had in fact used it as a caravan site for upwards of four years and no notice for discontinuance under that section had ever been served upon him by the local planning authority. It follows that the respondent had under the existing legislation at the relevant date what may fairly be called a "vested right" to the use of this part of the field as a caravan site to which the local planning authority could only put an end by the procedure under section 26 of the Act of 1947, which would involve payment of compensation under the succeeding section.

Lord Wilberforce, at p 1169D, described the existing use right as "a form of vested right analogous to a right established by M prescription".

Under the Town and Country Planning Act 1971 any use of land begun without planning permission since the end of 1963 is now amenable to enforcement proceedings without limit of time. Four years remains the time-limit for the service of an enforcement notice relating to building development carried out without planning permission. The important status of a use of land begun at any time before the end of 1963 and continued ever since that date is recognised by the procedure under sections 94 and 95 and Schedule 14 which enable any person interested in the land, on proof of the relevant facts relating to that use, to obtain either from the local planning authority or from the Secretary of State on appeal what the

A Act calls an "established use certificate", which operates as a conclusive bar to protect the use, so long as it continues, against the operation of an enforcement notice.

If the local planning authority serve an enforcement notice within the four-year time-limit in respect of a building erected without planning permission or at any time in respect of a use of land begun without planning permission since the end of 1963 the notice, when it takes effect, may, of course, require the removal of the building or the discontinuance of the use without payment of compensation. If, however, the local planning authority seek to secure either the removal of a building or the discontinuance of a use of land at a time when the building or use is no longer amenable to enforcement proceedings, they may do so only by order under section 51 subject to
B all the safeguards which that section provides and subject also to the obligation to pay compensation under section 170. The measure of compensation is the damage which any person has suffered "by depreciation of the value of any interest in the land to which he is entitled" or "by being disturbed in his enjoyment of the land": section 170(2). Section 178(1) then provides, in relation to compensation for depreciation but not in relation to compensation for disturbance, that:

> The rules set out in section 5 of the Land Compensation Act 1961 shall, so far as applicable and subject to any necessary modifications, have effect as they have effect for the purpose of assessing compensation for the compulsory acquisition of an interest in land.

C In the light of "the well-known principle that a statute should not be held to take away private rights of property without compensation unless the intention to do so is expressed in clear and unambiguous terms" (see *Colonial Sugar Refining Co Ltd* v *Melbourne Harbour Trust Commissioners* [1927] AC 343, at p 359, *per* Lord Warrington) and in the light of the provisions of the Act of 1971 to which I have referred, it seems to me impossible to treat an established use under the Act of 1971 as being "contrary to law" within the meaning of rule (4) of the compensation rules. The right to such a use is aptly described by Lord Wilberforce as "analogous to a right established by prescription" and the legitimate status of the use is now expressly recognised by the statutory procedure for giving it the imprimatur of an established use certificate. If, on compulsory acquisition, a
D landowner is not entitled to compensation for the enhanced value of land attributable to an established use or for being disturbed in the enjoyment of such a use, it would often be cheaper for a local planning authority to acquire the land compulsorily than to secure the discontinuance of the use by order under section 51 of the Act of 1971. Moreover, if rule (4), as applied by section 178 of the Act of 1971 to the assessment of compensation under section 170, were construed as applicable to an established use as "contrary to law", this would produce the bizarre result that section 178 would operate to nullify any right to compensation conferred by section 170(2) in respect of damage suffered by depreciation of the value of land caused by discontinuance of an established use, but would have no
E effect on compensation for disturbance occasioned by the same discontinuance. If the legislature had intended that any distinction were to be made in the compensation payable under section 170 in respect of a discontinuance order under section 51 between, on the one hand, a use of land begun before July 1 1948 or with planning permission after that date and, on the other hand, a use of land begun between July 1 1948 and December 31 1963 without planning permission, it is surely inconceivable that they would not have done so in clear and explicit terms.

As Dillon LJ, in his dissenting judgment, aptly observed [1990] 1 WLR 845 at p 852E:

> Indeed it would be monstrous if the acquiring authority could avoid paying compensation for the disturbance, by closing down of the business carried on
> F by the claimants on the blue land, when that business, in so far as carried on on the blue land, could have legitimately been sold by the claimants to any other purchaser of the blue land.

The majority of the Court of Appeal felt constrained to reach an opposite conclusion by the authority of *LTSS Print & Supply Services Ltd* v *Hackney London Borough Council* [1976] QB 663. The question in that case was whether a landowner, being required by an enforcement notice to discontinue a use of land begun without planning permission since the end of 1963, was entitled to resume the use being made of the land immediately before the development which was the subject of the enforcement notice, if that previous use was begun before the end of 1963. The answer to that question depends on section 23(9) of the Act of 1971, which provides:

> Where an enforcement notice has been served in respect of any development G
> of land, planning permission is not required for the use of that land for the purpose for which (in accordance with the provisions of this Part of this Act) it could lawfully have been used if that development had not been carried out.

The Court of Appeal held that the previous use, although it had been immune from enforcement proceedings, was not a use for "the purpose for which (in accordance with this Part of this Act) [the land] could lawfully have been used" if the subsequent development had not been carried out. Your lordships have not been invited to overrule this decision, nor do I see any reason to do so. All three members of the court emphasised the importance of the words in parenthesis as justifying a narrow construction of the subsection and with this I have no quarrel. But in so far as the reasoning in the judgments goes
further and may suggest that a use of land begun without planning H
permission between July 1 1948 and December 31 1963 must be treated for all purposes and in all contexts as unlawful I am unable to agree with it. Section 23(9) is concerned solely with the question whether a use voluntarily discontinued and unlawfully supplanted by another materially different use, which is then forcibly discontinued, may be resumed (involving another material change) without planning permission. Looking at the underlying policy of the Act there is good reason why the legislature should have wished to authorise such resumption only of uses lawfully begun before July 1 1948 or thereafter with planning permission. This supports the narrow construction which the court adopted, but the words
"contrary to law" in rule (4) of the compensation rules appear in a J
totally different context and relate to a totally different subject-matter. To hold that these words apply to an established use under the Act of 1971 leads, for the reasons that I have sought to demonstrate, to manifest injustice and is quite unacceptable. Again in agreement with Dillon LJ, I would answer the second question of law in favour of the claimants and allow the cross-appeal.

The effect of allowing the cross-appeal is simply to restore Mr Wellings' valuation of the uncovered part of the blue land at £17,500. The effect of allowing the appeal, however, is that compensation for disturbance will have to be reassessed on the basis that the claimants are entitled to be compensated for the effect of the acquisition upon their business in relation only to the blue land.

I would, accordingly, set aside the orders of the Court of Appeal K
and the Lands Tribunal and remit the case to the Lands Tribunal to redetermine the amount of compensation in accordance with this opinion. I think the fair result as regards the costs of the appellate proceedings is that there should be no order. The costs before the Lands Tribunal remain in the discretion of the tribunal.

LORDS TEMPLEMAN, GRIFFITHS, ACKNER and LOWRY agreed with the speech of Lord Bridge of Harwich and the order proposed by him and did not add anything.

House of Lords L

February 14 1991

(Before Lord MACKAY OF CLASHFERN LC, Lord KEITH OF KINKEL, Lord BRIGHTMAN, Lord OLIVER OF AYLMERTON and Lord LOWRY)

HERTFORDSHIRE COUNTY COUNCIL v OZANNE AND OTHERS

Estates Gazette April 6 1991

[1991] 13 EG 157

Compensation for compulsory acquisition of land — Value of M
"ransom strip" — Third stage in this litigation, Lands Tribunal, Court of Appeal, House of Lords — Whether rule (3) in section 5 of Land Compensation Act 1961 applied to the valuation of the strip so as to confine compensation to agricultural value — Agricultural value £5,500, award of Lands Tribunal £1.24m — Present appeal, by acquiring authority, related to the application of rule (3), which the Lands Tribunal and the Court of Appeal held not to apply — Appeal to House of Lords not concerned with that part of decision of Court of Appeal which remitted case to Lands Tribunal to identify the relevant scheme — House of Lords

A **agreed that the provisions in rule (3) excluding purposes to which the land could be applied only in pursuance of statutory powers did not apply to the land acquired — The reason was that the statutory powers relied on by the council for the application of rule (3), namely to stop-up an existing highway, did not relate to the use of the land acquired**

This litigation arose from the compulsory acquisition of a strip of land of 1.605 ha used for agricultural purposes adjoining the south side of Thorley Lane, Bishop's Stortford — North of Thorley Lane there existed a substantial area allocated for residential development — The claimants B **(respondents to the appeal), the owners of the strip, contended that it was a "ransom strip", required for the purpose of realigning and improving Thorley Lane as a prerequisite to the full development of the large development area to the north of the lane — The value of the strip on this basis was claimed to be £1.24m and this was in fact the amount awarded by the Lands Tribunal — The tribunal found that in the imaginary no-acquisition world with which one must grapple there existed a market for the land as a "ransom strip" — The tribunal had considered that it was unnecessary to go into the submissions on law or fact as to the existence and extent of any scheme —** C **The Court of Appeal disagreed with this conclusion — They considered that it was impossible to say that in the imaginary no-acquisition world the land had a value as a ransom strip unless the relevant scheme were identified — The court accordingly remitted the case to the tribunal to examine the issue as to what was the scheme and to consider any matters consequential on its identification — As the Lord Chancellor mentioned in his speech, the House of Lords was not called upon to adjudicate on the merits of this part of the case**

The only matter with which the House was concerned was the appellant council's submission that part of rule (3) in section 5 of the Land Compensation Act 1961 applied to the D **valuation of the respondents' strip of land — The part of rule (3) which was claimed to apply provided that "the special suitability or adaptability of the land for any purpose shall not be taken into account if that purpose is a purpose to which it could be applied only in pursuance of statutory powers" — It was the appellants' contention that the enhanced value referred to in the rule should accordingly be disregarded in the present case — The steps in the appellants' argument were (1) that the land in question had an enhanced value over its agricultural value because of its special suitability or adaptability for the purpose of providing a realignment of** E **Thorley Lane; (2) that, in order that it should be used as such a realignment, it was necessary that part of Thorley Lane should be stopped-up; (3) that, as Thorley Lane was a public highway, such stopping-up required the exercise of statutory powers; and (4) that, as a result, the special suitability or adaptability of the land for the alignment of Thorley Lane should be disregarded — It followed that the compensation should be assessed at agricultural value, namely £5,500**

In the opinion of the House of Lords, however, there was a fatal flaw in this argument — The correct construction of the part of rule (3) on which the acquiring council relied was that F **the statutory powers referred to must be powers enabling the person who is entitled to use the land to apply it to the purpose in question; and the statutory powers must be necessary to enable such person to use the land for that purpose — Statutory powers not related to the use of the land acquired could not form a basis for the application of this part of the rule — It was accepted that no part of the existing Thorley Lane which could be affected by any such stopping-up order lay within the land acquired from the claimants — It followed that statutory powers conferred on the Secretary of State to order the stopping-up of a highway on land which was not part of the land acquired could not justify the application of the rule to the land acquired — The council's submission** G **therefore failed and their appeal must be dismissed**

The following cases are referred to in this report.

Cedars Rapids Manufacturing & Power Co v *Lacoste* [1914] AC 569, PC
Hampson v *Department of Education and Science* [1990] 3 WLR 42; [1990] ICR 511; [1990] 2 All ER 513, HL
Horn v *Sunderland Corporation* [1941] 2 KB 26; [1941] 1 All ER 480, CA

This was an appeal by Hertfordshire County Council from a decision of the Court of Appeal allowing the council's appeal from a decision of the Lands Tribunal (Mr C R Mallett FRICS) awarding compensation in the sum of £1.24m to the claimants, P J Ozanne, M J Wilson and Rothschild Trust Co (CI) Ltd, trustees of the Boyd H Gibbins Family Trust, in respect of the compulsory acquisition of a strip of land, of 1.605 ha adjoining the south side of Thorley Lane, Bishop's Stortford. The decision of the tribunal was reported at [1988] 2 EGLR 213; [1988] 36 EG 195 and 37 EG 123 and the decision of the Court of Appeal at [1989] 2 EGLR 18; [1989] 43 EG 182.

Michael Rich QC and John Howell (instructed by Mr W J Church, County Secretary, Hertfordshire County Council) appeared on behalf of the appellant council; Matthew Horton QC and Sebastian Head (instructed by Berwin Leighton) represented the respondent trustees.

In his speech, **Lord Mackay of Clashfern LC** said: This is an appeal by the Hertfordshire County Council ("the council") from an J order of the Court of Appeal dated August 2 1989 allowing the council's appeal from a decision of the Lands Tribunal given on March 2 1988 whereby the council was ordered to pay £1,240,000 in compensation to the respondents ("the claimants") but remitting to the said Tribunal the issue as to what was the scheme and any matters consequent upon the Tribunal's identification of it. This appeal proceeds by virtue of leave granted by this House on December 14 1989.

The council is the highway authority for the County of Hertford on whose behalf the East Hertfordshire District Council (Thorley Lane, Bishop's Stortford) Compulsory Purchase Order 1976 was made. That order authorised the compulsory purchase of 1.605 ha of land adjoining the south side of Thorley Lane, Bishop's Stortford, the K property of the claimants. The compulsory purchase order was made under section 214 of the Highways Act 1959 and section 22 of the Land Compensation Act 1973 and the purpose for which the purchase was authorised was described as (a) the construction of a new highway from the existing junction of Thorley Lane and the London-Norwich Trunk Road A11 to a point 123m west of the junction of Thorley Lane and Pynchbek at Bishop's Stortford in the district of East Hertfordshire in the County of Hertford; (b) the construction of new highways to connect the above-mentioned highway with the existing road system at Bishops Avenue and Pynchbek at Bishop's Stortford; (c) the improvement of Thorley Lane and Oxcroft, Bishop's Stortford; and (d) the mitigation of the L adverse effect which the existence or use of the new highways proposed to be constructed and improved at (a), (b) and (c) will have on the surroundings of the highways.

The claimants referred the question of the determination of the amount of compensation payable to them on the acquisition of the said land to the Lands Tribunal. The agricultural value of the said land was agreed to be £5,500. The claimants contended that the said land was a "ransom strip". They argued that the said land was required in order to enable residential development of a substantial area lying to the north of Thorley Lane which has been referred to conveniently as the Thorley Development Area. The claimants contended that, since the said land was necessary in order to enable the development of the Thorley Development Area to proceed, its M value was considerably enhanced. By a calculation relating to the difference between the value of the Thorley Development Area if development was possible and compared with its value if no development was possible the claimants contended that the said land should be valued at £1,240,000, which appears a remarkably large figure for the amount of land in question, particularly as the planning permission granted in respect of the Thorley Development Area does not appear to have required as a condition that an access from the south of the development area should be made available over the said land. The Lands Tribunal determined compensation in the sum of £1,240,000, being Mr Mallett's assessment of the value of the land under the terms of the relevant statutes.

A The council appealed to the Court of Appeal. There the council had two arguments. The first was that the Lands Tribunal had not identified the scheme underlying the acquisition and, therefore, had failed to identify, as it was necessary to do, the extent to which the value of the said land was affected by the scheme. The Court of Appeal accepted this argument and remitted the case to the Lands Tribunal, on the issue as to what was the scheme, and any matters consequent upon the Tribunal's identification of the scheme. They ordered the costs in the Lands Tribunal to be costs in the cause and three-quarters of the council's costs in the Court of Appeal to be paid by the respondents.

On the merits of this part of the case your lordships are not called
B upon to adjudicate, although, of course, the question of costs in the Court of Appeal may be affected by the decision of this appeal to your lordships' House.

The second argument advanced by the council in the Court of Appeal was the only argument with which your lordships are concerned. If it were to succeed, there would be no need for the case to go back to the Lands Tribunal and the compensation could be determined as the agricultural value of the said land.

The assessment of compensation for the compulsory acquisition of the said land is subject to the provisions of section 5 of the Land Compensation Act 1961 which provides:

> Compensation in respect of any compulsory acquisition shall be assessed in accordance with the following rules:
> C (1) No allowance shall be made on account of the acquisition being compulsory:
> (2) The value of land shall, subject as hereinafter provided, be taken to be the amount which the land if sold in the open market by a willing seller might be expected to realise:
> (3) The special suitability or adaptability of the land for any purpose shall not be taken into account if that purpose is a purpose to which it could be applied only in pursuance of statutory powers, or for which there is no market apart from the special needs of a particular purchaser or the requirements of any authority possessing compulsory purchase powers: . . .

The only provision relied upon by the council before your lordships is that:

> the special suitability or adaptability of the land for any purpose shall not be
> D taken into account if that purpose is a purpose to which it could be applied only in pursuance of statutory powers.

Counsel made it absolutely clear that, having considered the matter, the council's argument was based solely on the passage I have just quoted.

The statutory powers which the council say are relevant are the powers to stop-up an existing highway which are contained in sections 212 and 219 of the Town and Country Planning Act 1971 (sections 249 and 255 of the Town and Country Planning Act 1990). No order under either of these provisions had been produced and your lordships were informed by counsel for the council that it has not established that any order was made.

E The argument for the council was that the said land could be used for the realignment of Thorley Lane only if there were a stopping-up order by the Secretary of State in respect of the existing Thorley Lane. It is accepted that no part of the existing Thorley Lane that could be affected by any such stopping-up order lies within the said land acquired from the claimants.

In elaborating their contention, the council argued that it could not be disputed that the said land had an enhanced value over its agricultural value only in respect of its special suitability or adaptability for the purpose of providing a realignment of Thorley Lane. Further, that in order that it should be used as a realignment of Thorley Lane it was necessary that part of Thorley Lane should be stopped-up and that, since Thorley Lane was a public highway, such
F stopping-up required the exercise of statutory powers. On this basis, it was claimed that the part of rule (3) relied upon required the special suitability or adaptability of the said land for use for the realignment of Thorley Lane to be disregarded in assessing the compensation for the compulsory acquisition of the said land and, therefore, required the compensation to be determined at agricultural value, namely £5,500.

In support of this argument, the council submitted that all compensation for compulsory acquisition is statutory; that the Acquisition of Land (Assessment of Compensation) Act 1919, in which rule (3) was originally enacted and which is now consolidated in the Land Compensation Act 1961, was passed for the purpose of mitigating the evil of excessive compensation; that special suitability or adaptability of the land for a purpose for which there is no market G apart from the special needs of a particular purchaser or requirements of any authority possessing compulsory purchase powers are dealt with in the part of rule (3) not relied upon here; and that it is necessary to give some other subject-matter to the part of the rule relied upon from the subject-matter of the part not relied upon, if a sound construction of the part relied upon is to be made.

It was submitted that it was legitimate, in order to arrive at a sound conclusion upon the scope of the part of the rule relied upon, to look at cases decided before Parliament passed the Act of 1919 and that in doing so the case of *Cedars Rapids Manufacturing & Power Co* v *Lacoste* [1914] AC 569, a decision of the Privy Council on appeal from the Superior Court of Quebec, provided an example of the class H of case which Parliament had in mind in enacting the part of rule (3) relied upon by the council. Finally, it was submitted that the purpose in question in rule (3) must be one to which the land being acquired can be applied and whether the land can be so applied only in pursuant of statutory powers must be determinative of the application of the first part of the rule.

In opening the appeal, counsel did not refer to the second report of the committee dealing with the law and practice relating to the acquisition and valuation of land for public purposes (1918) (Cd 9229) usually referred to as the "Scott Report" because its recommendations had not been implemented in the Act that followed in 1919.

Counsel for the claimants replied that the statutory powers J referred to in the part of the rule relied upon by the council must be those which empowered the application of the land taken to the purpose for which it was specially suitable or adaptable and that the powers must be powers granted to the person thereby enabled to apply the land taken to the purpose for which it is specially suitable or adaptable. He further submitted that the stopping-up of Thorley Lane, in exercise of powers of the Secretary of State to make a stopping-up order, could affect rights of the public only in respect of those parts of the land comprising Thorley Lane affected by the order, at the time when it took effect, and that this in no way affected or empowered any uses to which the said land acquired from the claimants could be applied.

K Counsel for the claimants submitted that the construction of the rule relied upon was clear and that accordingly it was not necessary to examine the history of the process by which it had first come to be enacted, but by way of background he referred your lordships to the Scott Report. He also pointed out that the construction advanced by the council would appear to rule out from consideration purposes for which planning permission would be required or building regulation consent might be required under the Building Act 1984, and he gave a variety of illustrations of other types of consent granted under statutory powers as essential to particular uses of land which would fall to be excluded from account if the council's construction were to prevail.

In the course of his reply, counsel for the council referred to the L speech of my noble and learned friend Lord Lowry in *Hampson* v *Department of Education and Science* [1990] 3 WLR 42, particularly to the paragraph beginning at p 49 between A and B.

A comparison of the Act of 1919 and in particular the rules that it enacted for assessing compensation with the pre-existing law makes it clear that the general intention of Parliament in 1919 was to direct the omission from compensation of certain items for compulsory acquisition which under the previous law had been included. This view is strongly supported by the opinion of Scott LJ, as he had by then become, in *Horn* v *Sunderland Corporation* [1941] 2 KB 26, at p 40, where he said:

> The main object of the Act of 1919 was undoubtedly to mitigate the evil of M excessive compensation which had grown up out of the theory, evolved by the courts, that because the sale was compulsory the seller must be treated by the assessing tribunal sympathetically as an unwilling seller selling to a willing buyer.

To go further and seek to suggest that the meaning of particular provisions adopted by Parliament in 1919 could be discovered by identifying the particular cases in the previous law that these provisions were intended to reverse appears to me to amount to speculation upon which it is not profitable to enter in the present appeal.

I regard the language in question as sufficiently plain to reach a clear conclusion upon its application to the present case. The special suitability or adaptability of the land for any purpose is directed to be left out of account if that purpose is a purpose to which it could be applied only in pursuance of statutory powers. This is expressed in the passive voice but the context shows that the application referred to is by a person using the land and, therefore, the statutory powers in question must be powers enabling a person entitled to use the land to apply it to the purpose in question and since the purpose in question is one to which the land could be applied *only* in pursuance of the statutory powers the statutory powers must be necessary to enable such person to use the land for that purpose. I do not see how statutory powers not related to the use of the land acquired could form a basis for the application of this part of the rule.

Therefore, I consider that statutory powers conferred upon the Secretary of State to order the stopping-up of a highway on land which is not part of the land being acquired could not form the basis of the application of this part of the rule to the land acquired. Since the only statutory powers here relied upon by the council are the statutory powers of the Secretary of State to stop-up parts of Thorley Lane, I consider that the council's argument must fail.

The construction of the relevant part of the rule which I have adopted is consistent with the approach taken by my noble and learned friend Lord Lowry to the construction of the words "in pursuance of" in a different enactment in *Hampson.*

As I said earlier, in the submissions for the council great stress was laid on the decision of the Privy Council in *Cedars Rapids Manufacturing & Power Co* v *Lacoste* [1914] AC 569 as illustrative of the type of case with which the part of the rule founded upon was intended to deal by reversing the committee's decision. Cedars Rapids was incorporated by a statute of the Parliament of Canada of 1904 and it was thereby empowered to construct and develop water powers in or adjacent to the St Lawrence river in a certain parish and to take by way of expropriation lands within that parish actually required for those purposes. The respondents in the appeal owned, within the limits of the Cedars Rapids Company's Act, two islands in the river and certain rights over a promontory of land. One of the islands was at the head of the Cedars Rapids, the promontory was at the foot of the rapids and the second island was between. There was a fall of some 28 ft in all from the head to the foot of the rapids. The respondents' rights upon the promontory consisted of a reservation of a road and the right to erect a mill with the right to use the power of the water flowing by and over the land. The question before the Privy Council related to the basis upon which compensation for the two islands and the respondents' rights in the promontory should be determined. Giving the judgment of the board, Lord Dunedin said, at pp 579-580:

> The real question to be investigated was, for what would these three subjects have been sold, had they been put up to auction without the appellant company being in existence with its acquired powers, but with the possibility of that or any other company coming into existence and obtaining powers. It is on account of the latter consideration that their Lordships, while unable to accept the judgment under appeal, are also unable to restore the judgment of the arbitrators. Unfortunately, the appellants led no evidence except as to bare agricultural value. Now with regard to the Ile aux Vaches and the reserved water rights, it seems possible that there may be some value over and above the bare value. If the situation be naturally favourable to the establishment of power works like those of the appellants, then it is possible that the respondents and others might have been prepared to offer an enhanced value on this account, taking the chances of a situation in which they might or might not obtain the requisite Parliamentary powers to work out a commercial scheme. But the value emerging through a grant of such powers having been actually given cannot after the event be taken into account . . . These considerations, however, point to the possibility of something more being given for the subjects than the bare value; or in other words, that if they had been put up to auction as beforesaid, there was a probability of a purchaser who was looking out for special advantages being content to give this enhanced value in the hope that he would get the other powers and acquire the other rights which were necessary for a realised scheme.

In the passages that I have quoted, the Privy Council was recognising that the application of the lands to be acquired for use by a company obtaining the necessary statutory powers to work out a commercial scheme was a matter that should be taken into account in determining the compensation. It appears clear, from the facts, that the Privy Council made these observations in connection with lands and rights over lands which were required to be used in order to implement a scheme which required Parliamentary powers. The interpretation which I have placed upon the part of rule (3) relied upon in this case would have the effect that the enhancement of value referred to by the Privy Council in *Cedars Rapids* in the passage to which I have referred would not be taken into account in a case to which rule (3) of section 2 of the Act of 1919 and its statutory successor applied. I consider, therefore, that the council's submission was correct that the *Cedars Rapids* decision provides an illustration of the cases that Parliament has covered by the part of the rule founded upon, although wrong in contending that it applies to the present case. In *Cedars Rapids* the purpose giving rise to the enhancement of value, namely the use of the lands in question as part of a water power development of the type in question, could arise only where the appropriate statutory powers had been granted, whereas in the present case the land acquired could have been used for a highway without the exercise of any statutory power and certainly was not dependent upon the Secretary of State exercising any statutory power to stop-up any part of Thorley Lane. I should perhaps add that it is clear, from the passage that I have quoted from the decision in *Cedars Rapids,* that the obtaining of the requisite Parliamentary powers was to be borne in mind as providing opportunity for realisation of the enhanced value. Even after the company had obtained the necessary statutory powers, they also required the consent of the Dominion Government to erect works in the bed of the river and to abstract water from it, but the provisions of rule (3) do not refer to this. This I think goes to emphasise the distinction referred to by counsel for the claimants when he pointed out that, if the present case was covered by the rule, it was very difficult to see why the rules should not also cover a purpose to which a piece of land could be put only after obtaining some particular statutory consent such as planning permission, consent under the Building Acts, or the like. It is clear from the modern statutory provisions governing compensation for the compulsory acquisition of land that the question of what types of development would receive planning consent is highly relevant to the determination of compensation for compulsory acquisition and any construction of the provision founded upon which resulted in any enhancement of the value of a piece of land resulting from its use for a purpose which required planning permission being disregarded would be absurd.

The Lands Tribunal came to the same conclusion as I have done on the submission of the council which has been the subject of appeal to your lordships' House. Mr Mallett said [1988] 2 EGLR 213 at p 216:

> Clearly the first limb of the provision of rule (3) cannot apply in that the special suitability or adaptability of the land can be realised other than by the use of statutory powers.

In the Court of Appeal Mann LJ dealt with the argument in this way [1989] 2 EGLR 18, at p 20:

> Mr Rich relied upon the first limb of the rule, and put the matter in this way: the Thorley Lane improvement included not only the construction of a new carriageway but also the closure of certain parts of the existing lane (none of which were on the reference land). That, he said, could be achieved only in the exercise of statutory powers, the exact nature of which I do not particularise because there may be doubt as to whether the proper procedures were subjected to adherence.
>
> Undoubtedly the reference land was locationally suitable for the improvement, and it may even be taken that it has "a special suitability". However, I am not persuaded by the argument.
>
> Mr Horton countered it by saying that private developers can and do so construct roads or road improvements upon land which they own, subject only to the obtaining of planning permission. He also pointed out that developers frequently have to stop-up existing highways and they may be authorised to do so at their own instance under section 209 of the Town and Country Planning Act 1971. That being so, and bearing in mind that the statutory closure was not upon the reference land, I do not see, in common with the tribunal (albeit for different reasons) that rule (3) applies.

The order for costs made in the Court of Appeal no doubt reflects the fact that the Court of Appeal decided against the council on this argument, although in their favour on the other argument which resulted in the Court of Appeal deciding to remit the case to the Lands Tribunal. Since, in my opinion, the result on the argument adduced to your lordships is the same as it was both in the Lands Tribunal and in the Court of Appeal, the decision of the Court of Appeal as to costs should stand, but the council having failed in their appeal to this House must pay the claimants' costs here.

Lords Keith of Kinkel, Brightman, Oliver of Aylmerton and **Lowry** agreed that the appeal should be dismissed, and did not add anything.

For a further case on this subject see p 256

ESTATE AGENTS

A

Queen's Bench Division

July 17 1990

(Before His Honour Judge LEVER QC, sitting as a judge of the High Court)

MICHAEL ELLIOTT & PARTNERS LTD v UK LAND PLC

Estates Gazette March 2 1991

[1991] 08 EG 123

B **Estate agents — Claim by agents against purchasers for commission for services in finding properties — Agents not retained by vendors — Portfolio of properties in various areas — Difficulties caused by arrangements made through intermediaries — Lengthy review of evidence — Different routes by which plaintiff agents might claim entitlement to remuneration, direct contract, contract through other agents, by intermediate agents' ostensible authority, by ratification on the part of defendants, or, as a last resort, by a *quantum meruit* claim — Lack of clarity in arrangements leading to misunderstanding — Decision in favour of plaintiffs for a modest "reasonable fee"**

C **The plaintiffs in the present case, estate agents, surveyors and property consultants, claimed against defendants, owners and dealers in commercial properties, an introduction fee of 1% on the purchase price of the properties introduced to the defendants — The purchase price of the properties in question, a portfolio of properties situated in Plymouth, Hemel Hempstead, Welwyn Garden City and Soho Square, London, was £22.3m — At the rate of 1% the fee would have been £223,000 (plus VAT) — In the result the judge decided that the plaintiffs were entitled to a fee of one quarter of the amount claimed**

D **The main figures in the litigation were three, the plaintiff agents, another firm (referred to in this headnote as "NB") described as retained or in-house surveyors and agents for the defendant company, and the defendants — The history of the events which ended in court was described in great detail in the judgment — The judgment makes it evident that one of the difficulties was a curious lack of clarity and business precision in the negotiations and arrangements, leading inevitably to misunderstanding and dispute**

After a review of the evidence and submissions the judge's finding was as follows:

E **(1) there was no evidence of a concluded contract, made either directly between the parties or through the agency of NB, to pay the plaintiffs a commission of 1% on the purchase price of the portfolio as claimed;**

(2) there was, however, evidence that the plaintiffs were entitled to be paid a reasonable fee for their services, either —

(a) by an express contract made through the agency of NB to act for the defendants, with an implication that the plaintiffs would be paid a reasonable fee for their services, or

(b) by NB having been clothed by the defendants with ostensible authority to agree that a reasonable fee would be paid to the plaintiffs, or F

(c) by the defendants having, through acquiescence or inactivity, ratified the arrangements made by NB with the plaintiffs, or

(d) on the basis of the principle of *quantum meruit*;

(3) a reasonable fee for the plaintiffs' services, whether on the basis of contract or *quantum meruit*, was 0.25% of the purchase price of the portfolio

Accordingly judgment was given in favour of the plaintiffs, on the basis of a fee of 0.25% of the purchase price of the portfolio (£22.3m), namely £55,750, plus VAT at 15%, £8,360, ie £64,110 in all G

The following cases are referred to in this report.

Armagas Ltd v *Mundogas SA* [1986] AC 717; [1985] 3 WLR 640; [1985] 3 All ER 795; [1985] 1 Lloyd's Rep 1, CA

Batis Maritime Corporation v *Petroleos del Mediterraneo SA* [1990] 1 Lloyd's Rep 345

In this action the plaintiffs, Michael Elliott & Partners Ltd, claimed against the defendants, UK Land plc, the sum of £256,450, being 1% of a purchase price of £22.3m plus VAT of £33,450, by way of agreed commission for the plaintiffs' professional services in relation to the purchase of properties by the defendants. H

Andrew Onslow (instructed by Jeffrey Green Russell) appeared on behalf of the plaintiffs; Dominic Kendrick (instructed by Russell-Cooke Potter & Chapman) represented the defendants.

Giving judgment, JUDGE LEVER QC said: In the spring of 1989 Regent Crest plc, who are owners of and dealers in commercial property on a very large scale, were minded to sell a number of their holdings. The value of these properties, which were of widely ranging types, situated at four sites in Plymouth, Hemel Hempstead, Welwyn Garden City and Soho Square, London W1, being in the region of £25m. It followed that the number of potential purchasers was limited, and if a sale was concluded, it would in all probability be to a company operating in the same field who would probably acquire them on behalf of investment clients or for individual development. J The properties have been throughout described in investment terms as a portfolio.

So it transpired that the defendant company, who themselves trade in land and in particular commercial property, frequently on behalf of clients who wish to invest in or develop it, agreed to buy the portfolio for that sum. In the ultimate event, one of the properties, namely a retail warehouse development in Plymouth, was withdrawn from the portfolio and in the deal eventually concluded the defendants purchased the remaining three at a revised price of £22.3m.

The plaintiff company, who are surveyors, estate agents and consultants in this field, were one of the intermediaries concerned in K this transaction and claim against the defendants the sum of £256,450, being as to £223,000, 1% of the eventual purchase price paid by the defendants, plus £33,450 VAT payable on the transaction. This sum, they claim, represents the fee agreed with the defendants by way of a 1% commission earned by the plaintiffs for their professional services in the transaction.

They assert alternatively — and although the case is not specifically so pleaded in a specially endorsed writ, no point has been taken in this respect by the defendants — that if no agreement to pay the specific sum of 1% is to be found in the contract between the

A parties, then it is an implied term of the agreement that the defendants are bound to pay them a reasonable fee under the contract and that such a reasonable fee would indeed be 1% of the purchase price.

In the further alternative they contend that if no binding contract existed between the parties, then they are entitled to the same sum *quantum meruit* for services which they have rendered to the defendants and from which the defendants have benefited.

The defendants, by their defence, deny that either directly or through their agents they concluded a binding agreement for the payment of commission in any sum to the plaintiffs, and in their pleaded defence deny the plaintiffs' right to recover any sum from them *quantum meruit*. Without prejudice to that denial they further
B assert that if *quantum meruit* is appropriate, then the true value of the plaintiffs' services is no more than what must be described in the context of this case as a nominal sum of £5,000.

As far as the defence in its pleaded form is concerned I should also say that by para 5 the defendants further contend that because the original four-property portfolio agreement of sale between Regent Crest and the defendants was in fact rescinded on or about July 13 1989, by which date the plaintiffs' involvement if any in the transaction had ended, and was thereafter replaced by the three-property portfolio sale in about September 1989, the plaintiffs were strangers to the contract eventually concluded and as such disentitled in any event to any fees by way of professional services in the matter.

C At the outset of the case Mr Kendrick, for the defendants, realistically in the light of the evidence if I may say so, abandoned this contention *in limine*, and the contents of para 5 of the defence and extensive particularisation about its contents have played no part in the issues I am called upon to resolve.

One of the more striking features of this case, apart from the remarkably brief timescale — virtually all the central events spanning a period of hardly more than two weeks in April 1989 — is the brevity of really significant documentary evidence. It is also notable that in respect of many of these events and documents there is a large area of agreement between the parties and their witnesses as to what was said and done by the various principal actors in the matter. In these circumstances, it is convenient to set out the history of the matter
D primarily by way of an analysis of the evidence of the plaintiffs' first and main witness, Andrew Reilly, to which I now turn.

Mr Reilly, 28 years of age at the time and an associate of the Royal Institution of Chartered Surveyors, is now a director of what is now the plaintiff company, Michael Elliott. He was, in my judgment, a patently competent but — and I do not use the word in any pejorative sense — inexperienced practitioner. He was primarily involved in investment agency work. He told me that his partner, Jeremy Charles [ARICS], had a telephone call from a firm of estate agents and property consultants by the name of Lanyon Fleming early in April 1989; that Charles told him that Lanyon had offered them these properties to act as an intermediary in their sale in an endeavour to
E find a suitable purchaser for Lanyon Fleming's client, Regent Crest. At first, he said, we only had a rough outline of the proposition, and the general tenor of Lanyon Fleming's approach was that the properties would be offered, as he put it, very discreetly to a few agents and principals. He and his partners examined the proposition and put their mind to the question of potential purchasers, and thought of UK Land plc as being a prime candidate. At that time, he said, they were buying a lot of portfolios, and they were one of the more aggressive purchasers of property in what was then a buoyant market.

They had had previous dealings with UK Land plc, but were not closely connected with them in any way, and in fact I think they had only been involved in one entirely unconnected transaction. He told
F me that the defendants were primarily traders in property, and to the end of making the approach he contacted a young man called Simon Marriott [ARICS] of Nelson Bakewell. It is significant in this case, in my judgment, that Marriott and Reilly were friends and contemporaries. Marriott was employed by Nelson Bakewell and Nelson Bakewell were variously described as the retained or in-house surveyors and agents for UK Land. The fact of the matter is that the defendants and Nelson Bakewell had no particular contractual relationship with each other, but I accept, as I was told by members of both firms, that the latter commonly used the former's services in the many transactions in which they were involved.

It was in these circumstances that the two young men met for a drink about April 11, and Reilly mentioned to Marriott the availability of these properties in a later telephone call. At that point G
Marriott asked Reilly to let him have details on their behalf.

Thereafter, events moved with exceptional speed. Armed with the information he had from Reilly, Marriott, on behalf of Nelson Bakewell, wrote to inform the defendants of the existence of a portfolio which, as he rightly surmised in the event, would be of great interest to them. The letter in question has played an important part in the case. It is necessary to examine its contents.

By an interesting preliminary the subscription to the letter is the initial "CJEA/SCM", which I later learnt indicated that, although "SCM", Simon Marriott, was dealing with the matter, he was doing so under the general authority and supervision of Charles Allison [ARICS], a senior partner in Nelson Bakewell.

The letter in fact was addressed to Mr Barclay of UK Land plc, who H
has played no direct part in the case but was apparently Marriott's immediate source of contact, and told Barclay that he, Marriott, had just received details of a Regent Crest plc portfolio "which I am considering on your behalf". The letter continues by stating what the portfolio consists of and in its penultimate paragraph informs UK Land of the following:

> This portfolio has been introduced to me through Michael Elliott & Partners who require a fee if we are successful in this portfolio acquisition.

Marriott, in passing, told me in his evidence that he was aware at this time that Michael Elliott were not the retained agents of Regent Crest.

That letter was sent to the defendants on April 13 1989 and a copy J
of it was forwarded to Reilly himself. The weekend passed, and on April 17 Marriott visited Reilly in the morning. They discussed the properties in some detail and later on that day, Reilly told me, Marriott phoned him and said that his clients wished to proceed with the transaction, and they discussed the relevant timescale.

It was at this stage that Reilly ascertained from Marriott what the written conditions would be, and as a result of that caused to be written, under the signature of Mr Charles, a partner in the firm of which Reilly was at that time an employee, a letter to Richard Lanyon [ARICS] of Lanyon Fleming, apparently a personal acquaintance of Charles (hence the personality of the author and the recipient) in which he set out an offer on behalf of what he described as "my
clients, UK Land plc, to purchase the portfolio for the sum of £25m", K
and also set out particulars in the form of description of the properties, that the deal was subject to contract, subject to survey, timescale, nominated solicitors, and some general background information.

There appears in the correspondence, as he told me, a letter which he wrote to Marriott on the same day, April 18, enclosing a copy of the letter which he had written and raising for the second time the question of commission. On this occasion he told Marriott:

> I wish to take this opportunity in confirming that this is an instance where we are acting jointly with yourselves, and we would therefore be looking to you for a moiety of your fees should this proceed towards a successful acquisition.

Then, in a perfectly legitimate attempt to involve his own firm further L
in the matter, to their obvious potential profit, he added:

> I would also mention that we would be pleased to have a further joint involvement should UK Land wish to sell on any of the properties within the portfolio.

He assumed at that time that Nelson Bakewell, he said, would pay Michael Elliott's fees; and indeed, asked by his counsel, if the matter had ended there and then, and there had been a question of litigation of the matter, that Nelson Bakewell would have been the appropriate defendants because it was to them and solely to them that he was looking for his fees. He saw it, he said, as a joint agency at that time and he did not in fact know what Nelson Bakewell's fees would be,
but he surmised that they would be agreed on a percentage basis with M
their principals, UK Land.

On April 19 he went to a meeting at Lanyon Fleming's, and it is accepted — although Lanyon Fleming have played no part in this case in any direct sense — that the document which appears in the papers is an accurate minute of a meeting that took place between himself and Lanyon Fleming. The memorandum concludes:

> After the meeting AR

— that is Andrew Reilly —

> was going on to see UK Land and Nelson Bakewell and would be talking to Lanyon Fleming if there was any further questions, disagreements etc. By Monday everything should be worked out for an engrossment on Tuesday or Wednesday [of the next week].

A He agreed that it was very strikingly the case that he had no direct contact with UK Land before the exchange of contracts, and on the question of agency or contract this is an interesting but accurate point, because his assertion to that effect was confirmed by everybody else who gave evidence on the relevant point. It is also interesting and true that it appears to have been the case that Nelson Bakewell had no direct contact with either Regent Crest or Lanyon Fleming in the negotiations.

"I then," he said, "had a meeting with Nelson Bakewell and told them what had appeared in the memorandum of the meeting" to which I have just referred. Nothing specific was said about the fees. The relevant document represents Lanyon Fleming's response and B their confirmation that their clients, Regent Crest, were prepared to sell the property to the defendants. Consequently a contract was concluded between those two for that sale on that date. There is a further letter not of major significance, again from Lanyon Fleming, and he forwarded that. His next problem was to tackle the question of how long was going to elapse between exchange and completion,and in that respect Mr Reilly did a certain amount of work, and the relevant correspondence represents the solicitors' correspondence in that context.

On April 20 or 21 he and Marriott actually visited two of the properties, at Hemel Hempstead and Welwyn Garden City, and as appears from his desk diary further communications took place with Lanyon's and Marriott on these dates.

C In the course of these discussions — it is important in relation to my judgment in this case — "Marriott told me," he said, "that his firm did not want Michael Elliott to be involved in the disposals, and our fee would relate solely to the introduction." I was told by Mr Reilly, and indeed by virtually all the other witnesses in the case, that agency fees in these matters can relate sometimes to what I would describe as acquisitions — that is, the introduction of a property to a potential purchaser who acquires it — and/or disposals, the agency in those circumstances disposing to further purchasers the property so acquired. It appears that sometimes fees are paid on acquisitions; sometimes fees are paid on disposals; but frequently, as is the case with Nelson Bakewell in relation to the work they carried out for the defendants, an agent will be paid a fee which will cover both D acquisition and disposal, the fee by convention sometimes being divided.

"Marriott", he said, "asked me what our introduction fee would be. I told him 1%. Marriott said, OK, he would clear it with his clients." A couple of days later, Reilly told me, Marriott told him that UK Land were happy to act on that basis.

I turn now to the letter which, says Reilly, came to be written against that background. In that letter for the first time Marriott communicates with Mr Colin Tatt. Mr Tatt is a director of UK Land who played a major role in this transaction, and indeed is one of the leading actors in this case, because it was he who was concerned, if anybody was, with the question of commission which might or might E not be earned by the plaintiffs. Mr Tatt, as chairman, was the recipient of Marriott's letter of April 21, and Marriott wrote to him at UK Land's offices in Kensington:

> Dear Colin, I am pleased to report that the solicitors have now received the papers on the Regent Crest plc property portfolio and are proceeding apace with . . . inquiries and amending the contract.

He told Tatt about the proposed timetable in these matters and concluded his letter:

> I would add that Michael Elliott & Partners, who introduced the deal to us, have requested a 1% introductory commission for their work.

He sent a copy of that same letter to Reilly.

The only further discussion of fees that Reilly said took place was F that he asked for written assurance, or written confirmation, that these fees had been agreed. "We had got", he said, "numerous verbal, oral assurances from Allison [C J E Allison ARICS, a director at Nelson Bakewell] and Marriott to the effect that there would be no problem in the matter." After April 21, apart from certain liaison work between Lanyon Fleming and Nelson Bakewell, his firm did very little in the matter. On about May 3 he had a conversation with Marriott with a further request for written confirmation of the fee, and Marriott said that he need have nothing to worry about.

It was in these circumstances that he received a letter from Marriott indicating what he was minded to do. The letter said:

> Dear Andrew, I take pleasure in enclosing correspondence from me to UK Land plc on April 13

— that was the original letter saying that the portfolio had been G introduced by Elliott's who require a fee —

> and April 21

— the letter which said that Elliott's requested a 1% introductory commission —

> confirming your involvement in the acquisition of the above. As you are aware completion is set for July 3 after which you should submit your account direct to UK Land. Yours sincerely, Simon Marriott.

He said he had sent a copy of that letter to Mr Tatt.

This was indeed the sort of reassurance that Marriott was legitimately able to give, and it was as a result of this that a fee note was finally sent to the defendants by the plaintiffs. When the fee note arrived, it received a curt and dusty answer, described in various H terms by members of the plaintiff company as a bitter blow, an ugly shock. The letter simply stated:

> Dear Mr Reilly, Regent Crest Portfolio. Thank you for your letter of 27th June enclosing a fee note. I understand that you were joint agents with Nelson Bakewell and I suggest that you should agree your share of the fees directly with them. Yours sincerely, C.C. Tatt.

He said that, because we had agreed with Nelson Bakewell that we would take a 1% commission, this letter, of course, as I have said, was a bitter blow.

The matter was in fact taken out of his hands, I conclude, by senior members of the firm and in particular Mr Shipman [ARICS], to whose evidence I will turn in due course, in pursuing the matter with Mr J Tatt.

Under cross-examination Mr Reilly stated that he regarded the appointment of Michael Elliott as dating from the meeting of April 17 at which Marriott had told him that his clients were indeed going ahead, and which led Reilly to write the letter of offer to Lanyon Fleming. He agreed that introducing agents can often share a fee with the main agent; that they can make a separate claim, and in the case of a separate claim the principal might not even know of the same.

As to the obvious point that he had nothing in writing to support his claim from the defendants he said, simply and plainly, that "I took Marriott's and Allison's word for it". He himself had not pursued the matter after the letter from Mr Tatt and left the matter entirely to Shipman. K

There was a certain amount of evidence as to what the plaintiff firm had done by way of services carried out on a *quantum meruit,* based on the amount of work in terms of hours worked, documents produced, but, in my judgment, that is essentially irrelevant in this case. The question of agency and the possibility of a fee was primarily concerned with the question of an introduction.

Well, I turn to analysis of what Marriott himself says he said and did in the matter. Mr Marriott produced early in his evidence a memorandum of the fee structure agreed in broad terms between Nelson Bakewell and UK Land plc at whose service he primarily worked during his two years with Nelson Bakewell. This memorandum is in the form of a document emanating from Phillip Nelson [ARICS], presumably the senior partner of Nelson Bakewell, L to the directors of UK Land plc. It states plainly enough that this is intended as a guide only and will be subject to ratification for each job undertaken and/or completed at the weekly board meetings, indicating as was the fact that UK Land and Nelson Bakewell were working in close and frequent co-operation. However, subject to that the terms were simple enough. On acquisitions introduced by Nelson Bakewell, 0.5% of the purchase price plus VAT was to be their commission. In the cases of acquisitions dealt with but not introduced by Nelson Bakewell, a valuation of 0.1% to 0.25% of the value, exclusive of VAT, dependent on work involved and *quantum* value.

For the sake of completeness I should add that the second M paragraph deals with disposals and states that 1.5% of the sale price exclusive of VAT and extraordinary marketing expenses shall be payable. In other words, in summary, that Nelson Bakewell, if they dealt with both acquisitions and disposals, would earn commission of 2% on the entire transaction.

What were the circumstances in which they came to work together? "Reilly and I," Marriott said, and he put the matter bluntly, "are mates. We have worked together and our firms have had much closer contact since we joined respectively." His evidence as to how the matter began was essentially in agreement with that of Mr Reilly. "Reilly rang with the details", he said. "We met at a pub. He told me he was not retained by the vendors and that he would want a fee. Mr

Allison was my immediate superior and I went back to the office and told Allison. It was the next day, April 13, that I wrote to Barclay at UK Land" — the letter which I have already quoted. "I pointed out, obviously enough, that Michael Elliott would require a fee", and then he said, "We got cracking" — a fair description of the speed with which the matter then moved: reports from the clients, instructing solicitors, inspecting the properties, finding out all about them.

When I received the letter demanding or requesting a moiety of "your fees", I did indeed tell Reilly

— and he was supported in this by Mr Allison —

that it was not Nelson Bakewell's general practice, and it was not going to be their practice in this case, to split fees. They were solely involved in the acquisition. I suggested to him that he might consider a fee payable by UK Land solely for the acquisition. I discussed that with Allison. Reilly raised the question of 1 per cent in answer to my question, "What would your fee be solely on the acquisition?" I said, "Very well, I'll write to our client and put it to him". Allison himself thought that 1 per cent was reasonable, so it was in those circumstances that I wrote the reassuring letter to Reilly. There was no response from Mr Tatt, and I simply understood the fee had been agreed. However, later in conversation with Mr Allison, Mr Allison specifically told me that Tatt had indeed agreed this fee, and I passed that information on to Reilly.

That was a critically important piece of evidence, but I should say straight away that it was met immediately in cross-examination by an equally important piece of evidence, and that is this: that during the course of these proceedings Mr Marriott had made a witness statement to both the defendants' and the plaintiffs' solicitors. As a qualified professional man he would, of course, understand the importance of the forging of an effective contractual chain in this case, and that evidence that he had given in court could be a critical link in that chain. There is no mention in either of these statements of his having been told by Allison that Tatt had told Allison of a fee agreed in these terms.

Jeremy Charles is the senior partner in Michael Elliott; Mr Shipman is another; and they were, if I may so describe them, the plaintiffs' subsidiary witnesses, not having played any significant part in the deal itself. It was indeed a telephone call, Mr Charles said, from Richard Lanyon that set this whole transaction in motion. He discussed it with Shipman and with Reilly. They thought of UK Land. They had had dealings with UK Land in 1986 and 1987, to which I will have cause to refer, and "we thought", he said, "it would be ethical to take the matter through Nelson Bakewell" who he well knew were the defendants' retained or in-house agents.

There was drawn to Mr Charles' attention in cross-examination the circumstances of the previous dealings between the plaintiffs and the defendants in 1986-87 in a matter which I will describe as the Woolf portfolio in which the plaintiffs had, apparently as a link in a chain of agents, offered the defendants an introduction to the purchase of a portfolio, which was eventually carried through at a price of £14.5m. The acquisition fee on the transaction (roughly three-fifths of the value of the present one) was in fact agreed between Mr Tatt of UK Land not in percentage terms but in the sum of £50,000. That sum was in turn divided between a number of introducing agents, so that the eventual sum earned by the plaintiffs from the acquisition alone was no more than a fifth of £50,000, namely £10,000.

All these matters are self-evident from the separate bundle of correspondence under the heading of "the Woolf bundle", which was placed before Mr Charles, and containing, interestingly enough, a specific written confirmation of the £50,000 fee on the acquisition to the plaintiffs and their fellow agents in the chain. For completeness, I should say that the evidence came from Mr Tatt as well as Mr Charles that Michael Elliott also earned I think the sum of £184,000 on sales commission. In other words, they benefited by acting on acquisitions and disposals in the sum of £194,000.

That introduction was an exclusive introduction — a topic to which I will turn later in this judgment.

Finally for the plaintiffs I heard from Mr Shipman, who came into the picture on receipt of Mr Tatt's letter. Mr Shipman was, as he put it very vividly, extremely sickened by this letter. He telephoned Tatt, and he made a memorandum of his telephone call after it had been made, and I should say that although Mr Tatt had no specific recollection of the terms of the telephone call, he had no reason to deny that Mr Shipman had recorded its terms accurately. Mr Shipman wrote:

I spoke to Colin Tatt, Chairman of UK Land, Monday morning approximately 11.30 am. I inquired what Colin Tatt intended us to do with our fee invoice. He advised me that this deal was a loss-making situation for a trade-on, and that we were due a moiety of Nelson Bakewell's commission. He only paid them a couple of thousand pounds for writing the report and acquisition advice. He offered us the sale of the lot in Hemel Hempstead and said he would hardly be paying Nelson Bakewell any form of fee on the disposal of the properties. He advised me to sort out our position with Nelson Bakewell.

Naturally, Mr Shipman said this was unacceptable, and effectively no more is to be said of the plaintiffs' case than that the specially endorsed writ followed within a month or two.

He amplified that memorandum to some extent and said he was astonished by Tatt's attitude. Tatt said that Nelson Bakewell were only earning a very small fee — £4,000. Shipman indicated he could barely believe his ears, and said, "You're only offering us £2,000 then?" "Correct", replied Tatt. The sum, Shipman indicated, was derisory and he felt that Tatt's tactic in fact was to offer instructions for the opportunity of acting on disposals by way of some form of inducement or sweetener.

What had he considered the position to be?

I took Allison's word for it

— quoting the word that Marriott said that Allison had given —

He is a highly qualified man of some standing in the profession. It is true I did not put our case in writing to Tatt. I simply thought that the deal was off, the disposal deals were off, and there was only one thing to do and that was to resort to the present litigation. I didn't want to be involved in any further discussions. I didn't want to reflect on Allison's integrity by asking him for confirmation in black and white. That was why it had not been done at an earlier stage.

The defence case was put in succinct form by Mr Allison and Mr Tatt, and I deal first of all with the evidence of Mr Allison who, I remind myself, was responsible for all the letters written by Marriott, even though he very honestly agreed that he may have given Marriott a free hand to the extent of not ensuring that every letter complied to the last syllable with what he would have written before he allowed it to be posted. Mr Allison is the director in charge of the property investment department at Nelson Bakewell. He told me, in a series of plain and simple propositions, first that Nelson Bakewell had no authority to agree the payment of commission from UK Land during the course of their work as UK Land's agents. Second, that nobody from UK Land had told him that they had agreed to pay 1% or any commission at all to Elliott's. Third, that he never told Marriott that the fee of 1% had been agreed by Tatt.

As far as his firm were concerned, the transaction which had appeared to be potentially very profitable turned out to be the loss-maker that Mr Tatt described when he discussed the matter with Mr Shipman. In those circumstances, as a matter of obvious professional prudence and anxiety to ensure that their future relations were harmonious, Nelson Bakewell totally waived their acquisition fee and dramatically reduced their disposal fee to the point that the present position, which still obtains, is that Nelson Bakewell are going to be paid or have been paid 0.25% on the disposal of the remaining properties.

Under cross-examination Mr Allison said that he simply had no recollection of the letter requesting a moiety. He was never aware, he said, that a one-off acquisition fee, an introductory fee as opposed to the moiety, was ever discussed. He could not remember if he had ever seen the letter by Michael Elliott requesting a 1% introductory commission. He conceded that he must have approved it. He could not remember discussing the matter with Marriott. He presumed that a letter written again under his name, or under his aegis, was indeed to reassure Michael Elliott. He said, and I accept that he was correct, "I do not think I ever discussed Elliott's fee with Tatt before Elliott's sent their fee note."

An interesting letter, however, was written by Mr Allison. This, of course, was when the matter was moving rapidly towards this court, and was written on September 22 1989. Allison wrote to Tatt on the subject of Elliott's fees. "Andrew Reilly", he told Tatt, "of Michael Elliott & Partners telephoned me wondering what the position was regarding any fees which may be due to him for introduction for the above." The question of the rescinded contract was then canvassed; I pass over that. "From the enclosed correspondence" — and he enclosed the relevant letters — "you will see that no actual agreement was made regarding fees, and I have suggested that Andrew Reilly speaks to you direct."

A He said that as far as he himself was concerned he would have charged 0.5% if Lanyon Fleming had come direct to him; that throughout he simply expected UK Land to pay Elliott.

Mr Tatt gave evidence in equally plain terms. He is the chairman of UK Land. He described Nelson Bakewell as their main agents but not tied to them in any professional sense. Mr Tatt told me that it was his policy to negotiate a separate fee with the first introducer, if I can so describe them — the introducer who stood in the position of the plaintiffs in this case. Nelson Bakewell had no authority to agree fees, as he put it, off their own bat. After the letter from Mr Marriott he said he would have expected Michael Elliott to get in touch direct with him.

B Following Marriott's assertion, I would add, that Michael Elliott's who made the introduction to us have requested a 1 per cent introductory commission. Nobody ever told me that 1 per cent had been agreed. I would never pay 1 per cent because, of course, I have my obligations to Nelson Bakewell. We do not want to have to pay a total of more than 1 per cent on the acquisition.

As he put it, 0.5% under the agreement set out in the correspondence to Nelson Bakewell and another 0.5% to play around with. "Had Michael Elliott approached me, I would have negotiated a fee in this case. I would have put it in the region of £5,000 to £20,000 because the introduction was not exclusive." It will readily be appreciated that that is something like one-fortieth to one-tenth of Elliott's claim.

He repeatedly said that he would never have agreed 1% and stated — and in my judgment this is a very important piece of evidence C indeed — "If Elliott's had picked the phone up, I would have agreed a fee with them". It was an assertion that he made on more than one occasion, and in robust terms. He said he simply did not remember the letter with the request for 1%. "As far as I am concerned, the terms of commission are always put in writing. It's my standard practice"; and indeed, as I have already indicated, he was able to point to the Woolf portfolio bundle where he had done exactly that. "As far as Shipman's telephone call is concerned, I have no recollection of that, but I have no reason to quarrel with Mr Shipman's account", Mr Shipman's memorandum. "I certainly don't deny that it happened."

In Woolf, of course, he paid a £50,000 introductory fee. True, it D had to be divided; but, contended Mr Tatt,

as far as I am concerned the key to a generous commission, or the opposite thereof, is the degree of exclusivity in the introduction. I am prepared to pay,

he said, putting the matter in broad terms,

good money for an exclusive opportunity to acquire a property. If at the other end of the scale I feel that the introducer is simply passing on market-place gossip which can be acquired by anybody in this relatively narrow field of operations, the few men who are operating in this field, why then I am not prepared to pay a large fee.

He stated that he was astonished, or words to that effect, when a bill for £¼m landed on his desk, when, as he made the point, there had been no previous correspondence between himself and the person E who was making the claim. "I did look at the correspondence", he said, "when that bill arrived, and as a result of that I wrote the letter to Mr Reilly. At the time that I wrote that letter" — which states, of course, "you are joint agents with Nelson Bakewell" — "what I had in mind was the correspondence that I had seen earlier in which Michael Elliott were talking of a moiety to be shared between the two of them." He said this particular deal, in fact, was far from exclusive. Regent Crest were offering it to agents, and apparently directly to other potential purchasers. "I agree", he said, "that by the time of Shipman's telephone call the sky had darkened", and that by implication he would have to look with great care at what he was going to have to pay out in what was already a loss-making situation.

He confirmed that Nelson Bakewell were being paid only 0.25% F on a fraction of the portfolio, the amount in cash terms amounting to £37,500. "If", he said, "fees are not agreed, those eventually paid by me are usually very much dependent on the result of the deal", and it was he who introduced into the case for the first time the concept of renegotiation; in other words, the agreement of a fee with the implication that it would be dependent on the success or failure of the deal, and that the agent, as Nelson Bakewell had done in this case, would be prepared, in the event of the transaction not being profitable, to renegotiate his normal fees on a more modest basis.

He said that if there had been further introducers in previous cases they would generally get half of Nelson Bakewell's fee. In his experience he could never recall a case where he had paid an introducer more than the main agent — although it should be added, for sake of completeness, that he also said: "I must have done it on G some occasion, but I have no specific recollection of it".

This, then, is the evidence on which I am called to resolve the issues posed at the outset of this judgment. The answer to the question, "was there a direct contract between the parties that the defendants should pay the plaintiffs an agent's commission of 1% on this sale?" is plain. Indeed, it was effectively conceded by Mr Onslow, for the plaintiffs, in the course of his closing submissions. No such contract was made. There is a complete absence of evidence that UK Land in writing, or orally by Mr Tatt or Mr Barclay — to whom the original offer was made by the letter of April 13 — or by conduct, agreed Reilly's 1% proposition forwarded to them by Marriott on April 21, by which date the introduction and UK Land's offer to purchase had H already been made.

The second question that falls to be decided is, therefore, did the defendants, through the agency of Nelson Bakewell, agree to pay to the plaintiffs a fee of 1%? If such an agreement was made, its existence can only be spelled out from the evidence of Messrs Reilly, Marriott and Tatt, and in particular that of Mr Marriott. Can the binding links of contract through an agent be established from what they said, did and wrote?

As far as the defendants were concerned, I have the evidence of Mr Allison and Mr Tatt that Nelson Bakewell had no general authority on behalf of the defendants, let alone specific authority, to fix an appropriate sum or percentage by way of commission. Mr Onslow contends that such an agreement came into existence on April 17 J when Marriott asked Reilly to write to Lanyon Fleming making UK Land's offer, and that the term of the contract that the fee should be 1% is to be found in the evidence of Mr Marriott. He further contends, in an argument which I confess I had difficulty in following, that ratification for this is to be found in the fact that when Allison forwarded to Tatt a copy of the correspondence in the matter, comprising the letters of April 13, April 21 and May 10, Tatt on behalf of the defendants made neither comment nor complaint.

In my judgment, it is quite impossible for the court to read Tatt's silence as acquiescence to Reilly's proposal of a fee of 1%.

What are the facts of the matter? In my judgment, they can be decided only in the light of the whole background of the case, which is K very significant. This is a small world of professional men in which virtually all the principal actors in the case knew each other well. Indeed, I think the only two witnesses in the case who were not personally acquainted were Reilly and Tatt. Virtually all the correspondence in the case is in Christian name terms. In particular, Reilly and Marriott had been at school together. They had maintained a firm friendship since they had left school and had embarked on parallel careers. In such a small world it is by no means uncommon, and it should be said it is not an unattractive characteristic in itself, not to be too precisely demanding that agreements should be reduced to writing, particularly in relation to sums of money that one party may be liable to pay to another; the L underlying assumption being that, as between gentlemen and men of honour, such questions can be resolved without disagreement or, still less, rancour. Unhappily, there are occasions when misunderstandings do occur, and in my judgment this was exactly one such.

I find the facts to be as follows. First, Reilly assumed, and was entitled to assume, that if the plaintiffs made the introduction and an acquisition followed, they would be entitled to a fee. Marriott, and for that matter Allison, who approved Marriott's letters to Barclay, were of the same mind. Nelson Bakewell knew that Lanyon Fleming and not the plaintiffs were Regent Crest's retained agents. Reilly's initial proposal was that the plaintiffs should receive one-half of Nelson Bakewell's fee from Nelson Bakewell as principals. Why? M Because he, not unnaturally, by the custom of the market — as evidenced in the Woolf portfolio case and other similar cases cited by later witnesses, to whose evidence I shall turn, in the persons of Messrs Owen and Roberts — hoped that his firm would make further profits from fees earned by way of agency work in disposals as opposed to acquisitions.

When Marriott accurately informed him that Nelson Bakewell would not agree to such an arrangement, his second proposal was that forwarded by Marriott to Tatt, namely an introductory commission of 1%; Mr Tatt simply did not respond. I have no doubt that his failure to do so stemmed from the fact that he was preoccupied with the problems of carrying through a £23m deal as

A opposed to striking an ancillary agreement in a sum involving one-hundredth, or less, of that sum.

Reilly and Marriott were thus placed in a delicate position. The former pressed the latter to obtain written confirmation of the fee from Tatt, and Marriott did his best to reassure Reilly that all would be well, in particular by writing a letter of reassurance and sending a copy to Tatt, reminding the latter of the plaintiffs' 1% proposal in the enclosure. By this time (by the time of the letter of reassurance) three weeks had passed. I have no doubt that Marriott did raise the matter with Allison who, he correctly appreciated, as a senior partner with the firm, would be more likely to have Tatt's ear; and although in evidence Allison was somewhat vague on the point, he, Allison, may B have said words to the effect to Marriott, "Tell him not to worry; there'll be no problem over fees." In other words, the sort of response that I have postulated as typical of dealings between businessmen who are also friends.

What I have to decide is whether Allison ever said to Marriott words to the effect that Tatt would pay a 1% fee. Marriott says that Allison did; Allison says he did not. I have no doubt the latter is correct.

First, I accept that Allison knew he had no authority to bind the defendants to such a fee. He knew that Tatt was very much his own man. Second, it would have involved his authorising a fee twice the size of his own; an unlikely enough proposition in itself. Third — and it is necessary to remind oneself we are speaking of a sum of about C £$\frac{1}{4}$m — although much of what transpired in this case took place on a very informal basis, this would have been taking informality to a quite extraordinary degree. Fourth, no note or memorandum of that alleged agreement was ever made by anyone. Finally, I remind myself of the letter in which Allison wrote to Tatt that no agreement had been made. If Allison had made an agreement, this would involve either an extraordinary aberration, forgetfulness, or, alternatively, the crudest and most fraudulent lie to Tatt on his part. I do not think he was guilty of either.

I do not think that Mr Marriott deliberately lied in the witness box. I think that as far as he was and is concerned, this is very much a case — as I said during the course of the submissions — of the wish being D father to the thought. In other words, he has allowed a perfectly natural desire not to let his friend, Reilly, down by failing to do what the latter asked to translate itself into a conviction that what was probably an easy and casual assurance by Allison that everything would be all right meant what he now wants it to mean, and that he read into Allison's words something far more precise than the latter intended. I am conclusively reinforced in this view by the striking piece of evidence, to which I have referred, emerging in his cross-examination. Before the trial he had given statements to the solicitors. These statements were at some points carefully redrafted by him. In neither of them, as I have said, is there any reference to the conversation with Allison from which he now claims he was authorised to confirm a 1% fee with Reilly. Further comment on this E point is unnecessary.

In my judgment, this is a critical finding of fact, because it renders unnecessary any analysis of whether Nelson Bakewell had any form of authority, express, implied or ostensible, to conclude the specific agreement of a commission of 1% contended for by the plaintiffs. My finding is that no agreement, either express or via an agent, was made between the parties for the payment of a fee of 1% commission on this transaction.

I now address myself to the question, was there a binding agreement that in the event of an acquisition the defendants would pay the plaintiffs for their professional services, containing an implied term that the services would be paid for at a reasonable rate?

F It is possible to contend for this conclusion via four possible routes. Canvassed least in the submissions of counsel was the most direct of these routes, namely that the agreement was in fact made between the plaintiffs and the defendants direct, even though the contact between themselves was solely through a third party until the submission of the plaintiffs' fee note. The other possibilities argued before me are that the agreement was reached through the agency of Nelson Bakewell, either as actual agents or ostensible agents, or in the further alternative by their ratification of what Nelson Bakewell did.

It is important to bear in mind that the question to be answered is not whether Nelson Bakewell had any real or ostensible authority to agree a reasonable fee, whatever that might be. The question is, did they possess the actual or ostensible authority to agree that their principal would pay a reasonable fee, the negotiation of the amount G of which would follow in the normal course of events?

I can deal with the question of actual authority briefly. Actual authority is either express or implied. It seems to me that, despite Mr Onslow's attractive submissions on the question of agency, his arguments were more properly directed to the question of ostensible rather than actual authority. In relation to express actual authority the evidence is simply that of Mr Allison and Mr Tatt, the agent and the principal in question. They said simply that their practice was that Mr Tatt should agree fees and not Mr Allison. There is no reason to disbelieve them. Implied authority is given to an agent where he is given broad powers to deal with incidental matters, or to do what is usually and normally done in a particular trade or profession. In my H judgment, this is not such a case. Such an authority cannot arise where it is expressly understood between principal and agent, as it was here, that the former and not the latter would deal with the fees of parties acting as intermediaries in transactions such as the present.

Any remark that might have been made — and I stress the word "might" — by Allison to Marriott and passed on to Reilly to the effect that the latter need not worry about his fees, and that there was no problem in that direction, falls far short of his exercising an implied authority to bind his principal to the payment of a reasonable fee.

However, I do hold that it is the case here that the defendants in the person of Mr Tatt always considered themselves contractually bound to pay the plaintiffs a reasonable fee for their services. If I am wrong J in this, then I hold in the alternative that Nelson Bakewell had ostensible authority to agree such a fee, and did so. If, yet again, I am in error, I hold that the defendants ratified Nelson Bakewell's acts in purporting to agree that a reasonable fee would be paid by their subsequent actions.

Critical to all these findings is an analysis of what Mr Tatt said and did, and equally importantly did not say in the witness box. It was stated several times in the course of the evidence, and not just by protagonists of the defendant company, that UK Land plc had over the years won a reputation as "good payers". Never once in his evidence did Mr Tatt suggest that he did not accept an obligation that the plaintiffs would be paid. He quite candidly admitted that the K letters of April 21 and May 10 were documents of which he had no recollection. As he saw the matter, and I believe he still sees it, the first time that the question of the plaintiffs' fees came to his mind was when the plaintiffs' fee note in what he considered a grossly exorbitant sum reached his desk at the end of June 1989 — two months after their active involvement in the matter had ceased. His initial response was to write the letter which said "You must agree your fees direct with Nelson Bakewell"; but, as he stressed in evidence, he had responded in this way because of his recollection of Reilly's original moiety proposal, rather than after perusal of the subsequent correspondence. He stated quite firmly, and to me convincingly, that after the 1% proposal of April 21 and confessing that he had not consciously directed his mind to it, that he would have L expected the plaintiffs to get in touch with him direct (and I emphasise and underline the word "direct") to negotiate a mutually agreeable fee.

In my opinion, as I have already indicated, the most important single piece of evidence in this case was Mr Tatt's assertion, which I find entirely convincing and I quote it again, "Elliott's only had to pick the phone up and I would have agreed a fee with them."

A great part of the significance of that remark lies in implication of the time of which he was speaking, namely before the fee note arrived. By that time, the time the fee note arrived, the deal as far as UK Land were concerned was beginning to look a far less lucrative proposition than Mr Tatt originally hoped: see Mr Shipman's memorandum accepted by Mr Tatt as an accurate record of their M conversation of about June 30.

I have to consider, however, what was in the minds of the parties at the time a contract, if any, was made. When Nelson Bakewell provided the material for Reilly to write the plaintiffs' offer letter it is patent they could have done so only on the express authority of the defendants. Mr Tatt, and/or his colleague Mr Barclay, the recipient of the original letter, must have known at the time (although the matter was clearly not uppermost in their minds, as I have already said) that the plaintiffs were acting for him in so doing. Contracts were indeed exchanged very soon after. What was his normal practice in such a situation? He told me at the very outset of his evidence. Nelson Bakewell were the agents he normally used. If, he said, a firm

A other than Nelson Bakewell effected the actual introduction he would expect to negotiate a separate fee with them. In other words, we returned to his assertion that Michael Elliott & Partners only had to pick the telephone up.

It is, therefore, my judgment that the plaintiffs were expressly engaged by Nelson Bakewell to act for the defendants on April 17 or 18; that it was an implication of this agreement that the defendants would pay a reasonable fee for their services; that Mr Tatt, on behalf of UK Land, instinctively accepted, though he may not have given voice to his thoughts, or even troubled to analyse the precise position in law, that the plaintiffs were entitled to look to him for a fee which he would negotiate in a reasonable sum. By the time his conversation B with Shipman took place, he made the suggestion that the plaintiffs should look to Nelson Bakewell; events had moved on a long way.

If I am wrong in this, I hold alternatively that the circumstances of this case were such that the defendants clothed Nelson Bakewell with ostensible authority to agree that a reasonable fee — not, I should emphasise, any particular reasonable fee — would be paid to them by the defendants.

I take the general principle of such authority from the text of *Bowstead on Agency*, 15th ed at p 284:

> Where a person by words or conduct represents or permits it to be represented that another person has authority to act on his behalf, he is bound by the acts of such other person with respect to anybody dealing with him as an agent on the faith of such representation, to the same extent as if such other person had C the authority that he was represented to have, even though he had no actual authority.

Every such case must turn very much on its own facts. At one point in his submissions Mr Kendrick, for the defendants, stated that in this particular case he was indeed prepared to argue that the contract was simply one between the plaintiffs and Nelson Bakewell as principal. But in my view, quite correctly, he abandoned the attempt to do so *in limine*.

All the evidence points to the conclusion that Nelson Bakewell, who are estate agents, were accepted as exactly that by everyone concerned throughout, after Reilly's abortive attempt at the outset to negotiate a fee with them direct. It is, however, true, as Mr Kendrick D says, for the law to tend to protect third parties by giving them rights against the agent rather than the principal where the agent acts without authority. It is also true that ostensible authority cannot be held to exist generally in most cases, and in each case the third party must trust the warranty of an agent that a particular authority has been obtained, on which topic I was referred to the observations of Robert Goff LJ in the case of *The Ocean Frost** [1986] AC 717.

What in fact was the position here? Although we have no direct evidence as to what happened between the defendants and Nelson Bakewell before Reilly's letter of offer was written, it is a fact that UK Land knew when they made their offer to purchase that the introduction had arrived via the plaintiffs and that in the normal course of events their offer would return via the same route. Mr E Onslow is absolutely correct, in my view, in saying that as it was plainly the case that UK Land were giving Nelson Bakewell authority to accept the introduction, the inevitable implication must be that UK Land could not expect Nelson Bakewell to use the intermediary services of the plaintiffs without cloaking them with either actual or ostensible authority to tell the plaintiffs that they would be remunerated for their services in a reasonable sum.

UK Land provided Nelson Bakewell with all the material to allow the plaintiffs to make the offer; they knew the source of the introduction, and they permitted their representative to deal with the plaintiffs in all material matters, save as to the actual *quantum* of the fee in question. I have no doubt that they clothed Nelson Bakewell with ostensible authority to permit Marriott or Allison to declare that F a reasonable fee would be payable.

When Marriott gave Reilly the assurance that there would be no problem over fees, and in particular when he wrote the various letters to Tatt on the subject, it was plainly his understanding that a fee of some sort would be payable, and by all its implication that such a fee would be reasonable. Allison, in approving these letters, was clearly of the same mind. The entire picture is of men clothed with ostensible authority to say just that, and Tatt could see that picture himself throughout by virtue of his receipt of these letters or copies of them.

Finally, ratification. Once more, I take the general principles set out in *Bowstead on Agency* and in particular articles 13, 16 and 17 at p 51 of that volume. It is unnecessary for me to recite them here. G

Mr Kendrick is right to emphasise that there is no unequivocal, positive conduct by the defendants to evidence ratification. What do the defendants know at all material times? First, that the introduction had come from the plaintiffs; second, that the offer to purchase was returning via the plaintiffs; third, from the outset that the plaintiffs were expecting a fee (Marriott's letter to Barclay). What is there that points in the direction of ratification? Well, first of all it must be said that it is the plain and simple fact that the defendants bought the land. Second, they neither commented on nor still less complained about the use of the words "my clients" by the plaintiffs in their letter to Regent Crest's agents as descriptive of them. Third, they never commented nor still less complained about Nelson Bakewell's various reminders that the plaintiffs would require a fee. H In summary, they made absolutely no comment on the relevant letters.

By reason of this undisputed evidence I am prepared to hold that if, contrary to my earlier findings (a contract to pay a reasonable fee was made either directly between the plaintiffs and the defendants, or the plaintiffs and Nelson Bakewell acting as the defendants' ostensible agents), there was no contract so concluded, on the facts as I have previously found them the defendants ratified what had been done by Nelson Bakewell.

They did not say or write a single word to suggest that the acts which the plaintiffs were carrying out on their behalf and to their benefit were unauthorised. In the circumstances of this case, I infer J ratification from the acquiescence or inactivity of the principal, to quote the words of *Bowstead* at p 66. Such acquiescence, it seems to me in this case, was total.

Finally, the plaintiffs claim *quantum meruit*. If, they contend, there was no contractual nexus between themselves and the defendants, they have nevertheless rendered services to the defendants, who have freely accepted and not rejected them, with the knowledge that they could not possibly be gratuitous, and the defendants have incontrovertibly benefited from these services. Consequently, say the plaintiffs, they are entitled to be paid for these services *quantum meruit*.

Although he did not formally concede the point, Mr Kendrick realised that this was virtually an impossible claim to contest in K general principle, and in my judgment, in the light of the contents of the relevant letters, and the simple central fact in this case that the defendants did in fact enter into a contract which they saw at the time as potentially to their great benefit as a result of the part played by the plaintiffs, he is right. If no single one of my findings in law to this point of my judgment is correct, I hold that the plaintiffs are entitled to be paid a sum *quantum meruit* in this case. Such a sum is one which is reasonable and proper in the particular circumstances in which the obligation to pay arises, and represents no more and no less than the reasonable implied fee which is due from the defendants if, as I have held, they are liable in contract to the plaintiffs.

This is an important point in the present case. This *quantum meruit* L is not, unhappily, to be assessed in the straightforward terms that might be appropriate for the services of a tradesman or artisan (say, X days' work at Y pounds per day), or a professional man, say a solicitor or an accountant, who might charge for X hours' work at Y pounds per hour. In this case the plaintiffs wrote no more than half a dozen letters, and at most, primarily through Mr Reilly, devoted probably less than 100 hours of their time to the defendants' affairs. It is agreed on all sides that the simple fact that the plaintiffs introduced the defendants to a £25m deal must be the starting point of my inquiry.

Mr Kendrick places reliance on a recent case of *The Batis** reported at [1990] 1 Lloyd's Rep 345, and in particular a passage at p 352 in column 2. But I do not derive significant assistance from the M judgment of Hobhouse J in that case, because his lordship was there concerning himself with a situation fundamentally different from that obtaining in the present case, namely the appropriate *quantum* recoverable where circumstances have vitiated a contract already existing between the parties, and one party has gone on to behave in a manner forced on him by circumstances to the benefit of the other. In my view, the correct approach to *quantum meruit* in this case is indeed to find the answer to the question, had the parties agreed a reasonable fee at the material time — that is, in the last fortnight of April 1989 — what would that fee have been?

*Editor's note: This refers to *Armagas Ltd* v *Mundogas SA*.

*Editor's note: This refers to *Batis Maritime Corporation* v *Petroleos del Mediterraneo SA*.

Each party called an expert in these matters to support their contentions. I intend no discourtesy to these gentlemen if I say that I found their evidence of limited assistance. Mr Nicholas Owen [FRICS], the chairman of Herring Son & Daw, and Mr J W Roberts [FRICS], a former partner in Jones Lang Wootton, are both chartered surveyors with highly impressive experience in the work of estate agents dealing with commercial property. By a multitude of examples they demonstrated to me essentially that the commission on a one-off transaction, as this was, amounted to no more and no less than, as Mr Tatt himself said in evidence, what the market would bear. The examples of commissions actually paid on purchases analogous to a greater or lesser degree with the present one range from 1.5% at one extreme down to a few thousand pounds at the other.

I do not think I can reach the correct conclusion by deciding that any particular one of the numerous examples cited to me is especially analogous to the facts of the present case; because, as both Mr Owen and Mr Roberts told me, many factors come into play in each case. First, the relationship between the principal purchaser and the agent. Where, as Nelson Bakewell stand to the defendants, and Mr Owen's company stand to organisations like Crusader Insurance, as retained agents, a generous commission usually seems to be catered for by express agreement. Second, the exclusivity of the introduction, contended by the defendants to increase and the plaintiffs not to affect the fee. Third, the question of the size of the fee in absolute terms. Mr Roberts strongly contended that a fee should be assessed at the end of the day in pounds and pence rather than by a percentage when the amount involved was as large as it is in the present case. Fourth, the proportion of the fee of the introducer to that of the main agent. I deduce from Mr Owen's and Mr Roberts' evidence that this will seldom be as great, and even more rarely greater. Finally, the question of renegotiation. This is an important point.

Mr Owen unequivocally stated that he would not expect a correct fee, as he described it, to be renegotiated. As it so happened, the purchase of this portfolio by the defendants resulted in a loss and not a profit. Nelson Bakewell waived the 0.5% acquisition fee which would normally have been theirs, and they have agreed to accept a greatly diminished fee from that laid down in the original memorandum; as I have said, at 0.25% commission on disposals, this means that they will receive in this case no more than a total of some £35,000 or £37,500 from their principals, the defendants. But Nelson Bakewell as the defendants' retained or in-house agents have, to quote Mr Kendrick's word in argument, "tied their wagon to the star of the defendants' success in the property market". They are the defendants' advisers. In my view, renegotiation of fees for a retained agent concerned with both acquisition and disposal is one thing; the fee payable to an agent who simply introduces a property, and performs no role in its evaluation or subsequent disposal, is quite another.

It follows that I decline to approach the question of *quantum* by reference to the potential success or, as it transpired, actual failure of the defendants' venture in purchasing this portfolio.

The crux of Mr Owen's evidence was his opinion that 1% would be the correct fee in this case; or, as he somewhat confusingly modified it, an expectation from which people would start unless negotiated otherwise. He conceded that skill and knowledge in the property market would relate to the activities of Nelson Bakewell rather than the plaintiffs in the present case and that the essential value of the plaintiffs' services lay in the introduction.

Mr Roberts laid great emphasis on the criterion of exclusivity, namely whether the potential purchaser was being offered an exclusive chance to purchase or whether, as I have said, at the opposite end of the scale the agent was simply passing on market-place gossip that a particular property was coming on to that market.

I found this an attractive argument in terms of both logic and what I learnt from the experts to be commercial practice.

The other central element in Mr Roberts' evidence was that where, as in this case, one agent introduced to another, it is common to accept half a global fee. In the present case such a calculation would have resulted in the plaintiffs receiving 0.25%, being half of the conventional fee payable by the defendants to Nelson Bakewell in such a case. Such a figure would have resulted in the fee which Mr Roberts robustly described as "bang on" for a reasonable fee in the present case. He derided the idea of an acquisition fee of £¼m. Nobody, he asserted, would ever pay that for an introduction *simpliciter*.

A striking corollary to his evidence was the fact that 0.25% was exactly what, unknown to him and unknown to Reilly (for I accept Reilly's evidence on this point), the plaintiffs were in fact in their very first letter asking Nelson Bakewell to pay. Further, he made the point that for their conventional 0.5% Nelson Bakewell would have the task, a heavy one, of evaluating these proposed acquisitions.

Narrowing the issues, I discount Mr Kendrick's submissions that the plaintiffs were entitled to no more than £17,500, or £35,000, as being figures calculated on the basis of Nelson Bakewell's actual remuneration, for the reasons I have already adumbrated. At the other end of the scale, I accept Mr Roberts' evidence and reject that of Mr Owen by holding that a fee of 1%, twice that of the main agent, is inappropriate in a case of one-off introduction such as the present. This effectively reduces my range of choice to a figure of 0.25% or 0.5% of the purchase price, and I am acutely conscious of the fact that in cash terms this amounts to a difference of no less than £55,000 or thereabouts.

For the plaintiffs it was argued that Mr Owen provided a number of examples in which the market rate was demonstrated significantly to exceed 0.25%. But, in my view, the individual features of these examples fail to produce a coherent picture. I give full weight to Mr Onslow's reliance on the fact that the Woolf portfolio deal produced a net payment of £194,000 in all for the plaintiffs on a £14.5m deal. Against this it must be said in the last analysis that the plaintiffs' acquisition fee in the Woolf case was a mere £10,000.

However, as between these two figures I found the submission of Mr Kendrick the more convincing. This was not an exclusive introduction. In my opinion, this factor mitigates against a high fee and I accept what Mr Tatt and, more particularly, Mr Roberts say on the point. No negotiations of price were involved, a simple but important factor that is not to be overlooked.

The most compelling argument, however, in my judgment, derives from the actions of the parties themselves. When the plaintiffs made the introduction, they proposed, by implication, a fee of 0.25% — a moiety of Nelson Bakewell's normal fee. Looked at from the defendants' point of view, Nelson Bakewell and the defendants, by the terms of their memorandum on fees, envisaged a payment of 0.25% to Nelson Bakewell where they were not entitled to an introducing fee, with the inference that what they were not earning was the other 0.25% which they would have earned had they made the introduction.

In summary, balancing as best I can the many factors involved in reaching such a conclusion, I hold that the plaintiffs are entitled to succeed on their claim: first, on the basis that the defendants agreed directly to pay them a reasonable fee for their services; in the alternative, that the defendants clothed Nelson Bakewell with ostensible authority to agree such a fee; in the further alternative, that the defendants ratified the acts of Nelson Bakewell in agreeing such a fee; and in the yet further alternative, *quantum meruit*. Such a reasonable fee in the judgment of this court is 0.25% of the sale price of the properties involved, namely £55,750 plus 15% VAT, which by my calculation is £8,360; £64,110 in all. I give judgment for the plaintiffs in that sum.

I would not wish to part with this case without thanking both counsel, Mr Onslow and Mr Kendrick, for the moderate, attractive and forceful way in which they have presented their cases. Their clients have reason to be grateful for their services.

Queen's Bench Division

February 1 1991

(Before Mr Justice MACPHERSON)

KNIGHT FRANK & RUTLEY v RANDOLPH

Estates Gazette June 22 1991

[1991] 24 EG 164

Estate agents — Commission — Property introduced by plaintiff agents but vendor refused to pay agents' share of commission — Allegation by vendor that a higher price could have been obtained but for agents' negligence or breach of contract — Vendor, in fact, achieved a very good price, £925,000, when the guide price was thought to be £900,000 —

A **Vendor's evidence unsatisfactory — Litigation which should have been avoided — Criticism of agents and staff not justified — No breach of duty of care — Judgment for agents for commission due to them**

The case concerned the sale of a property known as Rush Manor, Wallingford, Oxfordshire — There was no question that the plaintiff agents had introduced the purchaser who bought the property, but the defendant vendor refused to pay the plaintiffs the share of commission due to them — The plaintiffs had acted jointly with local agents, whose share of commission had been paid by the defendant — The defendant B **had been hoping for £1m for the property and, in fact, sold it to a Mr Potter for £925,000 — He claimed to be entitled to withhold the plaintiffs' commission on the ground that there was "every prospect" of a higher price being achieved but for the plaintiffs' failure — There was a counterclaim by the defendant in support of his allegation**

The judge found that the defendant's evidence was distorted by irritation which had clouded his judgment and produced unjustifiable criticism of the plaintiffs — Disregarding irrelevant matters, the defendant's main complaint was that a representative of the plaintiffs had failed to keep him fully informed of the interest and attitude of the C **Mr Potter who eventually purchased the property — Mr Potter was an astute businessman who played his cards close to his chest — He had given the plaintiffs to understand that he would not go above £805,000 — When the plaintiffs became aware that another bidder, negotiating privately with the defendant, had made an offer of £875,000, they did not think that Mr Potter would be interested at that sort of figure — There were, however, strange developments — The defendant made a deal with the other bidder, a Mr Pither, for the sale of the property to him for £900,000, but this deal was later "unscrambled" — Apparently "the astute Mr Potter"** D **persuaded the defendant to retreat from the private deal and accept Mr Potter's own bid for £925,000**

The judge's conclusion was that there was no substance in the allegation against the plaintiffs' representative — She had acted reasonably throughout and with all due care — The defendant was kept well informed of Mr Potter's interest — The defendant had got a good price for the property in a sparse market — The plaintiffs were entitled to their commission and earned a commendation from the judge, who was "impressed by the moderation and care of their witnesses" — Judgment for plaintiffs on claim and counterclaim

E The following cases are referred to in this report.

Dunton Properties Ltd v *Coles Knapp & Kennedy Ltd* [1959] EGD 221; (1959) 174 EG 723, CA
Smith v *Burns* The Times February 1 1991

In this action Knight Frank & Rutley claimed against Denys Randolph their share of commission due on the sale of the defendant's property at Rush Manor, Wallingford, Oxfordshire. The arrangement was that the commission of 3% was to be shared between the plaintiffs and Adkin Commercial, of Oxford, in the proportion of $1\frac{3}{4}\%$ and $1\frac{1}{4}\%$ respectively. The defendant had paid the amount due to Adkin Commercial but not the amount due to the F plaintiffs.

David Jeremy (instructed by Corbald Rigby & Co) appeared on behalf of the plaintiffs; Alan Steynor (instructed by Lewis Silkin) represented the defendant.

Giving judgment, **Macpherson J** said: Since I do not believe in unnecessary suspense I say at once that the plaintiffs succeed in this case.

In this judgment I shall refer to the plaintiffs, Knight Frank & Rutley, as KFR and to the defendant (who is plaintiff on the counterclaim) as Mr Randolph.

On August 15 1989 Mr Randolph agreed to sell the property known as Rush Manor, at Wallingford, to Mr David Potter. In fact the property was jointly owned by Mr and Mrs Randolph and sold, I believe, to Mr and Mrs Potter. Neither Mrs Randolph nor Mrs Potter G has given evidence, although both of them figure in this unfortunate story.

The sale was admittedly the result of the introduction by KFR of Mr Potter to the property and its owners. KFR acted, in fact, in concert with other local agents called Adkin Commercial, represented in this case by Mr Harper. The deal with the agents was that Mr Randolph would pay 3% plus VAT and expenses should there be a fruitful introduction. The commission was to be shared between KFR and Adkin as to $1\frac{3}{4}\%$ to KFR and $1\frac{1}{4}\%$ to Adkin.

Mr Randolph was hoping for the magic £1m for his property. But throughout this relevant time the guide price was indicated to be £900,000. So that price was eventually exceeded by £25,000. In my judgment, £1m was never on the cards. H

There is no doubt that KFR did the lion's share of the joint agents' work. They incurred virtually all the expenses during the sale period. Mr Stephen Lambert [FRICS], in Oxford, and Miss Nicky Althaus, in London, were engaged in the activity, and so was Mr William Yates [FRICS], the head of the residential division of the firm, and others.

KFR produced the glossy brochures which are familiar to those interested in expensive property. They did all the usual circulation and advertising. All this can be seen in the uncontested part of Mr Yates' evidence, which was put in and proved in statement form. This practice was sensibly followed for all the witnesses as they were called. No extensive reference by me to the papers and evidence is necessary. J

Adkin have been paid their $1\frac{1}{4}\%$, £13,296.88, with the thanks and congratulations of Mr Randolph. But this action arises because Mr Randolph has refused to pay KFR. In a nutshell, he says that he is entitled to withhold payment because "there was every prospect" of a higher price having been achieved but for KFR's breach of contract or negligence. Mr Randolph in his proof says that he believes that he could have achieved "possibly as much as £50-£100,000 higher". I feel bound to say that this assertion is fanciful and has no support from the evidence.

I am in this context disappointed to have to say that I found Mr Harper's evidence, on behalf of Adkin Commercial, on this aspect of the case disingenuous. He does seek to say that he believes that if there had been competitive bidding the price could have been as high K as £975,000. But I find as fact that on September 8 1989 he said to his fellow agent that £925,000 was "a bloody good figure" and that the sale to Mr Potter was "one of the luckiest sales of the year". He also told his client that the figure was "very satisfactory . . . in the light of current market conditions". Later he repeated to Mr Lambert by letter his view that the price that was eventually achieved was a very good one, taking into account the property and the market.

Of course, Mr Harper's position is a difficult one. He was joint agent with KFR. Mr Randolph castigates KFR, yet Mr Harper has been paid in full. Mr Harper did somewhat half-heartedly say that KFR might bear some blame for a failure of communication. But my strong impression was that his heart was not in that criticism. He said L twice that what he thought would happen after the protest and storm died down was that Mr Randolph would pay up to KFR and the whole matter would be over. In my judgment, it is a pity that this did not happen. Very large cost has been incurred in litigation which should, in my judgment, have been avoided. But the trouble is that sometimes hindsight and justification rule the head. I believe that this is what happened in this case to Mr Randolph, helped perhaps unwittingly by Mr Potter, who has given evidence in Mr Randolph's support.

I say at once, however, that I am sorry that Mr Jeremy, counsel for the plaintiffs, impugned the good faith of Mr Potter and indeed, to some extent, of Mr Randolph. Both of them have, as it seems to me, warmed to the task of fighting this case, to an extent which has led to exaggeration; and the prospect of more money does distort views. M But I do not find that there has been any purposeful or devious scheme afoot. Both Mr Randolph and Mr Potter are highly successful and honourable businessmen. And I believe that their temporary irritation (to be seen in the early letters and attitude of themselves and their wives) has been unnecessarily prolonged and has indeed clouded their good judgment, but no more than that.

Mr Potter has, of course, no financial axe to grind. He achieved this property at £925,000 and is living in it. But certainly in some respects his contentions are not made out by his evidence, as I will shortly demonstrate. For example, his mistaken belief that his secretary passed on his message from the beach in Portugal on July 20 or 21 1989 has wrongly coloured his part in this case.

As to Mr Randolph, he too has, I fear, to a considerable extent allowed irrelevance and unjustifiable criticism of KFR to infect his approach to the heart of this case. For example, he has said that he believes that the change of name of this property from "The Cottages" to the more impressive "Rush Manor" was bad for the sale, and he plainly seeks to blame KFR for this. KFR did say that "The Cottages" was not perhaps a good selling name, but it is palpable that the idea of "Rush Manor" came from Mrs Randolph. She researched the Victorian history of the Liberty of Clapcot and said to KFR: "I am sure you will agree that the house should be called Rush Manor. Had I known in 1965 that which I now know it would not have been called The Cottages, but all that is past." In the circumstances any criticism about the name of the house cannot possibly be put at the door of KFR.

Nor, in my judgment, has the failure to achieve the agreed *Country Life* advertisement date any bearing on the case, apart from the fact that it was (as Mr Yates accepted) an error which should not have happened. But I expect that even in Mr Randolph's business life he must have seen such mistakes made by staff. And KFR at once accepted that it was their mistake, entered an alternative *Country Life* advertisement, and paid for two separate advertisements in leading newspapers as well.

Yet this complaint appeared in the pleadings and still troubles Mr Randolph two years later, although his pleadings accepted, when particulars were sought of him, that this error did not cause any loss to him but "was one of a number of factors which caused the defendant to lose confidence in the plaintiff". Loss of confidence is not a cause of action.

Without wishing to waste time on irrelevant matters, I am also convinced that Mr Randolph's criticism about the projected splitting of the property is unjustified. Mr Lambert told me that it went ahead for a time because of a client's inquiry and was simply part of KFR's efforts to achieve the best possible price. Certainly Mr Randolph never vetoed the idea at the time. I do not believe that his irritation over this matter is reasonable. In parenthesis, the joint agent Adkin did not consider the proposal to be a waste of time either.

The response to the offer for sale was disappointing. People liked the site but not the house. All options were followed, including for a time the split, which had been an idea emanating from a client's inquiry.

In the witness box Mr Randolph also slipped in a criticism about the availability of those dealing with the matter of KFR. He has not any good word for them. I believe that these criticisms have fed upon themselves and are unjustified.

I turn, then, to the heart of the case. As often happens, the heart can now, at the end of the day, be clearly seen. The main issues are in a narrow compass. The duty of the agents in law is to act with all reasonable care and skill and to do their job properly as professionals. Specifically, as is set out, for example, in *Dunton Properties Ltd* v *Coles Knapp & Kennedy Ltd* (1959) 174 EG 723:

> The duty of an agent engaged . . . upon the sale of a property must be in broad terms, as Mr Bagnall stated, to do all that they can as professional men to get for their principals the best possible price. It is not in issue and has been decided, that if before any binding bargain is made it had come to the notice of the defendants that another purchaser was available offering a higher sum of money they would have had so to inform their principals . . .

(see Lord Evershed MR, p 723).

The general duty is usefully and broadly set out in *Bowstead on Agency*, 15th ed p 144:

> An agent must exercise such skill and care and diligence in the performance of his undertaking as is usual or necessary in or for the ordinary or proper conduct of the profession or business in which he is employed or is reasonably necessary for the proper performance of the duties undertaken by him.

The question here is whether Miss Althaus made culpable mistakes and failed in her duty in two directions between July 17 and July 24 1989. First, as pleaded in the defence, para 9(b), in failing to pass on relevant information to Mr Randolph. I stress the way in which the case was originally pleaded; not because I wish or intend to be over-formal or rigid about the case or the pleadings, but because it is, in such a case as this, important to see where the original thrust of the case aimed. An amendment was made to the pleadings after the speeches were ended. It is unusual even to contemplate such late amendment: see *Smith* v *Burns, The Times* February 1 1991. But since the alternative case (pleaded now as para 9(c)) was canvassed in evidence, it seemed to me right to allow that second interlocking seam to be followed. Thereby Mr Randolph says that Miss Althaus failed properly or fully to respond to a request made for information about this sale made by Mr Potter's secretary on July 19 1989. The full allegation can be seen in the amended subparagraph.

The course of the proceedings connected with the sale from the moment the agency was agreed until the end can be seen in the full bundle of documents and in the statements of the witnesses. There are, however, some important factual findings which I make having seen and heard the witnesses. In my judgment, these findings, together with my conclusions as to Miss Althaus' actions in the light of all the circumstances, tested against the standard set by the law, decide this case in the plaintiffs' favour.

First and foremost, I do not accept Mr Randolph's evidence that until the receipt of Mrs Potter's letter dated July 27 he "had no idea that Mr Potter was interested in the property nor had I realised that he had submitted a bid". I reject his evidence that he was "not notified of Mr Potter's interest in the property in June or July . . .". It is palpable both from KFR's witnesses and from the documents that Mr and Mrs Randolph were told about Mr and Mrs Potter's interest on a number of occasions. If indeed Mrs Randolph did not tell Mr Randolph about her telephone calls, that cannot be the fault of KFR. Indeed, I am sure that both Mr and Mrs Randolph were told all that the telephone notes record. These were all fully verified in evidence. The result is that I find as fact that the Randolphs must have known the following facts:

(a) that Potter was interested. He had been twice to see the property. His interest was clearly passed on to the Randolphs;

(b) that Potter had bid only £765,000 and then later £805,000, but had said that he was not interested at £900,000 and had said that he would "not be budged" from his offer;

(c) that Potter would not increase his bid, but did wish "to be kept informed";

(d) that by July 10 Mr Pither was well and truly on the scene and was the most interested viewer and the only bidder who approached £900,000 at any time until July 27.

Furthermore, I am wholly satisfied from Mr Potter's evidence that, until he heard from his secretary on July 19 or 20 that which Miss Althaus told her, he played his cards close to his chest. That is what he told Mr Lambert and accepted readily in the witness box. He played a waiting game, and as it turned out he waited nearly too long. He did say that £805,000 was a figure at which he would stick, and I find that Miss Althaus wholly believed this to be the position. If it was not the position I do not see how Miss Althaus can be blamed for believing that it was and that Mr Potter was not and would not be interested at £900,000. He did say that he wished to be kept informed. But at no time did he say that which Mrs Potter set out in her letters, or that which Mr Potter set out in his proof, namely that Mr Potter "insisted that no sale should be agreed without seeking a further offer from us".

Mr Potter accepted readily in the witness box that he did not add that he must be allowed to respond to any higher offers that might be made, nor in evidence did he assert that he had indicated "very strong interest and that we were prepared to bid higher as and when necessary". Certainly he never transmitted that to Miss Althaus, or his private view that he might go up to £950,000 or £975,000. Any such intention was, in my judgment, totally invisible on the back of the cards which were held close to his chest. He is not to be blamed for this, of course. He was a buyer; and negotiation can be like a game of poker. But the result was, I am perfectly satisfied, to cause Miss Althaus to believe that he was not a keen bidder above his highest recorded bid of £805,000. It is with that finding in mind that I approach Miss Althaus' actions between July 17 and 24.

By that date Mr Pither was well on the way to the agreement that he made privately with Mr Randolph. By July 17 the offer was £875,000, and Mr Randolph indicated that he would take over the reins and did not require KFR to be present. Mr Yates was not happy about this, but Mr Randolph told him that he would carry on alone. We know that he made an agreement at £900,000 at about 10 am on Monday July 24. Completion was to be in December because Mr Pither could not produce the money until then.

The notes show that Miss Althaus and all concerned knew about the £875,000 bid by July 17 or 18 1989. Miss Althaus did not inform Mr Potter. This she accepted was in hindsight probably wrong. But her attitude, and I accept her evidence about this, was that Mr Potter would not be interested at that figure. In any event, Mr Potter's secretary, Miss McGowan, telephone call on July 19 overtook any such error, because on that day Miss McGowan rang and asked if

A there were any developments. Mrs Potter says that the precaution had been taken of "asking my husband's secretary to contact the agents, say we were away, but contactable, and re-express our very keen interest in the house". This does not square with the note of her actual request for "any developments". Miss McGowan was not called to give evidence.

It should of course in parentheses be noted that Mrs Potter's letters were written after July 24 when she and her husband came out into the open and registered for the first time, as I find, their great keenness to buy. Furthermore, the letters were aimed at persuading Mr Randolph to unscramble the sale to Mr Pither. This was, as we know, later achieved. It was not, I suppose, a really bad gazump on B Mr Pither, who was told that completion must be advanced, thus making purchase by him impossible. But he did necessarily pull out, leaving the field free for Mr Potter's accepted bid of £925,000.

I do not accept that Miss Althaus can be justifiably criticised for giving the information that she did to Miss McGowan, in the manner and words set out. She did not think that the request for information was keen or urgent. She did not regard the waiting Mr Potter as likely to come in over the keen Mr Pither. And I am absolutely sure that Miss McGowan never came back to Miss Althaus with whatever orders Mr Potter gave her from the beach in Portugal. Again I say that Miss McGowan has not been called; and Miss Althaus' note of July 26 1989 (Wednesday) is plainly genuine. Mr Lambert saw it very soon afterwards when he went over Miss Althaus' file.

C Of course, if Mr Potter had said to her (by himself or through Miss McGowan) that he was going to make a competing bid on the 24th, that would have been something which had to be passed on to Mr Randolph: see *Dunton* v *Coles*. But I am sure that such an intention was not transmitted to her. And, of course, Mr Potter made no personal contact with anybody else until it was too late — or seemed to be too late — on Monday July 24 or thereafter, although he was in the United Kingdom by Friday morning, July 21.

As to that contact by Mr Potter, I believe that he may unsuccessfully have tried to make contact on July 24. But I am satisfied, because of the dated notes, that the successful contact was made on July 25, perhaps by Mrs Potter, when Miss Althaus was fully and otherwise pleasurably engaged at Buckingham Palace. Nothing D very much turns on this part of the case in any event. But those are my conclusions.

It may well be because he was unwell and had to attend two hospitals at once on his return that Mr Potter did not press his suit before Monday, July 24.

I wholly acquit Miss Althaus of negligence or breach of her duty of care in that which she did *vis-à-vis* Mr Potter.

Again I stress that in fact Mr Potter achieved what he regards as a good deal, because he did persuade Mr Randolph to unscramble the deal with Mr Pither. And his offer of £925,000 was snapped up with understandable alacrity by Mr Randolph, who thus gained £25,000 above the price which he had also gladly accepted on his own account E from Mr Pither, who was the only other interested bidder.

That disposes of all but the originally pleaded case. In this context I note that in evidence Mr Randolph was in fact perhaps keener upon the amended point than the original plea. This may well be because, as para 9(a) shows, the plea is based upon the unsustainable allegation that Mr Randolph was

> told by the Plaintiffs that the only potential purchaser interested in acquiring the property for a price in the region of £900,000 was a Mr Pither. This was incorrect. Mr and Mrs Potter, the eventual purchasers, had told the plaintiffs that they were very interested in the property, that they were cash buyers, and that they wished to be kept closely informed and that no sale should be agreed without a further offer being sought from them. The plaintiffs failed to pass this information to the defendant. On July 19 1989 the Potters' secretary telephoned the plaintiffs and reiterated that the Potters were still interested F and wished to be kept informed. The Plaintiffs failed to pass this information to the Defendant.

I have already indicated that the suggestion that Mr and Mrs Potter were "very interested" and that they should effectively be given the "last refusal" is not sustained at all by the evidence. All that the Potters said and offered up to July 17 was, as I have found, transmitted to the Randolphs. If that information went, so to speak, over the top of their heads, that is not KFR's fault. Indeed it is important to note that Mr Harper himself agreed that the earlier bid G and information about the Potters was passed on to Mr and Mrs Randolph.

The final question as to liability, then, is whether Miss Althaus should have transmitted the information about Miss McGowan's call to Mr Randolph. The highest, in my judgment, that this can be put is that with hindsight and in some circumstances it might have been better if the call had been passed on. But, as I have already said, I do not blame Miss Althaus for believing that the Potters were cool, if not cold, and for not being much affected by the anodyne request of Miss McGowan as to "any developments".

It must also be noted that if Mr Randolph's belief or knowledge was as he said that it was, namely that he had never ever heard of the Potters, he would not have been likely to pause in his negotiations H with Mr Pither, whose bid was £95,000 up on Mr Potter's apparently final offer and with whom he wished to personally to go ahead.

The duty of Miss Althaus was, in my judgment, to act reasonably and with all due care. I am not prepared for a moment to find that she failed in her duty or that she could possibly be described as being negligent in failing to pass on the content of Miss McGowan's call and her response to it. Furthermore, no later call came back from Miss McGowan. Miss Althaus could not possibly know that Mr Potter was in Portugal. He was in fact contactable and could himself have telephoned if he were keen enough so to do. Silence until the end of the week reinforced Miss Althaus' view that there was nothing further to report. She was justified, in my judgment, in that J conclusion.

Once again I stress also that Mr Randolph was not "landed" with Mr Pither's bid. He was able to extract from that agreement, and then he chose to open up the field and start fresh, competitive bidding. He went at once for Mr Potter's good bid of £925,000. If he was allowed to contract with Mr Pither when he would not have done so but for Miss Althaus' failure, he was able in fact to retrieve the situation.

I saw Miss Althaus in the witness box. She is younger than the professional men in the case. But I was impressed by her evidence and by her attitude. If she thought for a moment that she was in breach of her duty to Mr Randolph I believe that she would have accepted blame. Neither she nor the two very experienced partners in KFR, Mr Yates and Mr Lambert, thought that she had erred. Nor, I believe, in K his heart of hearts did Mr Harper. Nor do I.

In addition, and for good measure, I repeat that, in my judgment, £925,000 was a good price for Rush Manor. It was not a popular house among the many viewers. There were in truth very few bidders. Even if a competition should have begun between Mr Pither and Mr Potter I am most doubtful whether £925,000 would have been exceeded. Mr Pither's offer was subject to contract and in particular to survey. I have seen his survey report. His surveyor was "deeply concerned that this particular property is unsuitable for your individual needs," and he thought "the overall transaction unattractive". With that albatross round his neck, and particularly with Mr Randolph's insistence upon October completion, which L made the deal financially impossible for Mr Pither, I do not see how he could be considered a likely competitor against the astute Mr Potter, who came in with a well-timed "ace" at £925,000. I repeat my reference in this context to Mr Harper's repeated and considered views given at the time that £925,000 was a very good price. He was Mr Randolph's favoured agent. Even if there were a breach of duty — which I do not believe there was — I would have awarded no damages in this case.

I end simply by repeating that, in my judgment, there are no knaves or villains in this case at all. Unfortunately, I believe firmly that Mr Randolph's irritation about irrelevant matters allowed him to embark upon this refusal to pay KFR's fees. And some exaggeration of the true position and perhaps some fatal lack of communication by M Mr Potter with his own secretary have allowed him to continue his case. I am sure that following Mr Harper's views he should have made a short protest and then paid up. That he will now have to do.

As to KFR, I was impressed by the moderation and care of their witnesses, and I find no case proved against them.

There will therefore be judgment for KFR on the claim and counterclaim.

A
Court of Appeal

October 16 1990

(Before Lord Justice FOX and Lord Justice STOCKER)

HYDE HOUSING ASSOCIATION LTD v HARRISON

Estates Gazette February 23 1991

[1991] 07 EG 131

Housing Act 1985, Schedule 1, para 3(1) — Whether a person occupying a flat under a licence from a housing association B pending development was a secure tenant — Para 3(1) provided that a tenancy is not a secure tenancy if the dwelling-house in question is on land which has been acquired for development and the dwelling-house is used by the landlord, pending development of the land, as temporary housing accommodation — Whether, in order to give rise to this disqualification, the land must have been acquired for development by the landlord or could have been acquired by a third party — In the present case the landlord was a housing association but the land had been acquired for development not by the association but by the Department of Transport — C The Court of Appeal, reversing the decision of the county court judge, decided that para 3(1) applied and that the respondent did not have a secure tenancy

The facts were that the appellant association had granted to the respondent what was described as a temporary weekly licence to occupy a flat pending development, at a weekly occupation charge of £16.30, subject to 28 days' notice of termination — After the respondent had fallen into arrears the association gave notice terminating his occupation — The issue in proceedings for possession was simply whether the respondent had a secure tenancy — This was purely a question D of construction of para 3(1) of the schedule to the 1985 Act — The county court judge decided that the respondent did have a secure tenancy and dismissed the association's action for possession — He took the view that, as a matter of construction, the provision contemplated that the person who was landlord would be the same person as had acquired the land for development

The Court of Appeal took the opposite view — They held that the ordinary meaning of para 3(1) in the English language did not require the landlord to be the person who had acquired the land for development — The paragraph imposed two quite E distinct requirements — The first was that the dwelling-house must be on land which had been acquired for development — The second was that the dwelling-house was used by the landlord, pending development of the land, as temporary housing accommodation — These two requirements were satisfied here — Hence the tenancy was not a secure tenancy and the landlord was entitled to possession — In the view of the Court of Appeal the construction for which the respondent contended involved writing into para 3(1) words which were not in the statute, namely the words "by the landlord" after "acquired for development"

The court rejected a number of other arguments on F construction put forward on behalf of the respondent — On the question of policy they saw no reason why the construction for which the appellants contended should not have been in accordance with Parliament's intention — It was undesirable that land should be sterilised and removed from the housing stock pending development — The case of *Attley* v *Cherwell District Council,* which had been cited, was distinguished — Appeal allowed

The following case is referred to in this report.

Attley v *Cherwell District Council* (1989) 21 HLR 613 G

This was an appeal by Hyde Housing Association Ltd from a decision of Judge James, at Woolwich County Court, whereby he dismissed the association's action for possession of a flat at Wickham Lane, London SE2, occupied by the defendant (the present respondent), Mr Danny Harrison.

Nicholas Vineall (instructed by Evans Butler Wade) appeared on behalf of the appellants; Bryan McGuire (instructed by Powell & Co) represented the respondent.

Giving judgment, **FOX LJ** said: This is an appeal by the Hyde Housing Association Ltd ("Hyde") from a decision of His Honour Judge James in the Woolwich County Court, dismissing Hyde's H action for possession of premises consisting of a flat at Wickham Lane, London SE2.

Hyde is a charitable housing association registered under the Industrial and Provident Societies Act 1965 and the Housing Act 1976.

Hyde holds the premises under an agreement from the Department of Transport to use the premises for temporary housing for those in need. That agreement was made in October 1984. Previously, in April 1984, the Department of Transport had acquired the premises as part of the land required for a road development scheme.

In November 1984 Hyde granted to the defendant, Mr Harrison, what was described as a temporary weekly licence to occupy the J premises pending development, in consideration of the payment by the defendant of an occupation charge of £16.30 per week. The licence contained provisions that the defendant would pay the charge promptly and would vacate the premises on 28 days' notice by Hyde to vacate. The defendant became in arrears with his payments. At the hearing it was agreed that he owed some £803 to Hyde. Notice was given by Hyde determining the licence or other right to occupy. The judge recorded that there was no dispute about that. The issue is whether the defendant had a secure tenancy. That depends upon the provisions of para 3 of Schedule 1 to the Housing Act 1985.

Para 3 is in the following terms:

> (1) A tenancy is not a secure tenancy if the dwelling-house is on land which K has been acquired for development and the dwelling-house is used by the landlord, pending development of the land, as temporary housing accommodation.
>
> (2) In this paragraph "development" has the meaning given by section 22 of the Town and Country Planning Act 1971 (general definition of development for the purpose of that Act).

It is common ground that the premises were "acquired for development" by the Department of Transport within the meaning of para 3. It is also common ground that the premises have been used by Hyde at all material times for temporary housing. The issue is whether, upon the proper construction of para 3, the land must have been "acquired for development" by the landlord (that is to say, by Hyde in the present case) or whether the acquisition may be by a third party (the Department of Transport in this case).

The judge held that "an ordinary reading of para 3 would normally result in the reader assuming that the person who was the landlord would be the same person as had acquired the land for development". Accordingly, the judge held that there was a secure tenancy and dismissed the action.

In my opinion, para 3(1) of Schedule 1, read according to its ordinary meaning in the English language, does not require that the landlord be the person who has acquired the land for development. The paragraph, it seems to me, imposes two distinct requirements. The first is that the dwelling-house must be on land which has been acquired for development. The second is that the dwelling-house is used by the landlord, pending development of the land, as temporary housing accommodation. If these two requirements are satisfied, the tenancy is not a secure tenancy.

There is, in my view, no requirement, express or implied, that the acquisition should have been by the landlord. The statutory language is simply "if the dwelling-house is on land which has been acquired for development". There is nothing in that language which requires that the developer should be the landlord. It is only in relation to the second requirement that the landlord is introduced at all.

The construction which I have adopted seems to me to be the ordinary meaning of the language used and does not involve extending or rewriting it in any way.

The construction contended for by the respondent, on the other hand, involves, in effect, the rewriting of para 3(1) to read:

> A tenancy is not a secure tenancy if the dwelling-house is on land which has been acquired for development *by the landlord* and the dwelling-house is used by the landlord, pending development of the land, as temporary housing accommodation.

As to the language of the paragraph, the respondent advances the following contentions:

(1) The "landlord" is the only person expressly referred to in the paragraph. That is true, but it is not a compelling reason for saying that what I have referred to as the first requirement should be given a meaning which its language does not bear. The fact is that the draftsman did not introduce a reference to the landlord into the first requirement.

(2) There is no comma after the words "acquired for development". I do not think that there needs to be a comma to justify the interpretation which I have adopted.

(3) The draftsman was merely avoiding the inelegance of using the words "the landlord" twice over. No doubt the draftsman might try to avoid inelegancies, but not in such a way as to distort or obscure the meaning.

(4) The draftsman could have put the matter beyond doubt, if he really intended that the paragraph should have the meaning contended for by Hyde by, for example, setting out the paragraph in a different form:

> A tenancy is not secure if —
> (*a*) the dwelling-house is on land which has been acquired for development, and
> (*b*) the dwelling-house is used by the landlord pending development of the land, as temporary housing accommodation.

It seems to me that the paragraph, as enacted, provides exactly the same result as that wording, and I do not see why the draftsman should not have supposed so.

(5) The words falling between "acquired for development" and "used by the landlord" namely "and the dwelling-house is" are to make it clear that it is the dwelling-house and not the land which must be used pending development for temporary housing accommodation. There is no grammatical reason, it is said, to separate the terms "acquired for development" and "used" or to treat them as referring to different persons. The reason, it seems to me, is because "used" is followed by the words "the landlord", and there are no words specifically identifying the acquirer for development. The language, so far from identifying the landlord as the necessary acquirer, leaves the identity of the acquirer entirely at large.

It is said that the purpose of the Act and of the schedule is to regulate agreement between landlord and tenant. Accordingly, so it is said, the strong assumption is that unless otherwise stated the parties being referred to were landlord and tenant. I do not feel able to accept that. Let it be assumed that the purpose is to regulate relations between landlord and tenant. The question in the present case is the identification of the circumstances in which they are to be regulated. It is not unreasonable that the relationship between the landlord and the tenant should be regulated, in a particular way as to security of tenure, if the land has been acquired for development whether by the landlord or by a third party, and the tenant's occupancy was on a temporary basis pending development.

As to the context in which para 3 appears, it is no doubt true that there may be overlapping between the provisions of para 3 and para 6. But the two paragraphs, though they both are specifying circumstances in which a tenancy is excluded from the category of secure tenancies, impose quite different requirements for the exclusion. Further, para 6 is dealing with what may be called "three-party" situations, ie a superior lessor, a landlord and a tenant — the first letting to the second, and the second to the third. Para 3, however, extends also, in my view, to "two-party" situations, ie where the acquirer grants the tenancy direct to the occupier of the house. On the respondent's construction, para 3 is always limited to the two-party situation.

It is contended that, if Hyde's construction of para 3 is correct (ie that the acquirer need not be the landlord), the tenant will never be sure of his position, since para 3 contains no provision for the insertion in the lease of any express terms in the tenancy agreement or in the agreement between the acquirer and the landlord as to intention to develop. But para 6 equally contains no provision as to the contents of the tenancy of the house to the occupying tenant (as opposed to the landlord). But in any event, it would normally be in the landlord's interest that the tenancy to the occupying tenant should contain provisions enabling the landlord to obtain possession when the developer requires it, since the inclusion of such provisions will reduce the risk of disputes and difficulties between the landlord and the tenant when possession is required for development.

In general I see no reason why, as a matter of policy, Parliament should not have intended para 3 to have the meaning contended for by Hyde. It is undesirable that land should be sterilised and removed from the housing work pending development. It is desirable to encourage its use for housing, provided that that does not impede the development. But there may well be cases where the developers do not wish to assume the burdens of dealing with residential tenants and the day-to-day problems regarding houses but would be quite content to allow temporary use under a tenancy agreement to a third party who would assume these burdens and provided that the tenancy ensured the availability of the land when required for the development.

We were referred to the decision of the Court of Appeal in *Attley* v *Cherwell District Council* (1989) 21 HLR 613. I do not, however, think that the present point really arose in that case. Cherwell District Council were seeking to obtain possession of a dwelling-house and relied on para 3. The land was originally acquired by Banbury Borough Council but vested in the Cherwell District Council in the local government reorganisation of the early 1970s. The court held that the effect of the Local Government Act 1972 was that Cherwell District Council themselves were to be treated as having acquired the land for development. It was therefore a case where the plaintiff landlord had acquired the relevant land for development.

In the circumstances I would allow the appeal.

STOCKER LJ agreed and did not add anything.

The appeal was allowed: possession to be given forthwith. No order was made as to costs. Leave to appeal to the House of Lords was refused.

A

Court of Appeal

May 18 1990

(Before Lord Justice PURCHAS, Lord Justice RALPH GIBSON and Lord Justice FARQUHARSON)

DUDLEY METROPOLITAN BOROUGH COUNCIL v BAILEY

Estates Gazette March 16 1991

[1991] 10 EG 140

B **Housing Act 1985, Part IV, section 83 — Secure tenancies — Proceedings for possession — Prescribed form of notice of proceedings — Grounds for possession order — Particulars of ground — Whether notice given by council in accordance with prescribed form — Whether proper particulars given — Notice stated that possession was sought on ground that rent lawfully due had not been paid — Figure stated as arrears of rent in fact included amounts for rates and water rates — Appeal from decision of county court judge, on a preliminary point of law, holding that council's notice was defective and that court had no jurisdiction to entertain proceedings — Appeal allowed**

C **The tenant, respondent to the appeal, held a weekly secure tenancy the rent of which was not inclusive of rates, although the tenant was under a separate obligation to pay rates and water rates to the council — The council had apparently for some time followed the practice of stating "rent" in a notice under section 83 of the Housing Act 1985 as including sums due for rates and water rates under the terms of the tenancy — It was accepted for the purpose of the preliminary point that in the notice the arrears of rent were said to be £145.96 when the sum in fact due for rent was £72.88**

The two questions before the county court judge were (1) D **was the form of notice issued by the council a form prescribed by the relevant regulations or in a form substantially to the same effect, and (2) did the form give sufficient and proper particulars in compliance with section 83(2)(*c*) — The judge answered "no" to both questions — As regards the first, he considered that the wording on the council's form "The reasons for taking this action are . . ." departed too far from the prescribed words "Particulars of each ground are as follows . . ." — He thought that the council's words suggested an explanation of their motives for taking action rather than a full explanation of the ground on which reliance was placed — As regards the second question, the sum shown** E **in the particulars for rent was £145.96, whereas the sum due for rent was £72.88 — The particulars were therefore, in the judge's view, factually incorrect and misleading to the tenant, bearing in mind the number of state benefits, including housing benefit, with which tenants had to cope and which might affect their obligations in regard to rent — The council had misunderstood the law**

On appeal the court refused to accept as correct the judge's finding that the notice was not in a form substantially to the same effect as the prescribed form — The court also rejected a new point raised on behalf of the tenant that the Secure F **Tenancies (Notices) Regulations 1987 were *ultra vires* the 1985 Act by including the words "or in a form substantially to the same effect", as the Act required the notice to be strictly "in a form prescribed by regulations" — This challenge which, if sound, would have caused widespread havoc among regulations, was shown to be fallacious — There is no prescribed form until the regulations are made and the form then prescribed is eg, that "specified in Part I of the Schedule or in a form substantially to the same effect" — As regards the error in the particulars, the court's view was that it did not invalidate the notice and deprive the judge of jurisdiction, although it might well affect the action which the judge would take on the merits of the case — The requirement of** G **particulars is satisfied if the landlord had in good faith stated the ground relied on and supplied particulars of that ground, giving in summary form the facts he intended to prove in support of the stated ground — It would be an impossibility to impose any effective requirement of objective accuracy in the particulars — The case of *Torridge District Council* v *Jones*, cited on behalf of the tenant, was distinguishable — Appeal allowed and case remitted to the county court**

The following case is referred to in this report.

Torridge District Council v *Jones* (1985) 18 HLR 107; [1985] 2 EGLR 54; 276 EG 1253, CA H

This was an appeal by the landlords, Dudley Metropolitan Borough Council, from a decision of Judge Stuart-White, at Stourbridge County Court, holding that the court was deprived of jurisdiction under section 83 of the Housing Act 1985 by reason of an error in the notice of proceedings for possession. The case related to a dwelling-house at 117 Byron House, Avon Road, Halesowen, let under a secure tenancy to the tenant, Carlos Bailey, the defendant in the proceedings and the respondent to this appeal.

Jeremy Sullivan QC and Miss Julia Macur (instructed by the solicitor to Dudley Metropolitan Borough Council) appeared on behalf of the appellants; Stephen Sedley QC and Andrew Tucker J (instructed by Silks, of Netherton) represented the respondent.

Giving the first judgment at the invitation of Purchas LJ, **Ralph Gibson LJ** said: This is an appeal from the order made on November 2 1989 by His Honour Judge Stuart-White in Stourbridge County Court on a preliminary point of law in a possession action brought by Dudley Metropolitan Borough Council ("the council") against Carlos Bailey ("the tenant") for possession of a dwelling-house, 117 Byron House, Avon Road, Halesowen. The learned judge held that the notice of proceedings for possession served by the council on the tenant pursuant to section 83 of the Housing Act 1985 was defective because it was not substantially in the form specified in the regulations and because the particulars given were misleading; and K that, therefore, the court had no jurisdiction to entertain the proceedings for possession. The requirement of a notice of proceedings under section 83 is part of the secure tenancy provisions contained in Part IV of the Housing Act 1985.

It is necessary to set out the relevant statutory provisions. It is not in issue that the tenant held a weekly secure tenancy.

> Section **82.**—*Security of tenure*
> (1) A secure tenancy which is either —
> (*a*) a weekly or other periodic tenancy . . .
> cannot be brought to an end by the landlord except by obtaining an order of the court for the possession of the dwelling-house . . .
> **83.**—*Notice of proceedings for possession or termination*
> (1) The court shall not entertain — L
> (*a*) proceedings for the possession of a dwelling-house let under a secure tenancy . . .
> unless the landlord has served on the tenant a notice complying with the provisions of this section.
> (2) The notice shall —
> (*a*) be in a form prescribed by regulations made by the Secretary of State,
> (*b*) specify the ground on which the court will be asked to make an order for the possession of the dwelling-house or for the termination of the tenancy, and
> (*c*) give particulars of that ground.
> (3) Where the tenancy is a periodic tenancy the notice —
> (*a*) shall also specify a date after which proceedings for the possession of the dwelling-house may be begun, and
> (*b*) ceases to be in force twelve months after the date so specified; . . . M
> **84.**—*Grounds and orders for possession*
> (1) The court shall not make an order for the possession of a dwelling-house let under a secure tenancy except on one or more of the grounds set out in Schedule 2.
> . . .
> (3) The court shall not make such an order on any of those grounds unless the ground is specified in the notice in pursuance of which proceedings for possession are begun; but the grounds so specified may be altered or added to with the leave of the court.

The relevant regulations made under section 83(2) are the Secure Tenancies (Notices) Regulations 1987 (SI 1987 no 755). Para 2(1) is as follows:

> The notice to be served on a secure tenant under section 83 of the Housing Act

A 1985 before the court can entertain proceedings for possession of a dwelling-house let under a secure tenancy which is a periodic tenancy, shall be in the form specified in Part I of the Schedule to these regulations, or in a form substantially to the same effect.

Part I of the Schedule specifies a form in five numbered paragraphs as follows:

> This notice is the first step towards requiring you to give up possession of your dwelling. You should read it very carefully.
> 1. To . . . [name(s)] of secure tenant(s)
> 2. The [name of landlord] intends to apply to the Court for an order requiring you to give up possession of . . . (address of property).
> 3. Possession will be sought on Ground(s) . . . of Schedule 2 to the Housing Act 1985 which reads:
> B [Give the text in full of each Ground which is being relied on].
> 4. Particulars of each ground are as follows:
> [Give a full explanation of why each Ground is being relied upon].
> 5. The Court proceedings will not be begun until after . . . [give the date after which Court proceedings can be brought].

The specified form includes notes of advice and explanation following each numbered paragraph; for example, following para 1 there are the words:

> If you need advice about this notice, and what you should do about it, take it as quickly as possible to a Citizens Advice Bureau, a Housing Aid Centre, or a Law Centre, or to a solicitor . . .;

and under para 4 the following appears:

> Before the Court will grant an order on any of the Grounds 1 to 8 or 12 to 16 (in
> C Schedule 2 to the Act of 1985), it must be satisfied that it is reasonable to require you to leave. This means that, if one of these Grounds is set out in paragraph 3 of this notice, you will be able to argue at the hearing in Court that it is not reasonable that you should have to leave, even if you accept that the Ground applies.

The facts in this appeal are as follows. The tenant went into occupation in September 1988. The council provided a tenant's handbook containing details of the conditions of tenancy and information about a tenant's rights and obligations. It is sufficient to say that, for the purposes of the preliminary point, it is accepted that it was an obligation of the tenant to pay rates and water rates to the council, but the obligation to pay rent was an obligation distinct from that relating to rates. This was not an inclusive rent.

D The council served on the tenant the "Notice of Seeking Possession" on January 18 1989. The heading of the document and the certificate of service and signature can all be omitted. The effective part of the notice was as follows:

> NOTICE OF SEEKING POSSESSION
> Housing Act 1985, Section 83
> THIS NOTICE IS THE FIRST STEP TOWARDS REQUIRING YOU TO GIVE UP POSSESSION OF YOUR DWELLING. YOU SHOULD READ IT AND THE ACCOMPANYING NOTES PRINTED OVERLEAF VERY CAREFULLY.
> 1. To: Mr Carlos Bailey.
> The tenant(s) of 117 Byron House, Avon Road, Halesowen.
> 2. The Borough Council of Dudley intends to apply to the Court for an Order
> E requiring you to give up possession of:
> 117 Byron House, Avon Road, Halesowen.
> 3. Possession will be sought on Ground 1 (ONE) of Schedule 2 to the Housing Act 1985 which reads:—
> RENT LAWFULLY DUE FROM THE TENANT HAS NOT BEEN PAID OR AN OBLIGATION OF THE TENANCY HAS BEEN BROKEN OR NOT PERFORMED.
> 4. The reasons for taking this action are:
> RENT DUE TO COUNCIL HAS NOT BEEN PAID IN THAT AS AT THE 16TH DAY OF JANUARY 1989 YOU ARE IN ARREARS TO THE SUM OF £145.96.
> 5. The Court proceedings will not be begun until after:— 20th February 1989.
> Dated this 18th day of January 1989.

The notes of advice and explanation contained in the specified form were served with the notice on the tenant in a separate printed sheet. No submission has been directed to that part of the notice.

F Particulars of claim were issued on April 25 1989. It was alleged that the weekly rent was £34.78; that the tenant had made default in payment of rent; that a notice had been served under section 83; that the arrears of rent to the date of the particulars of claim were £250.80 and sums due for rent from that date until the date of the hearing on June 6 1989 would be £173.90.

On June 6 1989 the tenant did not appear and a suspended possession order was made. The council did not ask for an immediate order. On June 27 1989 the tenant applied for the suspended order to be set aside. He had by that time obtained the advice and assistance of the Dudley Law Centre. His complaint was that the council had not given details as to how the money claim of the council had been calculated. If such details had been given he would have fully defended the proceedings on the ground that the sum claimed as rent G was not "rent" properly so called but consisted of "unrebated elements of general rates and water rates". He claimed to have a good defence in law. The Dudley Law Centre, on his behalf, explained his main contention by reference to the fact that the tenant was "on income support". Under Regulation 61 of the Housing Benefit (General) Regulations 1987 (SI 1987 no 1971) it is provided that:

> The amount of a person's maximum housing benefit in any benefit week shall be —
> (a) 100% of his eligible rent calculated on a weekly basis in accordance with Regulations 69 and 70 . . .; and
> (b) 80% of his eligible rates calculated on a weekly basis in accordance with those Regulations . . .

It was contended that the only liability of the tenant was for 20% of H general rates and water rates.

On July 27 1989 the suspended possession order was set aside by the judge. The correctness of that decision is not challenged. The council on August 8 1989 served amended particulars of claim. There was added an allegation that:

> In breach of the tenant's obligation under the terms of the tenancy the tenant has failed to pay . . . the sums due in respect of rates and water rates . . .

The details of the money claim were amended so as to assert that the arrears of rent were £72.88 and rates and water rates £154.90. The defence, in addition to disputing the sums claimed, alleged that the council's notice did not comply with the requirements of section J 83(2)(*c*) of the 1985 Act.

The proceedings came for trial on November 2 1989. Again the council were not seeking an immediate order for possession. It was decided to determine the validity of the council's notice as a preliminary point. The council had for some time followed the practice of stating "rent" in a section 83 notice as including sums due for rates and water rates under the terms of the tenancy. It was conceded for the council that in the notice the arrears of rent were said to be £145.96 when the sum due in fact for rent was £72.88. It was accepted for the purposes of the preliminary point that it was an obligation of the tenant under the tenancy to pay rates and water rates and that the balance of the sum of £145.96 was owing by the tenant in respect of rates and water rates. K

Two questions were raised for decision by the learned judge. The first was whether the form of notice used by the council was, as required by section 83(2), "in a form prescribed by regulations made by the Secretary of State", or in a form substantially to the same effect. The second question was, if the form did comply with section 83(2) by being in the prescribed form, whether the form gave sufficient and proper particulars to comply with section 83(2)(*c*).

The judge answered both questions adversely to the validity of the notice. As to the first question, he held that the council's notice was not in a form substantially to the same effect as the form specified in the regulations because, instead of a paragraph reading "Particulars of each ground are as follows . . .", followed by an explanation of L why each ground was being relied upon, the council's form had a paragraph headed "The reasons for taking this action are . . .". What was required, both by section 83(2)(*c*) and by para 4 of the specified form, was that a full explanation should be given of the ground upon which reliance is placed, but the council's form, in the judge's view, explained their motives for bringing the proceedings for possession and that was a quite different matter.

As to the second question, the judge's reasoning was as follows. The sum specified in the notice as "rent due and not paid", was a sum comprising a total owed for rent, general rates and water rates. It was conceded that the sums due for rates were not arrears of rent. The sum shown for rent was £145.96, but in fact the sum due for rent was £72.88. The particulars were, therefore, not sufficient and proper so M as to comply with section 83(2)(*c*). First, the particulars were incorrect to the factual knowledge of the council, although no doubt there was a misunderstanding of law. Second, the whole purpose of giving particulars is to enable the tenant to understand the case which he has to meet. Since many tenants, including this tenant, are "in receipt of various state benefits, including housing benefit, the effect of which was that it was difficult for the tenant to know what was his true position in relation to his obligation to pay rent", particularly where the rent is paid directly by the Department of Social Security, it is important that the particulars given should not be misleading. The particulars given to this tenant were misleading because the council misunderstood the law and they could not fall back on a mistake of law. There was, accordingly, no jurisdiction in the court to entertain the proceedings.

The learned judge emphasised the distinction, as he saw it for these purposes, between error of law and error of fact in the following passage:

> If what had happened had been that they had been mistaken as to the facts and they had been shown, after a hearing and evidence given on both sides, to be mistaken as to the facts then perhaps different considerations might have arisen, but they were not mistaken as to the facts; they knew perfectly well what the arrears of rent were and they asserted on the face of that document . . . that the arrears of rent were that which they knew they were not.

It should be emphasised that the judge was not referring to any conscious statement by any officer of the council of an assertion in the notice which that officer knew to be untrue. The judge was referring to information within the files and papers of the council and available to the officers of the council.

For the appellant council, Mr Sullivan submitted that the judge was wrong to hold that the form of the notice was not in a form substantially to the same effect as the form specified in the schedule. For the respondent, Mr Sedley raised the contention, which was not made before the judge, that the words in para 2 of the regulations "or in a form substantially to the same effect" were introduced as licence to depart from the prescribed form for which there is no authority in the statute. Mr Sedley did not develop any argument based upon substantial departure from the form for the purposes of supporting the judge's ruling if the *ultra vires* point failed.

The argument on *vires* was that the statute required the notice to be "in a form prescribed by regulations". It was intended to impose a strict requirement as a precondition for the entertaining of the proceedings for possession. A licence to depart from the prescribed form must be found, it was said, in the statute and could not be introduced by means of the regulations. I do not accept this submission. There is no prescribed form until the minister has made the regulations by statutory instrument under section 83(6). The minister has prescribed a form which is that "specified in Part I of the Schedule or in a form substantially to the same effect". I have no doubt that the section gives to the minister power to do what he did and I respectfully think that the method adopted was wholly in accordance with the intentions of Parliament.

As to the judge's finding that the notice was not substantially to the same effect as the form in the schedule, I cannot accept it as correct. I share the judge's surprise that anyone should have thought it useful or necessary on this point to depart from the words of the form in the regulations, but I have no doubt whatever that the form of this notice was substantially to the same effect. In para 3 it is stated that possession would be sought on ground 1, ie that rent was lawfully due from the tenant. In ground 4 the reasons for taking this action were stated as: "arrears of rent unpaid in the sum of £145.96". The specified form called for "particulars of each ground" to be stated and the note thereunder called for "a full explanation of why each ground was being relied upon". The council gave the reasons for taking proceedings for possession on the ground of failure to pay rent by stating that the tenant had failed to pay £145.96 due from him as rent. That, in my judgment, is giving particulars of the grounds upon which the council were relying and gives a full explanation as to why the ground was relied upon. No other point was taken in support of the contention that the notice was not in a form substantially to the same effect as that specified in the schedule.

On the second question, Mr Sullivan's submission was that error in the particulars does not invalidate the notice so as to deprive the county court of jurisdiction under section 83 of the Act, although error in the particulars in the notice would be relevant to, and in a particular case might be decisive of, the question whether the court would make a possession order under section 84(2), and, if an order should be made, whether it should be suspended and on what terms. The requirement in section 83(2)(*c*) is that the council shall give particulars of the ground specified in the notice on which the court would be asked to make an order for possession. The council had in good faith done that and they had therefore complied with the requirement. The statute did not require for the validity of the notice any degree of conformity between the particulars given and the facts as objectively determined by the court.

Mr Sedley, for the tenant, submitted that the purpose of section 83 is to ensure that the tenant is given, by accurate particulars, a chance to remedy any breach, if he can, and so make proceedings unnecessary. Reference was made to the judgment of Oliver LJ in *Torridge District Council* v *Jones* (1985) 18 HLR 107 at p 114 where he said that the required particulars "must be nothing short of a specification of the amount which is claimed as being in arrears".

The power given by section 84(3) for the court to permit alteration of or addition to the ground specified in a notice emphasised, it was said, the rigour of the principal requirement. Further, the differences in law between an obligation to pay rent and an obligation to pay rates or water rates, whether directly under the Rating Acts or to the landlord as an obligation of the tenancy, are real and the consequences of breach may be different. It was necessary, therefore, in order to fulfil the statutory purpose, to differentiate between those liabilities in the notice. The position of a tenant on income support, and the complication of the regulations, which defies his entitlement and the method of payment of benefits, is such that accuracy of the particulars must be regarded as central to the statutory purposes. It would be open to the court to disregard minor errors as *de minimis*. The purpose of the requirement, strictly construed, could be seen as disciplinary and as enforcing the greatest efforts to achieve accuracy. It would, said Mr Sedley, be disastrous and wholly contrary to the purpose of Parliament if a serious error in a notice, whereby the liability of the tenant was overstated, should cause the tenant to be dismayed and to give up possession without any contest.

For my part I cannot accept the conclusion to which Mr Sedley's submissions were directed. On the construction of section 83(2), considered in its context as part of the provisions for the right of secure tenants, I think that the judge was wrong. As Mr Sullivan submitted, the requirement is that the landlord give particulars of the ground specified by him on which the court will be asked to make an order for possession. The question is whether, at the date of the notice, the landlord has in good faith stated the ground and given the particulars of that ground. The requirement of particulars is satisfied, in my judgment, if the landlord has stated in summary form the facts which he then intends to prove in support of the stated ground for possession. Error in the particulars does not, in my judgment, invalidate the notice although it may well affect the decision of the court on the merits. It would be impossible, I think, to impose any effective requirement of objective accuracy in the particulars and there is no need to try to derive from the words of the section any such requirement. The protection given by the notice provisions is effective without any such additional requirement. The warning notes, which are required to be part of the form of the notice, should serve to cause a tenant to get advice if the tenant does not understand his position. It must be rare indeed for a tenant not to know roughly what he owes and what he has paid. An example raised by Farquharson LJ served well, I think, to test Mr Sedley's submissions. It was asked what the position would be if the ground of possession specified in a notice were ground 2: ie conduct which is a nuisance or annoyance to neighbours. The particulars given might include allegations of grave misconduct based upon statements made by three neighbours. The statements of two of them might be accurate and fully justify the making of an order for possession. The statement of the third neighbour might be demonstrably untrue. Provided the particulars were given in good faith by the council, what warrant is there for supposing that Parliament intended the court to be deprived of jurisdiction to hear the case and to require the council to start again with a new notice?

It is to be noted that the power of amendment of the notice given by section 84(3) is directed to altering or adding to the grounds specified in the notice and not to the particulars. The existence of that power of amendment is, I think, of no significant help in determining the issue as to the requirement of objective accuracy under section 83(2)(*c*). The power is primarily directed, in my judgment, to circumstances such as where a landlord, who may have served a notice on the ground of non-payment of rent, wishes to raise for determination in the same proceedings an issue, for example, as to breach of covenant for repair when the court can see that the tenant will not be disadvantaged by having it included in the proceedings without a prior section 83 notice specifically directed to it.

As to the decision of this court in *Torridge District Council* v *Jones*, cited above, no support can be derived from it for Mr Sedley's submissions. In that case, the landlord had stated under para 4 of the notice that "the reasons for taking this action are non-payment of rent", but no particulars whatever were given of the rent alleged to have been unpaid. Oliver LJ had no occasion to consider the need for

A objective accuracy in any statement of the amount due because there was no such statement. The landlord had failed to state what was claimed as being in arrears and had therefore failed to comply with the requirement in section 83(2)(*c*).

Questions were raised in argument as to a notice which might be, as it was called, "bad on its face", such as a notice asserting, by means of computer error, or computer operator's error, arrears of rent of £1m as due from a council tenant on income support. For my part, I think that such a notice would be demonstrably defective because it is, at this date, impossible to suppose that the council at the date of such a notice intended in good faith to prove those arrears of rent in support of the claim to possession. Where, as on the assumed facts in this case, the council have made an honest error as to the sums owing by the tenant for rent, and as to the description properly to be applied in law to different parts of the money owed by the tenant, the notice is B not by reason of that fact invalidated, although, of course, the notice, unless altered with the leave of the court under section 84(3) states only the ground of non-payment of rent.

I would therefore allow this appeal, set aside the order of the learned judge and remit the possession proceedings for trial in the Stourbridge County Court.

Purchas and **Farquharson LJJ** agreed and did not add anything.

The appeal was allowed, the order below set aside and the matter remitted to Stourbridge County Court for possession proceedings to take place; no order was made as to costs; legal aid taxation was ordered. Leave to appeal to the House of Lords was refused.

LANDLORD AND TENANT
GENERAL

Court of Appeal

October 19 1990

(Before Lord Justice DILLON and Lord Justice RALPH GIBSON)

HOLDOM v KIDD AND OTHERS

Estates Gazette January 19 1991

[1991] 02 EG 163

Restrictive covenant — Restriction of use of land to agricultural purposes — Land subject to covenant used to provide accommodation for caravans or mobile homes occupied by gypsies — Interlocutory injunction sought by person entitled to benefit of covenant to restrain use in breach of it — Prospect of obtaining planning permission for the forbidden use followed by application to Lands Tribunal under section 84 of the Law of Property Act 1925 to discharge or modify restriction — Whether judge in exercise of his discretion should have granted the injunction or should have refused it in order to keep the gypsies and their families on the site pending the result of proceedings before the planning authority and the Lands Tribunal — Judge held to have been entitled to grant the injunction

The covenant at the centre of these proceedings restricted the use of the subject land to agricultural purposes only and forbade the erection of residential buildings on any part of it — The defendants (present appellants) were assignees of the original covenantor's land and thus bound by the covenant — The plaintiff (present respondent) was entitled to the benefit of the covenant — The defendants had placed on their land concrete slabs for the placement of caravans or mobile homes belonging to gypsies; and gypsies and their families were living in the accommodation thus provided — There was an established breach of the covenant on which the plaintiff had the right to rely — The county court judge had granted him an interlocutory injunction enforcing the restriction until judgment or further order — The defendants appealed

The main point in the appeal was the possible effect on the exercise of the judge's discretion of the proceedings taken by the appellants with a view to obtaining planning permission for the gypsies' residential caravans and then, if such permission was obtained, to apply to the Lands Tribunal under section 84 of the Law of Property Act 1925 for the discharge or modification of the restrictive covenant — The judge had said that if he granted the interlocutory injunction there would be nothing to prevent the appellants from going forward with the planning appeal and, if so advised, the application to the tribunal — The appellants submitted that this was a misdirection — They argued that the judge should have refused the injunction so as to keep the gypsies and their families on the site for the hearing of the planning appeal — The appellants should be allowed to present their case from the advantageous ground of present occupation

The Court of Appeal rejected this submission — It was the wrong approach — The judge was entitled to take the view that the matter to be decided on planning considerations was whether the land should or should not be used for the residential accommodation of gypsies, not whether particular gypsy families who happened to be there should be preserved from disturbance — The grant of the injunction would not prevent the appellants from going through with the planning appeal or the application to the tribunal, if so advised — The judge had exercised his discretion correctly — Appeal dismissed

The following case is referred to in this report.

Richardson v *Jackson* [1954] 1 WLR 447; [1954] 1 All ER 437; (1954) 7 P&CR 345

This was an appeal by the second, third and fourth defendants, Anthony Gaskin, Horace Gaskin and Charles Randall, from the decision of Judge Mellor, at Norwich County Court, granting the plaintiff, John Holdom, an interlocutory injunction against the breach of a restrictive covenant affecting an area of land at Attleborough, Norfolk. The first defendant in the action, Andrew Peter Kidd, was not represented and took no part in the proceedings.

Barry Payton (instructed by Cunningham John & Co, of Thetford, Norfolk) appeared on behalf of the appellants; John Holt (instructed by Greenland Houchen, of Attleborough, Norfolk) represented the respondent.

Giving judgment, **DILLON LJ** said: This is an appeal by the second, third and fourth defendants in this action against an order of Judge Mellor made in the Norwich County Court on June 26 1990 whereby the judge granted the plaintiff, Mr Holdom, an interlocutory injunction until judgment or further order, enforcing against the second, third and fourth defendants a restrictive covenant in a conveyance of certain land.

The conveyance in question is dated March 27 1980 and was a conveyance by the plaintiff, Mr Holdom, and his then wife (whose interest he has apparently acquired) to a Mr Bartholomew (called the purchaser), who was the predecessor in title of the second, third and fourth defendants. The covenant is in clause 3 of the conveyance. The conveyance is of an area of land near Attleborough in Norfolk, comprising $2\frac{1}{2}$ acres as delineated on a plan, and by clause 3 the then purchaser, Mr Bartholomew, covenanted as follows:

> 3. For the benefit and protection of the adjoining and neighbouring property retained by the Vendors known as Potmere Cottages Hargham Road Attleborough aforesaid or any part or parts thereof and so as to bind so far as may be the property hereby conveyed into whosoever hands the same may come the Purchaser hereof covenants with the Vendors that the purchaser and the persons deriving title under him will at all times hereafter observe and perform the restrictions and stipulations set out hereunder
>
> (a) To use the land hereby conveyed for agricultural purposes only such agricultural purposes to be defined as in section 290(1) of the Town and Country Planning Act 1971 . . .
>
> (b) Not to erect any residential building upon the land hereby conveyed or any part thereof.

That restrictive covenant was duly registered and it is not in doubt that it binds the defendants as purchasers of part of the land conveyed. The words used were effective to annexe the benefit of the covenant to the then vendor's retained land, known as Potmere Cottages, and every part thereof. Consequently it matters not that part of the retained land had been sold off, because the plaintiff retains a part, and in fact the part nearest to the land which the second, third and fourth defendants have acquired. Therefore, we have a position where the covenant binds these defendants and the plaintiff is entitled to the benefit.

The defendants have placed on their land concrete slabs for taking caravans or mobile homes belonging to gypsies. Apparently there are approximately 10 plots and 17 caravans or mobile homes. The gypsies, though travellers and not always there, do work on local farms when they are there and it is said therefore that if the man of the family is working on a local farm, the land where the caravan is situated is indeed being used for agricultural purposes within the meaning of the covenant. That has not been pressed very hard in this court by Mr Payton. It seems to me obvious that if you have a housing estate and one house is occupied by an agricultural labourer who works on a nearby farm, another house is occupied by a bank clerk and a third house is occupied by a railway porter you cannot say that part of an estate is being used for agricultural purposes, part for banking purposes and part for railway or transport purposes. I take the position to be, therefore, that there is an established breach of the covenant and that the plaintiff has the benefit and the right to enforce. It was suggested that he might have lost the right to enforce because he has allowed a friend to carry on activities on his land in the way of dismantling vehicles or repairing vehicles, but I do not see that there is any arguable basis that that amounts to a general relaxation or release of the covenant which would authorise this quite different user of land subject to the covenant, which is very close indeed to the vendor's retained land.

The real point in this case is that the defendants say, through Mr Payton, that they have made an application for planning permission for use of the land for accommodating gypsies' caravans and if and when they get planning permission they will apply to the Lands Tribunal under subsection (9) of section 84 of the Law of Property Act 1925 to have the restrictive covenant wholly or partially modified or discharged so as to permit continued user by the gypsies.

It is not in doubt that the judge in such cases has a discretion to stay the enforcement of a covenant pending application to the Lands Tribunal if it appears that an application to the Lands Tribunal would have some chance of success. The notes in *Halsbury*, to which we have been referred, cite the case of *Richardson* v *Jackson* [1954] 1 WLR 447, in which a stay was granted. It is equally not in doubt that there would be no purpose at all for these appellants to seek to apply to the Lands Tribunal until they have got, if they can, planning permission for the desired use of the land. The Lands Tribunal is required to take account of the planning position in relation to the land, the subject of any application, and would obviously refuse any modification of the covenant to permit a user which was contrary to planning law.

At the time of the hearing before the judge in the Norwich County Court the position was that the application for planning permission by the defendants was pending before the local planning authority. Since the judge's order, the planning application has been refused by the local planning authority, on July 3 1990, and an appeal to the Secretary of State for the Environment against that refusal has been lodged. We are told that a public inquiry before an inspector has been appointed to commence on February 14 1991 and that the decision of the Secretary of State may be expected within perhaps a minimum of six to eight weeks after the inquiry has been concluded. There would then, of course, before it was lawful to use the land for gypsies, have to be a successful application to the Lands Tribunal.

The judge dealt with this in his judgment in this way:

> Thirdly . . . Mr Payton asks me to bear in mind that an injunction is an equitable remedy and it is in many ways the equivalent of a final judgment (although not necessarily so). The argument is that the defendants are Romany gypsies. It is universally known that they have very real difficulty in finding sites. They have been given special favour in the planning system to find sites. They have made application for planning permission. As of today it is possibly not entirely clear but quite probably it appears that the local planning authority either have or are just about to refuse their application. That, however, would not be the end of the story because it is open to them to appeal and Mr Payton tells me they will undoubtedly appeal . . . and it is anticipated there will be a public inquiry.

He suggested that there should be evidence from a planning expert as to the likelihood of success and the judge pointed out that that would not conclude the matter because of the restrictive covenants. He went on:

> It is said, however, that if the planning appeal were resolved in their favour, they would wish to apply to the Lands Tribunal under s84(5) of the Law of Property Act 1925 . . . to discharge or modify the covenant to accord with such planning permission as may be granted.

The judge then said:

> It seems to me to follow as a matter of principle that I must approach the case on the basis that there is, at least, a prospect that if all the procedures were followed there is a chance that planning permission would be granted and that the restrictive covenant may be modified or discharged. If I were to make the order sought, there is nothing to prevent the defendants from going through with a planning appeal and if so advised going to the Lands Tribunal. I must bear in mind that if I make the order, I do not deprive them of their further rights in the land although in practical terms it would affect their homes and the homes of their families.

The judge recognised that the case was a difficult case in the sense that it involved people who have their families on site, who have difficulties in finding accommodation, and in the exercise of his discretion he decided to grant the injunction.

Mr Payton says that the judge has seriously misdirected himself in saying that if he makes the injunction "there is nothing to prevent the defendants from going through with the planning appeal and if so advised going to the Lands Tribunal". He says that that is an unrealistic view for the judge to take in that, if the injunction is granted, there will be great difficulty in succeeding with the planning appeal and, as I have already indicated, if they do not succeed on the planning appeal there is no basis for going to the Lands Tribunal. So, says Mr Payton, the judge has erred in principle and should have refused to grant any injunction so as to keep the gypsies and their families on the site for the hearing of the planning appeal and, if circumstances fit, the subsequent application to the Lands Tribunal as persons in occupation whose occupation the court has felt should not be disturbed. In other words, the judge should have given the defendants that advantage, in effect presenting something which is nearly a *fait accompli* to the Secretary of State and to the Lands Tribunal that these people, although not lawfully there, should not be disturbed.

That seems to me a completely wrong approach in the context of this case. I think the judge was fully entitled to recognise that granting the injunction and enforcing it does not prevent the defendants from going through with the planning appeal and, if so advised, going to the Lands Tribunal. They are fully entitled to say that they had gypsies there. While the covenant stands they cannot have gypsies there, but if planning permission is granted gypsies will be available to go there and if planning permission is granted they will have a prospect of success in having the covenant released. It seems to me that the judge was entitled to take the view that the function of the minister on the planning appeal was to decide, on planning considerations, whether this land should or should not be usable for accommodating gypsies and not to consider, on considerations of humanity, whether this land should or should not be used to accommodate particular gypsy families who happen to be there now.

It follows, in my view, that there is no basis for interfering with the judge's exercise of his discretion. Accordingly, I would dismiss this appeal.

Ralph Gibson LJ agreed and did not add anything.

The appeal was dismissed with costs. Legal aid taxation was ordered.

Chancery Division

July 16 1990

(Before Mr Justice WARNER)

HOWARD AND ANOTHER v MIDROME LTD AND OTHERS

Estates Gazette January 26 1991

[1991] 03 EG 135

Landlord and Tenant Act 1987, Part II — Appointment of managers by the court — Application by motion for interim order appointing a company representing tenants to manage flats — Management by landlords "simply non-existent" — One matter requiring urgent attention was leaking roof — Judge satisfied within the terms of section 24(2)(*a*)(i) and (ii) that landlords were in breach of obligations owed to tenants and that the circumstances were such that the breaches were likely to continue — Management company appointed — Further proceedings to go to county court

A **The judge was also satisfied that it would not be "just and convenient" to use the words of section 24(2)(*a*)(iii) or section 24(2)(*b*) to allow the present conditions in regard to management to continue pending the trial of the originating summons between the parties — Hence the need for these interlocutory proceedings — According to affidavit evidence the damage caused by penetration of water into the subject building was getting worse — There was no suggestion from the defendant landlords, Midrome Ltd, that, if the judge refrained from making the order now sought, they would take steps to remedy the condition of the roof or indeed take any**
B **other steps towards the proper management of the property**

In view of the judge's conclusions on the substance of the application, the points raised on behalf of the defendants were necessarily of a procedural and technical nature — They were as follows:

(1) The first point was that the preliminary notice under section 22 of the 1987 Act was defective in not containing a statement that the tenants would not make an application under section 24 if the landlords complied with the specified requirements — This criticism appeared to imply that all the matters relied on by the tenants were capable of remedy, which was not the case here — In any case the judge decided to
C **exercise the dispensing power given him by section 24(7) to overlook this defect, if it was a defect**

(2) It was next objected that the tenants' notice under section 22 did not sufficiently specify the steps to be taken by the landlords to remedy certain of the defects — The judge held that in fact the specification given was sufficient

(3) It was objected that there were certain matters which were capable of being remedied but which the tenants' notice did not require the landlords to remedy — These, however, related to insurance which could not be obtained and to the payment of money which the landlords had no intention to
D **pay — If there were defects here, the judge again relied on his dispensing power under section 24(7)**

(4) It was suggested that a specified period of 28 days for compliance was not in the circumstances reasonable — The judge rejected this criticism on the ground that, whatever period was specified, the landlords had no intention of complying

(5) It was contended that the requirement of RSC Ord 97, r15 had not been complied with, as the "qualifications" of Croftbride Ltd, the proposed manager, had not been stated — The judge rejected a suggested implication that the manager
E **must have a professional qualification; the provision merely asked for qualifications to be stated if there were qualifications**

(6) It was suggested that the applicants should be debarred by delay from obtaining interlocutory relief, the notice having been served in August 1989, the originating summons in December 1989 and the present motion launched in July 1990 — The judge, however, decided that it was "just and convenient" to make the order sought notwithstanding the delay; another winter was on the way and the physical deterioration and the lack of management continued

F **The judge accepted a submission that the words "profits" and "other moneys" should be omitted from the wording of the powers to be conferred on Croftbride Ltd, so that the wording should read "receive or levy all rents and service charges payable by the tenants"**

The judge made an order appointing Croftbride Ltd manager for an interim period before the trial of the action, subject to Croftbride Ltd providing security to the satisfaction of the master, the costs of the application to be costs in the cause — The judge ordered further proceedings to be transferred to the appropriate county court

No cases are referred to in this report.

This was an interlocutory motion for an interim management order under section 24 of the Landlord and Tenant Act 1987. The G
main application was for an acquisition order under Part III of the Act. The plaintiffs were the tenants' representative, Catherine Howard, and Croftbride Ltd, the proposed management company. The defendants were the landlords, Midrome Ltd, and Thomas Collins and Nadir Craft. The premises which were the subject of the proceedings were at Berry Lodge, Crouch Hill, London.

Philip Waller (instructed by Portner & Jaskel) appeared on behalf of the plaintiffs; Barry Denyer-Green (instructed by David Wise) represented the first defendants.

Giving judgment, WARNER J said: Having regard to the lateness of
the hour I will state my reasons for the order I am going to make as H
shortly as I can. I will make the order sought by the plaintiffs with, however, in the light of Mr Denyer-Green's submissions, the omission of some words from the provisions relating to the powers that are to be exercisable by Croftbride Ltd as manager of the property.

Shortly stated, the need for the appointment of a manager of the property arises first because of the leaking roof. It has, of course, been leaking for some time, but, as was pointed out by Miss Howard in her affidavit, another winter is going to occur in a very few months' time, and I infer from what she says in the last of her affidavits that the damage caused by penetration of water into the building is getting worse. Second, it seems to me clear from the evidence that the
landlord, Midrome Ltd, has no intention of doing anything at all J
about that roof unless paid in advance to do it. Third, the evidence also points to management of the property by or on behalf of Midrome Ltd being now simply non-existent. As long ago as October 1988 it was stated on behalf of Midrome Ltd in a letter to one of the tenants at the property that Midrome Ltd was not equipped to cope with the management due to the difficulty it had had in securing the services of suitably qualified staff.

It would not be just and convenient within the terms either of section 24(2) (b) or of section 24(2)(a)(iii) of the Landlord and Tenant Act 1987 to let the present conditions as regards the management of the property continue pending the trial of the originating summons. I am satisfied within the terms of section 24(2)(a)(i) and (ii) that the K
landlord is in breach of obligations owed by it to the tenants, in particular the failure to keep the roof in proper repair, and that the circumstances by virtue of which the landlord is in breach of those obligations are likely to continue.

Mr Denyer-Green himself, on behalf of Midrome Ltd, attributed the situation to three difficulties faced by Midrome Ltd. One was its difficulties with previous agents and builders. Another was its difficulties over insurance. The third was its inability to require payments in advance, which he stressed. There is no suggestion from Midrome Ltd or Mr Antonelli, who as a director has sworn evidence on its behalf, that, if I refrain from making this order, Midrome Ltd will take steps to remedy the condition of the roof or indeed any other
steps towards the proper management of the property. L

Mr Denyer-Green, who has said everything that could possibly be said in opposition to the motion, took four points, apart from relying on the difficulties facing Midrome Ltd. One of those points was that, so he submitted, the notice served by the plaintiffs on August 3 1989 was defective, so defective that it could not be relied upon under the Act. However, Mr Denyer-Green, quite properly, drew my attention to section 24(7), which says:

> In a case where an application for an order under this section was preceded by the service of a notice under section 22, the court may, if it thinks fit, make such an order notwithstanding —
> (a) that any period specified in the notice in pursuance of subsection (2)(d) of that section was not a reasonable period, or
> (b) that the notice failed in any other respect to comply with any M
> requirement contained in subsection (2) of that section or in any regulations applying to the notice under section 54(3).

The first point taken by Mr Denyer-Green under this head was that the notice did not contain a statement in accordance with the second limb of section 22(2)(b). By virtue of that provision a notice under section 22 must, if para (d) of subsection (2) is applicable, state that the tenant will not make an application for an order under section 24 if the landlord complies with the requirements specified in pursuance of para (d). Para (d) provides that where the matters that would be relied upon by the tenant for establishing the grounds upon which the court would be asked to make an order under section 24, if an application for that order were pursued, are capable of being remedied by the landlord, the notice must require the landlord, within such reasonable period as is specified in the notice, to take such steps for the purpose of remedying them as are so specified.

Here, Mr Denyer-Green pointed out, the notice did not in terms contain a statement to that effect, that is to say a statement that the tenant would not make an application for an order under section 24 if the landlord complied with the specified requirements. That seems to me to be a highly technical point. Possibly it was met by Mr Waller's retort that the second limb of para (b) of section 22(2) does not apply unless all the matters relied on by the tenant are capable of being remedied, which is not the case here. At all events, as I say, it is a highly technical point, and I think it would be just for me to disregard it in exercise of the power under section 24(7).

Then Mr Denyer-Green objected that the steps to be taken by the landlord to remedy certain of the defects were not specified in the notice. I do not think that that is right. The notice required the landlord to remedy the matters specified in paras 1(i), 1(ii) and 3(iv) of Schedule 2. Para 1(i) relates to the failure to repair the roof and gutters of the property, and under the particulars in that part of Schedule 2 there is this paragraph:

> Full particulars of the defects in the roof and guttering and of the work required to remedy them appear from the reports of Peter M Olley and Associates, dated January 13 1989, a copy of which has been served upon the landlord.

That is so, a copy of that report had been served. Para 1(ii) relates to the failure to decorate the exterior of the property as required by the terms of the leases, and in particular to paint the exterior parts of the property at least once in every three years, contrary to the said terms. The particulars under that simply state that the exterior of the property was last decorated in 1984. It is difficult to see how the tenants could be more specific than that, because the terms of the leases were not more specific.

Third, Mr Denyer-Green objected that there were in the schedule certain matters relied upon by the tenants which were capable of being remedied but which the notice did not require the landlord to remedy. One of those was the failure by the landlord to insure. That, as Mr Denyer-Green himself rightly accepted from me during the course of the argument, was a somewhat hollow complaint in view of Mr Antonelli's evidence that it is not possible (for reasons that I need not go into) for Midrome Ltd to insure the property. There is really no alternative at present to the property being insured by the tenants, as it is. The other matter that Mr Denyer-Green mentioned under this heading was the repayment of £1,094.67. Since there is absolutely no intention on the part of Midrome Ltd to repay that amount, that complaint also is a little hollow. So I think that, if those are technically defects in the notice, I should and I do waive them under section 24(7).

Then Mr Denyer-Green submitted that 28 days, which was the period specified in the notice, was not in the circumstances a reasonable period. Indeed, he said, Midrome Ltd could not comply in that time, so far, at all events as the external painting of the property was concerned, with the requirements of section 20(3) of the Landlord and Tenant Act 1985. That seemed to me at one time to be a valid objection. In fact when I first saw this notice I thought that 28 days was an unreasonably short period in the circumstances. However, after some hesitation, I have come to the conclusion that that objection too I should waive under section 24(7), because it is quite plain from the evidence that, whatever period had been specified, Midrome Ltd had no intention of complying and would not have complied. It would be unrealistic to hold that that requirement, namely the requirement that the period specified in the notice should be reasonable, should in the circumstances of this case result in the plaintiffs being denied the remedy to which they are otherwise entitled.

Mr Denyer-Green's next contention rested on the terms of Ord 97, r15 of the Rules of the Supreme Court which, among other things, require an application for an order under section 24 of the Act of 1987 to state "the name, address and qualifications of the person it is desired to be appointed manager of the premises". Here the originating summons, while it states the name of Croftbride Ltd and its address, 8 Welbeck Way, London W1M 7PE, does not of course state its qualifications. Its memorandum and articles of association are in evidence. Under the objects clause in its memorandum it has as its principal objects to undertake the management and administration of flats etc and to manage any land and buildings or other property. But Mr Denyer-Green's point is that Ord 97, r15 points to the need for the person appointed as manager of premises to have professional qualifications. I do not think that that is right. I do not think that the order is intended to restrict the discretion that the court has under the Act. It merely requires the court to be informed of such qualifications as the person concerned has, and obviously if it has none of a professional character none can be stated. What matters for present purposes is that the tenants, whose creature Croftbride Ltd is, have shown that they are likely to be capable of managing this property. They have provided for it to be insured; they have had these reports from Peter M Olley & Associates; and I understand that they have obtained estimates for repairs to the roof, and so on.

Before I deal with Mr Denyer-Green's next point I will deal with his last point, which was that the delay which the plaintiffs have allowed to elapse should debar them from obtaining interlocutory relief. They served their notice, as I have mentioned, in August 1989; the originating summons was issued in December 1989; and this motion was launched in July 1990. I have already said that in my view it is just and convenient to make the order sought. I think that that is so notwithstanding the delay, because another winter is on its way, the deterioration in the property is continuing, and so is the lack of management of the property.

Objections were raised by Mr Denyer-Green to the wording of the powers which the notice of motion seeks to have conferred on Croftbride Ltd. Those objections were, first, to the use of the words "profits" and "other money" in the provision for Croftbride Ltd to "receive or levy all rents, profits, service charges and other moneys payable by the tenants of the said premises under their respective leases." As I think Mr Waller was inclined to concede, Croftbride Ltd will not really need the words "profits" and "other moneys" in that provision, and it would be better if they were omitted. So it would read: "receive or levy all rents and service charges payable by the tenants".

Then Mr Denyer-Green objected to the provision in para 8 for Croftbride Ltd to "receive such remuneration from the Defendant or otherwise as the court may think fit". I am not at all sure that in the circumstances of this case, bearing in mind that Croftbride Ltd is the alter ego of the tenants and that it is being appointed only for an interim period until the trial, it is necessary or even proper for any provision for its remuneration to be included at all. My inclination would be to omit that altogether, but I have not heard Mr Waller on it, and if he wishes to address me on it I will hear him. Otherwise I will make an order in the form that I have indicated.

Croftbride Ltd appointed manager of the premises pending trial of the action, subject to providing security to the satisfaction of the master; costs of the motion to be costs in the cause; further proceedings transferred to the appropriate county court.

Chancery Division

July 20 1990

(Before Mr Justice KNOX)

VAUX GROUP PLC v LILLEY AND ANOTHER

Estates Gazette February 2 1990
[1991] 04 EG 136

Landlord and Tenant Act 1927, section 19(1)(*b*) — Qualified covenant against assignment — Building lease for more than 40 years with more than seven years of the term unexpired — Provisos requiring lessees to obtain, if so required by the lessors, an acceptable guarantor and a direct covenant by the assignee with the lessors to observe and perform the covenants and conditions of the lease — Whether effect of section 19(1)(*b*) was to relieve lessees of the obligations contained in the provisos — No existing authority on the true construction of section 19(1)(*b*) — Held by Knox J that the obligations in regard to a guarantor and a direct covenant by the assignee were not struck down by the 1927 Act

The parties to these proceedings were tenants and landlords under four underleases which related to premises connected with the Swallow Hotel, Newgate Street, Newcastle upon

A Tyne — The subject underlease, which had included a covenant to erect buildings which were to constitute an hotel, was for a term of 99 years less three days from August 25 1967 — It was thus common ground that section 19(1)(*b*) of the Landlord and Tenant Act 1927 applied *inter alia* to any assignment — The relevant covenants in the underlease were not to assign the demised premises without the consent of the underlessors; upon any assignment to obtain, if required by the latter, an acceptable guarantor for any private limited company; and to obtain a direct covenant by the assignee with the underlessors

B The present proceedings arose because the plaintiff underlessees claimed to be entitled to assign to WHL Hotels Ltd without being under an obligation to provide an acceptable guarantor — The plaintiffs' originating summons accordingly sought declarations that they were entitled to assign the underlease (a) without the consent of the defendant underlessors, (b) without obtaining any guarantor and (c) without obtaining a direct covenant by WHL Hotels Ltd with the defendants — There was, of course, no contention with regard to (a)

The plaintiffs submitted that section 19(1)(*b*) was a much
C more far-reaching provision than section 19(1)(*a*) — Section 19(1)(*b*), it was suggested, showed a discernible statutory policy designed to confer freedom of assignment, so that anything which inhibited such freedom was repugnant to the policy and was intended to be struck down — The plaintiffs also submitted a somewhat more philosophical doctrine, namely, that the congeries of obligations, including the specific covenant not to assign without consent, the provisos as to a guarantor and a direct covenant by the assignees, together with other associated provisions, should all be regarded as parts of one covenant, "which made a seamless whole" — The idea appeared to be that this one seamless
D covenant was in its entirety subject to section 19(1)(*b*) and as such enjoyed a dispensation from due performance — Neither of these submissions found favour with the judge

The judge, however, derived some assistance from the authorities on the construction of section 19(1)(*a*), such as *Adler* v *Upper Grosvenor Street Investment Ltd* and *Bocardo SA* v *S&M Hotels Ltd*, which vindicated, as against dicta in *Greene* v *Church Commissioners*, the continued validity of a requirement for a tenant to offer to surrender his term to his landlord as a condition precedent to the operation of the covenant against assignment without consent — The correct
E procedure was to look at the relevant clause and see whether the words introduced by the operation of the statute necessarily prevented effect being given to what the parties had agreed should be the bargain between them — Taking as an example the proviso about a guarantor, the obtaining of a guarantor was not a condition precedent to the assignment being effected; it was something to be carried out once there was an assignment — The judge therefore rejected the proposition that the introduction of the statutory words had the effect of destroying the bargain which the parties had made about the provision of a guarantor — Accordingly, the
F guarantor covenant was not struck down by section 19(1)(*b*) — The same reasoning applied to the proviso about securing a direct covenant by the assignee

Knox J touched briefly on the suggestion made by Danckwerts LJ in *Property & Bloodstock Ltd* v *Emerton* that the application of section 19(1) to an absolute covenant against assignment remained undecided

In the result the answers to the questions in the originating summons were that the plaintiffs were entitled to assign the underlease to WHL Hotels Ltd without licence or consent but were not entitled to do so without obtaining a guarantor or without obtaining a direct covenant by the assignees

The following cases are referred to in this report. G

Adler v *Upper Grosvenor Street Investment Ltd* [1957] 1 WLR 227; [1957] 1 All ER 229
Bocardo SA v *S&M Hotels Ltd* [1980] 1 WLR 17; [1979] 3 All ER 737; (1979) 39 P&CR 287; [1979] EGD 500; 252 EG 59, CA
Creer v *P&O Lines of Australia Pty Ltd* (1971) 125 CLR 84
Greene v *Church Commissioners for England* [1974] Ch 467; [1974] 3 WLR 349; [1974] 3 All ER 609; (1974) 29 P&CR 285, CA
Property & Bloodstock Ltd v *Emerton (No 2)* [1967] 2 All ER 839
Woolworth (FW) & Co Ltd v *Lambert* [1937] Ch 37

This was an originating summons whereby the plaintiffs, Vaux Group plc, sought declarations on the true construction of provisions in the underlease of the Swallow Hotel in Newgate Street, Newcastle
upon Tyne. The plaintiffs were the underlessees, the underlessors H
being the defendants, Arthur John Lilley and Basil Layton Mitchell.

Nigel Thomas (instructed by Slaughter & May) appeared on behalf of the plaintiffs; Jonathan Gaunt (instructed by Titmuss Sainer & Webb) represented the defendants.

Giving judgment, **Knox J** said: Part of an originating summons between Vaux Group plc as plaintiffs and Arthur John Lilley and Basil Layton Mitchell as defendants has been adjourned to be tried separately from the rest. The parties are tenant and landlords under four underleases between predecessors in title of the defendants and the plaintiff company, in some cases under a different name. The most important of those four underleases is one dated August 25
1967, made between Crossgate Properties Ltd and Vaux & J
Associated Breweries, as the plaintiffs were then called. That was accepted as being a building lease: there were terms in clause 2 whereby the lessee covenanted to erect buildings which were to constitute an hotel. That is not a matter which is in any way in dispute, nor indeed are any other facts in dispute in respect of the questions which have been adjourned for decision now. The term was for 99 years less three days from August 25 1967, and that means that it is common ground between the parties that the provisions of section 19(1)(*b*) of the Landlord and Tenant Act 1927 apply in relation to any assignment that is sought to be made.

The other relevant terms of this underlease are the covenant in
clause 3(17)(a) and (b), for which the sidenote to the lease says: "As K
to assignment and underletting." The covenant reads as follows:

> (a) not to assign underlet or part with the possession of part only of the demised premises save for the kiosk in the pedestrian mall for use as a barber's shop
>
> (b) not to assign underlet or part with the possession of the whole of the demised premises without the consent in writing of the lessor first obtained which consent shall not be unreasonably withheld in the case of a respectable and responsible assignee or underlessee proof of which is furnished to the lessor
>
> Provided Always that the demised premises shall not be sublet at a lower rent than the rent hereby reserved nor otherwise than upon the covenants and conditions herein contained and before granting such licence the lessor shall
> be entitled to be furnished by the lessee with particulars of the proposed L
> assignment or underletting and the lessee shall pay the lessor's solicitor's proper charges for every such licence. AND upon any assignment to obtain if the lessor shall so require an acceptable guarantor for any private limited company and a direct covenant by the assignee with the lessor to observe and perform the covenant and conditions of this underlease, and on the grant of any underlease out of this underlease to obtain
>
> (i) an unqualified covenant on the part of the underlessee that the underlessee will not assign underlet or part with possession of part only of the premises thereby demised (as aforesaid) and
>
> (ii) a covenant on the part of the underlessee that the underlessee will not assign underlet or part with possession of the whole of the premises thereby demised without obtaining the previous written consent of the lessor under this present lease and its superior lessor (such consent not to be unreasonably withheld) and to provide in any such underlease that any sub-underleases
> granted out of such underlease whether immediate or mediate shall contain M
> similar covenants
>
> Provided Always that in no event shall the lessee be entitled to assign underlet or part with possession of the demised premises or any part thereof prior to the completion of the erection and fitting out of the said hotel in manner aforesaid and the opening of the same for business.

The next covenant is one for the registration with the lessor within one month after execution of any assignment, transfer, underlease or devolution. I need not read it in full; it is a common form type of provision. The user was restricted to that of a high-class residential hotel, and there were numerous other covenants entered into by the tenant which do not, in my judgment, directly affect the question of construction which falls to be decided.

The underlease was in fact varied by two deeds which are not material and I need not take time in describing their effect.

The other three underleases are of smaller adjoining or abutting premises. The first is an underlease of a public house lately known as "Images"; that was dated April 16 1971. It was not a building lease; it was a lease of a constructed building for 99 years less three days from June 25 1969. The rent was a peppercorn and there was a provision (as indeed there was in relation to the first underlease of the hotel, which is now called the Swallow Hotel) with regard to the service charge. The provisions in the public house lease do contain a reference to the Swallow Hotel for the purposes of calculating and apportioning the subject-matter of the service charge. Otherwise there is no direct connection between the two. There are restrictions in that public house underlease on assignment and underletting which it is unnecessary to read in full. They are contained in clause 2(17)(a) and (b) and they very closely follow the terms of the underlease of the Swallow Hotel. There are differences, which are largely attributable to the fact that these were licensed premises.

The third of the four underleases was dated January 4 1980 and that concerned premises which have been used (and indeed were required only to be used by the terms of the underlease) as a generator. The land thereby underdemised was quite small (an area of 59 sq yds), and the term in that case was from December 25 1979 until August 21 2066. The rent was £50 per year variable upwards and has indeed been varied upwards.

Finally, the fourth of these underleases concerned a basement storage area. That was dated October 2 1987 and the landlord was in fact different, but nothing turns on that. The term there was to last until August 24 2066 and the initial rent was £130 which thereafter went up, and again there were provisions for service charge. Those premises were linked specifically in one respect with the underlease of the Swallow Hotel, because in the covenant not to assign underlet or part with possession of the whole of the demised premises (there being an absolute covenant not to do that in respect of part), the lessee was required at the same time to assign underlet or part with the possession as the case might be of the whole of the premises demised by the underlease of the Swallow Hotel. To that extent, therefore, the two had either not to be assigned or to be assigned together. In other respects, in common with the other three underleases besides the Swallow Hotel one, there were similar covenants not to assign without obtaining an acceptable guarantor, and the lessee was required to obtain covenants on the part of a further underlessee if there were an underlease out of this underlease.

The defendants now have been approached for this reason: that the plaintiffs claim that section 19(1)(*b*) of the Landlord and Tenant Act 1927 applies to the Swallow Hotel and has the effect of relieving the plaintiffs from the requirement of providing an acceptable guarantor, and it is whether that contention is correct that forms the principal subject-matter of the first question in the originating summons before me.

The originating summons seeks this relief in para 1: declarations that on the true construction of the underlease of the Swallow Hotel the plaintiff is entitled to assign the underlease to WHL Hotels Ltd (a) without any licence or consent from the defendants, (b) without obtaining a guarantor for WHL Hotels Ltd acceptable to the defendants or any guarantor, and (c) without obtaining a direct covenant by WHL Hotels Ltd as assignee with the defendants to observe and perform the covenants of the underlease. Question 2 of the originating summons raises questions which are not before this court, namely whether a refusal to grant a licence to assign to WHL Hotels Ltd was unreasonable, and other similar questions in relation to the refusal of a guarantor. Question 3 of the originating summons, which is before me, seeks declarations that on the true construction of the three underleases of the other three premises that I have detailed the plaintiff is entitled to assign those underleases to WHL Hotels Ltd (a) without seeking any licence or consent from the defendants, (b) without obtaining a guarantor for WHL Hotels Ltd acceptable to the defendants or any guarantor, and (c) without obtaining a direct covenant by WHL Hotels Ltd as assignee with the defendants to observe and perform the covenants and conditions of the said underleases.

There is unchallenged evidence, which does not in fact, I think, affect the matters that I have to decide, to the effect that the public house has been since August 1987 out of use and is now locked and boarded up, and that the generator building which is the subject of the third of the four underleases has been out of use since 1987 and is now empty and locked up. The last of the four premises, the basement, is currently used as a wine store for the Swallow Hotel. There is in evidence the amount of rent which is currently reserved and passing under the various underleases, and also the service charges that are payable under them. The rent for the hotel is by far and away the most substantial; it is currently £84,853. The other rents are much smaller. Similarly, the service charge in respect of the hotel is very much higher than the service charge payable in respect of the three other premises, which are all lumped together for the purposes of the evidence before me. Nothing, I think, in fact turns on the amount of the rent or the service charge. Equally, in my judgment, nothing turns on the correspondence that has passed between the parties, because it is well settled that events subsequent to the grant of an interest by a document such as an underlease cannot affect the true construction of the underlease itself, and I need not refer to that correspondence.

I turn, therefore, to section 19(1) of the 1927 Act. It reads as follows:

> In all leases, whether made before or after the commencement of this Act containing a covenant condition or agreement against assigning, underletting, charging or parting with possession of demised premises or any part thereof without licence or consent, such covenant, condition or agreement shall, notwithstanding any express provision to the contrary, be deemed to be subject
>
> (*a*) to a proviso to the effect that such licence or consent is not to be unreasonably withheld, but this proviso does not preclude the right of the landlord to require payment of a reasonable sum in respect of any legal or other expenses incurred in connection with such licence or consent; and
>
> (*b*) (if the lease is for more than 40 years, and is made in consideration wholly or partially of the erection, or the substantial improvement, addition to or alteration of buildings, and the lessor is not a Government department or local or public authority, or a statutory or public utility company) to a proviso to the effect that in the case of any assignment, under-letting, charging or parting with the possession (whether by the holders of the lease or any under-tenant whether immediate or not), effected more than seven years before the end of the term no consent or licence shall be required, if notice in writing is given to the lessor within six months after the transaction is effected.

I am told that the researches of counsel, which I have no reason at all to doubt, were exhaustive, show that there is no authority in English law which touches the true construction of section 19(1)(*b*). There is a considerable weight of authority in relation to section 19(1)(*a*), which has been the target of considerable expenditure of ingenuity by draftsmen over the years. Some of those efforts have been crowned with success and others not.

The main issue before the court is really one of approach to the construction and effect of section 19(1)(*b*). It was submitted to me on behalf of the plaintiffs by Mr Thomas that there was a discernible policy which is much more far-reaching than that in relation to section 19(1)(*a*) to be seen in the provisions of section 19(1)(*b*). It is, of course, clear that they both are preceded by the same initial five lines, and notably by the words "notwithstanding any express provision to the contrary". They have in common the legislative technique of introducing a proviso into the terms of the lease in question. Their difference, upon which Mr Thomas insisted, is that, whereas in 19(1)(*a*) the proviso was only to the effect that licence or consent was not to be unreasonably withheld, in 19(1)(*b*) the introduced proviso is to the effect that no consent or licence is to be required. This, it was submitted, is intended to confirm freedom of assignment on assigns so that anything which inhibits such freedom is repugnant to, and therefore struck down by, the provisions of section 19(1)(*b*) and the discernible statutory policy behind it.

The second main submission made on behalf of the plaintiffs was that the clause which I have read in the Swallow Hotel underlease in clause 3(17)(b) is all one covenant and that one should treat as within the terms of a single covenant not only the provision which on an assignment requires consent not to be unreasonably withheld but also the first proviso that there should not be a subletting at a lower rent than the rent thereby reserved or otherwise than upon the covenants and conditions, and the second one, that the lessor should be entitled to have particulars of the proposed assignment before any assignment was made, and the further requirement that, upon any assignment, the lessee should obtain, if the lessor should so require, an acceptable guarantor for any private limited company and a direct covenant from the proposed assignee. Those, it was submitted to me, were really all one covenant which made a seamless whole. There is in fact, for what it is worth, upper case or capital letters used in the word "and" which introduces the requirement that upon any assignment the lessee should obtain a guarantor and a direct covenant by the assignee, but I doubt whether that is anything upon which it would be safe to place weight.

I was referred, principally in connection with the authority which does exist concerning the ambit of section 19(1)(*a*), to the last of the decisions regarding attempts that successive generations of conveyancers have made to give a landlord an opportunity of obtaining a surrender of the lease when an application is made to them for leave to assign and there is a qualified covenant against such assignment so that section 19(1)(*a*) applies. In *Bocardo SA* v *S & M Hotels Ltd* [1980] 1 WLR 17 there was a lease which contained a clause that provided that if the tenants desire to assign they should first by irrevocable notice in writing to the landlords offer to surrender the lease to the landlords by deed absolutely and without any consideration. The issue that fell for determination by the Court of Appeal was whether that was effective, regard being had to section 19(1)(*a*) of the Landlord and Tenant Act 1927. The Court of Appeal held that it was, approving an earlier decision of Hilbery J, *Adler* v *Upper Grosvenor Street Investment Ltd* [1957] 1 WLR 227, and not applying certain *dicta* that had been uttered by Lord Denning MR and Sir Eric Sachs in *Greene* v *Church Commissioners for England* [1974] Ch 467. The Court of Appeal was influenced by the fact that the High Court of Australia, in *Creer* v *P & O Lines of Australia Pty Ltd* (1971) 125 CLR 84, had approved the decision of Hilbery J in *Adler* v *Upper Grosvenor Street Investment Ltd*. Megaw LJ gave his reasons for reaching the conclusion which he did at p 24 in the following terms:

> In the balance of conflicting arguments, I reach that conclusion substantially for three reasons. First, the respect due to the decision of the High Court of Australia, and the reasoning of the judgments therein; secondly (as is indeed a ground which clearly strongly influenced Menzies J in that case), the fact that the *Adler* decision has stood and has been acted on, for so many years; and thirdly (a ground which makes me feel able to place much greater reliance on the second ground than I should otherwise have done), the fact that I cannot see any good argument of policy for interfering, more than is essentially required by the words of the statute or by binding authority, with freedom of contract in respect of an agreement between the parties that the landlord should be entitled to the option of requiring a surrender of the lease, where the tenant desires to be freed from his obligation under the lease; bearing in mind that the legislature did not, in 1927, consider, and has not since considered, that policy requires infringement of freedom of contract between a potential landlord and a potential tenant validly and effectively to agree that there shall be no right of assignment at all, however reasonable the tenant's subsequent desire to assign may be.

In my judgment, that attitude should govern my approach to this particular problem under section 19(1)(*b*).

That is reinforced by the approach of the Court of Appeal in *F W Woolworth & Co Ltd* v *Lambert* [1937] Ch 37, at p 60 of which Greene LJ said of section 19(2) (which is concerned with the making of improvements with consent rather than assigning with consent but is otherwise in closely similar terms to section 19(1) of the same Act):

> It is, in my judgment, a misconception to treat the relevant subsection here as though it were a subsection conferring upon the court some remedial jurisdiction to relieve the lessee from the covenant or to modify it. It is in truth nothing of the kind; it is a statutory addition to the terms of a particular type of covenant and the proviso which the subsection mentions is to be read into the covenant. What the legal consequences of that may be in relation to any particular set of facts is a matter which will fall to be determined in the usual way. It is quite wrong to look upon this as something which enables the court to say that a particular set of terms or some particular conditions would be the reasonable thing for the lessor to impose in the circumstances. The practical question which falls to be decided in matters of dispute under this is simply the question aye or no, has the covenant been broken? Aye or No, does the covenant, or does it not, apply in the particular circumstances of this case?

That attitude was, in my judgment, re-echoed by the judgments in the Court of Appeal in *Bocardo SA* v *S & M Hotels Ltd*, in addition to that of Megaw LJ to which I have referred, where Lawton LJ in particular, at p 25, quoted part of what I have just read from Greene LJ in the *Woolworth* case and said of it:

> The same can be said about the proviso mentioned in subsection (1) with which this appeal is concerned.

So far as general approach is concerned, therefore, I prefer the arguments advanced on behalf of the defendants that the appropriate technique for the court to adopt is to regard this as a provision which requires the writing into the relevant clause of the words that the section contains.

On the second question, that is to say, the question whether this is truly to be treated as a single covenant or whether it is appropriate to look at the detailed effect of the covenant and see how it should be construed, read as a whole, it is, in my judgment, clear that, applying the technique which I have mentioned as laid down in the *Woolworth* case and in the *Bocardo* case, the insertion that falls to be made is one which, first, requires the first part of clause 3(17)(b) to have a limited operation. It is to be noted that section 19(1)(*b*) itself is hedged around with limitations. It operates only during the period of the lease before the last seventh of the term; it operates only where the lessor is not a government department or other public authority; and it operates only if notice in writing of the transaction is given to the lessor within six months after the transaction is effected. So that for all those three reasons it is not possible to say that, even where section 19(1)(*b*) clearly applies so as to modify the provisions that the parties have sought to put in their lease, the modification takes the effect of deleting the provision in question for all purposes and for the whole of the life of the lease. That, incidentally, is another difficulty in the path of an argument that there is a discernible statutory policy which requires total freedom for lessees in relation to the assignment of this particular type of building lease.

But in connection with the construction of the statutory words, in my judgment one has to look at the clause and see whether the words thus introduced by the operation of the statute do prevent effect being given to what the parties have otherwise agreed shall be the bargain between them. It was submitted to me that, although obtaining a direct covenant from an assignee and registering the transaction after it had taken place and obtaining covenants from the subtenant when there is a further underletting all fell outside the ambit of something that was prevented by the introduction of the statutory words, the requirement that a guarantor acceptable to the lessor should be procured was contrary to, and therefore nullified by, the introduction of the statutory words. In my judgment that is not a correct analysis of the situation. It seems to me clear that the assignment is one operation and the obtaining of a guarantor is not a condition precedent to its being effected; it is something that has to be effected once there is an assignment, but that, in my judgment, is not by any means the same thing. I do not, therefore, accept the proposition that the introduction of the statutory words involves a prohibition of giving effect to the bargain which the parties have reached that a guarantor should be provided.

In those circumstances it seems to me that the guarantor covenant is, as a matter of construction, not excluded by the terms of section 19(1)(*b*). I am encouraged in reaching that conclusion by a consideration of what the position would be if in fact there was no covenant regarding assignments and underletting at all. It seems to me plain beyond any sort of argument that section 19(1) would not apply at all in those circumstances because it starts with the words, "In all leases containing a covenant, condition or agreement against assigning, underletting, charging or parting with the possession of demised premises or any part thereof without licence or consent . . ." certain things shall take effect, and obviously, it seems to me, a lease that contained nothing at all on the subject of assignments and underlettings would not fall within the ambit of the subsection. In those circumstances section 19(1)(*b*) could not apply and, accordingly, a guarantee covenant such as the one that is included in this lease would, it would seem, necessarily be valid and effectual. This is yet another argument against there being discerned a statutory policy giving freedom of operation in all respects to lessees who fall within section 19(1)(*b*) as these lessees do.

I have not gone into the difference of opinion that is contained in two decisions of the Court of Appeal, both, I think, by way of *dicta*, in relation to the question whether section 19(1) applies to unqualified covenants against assigning. It is clear, of course, that qualified covenants (that is to say, covenants against assigning without licence or consent) are within the subsection, because those initial words that I read out include a reference to covenants, conditions or agreements against assigning, underletting, charging or parting with the possession of demised premises or any part thereof without licence or consent. But Danckwerts LJ in *Property & Bloodstock Ltd* v *Emerton* [1968] Ch 94, at the end of his judgment, specifically reserved his opinion on the point whether the provisions of section 19(1) of the Landlord and Tenant Act 1927 had no application to a covenant in simple terms against assignment or assignment without the landlord's consent of the property comprised in the lease, saying that he was by no means convinced that that was

A so. That is at pp 119-120. On the other hand, neither Megaw LJ, nor indeed counsel on either side, in the *Bocardo* case felt able to discern how an unqualified covenant could be embraced by the words of section 19(1). What Megaw LJ said on the score, at p 22, was:

> Neither of the parties before us asks us to give effect to the doubts on that point expressed by Danckwerts LJ in *Property & Bloodstock Ltd* v *Emerton*. It is hard to see how the words of section 19(1), "in all leases . . . containing a covenant . . . against assigning . . . demised premises . . . without licence or consent . . ." could fairly be construed as applying to leases which contain a simple covenant against assigning, with no reference whatever to "without licence or consent". I do not pursue that question because it is accepted before us that section 19(1) has no application where there is a prohibition of assignment.

B It is for that reason that I treat this question as being one on which there is no decision of the Court of Appeal binding upon me, in the one case because Danckwerts LJ clearly was speaking of something that he was not deciding, and in the other case because the matter was not argued and went by concession.

Nevertheless, it is of marginal significance to this case because it was accepted, rightly, I think, by Mr Thomas, on behalf of the plaintiffs, that unless he could establish that section 19(1) did apply to unqualified covenants against assignment, it would be virtually impossible to discern a statutory policy requiring total freedom against fetters on assignment in favour of lessees. There would be a necessary inconsistency between lessees with qualified covenants being totally free and lessees with unqualified covenants being C anything but. My view, for what it is worth, is that the opinion in passing of Megaw LJ is very much to be preferred. I, too, find it impossible to see how a lease with an unqualified covenant can be described as "a covenant not to assign without licence or consent". However, that is not an issue which is central to the decision which I reach, which is based on the policy of the Act not being the clear guide which would be needed to assign such a wide effect to section 19(1)(*b*) and on the conclusion that I have reached, that there is not here a single covenant but that the requirement for a guarantee is something which can, and legitimately should, operate independently of, although closely in connection with, the covenant against assignment.

D For those reasons, it seems to me that question 1 of the originating summons should be answered (a) "yes", but (b) and (c) "no". That means that the question in para 3 hardly arises, because it was not, I think, submitted that if I answered question 1 in the way which I propose, question 3 could be answered in a different way. But even if I had decided question 1 the other way, I would have been unpersuaded that the quite separate underleases, at a later date, being underleases which, it is rightly accepted, are not themselves within section 19(1)(*b*), could be, because of the reference at any rate in the last of the three to the Swallow Hotel, bound up in the same qualification. However, the point does not, in the circumstances, arise and I need not elaborate it further. I therefore propose to answer question 3 in the sense of "no" to all three of the subparagraphs.

E

Queen's Bench Division

August 9 1990

(Before Mr P J COX QC, sitting as a deputy judge of the division)

CARDIGAN PROPERTIES LTD v CONSOLIDATED PROPERTY INVESTMENTS LTD

Estates Gazette February 23 1991

[1991] 07 EG 132

F

Landlord and tenant — Re-entry by lessor into premises subject to 999-year lease — Premises sublet to London Borough of Camden for temporary housing — Alleged breaches by lessee of insurance covenant — Notice under section 146 of Law of Property Act 1925 — Action by lessee claiming that the lease had not been forfeited and that entry by lessor was unlawful — Whether lessee had been allowed a reasonable time to comply with section 146 notice — Discussion of waiver, forfeiture and equitable jurisdiction to grant relief — Re-entry by lessor held to be a trespass — Judgment for lessee

The subject premises were a block of residential flats, Scala House, Tottenham Street, London W1 — Plaintiff lessee held a lease of 999 years from 1976 subject to a substantial mortgage in favour of a Danish bank — The block had been sublet in 1987 to Camden and had fallen into a squalid condition through Camden's failure to observe its covenants and the activities of short-term occupiers and squatters — There had been a history of problems in regard to insurance — The lease contained a covenant by which the lessee was obliged to insure the premises in the names of the lessor, lessee and other interested parties — Scala House had been insured with Sun Alliance for a number of years, but in July 1989 it refused to renew the cover owing to the conditions of the occupation — A limited cover was obtained for a time and then cancelled by the plaintiff lessee — At this time it was trying without success to sell the leasehold interest and was in arrear with mortgage payments; the mortgagee had obtained a *Mareva* injunction to protect its interests — The mortgagee had obtained its own insurance but this cover extended only to its interest in the premises — The lessor, concerned about the insurance position, issued a writ in the Chancery Division for production of the current policy of insurance and these proceedings remained in being during the present action — Subsequently the lessor served on the lessee a notice under section 146 of the Law of Property Act 1925 complaining of a breach of the insurance covenant by failure to take out agreed insurance in the names of the lessor and lessee — After a warning that its patience was exhausted the lessor made a physical re-entry into the demised premises and remained in possession — The lessee then commenced the present action in which its claims as plaintiff were that its lease had not been forfeited, that the lessor's entry had been a trespass, and that the lessor should be ordered to quit the premises; alternatively, if the lease had been forfeited, the lessee asked for relief G H J K

In its action the plaintiff lessee submitted that the defendant lessor had failed adequately to specify particulars of the breach complained of; had made a physical re-entry only 10 days after service of the section 146 notice, thus not allowing the plaintiff a reasonable time for compliance; and had in any case waived any right to forfeit by taking proceedings in the Chancery Division with knowledge of the plaintiff's breach of covenant — The judge rejected the complaint about the adequacy of the notice but held that the defendant had not allowed a reasonable time for the plaintiff to comply with the notice — The defendant's entry into the premises was therefore a trespass — This was sufficient to decide the case in the plaintiff's favour — As, however, the matter might go further, the judge expressed his views on other points raised — He considered that by continuing the Chancery proceedings, thereby acknowledging the continuation of the tenancy, and having undoubted knowledge of the plaintiff's failure to insure, the defendant did waive its right to forfeiture — The judge also expressed the view that if the forfeiture had been held to be effective and if he had the necessary jurisdiction, he would have granted relief under the general equitable jurisdiction of the court L

In the result the plaintiff was granted a declaration that the lease was not forfeit, an order for possession and an inquiry as to damages; it was also awarded costs M

The following cases are referred to in this report.

Alghussein Establishment v *Eton College* [1988] 1 WLR 587, HL
Central Estates (Belgravia) Ltd v *Woolgar (No 2)* [1972] 1 WLR 1048; [1972] 3 All ER 610; (1972) 24 P&CR 103; [1972] EGD 648; 223 EG 1273, CA
Expert Clothing Service & Sales Ltd v *Hillgate House Ltd* [1986] Ch 340; [1985] 3 WLR 359; [1985] 2 All ER 998; [1985] 2 EGLR 85; (1985) 275 EG 1011 & 1129, CA
Fox v *Jolly* [1916] AC 1
Hick v *Raymond & Reid* [1893] AC 22

A *Shiloh Spinners Ltd* v *Harding* [1973] AC 691; [1973] 2 WLR 28; [1973] 1 All ER 90; (1973) 25 P&CR 48, HL

In this action the plaintiff, Cardigan Properties Ltd, lessee of Scala House, 21 Tottenham Street (off Tottenham Court Road), London W1, sought a declaration that its lease had not been forfeited and consequential relief against the lessor, Consolidated Property Investments Ltd. In the alternative the plaintiff claimed relief against forfeiture.

John Colyer QC and Martyn Zeidman (instructed by Arnold George & Co, of Ilford) appeared on behalf of the plaintiff; James Reid QC and Michael Driscoll (instructed by Hamlin Slowe) represented the defendant.

B

Giving judgment, **Mr P J Cox QC** said: In this action the plaintiff claims a declaration that its lease of Scala House, 21 Tottenham Street, London W1, is not forfeit and that the defendant has trespassed therein. In the alternative the plaintiff claims relief against forfeiture and for other relief as set out in the statement of claim.

The writ was issued on June 26 1990 and the trial of the action took place over four days commencing on July 26 1990. This may be a speed record for the progress of any civil action this century. The facts which are material to the issues in this case are not seriously in dispute and it is convenient to relate them in some detail in chronological order.

By a lease dated June 16 1976 ("the lease") the defendant demised
C the block of residential flats, Scala House, to Salut Co for a term of 999 years at a peppercorn rent in consideration of the payment of £770,000. The eight storeys of Scala House are constructed above and supported by the second storey of a building not forming part of the demised property.

By a sublease dated December 1 1987 Flowplace Ltd, the then lessee under the lease, granted Scala House to the London Borough of Camden ("Camden") for a term of two years at an annual rent of £618,000 for the purpose of providing temporary housing and requiring vacant possession at the end of the term on December 1 1989. Under the sublease Camden was required to yield up the premises at the end of the term in the same clean state and condition
D as at the beginning and to replace or repair broken fixtures and fittings.

The plaintiff, a company registered in Panama, the sole shareholder being Mr Ehab Debs, purchased the lease from Flowplace in July 1988 by a transfer dated July 13 1988 for the sum of £6,829,000. This purchase was substantially financed by a mortgage advance from a Danish Bank, "KD". Thereafter Mr Debs acted for the plaintiff under a power of attorney.

Scala House had been insured with Sun Alliance for a number of years and, when the plaintiff purchased, a new policy with Sun Alliance was negotiated. The proposal form, signed by Mr Debs, showed the mortgagee's requirements as to rebuilding costs to be
E £5,464,000. The schedule to this policy shows the policy-holder to be the plaintiff, the sum insured to be £5,464,000, and the annual premium to be £4,185.20.

In July 1988 the flats in Scala House were still in a reasonable state of internal repair, but because of the nature of the occupancy by Camden, involving short-term tenants and later on by squatters, there was a rapid deterioration internally. On October 23 1988 a fire occurred in one of the flats, doing damage to the value of over £30,000.

On April 26 1989 Sun Alliance wrote to the plaintiff's brokers indicating that, owing to the occupation of the premises, renewal of the policy on July 20 1989 would not be invited. The letter ended with these words:

F We trust this has provided you with adequate time to place the business elsewhere.

On May 16 1989 the plaintiff's brokers, David White, wrote by recorded delivery to the plaintiff's solicitor advising urgent action and recommended contact with specialist high-risk brokers. Thereafter Mr Debs attempted to obtain insurance cover for Scala House and was also trying to find a buyer for the property.

On August 29 1989 Kay International, acting for the plaintiff, wrote to the plaintiff's solicitors confirming that 21 Tottenham Street (that is Scala House) was held covered in the sum of £2,000,000 from August 2 1989 at an annual premium of £24,000. The cover was limited to fire, explosion and aircraft, and Mr Debs told me that he thought this was adequate to cover the cost of rebuilding. This must have been wishful thinking on Mr Debs' part, no doubt dictated by G
the slump in property values in late 1988 and early 1989 and the high cost of insuring this high-risk building.

On November 24 1989 Kay International wrote to the plaintiff's solicitor informing him that Mr Debs had cancelled the insurance and requesting that the mortgagee should be informed. Mr Debs, in evidence, explained that he did this because he thought he had found a purchaser whom he expected to insure the building, I think that day.

I should say this: so far as Mr Debs is concerned, I found him a truthful witness, but in the field of property management he was probably somewhat naïve, although he told me that he had had a number of property transactions apart from this one.

During November 1989 the defendant and National Car Parks
Ltd, its associate company, became anxious about the insurance H
cover for Scala House, and on November 30 1989 Mr Phillips, the insurance manager of National Car Parks, informed the plaintiff's solicitors that the defendant had arranged the necessary cover and would recoup the premium calculated on a daily basis from the plaintiff until the arrangement of satisfactory cover by the plaintiff in accordance with the terms of the lease. This cover was arranged with the Royal Insurance, who had previously surveyed the property. Royal Insurance, on May 21 1990, wrote to Mr Phillips referring to the property. The letter contains these words:

Our survey report is not too complimentary on this building. It would appear to have been very poorly maintained and that there is a lack of management
control on the part of the London Borough of Camden. They are in a very J
squalid condition with trash on the landings and stairs and damage to the fixtures such as doors and windows and lift machinery. There is 24-hour security at the front entrance and all the unoccupied flats (only five of the 34 flats are occupied) are sealed with steel doors to avoid squatters. Apart from this the security of these premises is negligible.

It is not clear from that letter at which date the survey report had been obtained, but it must have been well before May because, according to the evidence in the case, the last of the squatters were moved during the month of April.

By December 1989 the plaintiff had fallen into arrears with the mortgage repayments and KD, on January 16 and 17 1990, obtained *ex parte Mareva* injunctions against the plaintiff and Mr Debs. These
injunctions continue in force to the present day. K

In early December the plaintiff's solicitor, Mr Orchover, stressed to Mr Debs the urgent need to insure. KD had arranged for the insurance of Scala House with Preferred Insurance in early December to cover the period from December 22 1989 to December 31 1990 in the sum of £8,830,000 at a premium of £12,635. The plaintiff became aware of the existence of this insurance in February 1990. On January 19 1990 in a valuation report to KD Knight Frank & Rutley gave a vacant possession price on the open market for Scala House of £3,000,000.

At the end of 1989 the proposed purchaser of Scala House withdrew and Mr Debs continued his efforts to sell the property,
without success. This is not surprising with many of the flats L
unoccupied and about 15 still occupied by squatters.

On February 16 1990 the defendant issued a writ against the plaintiff in the Chancery Division for production of the current policy of insurance. These proceedings remain in being and give rise to a legal argument on waiver, to which I will refer later in this judgment.

Knowing that both the defendant and the KD Bank had obtained insurance cover for Scala House no serious effort was made by the plaintiff to obtain its own insurance cover until June 1990, as required by the lease. However, on May 24 1990 — by which time the last of the squatters had left — Mr Orchover sent a copy of the Preferred policy to the defendant requesting confirmation that the
Chancery proceedings for production would be discontinued. Such M
confirmation has never been forthcoming.

On May 31 1990 Mr Phillips wrote a letter on behalf of the defendant to Mrs Brown of Preferred Insurance. The fifth paragraph of that letter reads as follows:

Doubtless you will note the insured interest does not bear the name of Consolidated Property Investments Ltd as stipulated by the lease, but obviously in fairness you are not responsible for the omission. We have in our possession only a photostat copy of the policy wording and schedule but we would appreciate the opportunity to visit your offices to examine the original policy document and also, if possible, to view the proposal form (if any) and your survey of the risk. It is possible that we may have material information which may affect the acceptability of the building for insurance cover.

A Mrs Brown replied to that letter on June 6:

> Dear Mr Phillips,
> We acknowledge receipt of your letter. You are welcome to come to our offices to discuss this matter and we look forward to hearing from you at which time a mutually convenient appointment can be arranged.
> We are not in possession of the original policy document relating to the aforementioned premises. This of course is with our insured. Our insurance arrangements are confidential but the policy, as you have already established, only protects the interest of KD and affords no protection to Consolidated Property Investments Ltd.
> We are most interested in the comments contained in the fifth paragraph of your letter and would welcome the opportunity for discussion at your earliest convenience.

B This correspondence was not copied to Mr Orchover, who remained unaware of this development until these letters were disclosed in the course of these proceedings.

On Tuesday June 5 1990 the defendant served a section 146 notice on the plaintiff. Para 2 of the notice reads as follows:

> The lease contains covenants by the lessee with the lessor as follows . . .
> (2)(1) At all times during the said term to insure the demised premises and the building erected thereon and to keep the same insured in the name of the lessor and the lessee and in the name of other interested parties with a reputable insurance office or offices and at not less than the full replacement value thereof as agreed from time to time between the parties hereto. Provided that if the parties shall fail to agree upon the full replacement value thereof, then the matter shall be referred to the arbitration of an independent surveyor appointed for that purpose by the parties jointly in writing or, upon their
> C failure to agree upon such appointment within one month, then by an independent surveyor appointed for that purpose on the application of either party by the President for the time being of the Royal Institute (*sic*) of Chartered Surveyors. And in either case in accordance with the provisions of the Arbitration Act 1950 including architects' and surveyors' fees Against loss or damage by fire, explosion, tempest or aircraft other than hostile aircraft and all other risks commonly or usually covered by comprehensive "All Risks" policy, and pay all premiums for such insurance within 7 days after the same become payable and will produce to the lessor on demand policies of such insurance for the current year's premium.

Para 3:

> You are in breach of the covenants set out above, or one or some of them, in that you have failed to insure and keep the demised premises and the building
> D erected thereon insured in the name of the lessor and the lessee and in the name of other interested parties And/or pay all premiums for such insurance within seven days after the same become payable.

Para 4:

> We require you to remedy the said breaches insofar as they may be capable of remedy and to compensate the lessor in money for the commission thereof.

Para 5:

> On your failure to comply with this notice within a reasonable time, then it is the intention of the lessor to re-enter the demised premises and forfeit the lease and/or claim damages for your breaches of covenant.

This notice accurately reproduced the wording of clause 2(1) of the lease. At the time of service the plaintiff was in breach of the clause.

E On June 6 1990 Mr Orchover wrote to KD's solicitors. The letter is addressed to Messrs Hedleys:

> Dear Sirs,
> We write to confirm our telephone conversation today regarding the problem in connection with the insurance. We enclose herewith a copy of the notice received from the freeholders, from which you will see that the problem is that the insurance is not taken out in the names of the lessor and lessee. This is also holding up exchange of contracts because the insurance position is not totally satisfactory in that the freeholders are contending that we are in breach of the terms of the lease.
> Would you therefore get your client Bank kindly to contact the Preferred Insurance Company Ltd and ask them to forward a revised schedule showing that the insured names are (1) Consolidated Property Investments Ltd. as lessor; (2) Cardigan Properties Incorporated as lessee, and your client bank as
> F mortgagees (3). This will remedy the situation and we can then proceed with the transaction without further delay.
> Your early attention to this matter will be much appreciated.

It is clear from this letter that efforts to sell Scala House were continuing; that an exchange of contracts was being prejudiced by the insurance position, and that Mr Orchover was very anxious to correct the insurance position. On Friday June 8 1990 Mrs Brown of Preferred wrote to KD's solicitors. The letter is addressed to Hedleys and signed by Mrs Brown:

> We would confirm that on receipt of a written instruction from KD we would be able to elevate Consolidated Property Investments to the status of joint insured as required by the lease. At present we are prepared to revise the insurance as indicated above for a period of 14 days from today pending the visit of Mr Phillips of Consolidated to this office on Tuesday 12th June. Our G
> letter of 6th June to Mr Jenkins refers.
> With regard to Cardigan, we are unable to provide cover on the same basis to Cardigan Properties without receipt of their own instructions in this regard. Without this instruction we could only comply with Consolidated's request on receipt of written confirmation from Consolidated that if Cardigan also be elevated to status of joint insured, Consolidated would agree to pay such additional premium as the company at such time in the future may require in this regard. Once again, this cover would only be available at present for 14 days pending discussion with Mr Phillips.

It is apparent from this letter that Preferred was by no means certain of continuing to extend more than very limited cover after the meeting with Mr Phillips on Tuesday June 12.

On Monday June 11 Mr Orchover sent by fax a letter to Mrs H
Brown. It is marked "Urgent" by post and fax:

> Dear Mrs Brown,
> Your fax addressed to Mrs Jackson dated 8th June has been passed to me. I can confirm that we act for Cardigan Properties who are the lessees of the above property holding under a registered lease from the landlords Consolidated Property Investments being a lease of 999 years from 25th March 1976. We are therefore able to confirm instructions on behalf of our clients Cardigan Properties that they should be shown on the schedule of the insurance as having an interest as lessees of the property with Consolidated Property Investments being shown as the freeholders. We trust this is sufficient for your purposes because there is some urgency in obtaining the necessary revised schedule for confirmation of the insurance. Should there be anything further please do not hesitate to fax me by return.

Mrs Brown replied to this letter the next day: J

> Dear Mr Orchover,
> We are presently awaiting the instructions of our policy-holder in this connection. We will contact you again in due course.

and Mrs Brown signed that letter.

There was then a flurry of letters, mainly by fax from Mr Orchover, who plainly was mystified by his failure to persuade Preferred to add the names of the plaintiff and the defendant to KD's policy. These letters and replies cover the period between June 13 and 19.

One of these letters is a letter from the defendant's solicitors to Mr Orchover and I read that letter:

> Dear Sirs, K
> We thank you for your letter of 13th June received on 15th. Our client's patience is exhausted. The section 146 notice is crystal clear as to what needs to be done and your client has lamentably failed to do it. As you will be aware, even if your clients are not, there is a world of difference between an interest being noted on a policy and a policy in joint names. Our client's building is substantially at risk and in view of your client's non-compliance with the section 146 notice our clients are taking such further steps as may be necessary fully to protect their interest.

That letter was written nine days after the service of the notice and on a reading of that letter it casts doubt in my mind as to the defendant's understanding of the significance of Mr Orchover's efforts, or as to its knowledge of what had passed between Mr
Phillips and Mrs Brown. In the absence of any evidence from Mr L
Phillips one can only surmise about these matters. But what is clear is that the defendant by June 15 was contemplating steps to re-enter Scala House and to forfeit the lease. Such re-entry occurred on June 21 when the premises were entirely unoccupied and there is now no dispute between the plaintiff and the defendant as to the fact of re-entry and the defendant remains in possession pending the resolution of this case.

On June 22 1990 the plaintiff's solicitors wrote to the defendant:

> Dear Sirs,
> We refer to your letter faxed to us yesterday evening.
> We do not accept that your clients' purported forfeiture is valid or lawful and our clients will be seeking appropriate relief in the courts as already notified to
> you. M
> Your clients were fully aware that steps were being taken to have the insurance, which is in the names of KD, placed in the joint names of our respective clients. This coupled with the fact that your clients reversionary interest is not at risk, clearly demonstrates that quite apart from any other argument, your clients acted hastily and prior to the expiry of a reasonable period following the service of the section 146 notice.
> Entirely without prejudice to our clients' position we can confirm funds are immediately available for our clients to take out a new policy of insurance in their name jointly with the name of your clients in the insured sum of £4,000,000,

and so on.

Now section 146 of the Law of Property Act 1925, subsection (1), provides essentially as follows — and I omit many of the words:

> A A right of re-entry or forfeiture . . . for a breach of any covenant . . . in the lease shall not be enforceable, by action or otherwise, unless and until the lessor serves on the lessee a notice . . . specifying the particular breach complained of . . . and the lessee fails, within a reasonable time thereafter, to remedy the breach, if it is capable of remedy. . . .

Subsection (2) essentially provides that where the lessor is proceeding, by action or otherwise, to enforce such a right of re-entry or forfeiture, the lessee may apply to the court for relief . . .

By the amended statement of claim in this action the plaintiff claims, first, a declaration that the lease was not forfeit and that the defendant's entry was a trespass; in the alternative, relief against forfeiture; third, an injunction requiring the defendant to quit the
B premises; fourth possession; fifth an inquiry as to damages; and sixth, interest.

The plaintiff supports its case on four main grounds. First, the defendant failed adquately to specify the particular breach complained of in the section 146 notice, which *ipso facto* could not be relied upon as giving rise to a right of re-entry. Second, the defendant by re-entering the premises only 10 clear working days after the service of the section 146 notice did not allow reasonable time for the plaintiff to comply with the terms of the notice in all the circumstances of the case. Third, the defendant, with knowledge of the plaintiff's breach of covenant to insure, by issuing proceedings in the Chancery Division on February 16 1990 against the plaintiff,
C which proceedings are still continuing, has waived the forfeiture. Fourth, in the event of a finding by the court of a valid forfeiture of the lease, the plaintiff seeks relief under the equitable jurisdiction of the court.

I propose to consider each of these in turn in the light of the evidence and the relevant case law.

First, the adequacy of the section 146 notice. By clause 2(1) of the lease the lessee covenanted to "insure the demised premises . . . and keep the same insured in the name of the lessor and the lessee and in the name of other interested parties with a reputable insurance office . . . and in not less than the full replacement value . . . [to be] agreed from time to time between the parties . . . Against loss or damage by fire, explosion, tempest or aircraft . . . and all other risks commonly
D or usually covered by a comprehensive 'All Risks' policy." It is clear from these words that in order to comply with the terms of the covenant the lessee must reach agreement with the lessor as to the replacement value of the property. Furthermore it is common ground between the parties that the expression "a comprehensive 'All Risks' policy" is a contradiction in insurance terms and thus in my judgment, in order to give effect to the clause, it would be necessary for the lessee and the lessor to reach agreement as to the cover required.

In the case of *Fox* v *Jolly* [1916] AC 1, where the House of Lords was considering the sufficiency of a notice served under section 14(1) of the Conveyancing and Law of Property Act 1881, which was in
E precisely the same terms as section 146(1) of the Law of Property Act 1925, Lord Buckmaster said, at p 11:

> I can find nowhere in the section any words which cast upon the landlord the obligation of telling the tenant what it is that he must do. All that the landlord is bound to do is to state particulars of the breaches of covenants of which he complains and call upon the lessee to remedy them. The means by which the breach is to be remedied is a matter for the lessee and not for the lessor. In many cases specification of the breach will of itself suggest the only possible remedy.

Applying this clear guidance to the facts of the present case, I consider that the notice complied with the requirements of section 146(1) of the Act of 1925 as it plainly called upon the plaintiff to
F insure the property in accordance with the covenant in the lease.

Second, was a reasonable time allowed for compliance with the notice? I approach this question on the basis that all relevant facts relating to the particular case should be considered. As a consequence there is little help to be derived from the reported cases concerning such a notice. These in the main are concerned with breaches of repairing covenants where, depending upon the nature of the work to be done, times of up to a year or more have been considered to be reasonable. Mr Reid, for the defendant, pointing out the history of the plaintiff's failure to keep the premises insured, has cited two cases in support of the proposition that this history ought in some way to curtail the time to be allowed for the breach of covenant to be remedied.

The first of these cases is that of *Expert Clothing Service & Sales Ltd* v *Hillgate House Ltd* [1986] Ch 340. Slade LJ*, in his judgment, at p 358, said: G

> An important purpose of the section 146 procedure is to give even tenants who have hitherto lacked the will or the means to comply with their obligations one last chance to summon up that will or find the necessary means before the landlord re-enters. In considering what "reasonable time" to allow the defendants, the plaintiffs, in serving their section 146 notice, would, in my opinion, have been entitled to take into account the fact that the defendants already had enjoyed fifteen months in which to fulfil their contractual obligations to reconstruct and to subject the defendants to a correspondingly tight time-table running from the date of service of the notice, though, at the same time, always bearing in mind that the contractual obligation to reconstruct did not even arise until 29th June 1981, and that as at 8th October H
> 1982 the defendants had been in actual breach of it for only some ten days.

That case involved consideration of the tenant's failure to perform an obligation to reconstruct by the due date, which seems to me to be of a completely different nature from an obligation to insure. In considering the appropriate time for performance of the latter obligation I reject the submission that earlier conduct can properly be taken into account.

Mr Reid also referred to a case of *Alghussein Establishment* v *Eton College* [1988] 1 WLR 587. This case again involved the consideration of the obligation by a tenant to complete redevelopment or development by a particular date. My attention was drawn to the headnote, which conveniently sums up the effect of the opinion of Lord Jauncey: J

> Held, dismissing the appeal, that the well-known rule of construction, that there was a presumption that a party to a contract could not be permitted to take advantage of his own wrong as against the other party, applied, in the absence of an express provision to contradict the presumption, as much to a party who sought to obtain a benefit under a continuing contract on account of his breach, as it did to a party who relied on his breach to avoid a contract and thereby escape his obligations; that the terms of the proviso to clause 4 were not apt to displace that rule of construction and that, therefore, the plaintiffs were not entitled to invoke the proviso to obtain the grant of the lease.

Mr Reid argued that in some way I should apply the finding in *Alghussein*'s case to a consideration of the time adjudged to be reasonable for the performance of the covenant to insure. I find that K
submission unacceptable and accordingly I reject it.

The third case cited by Mr Reid was *Hick* v *Raymond & Reid* [1893] AC 22, a House of Lords' decision, where the House was considering what considerations applied in determining what was a reasonable time for the discharge of a ship's cargo. In his opinion, at p 32, Lord Watson said:

> When the language of a contract does not expressly, or by necessary implication, fix any time for the performance of contractual obligation, the law implies that it shall be performed within a reasonable time. The rule is of general application and is not confined to contracts for the carriage of goods by sea. In the case of other contracts the condition of reasonable time has been frequently interpreted; and has invariably been held to mean that the party upon whom it is incumbent duly fulfils his obligation, notwithstanding L
> protracted delay, so long as such delay is attributable to causes beyond his control, and he has neither acted negligently nor unreasonably.

In the same case, at p 29, Lord Herschell said:

> The only sound principle is that the "reasonable time" should depend upon the circumstances which actually exist. If the cargo has been taken with all reasonable despatch under those circumstances I think the obligation of the consignee has been fulfilled. When I say the circumstances which actually exist I, of course, imply that those circumstances, in so far as they involve delay, have not been caused or contributed to by the consignee.

Despite the submission of Mr Colyer that the case of *Hick* v *Raymond* had no bearing upon this present case, on a consideration of what is said in these judgments it does seem to me that these words are of assistance in the present case as to the way in which I should M
approach this question of reasonable time, and I propose to apply them.

Now on June 5 1990 when the notice was served on the plaintiff it was incumbent upon Mr Orchover, its solicitor, to take immediate steps to comply with the covenant. He was aware that both the defendant and KD had separate insurance cover on the property. In my judgment, in the prevailing circumstances, it was reasonable and prudent of him to endeavour (as he did) to negotiate with KD in order to get the plaintiff and the defendant added as insured names to the Preferred policy. He was aware of the terms of the covenant and that

*Editor's note: Also reported at [1985] 2 EGLR 85 p 90B.

the premises constituted a high risk so far as insurance was concerned. He was quite unaware that Mr Phillips, the defendant's insurance manager, had already queried with Mrs Brown of Preferred the acceptability of the premises for insurance, which made it most unlikely that Preferred would comply with Mr Orchover's request. It was not until about a week later that it should have become clear to Mr Orchover that he would have to take alternative steps on behalf of the plaintiff to comply with the notice.

As I have already indicated, due performance of the insurance covenant involved the securing of agreement between lessor and lessee as to the full replacement value of the premises. As to the reasonable time required for compliance with this covenant, I have heard the evidence of two expert witnesses: Mr Peters on behalf of the plaintiff and Mr Stratton on behalf of the defendant. I am satisfied that both these gentlemen were very experienced in this type of insurance cover. Each was cross-examined at some length and both impressed me as careful, honest witnesses.

Mr Peters, however, was more alert to the problems presented by the terms of the covenant and in expressing the opinion that four weeks would be a reasonable time for compliance he was plainly allowing for these. Mr Stratton held the view that two weeks or 10 working days would have been a reasonable time, but he seemed to take little or no account of the need for prior agreement between lessor and lessee as to rebuilding cost and the resultant need for surveys of the building, which was, of course, not a normal free-standing structure but dependent for support upon a structure below it. I also bear in mind that whereas Mr Peters prior to his involvement in this case had no prior dealings with any of the interested parties, Mr Stratton had in the past had business dealings with Mr Phillips through the National Car Parks company. Accordingly, where they differ as to the period of time I prefer the opinion of Mr Peters that a reasonable time would be four weeks. Even if I had preferred the 10 working-day period suggested by Mr Stratton, on the particular facts of this case, taking into account the time I find was justifiably spent by Mr Orchover in his abortive negotiations with Preferred, the defendant did not allow reasonable time for the plaintiff to comply with the notice; they re-entered the premises only 10 working days after service.

It follows that I find that the defendant's entry into these premises was a trespass.

Now I understand that this case is likely to go further and I have been invited to express my view upon the other points raised by the pleadings. So far as the waiver point is concerned, it is alleged by the plaintiff, in para 7(a), (b) and (c) of the amended statement of claim, that by continuing the action in the Chancery Division claiming a mandatory injunction for production of the policies of insurance in accordance with the lessee's covenant to produce such policies on the lessee's demand under clause 2(1) of the lease the defendant has waived its right to forfeiture. The writ claiming this injunction was served on the plaintiff on May 22 1990 and such proceedings were continuing up to the date of re-entry on June 21 1990. At the time the writ was issued I am satisfied that the defendant was aware that the plaintiff was in breach of the covenant to insure and that state of knowledge I find continued up to the time of the serving of the notice.

My attention has been drawn by Mr Colyer to the opening words in *Woodfall* under *Waiver of forfeiture*. This is the current edition [28th ed] of *Woodfall*, para 11906 at p 851:

> Courts of law always lean against forfeitures; therefore whenever a landlord means to take advantage of any breach of covenant or condition so that it should operate as a forfeiture of the lease, he must take care not to do anything which may be deemed an acknowledgment of the continuance of the tenancy, and so operate to waive the forfeiture. Merely lying by and witnessing the breach is no waiver; some positive act must be done.

Here it is argued by Mr Colyer for the plaintiff that although it is a very technical point it is none the less a good one because, by not discontinuing the action for the injunction and having undoubted knowledge of the plaintiff's failure to insure, the defendant has waived the forfeiture by the unequivocal act of commencing and continuing the proceedings. I have been referred to the words of Buckley LJ in *Central Estates (Belgravia) Ltd* v *Woolgar (No 2)* [1972] 1 WLR 1048 at p 1054 E:

> In my judgment the effect in law of an act relied on as constituting a waiver of a right to forfeit a lease must be considered objectively, without regard to the motive or intention of the landlord or the understanding or belief of the tenant.

I have considered this interesting point very carefully in the light of the arguments of Mr Colyer and also, of course, of Mr Reid, and I have come to the conclusion that by continuing those Chancery proceedings the defendant did in fact waive its right to this forfeiture.

Now in the event of the forfeiture being held to be valid is there a right of relief under the equitable jurisdiction of the court? My attention has been drawn to a number of authorities which make interesting reading, and it is clear that there is a sharp conflict of judicial opinion as to this. I propose to do no more than to express the view that I find it difficult to accept the proposition that section 146, in the absence of express words, is a complete code ousting an existing jurisdiction in the court to grant equitable relief in cases not covered by section 146. No doubt sooner or later the matter will be investigated and ruled upon at length by higher authority.

If I had such jurisdiction in this case I would have borne in mind the considerations mentioned by Lord Wilberforce in *Shiloh Spinners Ltd* v *Harding* [1973] AC 691 at p 725 between letters "D" and "F", where Lord Wilberforce observes:

> Failures to observe the covenants having occurred, it would be right to consider whether the assignor should be allowed to exercise his legal rights if the essentials of the bargain could be secured and if it was fair and just to prevent him from doing so. It would be necessary to consider the conduct of the assignee, the nature and gravity of the breach and its relation to the value of the property which might be forfeited. Established and, in my opinion, sound principle requires that wilful breaches should not, or at least should only in exceptional cases, be relieved against, if only for the reason that the assignor should not be compelled to remain in a relation of neighbourhood with a person in deliberate breach of his obligations.

If I had such jurisdiction in this case I would have granted relief against forfeiture because of the following facts:

First, the plaintiff's failure to insure was not deliberate but was brought about by a combination of circumstances beyond its control:

(a) The failure of Camden to comply with its covenant as the sublessee so that the premises fell into a state of serious squalor shortly after the plaintiff acquired the lease, which led to Sun Alliance's refusal to renew in July 1989.

(b) The existence of squatters in the building for five months until April 1990 after the sublease to Camden had come to an end compounded the problem by making refurbishment a virtual impossibility so that the whole became unoccupied and producing no income.

(c) The sharp decline in property values following the plaintiff's acquisition coupled with the heavy borrowing rendered it difficult for the plaintiff to find a purchaser at a reasonable price and to keep up the mortgage repayments.

(d) The constraints upon the plaintiff imposed by the *Mareva* injunction.

(e) The comparative inexperience of Mr Debs.

Second, the lease is virtually a freehold, being the unexpired portion for 999 years at a peppercorn rent for which the plaintiff paid £6,829,000.

Third, there was no evidence that the value of the defendant's reversion had been materially affected by the non-insurance.

Fourth, the effect of refusing relief would be to cast an enormous burden of debt on Mr Debs, who has personally guaranteed the mortgage to KD, and to result in a huge windfall for the defendant.

The terms upon which I would have granted such relief are: first, that the plaintiff should reimburse the defendant for the costs of insurance paid to date; second, that the plaintiff forthwith effects insurance in the joint names of the tenant, landlord and mortgagee, and that the plaintiff should pay the costs of this action. Of course, as I am not dealing in this particular case with the question of relief against forfeiture, it is my view that the plaintiff is entitled to judgment for the relief claimed in the statement of claim.

Judgment was given in favour of the plaintiff with costs. It was granted a declaration, as sought in its statement of claim, an order for possession and an inquiry as to damages.

A
Chancery Division

June 19 1990

(Before Mr Justice MILLETT)

ROPEMAKER PROPERTIES LTD v NOONHAVEN LTD (No 2)

Estates Gazette March 9 1991

[1991] 09 EG 125

Landlord and tenant — Forfeiture — Premises found to have B been used for purposes of prostitution — Although breaches of covenant had been established, the present judge, in his judgment of January 27 1989, granted the defendant owners of the premises relief against forfeiture on terms — One of the terms required by the judge was that the defendants should enter into a deed of variation of the user clause in order to exclude hostesses from the premises — Agreement on the wording of the deed caused difficulties and led to the present application to the judge under the liberty to apply — In requiring a deed of variation the judge was concerned to relieve the plaintiffs of the need to prove knowledge on the part of the defendants that the girls were offering their services C for sex

In the course of the discussions on the proposed deed of variation the original wording suggested by the landlords was objected to as being too wide and a modification suggested by the landlords was also criticised as involving the risk of forfeiture if any female visitor unknown to them were, privately and entirely for her own benefit, to proposition a customer — The tenants then, after some discussion and amendment, put forward a revised version and although the landlords were not entirely happy with this, as it involved proof of some knowledge on the part of the tenant or D management of the club of the activities of the girl in question, the judge decided to approve it as a term for granting relief — He pointed out that under the provisions of the lease the landlords had to prove knowledge not only of the presence of the hostesses but also of the fact that they were acting as prostitutes — Under the draft suggested by the tenants it would be sufficient to prove knowledge of the fact that the girls were acting as hostesses, that is to say providing their company, however innocently, for reward

The wording proposed by the defendant tenants was accordingly approved as satisfying the term of relief relating E to the deed of variation

The following case is referred to in this report.

Ropemaker Properties Ltd v *Noonhaven Ltd* [1989] 2 EGLR 50; [1989] 34 EG 39

This was an application by the parties, Ropemaker Properties Ltd and Noonhaven Ltd, under the liberty to apply granted by Millett J in his judgment delivered on January 27 1989 in the case of *Ropemaker Properties Ltd* v *Noonhaven Ltd* reported at [1989] 2 EGLR 50; [1989] 34 EG 39. The application concerned the deed of variation required by the judge as one of the terms of the relief from forfeiture F granted to Noonhaven Ltd.

Derek Wood QC (instructed by Linklaters & Paines) appeared on behalf of Ropemaker Properties Ltd; Jonathan Gaunt (instructed by Rubinstein Callingham Polden & Gale) represented Noonhaven Ltd.

Giving judgment, **MILLETT J** said: This is an application under a liberty to apply to resolve a difficulty which has arisen in carrying out the terms of an order made following a judgment which I delivered at the conclusion of the action in January 1989.

The action was a claim for forfeiture of a lease of West End premises for breach of the user covenant. The premises were used as hostess clubs. The user clause of the lease was contained in clause 3(12)(a) and (b) and was in these terms:

> (a) Not knowingly to use or permit the demised premises or any part thereof G to be used for any illegal or immoral purpose and that the demised premises shall be used only in accordance with the provisions of sub-clause (b) of this sub-clause (12) and for no other purpose which use shall not be construed to be a breach of this clause . . .
>
> (b) Not to use or permit the basement ground and first floors of the demised premises or any part thereof to be used for any purpose other than as a high class restaurant/night club to include music dancing and cabaret and gambling . . .

The premises were used as hostess clubs. The plaintiffs alleged that the hostesses were propositioning men for the purposes of prostitution. The defendants denied that they were doing so or that they themselves were aware of the fact and denied that they were in breach of the covenant. H

When the action began the plaintiffs were under the impression that the defendants were themselves managing the hostess clubs and must have known of the activities of the hostesses because it was so widespread. In fact it turned out that the clubs were managed by an undertenant in one case and a licensee in another, and the defendants attempted to distance themselves from the activities of the clubs.

I found that the clubs were used for the purposes of prostitution and that the defendants, if not actually aware, must have been taken to have been aware of those activities. I found that it was a case of wilful blindness.

I therefore found that breaches of covenant were established but I granted relief from forfeiture upon terms. One of the terms which I required was that the defendants should enter into a deed of variation J of the user clause in order to exclude hostesses from the premises. My intention in imposing that requirement was to make it easier for the plaintiffs to obtain forfeiture should there be any repetition. In particular, I was concerned to relieve the plaintiffs from the need to prove knowledge on the part of the defendants that the girls were offering their services for sex.

In the course of drafting the deed of variation, however, the parties' respective solicitors were unable to agree on its terms. This has given rise to a critical difference which was properly brought back to me because it is of considerable importance to the parties. The landlords' proposed clause originally read as follows:

> That there shall not at any time be present on the demised premises hostesses, K that is to say, female persons accompanying or offering to accompany members of or visitors to the club for payment commission or reward.

That is now recognised as being too wide because there would be a breach of such a covenant if a customer entered the club accompanied by (say) a girl from a legitimate escort agency which had nothing to do with the management of the club or the defendants and which did not provide girls for sex, the girl having arranged to accompany the customer for money. The landlords now put forward a modified clause under which the words "accompanying or" are omitted. The result of the amended clause is that it would not establish a breach of the covenant merely to prove that a girl from an escort agency was on the premises; it would be necessary to prove that L there were present on the demised premises at any time girls who *while on the premises* were offering to accompany members of or visitors to the club for payment, commission or reward. This would avoid the need to prove knowledge on the part of the defendants.

The defendants object because they claim that it would be unfair to put them at risk of forfeiture if any female visitor unknown to them but privately and entirely for her own benefit were to proposition a customer. The only way in which in practice the defendants could ensure that there was no breach of such a covenant would be to insist that no unaccompanied girls should attend the premises at all. Even that might not be sufficient because it would be impossible to be absolutely certain that waitresses and bar girls properly employed by the management of the club did not on occasion proposition a M customer.

Accordingly, the defendants have put forward a clause which, after some discussion and amendment, reads:

> That there shall not at any time be present on the demised premises hostesses, that is to say, female persons whose presence on the demised premises is permitted or suffered by the tenant or the management with a view to their offering to accompany members of the club or visitors to the club for payment commission or reward.

The landlords object to that form of wording because they say it reintroduces in the definition of the hostess the concept that there must be some knowledge on the part of the tenant or management of the activities of the girl in question. That is true, but, in my judgment, it is not inappropriate because the essential feature of a hostess is that her presence is integral to the way in which the club is run and that her services are provided by the club as one of the facilities offered to customers.

It is not correct to say, as the landlords submit, that the formula put forward by the tenants leaves the landlords in no better position than they were before. Under the clause in the present lease, the landlords were faced with two difficulties. First, they had to prove knowledge on the part not only of the management of the club but of the defendants themselves. Second, they had to prove knowledge not only of the presence of hostesses but of the fact that they were acting as prostitutes. The clause proposed by the defendants obviates both those difficulties. It will still be necessary for the landlords to prove some knowledge on the part of either the tenant or the management of the club, but it will be sufficient for them to prove knowledge of the fact that the girls are acting as hostesses, that is to say, providing their company, however innocently, for reward. I doubt that this will in practice cause any great problem should this activity recommence. It should be easy enough to detect any waitresses or other employees of the club sitting at tables with customers. That is not the normal activity of a waitress or bar girl and it would be extremely implausible should she sit with a customer unless there was something in it for her. That fact would be known to the management of the club and the inference would be obvious. Equally, the appearance of a number of girls unaccompanied at the beginning of the evening waiting for customers to arrive and then approaching them and sitting with them would also lead to an almost irresistible inference, and if such an inference can be drawn by the court that would only reflect the obviousness of the situation to the management of the club.

In my judgment, therefore, the form of clause proposed by the defendants is one which I think should sufficiently deal with the problem and I will be satisfied with it as a term of giving relief.

Court of Appeal

February 11 1991

(Before Sir Nicolas BROWNE-WILKINSON V-C, Lord Justice PARKER and Lord Justice NICHOLLS)

BILLSON AND OTHERS v RESIDENTIAL APARTMENTS LTD

Estates Gazette May 11 1991 and May 18 1991

[1991] 18 EG 169 and [1991] 19 EG 122

Landlord and tenant — Breach of covenant — Forfeiture — Relief against forfeiture — Peaceable re-entry on premises by landlords — Whether statutory provisions for relief under section 146(2) of Law of Property Act 1925 excluded because landlords had taken possession — Whether relief obtainable under inherent equitable jurisdiction of the court — Division of opinion in Court of Appeal — Majority opinion that tenants were not entitled to claim relief either under section 146(2) or under the old equitable jurisdiction of the court — Nicholls LJ, dissenting, held that the equitable jurisdiction was not excluded — Tenants' appeal dismissed

The tenants in the present case, in clear breach of covenant not to make alterations to demised premises without previous consent in writing, proceeded to make such alterations without even seeking the landlords' consent — After fruitless remonstrations the landlords served a section 146 notice on the tenants — The tenants' attitude continued to be unsatisfactory and one morning the landlords peaceably re-entered the premises and changed the locks — A few hours later, however, the tenants' workmen regained possession and subsequently completed the works — The landlords issued a writ claiming possession and succeeded before Mummery J — He held that there had been a breach of covenant; that the landlords had not waived the right to forfeit; that a reasonable period had elapsed between the service of the section 146 notice and the forfeiture by re-entry; that the re-entry constituted a lawful and valid forfeiture; and that the court had no statutory or equitable jurisdiction to grant relief from forfeiture — The tenants appealed — The main issue in the Court of Appeal was the question of the court's jurisdiction to grant relief from forfeiture

The views of the majority of the court (the Vice-Chancellor and Parker LJ) were as follows — Although the landlords had been in possession, following their peaceable re-entry, for only a few hours, this precluded the court under section 146(2) from granting such relief from forfeiture as could have been granted if they had enforced their rights by action instead of by self-help — This was because it cannot be said of a landlord who has obtained possession that he "is proceeding" to enforce his rights, in the words of section 146(2) — He *is* not proceeding but *has* proceeded and has obtained his object — The authorities for this interpretation were: *Rogers* v *Rice, Quilter* v *Mapleson* and *Pakwood Transport Ltd* v *15 Beauchamp Place Ltd*

The tenants argued that, if the court had no statutory jurisdiction to grant relief against forfeiture, it could nevertheless be granted under the old equitable jurisdiction — It was this contention that caused the most difficulty in the case — From the early 19th century until the decision of the House of Lords in *Shiloh Spinners Ltd* v *Harding* in 1972 it was thought that the courts had no inherent jurisdiction to grant relief from forfeiture for "wilful" breach of covenant other than the covenant to pay rent or some other sum of money — The *Shiloh Spinners* case showed that this belief was an error — However, it was not a landlord and tenant case and the House of Lords recognised that in certain areas the inherent equitable jurisdiction might have been implicitly removed by Parliament when it conferred statutory powers of relief — The crucial question was whether the old equitable jurisdiction had been extinguished by legislation in the case of wilful breach of covenant between landlord and tenant

On this point the views of the majority and those of Nicholls LJ differed — In three recent cases the tenant's claim to such equitable relief had been considered — In two, *Official Custodian for Charities* v *Parway Estates Developments Ltd* and *Smith* v *Metropolitan City Properties Ltd*, the claim was rejected — In one, *Abbey National Building Society* v *Maybeech Ltd*, it was accepted (the judge was Nicholls J) — In the present appeal the majority came reluctantly to the conclusion that the inherent equitable jurisdiction as between landlords and tenants to relieve against forfeiture from wilful breach of covenant (other than a covenant to pay rent) had been extinguished by reason of Parliament having legislated comprehensively in that field

Nicholls LJ, in his dissenting judgment, pointed out that the argument of the landlords in the present case meant that if the landlord chooses to effect the forfeiture by forcing his way into the premises he is in a more favourable position than if he had applied to the court for a possession order — "That cannot be right" — In view of the unsatisfactory consequences of such decisions as *Official Custodian for Charities* v *Parway Estates Developments Ltd*, Nicholls LJ suggested that the interpretation adopted in *Barrow* v *Isaacs & Son* and *Gray* v *Bonsall* should be preferred — He would have held that the court in the present case had jurisdiction to entertain the tenants' application for relief

In accordance with the views of the majority, the tenants' appeal was dismissed — Leave to appeal to the House of Lords was given

The following cases are referred to in this report.

Abbey National Building Society v *Maybeech Ltd* [1985] Ch 190; [1984] 3 WLR 793; [1984] 3 All ER 262; [1984] EGD 479; (1984) 271 EG 995

Barrow v *Isaacs & Son* [1891] 1 QB 417

Bracebridge v *Buckley* (1816) 2 Price 200

Burt v *Gray* [1891] 2 QB 98

Egerton v *Jones* [1939] 2 KB 702
Graf v *Hope Building Corporation* (1930) 254 NY 1
Gray v *Bonsall* [1904] 1 KB 601
Hill v *Barclay* (1811) 18 Ves 56
Howard v *Fanshawe* [1895] 2 Ch 581
Lovelock v *Margo* [1963] 2 QB 786; [1963] 2 WLR 794; [1963] 2 All ER 13, CA
Nash v *Earl of Derby* (1705) 2 Vern 537
Nind v *Nineteenth Century Building Society* [1894] 2 QB 226
Official Custodian for Charities v *Parway Estates Developments Ltd* [1985] Ch 151; [1984] 3 WLR 525; [1984] 3 All ER 679; (1984) 48 P&CR 125; 270 EG 1077, CA
Pakwood Transport Ltd v *15 Beauchamp Place Ltd* (1977) 36 P&CR 112; [1978] EGD 233; 245 EG 309, CA
Quilter v *Mapleson* (1882) 9 QBD 672
Rexdale Investments Ltd v *Gibson* [1967] 1 Ont LR 251
Rogers v *Rice* [1892] 2 Ch 170; (1892) 8 TLR 511
Sanders v *Pope* (1806) 12 Ves 282
Shiloh Spinners Ltd v *Harding* [1973] AC 691; [1973] 2 WLR 28; [1973] 1 All ER 90; (1973) 25 P&CR 48, HL
Smith v *Metropolitan City Properties Ltd* [1986] 1 EGLR 52; (1985) 277 EG 753
Thatcher v *C H Pearce & Sons (Contractors) Ltd* [1968] 1 WLR 748; (1967) 19 P&CR 682
Webber v *Smith* (1689) 2 Vern 103

This was an appeal by the tenants, Residential Apartments Ltd, from the decision of Mummery J granting the landlords, Hew Richard Dalrymple Billson, Charles George Findlay, John de Cardonnel Elmsall Findlay and Dr Robert Francis Ker, possession of premises at 17 Gledhow Gardens, Earl's Court, London SW5.

Michael Driscoll (instructed by Gouldens) appeared on behalf of the appellants); Robert Reid QC and Gordon Bennett (instructed by Frere Cholmeley) represented the respondents.

Giving judgment, **Sir Nicolas Browne-Wilkinson V-C** said: This appeal raises a number of points relating to alleged breaches of covenant by the defendants, as assignees of the residue of a term of 33 years from March 25 1964 granted by a lease dated July 15 1964. The property is 17 Gledhow Gardens, London SW5. The plaintiffs are the freehold reversioners who, as is now common ground, terminated the lease by peaceable re-entry.

The original lessees were two brothers, Edwin and Leslie Rogers, but the term became vested in Leslie Rogers alone. The lease is rather unusual. The parcels clause describes the property as having been "used and occupied for letting in furnished service rooms or service suites". The original lessees convenanted by clause (c) to carry out certain specified alterations to the property and further covenanted:

(f) Subject to the provisions of Clause (c) hereof not to make any alteration in or addition to the demised premises or any part thereof without the consent in writing of the Lessors first had and obtained
(g) To allow the Lessors and all persons authorised by them at all reasonable times to enter upon the demised premises or any part thereof and inspect or examine the same or the condition thereof
(h) [Not to contravene the planning laws] . . .
(m) During the personal occupation of the demised premises by the Lessees Edwin Henry Rogers and Leslie George Rogers or the survivor of them to use and occupy the demised premises for letting in furnished service rooms or service suites . . .
(s) So soon as the survivor of the said Edwin Henry Rogers and Leslie George Rogers cease personally to occupy the demised premises the consent to the letting of the demised premises in furnished service rooms or service suites shall cease until the Lessors shall upon application being made to them in writing give their written consent to the continuance of such user hereinbefore authorised by a successor in title of the said Edwin Henry Rogers and Leslie George Rogers . . .
(u) During the last seven years of the said term not to assign underlet or part with the possession of the demised premises or any part thereof without the previous consent in writing of the Lessors such consent not to be withheld in the case of a reputable and responsible assignee or underlessee being proposed by the Lessees.

On May 3 1989 Leslie Rogers assigned the residue of the term to the defendants in consideration of the payment of £280,000. The defendants planned to use the property for letting in self-contained furnished suites, a proposed user which would manifestly breach covenant (s) unless the plaintiffs' consent was obtained. In pursuance of their desire, on the day after the completion of the assignment, May 4 1989, the defendants embarked on major works of reconstruction in order to convert the furnished accommodation into self-contained flatlets. This involved major structural work as well as comprehensive refurbishment. The defendants neither sought nor were given the consent of the plaintiffs to such works which, on the face of it, involved plain breaches of covenant (f). The works were done at great speed, there being as many as 52 people working on the property from 8am to 8pm at one stage. The works were completed in mid-August 1989 at a cost to the defendants of £260,000 to £375,000.

During the course of the negotiations for the assignment of the term in 1988 and early 1989, the plaintiffs' solicitors made it absolutely clear to the defendants that: (i) the defendants could not use the property for letting in furnished suites without the plaintiffs' consent; (ii) the plaintiffs' written consent to alterations was required and was not subject to any proviso as to reasonableness; (iii) the plaintiffs would consider consenting to the continued use of the property for furnished lettings but subject to certain specific conditions, including a condition that:

No alterations shall be carried out without our Clients' consent and any alterations will be strictly subject to the Trustees' surveyors being satisfied as to the proposed work. A formal licence will be required and the works will be supervised by the Trustees' building surveyor. No lettings may commence until all works have been completed to the satisfaction of the Trustees' surveyors.

One of the mysteries of the case is that, despite that clear warning, the defendants proceeded to start the extensive works I have mentioned without asking for any consent from the plaintiffs either to carry out the alterations or to use the property for furnished lettings. At least part of the explanation appears to be that the defendants and their then solicitors, Piper Smith & Basham (who are not acting for the defendants in these proceedings), were in a state of some confusion as to the applicable law. Under section 19(2) of the Landlord and Tenant Act 1927 any covenant against making alterations which are improvements without the consent of the landlord is subject to a proviso that the landlord's consent shall not be unreasonably withheld. In Part I of the same Act there are different provisions relating to improvements by tenants. If a tenant serves on the landlord notice of his intention to make improvements, together with a specification and plan showing the proposed improvements, and the landlord objects, the court can certify the improvements to be proper and the tenant can thereafter lawfully effect such improvements: section 3(4). The defendants seem to have become hopelessly confused between these two sets of provisions. They never asked for the landlords' consent under section 19(2); they never provided a specification of the proposed works or applied to the court under section 3(4).

By a letter dated May 3 1989 the defendants informed Frere Cholmeley (the plaintiffs' solicitors) of the assignment to the defendants and of immediate steps being taken to "restore the property to the decorative order and condition required by the lease" in the course of which internal improvements would be effected. "As a matter of courtesy" they supplied them with various plans of the proposed works in a reduced form and confirmed that compensation for those matters would not be sought by their clients at the end of the lease in accordance with section 19(2) of the Landlord and Tenant Act 1927.

On May 17 1989 Frere Cholmeley wrote to the defendants pointing out that the letter dated May 3 had, in fact, been posted on May 11 and had been received by them only on May 16. They drew attention to the covenant against alterations and said that no alteration or addition should be carried out until formal written licence had been obtained and that work of this nature must cease immediately. In subsequent correspondence, Frere Cholmeley unsuccessfully made repeated attempts to persuade the defendants to submit an application for written consent to the alterations in the normal way. In addition, visits were made to the site by the plaintiffs' surveyors to ascertain what works were being carried out to the property. The surveyors requested specifications, calculations, justifications and full-size working drawings, but these, though promised, were never provided. Instead, the defendants pressed on with the work. Serious reservations were expressed by the plaintiffs' structural engineer about the works.

On July 4 1989, a section 146 notice was served on the defendants. It referred, *inter alia*, to the covenant in clause (f) of the lease and specified the alleged breach in these terms:

The lessee has made and continues to make alterations to the premises for which the Trustees' consent has not been sought or given.

The plaintiffs then requested the defendants "within a reasonable time to remedy the said breaches in so far as they are capable of remedy".

In the letter accompanying the notice, Frere Cholmeley expressed the view that the defendants' breaches were irremediable but added that, as the plaintiffs were anxious to establish precisely what had been done to the property, their surveyors wished to see all working drawings, specifications and other documents from which the scope and extent of the works could be accurately determined. They also reminded the defendants that consent had not been sought under clause (s) of the lease, and would not be given, to the use of the premises as "furnished service rooms or service suites". Reference is made to the general prohibition in clause (n) against the use of the property for any trade or business whatsoever.

On July 10 1989, Piper Smith & Basham sent to Frere Cholmeley what they described as "a counternotice". That counternotice alleged that the works were "improvements adding substantially to its value and accordingly the landlord is restrained from objecting thereto by virtue of section 19 of the Landlord and Tenant Act, 1927". It was contended that no remedy or damages were appropriate and the works were, in any event, remediable by reinstatement. It was also contended that the plans in respect of the proposed works had been supplied with the letter of May 3 and, to the extent that approval was not given in respect of those plans, "consent is being unreasonably withheld by the landlord". The letter concluded with a request for the section 146 notice to be withdrawn and, if that was not agreed, the landlords were required to take action through the courts.

Thereafter the defendants continued to press on with the works at the property. The plaintiffs' advisers concluded that the service of the notice and the correspondence had no effect on the defendants. Arrangements were therefore made to re-enter the property.

On July 18 at 6 am agents for the plaintiffs peaceably re-entered the property, which was vacant, and changed the locks. By 10 am the same day workmen engaged by the defendants had retaken possession of the property, as the plaintiffs say unlawfully. The works were thereafter completed by the defendants.

The plaintiffs issued the writ in this action on July 19 1989 claiming possession. The defendants denied breach of covenant, denied that the plaintiffs had determined the lease by peaceable re-entry and counterclaimed for relief against forfeiture.

The action came before Mummery J, against whose judgment this appeal is brought. The issues before the judge were as follows:

(1) Have the defendants committed a breach of the lessees' covenant (f) "not to make any alteration in or addition to the demised premises or any part thereof without the consent in writing of the lessors first had and obtained"? The judge held that they had.

(2) If so, have the plaintiffs waived their right to forfeit the lease in respect of that breach? The judge held that they had not.

(3) Did a reasonable period elapse between the service of the section 146 notice on July 4 and the claimed forfeiture by re-entry on July 18? The judge held that a reasonable period had elapsed.

(4) Was the re-entry effected by the plaintiffs on July 18 a lawful and valid forfeiture of the lease? The judge held that it was.

(5) If the lease was forfeited by re-entry on July 18 1989, does the court have any statutory or equitable jurisdiction to grant the defendants relief from forfeiture on their counterclaim? The judge held that there was no such jurisdiction.

(6) If there were any jurisdiction to grant relief from forfeiture, ought the court to exercise its discretion and, if so, on what terms and conditions? The judge held that if, contrary to his view, he had jurisdiction to grant relief, he would have granted it.

The defendants appeal against the judge's decision on all the issues save (4) and (6) above. The defendants do not appeal against the judge's finding that there was a lawful and effective re-entry on July 18.

Of the issues which remain, issue (5) (jurisdiction to grant relief from forfeiture) is both the most difficult and the most important. But I must first deal shortly with the other points.

(1) *Has there been a breach of covenant (f)*

The defendants' case is that the alterations constituted improvements; that, consequently, under the Landlord and Tenant Act 1927, section 19(2), the covenant against effecting alterations without the consent of the landlord was subject to a proviso that such consent should not be unreasonably withheld; that, although the defendants never asked for such consent, the plaintiffs had made clear in the letter of May 17 1989 that their consent would never be given and that such blanket refusal constituted an unreasonable withholding of consent.

As at present advised, I would have thought this argument failed at the first step: how could alterations designed to improve the suitability of the property for furnished letting constitute an "improvement" when the tenant had no right to use the property for such lettings? This point was not argued below and I express no concluded view on it.

The judge held that the argument failed, since the letter of May 17 1989, read in context, was not a blanket refusal. I agree. Before May 17 the plaintiffs had made it clear that they could not consider whether to give consent until plans and specifications had been provided. Before the assignment of the term to the defendants, the letter of September 1 1988 had indicated that consent to alterations might be given. In the letter of May 3 1989 the defendants did not ask for consent to make the alterations: they say they were informing the landlords as a matter of "courtesy". The letter of May 17 from the plaintiffs contains the following passages:

> Please note that no alteration or addition should be carried out until formal written Licence has been obtained and in the event that any work of this nature has commenced, it must cease immediately.

After referring to the proposed user of the property the letter ended:

> Our instructions will be to enforce any breach of the user covenant and there can be no question of Licence for Alterations being given.

These are the words relied upon as being an unequivocal refusal of consent to making alterations.

I agree with the judge that, read in that context, these words do not amount to an unequivocal refusal of consent to alterations. Given the extremely confusing nature of the defendants' letter of May 3, the words relied upon are at least as consistent with an assertion that no licence has in the past been given as with meaning that no licence will in the future be given. This makes it consistent with the earlier passage which I have read from the letter of May 17, which is inconsistent with a blanket refusal since it refers to the possibility of formal written licence being obtained in the future. The letter of May 17 was certainly not subsequently treated as having been a blanket refusal. For example, in a letter of June 1 1989 from the plaintiffs' solicitors to the defendants' solicitors, the plaintiffs' solicitors say "that our clients have not consented to the carrying out of work of alteration and will not do so until they have had a proper chance to consider that work".

In my judgment, it is impossible to spell out of this correspondence a blanket refusal to grant consent to the alterations in any event.

(2) *Waiver*

The defendants contend that the plaintiffs have waived their right to forfeit the lease on the grounds of the breach of covenant (f). Ultimately, their contention is based on a letter of June 30 1989 from the plaintiffs' solicitors which contains the following passage:

> Our Clients' surveyors are of the opinion that the works are not being carried out in a proper fashion, in consequence of which there is a serious danger of damage to our Clients' reversion. Your Clients have failed to comply with their direct requests and we must therefore inform you that our Clients are not prepared to allow your Clients to continue carrying out work without Licence for Alterations. Your Clients are in breach of covenant and we are instructed to require you to let us have an undertaking that your Clients will cease carrying out work to the property until they have submitted the required details to Pollock & Buchan and have received formal authority to proceed. Unless we have the undertaking requested by 12 noon on Monday, we will be issuing a Summons seeking an injunction against your Clients.

The defendants contend that this letter is an unambiguous election, with full knowledge of the breaches, to enforce the covenants in the lease by action and therefore constitutes a waiver of the breaches. The judge rejected this argument on two grounds. First, he held that, even if the letter constituted a waiver of any breaches made before June 30 1989, further works were done between June 30 and the service of the section 146 notice on July 4, as to which further works there was no possible waiver; second, he held that the letter of June 30 did not contain an unequivocal waiver.

As to the judge's first ground, Mr Driscoll, for the defendants, submitted that the weight of the evidence was against the judge's finding that further works had been done between June 30 and July 4. He took us to passages in the evidence which, he said, indicated that all structural alterations had been finished by June 30 and that only works of fitting out remained to be done thereafter. I am far from satisfied that the evidence showed any such thing: a letter from the landlords' agents dated June 27 1989 suggests that at that date alterations to the plumbing and drainage were still being done. More important, the defendants pleaded in para 10 of the defence:

A It is admitted that the Defendant continued to carry out the said works of alteration and improvement after service of the said notice but the majority of those works had been completed before service of the said notice.

In the light of this pleading, the exact date when any particular works were done was not in issue and, in my judgment, the judge was fully entitled to hold as he did that "the alterations to the property were being made as part of a continuing programme of process of alteration throughout the whole 14-week period commencing May 4 1989 and continuing down to August including the period between the letter of June 30 and the section 146 notice of July 4".

(3) *Did a reasonable time elapse between the service of the section 146 notice and the forfeiture?*

B Section 146(1) provides that a right of re-entry or forfeiture for breach of covenant is not to be enforceable until:

(1) a notice has been served specifying the breaches and, if such breaches are remediable, requiring them to be remedied, and

(2) "the lessee fails within a reasonable time thereafter, to remedy the breach".

The defendants contend that the breach was remediable by reinstating the property to its condition before the alterations; that such reinstatement would take some four months; therefore, re-entry on July 18 (ie only 14 days after the service of the section 146 notice) had not allowed a reasonable time.

The judge held, first, that since the alterations had been started C without *prior* consent of the plaintiffs the breach was irremediable. Second, he held that even if he were wrong on the first point, remedying the breach would consist not in doing the works of reinstatement but in stopping the works, submitting the necessary plans and specifications and then awaiting the giving or unreasonable withholding of consent.

I prefer to express no view on the judge's first ground for decision, beyond expressing some doubt as to whether he was right in holding that the breach was irremediable. In my judgment, even if the breach were capable of remedy the defendants' actions in continuing to press on with the work in defiance of the warnings and in defiance of the section 146 notice demonstrated that the defendants had no intention to remedy the breach at all. It is to be noted that the section 146 notice D itself does not have to limit a time within which the breach is to be remedied. All that the statute requires is that a reasonable time to remedy the breach must elapse between service of the notice and the exercise of the right of re-entry or forfeiture if the actions of the lessee make it clear that he is not proposing to remedy the breaches within a reasonable time, or indeed any time. In my judgment, a reasonable time must have elapsed for remedying the breaches once it is clear that he is not proposing to take the necessary steps to remedy the breach, but is committing further breaches. If this were not the case, what would the landlord's rights be if the defendant continues to commit the very breaches complained of by the section 146 notice after the date of its service? If he were to take proceedings to restrain further breaches of covenant, he would subsequently be faced with the E contention (such as that advanced by the defendants based on the threat of proceedings in the letter dated June 30 1989) that the landlord had waived his right to forfeit by seeking to enforce the covenants. The only effective remedy of a landlord, faced with intransigent behaviour such as that of the defendants in the present case, must be to forfeit the lease on the grounds that, whatever time was allowed, the defendant was showing no intention of remedying the breach at all.

In my judgment, given the stance adopted by the defendants in response to the section 146 notice, a reasonable time did elapse for the defendants to stop doing the work and either apply for the necessary consents or put forward proposals for remedying the breach.

F (5) *Jurisdiction to relieve against forfeiture*

(a) Under section 146(2)

On the judge's finding, the plaintiffs forfeited the lease by peaceable re-entry on July 18 1989. They were in possession for only four hours before the property was re-occupied by the defendants. Yet, it is said, this period of four hours' possession precludes the court from granting such relief from forfeiture as could have been granted had the plaintiffs enforced their rights by action instead of by self-help.

This contention flows from the wording of section 146(2), which confers the statutory jurisdiction to relieve from forfeiture "where a lessor *is proceeding*, by action or otherwise, to enforce such a right of re-entry or forfeiture". It has been established by Court of Appeal decision for nearly 100 years that once a landlord has obtained G possession of the premises under order of the court, the statutory right to relieve the tenant against forfeiture is lost, since it cannot thereafter be said that the lessor "is" proceeding to enforce his rights: he *has* proceeded: *Rogers* v *Rice* [1892] 2 Ch 170; *Quilter* v *Mapleson* (1882) 9 QBD 672.

Mr Driscoll, for the defendants, contends that those cases do not cover a case, such as the present, where the landlords have obtained possession not under order of the court but by peaceable re-entry. He submits that, in this case, the landlords, having again lost possession, are proceeding to enforce their right of re-entry by seeking possession under order of the court.

I have great sympathy with this contention and, were it not for authority, would uphold it. H As it seems to me, a lessor "is proceeding to enforce" his right of re-entry whenever he has to rely on such right in order to obtain or retain possession of the demised premises. If a lessor has obtained possession by way of execution of a court order and the former tenant thereafter evicts him or seeks to obtain possession by proceedings, the lessor does not need to rely on his right of forfeiture or re-entry at all: he relies solely on the order of the court which establishes finally that he is entitled to possession against the former tenant. But, where the lessor has not obtained an order of the court, in any proceedings between him and the former tenant the lessor is forced to rely on his right of re-entry to justify his possession. It does not seem to me an undue straining of language to say that, by such reliance, he is "proceeding . . . otherwise to enforce such a right J of re-entry or forfeiture".

But, in my judgment, whether the House of Lords may think such reasoning is correct, it is not open to us in this court so to hold. In *Quilter* v *Mapleson* all the members of the Court of Appeal expressed the view that under the statutory predecessor of section 146(2) (section 14 of the Conveyancing and Law of Property Act 1881) once there has been a re-entry, the statutory jurisdiction to relieve against forfeiture has gone. In *Rogers* v *Rice* those general dicta were adopted and applied in a case where relief was sought after a judgment for possession had been executed. In *Pakwood Transport Ltd* v *15 Beauchamp Place Ltd* (1977) 245 EG 309 the Court of Appeal held that a tenant could apply for relief against forfeiture as soon as a section 146 notice was served and before the landlord had K started proceedings. Their ground for so holding was that, in a case of peaceable re-entry, the right to relief against forfeiture would otherwise be illusory, since the tenant had no right to claim relief after peaceable re-entry. In my judgment, it was an essential step in the reasoning in *Pakwood* that, after peaceable re-entry, there is no statutory right to relief from forfeiture. Accordingly, that proposition was part of the *ratio decidendi* of the case and it is binding upon us.

(b) Under the inherent jurisdiction of the court

Mr Driscoll submitted that, even if there is no statutory jurisdiction to grant relief against forfeiture, relief can be granted under the old equitable jurisdiction of the court. L

There is no doubt that, before the intervention of the legislature, courts of equity did have some jurisdiction to relieve tenants against forfeiture for breach of covenant. The exact scope of that jurisdiction is a matter of doubt which I will consider shortly hereafter. But, until the decision of the House of Lords in *Shiloh Spinners Ltd* v *Harding* [1973] AC 691, the exact scope of such jurisdiction in relation to forfeiture of leases had not been an issue for over 150 years. There was no doubt that the courts of equity had jurisdiction to relieve for non-payment of rent. However, in relation to breaches of other covenants in the lease, from the early 19th century until 1972 it was thought that there was no equitable jurisdiction to relieve tenants from forfeiture for wilful breach of such covenants. M

Shiloh Spinners was concerned with relief against forfeiture for breach of a covenant to fence and support buildings. That covenant was entered into between the assignor and the assignee of a lease, ie the case was not concerned with covenants as between landlord and tenant. The House of Lords held that, although it had been assumed not to be the law, equity did, in fact, have jurisdiction to relieve against forfeiture for wilful breach of covenant other than a covenant for the payment of money "where the primary object of the bargain is to secure a stated result which can be effectively attained when the matter comes before the court, and where the forfeiture provision is added by way of security for the production of that result".

Although *Shiloh Spinners* was not itself concerned with covenants as between landlord and tenant, it has given rise to three decisions in which tenants, who were not within the statutory jurisdiction to relieve against forfeiture, have sought to rely on the old equitable jurisdiction. In the present case, the defendants submit that they can be given relief in equity, since it is established that in equity relief could be given even after the landlord had peaceably re-entered: *Webber* v *Smith* (1689) 2 Vern 103; *Nash* v *Earl of Derby* (1705) 2 Vern 537.

In each of the cases since *Shiloh Spinners*, the dispute has been whether the equitable jurisdiction to relieve from forfeiture for wilful breach of covenant in a lease has survived the legislation giving the statutory right to such relief, or whether such legislation has impliedly extinguished the old equitable jurisdiction in relation to leases. That is the prime question in the present case. But before turning to it I must deal shortly with the nature of the old equity jurisdiction.

The old equity jurisdiction

For present purposes, equity has exercised jurisdiction to relieve against forfeiture in three categories of case, namely:

(a) under the general principles of equity that the court will relieve a party from the consequences of fraud, accident or mistake;

(b) in cases of non-payment of rent or other sums certain;

(c) for wilful breach of covenants, other than covenants for the payment of rent, where adequate compensation can be made for the breach.

I will deal with each of these in turn.

(a) *The general equity jurisdiction*

The general jurisdiction of equity to relieve from the consequences of fraud, accident and mistake applied as much to leases as to any other legal relationship. The ambit of this jurisdiction is narrowly confined by authority. In particular, relief on the ground of mistake is not granted where such mistake is due to the negligence of the party mistaken or of his solicitor: *Barrow* v *Isaacs & Son* [1891] 1 QB 417. It is not suggested in the present case that the defendants are entitled to relief on the grounds of mistake. In my judgment, in an appropriate case relief can still be granted in this category of case even where the fraud, accident or mistake arises in the context of a lease. In *Barrow* v *Isaacs & Son* the Court of Appeal recognised that there was such jurisdiction in relation to the breach of a covenant against underletting but declined to hold that the mistake there made was a relevant one. This case was quoted in *Shiloh Spinners* with approval.

(b) *Non-payment of rent*

There is no question of the equitable jurisdiction having been impliedly removed. Save in one instance, relief against forfeiture for non-payment of rent has been regulated by Parliament quite separately from relief for breach of other covenants. The statutory history was fully deployed before us and can be summarised as follows.

The Landlord and Tenant Act 1730 recited that landlords had been adversely affected by the courts of equity granting relief from forfeiture for non-payment of rent long after the lease had been forfeited at law. Section 2 of the Act provided that any bill for relief from forfeiture had to be filed within six months after judgment for execution was recovered in the common law courts, failing which the equitable jurisdiction to relieve was barred. If the bill in equity was filed, the tenant was bound to pay into court the full amount of arrears of rent and costs. Moreover, under section 4, if the tenant tendered such arrears of rent and costs, such tender stayed the proceedings at common law. Finally, section 4 provided that, if relief was given in equity, the original lease revived without the necessity for the grant of a new lease.

The 1730 Act was repealed and in effect re-enacted by sections 210-212 of the Common Law Procedure Act 1852. Unless all arrears of rent and costs are paid within six months of the order for possession at law, all rights to relief at law and in equity are barred. Section 211 recognises the continuation of the right to apply for relief in equity but re-enacts the requirement that the tenant's right to relief in equity is to be barred unless he pays into court the arrears and costs claimed. Section 212 re-enacts the provisions whereby proceedings at law are stayed if payment or tender of arrears and costs is made. The 1852 Act is still the basic statute regulating relief from forfeiture for non-payment of rent.

By the Common Law Procedure Act 1860 the jurisdiction of courts of equity to relieve against forfeiture for non-payment of rent was conferred on the common law courts.

By the County Court Amendment Act 1856, the County Court Acts 1934 and 1959 and, currently, sections 138 and 139 of the County Courts Act 1984, there is conferred on the county court jurisdiction, in effect, to relieve against forfeiture for non-payment of rent on payment within a specified period of rent arrears and costs, failing which all rights to relief are barred. Section 139(2) of the 1984 Act (re-enacting section 191(3) of the 1959 Act) expressly confers on the county court power to relieve against forfeiture for non-payment of rent where the lease has been forfeited by peaceable re-entry. The application for relief has to be made within six months from the date of peaceable re-entry.

It has been decided by authority that both the High Court (under its inherent equitable jurisdiction) and the county court (under section 139 of the 1984 Act and its predecessors) can relieve against forfeiture effected by peaceable re-entry after such re-entry has occurred and without limit of time: *Howard* v *Fanshawe* [1895] 2 Ch 581; *Thatcher* v *C H Pearce & Sons (Contractors) Ltd* [1968] 1 WLR 748; *Lovelock* v *Margo* [1963] 2 QB 786.

In summary, the basic jurisdiction to relieve from forfeiture for non-payment of rent is the old equitable jurisdiction; that equitable jurisdiction (and the statutory application thereof in the county courts) is subject to a separate code of statutory provisions modifying and limiting the equitable jurisdiction in certain respects, particularly in relation to time; however, the old equitable jurisdiction to relieve, without limit of time, continues to apply where there has been a forfeiture by peaceable re-entry.

(c) Breach of covenant other than for payment of rent

At the beginning of the 19th century there were two conflicting views as to the scope of the equitable right to relieve from forfeiture where (i) the breach was "wilful" (ie not capable of being relieved against on the ground that the breach was due to fraud, accident or mistake), (ii) the covenant was inserted primarily to achieve a stated object (eg repair or non-assignment), (iii) the forfeiture provision was primarily intended as a means of securing the attainment of that object, and (iv) by imposing the necessary terms the court could broadly achieve the attainment of that object without insisting on a forfeiture. Lord Erskine LC asserted the existence of jurisdiction to relieve in such a case: *Sanders* v *Pope* (1806) 12 Ves 282; Lord Eldon LC rejected such jurisdiction: *Hill* v *Barclay* (1811) 18 Ves 56.

In *Shiloh Spinners* (at p 722G) Lord Wilberforce stated that Lord Erskine's view was "poorly regarded in Lincoln's Inn" and that Lord Eldon's view came to be accepted as correct. I would venture to add that the decision of the Court of Exchequer in *Bracebridge* v *Buckley* (1816) 2 Price 200 appears to have decided that the narrow view of Lord Eldon was correct and that either there was no equitable jurisdiction to relieve in such circumstances or, if such jurisdiction theoretically existed, that it would not be exercised. It has not been explained how, in 1816, the Court of Exchequer came to be ruling upon equitable doctrines.

The narrower view which triumphed was based on two grounds: first, that equity had no general right to relieve a man from his bargain; second, that in the case, for instance, of a breach of covenant to repair, the court was unable to supervise the carrying out of the repairs which it would necessarily impose as a condition of relief.

From the early 19th century until *Shiloh Spinners* in 1973 it was thought that the court had no inherent jurisdiction to grant relief from forfeiture for "wilful" breach of covenant other than the covenant to pay rent or some other specified sum of money. This is a point of central importance in the present case. The decision in *Shiloh Spinners* disclosed, retrospectively, that the view so held for 150 years had been erroneous. The court had originally had an inherent jurisdiction to relieve in such cases (though such jurisdiction was to be exercised with great caution). However, the House of Lords recognised that, in certain areas, such inherent jurisdiction might have been implicitly removed by Parliament in conferring on the court statutory powers of relief.

In *Shiloh Spinners* it was argued that, although the covenant and forfeiture provision there in question did not arise as between landlord and tenant, the statutory provisions for relief from forfeiture as between landlord and tenant contained in section 146 of the Law of Property Act 1925 and its predecessors had impliedly removed the whole equitable jurisdiction to relieve in such cases. The

A House of Lords rejected this argument. After mentioning some of the statutes dealing with relief in cases between landlord and tenant, Lord Wilberforce said (at p 725 at A-C):

> In my opinion where the courts have established a general principle of law or equity, and the legislature steps in with particular legislation in a particular area, it must, unless showing a contrary intention, be taken to have left cases outside that area where they were under the influence of the general law. To suppose otherwise involves the conclusion that an existing jurisdiction has been cut down by implication, by an enactment moreover which is positive in character (for it amplifies the jurisdiction in cases of leases) rather than negative. That legislation did not have this effect was the view of Kay LJ in *Barrow* v *Isaacs & Son* . . . when he held that covenants against assigning — excluded from the Conveyancing Act 1881 — were left to be dealt with B according to the ordinary law.

The question, therefore, is whether the legislature has stepped into the area of relief against forfeiture for wilful breach of covenant (other than to pay rent) as between landlord and tenant so as to have excluded the old equitable jurisdiction to relieve in that area.

Statutory history

By the middle of the 19th century it had been frequently decided that equity had no jurisdiction to relieve against forfeiture for wilful breach of covenant to insure. The legislature, therefore, enacted section 4 of the Law of Property and Trustees Relief Amendment Act 1859, which provided that a Court of Equity should have power to relieve against forfeiture in such a case, subject to strict limitations. C The words "a Court of Equity shall have power to relieve" necessarily implies that Parliament considered that it was conferring on the Court of Equity a jurisdiction which it otherwise lacked. This provision was repealed and re-enacted by the Common Law Procedure Act, 1860, section 2.

A much wider statutory power to relieve against forfeiture in such cases was conferred by the Conveyancing and Law of Property Act 1881, section 14, which is the basis of the 1925 Act. It contains most of the features of section 146, namely, the requirement of the landlord to serve a notice before forfeiting the lease and a power for the court to grant relief "where a lessor is proceeding by action or otherwise" to enforce a forfeiture. However, the 1881 Act did not extend to covenants against assignment and underletting or to certain D covenants in mining leases; nor did it contain any provision for relief against forfeiture on bankruptcy. Section 14(8) expressly provided that the section did not apply to covenants for non-payment of rent. The express provisions for relief in cases of insurance covenants contained in section 2 of the Common Law Procedure Act were repealed.

The Conveyancing and Law of Property Act 1892, section 2, for the first time conferred a limited power on the court to relieve against forfeiture for bankruptcy or insolvency (similar to that now contained in section 146(9) and (10)). Further, by section 4 jurisdiction was conferred on the court to relieve a subtenant of the whole or part of the demised premises, it having been held in *Burt* v E *Gray* [1891] 2 QB 98 that, in a case of breach of covenant to repair, relief could not be granted to a subtenant of part of the demised premises under section 14 of the 1881 Act.

Finally, section 14 of the 1881 Act and sections 2 and 4 of the 1892 Act were repealed and re-enacted with amendments in section 146 of the Law of Property Act 1925. The most notable amendment was the inclusion for the first time of power to relieve against forfeiture for breach of covenant to assign. The provisions of section 146 are too well known to require me to set them out in full. The relationship between the subsections of section 146 and their statutory predecessors is as follows:

F

Subsections of 146	*Predecessor*
(1)	1881 Act, section 14(1)
(2)	1881 Act, section 14(2)
(3)	1892 Act, section 2
(4)	1892 Act, section 4
(5)	1881 Act, section 14(3)
(6)	1881 Act, section 14(4)
(7)	1881 Act, section 14(5)
(8)(ii)	1881 Act, section 14(6)(ii)
(9) and (10)	1892 Act, sections 9(2) and (3)
(11)	1881 Act, section 14(8)

It will be apparent that, hitherto, Parliament had treated relief against forfeiture for non-payment of rent wholly separately from relief against forfeiture for breach of other covenants. However, it was held in *Gray* v *Bonsall* [1904] 1 KB 601 that section 4 of the 1892 Act applied so as to permit relief to be given to undertenants not only G against breach of covenant other than for payment of rent but also against breach of covenant to pay rent. This decision accounts for the one exception to the rule that in dealing with relief against forfeiture Parliament treated separately breach of covenant to pay rent and breaches of other covenants. As a result of the decision in *Gray* v *Bonsall*, section 146(4) of the Law of Property Act 1925 expressly applies to relief to undertenants for breach of covenant to pay rent; the rest of section 146 applies to covenants other than for payment of rent: see section 146(11).

Decisions since Shiloh Spinners

The House of Lords having opened up the possibility that the old equitable jurisdiction to relieve against forfeiture for wilful breach H was wider than had previously been thought, three cases have recently been decided in which the tenant claimed that, although there was no statutory jurisdiction to relieve him, relief could be granted to him under the court's old inherent equitable jurisdiction.

In *Official Custodian for Charities* v *Parway Estates Developments Ltd* [1985] Ch 151*, the tenant company had gone into liquidation. The tenant company failed either to sell the demised property or to start proceedings for relief under section 146 within a year of the liquidation. Accordingly, relief under section 146 was precluded by section 146(10). The tenant claimed to be entitled to relief against forfeiture under the inherent jurisdiction, but this was rejected by the Court of Appeal. Dillon LJ, after quoting the passage J from Lord Wilberforce's speech in *Shiloh* which I have set out above, held that the legislature had "stepped into" the area of law concerning relief against forfeiture in the event of bankruptcy and that, accordingly, any inherent equitable jurisdiction in that area was excluded. He pointed out that there was no point in section 146(10) imposing a one-year time-limit if the court was to have an inherent jurisdiction exercisable at any time, both before and after the expiry of that year, to relieve from forfeiture. However, I do not understand him as having decided that section 146 as a whole is inconsistent with there being a continuing equity jurisdiction to relieve against forfeiture for wilful breach of covenants other than payment of rent.

On the same day as the *Parway* decision, Nicholls J (as he then was) gave judgment in *Abbey National Building Society* v *Maybeech Ltd* K [1985] Ch 190†. The building society was mortgagee of a lease, which, in addition to a covenant for the payment of rent, contained a covenant to pay maintenance contributions. The tenant was in arrears with the payment both of rent and the maintenance contributions. The landlord forfeited the lease and obtained possession under order of the court, without notifying the building society of what he was doing. When the building society discovered the position, it applied for relief against forfeiture under the inherent jurisdiction of the court. The judge granted such relief. It should be noted that, in relation to non-payment of contributions, relief could not be given under the provisions relating to non-payment of rent: relief could be granted only under section 146. Since the landlord had L obtained possession, such relief under section 146 was no longer obtainable. Broadly, the judge held that the legislature had not "stepped into" the whole area of relief against forfeiture for breach of tenants' covenants and that the old equitable jurisdiction to relieve from failure to pay a sum of money (which was admitted) could still be exercised. I will consider his reasoning further below.

The third case is the decision of Walton J in *Smith* v *Metropolitan City Properties Ltd* [1986] 1 EGLR 52. The landlords had peaceably re-entered for breach of repairing covenants. The tenant thereafter applied for relief against forfeiture under the inherent jurisdiction. The judge applied Lord Wilberforce's test to section 146 as a whole and held that the equitable jurisdiction to relieve for breach of covenant as between landlord and tenant (other than for payment of M rent) no longer existed. He held that the reasoning applied in *Parway* to subsections (9) and (10) of section 146 applied equally to the rest of the section and declined to follow the *Abbey National* decision.

We therefore have to decide which of these two views is correct.

Conclusion

I have found this a very difficult case to decide, not least because common sense and justice both require that the tenant's right to relief ought not to depend upon whether the forfeiture is enforced by action or by peaceable re-entry: the tenant ought to have broadly the same rights whichever procedure the landlord adopts. That is

*Editor's note: Also reported at (1984) 270 EG 1077.
†Editor's note: Also reported at (1984) 271 EG 995.

A particularly the case here, where the plaintiff landlords obtained possession for only some four hours. However, I have reluctantly come to the conclusion that the inherent equitable jurisdiction as between landlord and tenant to relieve from forfeiture for wilful breach of covenant (other than a covenant for the payment of rent) has been extinguished by reason of Parliament having legislated comprehensively in that field.

First, the question is whether Parliament has impliedly shown an intention to exclude the inherent jurisdiction. In one sense it is impossible logically to find any such intention: since, at the time all the legislation was passed, no one (including Parliament) thought that the courts had any such jurisdiction, Parliament could not have intended to exclude such jurisdiction. However, in my judgment,
B that is to take too narrow a view. Since, at the relevant times, no one thought that there was an inherent jurisdiction, the intention of Parliament in legislating must have been to lay down a comprehensive and exclusive code setting out the circumstances in which relief could be given. If anyone had asked in 1859 (when the statutory power to give relief against forfeiture for failure to insure was introduced) whether Parliament intended that relief could also be given in cases of wilful breach of covenants to repair or against assignment, the answer must surely have been "no". If similar questions were asked in 1881 concerning relief against forfeiture for bankruptcy or for wilful breach of a covenant against assignment, the answer would surely have been the same. In 1925 could
C Parliament have contemplated that it was leaving a jurisdiction to relieve against forfeiture for wilful breach of a covenant against assignment committed before 1925 or for wilful breach of certain covenants in mining leases? (See section 146(8)). The subsequent disclosure of the true position by the House of Lords in *Shiloh Spinners* cannot alter the effect of legislation passed in ignorance of the true state of the law.

Second, as I read him, Lord Wilberforce in *Shiloh Spinners* was of the view that in the area of forfeiture for breach of covenant between landlord and tenant, Parliament had stepped in and laid down an exclusive code, save to the extent that equity retained its jurisdiction under category (1) above, ie in cases of fraud, accident or mistake. Lord Wilberforce said (at p 724A-B):

D
> As regards covenants to pay rent, in spite of Lord Eldon LC's reservations, the matter has, subject to qualifications which need not be discussed, been taken over by statute, first by 4 Geo 2 c 28, then by later Acts leading up to the Law of Property Act 1925. The same is true of covenants to insure and other covenants in leases.

If, in Lord Wilberforce's words, Parliament has "taken over" in the area of covenants between landlord and tenant, Parliament must, by definition, have stepped in "with particular legislation in a particular area" (see *per* Lord Wilberforce at p 725A). In my judgment, Lord Wilberforce, though not directly concerned with the question, was expressing the view that the statutory jurisdiction to relieve in this field was now exclusive.

E Third, there are decisions of this court that once a landlord has re-entered there is no right to relief: see *Egerton* v *Jones* [1939] 2 KB 702; *Pakwood (supra)*. Although the possible existence of the inherent jurisdiction was not argued in those cases, it cannot fairly be said that those decisions were reached *per incuriam*.

Fourth, I agree with Walton J that the approach to section 146(9) and (10) adopted by this court in *Parway* is equally applicable to the rest of the section. If Parliament intended the court to be able to grant relief even when the landlord is *not* proceeding to enforce his rights but has enforced them, what was the reason for including in section 146(2) and (4) the limitation that the powers of relief could be exercised only when the landlord "is" proceeding?

What were the factors which led Nicholls J in the *Abbey National*
F case to hold that the inherent jurisdiction remained? As I read the judgment there are three reasons. First, at p 201E-H he relied on the fact that, in cases of relief from forfeiture for non-payment of rent, it has been held that the court continues to have its inherent jurisdiction to relieve at any time where there has been a peaceable re-entry, notwithstanding the time-limits specified in the Common Law Procedures Act 1852. But, in my judgment, analogies drawn from the non-payment of rent cases are dangerous. As I have sought to show, Parliament has treated non-payment of rent cases on a completely different basis from breaches of other covenants. In relation to relief from forfeiture for non-payment of rent, Parliament has expressly recognised the continuing existence of the equitable jurisdiction. In such cases, all relief against forfeiture is still given under the inherent jurisdiction. Parliament's intervention in G
that area has been limited to providing limits on the time and circumstances when such relief can be granted by the court under its inherent jurisdiction and a statutory bar to obtaining an order for possession at all by means of the tender of arrears and costs. Therefore, in non-payment of rent cases, unless the case falls within an express statutory limitation on the inherent jurisdiction, the court's jurisdiction remains untrammelled and unfettered. In section 191(3) of the County Courts Act 1959 Parliament has expressly provided that in cases of peaceable re-entry the county court is to have the same jurisdiction as the High Court (ie the inherent jurisdiction) and there is no doubt that the High Court's inherent jurisdiction survived peaceable re-entry. To argue from the separate H
code applying to covenants to pay rent to the code regulating relief from forfeiture for breach of other covenants (which contains no express or implied reference to the inherent jurisdiction at all) is, in my judgment, not right.

Second, Nicholls J relied (at p 202H) on the decision in *Gray* v *Bonsall* [1904] 1 KB 601, where the question was whether jurisdiction to relieve for non-payment of rent existed only under section 1 of the 1860 Act or was also exercisable under section 4 of the 1892 Act. The Court of Appeal held that both statutory jurisdictions were exercisable. Again, that case was dealing with non-payment of rent, a dangerous analogy. More important, it was a case on the co-existence of two statutory jurisdictions: there was no question of the interaction between an express statutory provision and the inherent J
jurisdiction.

Third, Nicholls J relied on *Nind* v *Nineteenth Century Building Society* [1894] 2 QB 226. Again, in my judgment, that case was purely concerned with the question of statutory construction as to whether two statutory provisions could co-exist. It did not touch upon the interaction between statute and the inherent jurisdiction.

I have had the advantage of reading the judgment in draft of Nicholls LJ. I fully agree with his powerful exposition of the injustice and undesirable consequences of reaching the conclusion that, in cases far more meritorious than this, the tenant cannot be relieved from forfeiture for wilful breach of covenant in the case of peaceable re-entry. But, in my judgment, it is not our function to fill the gaps in the statutory framework by judicial decision. The remedy lies, in my K
view, first in asking the House of Lords to consider whether decisions in *Rogers* v *Rice* and *Pakwood* are correct and, if they are, second, for Parliament itself to correct the position.

On the more technical aspects, the difference between my view and that of Nicholls LJ comes down to the decision in *Barrow* v *Isaacs*. In that case the Court of Appeal undoubtedly held that the general equitable jurisdiction to relieve from the consequences of fraud, accident and mistake (my category (a) above) still applied to breach of covenants against assignment in leases even though relief for breach of such covenants was expressly excluded from the statutory jurisdiction to relieve conferred by the 1881 Act. But that is not the point at issue in this case. In my judgment, the general equitable L
jurisdiction to relieve from the consequences of fraud, accident or mistake has never been affected by the statutory provisions and is not affected by section 146. Say a landlord were by fraud to procure the tenant to commit a breach of covenant and then forfeit the lease and obtain possession. Can it be doubted that the court would relieve against the forfeiture on the grounds of fraud even though the statutory jurisdiction to grant relief was no longer exercisable? In my judgment, the general equitable jurisdiction to relieve from fraud, accident or mistake remains totally unaffected.

The question is whether the statutory right to relieve in cases falling within my category (c) (ie "wilful" breach of covenant) has impliedly extinguished the equitable jurisdiction to relieve in such circumstances. As I have said, everyone at all material times thought M
that there was no such jurisdiction, its existence only having been discovered retrospectively by the decision in *Shiloh Spinners*. On that point, in my judgment, *Barrow* v *Isaacs* casts no light, since it was concerned only with the general jurisdiction to relieve from the consequences of mistake.

I therefore conclude, with regret, that there is no longer any inherent jurisdiction to relieve from forfeiture for wilful breach of covenant as between landlord and tenant, save in relation to covenants for non-payment of rent. I would therefore dismiss the appeal.

Agreeing that the appeal should be dismissed, **Parker LJ** said: In this judgment I shall deal with only two of the many questions arising

A on this appeal. On all other questions I agree with and have nothing to add to the judgment of the Vice-Chancellor, which I have had the opportunity to read in draft. On the two questions I also agree with his judgment, but with trepidation I venture a contribution of my own.

Question 1

Where:

(a) a lessee is in breach of a covenant in a lease other than a covenant to pay rent;

(b) the lease contains a stipulation or proviso for re-entry or forfeiture for breach of that condition;

(c) the pre-conditions imposed by section 146(1) of the Law of B Property Act 1925 for the enforcement of such right of re-entry or forfeiture by action or otherwise have been fulfilled;

(d) the landlord has without resort to action exercised such right and forfeited the lease by lawfully retaking possession of the premises;

has the court any jurisdiction to grant relief against forfeiture under section 146(2) on a claim first raised by the lessee after such forfeiture has been effected?

The power arises under that subsection "Where a lessor is proceeding by action or otherwise to enforce such a right of re-entry or forfeiture".

It is settled law that, where a landlord has proceeded by action, the C jurisdiction to entertain a claim for relief under the subsection ceases when the landlord has obtained and executed judgment, ie retaken possession under the order of the court. Thereafter he is no longer proceeding to enforce the right. He has enforced it. The same must therefore, in my judgment, follow when the landlord has, without resort to the courts, fully and lawfully enforced his right. He is then no longer proceeding to enforce it. He has obtained everything which he could obtain, whether by action or by other means.

I do not find either result surprising. If the right to claim relief thereafter had been intended by Parliament, it appears to me that Parliament would surely have limited the time during which such a claim could be subsequently brought. If the lessee could not claim relief unless and until the landlord had exercised his right, different D considerations would apply, for he would then be exposed to a landlord exercising his right without protection. But this is not the case. He may protect himself by bringing an action for relief as soon as notice under section 146 has been served. (*Pakwood Transport Ltd* v *15 Beauchamp Place Ltd* (1977) 245 EG 309 *per* Buckley LJ at p 311 and Goff LJ at p 313) and he has a period in which to do so because the lessor cannot act until a reasonable time has elapsed.

Question 2

If, in the above circumstances, the lessee's claim for relief cannot be entertained under section 146(2), has the court nevertheless jurisdiction to entertain the claim pursuant to an inherent power to grant relief in equity?

E This in truth embraces two questions, namely (a) did such a power exist prior to the enactment of the section; (b) if so, was such power excluded by the section?

These two subsidiary questions are closely interrelated. As to the first, in his speech in *Shiloh Spinners Ltd* v *Harding* [1973] AC 691 Lord Wilberforce considered the history and extent of the equitable right to grant relief against forfeiture of property at pp 722-724. The appeal in that case concerned not covenants in a lease but the question of the power to grant relief for breach of covenant in other cases. At p 723 Lord Wilberforce said:

> . . . no decision in the present case involves the establishment or recognition directly or by implication of any general power — that is to say, apart from the F special heads of fraud, accident, mistake or surprise — in courts exercising equitable jurisdiction to relieve against men's bargains. Lord Eldon LC's firm denial of any such power in *Hill* v *Barclay* does not call for any revision or review in this case. Equally there is no need to qualify Kay LJ's proposition in *Barrow* v *Isaacs & Son* (cited above). I would fully endorse this: it remains true today that equity expects men to carry out their bargains and will not let them buy their way out by uncovenanted payment. But it is consistent with these principles that we should reaffirm the right of courts of equity in appropriate and limited cases to relieve against forfeiture for breach of covenant or condition where the primary object of the bargain is to secure a stated result which can effectively be attained when the matter comes before the court, and where the forfeiture provision is added by way of security for the production of that result. The word "appropriate" involves consideration of the conduct of the applicant for relief, in particular whether his default was wilful, of the gravity of the breaches, and of the disparity between the value of the property of which forfeiture is claimed as compared with the damage caused by the breach. G

This, on the face of it, is a plain statement (with which the other members of the judicial committee agreed) (i) that apart from the four cases mentioned (none of which is relevant on this appeal) there is no general power to relieve against men's bargains; but (ii) that courts of equity have and always have had power to relieve in the case of breach of covenant or condition where the primary object of the bargain is to secure a stated result which can effectively be attained when the matter comes before the court and where the forfeiture provision is added by way of security for the production of that result; (iii) that wilful or conscious breach of covenant was not a bar to relief but a factor to be taken into consideration when deciding whether relief should be granted. H

Lord Wilberforce thus dispelled what had for a long time been accepted, namely that, save in the case of a covenant to pay rent and any other covenants specially dealt with, no relief would be granted in the case of a wilful breach. Later, however, at p 725E-F he made it plain that, although wilful breach was not an absolute bar, it was only in exceptional cases that, if the breach was of this nature, relief should be granted, "if only for the reason that the assignor should not be compelled to remain in a relation of neighbourhood with a person in deliberate breach of his obligations".

Although in that case the reference was to the assignor, the reason appears to me to apply with equal force to a case of lessor and lessee.

I turn now to the second subsidiary question. J

Having dealt with the historical position and stated the existing law, Lord Wilberforce continued at p 724A:

> Both as a matter of history and by the nature of things, different considerations apply to different covenants. As regards covenants to pay rent, in spite of Lord Eldon LC's reservations, the matter has, subject to qualifications which need not be discussed, been taken over by statute, first by 4 Geo 2 c 28 then by later Acts leading up to the Law of Property Act 1925. The same is true of covenants to insure and other covenants in leases.

In my view, this is a clear statement that, in the case of covenants between lessor and lessee, relief against forfeiture is now regulated wholly and exclusively by statute. I can attribute no other meaning to the words "taken over".

That this is the meaning which Lord Wilberforce intended is, it K appears to me, reinforced by a later passage where he said at p 724G:

> Secondly, a point of more difficulty arises from the intervention of Parliament in providing specific machinery for the granting of relief against forfeiture of leases: see Law of Property (Amendment) Act 1859 (22 & 23 Vict c. 35), Common Law Procedure Act 1852, Law of Property Act 1925, Leasehold Property (Repairs) Act 1938 and other statutes. This, it is said, negatives an intention that any corresponding jurisdiction should exist outside the case of leases. I do not accept this argument. In my opinion, where the courts have established a general principle of law or equity, and the legislature steps in with particular legislation *in a particular area*, it must, unless showing a contrary intention, be taken to have left cases *outside that area* where they were under the influence of the general law.

(*emphasis supplied*) L

This proceeds on the basis that, although in the case of leases (the area in which the legislature had stepped in) equitable relief could not be granted, it does not follow from section 146 that in other areas equitable relief was also excluded.

On the face of section 146 the exclusion of any pre-existing law, save in the case of re-entry or forfeiture for non-payment of rent, follows from subsection (11).

With exceptions which do not matter, the section is dealing with and regulating re-entry and forfeiture for breach of *any* covenant or condition in a lease. By specifically saving in subsection (11) the law relating to re-entry or forfeiture or relief in case of non-payment of rent, there is a clear inference that in other cases existing law, whatever it may be, is excluded. M

It was, however, sought, by reference to statutory history and the case law at the time of the enactment and its predecessors to contend that that result did not flow.

Such an argument appears to me to involve the proposition that Lord Wilberforce did not mean what he said or introducing a qualification on what he said which he did not himself think fit to mention. I cannot accept this.

Quite apart from this, however, the argument appears to me to fail. Its essence is that the inference which would otherwise flow from the saving of the old law in the case only of non-payment of rent cannot be drawn. Parliament must, it is said, be presumed to have proceeded at the time of the enactment of the ancestor of subsections

A (1), (2) and (11) of section 146 — subsections (1), (2) and (8) of the Common Law Procedure Act 1881 — on the basis and in the belief that there was no existing law to exclude in other cases. This belief continued thereafter through the statutory stages leading up to section 146. Indeed it continued thereafter until the decision of the House of Lords in the *Shiloh Spinners* case in 1972. It follows, the argument goes, that Parliament cannot have intended to exclude old law which it believed not to exist. This consequence does not, in my view, follow. If it was believed that there was no relevant old law the parliamentary intent must, as it seems to me, have been the positive intent to create and thereafter maintain a comprehensive and exhaustive statutory position. I am unable to accept that the result of
B the 1972 decision should be that Parliament wholly failed in 1881 and thereafter to achieve this. I would dismiss this appeal.

I should perhaps add, lest silence on the point should be taken to indicate concurrence with the view taken by the learned judge, that, had there been jurisdiction to grant relief, I am far from sure that I would have taken the same view. This was not only a case of deliberate and flagrant breach but one in which there were many aggravating circumstances. To have granted relief in such a case, on the sort of conditions on which alone it could have been granted, would therefore appear to me to be permitting the lessee to buy his way out by uncovenanted payment and forcing the lessor to live in neighbourhood with a lessee who had no respect whatever for his contractual obligations. That both considerations militate against a
C grant is clear from the first of the above citations from Lord Wilberforce's speech in the *Shiloh* case and to grant relief in such circumstances would be likely to encourage lessees to disregard their obligations.

Dissenting, **Nicholls LJ** said: On July 18 1989 the landlords carried out a "dawn raid" at 17 Gledhow Gardens, London SW5. On May 3 Residential Apartments Ltd had bought the residue of a lease, due to expire in June 1997. The company had paid £280,000 for the lease. The company, through a host of workmen, proceeded at once to carry out extensive and expensive works to the house. This was done in blatant disregard of the covenant in the lease against making alterations without the landlords' consent. There was not even a request for consent. Letters of protest from the landlords' solicitors
D had no effect. On July 4 the landlords served on the company a notice under section 146 of the Law of Property Act 1925, calling on the company to remedy the breaches in so far as they were capable of remedy. That also had no effect. The work continued. So on the morning of July 18, just before six o'clock, the landlords' agent attended at the property with a locksmith. They turned up at this very early hour in order to be there before the tenant's workmen arrived and while there was no one in the property. The agent and the locksmith entered through the back door and changed the locks and fixed notices on the front and back doors. Then they left. They were there for less than an hour. When the landlords' agent returned later, at about 10 am, he found that, meanwhile, the tenant's workmen had
E arrived and gained access and resumed their work.

The principal question raised by this appeal is whether this re-entry by the landlords had the effect of terminating the court's power to grant the tenant relief against forfeiture. Mummery J held that it did. The judge went on to decide that, if the court had jurisdiction, he would have granted relief from forfeiture, on stringent conditions. Although the tenant had acted in deliberate breach of its obligations, it would be inequitable to refuse relief, since that would be to deprive the tenant of all benefit of its very substantial expenditure on the acquisition and improvement of the property. That would be unfairly disproportionate to the breach of covenant and to the tenant's conduct.

The main thrust of the landlords' case can be stated shortly. As
F reaffirmed by the House of Lords in *Shiloh Spinners Ltd* v *Harding* [1973] AC 691 at p 723, courts of equity have a jurisdiction, in appropriate and limited cases, to relieve against forfeiture where the primary object of the parties' bargain was to secure a stated result which can effectively be attained when the matter comes before the court and where the forfeiture provision was added by way of security for the production of that result. In section 146(2) of the Law of Property Act 1925 Parliament provided that the court should have power to grant relief against forfeiture, but only if the tenant applies for relief before the landlord has enforced his right of re-entry by actual re-entry: see *Rogers* v *Rice* [1892] 2 Ch 170. That limitation on the statutory power is inconsistent with the court's retaining, outside the statute, a concurrent, open-ended power to grant relief. Therefore, by necessary implication the statutory power in section G
146(2) has excluded the equitable jurisdiction.

At first sight this simple approach is compelling. But it leads to a wholly unacceptable conclusion. Take this very case. A landlord serves a section 146 notice, which is not complied with. So the landlord is entitled to forfeit the lease. According to the argument of the landlords in the present case, if the landlord chooses to effect the forfeiture by forcing his way into the premises, he is in a better position than if he had applied to the court for an order for possession. Had he applied to the court, the tenant would have applied for relief from forfeiture and the court would have granted relief if it were equitable to do so. But if he takes the law into his own hands, and without further warning to the tenant retakes possession
of the leased property, no application for relief from forfeiture can H
then be made. The court is powerless. The landlord is secure, even if he used physical force to obtain entry, provided he is careful to see that no one is actually on the property at the time and provided the premises are not let as a dwelling.

That cannot be right. Such a conclusion would be an incitement to all landlords to re-enter forcibly whenever they can do so. That would amount to the court's granting a charter for forcible entry. Lawyers would advise landlords that this course was not available where premises are let as a dwelling: see section 2 of the Protection from Eviction Act 1977. Further, that it would be a criminal offence, under section 6 of the Criminal Law Act 1977, to use or threaten violence
for the purpose of securing entry if the tenant, or someone else, was in J
the property and opposing the entry. But the lawyers would also have to advise landlords that, in the case of non-residential property, they should keep watch on the property. They should do this surreptitiously, so as not to alert their tenants to what they have in mind. They should then enter, by force if necessary, once they see that the coast is clear and there is no one on the property. In their own interests landlords should take this course, because thereby they would be better off than going to the court and seeking an order for possession. If they follow this hole-in-the-corner, self-help route they would thereby put an end to any claim by the tenant for relief. The stronger the tenant's claim for relief, the more important it would be for the landlord to keep away from the court and to pounce
on the property, in the evening or at the weekend, and change the K
locks and then sit back secure in the knowledge that the forfeiture is complete and that is an end of the matter.

There must be something wrong with an argument which leads directly to this conclusion. Whatever Parliament intended by section 146, it did not intend such a result.

Nor can it be right to encourage law-abiding citizens to embark on a course which is a sure recipe for violence. If a landlord enters business premises without warning out of business hours, violence is all too likely when the tenant arrives the next day to re-open his shop or offices and finds he is barred from entry. The policy of the law is to discourage self-help when confrontation and a breach of the peace
are likely to follow. If a tenant who is in breach of covenant will not L
quit but persists in carrying on his business despite the landlord's right of re-entry, the proper course for a responsible landlord is to invoke the due process of law and seek an order for possession from the court. But a landlord can hardly be expected to do so if, in terms of his legal rights, he will be severely prejudiced thereby. Nor can he be expected to respect, or even understand, a law which tells him that he should not resort to violence or force in such circumstances but tells him at the same time that if he does forcibly re-enter, his position in law will be better than if he invokes the court's process.

I turn therefore to consider whether this court is so imprisoned by the language of the statute and by previous decisions of this court that there is no escape from the conclusion for which the landlords
contend. M

An enabling provision

The structure of section 146 is simple. Subsection (1) precludes a landlord from enforcing, by action or otherwise, a right of re-entry unless the familiar form of notice has been served and the tenant has failed, within a reasonable time, to remedy the breach and make reasonable compensation. Subsections (2) and (4) enable a tenant or undertenant to apply to the court for relief. Subsection (5) contains some definitions. Subsections (8) to (10) provide for certain cases to which the section does not extend or apply. Subsection (11) provides in effect that, save for subsection (4), the section does not affect the law relating to re-entry or forfeiture or relief in case of non-payment

A of rent. The section overrides any stipulation to the contrary: subsection (12).

Thus the two subsections which enable applications for relief to be made, and relief to be granted, are subsections (2) and (4). These two subsections are expressed in positive, enabling terms. On their face they do not cut down or abrogate any other jurisdiction of the court to grant relief. Statutes are not to be construed so as to take away the jurisdiction of superior courts, unless they do so by express words or by necessary implication: see the authorities cited in *Halsbury's Laws of England,* 4th ed, vol 44, para 907. There are no express words to that effect in section 146. The question, therefore, is whether, nevertheless, subsections (2) and (4) do have that effect as a matter of B necessary implication.

Scope of the statutory power

The Law of Property Act 1925 was a consolidating statute. Section 146 is a re-enactment of section 14 of the Conveyancing and Law of Property Act 1881 and sections 2, 4 and 5 of the Conveyancing and Law of Property Act 1892. The only change in these provisions made by the 1925 property legislation was to repeal the exclusion from section 14 of covenants against assignment or underletting or the like: see the Law of Property (Amendment) Act 1924.

Subsections (2) and (4) of the 1925 Act, like their predecessor provisions in the 1881 Act and the 1892 Act, provide for application being made to the court for relief "where the lessor is proceeding, by C action or otherwise, to enforce . . . a right of re-entry or forfeiture . . .". The two subsections stipulate no cut-off date for the making of an application save in so far as the phrase "where the lessor is proceeding . . ." is properly to be regarded as identifying a period of time, with a beginning and an ending, within which alone the enabling provisions have operation. In *Rogers* v *Rice* [1892] 2 Ch 170 a landlord obtained possession pursuant to a judgment, and some days later the tenant applied for relief. This court held that the application was too late. The landlord's action, so far as it related to enforcing the right of re-entry, was at an end, and the landlord was not still "proceeding" to enforce his right of re-entry. Observations to the same effect had been made previously by members of this court in *Quilter* v *Mapleson* (1882) 9 QBD 672.

D In *Re Rexdale Investments Ltd and Gibson* [1967] 1 Ont LR 251, the Ontario Court of Appeal declined to adopt the same interpretation of the similar phrase in the (Ontario) Landlord and Tenant Act. That court preferred, in effect, to regard the phrase "where a lessor is proceeding . . ." as triggering but not terminating a tenant's right to apply for relief. In this country *Rogers* v *Rice* has been universally followed and applied. It is a decision binding on this court. That decision was given in a case where the right of re-entry had been enforced by action, but it is equally applicable where the right of re-entry has been enforced by actual re-entry without court action: see *Pakwood Transport Ltd* v *15 Beauchamp Place Ltd* (1977) 245 EG 309 at p 311, *per* Buckley LJ. The language of the phrase does not permit a distinction to be drawn between the two E cases, even though this produces the curious technical result that where a landlord brings possession proceedings the tenant has a period post-forfeiture when he can apply for relief, whereas when the landlord re-enters without having begun court proceedings the tenant has no such opportunity. In the former case, forfeiture occurs when the writ is served, whereas in the latter case the very act of forfeiture, namely taking possession, precludes any application for relief. More important, it is to be noted that this interpretation of the legislation produces a surprising distinction between forfeiture for non-payment of rent and forfeiture for breaches of other covenants. In rent cases, the tenant can apply for relief for up to six months after possession has been retaken, either under a court order or otherwise. F That is the period stated in section 210 of the Common Law Procedure Act 1852 or applied by equity by analogy. But the accepted construction of section 146(2) and (4) of the 1925 Act means that in all other cases the tenant is barred from applying for relief under the statute as soon as actual re-entry has occurred. It is difficult to see why there should be this inequality of treatment between rent and non-rent cases.

An exhaustive code?

In the recent decision in *Official Custodian for Charities* v *Parway Estates Developments Ltd* [1985] Ch 151 this court was concerned with an application for relief against forfeiture because of a tenant's liquidation. The application was made more than a year after the tenant went into liquidation, the lease not having been sold meanwhile. Thus the application was outside section 146 by reason of G the provisions of subsections (9) and (10). The court held that it had no jurisdiction to grant relief. Dillon LJ said (at p 165):

> The provisions of the subsections make no sense unless they are regarded as comprehensive within that particular area; in cases of certain types of tenancy listed in section 146(9) there is to be no relief at all, and in other cases the insolvent lessee has a year only to apply for relief, but if the leasehold interest is sold within that year, the time limit does not apply.

That was not how the statutory predecessors to section 146 were understood and interpreted at the time. The view that no relief at all could be granted in the cases excluded from the statutory power by section 146(9) is inconsistent with the approach adopted by this court in *Barrow* v *Isaacs & Son* [1891] 1 QB 417. In that case the question arose whether equity would relieve against the forfeiture of a lease for H breach of a covenant against underletting. Under section 14(6) of the 1881 Act such covenants were put on the same footing as conditions for forfeiture on bankruptcy: both were excluded from relief under the statute. So far as material, section 14(6) read:

> (6) This section does not extend —
>
> (i) To a covenant or condition against the assigning, under-letting, parting with the possession, or disposing of the land leased; or to a condition for forfeiture on the bankruptcy of the lessee, or on the taking in execution of the lessee's interest . . .

Kay LJ referred to the statutory provision and said (at p 430):

> . . . it is expressly provided that this does not extend to a covenant against assigning, under-letting, parting with the possession or disposing of the land leased. J
>
> Forfeiture for breach of this covenant is left to be dealt with according to the ordinary law and practice of Courts of Equity.

In *Gray* v *Bonsall* [1904] 1 KB 601 this court went further. In that case the court was not concerned with a covenant expressly excluded from relief under the statute. There the subject-matter for relief fell within section 4 of the 1892 Act, which was the predecessor to section 146(4). The court held that this did not exclude the court's relieving jurisdiction which existed outside the section. To some extent, *Gray* v *Bonsall* was a special case. Unlike subsection (2), subsection (4) of section 146 applies to non-payment of rent as well as other covenants. It would be surprising if this somewhat anomalous addition to the scope of subsection (4) had the consequence of wholly excluding K underlessees from resort to the established, ancient equitable jurisdiction to relieve for non-payment of rent. Parliament could not have intended that result. Romer L J (at p 606) said that it was clear that relief might be granted to the undertenants under either the Common Law Procedure Act 1860 or section 4 of the 1892 Act. Thus, in the context of the non-payment of rent, the court rejected the "exhaustive code" interpretation of section 4. But if this be correct, this makes the "exhaustive code" interpretation of section 4 difficult with regard to non-rent cases. To treat section 4 as non-exhaustive in rent cases but exhaustive in all other cases would be a strange construction of the section.

Is there an explanation for such a marked difference in approach between the *Parway* case on the one hand and *Barrow* v *Isaacs* and L *Gray* v *Bonsall* on the other hand? I think there is. The explanation lies in the changed perception of the circumstances in which a court of equity will exercise its jurisdiction to grant relief. When the statutes of 1881 and 1892 were enacted, the generally preferred view within the legal profession was the strict view of Lord Eldon LC in *Hill* v *Barclay* (1811) 18 Ves 56: see Lord Wilberforce's exposition in *Shiloh Spinners Ltd* v *Harding* [1973] AC 691 at p 722 *et seq*. Thus, when enacting those statutes, and the same would be true of section 146 of the 1925 Act, Parliament is not to be taken as having intended that the relevant sections impliedly superseded and excluded the residual, ancient jurisdiction of the Court of Chancery, which by then had become vested in the newly created High Court of Justice, for this reason: *as generally perceived at that time* this residual jurisdiction M seldom availed a tenant in non-rent cases. Thus there was nothing incongruous in the ancient jurisdiction and the new statutory jurisdiction existing side by side. For what it might avail him, a tenant could still seek to invoke the ancient jurisdiction even after 1881. For what it might be worth to him, the jurisdiction was still there and untouched by the new, enabling legislation, but it could not be expected to be of much help to him. Indeed, that is what happened in *Barrow* v *Isaacs*. However, in 1973 in the *Shiloh Spinners* case, the House of Lords reaffirmed, in more expansive terms, the right of courts of equity to grant relief against forfeiture. It is from this decision in 1973 that the incongruity perceived by Dillon LJ in the *Parway* case in 1985 springs.

How, then, does the matter now stand? If, as was understood to be the position when the relevant statutes were passed, section 146 and the predecessor sections were enabling provisions which did not supersede the existing jurisdiction of courts of equity, that must have remained the position. The proper interpretation of section 146, on the question whether that section constituted an exhaustive code or not, cannot have been changed by the reawakening of the court of equity's relieving jurisdiction in 1973. In this regard the section falls to be construed as at the date when it was passed. Either section 146, from the outset, wholly superseded the jurisdiction of the court of equity or it did not. In my judgment, for the reason I have sought to explain, at their inception the statutory provisions did not wholly supersede that jurisdiction. They did not do so expressly, nor did they do so as a matter of necessary implication. On this, faced with conflicting approaches adopted by this court, this court is entitled to choose between them. Given the manifestly unsatisfactory consequences to which the *Parway* interpretation of section 146 leads in some cases, I consider we are entitled to and should prefer the interpretation adopted in *Barrow* v *Isaacs* and *Gray* v *Bonsall.*

Equity follows the law

This is not to say that courts of equity should now grant relief without any regard to the statutory provisions. Equity follows the law, but not slavishly nor always: see Cardozo CJ in *Graf* v *Hope Building Corporation* (1930) 254 NY 1 at p 9. On this we have the benefit of guidance elsewhere in the field of relief from forfeiture. Section 210 of the Common Law Procedure Act 1852, which is still in force, limited to six months after judgment the period within which a tenant could apply for relief in the non-payment of rent cases to which that statute applied, namely where the rent was six months in arrears. Courts of equity have due regard to this statutory limitation in non-payment of rent cases where the statute does not apply: and in cases of forfeiture by peaceable re-entry, and in cases where possession has been taken under a court order where less than six months' rent was in arrears. In *Howard* v *Fanshawe* [1895] 2 Ch 581 at p 588, the landlord re-entered without the aid of the court. Stirling J said:

> . . . it does not follow that a Court of Equity would now grant relief at any distance of time from the happening of the event which gave rise to it. It appears to me that, inasmuch as the inconvenience of so doing has been recognised by the legislature, and a time has been fixed after which, in a case of ejectment, no proceedings for relief can be taken, a similar period might well be fixed, by analogy, within which an application for general relief in Equity must be made. A Court of Equity might possibly say that the action for relief must be brought within six months from the resumption of possession by the lessor.

These guarded observations were considered by Sir Jocelyn Simon P in *Thatcher* v *C H Pearce & Sons (Contractors) Ltd* [1968] 1 WLR 748 at p 755, another case of peaceable re-entry for non-payment of rent:

> . . . it seems to me to be contrary to the whole spirit of equity to boggle at a matter of days, which is all that we are concerned with here, when justice indicates relief. I think that a court of equity — and it is such jurisdiction that I am exercising now — would look at the situation of the plaintiff (sc the tenant) to see whether in all the circumstances he acted with reasonable promptitude. Naturally it would also have to look at the situation of the defendants [sc the landlords] to see if anything has happened, particularly by way of delay on the part of the plaintiff, which would cause a greater hardship to them by the extension of the relief sought than by its denial to the plaintiff.

Likewise with section 146. The concurrent equitable jurisdiction can be invoked only by those who apply with reasonable promptitude. What is reasonable will depend on all the circumstances, having due regard to the statutory time-limits. In the exercise of its jurisdiction a court of equity should apply, by analogy, the statutory time-limits as interpreted in *Rogers* v *Rice*, but not with a strictness which in all the circumstances would lead to a result Parliament could never have intended. Thus, a court of equity will entertain an application for relief by a mortgagee who knew nothing of the tenant's breaches and from whom the landlord deliberately concealed the existence of his forfeiture proceedings: *Abbey National Building Society* v *Maybeech Limited* [1985] Ch 190. On the other hand, the court will not entertain a tenant's application for relief against forfeiture on the ground of liquidation made three and a half years after the tenant went into liquidation, the property not having been sold within the statutorily prescribed period of one year: see the *Parway* case.

In the present case, the landlords re-entered by stealth. The property was not empty and unused. Workmen were carrying out alterations on behalf of the tenant. Without any warning, the landlords moved in on July 18. As was to be expected, the tenant ignored the new locks and the notices. The landlords issued their writ on the following day, and the tenant's counterclaim for relief was served a week later, on July 26. In my view, the tenant acted with promptitude, and the court can and should entertain its application for relief.

Shiloh Spinners case

I must mention one other point. In the *Shiloh Spinners* case Lord Wilberforce said (at p 724) regarding relief from forfeiture:

> As regards covenants to pay rent, in spite of Lord Eldon LC's reservations, the matter has, subject to qualifications which need not be discussed, been taken over by statute, first by 4 Geo 2 c 28 then by later Acts leading up to the Law of Property Act 1925. The same is true of covenants to insure and other covenants in leases.

I do not think that this passage is to be read as a statement by Lord Wilberforce that the statutory intervention in the field of leases has wholly superseded any jurisdiction that courts of equity otherwise would have. First, it is to be observed that Lord Wilberforce noted that the enactments relating to leases were positive in character and amplified the court's jurisdiction. He referred with approval to Kay LJ's observations in *Barrow* v *Isaacs*: see p 725 A-C. Second, the observations cited above must be read having in mind that, as regards covenants to pay rent, to this day the jurisdiction being exercised by the High Court when granting relief is that of the old Court of Chancery. Even today the jurisdiction does not stem from statute. The sole exception is section 146(4). In particular, section 210 of the Common Law Procedure Act 1852 imposes a time-limit on when relief may be given in some cases. It is not an enabling provision. The other statutory provisions regarding relief in rent cases are of a procedural character. Third, and most important, Lord Wilberforce's observations must be read in context. In this part of his speech and elsewhere, Lord Wilberforce was not concerned to examine in depth the position regarding leases. The covenants with which the House of Lords was concerned in the *Shiloh Spinners* case were not covenants in a lease. Lord Wilberforce traced the jurisdiction historically, and then put covenants in leases on one side because in those cases, unlike in the case of non-lease covenants, there had been statutory intervention. He was not concerned to consider such covenants further. Thus he did not direct his attention at the point in issue in the present case, and he is not to be taken as having expressed a view on the point.

For these reasons, for my part I would hold that the court has jurisdiction to entertain the tenant's application for relief. I would therefore proceed to hear the parties on whether relief should be granted and on what terms.

On the other points arising on the tenant's appeal, I agree with the reasoning and conclusions of Sir Nicolas Browne-Wilkinson V-C and have nothing to add.

The appeal was dismissed with costs. Leave to appeal to the House of Lords was granted. A stay was granted on terms.

Court of Appeal

January 18 1991

(Before Lord Justice GLIDEWELL and Sir Denys BUCKLEY)

CERIUM INVESTMENTS LTD v EVANS AND OTHERS

Estates Gazette May 25 1991

[1991] 20 EG 189

Landlord and tenant — Obligations of sureties — Sureties at fault in failing to ensure that landlords were notified of assignment and had the deed of assignment registered with them — Claim by sureties that, as a result of the failure to register, the licence to assign containing the sureties' obligations became null and void and their liabilities as guarantors ceased to exist — Appeal from decision of Sir Peter Pain (sitting as a High Court judge), who held that the sureties had an arguable defence and gave them unconditional leave to defend under Ord 14 — Court of Appeal held that sureties

were relying on their own breach to avoid their obligations — Court decided in favour of landlords without sending the case for trial

The problems in this case arose from the terms of a licence to assign business premises which included provisions for guarantee and suretyship — The sureties, defendants in the action, covenanted that the assignees of the lease would pay the rents and perform the obligations imposed on them by the lease — the sureties also covenanted that, in case of the assignees' default, they would make good all loss and damage arising from the breach — The licence also provided that it would become null and void if the assignment were not completed by a certain date and registered within a month thereafter — By registration was meant the production of the deed of assignment to the landlords — The assignment was duly completed in time but the registration was not — The assignees got into financial difficulties and rent became in arrear; receivers were appointed — Two actions were commenced by the plaintiff landlords, one for arrears of rent etc and the other for breach of other covenants — Applications were made for summary judgment under Ord 14, which came before Sir Peter Pain — He decided that the defendants had an arguable defence and granted them unconditional leave to defend — The submission which he accepted as arguable was that, as the licence became null and void on the failure to register in time, the result was that the defendant sureties did not become liable — The plaintiff landlords appealed against this ruling — In the Court of Appeal both sides invited the court to decide the issue one way or another rather than merely determine whether it should go for trial — The court agreed to do so following the guidance of Goff LJ in *European Asian Bank AG* v *Punjab & Sind Bank (No 2)*

The landlords, appellants in the Court of Appeal, took their stand on the well-established principle that a party is not entitled to avoid a contract otherwise binding on him by relying on his own wrong — Clause 3 of the licence imposed obligations on the respondent sureties which were breached when the time for registration had passed — By seeking to argue that the licence was void for non-registration the sureties were trying to escape from their obligations by relying on their own breaches — The respondent sureties, on the other hand, argued that the licence did not create any obligation on a surety until all the relevant conditions had been fulfilled — Thus, on this argument, since there was no registration the sureties were never under any obligation and it followed that there was no breach

The court accepted the appellants' submissions and rejected the respondents' argument — In the court's view that argument ignored the words of clause 4 of the licence, which provided that the licence was to become null and void *if* the assignment and registration were not completed by certain dates — This language indicated that initially from the date of its execution the licence was effective — If thereafter the assignment was not registered, the surety as well as the assignee was in breach — The surety, in saying that because there was no registration he was under no obligation, was seeking to rely on his own breach of covenant and thus on his own wrong

The court considered a respondent's notice served on behalf of one of the respondents but decided that the submission failed because it was not supported by the facts

Both of the landlords' appeals were allowed

The following cases are referred to in this report.

Alghussein Establishment v *Eton College* [1988] 1 WLR 587, HL
Doe d Bryan v *Bancks* (1821) 4 B & Ald 401
Eastern Counties Building Society v *Russell* [1947] 1 All ER 500
European Asian Bank AG v *Punjab & Sind Bank (No 2)* [1983] 1 WLR 642; [1983] 2 All ER 508; [1983] 1 Lloyd's Rep 611, CA
Junction Estates Ltd v *Cope* (1974) 27 P&CR 482
Lep Air Services Ltd v *Rolloswin Investments Ltd* [1973] AC 331; [1972] 2 WLR 1175; *sub nom Moschi* v *Lep Air Services Ltd* [1972] 2 All ER 393, HL
New Zealand Shipping Co v *Société des Ateliers et Chantiers de France* [1917] 2 KB 717; [1919] AC 1, CA and HL
Rede v *Farr* (1817) 6 M&S 121
Tottenham Hotspur Football & Athletic Co Ltd v *Princegrove Publishers Ltd* [1974] 1 WLR 113; [1974] 1 All ER 17; (1973) 27 P&C 101

These were two appeals by Cerium Investments Ltd against decisions of Sir Peter Pain, sitting as a judge of the High Court, whereby he rejected applications by Cerium Investments Ltd for summary judgment under RSC Ord 14 and gave unconditional leave to defend to Ernest John Evans, Richard John Ford and Stephen Brian Stubbs, defendants in two actions brought by Cerium Investments Ltd. The actions arose out of a lease of premises at 47 Nottingham Place, London W1, and an underlease of adjoining premises at 45 Nottingham Place.

Kim Lewison (instructed by Reid Minty & Co) appeared on behalf of Cerium Investments Ltd; Anthony Allston (instructed by Matossian & Co) represented Ernest John Evans; Christopher Coney (instructed by Comptons) represented Richard John Ford; Roderick Doggett (instructed by Mylles & Co) represented Stephen Brian Stubbs.

Giving judgment, **GLIDEWELL LJ** said: These are appeals against decisions of Sir Peter Pain, sitting as a judge of the High Court, who, on July 25 1990, after hearing two applications under RSC Ord 14 for summary judgment in two actions between these parties, gave unconditional leave to defend to the first, second and fourth defendants, Mr Evans, Mr Ford and Mr Stubbs; the third defendant, Mr Ridgeway, has taken no part in these proceedings from beginning to end, as I understand it.

The appeals raise a short, interesting but difficult point, and I would like to start by congratulating counsel and all who have been concerned in the matter for dealing with the point so clearly and so expeditiously. The hearing yesterday occupied no more than two and a half hours of argument. Four counsel were able to address us and, in that time, put their submissions with great clarity.

The facts of the matter are not in dispute. They can be summarised as follows. On October 4 1973 the plaintiff company, Cerium, granted a lease of premises, 47 Nottingham Place, London W1, to two gentlemen called Frischmann and Prabhu for 15 years from September 29 1973. Later, on January 13 1981, Cerium granted an underlease of the adjoining premises, 45 Nottingham Place, to Mr Frischmann and Mr Prabhu for the unexpired residue of a term of 21 years, which they (Cerium) held from December 21 1979 less one day. Both the lease and the underlease contained what for present purposes were identical covenants. Among the tenant's covenants were the following, quoting from the underlease. By clause 2(13) the tenant covenanted in the usual way not to assign or underlet without the previous consent of the landlord, which consent should not unreasonably be withheld. By clause 2(14) the tenant covenanted:

> Within one month of every assignment assent transfer or underletting of or relating to the demised premises or any part thereof to give notice thereof in writing with particulars thereof to the solicitors for the time being of the Landlords and Superior Landlords and produce to them such assignment transfer or underlease and to pay to the Superior Landlords a registration fee of Three pounds and to pay to the Landlords Solicitors a registration fee of Ten pounds in respect of such assignment transfer or underlease.

The process of production of the deed of assignment has been referred to in these proceedings as "registration" with the landlord.

On March 26 1987 two identical deeds were executed, each of them being called a licence. The parties to the deeds were the landlords Cerium, the tenants Mr Frischmann and Mr Prabhu, the assignee a company called the Harley Group Ltd, and as sureties the defendants to these two actions including Mr Ridgeway. Each licence achieved two separate objects. (I am now looking at the licence relating to 45 Nottingham Place.) First, by clauses 1 and 2 the deed granted a licence to the then tenants to assign and contained a covenant by the assignee that the assignee would thenceforth "pay the rents reserved under the Lease and observe and perform the covenants and the conditions on the part of the tenant contained in the Lease", and, second, by clause 3 the deed contained a form of guarantee and suretyship, the first part of which is in the following terms:

> The Sureties in consideration of the premises and the Licence having been granted at their request HEREBY COVENANT jointly and severally with the Landlord:—

A (a) That the Assignee will pay the rents reserved under the Lease and observe and perform the covenants and the conditions on the part of the tenant contained in the Lease and that in case of default in payment of such rents or in the performance or observation of such covenants as aforesaid the Sureties will pay and make good to the Landlord on demand all losses damages costs and expenses thereby arising or incurred by the Landlord . . .

There is a proviso to that which does not apply in the present case.

On April 1 1987 — that is to say some five days later — the lease and the underlease were duly assigned by Mr Frischmann and Mr Prabhu to the assignee company the Harley Group Ltd. That meant that in accordance with clause 4 of the licence another situation arose. This is the clause around which the argument in this case has revolved.
B Clause 4 provided:

This Licence shall become null and void if:

(i) the assignment of the Lease to the Assignee is not completed by 1st May 1987

(ii) such assignment is not registered in accordance with clause 2(14) of the Lease.

As I have just said, the assignment was completed on April 1. Therefore, in accordance with clause 2(14), it was required that the assignment should be registered within a further month — that is to say, before May 1 1987. That period passed without the assignment of either the lease or the underlease being registered. Harley Group Ltd fell into financial difficulties. In March 1988 Cerium obtained a
C default judgment against Harley Group Ltd for arrears of rent then unpaid. On June 7 1988 receivers of Harley Group Ltd's assets were appointed. The writs in the present actions were issued, first, on December 11 1989, the claim being for some £139,520-odd in respect of arrears of rent and insurance premiums with interest and, on April 18 1990, the claim being for damages for breach of other covenants in the lease and the underlease.

A number of arguments were advanced on behalf of the defendants before Sir Peter Pain. Most of those arguments did not succeed and have not been repeated before us. The argument which did succeed was that the failure to register the assignments in time or at all gave the defendants a defence. Put shortly, the argument was, first, the phrase "the licence" means the whole deed and that has
D been accepted in argument before us; second, on the failure to register, the licence (that is to say, the whole deed) in each case became null and void; and, third, therefore, either the defendants never were under any liability or, alternatively, they ceased to be under any liability on May 1 1987 when, by reason of non-registration, each of the licences became null and void.

The learned judge held that it was arguable that the defendants had such a defence, and so gave leave to defend. We, having heard the argument in full although within a short compass of time, have been invited by both sides to decide the issue one way or another. That invitation was based upon a proposition advanced by Goff LJ in this court in *European Asian Bank AG* v *Punjab & Sind Bank (No 2)* [1983] 1 WLR 642, at p 654C. I do not need to quote it, but it
E undoubtedly provides that in circumstances where there is no issue of fact and there is a straight issue of law such as the construction or effect of a document, then this court is in as good a position to decide the issue, having heard full argument, as would be the court of trial if leave to defend were given and there were a trial. On that basis we have agreed to decide the issue one way or another.

I should say, however, that by a respondent's notice Mr Coney for the second defendant, has raised another issue. That second issue does not arise if we rule in favour of the defendants on the first issue and in any case, if I understand it correctly, goes only to a part of the plaintiff's claim against his client. We have so far heard argument on only the first issue and this judgment is confined to that.

Mr Allston, for the first defendant, reminds us of the general rule
F that in interpreting a contract of suretyship, if there is doubt about the proper construction of the contract, it should be construed strictly. He relies upon a judgment of Hilbery J in *Eastern Counties Building Society* v *Russell* [1947] 1 All ER 500. At p 503C Hilbery J said:

The mortgage instrument, like any other written contract, must be taken to express what the parties intended to be the contract between them. It is not the function of the court to make a contract for the parties which would seem reasonable or which will more conveniently fit the circumstances which have supervened. The whole instrument, as it stands, must be construed so as to give effect to the intention of the parties discovered from the actual terms agreed by the parties and employed by them in the written instrument as expressing what they intend to agree. It must be borne in mind that the plaintiff society was the party putting forward the written instrument as G
expressing the terms on which they were willing to contract. If, therefore, there be a doubt about how an expression in the instrument is to be interpreted, it is to be construed *contra proferentes*. Moreover, the question here does not arise between the two principals to the contract, but between the society and the surety, and the terms employed in the contract, defining the surety's undertaking and expressing the terms on which he is to be freed from his undertaking, ought to be strictly construed. Rather will the court in case of doubt lean in his favour. Neither equity nor law will put a construction on the document which results in imposing on the surety any more than, on the strictest construction of the instrument, he must be said expressly to have undertaken, or so as to detract from the right given to the surety by the proviso defining the circumstances in which the surety is to be held discharged. If authority is needed for saying that the surety's contract has been said to be one *strictissimi juris*, I would refer to *Bacon* v *Chesney*, *Stamford, Spalding &* H
Boston Banking Co v *Ball* and *Blest* v *Brown* . . .

Then he referred to a passage in the judgment of Lord Westbury LC.

For my part I accept that as an entirely correct approach if and in so far as we are concerned with interpreting the contracts — that is to say, the two licences — and if and in so far as there is any doubt as to the proper interpretation of those contracts.

I summarise the submission of Mr Lewison for Cerium as follows. First, there is an established principle that a party is not entitled to avoid a contract binding on him, or indeed to secure a benefit under it, in reliance on his own wrong. For that principle he refers us to the recent decision of the House of Lords in *Alghussein Establishment* v *Eton College* [1988] 1 WLR 587. The sole speech was delivered by J
Lord Jauncey of Tullichettle, all the other of their lordships agreeing with him. Lord Jauncey said at p 591D:

My Lords, it is well established by a long line of authority that a contracting party will not in normal circumstances be entitled to take advantage of his own breach as against the other party.

He then cited passages from a number of authorities, *Rede* v *Farr, Doe d Bryan* v *Bancks*, the *New Zealand Shipping** case in the Court of Appeal [1917] 2 KB 717, and the same case on appeal to the House of Lords [1919] AC 1. He went on to say at p 594:

Although the authorities to which I have already referred involve cases of avoidance the clear theme running through them all was that no man can take advantage of his own wrong. There was nothing in any of them to suggest that the foregoing proposition was limited to cases where the parties in breach were K
seeking to avoid the contract and I can see no reason for so limiting it. A party who seeks to obtain a benefit under a continuing contract on account of his breach is just as much taking advantage of his own wrong as is a party who relies on his breach to avoid a contract and thereby escape his obligations.

Mr Lewison is of course relying on the first of those passages because this, he says, is a case where the sureties are seeking to rely on their own wrong to avoid the contract.

His second submission is that clause 3 of the licence imposed on each of the sureties two different kinds of obligations — namely, first, an obligation that the assignee would pay the rent reserved and observe and perform the tenant's covenants in the lease and, second, a covenant to indemnify the landlord against breach by the assignee of any of his obligations. For that dichotomy he refers us to the L
speech of Lord Reid in *Moschi* v *Lep Air Services Ltd* [1973] AC 331, the passage being at p 344G. Lord Reid said:

With regard to making good to the creditor payments of instalments by the principal debtor there are at least two possible forms of agreement. A person might undertake no more than that if the principal debtor fails to pay any instalment he will pay it. That would be a conditional agreement. There would be no prestable obligation unless and until the debtor failed to pay. There would then on the debtor's failure arise an obligation to pay. If for any reason the debtor ceased to have any obligation to pay the instalment on the due date then he could not fail to pay it on that date. The condition attached to the undertaking would never be purified and the subsidiary obligation would never arise.

On the other hand, the guarantor's obligation might be of a different kind. M
He might undertake that the principal debtor will carry out his contract. Then if at any time and for any reason the principal debtor acts or fails to act as required by his contract, he not only breaks his own contract but he also puts the guarantor in breach of his contract of guarantee. Then the creditor can sue the guarantor, not for the unpaid instalment but for damages. His contract being that the principal debtor would carry out the principal contract, the damages payable by the guarantor must then be the loss suffered by the creditor due to the principal debtor having failed to do what the guarantor undertook that he would do.

Mr Lewison's characterisation of the two forms of obligation imposed on the sureties by clause 3 of the licence are the two forms there described by Lord Reid although in the opposite order.

*Editor's note: *New Zealand Shipping Co* v *Société des Ateliers et Chantiers de France*.

A Third, Mr Lewison submits that the failure to register the assignment was both a breach of the relevant covenant in the lease and a breach by each surety of his obligation to ensure that the assignee observes and performs the covenants in the lease.

Fourth, these two breaches by the assignee and by the sureties were simultaneous, occurring when the time within which the assignee was required to register elapsed. Mr Lewison refers us again to other passages in the speeches in *Moschi* v *Lep Air Services*. Lord Diplock said at p 348D:

> It is because the obligation of the guarantor is to see to it that the debtor performed his own obligations to the creditor that the guarantor is not entitled to notice from the creditor of the debtor's failure to perform an obligation which is the subject of the guarantee, and that the creditor's cause of action B against the guarantor arises at the moment of the debtor's default and the limitation period then starts to run.

At G on the same page:

> It follows from the legal nature of the obligation of the guarantor to which a contract of guarantee gives rise that it is not an obligation himself to pay a sum of money to the creditor, but an obligation to see to it that another person, the debtor, does something; and that the creditor's remedy for the guarantor's failure to perform it lies in damages for breach of contract only. That this was so, even where the debtor's own obligation that was the subject of the guarantee was to pay a sum of money, is clear from the fact that formerly the form of action against the guarantor which was available to the creditor was in special *assumpsit* and not in *indebtatus assumpsit*.

One has to go back to the old forms of action for that explanation.

C Lord Kilbrandon said at p 359B:

> In the present case what the appellant guaranteed was "the performance by Rolloswin Investments Ltd of its obligations to make payments at the rate of £6,000 per week". This is the class of undertaking referred to in *Rowlatt on the Law of Principal and Surety*, 3rd ed (1936) at p 143. The learned author points out that as soon as a breach is committed of the duty, performance of which is guaranteed, the surety is immediately liable to the full extent of his obligation, and gives as "the reason for the rule that it is the surety's duty to see that the principal pays or performs his duty as the case may be".

Finally, Mr Lewison submits that in seeking to argue that the licence was void for non-registration, the surety was seeking to avoid the obligation imposed on him by clause 3 to indemnify by relying on his own breach of the first obligation in clause 3.

D To this argument Mr Allston for first defendant, with the support of Mr Coney and Mr Doggett for the second and fourth defendants, argues that the licence does not create any obligation on the surety until all the conditions in clause 4 have been fulfilled. Thus, since there was no registration, the sureties were never under any obligation and thus it follows not in breach of any obligation.

In my view, this argument ignores the initial words of clause 4 of the licence — "This licence shall become null and void if . . .". Those words make it clear that initially from the date of its being executed the licence was effective. In my judgment, the proper construction and effect of the licence was as follows. First, at the date of the execution of the licence the assignee gained a right to take an E assignment of the lease or underlease respectively. If he did not take the assignment — he was not, of course, obliged to do so — then no obligation on the surety would arise. Second, if, as happened, the assignee did take an assignment, then the sureties immediately came under the obligations imposed by clause 3 of the licence deed. Third, if thereafter the assignment was not registered within one month as required by the lease or underlease respectively, then the assignee was in breach of his covenants in the lease and underlease which he had taken on assignment and the surety was in breach of his covenant in clause 3 of the licence deed. Fourth, it follows, in my view, that if the surety seeks to say that because there has been no registration he is under no obligation, he is seeking to rely on his own breach of covenant and thus on his own wrong.

F For my part I conclude on this issue that Mr Lewison's argument is correct. For this reason I do not find it necessary to say anything about an alternative argument Mr Lewison based upon waiver. I would, therefore, allow the appeal but subject to the issue raised by Mr Coney in his respondent's notice.

Agreeing, **Sir Denys Buckley** said: As Glidewell LJ has said, we have been invited by the parties to determine the issues of law which arise in this case. Sir Peter Pain was not so invited. We are consequently in a different position from that which confronted him when the case was before him. Sir Peter, I think, quite rightly thought that he had no alternative to giving leave to the defendants to defend in order that the issues of law involved in the case should be resolved. Glidewell LJ in the judgment which he has delivered has indicated the G nature of those questions of law, and by the judgment which he has delivered he has resolved them in such a way that it must follow that the case is disposed of and there is no occasion to grant leave to defend to the defendants. Consequently, the position today is radically different from the position as it was before Sir Peter Pain. Sir Peter had, I think, no option but to grant leave to defend to the defendants, but now that the issues of law involved have been resolved there is no occasion to take the course which was the only course open, in my view, to Sir Peter.

I entirely agree with the judgment which my brother has delivered this morning. Accordingly, I agree that the appeal should be allowed, and I think it is unnecessary for me to add anything on the substance H of the legal argument. I am in entire agreement with Glidewell LJ in that respect.

Accordingly, I agree that the appeal should be allowed, subject of course to the outstanding question which has yet to be answered.

After hearing submissions on the respondent's notice, **Glidewell LJ** said: Mr Coney has just advanced with skill and — what I hope he will not regard me as discourteous in saying — ingenuity an argument which, appealing though it was, is about to fail. In order to deal with it I must go a little further into the facts of this matter, having already set out most of the relevant ones in my first judgment. I have already said that on June 7 1988 receivers of Harley Group Ltd were appointed. On June 27 1988 the receivers entered into an agreement J to sell the business, goodwill and assets including the two leasehold interests of Harley Group Ltd to a company then called Wilder Coe Secretarial Services Ltd. It was a consequence or perhaps an intention of the agreement that the purchaser company should adopt the name of the vendor company, and to that end the vendor company changed its name to another name which does not matter and Wilder Coe Secretarial Services Ltd changed its name to "The Harley Group Ltd" on September 20 1988. For the purpose of this argument those two companies have been conveniently referred to as "Harley Group 1" and "Harley Group 2".

Harley Group 2 then went into occupation of the premises or part of them. However, such rent as was paid thereafter to Cerium was paid by or on behalf of the receiver of Harley Group 1. There were no K direct payments by or on behalf of Harley Group 2. There were negotiations between the receiver and solicitors acting for Cerium designed to obtain Cerium's consent to a further assignment of the lease and the sublease respectively by him as receiver of Harley Group 1 to Harley Group 2. In the course of those negotiations new guarantors for the obligations in the lease and sublease if the assignment took effect were proposed, but those negotiations never achieved an agreement. The landlord (Cerium) therefore never gave its consent to the assignment of the lease or the underlease. There was thus no deed of assignment to Harley Group 2.

Mr Coney argues that at common law an assignee of a lease is liable under the tenant's covenants in the lease only for the period for which he holds the lease. L If he in turn reassigns validly to somebody else, then his obligations are limited to obligations arising during the period before the reassignment. So stated, Mr Lewison agrees that that is a valid proposition of law.

It follows, submits Mr Coney, that a surety of the obligations of a first assignee is liable on his guarantee only to seek to ensure that the first assignee undertakes the obligations in the lease and to pay compensation if the first assignee fails in those obligations, again for a failure which occurs during the period that the first assignee holds the lease. In other words, the surety is not liable for anything that happens after the second assignment has taken place.

Mr Coney drew our attention to a decision of McKenna J in *Junction Estates Ltd* v *Cope* (1974) 27 P&CR 482, which supports M that proposition. I do not, I think, need to cite the passage from the judgment to which he drew our attention, because, for my part, for the purposes of this argument I would be prepared to accept, without deciding it, that that is a proper contention in law as well. But where this argument falls down is on the facts. As I have said, there was no second assignment by deed. There was no consent by Cerium as landlord to any second assignment. If a situation arises between two parties where the parties agree that something shall be done and then put it into effect, but the transaction is never embodied in any conveyance or lease, it is the normal principle that nevertheless both will be bound by that which they agree to do. So if, for instance, as in the case of *Tottenham Hotspur Football & Athletic Co Ltd* v

A *Princegrove Publishers Ltd* [1974] 1 WLR 113, landlord and tenant agree that a tenant who has made an application for a new tenancy under the provisions of the Landlord and Tenant Act 1954 shall have by agreement a new tenancy for a year but that the new tenancy shall contain a clause excluding the tenant from making any further application under the 1954 Act for a further tenancy that binds the tenant even though no formal lease to that effect is ever entered into. But that is not the situation here. The situation here is that as between surety and landlord there were no negotiations at all, and as between Harley Group 1 and landlord the negotiations never achieved an agreement. Therefore Mr Coney cannot rely on any proposition which may derive from the situation that transpires when there has been an agreement.

B Moreover, he argues, perhaps with more enthusiasm than expectation, that one might arguably find an estoppel here, and he invited us to say that there ought to be leave to defend so that he could hereafter seek to suggest that there was an estoppel in Harley Group 2 going into possession and in the fact, as he asserts, that Cerium knew that Harley Group 2 were in possession.

With respect to him, I find that an unarguable proposition, since, whether or not Harley Group 2 were in possession, such rent as was paid during the period they were in possession was paid by and on behalf of the receiver for Harley Group 1.

Accordingly, as I have already said, in my judgment this argument fails because there is a total failure to show that there was any second
C assignment to Harley Group 2 or anything from which it could be held, if leave to defend were given, that the sureties had any protection arising out of any concepts of law such as those suggested by Mr Coney.

I would therefore rule against him on this argument. It follows, in my judgment, that the proper order is that the appeal succeed, that in the first action there be judgment for the plaintiffs against each of the present defendants for the sums claimed, and that in the second action there be judgments for Cerium, the plaintiffs, against each of the defendants for damages to be assessed.

SIR DENYS BUCKLEY agreed and did not add anything.

D *Both appeals were allowed with costs; costs below to remain undisturbed. Directions were given for the necessary calculations and assessment to be made.*

Court of Appeal

March 26 1991

(Before Sir Nicolas BROWNE-WILKINSON V-C, Lord Justice STOCKER and Lord Justice BELDAM)

DENETOWER LTD v TOOP AND OTHERS

E

Estates Gazette May 25 1991

[1991] 20 EG 194

Landlord and Tenant Act 1987, sections 12 and 13 — Right of first refusal — Reversion assigned by landlord — Failure to comply with statutory duty to offer tenants first refusal — Tenants served purchase notice under section 12 on the new landlords, the present respondents — Questions as to whether tenants were qualifying tenants, what interests tenants were entitled to acquire and as to validity of purchase notice — Appeal by tenants from decision of county court judge
F allowed — Jurisdiction of rent assessment committees under section 13 discussed

This appeal related to a pair of two-storey buildings at 71 Holden Road, London N12, called Brookdene — The tenants there hold long leases of their flats, eight in all, and the gardens which go with them, and are granted certain rights of way — Each lease includes in the demise of the flat the appropriate part of the external walls of the building — The garages are not included in the demise of the flats but most tenants hold one of the garages under a separate title — In November 1988 the original landlord transferred the freehold title to the reversion to Denetower, the present respondents, without having first complied with the requirement under section 1 of the 1987 Act G of offering the reversion to the tenants — On discovering the transfer the tenants served a purchase notice on Denetower under section 12, requiring the disposal to the tenants of the estate or interest that was the subject-matter of the previous disposal — Denetower then applied to the county court, seeking declarations that the tenants were not qualifying tenants, alternatively that the purchase notice was invalid, and in the further alternative that, if valid, the notice covered only the buildings — The county court judge held in favour of Denetower that the tenants were not qualifying tenants but indicated that, if necessary, he would have decided against H Denetower on the other points — The tenants appealed and Denetower cross-appealed

Whether tenants were "qualifying tenants"

Denetower contended that the tenants were not qualifying tenants on the ground that they were disqualified by section 3(2)(*a*)(ii), which excluded tenants where the demise included the flat "and any common parts of the building" — As the demise of each flat included part of the structure or exterior (defined as a common part in section 60) the tenants were not qualified — The Court of Appeal rejected this submission, holding that in order to be disqualified the demise must cover J the flat itself with its common parts and, in addition, some other common parts — Any other construction would produce absurd results

What were the tenants entitled to acquire?

It was submitted by Denetower that the premises to which the purchase notice applied consisted of buildings only, ie the physical structures consisting of the two blocks of flats — Accordingly, there was no power under section 12(3) to give a purchase notice relating to the gardens or roadway, as they were not part of a building — The court rejected this submission also — Building in this context included K appurtenances such as gardens and express or prescriptive rights over roadways and paths — Garages were not appurtenances, being held under a different title — As regards roadways and paths, however, the Vice-Chancellor suggested that a rent assessment committee, acting under section 13(1), might consider that the appropriate provision should be to grant the tenants perpetual rights rather than a freehold title

Validity of purchase notice

The court rejected a submission that the tenants' purchase notice was invalid because it purported to require Denetower to transfer to the tenants everything which Denetower had L acquired from the original landlord — As the tenants were not entitled to acquire all that land, the notice failed to comply with section 12(3) and was therefore invalid — The answer to this submission was that the purchase notice did not describe the property by reference to the registered title and it was expressly made "subject only to such modifications as are necessary or expedient in the circumstances" — The notice was entirely valid

Jurisdiction of court and of rent assessment committee

The question had been raised as to whether the county court (and therefore the Court of Appeal) had jurisdiction to M determine the points raised by Denetower — The jurisdiction of a rent assessment committee was set out in section 13(1) of the 1987 Act and the jurisdiction of the county court was by section 52 to hear and determine questions arising under Part I of the Act "other than questions falling within the jurisdiction of a rent assessment committee by virtue of section 13(1)" — It had been suggested that the validity of the notice was a matter falling within section 13(1) and therefore not within the jurisdiction of the county court — This was not correct — The rent assessment committee does not have jurisdiction to determine the validity of a notice served under

section 12 but only to determine questions arising under such a notice — The validity or otherwise of the notice served is a matter for the court to decide — The county court, therefore, had jurisdiction to decide on the validity of the notice — In order to decide the validity of the notice the court had to decide, *inter alia*, what the tenants were entitled to acquire — But, in default of agreement, it will be for the rent assessment committee to decide on the details of that issue in the light of the court's judgment

The following case is referred to in this report.

St Thomas' Hospital (Governors of) v *Charing Cross Railway Co* (1861) 1 John & H 400; 30 LJ Ch 395; 25 JP 771; 7 Jur NS 256; 9 WR 411

This was an appeal by tenants of two blocks of flats at Brookdene, 71 Holden Road, London N12, from the decision of Judge Goldstone, at Barnet County Court, that they were not "qualifying tenants" and thus not entitled to serve a purchase notice under section 12 of the Landlord and Tenant Act 1987 on their new landlords, Denetower Ltd, in pursuance of their right of first refusal under the Act.

Michael O'Sullivan (instructed by Moerans, of Edgware) appeared on behalf of the appellant tenants; Simon Brilliant (instructed by Wallace & Partners) represented the respondents.

Giving judgment, **Sir Nicolas Browne-Wilkinson V-C** said: This is an appeal from a decision of His Honour Judge Goldstone who, on the application of Denetower Ltd, declared that the appellants, who are tenants of the flats at Brookdene, 71 Holden Road, London N12, are not "qualifying tenants" within the meaning of section 3 of the Landlord and Tenant Act 1987 ("the Act"). There is also a cross-appeal by Denetower, which I will explain hereafter.

The Act was passed with a view to giving leaseholders of residential flats in a block of flats improved rights to control the upkeep and maintenance of the block as a whole. Part I of the Act confers on such tenants a right of first refusal when the landlord is proposing to dispose of his reversion. If the landlord assigns the reversion in breach of such right of first refusal, the tenants are given a right to acquire the reversion from the assignee. This case is concerned with that latter right, ie the tenants' right to acquire from an assignee from the original landlord.

The framework of Part I of the Act, in outline, is as follows. Under section 1(2), Part I of the Act is made to apply to premises which "consist of" the whole or part of a building which contains two or more flats held by "qualifying tenants" and if more than one-half of the flats in the premises are held by such qualifying tenants. Under section 1(1) the landlord cannot make "a relevant disposal" without serving on the qualifying tenants an offer notice under section 5, giving to the qualifying tenants a right of first refusal. Sections 5 to 10 contain elaborate provisions relating to the terms to be contained in the offer notice by the landlord, the acceptance of such offer or the making of a counter-offer by the tenants and other matters. If, for some reason, the right of first refusal is not taken up by the tenants, the landlord is then free for one year to dispose of the reversion on the same terms and at the same price as he has offered it to the tenants.

Sections 11 to 17 of the Act (which are the provisions directly in point in this case) deal with the position where the original landlord has, in breach of his obligations under section 1, disposed of his reversion to a new landlord. In those circumstances, section 12(1) provides as follows:

> the requisite majority of qualifying tenants of the constituent flats may, before the end of the period specified in subsection (2), serve a notice ("a purchase notice") on the new landlord requiring him (except as provided by the following provisions of this Part) to dispose of the estate or interest that was the subject-matter of the original disposal, on the terms on which it was made (including those relating to the consideration payable), to a person or persons nominated for the purposes of this section by any such majority of qualifying tenants of those flats.

It will be observed that, under section 12(1), the tenants' right is to acquire from the new landlord the property which was transferred to him by the original landlord. That transfer from the original landlord to the new landlord may well have included property in addition to the relevant block of flats. To deal with that contingency, section 12(3) provides as follows:

> a purchase notice —
> (a) shall, where the estate or interest that was the subject-matter of the original disposal related to any property in addition to the premises to which this Part applied at the time of the disposal —
> (i) require the new landlord to dispose of that estate or interest only so far as relating to those premises, and
> (ii) require him to do so on the terms referred to in subsection (1) subject to such modifications as are necessary or expedient in the circumstances;
> (b) may, instead of specifying the estate or interest to be disposed of or any particular terms on which the disposal is to be made by the new landlord (whether doing so expressly or by reference to the original disposal), provide for that estate or interest, or (as the case may be) for any such terms, to be determined by a rent assessment committee in accordance with section 13.

Brookdene consists of a pair of two-storey buildings facing on to Holden Road. Each building contains two ground-floor and two first-floor flats, making eight in all. Each ground-floor flat has two gardens, one at the front and one at the back, contiguous to the flat. Each first-floor flat has a garden which is not immediately adjacent to the building but situated behind the rear gardens of the ground-floor flats. A roadway runs between the two blocks leading to eight garages at the rear of the property. Finally, there is a small piece of land at the rear which is apparently unused and over which the tenants have no rights ("the unused land").

Each of the tenants holds a long lease of his or her flat and the garden which goes with it. Under the leases, each tenant is granted a right of way on foot either over a pathway leading to his flat and garden or over a narrow strip of the roadway between the buildings. It is common ground that each lease includes in the demise of the flat the appropriate part of the external walls of the building.

The garages are not included in the demise of the flats. Most of the tenants hold a long lease over one of the garages, but one of the garages is let to a non-tenant and the freehold of two of the garages has been sold to an outsider.

All the land, buildings, gardens, garages, roadway and unused land are registered at the Land Registry in one title described as "Brookdene, 71 Holden Road".

On November 8 1988, the original landlord transferred the freehold title to the reversion to Denetower Ltd, without first having complied with the requirements of section 1 of the Act by offering the reversion to the tenants. On discovering the transfer, the tenants, within time, served on Denetower a notice in the following terms:

> S12 Landlord and Tenant Act 1987
> Right of qualifying tenants to compel sale by new landlord
>
> To: Denetower Limited . . .
>
> We the requisite majority of the qualifying TENANTS of the constituent flats situate at Brookdene 71 Holden Road London N12 7DR HEREBY SERVE you with a NOTICE ("a purchase notice") requiring you to dispose of the estate or interest relating to the freehold reversion of Brookdene 71 Holden Road aforesaid TO THE QUALIFYING TENANTS NAMED BELOW ON THE TERMS on which the original disposal to you was made (including those relating to the consideration payable) SUBJECT ONLY TO such modifications as are necessary or expedient in the circumstances.

The notice was dated June 7 1989 and signed by all the tenants of both buildings.

On January 11 1990 Denetower Ltd applied to the Barnet County Court joining all the tenants as respondents seeking the following declarations:

> 1 A declaration that none of [the tenants] is a qualifying tenant of his or her respective flat at Brookdene within the meaning of section 3 Landlord & Tenant Act 1987 . . .
> 2 Alternatively, if it be found that any [tenant] is such a qualifying tenant, a declaration that a purchase notice served by the [tenants] and dated 7th June 1989 is invalid and of no effect.
> 3 In the further alternative, if it be found that the said purchase notice is valid, a declaration that the Applicants are only required to dispose of the two building(s) containing the [tenants'] respective flats and not the garages and grounds at Brookdene.

The judge held in favour of Denetower on the first declaration but indicated that, if it had been necessary for him to decide the points, he would have decided against Denetower on the second and third declarations. The tenants appeal against the judge's decision: Denetower cross-appeals on the subject of the second and third declarations.

The questions which accordingly arise for our decision are as follows:

1 Whether the tenants of Brookdene were "qualifying tenants" of their flats entitled to serve a purchase notice under section 12(3);

2 If so, whether the tenants are entitled to purchase only the buildings containing the flats, to the exclusion of, *inter alia*, the gardens and garages;

3 Whether the purchase notice dated June 7 1989 complied with the requirements of section 12(3).

1 *Are the tenants "qualifying tenants"?*

Under section 12(1), a purchase notice can be served only by the requisite majority of qualifying tenants. Denetower contends that the tenants are not "qualifying tenants" within the definition in the Act.

The term "qualifying tenants" is defined by section 3 of the Act for the purposes of Part I of the Act. It is common ground that the tenants are qualifying tenants unless they are excluded by the provisions of section 3(2)(a), which reads as follows:

> (2) A person is not to be regarded as being a qualifying tenant of any flat contained in any particular premises consisting of the whole or part of a building if —
> (a) he is the tenant of any such flat solely by reason of a tenancy under which the demised premises consist of or include —
> (i) the flat and one or more other flats, or
> (ii) the flat and any common parts of the building; or . . .

Section 3(2) has been amended by the Housing Act 1988. Such amendment was not retrospective so as to apply to this case. Section 60 of the Act provides that "'common parts', in relation to any building or part of a building, includes the structure and exterior of that building or part and any common facilities within it".

Denetower contends (and the judge held) that, since the lease of each flat included in the demise part of the exterior of the building, the demised premises include not only the flat but also "common parts of the building" within the definition in section 60. Therefore, it is said, the tenants fall within section 3(2)(a)(ii) in that the demised premises include both the flat and common parts of the building. Accordingly, it is said, none of the tenants is a "qualifying tenant" capable of serving a notice under section 12(1).

I do not accept these submissions. No one has been able to suggest what was the parliamentary intention lying behind the exclusion contained in section 3(2). If Denetower's submission is right, section 3(2) will have effectively emasculated the Act as originally enacted. The overwhelming majority of flat leases must include in the demise some part of the "structure" of the building in the form of load-bearing walls, whether internal or external. Moreover, in Part III of the Act (which provides for compulsory acquisition of the reversion), section 26 contains a similar definition of "qualifying tenants". Therefore, if the contention is correct, the protection of Parts I and III of the Act would be applicable only to the miniscule number of cases where a block of flats is let on terms which exclude structural walls, floors and ceilings from the demise. If Parliament has so clearly legislated, so be it. But in the absence of clear words I am very reluctant to reach such a conclusion.

In my judgment, the words of section 3(2)(a) are quite capable of bearing the meaning that, in order for the tenant to be excluded, there have to be two elements in the demise: first "the flat" and then, in addition, certain common parts. The words are "the flat *and* any common parts". The mere fact that "the flat" itself contains common parts is not enough: there has to be some element in addition to the flat which is part of the common parts.

It is, I accept, possible to construe the phrase in isolation as referring simply to the case where the demise contains two characteristics (as opposed to elements), *viz* the demise includes a flat and includes common parts, whether or not the latter form part of the flat. But, at best, this raises an ambiguity in the section which makes it legitimate to have regard to later legislation to explain that ambiguity. Para 2(1) of Schedule 13 to the Housing Act 1988 (which amends section 3(2) of the 1987 Act) describes section 3(2) as enacted as being a provision "which excludes persons having interests going beyond a particular flat". In my judgment, those words clearly show that in order to be excluded the tenant has to be entitled to something which is "beyond" the flat, ie not included in the flat.

2 *What are the tenants entitled to acquire?*

The basic right, conferred by section 12(1), is for the tenants to acquire the whole of the property transferred to Denetower, ie all the land comprised in the registered title. However, Denetower contends that section 12(3) applies, since "the estate or interest that was the subject-matter" of the disposal to Denetower related to property "in addition to the premises to which this Part applied at the time of the disposal". This requires one to identify the premises to which Part I of the Act applied. These are defined by section 1(2) which, so far as relevant, reads as follows:

> This Part applies to premises if —
> (a) they consist of the whole or part of a building; and
> (b) they contain two or more flats held by qualifying tenants; and
> (c) the number of flats held by such tenants exceed 50 per cent of the total number of flats contained in the premises.

Denetower contends that the word "consist" in its normal meaning means "comprise" or "consist only of". It therefore contends that the only "premises to which this Part applied" and to which the section 12 notice could relate were the physical structures consisting of the two blocks of flats as being the only "buildings" in which the flats were contained. Accordingly, it is said, under section 12(3) there was no power to give a purchase notice relating to the gardens, garages, roadway or the unused land, since they were not part of the building of which the relevant premises consisted.

To this the tenants have two answers. First, they submit that the requirement in section 1(2) that the premises must "consist of" a building does not mean that the premises must consist *only* of the building: the word "consist" can be used as meaning "include". I have some doubt whether this argument is correct, since the draftsman has shown in section 3(2)(a) that where he means to say "consist of or include" he does so in express terms.

However, it is unnecessary to decide this point, since I accept the tenants' second argument. They submit that the word "building" is not necessarily confined to the bricks and mortar of which the building is constructed. In *Governors of St Thomas' Hospital* v *Charing Cross Railway Co* (1861) 1 John & H 400 it was held that section 92 of the Lands Clauses Consolidation Act 1845 (which provided that the owner of land being compulsorily acquired could not be required to convey "a part only of any house, or other building or manufactory") required the purchase not only of the whole house but also of the gardens and appurtenances of the house.

In the present case, it would be to attribute to Parliament an entirely capricious intention if we were to hold that the tenants' right to purchase did not extend to the gardens and other appurtenances of the flats which are expressly or impliedly included in the demises of the flats to the tenants. In my judgment, we are not forced to adopt such an unreasonable construction, since it is a perfectly legitimate meaning of the word "building" that it includes the appurtenances of the building.

Denetower submits that there are other provisions in the Act which indicate that in this Act the word "building" does not include the appurtenances of such building. Under Part III of the Act, tenants can compulsorily acquire the reversion in certain circumstances. Under section 29(1) the court can make an order affecting "premises to which this Part applies". The premises to which Part III apply are defined by section 25(2) so far as relevant in the same way as section 1(2), ie as premises "consisting of" a building. Yet, section 29(4)(a) provides expressly as follows:

> An acquisition order may, if the court thinks fit —
> (a) include any yard, garden, outhouse or appurtenance belonging to, or usually enjoyed with, the premises specified in the application on which the order is made;

Denetower submits that section 29(4) shows that, apart from its provisions, yards, gardens, outhouses and appurtenances would not be included in the word "premises". While admitting the force of this point, in my judgment it is not decisive in construing such an ill-drafted, complicated and confused Act as this. In my judgment, section 29(4) was inserted in Part III out of an abundance of caution. Certainly there is no logical reason why the tenant who acquires under section 12 of the Act should not be entitled to acquire exactly the same property as could be acquired under Part III: yet there is nothing corresponding to section 29(4) in Part I of the Act.

I therefore reach the conclusion that the purchase notice under section 12 could have required Denetower to transfer not only the two buildings but also any appurtenances of that building.

What, then, are the appurtenances of the buildings? There can be no doubt that the gardens are included: they are expressly included in the leases of the flats which are part of the buildings. In my judgment, the garages are not appurtenances. The tenants enjoy no rights over the garages under or by virtue of their leases of the flats. The garages are held under quite separate leases; not every flatholder has a garage; the freehold of all the garages is not vested in Denetower. In those circumstances it is impossible to hold that the garages are

A appurtenant to the buildings. Similarly, the tenants of the buildings enjoy no rights over the unused land, nor is such land used in conjunction with the flats: therefore it is not an appurtenance of the building. As to the roadways and paths over which the tenants have either express or prescriptive rights of way, in my judgment they are appurtenances of the building. However, it may well be that a rent assessment committee acting under section 13(1) may reach the conclusion that the appropriate provision should be that the tenants should not acquire the freehold in the land of the roadway and paths but be granted perpetual rights of way over them.

3 *Was the section 12 notice valid?*

Denetower submits that the notice served on June 7 1989 required
B Denetower to dispose of the whole of the land comprised in the title registered at the Land Registry, ie buildings, gardens, garages, roadway and unused land. Since, as I have held, the tenants were not entitled to acquire all that land, it is submitted that the notice failed to comply with section 12(3) and was therefore invalid. If this argument is right, since the time for serving a section 12 notice has now expired, the tenants will have lost their rights completely.

The basis of this argument is that the notice, on its true construction, purports to require Denetower to transfer to the tenants everything which Denetower acquired from the original landlord, ie that the notice was a notice under section 12(1) not under section 12(3). I do not so construe it. First, the property is not
C described by reference to the registered title. The tenants are described as "the qualifying tenants of the constituent flats situate at Brookdene, 71 Holden Road . . .". This indicates that the notice, in referring to Brookdene, is referring to the buildings in which the flats are situate and not to the land comprised in the registered title. Next, the notice does not require (as would a notice under section 12(1)) the transfer to be made simply on the same terms as the original disposal: the terms are to be "subject only to such modifications as are necessary or expedient in the circumstances". Those latter words are a direct quotation from section 12(3)(*a*)(ii) and are appropriate only to a notice under section 12(3).

In my judgment, therefore, on its true construction the notice required the transfer of Brookdene (meaning the blocks of flats) plus
D their appurtenances on the same terms as the whole property was transferred to Denetower, but subject to the necessary modification to take account of the fact that the notice did not require the transfer of the whole property. As such, the notice was entirely valid. Given the complexity of the Act, a draftsman of a notice would be presented with an impossible task in seeking to specify in detail the exact property to be acquired. A notice in terms such as those contained in the present notice gives to the new landlord adequate notice of the claim. The details of the property to be acquired and the terms of the acquisition can in default of agreement be sorted out by the rent assessment committee on an application under section 13(1).

Jurisdiction

E In the course of argument, the question was raised whether the county court (and therefore this court) had jurisdiction to determine the points raised by Denetower. Under section 13(1) a rent assessment committee is given jurisdiction to hear and determine:

> (a) any question arising in relation to any matters specified in a purchase notice (whether relating to the nature of the estate or interest, or the identity of the property, to be disposed of or relating to any other terms on which the disposal by the new landlord is to be made); and
>
> (b) any question arising for determination in consequence of a provision in a purchase notice such as is mentioned in section 12(3)(b).

Under section 52 of the Act, the county court is given jurisdiction to hear and determine any questions arising under Part I of the Act
F "other than a question falling within the jurisdiction of a rent assessment committee by virtue of section 13(1) . . .". It was suggested that the validity of the notice was a matter falling within section 13(1) and therefore not within the jurisdiction of the county court.

In my judgment, the rent assessment committee does not have jurisdiction to determine the validity of a notice served under section 12, but only to determine questions arising under such a notice. The validity or otherwise of the notice served is a matter for the court to decide. Therefore the county court had jurisdiction to decide the validity of the notice. In order to decide on the validity of the notice, we have had to decide question 2 (above), *viz* what were the tenants entitled to acquire? But, in default of agreement, it will be for the rent assessment committee to decide on the details of that issue in the light G of this judgment.

I would therefore allow the appeal and declare that the tenants have at all material times been "qualifying tenants" within the meaning of Part I of the Act and that the notice dated June 7 1989 was a valid notice under section 12(3) of the Act.

Stocker and **Beldam LJJ** agreed and did not add anything of their own.

The appeal was allowed; declarations granted that the defendants have at all times been qualifying tenants within the meaning of Part I of the Landlord and Tenant Act 1987 and that the notice dated June 7 1989 was a valid notice under section 12(3) of the Act; plaintiff H *company to pay the defendants' costs here and below; application for leave to appeal to the House of Lords refused.*

Court of Appeal

February 6 1991

(Before Lord Justice STAUGHTON and Lord Justice McCOWAN)

SOOD v BARKER

J

Estates Gazette June 15 1991

[1991] 23 EG 112

Landlord and tenant — Breach of covenants — Use as a "public house beer shop" prohibited — Use without consent for purposes other than for sale or hire of videos restricted — Whether consent unreasonably withheld for change of use — County court judge decided all these questions adversely to tenant — Held by Court of Appeal, dismissing tenant's appeal, that there was ample evidence on which the judge could so decide, including his refusal to exercise his discretion to grant relief against forfeiture K

The plaintiff tenant held a nine-year term of shop premises — His lease prohibited, *inter alia*, the use of the premises for the trade or business of a "public house beer shop" and prevented without the consent of the landlord the use of the premises for any purpose other than the sale or hire of video tapes or equipment — The plaintiff, finding his video business falling off, obtained an off-licence and began to sell some wines and spirits and lager and Yorkshire bitter in cans — The landlord objected and served a notice under section 146 of the Law of Property Act 1925 requiring breaches to be remedied; failing which, the landlord stated that he would forfeit the L **lease — The tenant then commenced proceedings for a declaration that consent was being withheld unreasonably for the sale of intoxicating liquor — At the trial the county court judge held that there had been breaches of both the absolute and the qualified covenant, that consent had not been withheld unreasonably in respect of the latter, and that relief against forfeiture would not be granted — The tenant appealed**

The tenant contended that he was not in breach of the absolute covenant because the alcoholic liquor sold amounted only to a small percentage of his turnover and its display occupied only a small part of his shelf space — The judge had M **not accepted that evidence and from photographs shown to the Court of Appeal it appeared that there were four shelves containing intoxicating liquor of one kind or another, each perhaps 3m long — They may have comprised as much as a third or nearly a half of the shop — There was a possible query about the meaning of a "beer shop", but the old case of *Bishop of St Albans* v *Battersby* was authority that a beer shop was a shop where beer was sold for consumption off the premises — As to whether consent had been unreasonably withheld for the change of use, the landlord had asked eight questions to be answered in support of the tenant's**

application for consent — Some of these questions might have ranged too far but the tenant answered only one in part and the others not at all — In these circumstances it was not unreasonable for the landlord to refuse consent — The view was supported by the cases of *Fuller's Theatre & Vaudeville Co Ltd* v *Rofe* and *Isow's Restaurants Ltd* v *Greenhaven (Piccadilly) Properties Ltd* — On the final question, as to whether relief should be granted, the judge had given his reasons for refusal at considerable length and they included material which was perfectly adequate to justify refusal — Apart from these specific matters, the judge had found that he could not accept anything the tenant said in evidence; that there had been falsehood in the proceedings; and that points had been taken which could not really be regarded as put forward in good faith

In these circumstances it was within the judge's discretion to refuse relief and the appeal must be dismissed

The following cases are referred to in this report.

Fuller's Theatre & Vaudeville Co Ltd v *Rofe* [1923] AC 435; 128 LT 774; 39 TLR 236, PC

Isow's Restaurants Ltd v *Greenhaven (Piccadilly) Properties Ltd* [1970] EGD 82: (1969) 213 EG 505

St Albans (Bishop of) v *Battersby* (1878) 3 QBD 359

This was an appeal by Mr S C Sood, trading as Kumar Videos International, from the decision of Judge S M Willis, at Croydon County Court, dismissing an application for a declaration that permission was being withheld unreasonably by the landlord, Mr Paul Barker, for the sale of intoxicating liquor; and holding that the lease of the premises at 906 London Road, Thornton Heath, had been forfeited and granting an order for possession.

The appellant appeared in person; Michael Harington (instructed by Copley Singletons of Croydon) represented the respondent.

Giving judgment, STAUGHTON LJ said: This is an appeal from a decision of His Honour Judge Willis at the Croydon County Court. The story begins on October 20 1983 with a lease granted by Paul Anthony Barker (the defendant in this action and respondent to the appeal) to Kumar Videos International. It was for a term of nine years and related to premises at 906 London Road, Thornton Heath, which were a shop. There were covenants, and in particular clause 22, which reads as follows:

(a) That the demised premises shall not be used for any illegal or immoral purpose of any noisy noisome or offence

that does not seem to make sense

to the trade or business or as a public house beer shop betting shop or public place of amusement nor shall there be held or omitted any sale or auction or public meeting on the demised premises.

(b) Not without the consent of the landlord which shall not be unreasonably held in respect of trades not being of a competitive nature to that being carried on by the landlord to use the premises for any purpose other than for the sale and hire of video tapes and equipment provided that there shall not be implied any warranty or other obligation or representation on the part of the landlord that such is the permitted use for the purpose of the Town and Country Planning Act 1962 and 1968.

(Any apparent misprints in the clause are not mine but in the original.) There is also the usual proviso for re-entry in the case of breach of covenant.

Mr S C Sood, one of the plaintiffs, claims that he is the sole proprietor of, and trades as the tenant under the lease, Kumar Videos International. At one stage in the proceedings a Mr N C Sood was joined as co-plaintiff by order of the registrar. After the trial, the judge ordered that he should be deleted from the proceedings and that instead Mr R K Sood, a brother of Mr S C Sood, should be substituted, and that is the state of the action at this date.

In 1988, Mr Sood says, the video business was not doing too well. An off-licence nearby had closed down, and he applied for a licence to sell intoxicating liquor. His application was refused by the justices, but granted on December 15 by Croydon Crown Court. He then or thereabouts started to sell wines, spirits, lager in cans and Webster's Yorkshire Bitter in cans. On February 16 1989 there was a letter from the landlord's agents complaining that the use of the premises had been changed without consent and that this was a breach of the terms of the lease. They required Mr Sood to refrain forthwith from this unauthorised use.

There was then some correspondence, to which I shall return later. On May 17 1989 the solicitors of the landlord served a notice under section 146 of the Law of Property Act 1925, saying that there had been breaches of covenant in respect of the sale of alcoholic and other drinks, including beer for consumption off the premises, the premises had been used for the sale of food, and that they had been used as an employment agency. They required the breaches to be remedied and compensation in money. They said that, failing compliance within a reasonable time, it was the intention of the landlord to re-enter and forfeit the lease.

There was some further correspondence, and on June 13 1989 Mr S C Sood started proceedings in the Croydon County Court. He claimed a declaration that permission was being withheld unreasonably by the landlord to the sale of intoxicating liquor. On July 11 1989 the landlord's solicitors served a defence and counterclaim in which they claimed possession of the premises.

The case eventually came on for trial on July 30 and 31 1990, when Mr S C Sood was represented by his brother, Mr R K Sood, who eventually became a co-plaintiff, as he has been represented before us. The judge held that there were breaches both of clause 22(a) and clause 22(b), that the landlord's consent had not been withheld unreasonably and that there should be no relief against forfeiture. Accordingly, he held that the lease had been forfeited and granted an order for possession.

Notice of appeal has been given by Mr S C Sood alone. There are, as it seems to me, three issues. First, has there been a breach of clause 22(a) by the use of the premises as a beer shop? Second, has there been a breach of clause 22(b) by a change of use, and has the landlord's refusal of consent been unreasonable? Third, should there be relief against forfeiture?

I turn first to the question about clause 22(a) — use of the premises as a beer shop. The facts are that Mr S C Sood took to selling lager and Webster's Yorkshire Bitter in cans. He also sold wines and spirits. He said he did so on a small scale. He claimed that it was no more than 5% of his turnover as a whole and that the lager and Yorkshire Bitter were only half of that, about 2.5%. He said that there were only 3m of shelves in the shop devoted to lager, Yorkshire Bitter, wines and spirits as a whole.

The judge did not accept that evidence, or indeed anything else that was said on behalf of the plaintiffs, but he did not make an affirmative finding as to what percentage of the trade was in those commodities. We have been provided with some photographs, which the judge saw. They seem to show that there were four shelves containing intoxicating liquor of one kind or another, each perhaps 3 m long. They may have comprised as much as a third or nearly a half of the shop. We have no further information as to what the sales were.

The first point taken is that the lease contains the words "public house beer shop", without, as is unfortunately often the case, any punctuation. Mr Sood submits that that is all one noun. I am afraid that I cannot accept that argument. It seems to me that it is referring to two creatures, a public house and a beer shop. Next, it is argued that neither lager nor Webster's Yorkshire Bitter is beer. That is, to my mind, manifestly and plainly wrong. One can refer to "Yorkshire bitter", or "bitter beer" or "mild beer" or "Newcastle brown ale" for that matter. They are all beer. Then it is said that this only happened to a small extent. But I do not think that this was so trivial that it could be overlooked. It was argued that it was ancillary to the sale of videos, just as the sale of cigarettes and cigars in a restaurant might be said to be ancillary to the sale of food and drink. I cannot accept that argument either.

What did strike me initially as an arguable point is the meaning of the words "beer shop". It is not a term in current use, as far as I know, and it probably has not been for a number of years. This illustrates the vice of using old precedents which contain language which is no longer in current use. The new *English Dictionary* published by Collins first defines "beer house" as a tavern and then "beer shop" likewise as a tavern. That would make some sense of clause 22(a), because it would mean that it prohibited use of the premises as a public house where intoxicating liquors of all kinds were sold and an on-licence place where only beer was sold. If, on the other hand, "beer shop" meant a retail establishment for the sale of beer for consumption off the premises, one wonders why the clause did not also prohibit a wine shop, a gin shop or a whisky shop.

However, there is authority which shows this view is wrong. That is the case of the *Bishop of St Albans* v *Battersby* (1878) 3 QBD 359, where a man who was a grocer and baker also sold beer, and there was a covenant that he would not without the consent of the landlord use the premises as a beer shop, or public house, or any theatre, or public show, or exhibition. Cockburn CJ held that this was a breach of covenant, because a beer shop was a place where beer was sold for consumption off the premises. Mellor J agreed. After such a length of time, I do not think that it would be right for this court to depart from that interpretation of the word. Consequently, there was a breach of covenant in selling beer in this establishment.

The clause contains nothing about consent. Had it provided expressly that this was not to be done without the consent of the landlord, section 19 of the Landlord and Tenant Act 1927 would have had the effect that, whatever else the clause said, it was to be implied that the consent of the landlord was not to be unreasonably withheld. But as clause 22(a) contains no mention of the consent of the landlord, section 19 has no effect and the covenant is absolute.

I turn to the second question — whether there has been a breach of clause 22(b) in that the premises have been used for a purpose other than the sale and hire of videotapes and equipment without the consent of the landlord. For the reasons already given, I do not consider that the sale of intoxicating liquor was so minimal that it could be overlooked, or ancillary to the sale of videotapes. Accordingly, there has been a breach. But has consent been unreasonably withheld subsequently?

What happened was this. On April 13 1989 the landlord's solicitors wrote a letter, wrongly described as "without prejudice", and they said:

> To avoid unnecessary expense, inconvenience or unpleasantness, we are instructed by our client's managing agents to tell you that they are prepared to consider any reasonable application for licence to change the use of the demised premises. If you wish to make such an application you should deliver it to the office of [the landlord's agents]. The application should include a full statement of the following:
>
> 1 The use or uses to which the premises are now being put and will be put;
> 2 When such use or uses first started;
> 3 The full names of the proprietors of the business or businesses;
> 4 Copies of any off-licence and/or employment agency licence granted in respect of the business now being run at the premises;
> 5 The full addresses of any and all other properties at which you are carrying on the same or any other business or businesses;
> 6 The full names and addresses of the landlords of any and all such premises, whom our client's managing agents may contact to obtain references;
> 7 The names and addresses of the bank or banks at which you run accounts relating to the business or businesses, from whom also the managing agents may seek references;
> 8 The full trading accounts (including profit and loss account and balance sheet) of the business or businesses for the last three years or such shorter period as the business or businesses have been in operation.

They then say that this is to be answered within seven days.

Mr Sood says that they knew some of this information already, but not all of it. Some of those questions could very easily have been answered. None of them would have involved any great difficulty, although some involved confidential information which Mr Sood was not under any obligation to disclose to anybody and might wish to keep confidential unless the landlord had reasonable grounds for asking about it. On April 19 1989 there was a reply from Mr Sood which said:

> . . . this is to inform you that in pursuance of an off-licence granted to us for the sale of liquors at the above premises, we seek the landlords consent for the partial change of use.

So he had, in part at any rate, answered question 1. He purported to answer it entirely, but he might not have done so accurately.

None of the other questions was ever answered. On April 24 the managing agents wrote that the matter was in the hands of solicitors and it was important that Mr Sood respond to them promptly either directly or through his own solicitors, and they concluded:

> We shall be obliged, therefore, if you will respond in full to the enquiries raised in their correspondence to you.

There then came the section 146 notice, to which I have already referred. On May 19 Mr Sood wrote to the landlords' solicitors:

> We acknowledge receipt of your notice under s146 Law of Property Act 1925. However, we DO NOT accept the accusations made in the said notice. We have already applied to the managing agents to seek the Landlord's permission to sell intoxicating drinks off the premises, the use of which is not restricted under the terms of the lease anyway.

Then they refer to payment of rent. The landlord's solicitors replied on May 22. They challenged the contention that there was no prohibition on the use of the premises as an off-licence.

There it is. The judge held that the landlord was entitled to answers to all the questions asked before considering whether to grant consent. I quote from the judgment:

> The defendant is entitled to that information and the law is quite clear that where information is refused, as it has been here, it cannot be suggested that the landlord is being unreasonable in withholding his consent.

The judge did not refer to authorities for that proposition. We have been referred to two cases by Mr Harington. First, the case of *Fuller's Theatre & Vaudeville Co Ltd* v *Rofe* [1923] AC 435. That was a case of subletting without consent. The tenant had applied for consent but had refused to disclose the terms of the proposed sublease. His refusal was based on the argument that, as the landlord had unequivocally agreed to give his consent upon performance of a condition which had been fulfilled, there was no point in disclosing to him the terms of the sublease. Nevertheless, the Judicial Committee of the Privy Council held that the landlord was entitled to require to know the terms of the sublease and was not unreasonable in refusing consent until he had done so. One can quite see in general that, in the case of a covenant against subletting, the terms of the proposed sublease would be material for the landlord to know before he granted or refused consent.

The other case was *Isow's Restaurants Ltd* v *Greenhaven (Piccadilly) Properties Ltd* (1969) 213 EG 505. That was a case of both underletting and change of use. The proposed underlessee was to be a cinema club which was to show films of a kind not generally available to the public. The landlord wished to know what sort of films they were going to be and what record the proposed sublessee had of showing such films elsewhere. One can well see that those were matters which a landlord would reasonably wish to know before he granted or refused consent. Plowman J held that the information was for the greater part relevant.

I would not, for my part, accept the judge's proposition, if such it was, that a landlord is entitled to ask any questions he pleases and to refuse consent if he does not get an answer. However, one has to look at the questions asked in this case. Some of them were, I think, relevant to the question which the landlord had to decide — whether he should grant permission for these premises to be used not wholly as a video shop but also as a shop for selling intoxicating liquors. Others, as it seems to me, may well have been going too far. But Mr Sood did not answer any of them. In those circumstances, it seems to me that the landlord was not unreasonable in refusing his consent.

I should perhaps have mentioned earlier one argument of Mr Sood. This was that he had sent the landlord a copy of his application for a licence to sell intoxicating liquor, and the landlord had not objected to it. Whatever might have been the case if that were true, the judge did not accept that the letter was ever sent to the landlord. In fact, he concluded that it had not been.

So I come to the third issue, whether the court should grant Mr Sood relief against forfeiture under section 146 of the Law of Property Act 1925. Mr Sood, we are told, has a wife and two children, and this is his only livelihood. The business is not going well. We must take that into account, I suppose. He said that the shop next door had to close down. Hence he obtained a liquor licence. If he had given it up during the proceedings, he might have lost his chance, because two other shops close by also wished to have liquor licences; they might have been successful, and his licence might have been withdrawn. One remedy for that situation, which his brother and advocate described as "Catch 22", would have been to hasten on the proceedings in the Croydon County Court. Far from doing that, Mr Sood seems to have done the opposite.

The judge gave his reasons for refusing relief against forfeiture at considerable length. They cover two and a half closely written pages. In general, it seems to me that there was perfectly adequate material which would justify the judge in refusing relief. There are three points in the course of his discussion of the issue which have caused me some concern. He said that these were wilful breaches of covenant and, according to authority, that those should be relieved against by forfeiture only in exceptional cases. He observed that Mr Sood was still maintaining in his conduct of the action that he was entitled to act as he had done.

It does not seem to me that a tenant is precluded from asserting in court that he has committed no breach, and asking in the alternative for relief against forfeiture if he has committed a breach, provided always that that is done in good faith. If, on the other hand, a tenant puts up a case which is wholly unarguable and cannot be said to be put forward in good faith, then it may be that that is a powerful reason

A for refusing relief. Certainly some of the arguments advanced on Mr Sood's behalf in the Croydon County Court seem to me to come into the second category. The suggestion that the words "public house beer shop" comprise only one description of premises seems to me absurd, and even more absurd is the suggestion that neither lager nor Webster's Yorkshire Bitter is beer. I would find it difficult to accept that either of those arguments was put forward in good faith.

Second, the judge regarded as relevant the fact that Mr Sood never paid his rent on the due day but always at the last possible minute before he became in peril of forfeiture under the terms of the lease. I certainly agree that honest men should pay their debts on the day they are due. Unfortunately, in these times when people have come to realise the importance of cash flow, a great many of them do not do B that, but pay their debts only when there is a threat of some sanction in sight. I wonder whether it may be demanding too high a standard to say that a tenant who is always 20 days overdue but never 21 is to be reproached on that account, when he has an application for relief against forfeiture.

Third, there was an episode during the course of the proceedings when Mr Sood tendered the rent and the landlord or his solicitors or agents felt obliged to refuse it. Of course they had to, because the law states that acceptance of rent is a waiver of forfeiture. However, they proposed that Mr Sood should agree that it should be paid without prejudice so that there would be no waiver. Mr Sood declined to agree to that. For my part I would be inclined to think that he was exercising C his legal right. If the law is unfair to landlords in holding that in any and all circumstances an acceptance of rent is a waiver of forfeiture, that is a defect in the law, and Mr Sood cannot be reproached for taking advantage of it.

However, even without those points, it seems to me that there were ample grounds for the judge to refuse relief against forfeiture. They are set out at great length in the judgment and I do not repeat them. He found that he could not accept anything Mr Sood said in evidence. He found that there had been falsehood in the proceedings, that there had been undue delay, and that points had been taken which could not really be regarded as being put forward in good faith.

In those circumstances it was within the judge's discretion to refuse relief, and I would dismiss the appeal.

D

McCowan LJ agreed and did not add anything.

The appeal was dismissed with costs to be taxed if not agreed; costs order below to be varied.

Chancery Division

January 16 1991

(Before Mr Justice MILLETT)

E PRUDENTIAL ASSURANCE CO LTD v LONDON RESIDUARY BODY AND OTHERS

Estates Gazette June 25 1991

[1991] 25 EG 120

Landlord and tenant — Construction of agreement — Clause providing for tenancy of land to continue until the landlord, being the LCC in their capacity as highway authority, required it for road widening — Reversion passed to the GLC, then to the London Residuary Body and then to the second, third and fourth defendants in the present case, who had purchased the F land from the London Residuary Body — Effect of a notice to quit served on the tenants before the purchase — Was the clause as to the continuance of the tenancy repugnant to its nature? — Held, after discussion of authorities, that the notice to quit was valid

In this case the plaintiffs, the Prudential Assurance Co Ltd, raised issues as to the true construction of a tenancy agreement and the effect of a notice to quit purporting to determine it — By a memorandum of agreement in 1930 the LCC granted a tenancy, on a leaseback basis, to a tenant subject to a provision in clause 6 that the tenancy was to continue until the land in question was required by the council for the widening of Walworth Road, when it was terminable on two months' G **notice — As mentioned above, the reversion became vested in the London Residuary Body and then in the three other defendants to the present proceedings — The tenancy had become vested in the plaintiffs, the Prudential Assurance Co Ltd, who had sublet to the London Electricity Board — The GLC was a highway authority but the London Residuary Body was not — At some stage before the sale by the London Residuary Body they had served a notice purporting to terminate the tenancy under section 25 of the Landlord and Tenant Act 1954 — This notice was ineffective under the 1954 Act, but it was accepted that it was capable of taking effect as** H **a common law notice, provided that such a notice could be given by the London Residuary Body**

In the judge's view the agreement created a yearly tenancy, subject by clause 6 of the agreement to a restriction that it was to "continue until the land was required by the council for the purpose of the widening of Walworth Road", when at least two months' notice had to be given prior to the date of determination — The question then arose as to whether this restriction or condition was repugnant to the nature of a yearly tenancy — After considering a number of authorities the judge came to the conclusion that clause 6 did not exclude the tenants' right to determine the tenancy by notice and that the J **restriction placed on the landlords' right was not repugnant to such a tenancy — It was true that once the reversion had been transferred to the London Residuary Body, which was not a highway authority, it was no longer possible for the landlord to require the land for the purposes mentioned in clause 6 — What was the effect of that? It was clear that the council referred to in clause 6 must be both a highway authority, which required the land for the named purpose, and the landlord, which was capable of obtaining vacant possession by serving the requisite notice — The clause predicated a situation in which the landlord and the highway authority** K **were one and the same — In the judge's opinion, once they were different bodies clause 6 ceased to have any effect**

The effect of this construction was that the London Residuary Body had served a valid notice to quit and the second, third and fourth defendants, who had purchased from the Residuary Body, were entitled to possession

The following cases are referred to in this report.

Ashburn Anstalt v *Arnold* [1987] 2 EGLR 71; (1987) 284 EG 1375, CA
Breams Property Investment Co Ltd v *Stroulger* [1948] 2 KB 1; [1948] 1 All ER 758; (1948) 64 TLR 292, CA
Lace v *Chantler* [1944] KB 368
L *Midland Railway Co's Agreement, Re: Clay (Charles) & Sons Ltd* v *British Railways Board* [1971] Ch 725; [1971] 2 WLR 625; [1971] 1 All ER 1007; (1971) 22 P&CR 360, CA
Warner v *Browne* (1807) 8 East 165

In this case the plaintiffs, Prudential Assurance Co Ltd, raised a number of questions on the true construction and effect of a memorandum of agreement relating to land fronting 263-265 Walworth Road, London SE17, of which the plaintiffs were now the tenants and of which the reversion had been transferred from the first defendants, the London Residuary Body, to the second, third and fourth defendants, Barron Investments Ltd, Alan Moss Bayes, and Joan Estelle Bayes.

Paul de la Piquerie (instructed by the solicitor to the Prudential Corporation) appeared on behalf of the plaintiffs; Stephen Lloyd (instructed by Clifford Watts Compton) represented the defendants. M

Giving judgment, **Millett J** said: This case raises a number of questions on the true construction and effect of a memorandum of agreement dated December 19 1930 between the London County Council of the one part and Mr Samuel Nathan of the other part, which purported to grant a lease or tenancy of a piece of land fronting 263-265 Walworth Road. It appears that immediately prior to December 19 1930 Mr Nathan owned land fronting the Walworth Road and wished to redevelop the site, or part of it, by putting up a building upon it. The LCC, being the highway authority, contemplated the possible widening of Walworth Road and

A accordingly, on December 19 1930, acquired part of Mr Nathan's Land, consisting of the frontage of the premises to the Walworth Road, and on December 30 1930 leased it back to Mr Nathan, together with a right to put up a temporary building upon it. Without prejudging the questions I have to determine, it appears from the internal evidence of the document itself that the intention was that Mr Nathan should have the right to put up a temporary building on the frontage and to occupy it until such time as the LCC should determine to proceed with their proposal to widen the Walworth Road, whereupon he would have to give up possession. The new and permanent building which he intended to erect behind the temporary one would then have a frontage to the widened Walworth Road.

It is not in evidence whether similar transactions were entered into
B with adjoining premises on either side of the subject premises or further along Walworth Road.

Clause 1 of the memorandum of agreement is in the following terms:

> The Council hereby let to the Tenant and the Tenant takes from the Council the land (hereinafter called "the said Land") described in the Schedule hereto from 19th December 1930 at the rent of £30 per annum payable quarterly on the usual quarter days until the tenancy shall be determined as hereafter provided . . .

Clause 6 reads:

> The tenancy shall continue until the said Land is required by the Council for the purposes of the widening of Walworth Road and the street paving works
> C rendered necessary thereby and the Council shall give two months' notice to the Tenant at least prior to the day of determination when the said Land is so required and thereupon the Tenant shall give vacant possession to the Council of the said Land as hereinbefore provided.

The interest granted to the tenant by that document has become vested in the plaintiff, the Prudential Assurance Co Ltd, and in 1975 it sublet the premises subject to the memorandum of agreement, together with the freehold land behind it, to the London Electricity Board, which is still in occupation, at a substantial rent.

The proposal to widen the Walworth Road was never carried into effect and must, I apprehend, have been abandoned many years ago. In the course of time the LCC's rights and obligations became vested in their successor body, the Greater London Council, and a further
D statutory vesting occurred on the dissolution of the Greater London Council when its property, rights and interests were vested in the first defendant, the London Residuary Body. The LCC had, rightly or wrongly, conceived that it was not open to them to review the rent of £30 a year for the subject premises, which was fixed in 1930, or to determine the tenant's interest thereunder. The LCC and their successor, the GLC, were each the highway authority. The London Residuary Body, however, is not a highway authority. After taking advice they concluded that they could deal with the reversion on the footing that the interest created by the memorandum of agreement was terminable. Accordingly, on July 21 1988 the London Residuary Body sold the subject land to the second, third and fourth
E defendants, and the land was transferred to them by a transfer dated August 25 1988. Since that date the reversionary interest expectant on the determination of the interest created by the memorandum of agreement has been vested in the second, third and fourth defendants and, of course, ever since it became vested in the London Residuary Body it has been vested in a party other than a highway authority.

The question I have to decide is whether the memorandum of agreement created an interest which still subsists and which is not capable of being determined by the London Residuary Body or a successor in title not being a highway authority. Prior to the auction sale, the London Residuary Body purported to serve a notice under section 25 of the Landlord and Tenant Act 1954 determining the interest on December 19 1988. The notice was served on March 31
F 1988. It would have been a perfectly good notice if Part II of the Landlord and Tenant Act 1954 applied to the interest. However, since the plaintiff had sublet the whole of the land to the London Electricity Board and was not in possession of any part of it, it is plain that the land was not within Part II of the Landlord and Tenant Act 1954 at all. It is now common ground that the notice was not effective as a section 25 notice. The plaintiffs in fact served a counternotice claiming a new tenancy. Those proceedings have been transferred to the High Court and are before me, but it is common ground that they are a nullity and must be dismissed.

It is also common ground that the section 25 notice is capable of taking effect as a common law notice to quit and was apt to terminate an ordinary yearly tenancy if such a notice could be given by the London Residuary Body. The defendants rely upon it as a good G
common law notice. The parties are agreed as to the financial consequences which would ensue should I come to the conclusion that the notice was a valid common law notice to quit.

The defendants assert that the notice is a good notice to quit upon three alternative grounds. First, they submit that on its true construction the memorandum of agreement purported to demise the land for a fixed but uncertain term and that accordingly it created no legal estate in the land. Second, they submit that if a periodic tenancy was created, either by the memorandum of agreement itself or by Mr Nathan's entry into possession and payment of rent, then any restriction on either party's right to serve a notice determining the tenancy is void as repugnant to the nature of a periodic tenancy. They
submit that it was possible to envisage a situation, even in 1930, when H
it would become impossible for the landlord to serve such a notice, and that came about once the reversionary interest was vested in a body other than a highway authority. They submit also that in any event, on its true construction, clause 6 of the memorandum of agreement excludes any right of the tenant to give a notice determining the tenancy. Third, the defendants submit that on its true construction clause 6 restricts the landlord's right to serve notice to quit only if and so long as they are the relevant highway authority, and that any restriction on the right ceased when the land vested in the London Residuary Body.

The first question I must decide is whether, on its true
construction, the memorandum of agreement created a tenancy for a J
fixed term for, if it did, it is common ground that it infringes the requirement that the maximum duration of a tenancy for a fixed term must be known at its inception. If it created instead a periodic tenancy, it was obviously a yearly tenancy. Mr Lloyd, who appears for the defendants, submits that the structure and language of the memorandum of agreement are inconsistent with the creation of a periodic tenancy. The document does not purport to create a periodic tenancy and the words which it employs are not appropriate for that purpose. He points to the fact that the period of the demise is expressed as being from one given date until another: "*from* the 19th day of December 1930 *until* the Tenancy shall be determined as hereinafter provided". Moreover he points out that clause 6 itself
expressly provides that the tenancy shall continue until the said land K
is required by the council for the purposes of widening the Walworth Road: an express statement that the tenancy shall continue until a stated event. Mr Lloyd submits that the addition of the requirement that the council shall give two months' prior notice is merely machinery for enabling the date to be fixed by which possession shall be given.

I see the force of those submissions but I do not accept them. In my judgment, on its true construction, the memorandum of agreement was apt to create a periodic tenancy subject to a modification of the landlord's right to serve notice to quit. My ground for this conclusion is that the determining event is not, as it was in *Lace* v *Chantler* [1944]
KB 368, outside the control of either party. The determining event is L
the expiry of a notice to quit. Clause 1 of the memorandum of agreement so provides, "until the tenancy shall be determined as hereinafter provided", thus incorporating clause 6; and although clause 6 opens with the words, "The tenancy shall continue until the said land is required by the council for the purposes of the widening of Walworth Road" (which is also now outside the control of the landlord), it is apparent that in fact the tenancy continues until the requisite notice has been given and expires. It is not enough that the council should require the land for the stated purpose; they must state the fact in an appropriate notice, and then the tenancy continues until the notice expires. What brings the tenancy to an end is the expiry of the notice.

I am persuaded by Mr de la Piquerie that on the true construction M
of the memorandum of agreement it created a yearly tenancy subject nevertheless to determination by at least two months' notice by the landlord if, and only if, the land was required by the landlord "for the purposes of the widening of Walworth Road and the street paving works rendered necessary thereby". As he points out, clause 1 begins by granting a tenancy at a rent of £30 per annum payable on the usual quarter days. Had it stopped there, with nothing more, it would have created an express periodic tenancy. Had it stopped there and had clause 6 provided for the mode of determination, the words "the tenancy" in clause 6 would be construed as meaning "the yearly tenancy hereby created". The addition of the words "until the tenancy shall be determined as hereinafter provided" which link the

two clauses cannot change the nature of the tenancy created by the document. In my judgment there is nothing inconsistent with the language of the document to read it as if the word "yearly" appeared between "the" and "tenancy" wherever they appear. The fact that at first sight it appears to be determinable only by notice served by the landlord does not affect the matter for reasons which I shall consider in a moment.

If I am wrong about that and the memorandum of agreement creates a tenancy for a fixed term, it is agreed that the term is uncertain and that the document infringes the rule that the maximum duration of a fixed term of years must be known at the date of its creation. It would follow that any such fixed term was void, but there would still be a yearly tenancy which came into existence when Mr Nathan entered into possession of the land and paid the rent. That tenancy would be upon all the terms of the memorandum of agreement capable of applying to a periodic tenancy, and in so far as clause 6 modifies the rights of each party to give notice to quit and is not repugnant to a periodic tenancy it would continue to apply. Accordingly, in my judgment, the question whether this was a fixed-term tenancy which was void, thus letting in a periodic tenancy, or whether on its true construction the memorandum of agreement created an express periodic tenancy, does not determine the validity of the notice to quit, which has been served.

I therefore proceed on the basis that the memorandum of agreement created a periodic tenancy, the landlord's right to serve notice to quit being modified by clause 6.

The next question I have to decide is whether clause 6 is repugnant to the nature of a yearly tenancy. The first question that arises is whether the tenant's right to determine the tenancy by notice is thereby excluded. It is plain from the decision in *Warner* v *Browne* (1807) 8 East 165 that one cannot have a periodic tenancy in which one party is precluded from giving notice to terminate. Such a restriction is repugnant to the inherent nature of a periodic tenancy. As Mr Lloyd pointed out, clause 6 expressly provides that the tenancy "shall continue until the land is required by the Council for the purposes of the widening of Walworth Road" and the relevant two months' notice has expired. He invites me to read that clause, which does not provide for any corresponding notice on the part of the tenant, as excluding the tenant's right to serve a notice. A similar point came before the Court of Appeal in *Ashburn Anstalt* v *Arnold* [1987] 2 EGLR 71, where the language of the document was not dissimilar from the present in many respects. The agreement in that case contained the following provision:

> From and after September 29th 1973 Arnold shall be entitled as licensee to remain at the property and trade therefrom on the like terms save that it can be required by Matlodge Ltd to give possession on not less than one quarter's notice in writing upon Matlodge certifying that it is ready at the expiration of such notice forthwith to proceed with the development of the property and the neighbouring property involving *inter alia* the demolition of the property.

Until September 29 1973 it appears that Arnold had a tenancy at will or a licence at will on the payment of no rent, and thereafter was granted the rights I have quoted. The Court of Appeal held that Arnold had a tenancy and not a licence. The question arose whether Matlodge could bring Arnold's interest to an end by a notice to quit without desiring to develop the property. Having come to the conclusion that the agreement created a periodic tenancy, the Court of Appeal had to consider the question of repugnancy. The Court of Appeal dealt with the question of the tenant's right to determine the tenancy in this way:

> The plaintiffs say, however, that in this case . . . there is no provision for determination by Arnold & Co. It was said, therefore, that in the absence of notice by Matlodge, the term was uncertain in duration. We do not agree with that. As a matter of construction of the document, the possibilities are as follows: (i) Arnold & Co were not entitled to determine the arrangement at all. We reject that entirely. Bearing in mind that Arnold & Co were not required to pay any rent, such a construction is quite unreal in business terms; Arnold & Co were not obliged to occupy the premises and, if they did not occupy them, the outgoings would be nil or negligible, so there was no benefit to Matlodge in continuing the relationship.

In the present case there was a rent of £30, which may well have been a commercial rent in 1930, the time at which this document falls to be construed, and I cannot approach the matter in quite the same way. However, in my judgment, once one arrives at the conclusion that the document creates a periodic tenancy, the tenant's right of occupation as a periodic tenant must carry with it all the incidents of a periodic tenancy, and all the inherent rights, including the rights of either party to terminate the tenancy by giving the appropriate notice, save and in so far as they are expressly excluded by the document itself. In my judgment it is not a question of construing clause 6 in order to discover whether it confers a right of termination upon the tenant, but of construing the document as creating a periodic tenancy and considering whether the tenant's inherent right to terminate the tenancy by notice to quit has been excluded either expressly or by necessary implication. The landlord's right has been modified. The tenant's right has not and, in my judgment, it continues to subsist.

The next question is whether the restriction placed upon the landlord's right to terminate the periodic tenancy is repugnant to the nature of a yearly tenancy. As I have said, *Warner* v *Browne* is authority for the proposition that an absolute prohibition upon either party from giving notice to quit is repugnant to the nature of a periodic tenancy. *Breams Property Investment Co Ltd* v *Stroulger* [1948] 2 KB 1 is authority for the proposition that one can nevertheless suspend the right of one or other party from giving notice to quit for a fixed term certain, and that that is a valid condition which may be attached to a periodic tenancy without being repugnant to the nature of the periodic tenancy. In *Re Midland Railway Co's Agreement* [1971] Ch 725 the Court of Appeal went further and held that a fetter can be placed upon the right to determine a periodic tenancy for an uncertain period. In that case the agreement was for a half-yearly periodic tenancy. Clause 2 of the agreement provided for the termination by either party giving three months' written notice to quit to the other, subject to a proviso that the agreement should not be terminated by the landlord "until they should require the premises for the purposes of their undertaking". The Court of Appeal applied *Breams Property Investment Co Ltd* v *Stroulger* without distinguishing it. The Court of Appeal said at p 733:

> In *Breams Property Investment Co* v *Stroulger* . . . a curb on the lessors for three years unless they required the premises for their own use was upheld in this court, notwithstanding the earlier cases of *Warner* v *Browne* . . . and *Cheshire Lines Committee* v *Lewis & Co* . . . It follows that in a periodic tenancy a similar curb for 10, 20 or 50 years should not be rejected as repugnant to the concept of a periodic tenancy: and once the argument based on uncertainty is rejected we see no distinction in the present case.

I am bound by that decision to disregard any such distinction as may be derived from the fact that in the present case, as in that case, the period of suspension of the right to serve notice to quit was for an uncertain and indefinite period.

Mr Lloyd submits that in the present case, however, it must have been apparent from the outset that circumstances might occur by which it would be altogether out of the power of the landlord to serve a notice. That is a feature which is capable of distinguishing the present case from all previous cases. In every previous case where the landlord's right to serve a notice to quit has been suspended it has been suspended either for a fixed period or for an uncertain period, which the landlord himself could bring to an end by deciding to redevelop the property or use it for his own occupation and then serving the appropriate notice. In the present case, it is pointed out, it was always possible that the reversion could become vested in a body other than a highway authority and that, it is submitted, distinguishes the present case. A provision capable of depriving the landlord of the right to determine the tenancy, it is submitted, is repugnant to the nature of a periodic tenancy.

Before dealing with this submission it is, in my judgment, necessary to construe clause 6 and see how it applies on the transmission of the reversion to a third party. *Breams Property Investment Co Ltd* v *Stroulger* is authority for the proposition that a restriction on the landlord's right to serve a notice to quit runs with the reversion and accordingly it would, if valid, continue to bind successors in title to the LCC. It is not therefore submitted that the GLC was in any different position from the LCC. But when the land was transferred to the London Residuary Body it became impossible for the landlord, which was not a highway authority, themself to require the land for the purposes of widening the Walworth Road. It was not impossible, as Mr de la Piquerie submits, that one day the land would once again come into the ownership of a highway authority, which would then be again capable of serving a notice. But the circumstances in which the land should come again into the ownership of a highway authority which wanted the land for the purposes of road widening, without a compulsory purchase order which would extinguish the tenancy, rather than going through the

A procedure of clause 6, seems to me to be remote in the extreme and I would, for my part, disregard it.

The real question, in my judgment, is whether on its true construction the restriction imposed upon the landlord's right to serve a notice to quit by clause 6 endures once the landlord and the highway authority are no longer one and the same. It is clear that the council referred to inclause 6 must be both the highway authority which requires the land for the purpose of widening the Walworth Road and the landlord capable of obtaining vacant possession by serving the requisite notice. The clause predicates a situation in which the landlord and the highway authority are one and the same and, in my judgment, once they are different bodies, then clause 6 ceases to have any effect. Mr de la Piquerie submits that, if that were the case, Mr Nathan would have obtained no security for his interest and for the building which he proposed to erect upon it, since the very next day the council could have rendered his interest terminable by the B simple device of transferring the reversion to a third party. But that only raises a further question which has not been argued before me: whether there should be an implied term that the landlord would not assign the reversion to any party other than a highway authority. I can see the force of such a submission but, without deciding it, I have reached the clear conclusion that clause 6 is limited to the period during which the landlord and the highway authority are one and the same. It predicates a situation which no longer exists.

Accordingly, in my judgment, the London Residuary Body served a valid notice to quit and the remaining defendants are entitled to possession.

Declaration accordingly.

For further cases on this subject see p 261

LANDLORD AND TENANT
BUSINESS TENANCIES

Court of Appeal

July 25 1990

(Before Lord Justice NEILL, Lord Justice NICHOLLS and Lord Justice STAUGHTON)

ROMULUS TRADING CO LTD v HENRY SMITH'S CHARITY TRUSTEES (No 2)

Estates Gazette March 23 1991

[1991] 11 EG 112

Landlord and Tenant Act 1954, Part II, sections 30(1)(*f*) and 31A(1)(*a*) and (*b*) — Two Court of Appeal and three county court hearings required to elucidate the application of these provisions to a preliminary issue — Provisions analysed in depth by Court of Appeal at second hearing — Tenants not required to make advance election whether to claim a tenancy of the whole or a part only of the holding — Section 31A provides an escape route or fall-back position — After all the litigation the registrar's formulation of the preliminary issue was left unamended

The Kensington property which was the subject of the considerable litigation in this case consisted of an end-of-terrace house in Sumner Place with a rear extension on the ground and first floors along the flank wall in Old Brompton Road — The extension was reached from that road — Its position in relation to the entirety of the property explains some of the problems in the case — Following the termination of the lease of the property by the landlords' notice under section 25 of the 1954 Act the tenants applied for a new tenancy — The landlords objected, relying on section 30(1)(*f*) of the Act — The county court registrar ordered the trial of a preliminary issue as to the landlords' entitlement to oppose the grant, the registrar's statement of the issue following closely the wording of para (*f*) — The county court judge answered "no" to the question posed and the landlords appealed to the Court of Appeal — The Court of Appeal criticised the judge for lack of necessary findings and remitted the case to the county court to be tried by a different judge — Farquharson LJ in the Court of Appeal made some comments which caused questions later and were explained by the court in the present appeal — When the case got back to the county court a different judge refused to amend the registrar's formulation of the preliminary issue — Hence the present appeal from his order

The tenants sought directions which would enable the issue under para (*a*) of section 31A(1) to be decided first and the issue under para (*b*) to be decided second and separately — In this way, if the tenants failed on the para (*a*) point they would be able to express unqualified willingness to accept a tenancy of part of the holding (the Old Brompton Road part) and succeed under para (*b*) — The landlords' submission was that if the tenants wished to rely on para (*b*) they must make an advance election to abandon reliance on para (*a*) — It was suggested that they would not satisfy para (*b*) if they sought to rely primarily on para (*a*) and were willing to accept a tenancy of an economically separable part of the holding only if they failed to obtain a tenancy of the entirety under para (*a*) — The suggestion in regard to a duty to elect seemed to derive some support from the observations of Farquharson LJ in the first Court of Appeal hearing

The present Court of Appeal were in no doubt that the suggestion that the tenants needed to make an advance election was wrong — The true position was as follows — The relevant date for ascertaining the landlords' intention was the date of the hearing — If the primary question of the landlords' intention and need for possession was answered in favour of the tenants, that is, of course, the end of the landlords' opposition under the grounds in para (*f*) — If, however, the landlords were to succeed under para (*f*), read alone, Parliament has provided tenants with an escape route or fall-back position by section 31A(1)(*a*) and (*b*) — Before answering the primary question in favour of the landlord the court must go on to consider whether either para (*a*) or para (*b*) is satisfied — If para (*a*) is satisfied, the tenant will be entitled to a new tenancy of the whole holding — If para (*a*) is not satisfied, the tenant may still, under para (*b*), succeed in obtaining a tenancy of part — He will, of course, have to convince the court that he is willing to accept a tenancy of an economically separable part of the holding and that the other requirements of para (*b*) are satisfied

Having regard to this analysis of section 30(1)(*f*) and section 31A(1), the court decided that there was no justification for interfering with the wording of the preliminary issue in the form ordered by the registrar — Thus at the end of this fairly lengthy process the parties were due to go back to the county court for the matters to be decided in the light of the instruction now given

The following case is referred to in this report.

Betty's Cafés Ltd v *Phillips Furnishing Stores Ltd* [1959] AC 20; [1958] 2 WLR 513; [1958] 1 All ER 607; [1958] EGD 92; (1958) 171 EG 319, HL

This was an appeal by the tenants, Romulus Trading Co Ltd, from the order of Judge Phelan, in the West London County Court, whereby he refused to amend the form of a preliminary issue ordered by the registrar in proceedings under the Landlord and Tenant Act 1954 between the appellants and the respondent landlords, Henry Smith's Charity Trustees. The proceedings concerned the lease of a corner property comprising 13 Sumner Place and 47 Old Brompton Road in South Kensington, London SW7.

Michael Driscoll (instructed by Nabarro Nathanson) appeared on behalf of the appellants; David Neuberger QC (instructed by Denton Hall Burgin & Warrens) represented the respondents.

Giving judgment, **NEILL LJ** said: This is an appeal from the order dated March 16 1990 of His Honour Judge Phelan sitting in the West London County Court whereby he refused to amend the form of a preliminary issue which is due to be tried at a hearing beginning on September 24 1990.

It is the second occasion on which this matter has come before the Court of Appeal*.

*Editor's note: The earlier decision (by Mustill and Farquharson LJJ and Sir Michael Kerr) was given on November 30 1989 and reported at [1990] 2 EGLR 75.

The applicants in the proceedings (Romulus Trading Co Ltd, whom I shall call "the tenants") are presently holding over as the tenants of a corner property which comprises 47 Old Brompton Road and 13 Sumner Place in South Kensington under a lease which was granted to the tenants' predecessors in title by Henry Smith's Charity Trustees ("the landlords").

No 13 Sumner Place is a five-storey house at the end of a terrace. The flank wall overlooks Old Brompton Road. It has a rear extension on the ground and first floor only along that flank wall. The extension is known as 47 Old Brompton Road and is reached from that road. It is used by the tenants as a bureau de change. In the lease, dated November 21 1970 and granted for a term of 18 years and one quarter of another year from September 29 1969 by the landlords to the tenants' predecessor in title, there was a covenant by the tenants in clause 3(13) restricting the use of 47 Old Brompton Road and one rear room on the ground floor of 13 Sumner Place to that of a travel agency and offices in connection therewith or such other user as may be approved by the landlords, and that of the remainder of the premises, namely 13 Sumner Place, to that of a furnished and unlicensed guest house. The lease was assigned to the tenants on August 15 1979.

On May 12 1987 the landlords served a notice to terminate the lease on December 25 1987 in accordance with section 25 of the Landlord and Tenant Act 1954 ("the Act") which stated that the landlords would oppose an application to the court for a new tenancy on the ground specified in section 30(1)(*f*) of the Act. On May 27 the tenants notified the landlords that they would not be willing at the date of termination to give up possession of the property.

On September 9 1987 the tenants made an application to the West London County Court for a new tenancy in accordance with section 24 of the Act. On November 20 1987 the landlords served an answer repeating their objection to the grant of a new tenancy.

On January 29 1988 the registrar at the West London County Court made an order for the trial of a preliminary issue. The order provided:

> The following question to be tried as preliminary issue in this matter, namely whether the respondents are entitled to oppose the grant of a new tenancy to the plaintiffs on the grounds that on the termination of the current tenancy it is intended to demolish or reconstruct the premises comprised in the holding or a substantial part thereof, or to carry out substantial work of construction on the holding or part thereof, and this could not reasonably be done without obtaining possession of the holding.

The registrar also gave certain directions for the trial including directions as to expert evidence.

There then ensued some correspondence between the parties in which reference was made to a possible new lease restricted to 47 Old Brompton Road alone. It is sufficient to refer to the concluding documents in that correspondence. In the last paragraph of the fax dated September 16 1988 from the landlords' solicitors it was written:

> We note that you deny that our client has sufficient intent but that if we are successful then you ought to be entitled to a new lease of 47 Old Brompton Road. We assume that this is pursuant to the provisions of section 31A(1)(*b*). If that is the case then our client accepts that your client would be entitled to a new lease of that part.

There was a reply to that by the tenants' solicitors by fax on September 20 in which they say:

> We confirm your understanding in the last paragraph of your fax of September 16 is correct and note that your clients accept that our clients would be entitled to a new lease of 47 Old Brompton Road.

Then a few days later on September 26 1988 the solicitors for the tenants wrote again in relation to the fax of September 16:

> In respect of your final paragraph, we note that your clients accept that should your clients succeed in proving that they have the necessary intent (which is denied) then our clients are entitled under S31A(1)(*b*) to a new lease of 47 Old Brompton Road. Accordingly, our clients will not be preparing any expert evidence in relation to this aspect.

The preliminary issue was tried by His Honour Judge Oddie at about the end of 1988. He gave judgment on February 6 1989. Before I refer to the judgment, however, it may be convenient to set out the relevant sections of the Act. First section 30(1)(*f*) provides one of the grounds on which a landlord may oppose an application and it is in these terms:

> that on the termination of the current tenancy the landlord intends to demolish or reconstruct the premises comprised in the holding or a substantial part of those premises or to carry out substantial work of construction on the holding or part thereof and that he could not reasonably do so without obtaining possession of the holding.

Pausing there, it will be seen that para (*f*) may give rise to a number of issues which could include the following:

whether the landlords' intention is genuine;
whether the work to be carried out constitutes substantial work of construction;
whether the landlords could or could not reasonably do the work without obtaining possession of the holding.

I turn next to section 31A, which was inserted by the Law of Property Act 1969. That section, so far as is material, is in these terms:

> (1) Where the landlord opposes an application under section 24(1) of this Act, on the ground specified in paragraph (*f*) of section 30(1) of this Act the court shall not hold that the landlord could not reasonably carry out the demolition, reconstruction or work of construction intended without obtaining possession of the holding if —
>
> (*a*) the tenant agrees to the inclusion in the terms of the new tenancy of terms giving the landlord access and other facilities for carrying out the work intended and, given that access and those facilities, the landlord could reasonably carry out the work without obtaining possession of the holding and without interfering to a substantial extent or for a substantial time with the use of the holding for the purposes of the business carried on by the tenant; or
>
> (*b*) the tenant is willing to accept a tenancy of an economically separable part of the holding and either paragraph (*a*) of this section is satisfied with respect to that part or possession of the remainder of the holding would be reasonably sufficient to enable the landlord to carry out the intended work.

Finally, I should read section 32(1A), which was another addition to the Act made by the Law of Property Act 1969. That subsection is in these terms:

> Where the court, by virtue of paragraph (*b*) of section 31A(1) of this Act, makes an order under section 29 of this Act for the grant of a new tenancy in a case where the tenant is willing to accept a tenancy of part of the holding, the order shall be an order for the grant of a new tenancy of that part only.

At the trial before Judge Oddie the tenants accepted that the landlords intended to carry out a substantial amount of work and that the work could not be carried out under the terms of the existing tenancy. They argued, however, that the requirements of section 30(1)(*f*) were not met because some of the work was work of repair and most of the remainder was not structural work.

The judge considered the evidence relating to the work which was to be carried out and directed his attention both to the nature of this work and to the question raised by section 31A(1)(*a*) as to whether the landlords could reasonably carry out the work without obtaining possession of the holding and without interfering to a substantial extent or for a substantial time with the use of the holding for the purposes of the business carried on by the tenants. He stated his final conclusion in these terms:

> In my view, "construction must involve a structure, not merely refurbishment in ways not involving the structure of an existing building. Accordingly, I accept Mr Driscoll's argument. It follows, therefore, that in considering what amounts to the relevant work, I must disregard work which is not structural in that sense. In my judgment, on the facts of this case, the work of reconstruction or construction in the intended work includes that specified on p 4 of Mr Driscoll's argument. I would include the completion of the interior of the new accommodation so built as part of the construction. I consider it also includes demolition and replacement of the basement slab, the installation of staircases and the provision of new drainage and hot and cold water system and bathrooms where they did not exist before. If one disregards, as I must, the repair work, which I have already referred to, and the structural work, which I have listed above, then it seems to me that the rest of the work, although substantial in amount, really comprises work of refurbishment and not of demolition, reconstruction or construction works within the meaning of the Act. When one considers the nature of the interference described by Joyce and Hudson

— ie two of the experts who had given evidence —

> in the use of the holding, I conclude that there would not be substantial interference with its use for the purpose of the tenant's business. The answer to the question put in the preliminary issue is therefore "no".

It is not necessary to examine the judgment in detail, however, because on the subsequent appeal by the landlords to the Court of Appeal it was decided that the case should be remitted to the county court to be retried by another judge on the basis:

(a) that the judge had not made any finding as to whether the works which fell within para (*f*) of section 30(1) amounted to "works of reconstruction of a substantial part of the premises comprised in the holding" or to "substantial work of construction on part of the holding": *per* Farquharson LJ, and

A (b) that the judge had not identified the relevant works for the purpose of section 31A(1)(*a*) and had not then considered whether those works as distinct from the entirety of the works proposed constituted "interference to a substantial extent or for a substantial time with the use of the holding for the purposes of the business carried on by the tenants": also in the judgment of Farquharson LJ.

Nevertheless, it is right to draw attention to a short passage in Judge Oddie's judgment where he identified the issue which had to be decided under section 31A. At p 3 of his judgment he said:

> The tenant is willing to accept a tenancy of that part of the building known as 47 Old Brompton Road without 13 Sumner Place if it cannot obtain a new
> B tenancy of both. In particular, it is common ground that 47 Old Brompton Road is an economically separable part of the entire holding, and that possession of the remainder, namely 13 Sumner Place, would be reasonably sufficient for the landlord to carry out the intended work. And so the factual issue which will have to be resolved as the second main issue in the case [which] has to be decided by the court is whether the landlord could reasonably carry out the work without obtaining possession of 13 Sumner Place and without interfering to a substantial extent, or for a substantial time, with the use of the holdings for the purpose of a business carried on by the tenant. On this issue, the tenant contends that if the works involved were rephased, then they could be completed much sooner than the landlords say they could, with the tenants remaining in possession, and without interference to a substantial extent or for a substantial time with the use of the holding for the purposes of their business. Proposals were advanced by the tenants' surveyor as to how this
> C might be done. The landlord's case is that these proposals are quite unrealistic, so that the court could not hold that the landlord could not reasonably carry out the relevant works without obtaining possession of the holding.

It seems clear to me from this passage that the judge considered that the preliminary issue might raise questions under subsection (1)(*a*) of section 31A but would not raise any questions under subsection (1)(*b*) because

(i) the tenants were willing to accept a tenancy of 47 Old Brompton Road if they could not obtain a tenancy of the whole;
(ii) 47 Old Brompton Road was an economically separable part of the holding; and
(iii) possession of 13 Sumner Place would be reasonably sufficient to enable the landlords to carry out the intended work.

D When the case reached the Court of Appeal at the beginning of November 1989, however, subsection (1)(*b*) of section 31A had assumed a new significance. Counsel for the tenants sought to raise a preliminary point. He sought to argue, as I understand it, that the preliminary issue was bound to be decided in favour of the tenants because they were willing to take a tenancy of 47 Old Brompton Road alone. But this was a bad point for at least three reasons:

(1) because the preliminary issue had been argued below on the basis that subsection (1)(*b*) did not have to be considered;

(2) because subsection (1)(*b*) did not entitle the tenants to defeat the landlords' opposition to a new tenancy except to a limited extent; it was only a partial shield and could not lead to more than a new
E tenancy of part: see section 32(1A) of the Act, and

(3) because the issues before the Court of Appeal were whether the judge had reached the correct conclusions on the matters which he had considered, namely whether the requirements of section 30(1)(*f*) had been satisfied and how far the access and facilities which the tenants agreed to give under a new lease affected the position. Furthermore it is in the context of a preliminary point which did not begin to deal a knock-out blow to the landlords' appeal that one must consider what Farquharson LJ said about the matter in his judgment (with which Mustill LJ agreed). Farquharson LJ said*:

> At the outset of this appeal the respondent took a preliminary point arising under section 31A(1)(*b*). It appears that in correspondence between the
> F parties' solicitors in 1988, prior to the hearing before Judge Oddie, the respondent claimed to be entitled to a new tenancy of 47 Old Brompton Road on the basis that it was an economically separable part of the holding comprised in the original tenancy. This claim was conceded by the appellants' solicitors. Before this court Mr Driscoll on behalf of the respondent has argued that this concession meant that the judge was bound to answer the question put to him in the negative. The grounds for this argument were that even if the appellants succeeded in proving the matter required of them by section 30(1)(*f*) the fact that the tenant, ie the respondents, were willing to accept a new tenancy of 47 Old Brompton Road meant that the appellant landlords could not satisfy section 31A(1). Occupation by the tenant of 47 Old Brompton Road would not prevent the landlord carrying out the intended works on 13 Sumner Place. The fallacy in this argument was that in truth the respondent was only willing to accept a tenancy of 47 Old Brompton Road if it G could not obtain a tenancy of both 47 Old Brompton Road and 13 Sumner Place. It was a fall-back position. This meant that at the relevant time, namely at the date of the hearing, the respondent was not willing to take a tenancy of no 47. Accordingly, in my judgment, this point cannot be sustained.

It seems to me that, on a careful reading of what Farquharson LJ said, he meant no more than that he rejected the argument that section 31A(1)(*b*) provided a complete answer to the landlords' case. I do not believe that either he or the other members of the court were dealing with what matters might arise for examination on the retrial or were suggesting that the tenants had to elect whether to claim a tenancy of the whole or of a part only.

It is also to be remembered that the question of intent under section 30(1)(*f*) has to be judged at the date of the hearing and that evidence H of this intent can be admitted even though the intent was not formed in a provable form until after the hearing had begun: see *Betty's Cafés Ltd* v *Phillips Furnishing Stores Ltd* [1959] AC 20. I see no reason why a similar approach should not be adopted to evidence about the willingness of a tenant to accept a tenancy of part (for the purpose of section 31A(1)(*b*)) or, indeed, about his agreement to the terms of a new tenancy (for the purpose of section 31A(1)(*a*)).

On the hearing of an application under section 24(1) of the Act, where the landlord relies on the ground specified in section 30(1)(*f*) the court is required to consider that ground in the light of section 31A. In my judgment, this means that a tenant is entitled to put forward conditional arguments. For example, he can say that he J disputes the genuineness of the landlord's intention, or, if it is genuine, that the work satisfies section 30(1)(*f*), or, if section 30(1)(*f*) is satisfied, then he can say that the work can be carried out by means of the access and the facilities contemplated by section 31A(1)(*a*), or, if it cannot, that it can be done if the tenant were to accept a tenancy of a part only of the holding and that he is willing to do this.

Accordingly, in view of the conclusions which I have reached as to the meaning of what the Court of Appeal decided last November and as to the scope of what can be argued on the hearing of the preliminary issue, I would leave the issue in its present form.

Agreeing, **NICHOLLS LJ** said: On September 24 a judge sitting in the West London County Court is due to embark on a hearing, for the K second time, of a preliminary issue which was directed in these proceedings in a form commonly used when a landlord seeks to rely on ground (*f*) in section 30(1) of the Landlord and Tenant Act 1954. That form closely follows the wording of ground (*f*) itself. This formulation is none the less apt to embrace not only the issue raised by ground (*f*) read by itself but also the provisions of section 31A(1), because section 31A(1) is a statutory provision dictating the answer which, in prescribed circumstances, must be given by the court when ground (*f*) is in issue between the parties.

At the forthcoming hearing, as matters stand at present, the landlords intend to submit that if the tenants wish to rely on para (*b*) in section 31A(1), they must elect to abandon reliance on para (*a*). L They must elect, because if they seek to rely primarily on para (*a*), and they are only willing to accept a tenancy of an economically separable part of the holding under para (*b*), should they fail to obtain a tenancy of the whole under para (*a*), then they do not satisfy para (*b*). They do not satisfy para (*b*) because their willingness to accept a tenancy of part only is not unconditional.

That is the submission the landlords are proposing to make. It is in order to meet this point that the tenants seek directions which will have the effect of the para (*a*) issue being decided first and the para (*b*) issue being dealt with secondly and separately. In this way, if the tenants fail on the para (*a*) point, they will have an opportunity in court to express an unqualified willingness to accept a tenancy of part of the holding. Hence, if they fail on all else, they will succeed on para M (*b*). It is common ground that, leaving aside this election point, this is a case in which, if the tenants fail on everything else, they will be entitled to a new tenancy of part only of the holding under para (*b*).

I am in no doubt that the landlords' proposed submission, which has led to this appeal, is misconceived. In my view, there is no question of the tenants needing to elect as suggested. Neither the language nor the scheme of the statutory provisions seem to me to admit of room for doubt on this. When ground (*f*) is raised by a landlord, the primary question is whether the landlord has the requisite intention and, if so, whether he could not reasonably do the intended works without obtaining possession of the holding. The relevant date for ascertaining the landlord's intention is the time of

*Editor's note: See [1990] 2 EGLR 75 at p 76G.

the hearing: see *Betty's Cafés Ltd* v *Phillips Furnishing Stores Ltd* [1959] AC 20. If that primary question is answered in favour of the tenant, that is an end of opposition ground (*f*). But if, reading ground (*f*) alone, the landlord would succeed, Parliament has provided the tenant with a further possible escape route. That is found in section 31A(1). Thus, before answering the primary question in favour of the landlord, the court must go on to consider whether either para (*a*) or para (*b*) of section 31A(1) is satisfied. If the requirements of para (*a*) are satisfied, the tenant will be entitled to a new tenancy of the whole holding. So the judge will have to consider whether the tenant agrees to the inclusion in the terms of the new tenancy of terms giving the landlord access and other facilities for carrying out the work intended, and whether, if he does so agree, the other requirements of para (*a*) are satisfied.

I pause there. If the requirements of para (*a*) are satisfied, that will be an end to opposition ground (*f*); the tenant will succeed on para (*a*). But the matter does not rest there. A landlord is still not home. If para (*a*) is not satisfied, there is a further step which must still be taken. Para (*b*) provides a route by which a tenant, having failed to obtain a new tenancy of the whole holding, may still be entitled to succeed to a lesser extent by obtaining a new tenancy of part only. Accordingly, the tenant having failed on para (*a*), the court will need to consider para (*b*): is the tenant willing to accept a tenancy of an economically separable part of a holding and, if so, are the further requirements of para (*b*) satisfied? If they are, section 32(1A) will apply.

I can see no ground for thinking that Parliament intended that if a tenant seeks to rely on para (*a*), para (*b*) is not to be open to him as a fall-back position, or, I might add, vice versa. There are several features which point away from Parliament having so intended. First, it is open to a tenant to rely on para (*a*) even though his agreement to the inclusion of terms as to access and facilities is a fall-back position which arises only if he fails on ground (*f*) read alone. Thus a tenant may put a landlord to proof of his intention under ground (*f*) or he may dispute that the proposed works constitute demolition or reconstruction of the premises or a substantial part thereof. His attitude may be that, if he fails on these points, he is agreeable to the inclusion of terms as to access and facilities pursuant to para (*a*). I can see nothing in the Act to suggest that such an attitude by a tenant would put him outside para (*a*). If that is so under para (*a*), surely the position must be the same under para (*b*).

Second, this construction of the Act does not involve reading the phrase "is willing" in para (*b*) as applying to a case where a tenant is willing but only conditionally. As with the landlord's intention under ground (*f*), so with the tenant's willingness under para (*b*): this falls to be ascertained at the time of the hearing. What matters is how things stand at the conclusion of the case when the court comes to make its findings. If the tenant's attitude as established in evidence in the course of the hearing is that he is willing to accept a tenancy of part only should he fail, for example, on para (*a*), then when the court comes to decide para (*b*), the tenant's willingness will be unqualified and unconditional because the qualification or condition, namely failure of the tenant on the other issues, will by then have been satisfied.

Third, the contrary view of the construction of the statute would lead to an altogether extraordinary result. Even on the landlords' contention it would always be open to a court in this type of case to hear para (*a*) as a separate issue first and, having decided that issue against the tenant and having given judgment accordingly, to then go on and hear and determine para (*b*) as a separate issue. At that stage the tenant would be able to give evidence of an unqualified unconditional willingness to accept the tenancy of part only. So, if all the other ingredients of para (*b*) were present, the tenant would succeed on para (*b*). This is a highly unattractive conclusion. A riposte which is available to a tenant if the proceedings are carved into pieces and dealt with separately, but is not available if the proceedings are heard in their entirety in one piece in the normal way, would be nothing short of ludicrous. Parliament cannot have intended that the parties' rights in relation to the grant of a new business tenancy should be dependent on such procedural niceties.

The linchpin of the landlords' case is a dictum of Farquharson LJ on the previous occasion when these proceedings were before this court. For the reasons given by Neill LJ, I, too, do not think that that dictum was intended as a general proposition corresponding to the submission now being advanced by the landlords. The court had no occasion to consider such a general proposition and, read in the context of the issues before the court, Farquharson LJ is not to be understood as having enunciated any such far-reaching proposition.

Accordingly, in my view, the tenants' fears are groundless because the landlord's election submission is erroneous. Accordingly, I see no reason to make any alteration in the form of the issue which will come before the judge on September 24 1990.

Also agreeing, **STAUGHTON LJ** said: There are two problems, the general and the particular.

The general problem is whether, and if so when, a tenant has to elect to accept a tenancy of a separate part of a holding, pursuant to section 31A(1)(*b*) of the Landlord and Tenant Act 1954. Must he do this (i) in his application to the court for the grant of a new tenancy, or (ii) when he gives evidence at the trial, or (iii) in his final submissions, or (iv) after the judge has given judgment, or (v) after the determination of any appeal?

The answer, in my judgment, is that the tenant does not at any stage have to elect, in the sense of declaring that he wishes to have a new tenancy of part and does not wish to have a new tenancy of the whole. It is desirable that he should, from stage (i) onwards, make it clear that he wishes a new tenancy of part on the assumption that he cannot have a new tenancy of the whole — if that be the fact. But there is nothing in the statute which shows that he has to choose between those remedies, until a final determination has been made. If, for example, the case for a new tenancy of the whole fails at trial, the tenant can ask the judge to order a new tenancy of part but still preserve his full rights on appeal.

It is said that this is inconsistent with the judgment of Farquharson LJ (with whom Mustill LJ agreed) in the earlier proceedings. But there was no issue in those proceedings as to whether Romulus Trading Co Ltd should have a new tenancy of part, nor did it fall to be decided. I cannot read Farquharson LJ as expressing any view on whether they might subsequently be entitled to a new tenancy of 47 Old Brompton Road.

That being the answer to the general problem, it seems to me that there is no need to interfere with the preliminary issue ordered on January 29 1988. It will be for the judge at the trial to determine whether that issue does or does not include the question whether there should be a new tenancy of part. If it does not, and if the case for a new tenancy of the whole fails, he will then have to decide when he should try the alternative case that there should be a new tenancy of part — assuming that there is any issue as to that, which does not at present appear likely.

I would leave the preliminary issue untouched and leave the matter there.

Appeal dismissed.

Court of Appeal

December 11 1990

(Before Lord Justice FOX, Lord Justice BALCOMBE and Lord Justice STOCKER)

GURTON v PARROTT AND ANOTHER

Estates Gazette May 11 1991

[1991] 18 EG 161

Landlord and tenant — Landlord and Tenant Act 1954, Part II — Whether tenancy was business or residential — Dwelling-house and outbuildings converted to kennels — Groomage, breeding and kennelling of dogs carried on for a period of years — Appeal by landlords from decision of county court judge that the tenancy was a protected residential tenancy and that a notice under section 25 of the 1954 Act purporting to terminate the tenancy was void — *Cheryl Investments Ltd* v *Saldanha* considered — Appeal dismissed

The tenant in this case, the respondent to the appeal, had actually lived in the property as her home since 1939 but did not have a tenancy until 1974 — She started to take an interest in dogs in the 1960s, beginning by showing poodles and providing kennels for friends and relations — From that she

developed the activities of kennelling, grooming and breeding dogs until 1976, when she ceased to take an active part; and her son, who lived in a caravan nearby, continued to carry on for a time — The business came to an end about 1988 — Takings per annum from 1969 to 1976 were in the region of £3,000 — In 1988 the landlords gave the tenant notice to quit on the footing that she was a business tenant — In her originating application to the county court she sought a declaration that Part II of the 1954 Act did not apply to her tenancy; that her tenancy was a protected residential tenancy; and that the notice to quit was null and void — It was common ground in the county court that, if the letting to the tenant in 1974 was for occupation for the purposes of a business carried on by her, Part II of the 1954 Act applied, but, if it was not so let and occupied, the tenant was protected by the Rent Act 1977 — The judge decided that the tenancy was residential and the tenancy was so protected

In coming to this conclusion the judge applied the test stated by Lane LJ in *Cheryl Investments Ltd* v *Saldanha* — The judge held that in the present case the tenant occupied the property for the purposes of her residence only and that the running of the business on the property was merely incidental, something akin to a hobby — The question for the Court of Appeal was whether the judge was entitled to make that finding on the evidence before him — The judge had applied the correct test and the matter was peculiarly one within his ability to decide on the basis of evidence which he, not the Court of Appeal, had before him — There was no reason to interfere with his decision — Appeal dismissed

The following cases are referred to in this report.

Cheryl Investments Ltd v *Saldanha* [1978] 1 WLR 1329; [1979] 1 All ER 5; (1978) 37 P&CR 349; 248 EG 591, CA
Russell v *Booker* [1982] EGD 58; (1982) 263 EG 513

This was an appeal by Richard Parrott and Geoffrey Darrow, the landlords, trustees of the estate of the late E A Padfield, from a decision of Judge Paynter-Reece, at Romford County Court, holding that the tenancy of the respondent, Mrs Doris Lilian Gurton, of Wantzway Kennels, Upminster, Essex, was a protected residential tenancy, not a business tenancy subject to Part II of the Landlord and Tenant Act 1954.

Bernard Devlin (instructed by Hunt & Hunt & Houghtons, of Romford) appeared on behalf of the appellants; Steven Coles (instructed by Sackvilles, of Hornchurch) represented the respondent; Guy Newey was present as *amicus curiae*.

Giving the first judgment at the invitation of Fox LJ, BALCOMBE LJ said: This is an appeal by landlords from an order of His Honour Judge Paynter-Reece made in the Romford County Court on November 22 1989. By that order the judge directed that there be a declaration on behalf of the applicant, who is the respondent to this appeal (the "tenant"), that the tenancy relating to the property, at Wantzway Kennels, Hall Lane, Upminster, Essex, in the London Borough of Havering is not a tenancy to which Part II of the Landlord and Tenant Act 1954 applies; that the notice served by the respondents, the appellants in this court (the "landlords"), on January 27 1988 pursuant to section 25 of the Landlord and Tenant Act 1954 is null and void; and that the tenancy of the said property is a protected residential tenancy.

The tenant, Mrs Doris Lilian Gurton, in 1939, then aged 23, married Mr Arthur Gurton and took up residence at property then known as Emmanuel Cottage, but now known as Wantzway Kennels as described in the declaration made in the court below, and at the same time her mother-in-law, father-in-law and brother-in-law were already residing at the property. Her father-in-law died in 1945. A daughter was born to her in 1948 and a son, Graham, was born to her in 1951. In the late 1950s her mother-in-law died and in 1960 her husband died. Her brother-in-law died in the mid-1960s.

She has continued to live in the property from 1939 until today; until 1974 she paid no rent and had no tenancy.

In the mid-1960s she, with her son Graham, carried on a business at the premises, grooming, breeding and kennelling dogs and held a licence from the local authority to enable her to kennel dogs at the property. That business terminated early in 1988.

At some time before 1974 a Mr E A Padfield acquired the freehold of the premises, and in 1974 he granted to the tenant a tenancy of that property at a rent of 50 pence per week. There was no document evidencing or creating such tenancy.

In 1976 the tenant no longer played an active part in the business, she being then aged 60 and not in good health. Her son Graham, who lived in a caravan nearby, continued to carry on the business.

Mr Padfield died on January 12 1977, and the landlords are the trustees of his estate. On January 27 1988 they gave the tenant notice under the 1954 Act to terminate her tenancy on the basis that it was a business tenancy under Part II of the Act. In due course the tenant served a counternotice, and in April 1988 she issued an originating application in the Romford County Court seeking a declaration that her tenancy was not one to which Part II of the Landlord and Tenant Act 1954 applied; that the notice served on her by the landlords was null and void; and that her tenancy was a protected residential tenancy.

That was the application which came before Judge Paynter-Reece, with the result to which I have already referred.

Section 23(1) of the Landlord and Tenant Act 1954 is in the following terms:

> Subject to the provisions of this Act, this Part of this Act applies to any tenancy where the property comprised in the tenancy is or includes premises which are occupied by the tenant and are so occupied for the purposes of a business carried on by him or for those and other purposes.

In my judgment, the issue in this case is fairly and adequately summarised in the skeleton argument of the *amicus curiae*, for which I would like to say we are much indebted. Mr Newey, as *amicus curiae*, was appointed at a time when it was not known whether the tenant would be represented in this court, so we have had the benefit of his skeleton argument as well as argument on behalf of the tenant herself. Mr Newey's summary of the issue is whether, when let to the tenant in 1974, Wantzway Kennels was occupied for the purposes of a business carried on by her or "for those and other purposes". It is clear that whatever other reasons the tenant may have had for occupying the property, Wantzway Kennels had, at all material times, been her home — in fact since 1939. Accordingly — and this was common ground between the parties in the court below — if the property was not occupied for the purposes of the business within section 23(1) of the 1954 Act, the tenant is protected by the Rent Acts, and the judge so found.

If, on the contrary, Wantzway Kennels was, in truth, occupied for the purposes of a business when let to her, then the tenant will not have had the Rent Acts protection and Part II of the 1954 Act will have applied. If her tenancy initially fell within the 1954 Act, and not under the Rent Acts, then the tenant would seem no longer to have any statutory protection as a matter of strict law, since business user has now come to an end.

However, as I say, the matter seems to have been dealt with in the court below and in argument before us on the basis that if, at the time when the tenancy was granted, it was occupied for the purposes of a business, then it was a Landlord and Tenant Act business tenancy; but if it were not so occupied, then it was and still is a Rent Act protected tenancy.

I turn now to the judgment of the judge where he sets out in detail the evidence given by the tenant, to whom he refers as the "applicant". He said:

> I summarise the applicant's evidence. She told me that prior to her interest in dogs, which started in the mid-1960s, the outbuildings and the land on which they stood had been used for keeping chickens and a few pigs from time to time, mainly to provide food for the family, and this had been carried on by her brother-in-law. Her interest in dogs started at the suggestion of a friend. She bought a poodle and bred from it, selling the pups, and then started showing poodles. Her brother-in-law converted one of the huts into kennels to house the breeding dogs. She then started to kennel dogs for friends and relations, for which purpose she needed more kennels and the council licence. She had kennelling for 46 dogs but it was a seasonal activity, mainly summer holidays, and never had that number at any one time. She charged for this service and for grooming dogs £3 or £4 a week on average, and expanded the breeding side to eight or nine dogs over the years. She did not advertise in the press but used the name Wantzway Kennels in her handouts when showing dogs because she had to have a "prefix".

I should add that the judge had previously stated that there were two sign-boards advertising Wantzway Kennels and the services which she provided erected and kept in position on the property.

The judge continued:

A She had disclosed the accounts she had prepared for the Inland Revenue; these showed takings in the region of £3,000 per annum from 1969 to January 31 1976. She did not regard herself as being in business, rather it was a hobby which provided her with some money to help with living expenses and upkeep of the property.

She stopped playing an active part in 1976 due to her health, aged 60, but her son, who now lived in a caravan nearby, had carried on in his spare time. He had already been helping her for some years.

In 1974 Mr Padfield, with whom her family had always been on very good terms, visited her and suggested that for her security she should become a tenant and have a rent book. A rent of 10 shillings a week was agreed; he told her to get a rent book, which she did, and obtained two more when each became full. The rent books were produced. At the time the use of the property for kennelling and the like was known to Mr Padfield, who B thoroughly approved. Nothing specific was said by either as to whether the tenancy was to be business or residential but she was under the impression that she would be able to live on there until she died and that her security did not depend in any way on keeping on with the dog activities. Since early 1988, when the last council licence expired, no kennelling, breeding or grooming of any significance has been carried on by herself or her son, if at all.

Then the judge came to his first decision:

In my judgment those activities did constitute a business within the meaning of section 23(2). Wantzway Kennels was used as a business name, she worked on kennelling dogs and grooming dogs for profit and achieved a not insignificant turnover, which was shown in the business accounts. Although the applicant, in her evidence, tended to belittle the amounts earned and the work involved, I am quite satisfied on her own evidence, which generally I C found truthful, that she was conducting a business at the premises from the mid-1960s until 1988 . . .

There has been no cross-appeal against that finding of the judge, so that the issue which has been ventilated before this court was the other matter which the judge had to determine, namely: did she occupy the premises for the purposes of the business of kennelling, breeding and grooming dogs? She clearly did occupy those premises for the purposes of a residence, but did she also occupy them for business purposes?

The leading authority on this question, to which the judge below and this court have been referred, is *Cheryl Investments Ltd* v *Saldanha* [1979] 1 All ER 5* and I refer to the first finding at p 6:

Held — (i) Premises were occupied by a tenant "for the purposes of a business D carried on by him" within section 23(1) of the 1954 Act, only where the business activity on the premises was a significant purpose of the occupation of the premises or part of the reason for the occupation. Where, therefore, the business activity was merely incidental to residential occupation of the premises, and did not amount to a significant use of them, the premises were not occupied "for the purposes of a business" . . .

I would refer in particular to the judgment of Geoffrey Lane LJ at p 13 of the report. He had pointed out that section 17(1) of the Landlord and Tenant Act 1927 had referred to "holdings to which this Part of this Act applies are any premises held under a lease . . . and used wholly or partly for carrying on thereat any trade or business . . . ". He said that the alteration in the wording as affected by section 23 of the 1954 Act must have had a reason, and he E continued:

It is obviously a very fine distinction, but the words in section 23 seem to have been used in an attempt to make it absolutely clear that activities on the premises which are merely incidental to residential occupation do not bring the premises within the section although they may properly be described as using them for carrying on a trade or business. The businessman, for example, who takes work home in the evening which he does in a study set aside for the purpose may very well be using the premises partly for carrying on thereat a business, but he could scarcely be said to be occupying the premises for the purposes of a business, any more than the person who watches the television regularly every evening can be said to be occupying his house for the purpose of watching television.

Then come these very important words:

F It is only if the activity is part of the reason for, part of his aim and object in occupying the house that the section will apply.

Judge Paynter-Reece was referred to that case and he refers to it when dealing with this particular issue. He said:

I apply the test put forward by Lord Lane in the *Cheryl Investments* case to which I have referred; that is to say, was the business part of the aim and object of the applicant in occupying this property? I have come to the conclusion that it was not and never has been and that the applicant's occupation has not been for its purposes. I accept the submissions made by Mr Coles on this point. Although no one point is in any way conclusive, in my judgment, the totality of the points leads me to the view that she occupied this property for the purposes of her residence only and that the running of the business on the G property was merely incidental, something akin to a hobby, taking advantage of the facilities offered by the existing outhouses and other small amount of land, which might otherwise be used as a yard or garden, which was an adjunct of the house.

In particular, I am impressed by the fact that the applicant occupied this property purely as a residence, albeit merely under a licence from the landlord, for many years before starting the dog business, to provide herself with some extra money and a part-time activity. That business never became her aim and object in occupying the property; her aim and object was to continue to live in it as her home.

The judge then dealt with the view which Mr Padfield must have had in granting her the particular tenancy which he did. In my judgment, that part of his judgment can have no bearing on the question as to whether the tenant was occupying the premises for the purposes of a H business carried on by her; it was material to the question which was not substantially in issue before him, whether she had a Rent Act protected tenancy, and is covered by the case of *Russell* v *Booker* (1982) 263 EG 513. However, it is not necessary to refer to that because it was common ground between the parties that if the tenant had not, at the relevant date, been occupying the premises for the purposes of a business carried on by her, then she had a Rent Act protected tenancy, and the remainder of the judgment was, in my view, referable to that issue.

I turn, then, to deal with the question on this appeal: was the judge entitled to make the finding which he did on the evidence before him?

Mr Devlin, for the landlords, in this court has submitted a full and careful skeleton argument and has summarised in some detail the J evidence which was before the judge as to the extent of this business. However, I need not refer to it in detail because at the end of the day it seems to me that this was a matter peculiarly within the ambit of the judge's ability to decide on the basis of the evidence before him. It has often been said that this court suffers from the disadvantage of not being able to have what is sometimes called the "flavour" of the case because it does not hear the witnesses and it receives a somewhat attenuated picture on paper.

The question whether this tenant occupied this property for the purposes of a business carried on by her is to be decided by applying the test set out by Geoffrey Lane LJ in the *Cheryl Investments* case, which I repeat: "it is only if the activity is part of the reason for, part of his aim and object in occupying the house that the section will K apply".

Therefore, what Judge Paynter-Reece had to decide on the evidence before him was: what was the reason for this tenant's occupation of the premises? As he said, and I bear in mind, she had occupied the premises long before she commenced the business. What was the aim and object of her occupation?

It seems to me that this is a case where this court cannot and should not interfere with the judge's finding on the evidence, which he was entitled to make. I would not wish to disturb it, and therefore I would dismiss this appeal.

Fox and **Stocker LJJ** agreed and did not add anything. L

The appeal was dismissed with costs. Leave to appeal to the House of Lords was refused.

Court of Appeal

October 24 1990

(Before Lord Justice PURCHAS and Lord Justice STUART-SMITH)

DEMETRIOU v POOLACTION LTD AND ANOTHER M

Estates Gazette June 29 1991

[1991] 25 EG 113

Landlord and tenant — Landlord and Tenant Act 1954, Part II — Question as to status of tenancy for purpose of jurisdiction under 1954 Act — Continuity of business tenancy — Whether landlords under an express or implied obligation to repair premises — Circumstances in which an obligation to repair should be implied in a tenancy agreement — *Barrett* v *Lounova* discussed — No repairing obligation on landlords — Tenant held to have failed to establish claim to a business

*Editor's note: Also reported at (1978) 248 EG 591.

A **tenancy — Necessary continuity lacking — Landlords' appeal allowed**

The history of events leading to the present appeal was "a long and complex one going back over 20 or more years" — It is set out in detail in the judgment of Stuart-Smith LJ — The appeal raised a question of jurisdiction under the Landlord and Tenant Act 1954, the status of the tenancy having been open to doubt — At different times in its history the tenancy had been treated as a tenancy subject to the Rent Acts and held to be subject to section 32 of the Housing Act 1961 — The appeal was from a county court decision, which accepted that
B **the tenancy was subject to Part II of the 1954 Act and which granted the tenant a new lease of nine years at a rent of £13,000 pa but with a provision that the rent should be a peppercorn only until the landlords had put the premises in proper repair — The main issue was whether the tenant had the *locus standi* to apply for a business tenancy — This depended on whether the premises were occupied by the tenant at the material times and were so occupied for the purpose of the business carried on by him within the meaning of section 23 of the 1954 Act — The business which the tenant conducted was that of subletting rooms — A difficulty in the tenant's way was that the premises had not been occupied by any subtenants since**
C **1981, so that the essential element of continuity in the carrying on of a business in the premises was, on the face of it, lacking**

A central element in the county court judge's decision was his conclusion that the landlords were under an express or implied obligation to carry out repairs to the premises — The link between this conclusion and the question of the status of the tenancy and the court's jurisdiction was the finding that there had been a breach of the landlords' repairing obligation, rendering the premises virtually uninhabitable — Applying the principle laid down in *Morrisons Holdings Ltd* v *Manders Property (Wolverhampton) Ltd,* the judge held that the break
D **in continuity, due to the uninhabitable nature of the premises, was the fault of the landlords and so did not deprive the tenant of the protection given by the 1954 Act — The judge also appears to have held that, even if the landlords were not under an obligation to repair, the tenant's status as a business tenant was nevertheless preserved, but the reasoning under this head was difficult to follow**

The Court of Appeal held that there was simply no evidential material on which the judge could properly reach the conclusion that there was an express agreement by the landlords to repair — As to the alternative finding of an
E **implied agreement to repair, the court considered the circumstances in which it was justifiable to imply such a term — They considered in particular the judgment of Kerr LJ in *Barrett* v *Lounova* but decided that the circumstances in the present case were distinguishable from those in that case: there was, for example, no "correlative obligation" feature in the present case — The judge's finding of an implied obligation could not be sustained — The court also rejected the judge's view that the tenant's *locus standi* could be supported even in the absence of any repairing obligations on the part of the landlords — The tenant had failed to establish the necessary continuity required to qualify for the purpose of section 23 of**
F **the 1954 Act — Landlords' appeal allowed**

The following cases are referred to in this report.

Barrett v *Lounova (1982) Ltd* [1989] 2 WLR 137; [1989] 1 All ER 351; [1988] 2 EGLR 54; [1988] 36 EG 184, CA

Caplan (I&H) Ltd v *Caplan (No2)* [1963] 1 WLR 1247; [1963] 2 All ER 930

Cockburn v *Smith* [1924] 2 KB 119; [1924] All ER Rep 59

Hart v *Windsor* (1843) 12 M&W 68; (1843-60) All ER Rep 681

Morrisons Holdings Ltd v *Manders Property (Wolverhampton) Ltd* [1976] 1 WLR 533; [1976] 2 All ER 205; (1975) 32 P&CR 218; 238 EG 715, CA

Sleafer v *Lambeth Borough Council* [1960] 1 QB 43; [1959] 3 WLR 485; [1959] 3 All ER 378; (1959) 57 LGR 212, CA

Westminster (Duke of) v *Guild* [1985] QB 688; [1984] 3 WLR 630; [1984] 3 All ER 144; (1983) 48 P&CR 42; [1983] EGD 541; 267 EG 762, CA

This was an appeal by Robert Andrews (Estate Agencies) Ltd, the G second defendants in the action brought by the applicant (the present respondent), Varnavus Demetriou, claiming a new lease of premises at 51a Primrose Gardens, London NW3, under Part II of the Landlord and Tenant Act 1954. The appeal was against the order of Judge Dobry, at Bloomsbury County Court, granting the respondent a new tenancy and other relief. The appellants were the present reversioners, having purchased the freehold reversion from Poolaction Ltd. The appellants had applied to strike out the respondent's originating application on the ground that he had ceased to occupy the premises for business purposes.

Romie Tager (instructed by Lester Aldridge, of Bournemouth) appeared on behalf of the appellants; the respondent did not appear and was not represented. H

Purchas LJ said at the beginning of the proceedings: Before inviting Stuart-Smith LJ to deliver the first judgment, I wish to record that we have heard this appeal, which has been fully argued by Mr Tager for the appellants, in the absence of the respondent.

When the appeal was called on yesterday, Miss Suzanne Bell, who had been instructed by the respondent both at the hearing at first instance and during the preparation of the respondent's case on the appeal, informed the court that at a conference at 7 pm on the previous evening the respondent had informed her that he did not wish her to appear for him on the appeal.

The respondent had not arrived at court, so Miss Bell had no opportunity to confirm her position but was attended by a J representative of the respondent's solicitors. We refused an application made by Miss Bell for an adjournment to enable another counsel to be instructed, but delayed the hearing of the appeal in case the respondent had for some reason or other been delayed on his journey to court. He did not attend court yesterday, nor did he attend court today, nor has he so far as we know made any effort to contact the court to explain his absence.

I will now invite Stuart-Smith LJ to deliver the first judgment.

Giving judgment, **Stuart-Smith LJ** said: This is an appeal from an order of His Honour Judge Dobry sitting at Bloomsbury County Court and made on July 12 1989. It was made on the application of the tenant, Mr Demetriou, for the grant of a new lease of premises K consisting of the first, second and third floors at 51a Primrose Gardens, London NW3. The order was made under the Landlord and Tenant Act 1954 on the basis that the tenant had a business tenancy of those premises. The order was that the respondent landlords should grant a lease of nine years, that the rent payable should be £13,000 per year but that it should be a peppercorn rent until the landlords had put the premises into a state of proper repair and that there should be a rent review after three years from the grant of the tenancy and a second review three years thereafter. The landlords, who are effectively the second respondents, Robert Andrews (Estate Agencies) Ltd, now appeal to this court from that order. L

The principal issue both in this court and in the court below was whether the tenant had any *locus standi* to apply for a new tenancy. It was, therefore, an issue of jurisdiction. That depended upon whether the premises were occupied by the tenant at the material time and so occupied for the purposes of a business carried on by him within the meaning of section 23 of the Landlord and Tenant Act 1954. The business which the tenant conducted was of subletting rooms.

Integral to that question is the further question of whether, under the original lease, there was an express or implied obligation on the landlord to carry out repairs. The judge held that there was an express obligation to carry out structural and external repairs. Alternatively, he held that there was an implied obligation to that effect. The landlord appellants challenge those findings. M

The reason why that question was integral to the judge's conclusion was that the judge held that there had been a breach of that obligation which meant that the house at the material times was virtually uninhabitable. It had not, in fact, been occupied by the tenant's subtenants since 1981, but the learned judge held, applying the principle laid down in *Morrisons Holdings Ltd* v *Manders Property (Wolverhampton) Ltd* [1976] 2 All ER 205* that there had been sufficient continuity in the occupation of the premises by the tenant because the fault and the reason for the uninhabitability of the premises was the want of repair by the landlords.

*Editor's note: Also reported at (1975) 238 EG 715.

A The third issue that arises in this appeal is, assuming that the landlords were not under an obligation to repair: was the judge right in holding, as he appears to have done, that the tenant nevertheless occupied the premises for the purpose of his business and that on that basis also he had sufficient *locus standi* to apply for a new tenancy?

As Purchas LJ has pointed out, we did not have the advantage of argument by Miss Bell, who appeared for the tenant in the court below, but we did have available a skeleton argument provided by her and we have had the helpful argument by Mr Tager who, in accordance with his duties to the court, has drawn our attention to those points which are in favour of the tenant respondent.

The history of the matter is a long and complex one going back over
B 20 or more years, and it is necessary, I am afraid, that I set out in some detail that history.

It starts with an oral agreement in 1967 to grant a weekly tenancy of 51a Primrose Gardens, London NW3, between Woolmer Investments Ltd, on whose behalf Miss Ross, the director of that company, acted, as landlords of the respondent, Mr Demetriou, Mr Varnavus Demetriou, and Mr Chris Christoporou who is referred to as "Chris". The tenancy was taken in the name of Demetriou Varnavus. That was because the landlords wished to differentiate between these premises and other premises which they owned and which were let to one or other of those two partners. It appears that Mr Demetriou and Mr Christoporou were in fact taking the premises as a partnership, and the partnership name was for convenience
C called "Demetriou Varnavus". The rent was a modest one; indeed the final rent when the matter came before the judge was still only £13.32 per week. There had been increases of rent over the period of years reflecting only increases in rates, those increases having been effected by statutory notices of increase, which were appropriate to a Rent Act tenancy but not to a business tenancy.

In 1970 Miss Ross died, and in 1971 Woolmer Investments Ltd transferred the freehold reversion to Redspring Ltd. In 1974 the partnership between Demetriou and Chris was dissolved; and, on April 24 1974, Redspring Ltd (the then landlords) served a notice to quit on Demetriou Varnavus which expired on May 17 1974. That was not a Landlord and Tenant Act notice, and it is plain that the
D landlords from this time onwards, if not earlier, treated the premises as though they were subject to the Rent Acts. The landlords subsequently accepted rent from Mr Demetriou, the tenant, and a new rent book was supplied to him. It would therefore seem from that time onwards that he was the sole tenant of the premises.

In December 1974 Redspring Ltd transferred the freehold reversion to Pablan Ltd, who on September 28 1976 began proceedings against the tenant for £717 arrears of rent. I shall refer to those proceedings as "the 1976 proceedings".

In 1977 the tenant, Mr Demetriou, served an amended defence and counterclaim to those proceedings. The defence was one of set-off and in the counterclaim he claimed damages for breach of an implied repairing covenant, that covenant being implied under the provisions
E of section 32 of the Housing Act 1961.

On May 23 1977 a further notice to quit was served on the tenant. Again that was not a Landlord and Tenant Act notice.

On August 19 1977 the first respondents, Poolaction Ltd, completed the purchase of the freehold reversion from Pablan Ltd.

On December 9 1977 yet another notice to quit was served by Poolaction on the tenant. Again it was not a Landlord and Tenant Act notice.

On January 11 1978 Poolaction Ltd sought information from the tenant under section 139 of the Rent Act 1977 as to the details of lettings.

In January 1978 Pablan Ltd, who by this time had conveyed the
F freehold to their successors in title, obtained judgment for £2,626 for arrears of rent. There was a stay of execution on that sum pending trial of the counterclaim, that counterclaim being remitted to the registrar for report.

On March 14 1978 proceedings were issued by Poolaction Ltd against the tenant for possession on the grounds that the notice to quit had expired. I shall refer to those proceedings as "the 1978 proceedings". As I have indicated, those landlords also appear to have thought that the tenancy was one to which the Rent Acts applied, and the ground for seeking possession was not only the termination of the contractual tenancy but on the basis that the tenant no longer was entitled to the protection of the Rent Acts because he had sublet the whole of the premises.

On May 2 1978 Mr Demetriou served a defence and counterclaim. G
The defence relied upon security of tenure under Part II of the Landlord and Tenant Act 1954 and the counterclaim was for damages for breach of covenant to repair on the part of the landlords, again it being alleged that that covenant arose under the provisions implied by virtue of section 32 of the Housing Act 1961.

On April 20 1979 Poolaction Ltd served a notice under section 25 of the Landlord and Tenant Act 1954 on the tenant terminating the tenancy on October 31 1979. That was said to be without prejudice to their contention that in fact it was not a business tenancy at all because it was not within the 1954 Act.

On October 26 1979 His Honour Judge Pickering gave judgment in the 1978 proceedings and held that the Landlord and Tenant Act 1954
did apply to the premises and that the tenant was a business tenant. H
He adjourned the counterclaim and transferred it to the High Court, where it has never in fact been proceeded with to this day.

Throughout the period from about 1969 to 1979 the tenant had been complaining to his various landlords and their agents of want of repair of the premises. On no occasion was it alleged that the landlords are liable under an express repairing covenant. What is said, where the grounds are specified at all, is that there was a breach of the implied covenant, by virtue of section 32 of the Housing Act 1961.

On May 4 1979 the tenant served a counternotice saying that he was unwilling to give up possession on termination of the tenancy; and on
August 19 he issued an originating application for the new tenancy in J
these proceedings.

On January 23 1980 Poolaction Ltd served an answer not opposing the grant of a new tenancy.

On October 3 1980 judgment was given on the counterclaim in the 1976 action for £80. It will be borne in mind that that was for breach of the repairing covenant implied under section 32 of the Housing Act 1961.

On May 21 1981 the local authority served on Poolaction Ltd a notice pursuant to section 9(1A) of the Housing Act 1957 alleging failure to carry out many repairs to the house. Poolaction Ltd subsequently appealed from that order on the grounds that the tenant was a business tenant in control of the premises and was, therefore,
the proper person to be served with a notice under that section. That K
appeal was successful.

On July 20 1981 the local authority served a notice under section 26 of the Public Health Act 1961 on Poolaction Ltd on the ground of dangerous electricity supply. As a result of that notice the tenant served notices to quit on all his subtenants, leaving as an occupant of the premises only his part-time secretary, Miss Miqueax.

On November 5 1981 the local authority served a further notice pursuant to section 9(1A) of the Housing Act on the tenant on the basis that he was the person in control of the premises.

On February 4 1982, Mr Demetriou having appealed to this court, the Court of Appeal in the 1976 proceedings dismissed his appeal
and, therefore, affirmed the judgment that Mr Demetriou was L
entitled only to £80 on his counterclaim based on section 32 of the Housing Act.

On August 16 1984 a contract was entered into by which the freehold reversion of the premises was to be sold by Poolaction Ltd to a Mr D Davis, who assigned the benefit of that contract to the appellant company, Robert Andrews (Estate Agencies) Ltd. That contract was completed on September 13 1984, and on October 30 1984 the appellants applied to be joined as second respondents in these proceedings. That application was opposed by Mr Demetriou. He had registered a caution against title to the property which prevented the registration of the freehold title being made for some two years.

On November 25 1984 the section 9(1A) notice which had been M
served on the landlord was withdrawn by the local authority; and in 1984 Miss Miqueax died.

It was not until June 1986 that the appellants were joined as second respondents in these proceedings; and in January 1988, the matter having proceeded no further, the appellants applied to strike out the tenant's originating application on the ground that he had ceased to occupy the premises for business purposes. That matter was referred to the judge on the hearing of the application and, as I have said, the judge made his order on July 12 1989.

The judge held as follows. First, as I have already indicated, that the original letting to Mr Demetriou and Mr Christoporou was an oral one. Second, that there was an express term that the landlord was

A liable to repair the structure and outside of the building. Third, and alternatively, that there was an implied term to the same effect. It should be noted that, despite what was said in correspondence and indeed in the two sets of proceedings to which I have referred in the counterclaims to those proceedings, that finding was not based on section 32 of the Housing Act 1961 but was of a more limited nature. Fourth, the judge held that the tenant had done all that he could to retain occupation since 1980 or 1981. He relied on the principle set out in the case of *Morrisons Holdings Ltd* v *Manders Property (Wolverhampton) Ltd* [1976] 1 WLR 533 to the effect that events beyond the control of the tenant, namely the failure of the landlords to carry out the repairs, had led to the lack of occupation of the premises for business purposes and that that was something which B did not destroy the tenant's *locus standi* to bring the proceedings. The judge expressed it in this way:

> . . . the applicant has done all that he could in the circumstances to retain occupation since 1980/1981 and resume his business although the facts are somewhat different, the case falls fairly and squarely within the *Morrison* principle and indeed is if anything stronger.

He appears also to have held that irrespective of the *Morrisons* principle the tenant remained in occupation for the purpose of his business, assuming there was no express or implied repairing obligation on the landlords.

On those findings the learned judge concluded that since there was no ground of opposition to the grant of a new lease under section 30 C of the Landlord and Tenant Act 1954 it was simply a question of fixing the terms of the new lease. He also held that the repairs which would have to be carried out by the landlord pursuant to the repairing obligation which he had found rested upon him would cost some £100,000 and that, once those repairs had been done, before the tenant could resume his business of subletting rooms he would have to expend the sum of between £25,000 and £30,000 on complying with the requirements of the fire regulations in relation to a building in multiple occupation, and in refurbishing the premises and refurnishing them generally.

Mr Tager, first, submits that the judge's finding that there was an express covenant to carry out those repairs was wrong. Second, that the alternative finding that there was an implied covenant was also D wrong. Third, consequently the principle in *Morrisons Holdings* does not apply. Fourth, that in the absence of sufficient continuity of occupation for the purposes of carrying on his business the tenant was unable to apply for a new lease.

Dealing with the argument on the express agreement, the learned judge said in para 5.3 of his judgment:

> . . . It is most likely that the repairs were discussed at the initial meeting and subsequent correspondence confirms this.

Mr Tager points out that there was really no likelihood that repairs were discussed at that meeting and there was no evidence from those who had taken part in the meeting of any such express agreement.

Mr Demetriou at the time of that meeting in 1967 did not speak E English and he did not purport to give any evidence that the question of repairs had been discussed at that meeting, still less that it was agreed that the landlords should bear that burden. Indeed, in the light of his evidence about his inability to speak English he could not do so. He called as a witness Mr Christoporou who had conducted the negotiations on behalf of the tenants with Miss Ross in 1967. The burden of his evidence was that he could not remember anything about it at all. He could not remember whether the question of repairs had been discussed and again, still less, that there had been an obligation on the landlords.

The learned judge, in para 5.7 of his judgment, said:

> I accept that neither Mr Demetriou nor Mr Chris can now remember what F exactly was said about repairs at the time.

That, in my judgment, is not a wholly accurate reflection of what either of those witnesses said. In fact they said that they could not remember anything about it, not "exactly what was said".

It is significant, in my judgment, that at no stage throughout the correspondence, when allegations of non-repair were being made, is there any suggestion that there was an express agreement between the parties that the landlords should shoulder that obligation. In the two sets of proceedings to which I have referred it is quite plain that reliance is based upon section 32 of the Housing Act and not upon any express obligation.

In my judgment, there is simply no material upon which the learned judge could reach the conclusion that there was an express agreement to that effect. Not only is there no such evidence but it G seems to me to be wholly contrary to the contentions which were being put forward on behalf of Mr Demetriou either by himself or by his legal advisers. In my judgment, therefore, the judge's conclusion on that point cannot stand.

Turning to the question of the implied term, the judge seems to have reached his conclusion on the basis that throughout the period, at any rate to 1979 until Judge Pickering's judgment, various landlords did not dispute their liability to carry out repairs. But Mr Tager makes a number of submissions in relation to that point. First, he says that the express willingness to carry out repairs, although they were not done, sometimes at any rate relates to matters which were not structural or external, which was the extent of the repairing covenant found by the judge — for example, gas and electricity H installations. Those would be matters which would be covered by a covenant under section 32 but not by the one which the learned judge found.

Second, that in so far as the landlords appeared to accept an obligation, they were acting, Mr Tager submits, under the mistaken belief that the tenant was a Rent Act tenant and that section 32 of the Housing Act applied to the tenancy. That manifestly was their belief until the judgment of Judge Pickering. Since that time, apart from a "without prejudice" offer to carry out certain electrical repairs if arrears of rent were paid (see a letter of August 8 1984), I am unable to find that anything approaching an acceptance of liability to carry out repairs was given by the various landlords. On the contrary, as it J seems to me, the landlords rejected liability under the section 9(1A) Housing Act notices. The local authority served those notices on the tenant, who at no time that I can see alleged that the landlords were liable for those, by correspondence or by proceeding with his counterclaim in the 1978 action or by starting fresh proceedings seeking to enforce any such covenant.

The question whether an obligation on a landlord to repair should be implied into a tenancy agreement was considered recently by this court in *Barrett* v *Lounova (1982) Ltd* [1989] 1 All ER 351,* and it is fair to the learned judge to say that although the case had been reported by the time the matter was argued before him he was not referred to this decision. That case related to the lease of a house which was entered into before section 32 of the Housing Act 1961 K came into operation. There was, therefore, no implied obligation by statute. There was a provision in the tenancy agreement of an express covenant by the tenant to keep the inside of the premises in repair, but the tenancy agreement was entirely silent as to the obligation to repair the outside. The court held that in the circumstances of that case it was right to imply an obligation on the part of the landlord to keep the exterior of the premises in repair. I refer to the judgment of Kerr LJ at p 354F-G where he says, referring to the judge's analysis:

> . . . Those two ways of putting the test whether or not a term should be implied, sometimes referred to as "the officious bystander test" and "the business efficacy test", are of course correct. But whether or not, on applying those tests, the implication falls to be made is not easy, and the authorities are L of no direct assistance.
> The landlord relied strongly on a well-known passage in *Woodfall on Landlord and Tenant* (28th edn, 1978) para 1-1465, in the following terms. "In general, there is no implied covenant by the lessor of an unfurnished house or flat, or of land, that it is or shall be reasonably fit for habitation, occupation or cultivation, or for any other purpose for which it is let. No covenant is implied that the lessor will do any repairs whatever . . .".

Kerr LJ then referred to the authorities of *Hart* v *Windsor* (1843) 12 M&W 68; (1843-60) All ER Rep 681; *Cockburn* v *Smith* [1924] 2 KB 119; [1924] All ER Rep 59; *Sleafer* v *Lambeth Borough Council* [1959] 3 All ER 378; [1960] 1 QB 43, and at p 356F-G he refers to the case of *Duke of Westminster* v *Guild* [1984] 3 All ER 144; [1985] QB 688 and cites from the judgment of Slade LJ:

> We do not question the correctness of those decisions on their particular facts, M or doubt that in some instances it will be proper for the court to imply an obligation against a landlord, on whom an obligation is not in terms imposed by the relevant lease, to match a correlative obligation thereby expressly imposed on the other party. Nevertheless we think that only rather limited assistance is to be derived from these earlier cases where obligations have been implied.
> Then he referred to the proposition in *Woodfall* which I have read.
> So it follows that a repairing obligation on the landlord can clearly arise as a matter of implication. But that leaves the question already mentioned, which I find difficult and on the borderline, whether the terms and circumstances of

---

*Editor's note: Also reported at [1988] 2 EGLR 54.

> A this particular lease enable such an implication to be made. As to that, although I have not found this an easy case, I agree with the conclusion of the recorder. In my view the clue lies on what Slade LJ referred to as a "correlative obligation", in this case one which is correlative to the express covenant by the tenant to keep the inside and fixtures in good repair, order and condition. The considerations which lead me to that conclusion are the following. It is obvious, as shown by this case itself, that sooner or later the covenant imposed on the tenant in respect of the inside can no longer be complied with unless the outside has been kept in repair. Moreover, it is also clear that the covenant imposed on the tenant was intended to be enforceable throughout the tenancy. For instance, it could not possibly be contended that it would cease to be enforceable if the outside fell into disrepair. In my view it is therefore necessary, as a matter of business efficacy to make this agreement workable, that an obligation to keep the outside in repair must be imposed on someone.
> B For myself, I would reject the persuasive submission of counsel for the landlord that both parties may have thought that in practice the landlord (or possibly the tenant) would do the necessary repairs, so that no problem would arise. In my view that is not a businesslike construction of a tenancy agreement. Accordingly, on the basis that an obligation to keep the outside in a proper state of repair must be imposed on someone, three answers are possible.

He then goes on to consider the three possible solutions, namely that the obligation is on the tenant, that it is a joint obligation, or that it is on the landlord, and concludes that it has to be on the landlord.

Mr Tager submits that that case is distinguishable from the present because there was no obligation here on the tenant and, consequently, there is no correlative obligation on the landlord.
C There was nothing, he submits, in the circumstances of the letting that made it necessary to imply a term to give business efficacy to the contract. It was a weekly tenancy granted to a partnership, one of whose members was a builder or at any rate did building work. The tenant, Mr Demetriou or the partnership, at that stage would themselves have been under an obligation under section 32 of the Housing Act *vis-à-vis* their own subtenants. In those circumstances, Mr Tager submits that there is nothing which makes it necessary to give business efficacy to this contract that there should be an obligation imposed upon the landlord, and that the general position at common law applies.

The learned judge in dealing with this matter put it in this way:

> D . . . .alternatively and in any event it must be inferred from the nature of the letting and the conduct of the parties that such obligation by the landlord is to be necessarily implied on the facts. The only alternative construction would be that no one had the obligation, and letting of premises which are not repaired is, as events have proved, not a practical proposition.

In as much as the learned judge was seeking to rely upon the subsequent conduct of the parties to draw the inference that that obligation was to be implied into the original contract, in my judgment he was in error. It is not permissible to look at the subsequent conduct of the parties in construing the contract at the time that it was made. Moreover, it is a phenomenon, certainly known at common law, that there may be situations in which there is no repairing obligation imposed either expressly or impliedly on
E anyone in relation to a lease.

That may especially be so in a case such as this where it is common ground that the case was not one to which section 32 of the 1961 Act applied. The reason why that was common ground is because under the provisions of that Act, which are now re-enacted in the Landlord and Tenant Act 1985 and are to be found in section 11 of that Act (which is the equivalent of section 32 of the Housing Act 1961) and section 16 of the Landlord and Tenant Act 1985 (which is equivalent to section 32(5) of the previous Act), that section applies only to the lease of a dwelling-house (I refer to para (*b*)), which means a lease by which a building or part of a building is let wholly or mainly as a private residence. It is plain, on the judge's findings, that these premises were let not wholly or mainly as a private residence but for
F the purpose of subletting as rooms. Therefore, it was common ground that section 32 of the Housing Act did not apply and, in my judgment, in those circumstances it is not easy to imply a term into a business tenancy such as this in the way the learned judge sought to do.

I have come to the conclusion that the judge's finding, in the alternative, that there was an implied obligation on the landlords cannot be sustained. That being so, it follows, in my judgment, that the principle in the *Morrisons Holdings* case does not apply.

That is not, however, the end of the matter, because it does appear, as I have indicated, that the learned judge dealt with it on an alternative basis.

In reciting Miss Bell's submissions the learned judge said (at para 14.1 of his judgment) that the submissions made on his behalf were: G

> (1) Mr Demetriou was in fact living there since 1982 along with his mother and Miss Miqueax, he stored furniture there and used one room as his office;
> (2) It was his intention to resume business of subletting as soon as repairs were done and he was ready and willing at all times so to do.
> If this be correct as a fact, then as a matter of law the tenant would have the necessary *locus standi*: . . .

He then went on to deal with Miss Bell's alternative submissions, namely those based upon the *Morrisons Holdings* case.

The learned judge did not expressly accept those submissions which I have just read out. But, by implication, he appears to have done so, because he went on to consider, in the alternative to his finding that there was an express or implied repairing covenant, what the position would be if there was no such repairing liability, and at H
para 2, on p 42, he said:

> If, contrary to my conclusion there was no repairing liability on the landlord, it could only be introduced into the new lease in accordance with section 34 and its construction by the House of Lords in *O'May* v *CLRP* [1983] AC 726.

And he went on to impose such an obligation. It seems to me, therefore, that the learned judge put his judgment on that alternative basis.

Mr Tager submits that that was a false basis upon which the judge could found himself, because he says that the tenant was entitled to a new tenancy only if, throughout the period from the application for a new tenancy till the time when the order of the court was made, he was a tenant of the holding to which the 1954 Act applied. In 1981, when J
he gave notice to quit to his subtenants who never returned, or at any rate at some time thereafter but prior to the judge's determination in this case he ceased to be such a tenant. He relied on the authority of *I&H Caplan Ltd* v *Caplan (No 2)* [1963] 1 WLR 1247. At p 1254 Cross J said:

> . . . .The only sort of tenant who is given by section 24(1) a right to apply for a new tenancy is a tenant under a tenancy to which the Act applies. And if the tenant ceased to occupy the property for business purposes after the service by the landlords of a statutory notice but before he himself issued his summons asking for a new tenancy, then, as it seems to me, the landlord could clearly apply to have the summons dismissed on the ground that the tenant had no locus standi to apply to the court. Section 29(1) is telling the court how to deal with an application under section 24(1), and that, as I have said, is and must be K
> an application by the tenant under a protected tenancy. Section 29(1) and section 32 appear to me clearly to contemplate that if and when the terms of the new tenancy come to be settled, there will inevitably be some property to be comprised in it. There may be disputes as to the extent of that property, but the wording of those two sections, 29(1) and 32(1), seem to me to be totally inconsistent with the notion, which is really implicit in the tenant's argument here, of the judge solemnly sitting down to settle the terms of the tenancy and then suddenly finding at the end of the day that he cannot make any effective order because there is no holding at all. As I see it, it really is a continuing condition of the tenant's right to a new tenancy that he should be throughout the proceedings tenant under a tenancy to which the Act applies. If this condition ceased to be fulfilled at any time, then I think that the landlord can apply to have the tenant's summons dismissed just as he could if the tenancy was not protected by the Act when the application was made. For my part, I L
> reach that conclusion on a consideration of the wording of the Act itself, but I must say that I am confirmed in my view by the fact that the contrary view would, I think, lead to some astonishing results which I can hardly think would be consistent with the policy underlying it.

On the facts of that case the learned judge held that there was no lack of continuity in the tenant's occupation. At p 1260, after finding the facts, he says:

> . . . .what is the legal consequence? I think it is quite clear that a tenant does not lose the protection of this Act simply by ceasing physically to occupy the premises. They may well continue to be occupied for the purposes of the business although they are de facto empty for some period of time. One rather obvious example would be if there was a need for urgent structural repairs and the tenant had to go out of physical occupation in order to enable them to be
> effected. Another example would be that which the Court of Appeal had to deal with in *Teasedale* v *Walker* [1958] 1 WLR 1076. That was a case where premises were only occupied during seasonal periods: they were closed and empty in the winter and only used in the summer. On the other hand, as the Court of Appeal pointed out in *Teasedale* v *Walker*, a mere intention to resume occupation if you get a new tenancy will not preserve the continuity of business user if the thread has once been definitely broken. I think one must observe, however, that in *Teasedale* v *Walker* the discontinuance of business occupation was before the tenant applied to the court for a new tenancy. She had let someone else occupy it for business purposes. He had gone out of occupation and given her back the key and the premises were empty when the summons was issued. In those circumstances the Court of Appeal considered that her expressed intention to start again herself, if and when she got a new tenancy, was not enough. M

A Then he goes on to consider in what circumstances it might be sufficient.

That decision was approved by this court in the *Morrisons Holdings* case [1976] 1 WLR 533 at p 540. After citing those two passages from Cross J's judgment in *Caplan* Scarman LJ said:

> I respectfully agree with the view of the law expressed by Cross J in the two passages to which I have referred. I would put it in my own words as follows: in order to apply for a new tenancy under the Act the tenant must show either that he is continuing in occupation of the premises for the purposes of a business carried on by him, or, if events over which he has no control have led him to absent himself from the premises, if he continues to exert and claim his right to occupancy.

Mr Tager submits that on the hypothesis that there was no
B repairing covenant on the landlords, the tenant really here has done nothing to put into effect an intention to carry on his business by subletting. He has asserted against some landlords that they ought to have done repairs, but he has never proceeded with his counterclaim or started proceedings to enforce what he believed, if he did in fact have a contractual right. Had he done so, he would have received the answer given by this court, namely that he had no such right, and he would have received that answer long since. He would then have been in a position where, in order to carry on his business, he himself would be forced to expend an enormous sum of money, namely £100,000 on carrying out repairs and some £25,000 on carrying out the works to satisfy the fire regulations and to refurbish the premises. C
It is quite plain on the evidence that not only does Mr Demetriou not have the money to afford sums of that kind or anything like them but it would be quite absurd for him to expend those sums of money. It seems to me that, whatever his professed intention may have been, in the light of the finding that there is no express or implied repairing covenant it was quite impracticable for him to carry on his business. Indeed, he has in fact made no attempt to comply with the section 9(1A) notice. There has been no subletting of any of the rooms since 1981 and even allowing a reasonable time in which to put the premises into a state where he could carry on his business, in my judgment that period had long since expired before the matter came before the judge.

In my judgment, therefore, the tenant had no *locus standi* to make D
an application for a new lease and his application should have been refused. I would therefore allow this appeal.

PURCHAS LJ agreed and did not add anything.

The appeal was allowed with costs below and in the Court of Appeal, the order not to be enforced without the leave of the court. The order below was set aside and the application for a new tenancy dismissed, without prejudice to the appellants' pending application for an interim rent.

LANDLORD AND TENANT
LEASEHOLD REFORM

A
Court of Appeal

November 14 1990

(Before Lord Justice DILLON, Lord Justice STOCKER and Lord Justice BINGHAM)

WOODRUFF v HAMBRO AND OTHERS

Estates Gazette March 30 1991

[1991] 12 EG 63

Leasehold Reform Act 1967, section 1(4A) and section 23 —
B **Surrender of existing long lease and grant of new long lease — Attempt to secure lower rateable value in consequence of improvements — Proper construction of section 23(3) of 1967 Act — Distinction between surrender and agreement to surrender — Relief under section 23(3) not available where agreement to surrender consummated by performance — Leaseholder held to have no *locus standi* to apply for reduction of rateable value in order to qualify for enfranchisement — Appeal from decision of county court judge dismissed**

The premises concerned in this case consisted of a house divided into two maisonettes — One of the maisonettes was
C **occupied by a statutory tenant, the other by the present appellant — The lease of the whole house was for a term of 99 years from June 24 1891 — It was a house within the meaning of the Leasehold Reform Act 1967 — The maisonette occupied by the appellant was not a house — In 1976 the unexpired residue of the lease of the whole house was assigned to the appellant, subject to the statutory subtenancy — In 1985 the appellant entered into an agreement with his landlords, the respondent trustees, by which on the completion date he would surrender to the respondents the existing lease of the whole house, subject to and with the benefit of the**
D **subtenancy, and the respondents would grant to him a new lease of his maisonette alone — The consideration for the grant of the new lease (for a term of 60 years) was to be the surrender of the existing lease plus the payment of a premium**

In 1988 the appellant served on the respondents two notices which led to the present proceedings — One was a notice under the 1967 Act claiming the freehold of both maisonettes and the other was a notice claiming a reduction in the rateable value of the premises because of improvements which he had carried out — The landlords objected and the appellant made an originating application to the county court under section 20 of
E **the 1967 Act — The county court judge decided that it was so clear that the appellant did not have the *locus standi* to proceed that there was no point in the matter going to trial and he dismissed the application on the preliminary issue — The appellant appealed**

The Court of Appeal agreed with the county court judge — The appellant had in effect to overcome two obstacles — The first was to get the rateable value, as it stood in 1975 at the date of the agreement to surrender, reduced to the statutory limit for enfranchisement — The second was to get the surrender set aside under section 23(3) of the 1967 Act — However, examination of section 23(3) of the Act showed a fatal F
objection to the appellant's scheme — On its proper construction the relief afforded by section 23(3) applied to an agreement to surrender, not to a surrender which had been completed and taken effect, to an executory, not an executed, surrender — In 1985 the surrender took effect as part of the package which included the grant of a new lease — As the judge had concluded, there was no point in the application going to trial — Appeal dismissed

The following cases are referred to in this report.

Alnatt London Properties Ltd v *Newton* [1984] 1 All ER 423; (1982) 45 P&CR 94; 265 EG 601, CA G

Hennessy's Agreement, Re [1975] Ch 252; [1975] 2 WLR 159; [1975] 1 All ER 60; (1974) 29 P&CR 135

Joseph v *Joseph* [1967] Ch 78; [1966] 3 WLR 631; [1966] 3 All ER 486, CA

Westminster (Duke of) v *Oddy* (1984) 270 EG 945

This was an appeal by Charles Christopher Woodruff against a decision of Judge Phelan, at the West London County Court, dismissing on a preliminary issue an originating application seeking certain declarations in regard to improvements under section 20 of the Leasehold Reform Act 1967. The premises concerned were 26 and 26A Yeoman's Row, off Brompton Road in Kensington, London SW3. The respondents were Jocelyn C Hambro and others, trustees of Henry Smith's Charity. H

Martin Griffith (instructed by Winckworth & Pemberton) appeared on behalf of the appellant; Paul Morgan (instructed by Denton Hall Burgin & Warrens) represented the respondents.

Giving judgment, **DILLON LJ** said: This is an appeal by Mr Woodruff against a decision of Judge Phelan in the West London County Court on July 21 1989 whereby on a preliminary issue he dismissed a certain originating application which Mr Woodruff had made to that court on the basis that Mr Woodruff had no *locus standi*.

The matters in issue concern the Leasehold Reform Act 1967 as amended and certain premises known as 26 and 26A Yeoman's Row, J
off Brompton Road in Kensington. The premises 26 and 26A are a house divided into two maisonettes. One of the maisonettes, 26A, is occupied by a statutory tenant, a Mr Montalto; 26 is occupied by Mr Woodruff.

It is common ground that the house as a whole, comprising the two maisonettes, is a house within the meaning of that word in the 1967 Act but that the single maisonette, no 26 on its own, is not a house.

There was a lease of the premises, which subsequently became known as 26 and 26A, on March 19 1898 by the then trustees of Henry Smith's Charity to a Mr William Collbran for a term of 99 years from June 24 1891. The present trustees of Henry Smith's Charity are the present landlords of the premises and the respondents to this appeal.

On April 27 1976 the unexpired residue of that lease of the whole K
house was assigned to Mr Woodruff, subject to the subtenancy of Mr Montalto of no 26A. Mr Woodruff has been in occupation of no 26 ever since that assignment.

On September 10 1985 Mr Woodruff entered into an agreement with the trustees (the landlords) that on the completion date he would surrender to the trustees the existing lease, that is to say the lease of March 19 1898, subject to, and with the benefit of, the tenancy of no 26A of Mr Montalto and the trustees should grant and he would accept a new lease of no 26 alone. The consideration for the grant of the new lease was to be the surrender of the existing lease and the

A payment by Mr Woodruff of a premium for the new lease of £200,000. That agreement was completed on September 24 1985 and the new lease was then granted to Mr Woodruff. It was granted for a term of 60 years.

There were various interim matters in that it appears, though these matters are not directly relevant to the issues we have to decide, that Mr Woodruff, with leave, carried out various works to no 26 and he also raised various mortgages on the security of his new lease of no 26. But the next relevant step was that on November 9 1988 Mr Woodruff served two notices on the trustees. One was a notice under the 1967 Act of his claim as tenant of 26 and 26A Yeoman's Row, not just no 26, to have the freehold of those premises under the Act, and
B the other was a notice that he had carried out various improvements to 26 and 26A Yeoman's Row and required the landlords to agree to a reduction in the rateable value of the premises for the purposes of the 1967 Act. The landlords gave notice objecting and consequently on January 24 1989 Mr Woodruff launched an originating application in the West London County Court, which is the application which Judge Phelan dismissed on the preliminary issue. By it he sought, pursuant to section 20 of the 1967 Act, an order in the following terms:

1 A Declaration that Schedule 8 to the Housing Act 1974 applies to the improvements specified in the First Schedule to his notice in relation to improvements dated November 9 1988.

2 A Declaration that the works specified in the Notice were involved in
C those improvements.

3 A Declaration that he or his predecessors as tenants had made the improvements.

4 A Declaration that he or his predecessors' contributions amounted to the whole cost of the improvements.

He then set out as the grounds on which he claimed to be entitled to the order that he wished to invoke the powers of the county court under section 23(3) of the 1967 Act in respect of his agreement to surrender the lease of March 19 1898, but before doing so he had sought a reduction in the rateable value of the premises pursuant to Schedule 8 to the Housing Act 1974.

It is convenient therefore to turn to the 1967 Act. Section 1 provides by subsection (1) that Part I of the Act:

D shall have effect to confer on a tenant of a leasehold house, occupying the house as his residence, a right to acquire on fair terms the freehold or an extended lease of the house and premises where —

(*a*) his tenancy is a long tenancy at a low rent and

— subject to certain subsections to which I need not refer —

the rateable value of the house and premises on the appropriate day is not (or was not) more than

— a specified sum which again I need not go into —

and

(*b*) at the relevant time (that is to say, at the time when he gives notice in accordance with this Act of his desire to have the freehold or to have an extended lease, as the case may be) he has been tenant of the house under a long tenancy at a low rent, and occupying it as his residence, for
E the last three years or for periods amounting to three years in the last ten years . . .

It is not in doubt that if the rateable value qualification had been satisfied Mr Woodruff would have had the right to acquire the freehold or an extended lease if he had given notice before he entered into the agreement to surrender the existing lease. The apparent difficulty at that stage was that the rateable value of 26 and 26A as a whole was above the prescribed statutory limit. Subsection (4A) of section 1 of the 1967 Act, introduced by amendment, provides as follows:

Schedule 8 to the Housing Act 1974 shall have effect to enable a tenant to have the rateable value of the house and premises reduced for purposes of this section in consequence of tenant's improvements.

F Schedule 8 to the 1974 Act provides by para 1:

(1) Where the tenant, or any previous tenant, had made or contributed to the cost of an improvement on the premises comprised in the tenancy and the improvement is one to which this Schedule applies, then, if the tenant serves on the landlord a notice in the prescribed form requiring him to agree to a reduction under this Schedule, their rateable value as ascertained for the purposes of section 1 of the Leasehold Reform Act 1967 shall be reduced by such amount, if any, as may be agreed or determined in accordance with the following provisions of this Schedule.

(2) This Schedule applies to any improvement made by the execution of works amounting to structural alteration extension or addition.

There are then various supplementary provisions for working out whether the tenant is entitled to any, and if so what, reductions.

Section 23 of the 1967 Act provides as follows, so far as material: G

(1) Except as provided by this section, any agreement relating to a tenancy (whether contained in the instrument creating the tenancy or not and whether made before the creation of the tenancy or not) shall be void in so far as it purports to exclude or modify any right to acquire the freehold or an extended lease or right to compensation under this Part of this Act, or provides for the termination or surrender of the tenancy in the event of a tenant acquiring or claiming any such right or for the imposition of any penalty or disability on the tenant in that event.

(2) Subsection (1) above shall not be taken to preclude a tenant from surrendering his tenancy, and shall not —

(*a*) invalidate any agreement for a tenant to acquire an interest superior to his tenancy or an extended lease on terms different from those provided for by this Part of this Act; or

(*b*) where a tenant has given notice of his desire to have the freehold or an H extended lease under this Part of this Act, invalidate any agreement between the landlord and the tenant that that notice shall cease to be binding or any provision of such an agreement excluding or restricting for a period not exceeding three years the right to give a further notice of either kind with respect to the house or any part of it; or

(*c*) where a tenant's right to compensation has accrued, invalidate any agreement as to the amount of the compensation.

(3) Where —

(*a*) a person, being entitled as tenant of a house to acquire the freehold or an extended lease under this Part of this Act, enters into an agreement without the prior approval of the court for the surrender of his tenancy, or for the acquisition by him of an interest superior to his tenancy or of any extended lease; or

(*b*) a tenancy having been extended under this Part of this Act, the tenant, J on the landlord claiming possession for purposes of redevelopment, enters into an agreement without the prior approval of the court for the surrender of the tenancy;

then on the application of the tenant the county court or any court in which proceedings are brought against him on the agreement may, if in the opinion of the court he is not adequately recompensed under the agreement for his rights under this Part of this Act, set aside or vary the agreement and give such other relief as appears to the court to be just, having regard to the situation and conduct of the parties.

The person who can apply for relief under subsection (3) of section 23 is a person being entitled as tenant of a house to acquire the freehold or an extended lease under the 1967 Act. That, of course, refers back to the qualifications in section 1 of the 1967 Act, but Mr K Griffith, for Mr Woodruff, submits that that does not import the condition as to rateable value and in that regard reliance is placed on certain *obiter dicta* of Eveleigh LJ in the case of *Duke of Westminster* v *Oddy* (1984) 270 EG 945. They are to the effect that if the conditions which would enable a reduction to be obtained are there, the tenant can claim the freehold or an extended lease even though he has not yet established a notional rateable value below the limits. But Eveleigh LJ was not concerned with a case where there has been a purported surrender of the relevant lease. Mr Woodruff seeks in effect to overcome two hurdles. First, to get the rateable value as it prevailed at the time in September 1975 of the agreement for surrender reduced because of improvements to within the statutory limit, and then to get the surrender set aside under section 23(3) of the 1967 Act. Unless he L can achieve both of those objectives he gets nowhere.

Section 23(3) bears close parallel to section 38 of the Landlord and Tenant Act 1954, which is in Part II of that Act. That provides:

(1) Any agreement relating to a tenancy to which this Part of this Act applies (whether contained in the instrument creating the tenancy or not) shall be void.

— then there is an exception for applications to the court —

in so far as it purports to preclude the tenant from making an application or request under this Part of this Act or provides for the termination or the surrender of the tenancy in the event of his making such an application or request or for the imposition of any penalty or disability on the tenant in that event.

M That has plain parallels with the wording in subsection (1) of section 23 of the 1967 Act.

It has been held under section 38 that an agreement to surrender a lease or tenancy is caught by section 38: see *Joseph* v *Joseph* [1967] Ch 78 and *Alnatt London Properties Ltd* v *Newton* [1984] 1 All ER 423*, both decisions of this court. But under section 24(2) in Part II of the 1954 Act a tenancy does not come to an end by an actual surrender.

There is a further parallel in Part I of the Landlord and Tenant Act 1954, which is concerned with residential property. That provides by section 17:

*Editor's note: Also reported at (1982) 265 EG 601.

The provisions of this Part of this Act shall have effect notwithstanding any agreement to the contrary: Provided that nothing in this Part of this Act shall be construed as preventing the surrender of a tenancy.

It was consequently held that, though an actual surrender was permissible, an agreement to surrender the tenancy in the future was unenforceable: see *Re Hennessy's Agreement* [1975] Ch 252, a decision of the Vice-Chancellor, Sir Anthony Plowman.

The wording of section 23 is, of course, different because there is no counterpart to subsection (3) in the 1954 Act. The draftsmanship is not perhaps ideal, but it seems to me clear that what is achieved by section 23 is that certain classes of agreement — an agreement without the prior approval of the court for the surrender of the tenancy or for the acquisition by the tenant of an interest superior to his tenancy, or of any extended lease — are taken out of subsection (1) and therefore not rendered void (as but for subsections (3) and (2) they would be) by section 1. Instead they are to be voidable on the tenant's application under subsection (3). But subsection (2) of section 23 makes it plain that subsection (1) is not to be taken as precluding a tenant from surrendering his tenancy. That is in line with the provisions under the 1954 Act and it seems to me that it necessarily follows that in section 23 Parliament has preserved the same distinction as is to be found in the 1954 Act, as interpreted by the courts, between the actual surrender and the agreement for surrender. I do not think that subsection (3), where it refers to "an agreement . . . for the surrender of his tenancy" and provides that the court may on application "set aside or vary the agreement and give such other relief as appears . . . to be just", is contemplating setting aside or overriding an executed surrender of the lease or a surrender which has taken effect by operation of law and is no longer executory. So equally where there is reference to an agreement for an extended lease or the acquisition of an interest superior to his tenancy, I do not think the powers under section 23(3) can be exercisable once the extended lease has been actually granted or the superior interest has been actually acquired.

Therefore it must follow that even if the dicta of Eveleigh LJ in *Oddy*'s case are correct, an investigation into the improvements and their effect on the rateable value of the house, of which Mr Woodruff is no longer the tenant in any shape or form as a whole house, can serve no useful purpose at all. Consequently there is no point in this application going for trial and I agree with the judge that it should be dismissed. I would dismiss this appeal.

Agreeing, **Stocker LJ** said: For my part I should be content to rest my conclusion on the simple proposition that as a matter of construction section 23(3) of the Act is concerned solely with an agreement to surrender and not with the surrender itself once it has taken place in accordance with the agreement. The relevant words in section 23(3) "Where . . . a person . . . enters into an agreement . . . for the surrender of his tenancy" are plain and unambiguous. They relate to an agreement, not to a surrender which takes place thereafter in accordance with that agreement. If the draftsman of the Act had intended the subsection to apply to an actual surrender of the tenancy, in addition to an agreement to do so, it would have been a very simple matter to have included the surrender as well as the agreement for it in the wording of the subsection. The consequences of an actual surrender may well be irreversible, or virtually so, particularly, as in this case, if it is followed by the grant of a new lease. I cannot accept that the draftsman can have overlooked such potential consequences and would have expressly included actual surrender as well as an agreement to surrender if he intended the subsection to have that effect. I find it impossible to construe the subsection so as to extend the meaning of the words "agreement to surrender" to include the subsequent surrender itself. Accordingly, the appellant is not entitled to rely upon the provisions of section 23(3), with the consequence that the judge correctly held that he had no *locus standi* to apply for a variation of the rateable value of the house reduced in accordance with the provisions of section 1(4A). In the light of this conclusion, I do not find it necessary to consider other issues and arguments which have been raised on this appeal. I agree, therefore, that this appeal should be dismissed.

Also agreeing, **Bingham LJ** said: The 1967 Act intends to confer on tenants rights against their landlords which they did not enjoy at common law. It follows that the Act necessarily encroached on the common law rights of landlords. In deciding in any doubtful case how extensive are the tenant's rights and the encroachment on the rights of the landlord, the task of the court is to put a fair construction on the Act, bearing in mind that Parliament intended the Act to work to the advantage of tenants. In so far as there is a lacuna or ambiguity in the Act, it is proper to bear the purpose of the Act in mind in endeavouring fairly to reflect the intentions of Parliament.

I am persuaded, contrary to my first impression, that section 23(1), if it stood alone, would embrace an agreement for the surrender of a lease. So much seems to follow by analogy from *Joseph* v *Joseph* [1967] Ch 78. It therefore follows, as a matter of construction and logical necessity, that subsection (2) of section 23 must apply to an agreement to surrender.

The crux of this appeal turns on the proper construction of section 23(3), the terms of which Dillon LJ has cited. Mr Griffith, for Mr Woodruff, submits that the court enjoys the discretionary power conferred by that subsection if a tenant qualified to acquire the freehold enters into an agreement without the prior approval of the court for the surrender of his tenancy, notwithstanding that the agreement has been performed and the surrender made, and notwithstanding that, as has happened here, a premium of £200,000 has been paid and a new lease granted. All these matters, subsequent to the agreement, are, Mr Griffith submits, relevant to the exercise of the court's discretion and not to the existence of its jurisdiction.

This submission involves the conclusion that in subsection (3)(*a*) the draftsman intended "agreement . . . for the surrender" to include a surrender. The submission also involves the conclusion that when the draftsman used the words "set aside or vary the agreement and give such other relief as appears to the court to be just", he intended to give power to set aside a surrender as well as an agreement to surrender. I find it inconceivable that any Parliamentary draftsman who sat down to draft subsection (3) would not have made reference to surrenders as well as agreements to surrender if he had intended the subsection to have that extended effect. As it stands, the meaning of the subsection is, in my view, clear and, even construing the subsection bearing in mind the Act's benevolent intentions towards tenants, I cannot attribute to it the meaning contended for. This clear impression is reinforced when one notes that this extraordinary statutory jurisdiction is conferred not only on the county court but also on any court in which proceedings are brought against the tenant "on the agreement". That in itself is a clear indication, in my opinion, that the agreement has not been consummated by performance. In agreement with the learned county court judge and Dillon and Stocker LJJ, I conclude that the appellant is unhorsed at this first formidable obstacle and I prefer to express no opinion on other obstacles further down the course.

The appeal was dismissed with costs.

Court of Appeal

November 27 1990

(Before Lord Justice SLADE and Lord Justice RALPH GIBSON)

DYMOND v ARUNDEL-TIMMS AND OTHERS

Estates Gazette June 1 1991

[1991] 21 EG 107

Leasehold Reform Act 1967 — Application for freehold by leaseholder — Whether at the relevant time the leaseholder had been occupying the premises as a residence for the last three years or for periods amounting to three years in the last 10 years — Whether only a main residence — Question as to extent of occupation of "other residence" within the meaning of para 6(1)(*d*)(ii) of Schedule 3, Part II, to 1967 Act — Allegations of misrepresentation or concealment of material facts on the part of the leaseholder — County court judge's dismissal of application upheld by Court of Appeal — Order made under section 20(6) of 1967 Act making void any further notice claiming enfranchisement within five years beginning with order

The conflict and litigation in this case arose from the fact that in addition to the leasehold premises which the applicant had purchased with a view to enfranchisement, the applicant

A **had continued to own, and in some degree to occupy, a cottage — The leasehold premises consisted of a shop, which was let, with a residential flat or maisonette above — In her notice of claim to the freehold, in the standard form under the Leasehold Reform (Notices) Regulations 1967, the applicant stated that she had occupied the maisonette as her only residence since the date of its acquisition in 1984 — In reply to the question in the form of notice as to the applicant's occupation of any "other residence" she answered "none" — The freeholders challenged her claim on the grounds that she had not occupied the maisonette as her only or main residence**
B **for a period of three years and that she was not in such occupation on the date of her notice, February 18 1988 — The applicant issued an originating application in the county court applying for enfranchisement and in answer the freeholders denied her claim — They requested further particulars relating to places occupied and periods of occupation during the relevant period, to which the applicant replied with references to foreign holidays and occasional nights spent with a friend, but made no reference to any occupation of the cottage — At the trial the respondent freeholders obtained leave to allege misrepresentation or concealment of material facts on the part of the applicant**

C **Apart from the question of misrepresentation or concealment, the county court judge dealt with four main issues — He accepted that at the date when the applicant served her statutory notice, February 18 1988, she had been occupying the maisonette as her only or main residence for the requisite period — Her notice, however, failed to comply with para 6(1)(*d*)(ii) of Schedule 3 to the 1967 Act by omitting to state that during the relevant period she had another residence, namely, the cottage — This omission rendered the notice invalid unless the failure was merely an "inaccuracy in the particulars required" within the meaning of para 6(3) of**
D **Schedule 3 — The applicant argued that the failure to mention the cottage was such an inaccuracy and that the validity of the notice was saved by para 6(3) — The judge, however, who took an unfavourable view of the applicant's evidence at the trial, refused to accept that the omission was a mere "inaccuracy" — He held that the omission was deliberate — Hence the notice was invalid — The judge's finding was that the notice was "supported by concealment or misrepresentation of material facts" — He then ordered, in accordance with section 20(5) and (6) of the 1967 Act that any further notice by the applicant claiming enfranchisement would be void if given in the next five years**

E **After reviewing the judge's reasoning and conclusions, the Court of Appeal held that, unless there were grounds for upsetting the judge's finding of misrepresentation or concealment of material facts, there was no justification for disturbing his decision — In the court's view, there was ample material for his finding that the applicant's notice was supported by misrepresentation or concealment — The judge's order under section 20(6) of the 1967 Act followed upon that finding — Appeal dismissed**

The following cases are referred to in this report.

F *Cresswell* v *Duke of Westminster* [1985] 2 EGLR 151; (1985) 275 EG 461; 25 RVR 144, CA

Poland v *Earl Cadogan* [1980] 3 All ER 544; (1980) 40 P&CR 321; [1980] EGD 769; 250 EG 495, CA

This was an appeal by Mrs Wendy Denise Dymond from a decision of Judge Peck, at Barnstaple County Court, dismissing her application under the Leasehold Reform Act 1967 to acquire the freehold of a property at 17 Bridgeland Street, Bideford, Devon. The respondents to the application and the appeal were the trustees for the time being of the Bideford Bridge Trust.

Kirk Reynolds (instructed by Peter Peter & Wright, of Bideford) appeared on behalf of the appellant; Simon Monty (instructed by Bazeley Barnes & Bazeley, of Bideford) represented the respondents.

Giving judgment, **Slade LJ** said: This is an appeal by Mrs Wendy G
Denise Dymond from a judgment of His Honour Judge Peck given in the Barnstaple County Court on March 26 1990. By his order he dismissed her application, made in reliance on the Leasehold Reform Act 1967, to acquire the freehold of a property known as 17 Bridgeland Street, Bideford, Devon ("the house"). She wished to acquire it from the respondents, who were the freeholders of the house and are the present trustees of the Bideford Bridge Trust.

I find it convenient to begin this judgment by referring to certain relevant provisions of the Leasehold Reform Act 1967, as amended ("the 1967 Act"). Section 1(1), so far as material, provides as follows:

> This Part of this Act shall have effect to confer on a tenant of a leasehold H
> house, occupying the house as his residence, a right to acquire on fair terms the freehold . . . of the house and premises where —
>
> (*a*) his tenancy is a long tenancy at a low rent and . . . the rateable value of the house and premises on the appropriate day is not (or was not) more than £200 . . ., and
>
> (*b*) at the relevant time (that is to say, at the time when he gives notice in accordance with this Act of his desire to have the freehold . . .) he has been tenant of the house under a long tenancy at a low rent, and occupying it as his residence, for the last three years or for periods amounting to three years in the last ten years . . .

In the present case it is common ground that Mrs Dymond's tenancy of the house satisfied condition (*a*) and that at the relevant time, when she gave notice of her desire to have the freehold, she had been tenant of the house under a long tenancy at a low rent for the last three years. J
It is, however, in dispute whether at the relevant time she had been occupying it as her residence for the last three years or for periods amounting to three years in the last 10 years.

Section 1(2) defines what is meant by the phrase "occupying the house as his residence". So far as material it provides:

> In this Part of this Act references, in relation to any tenancy, to the tenant occupying a house as his residence shall be construed as applying where, but only where, the tenant is, in right of the tenancy, occupying it as his only or main residence (whether or not he uses it also for other purposes); but —
>
> (*a*) references to a person occupying a house shall apply where he occupies it in part only; . . .

Schedule 3, Part II, para 6(1) of the 1967 Act requires that a K
tenant's notice under Part I of his desire to have the freehold of a house and premises shall be in the prescribed form and shall contain a number of specified particulars, including in subpara (*d*):

> the periods for which since the beginning of the preceding ten years and since acquiring the tenancy the tenant has and has not occupied the house as his residence, together with the following additional particulars about the periods for which during that time he has so occupied the house, that is to say, —
>
> (i) what parts, if any, of the house have not been in his own occupation and for what periods; and
> (ii) what other residence, if any, he has had for what periods, and which was his main residence.

Para 6(3), so far as material, provides: L

> The notice shall not be invalidated by any inaccuracy in the particulars required by this paragraph or any misdescription of the property to which the claim extends; . . .

Section 20 of the 1967 Act provides for certain sanctions. Subsection (5) provides:

> Where a person gives notice of his desire to have the freehold . . . and the notice either is set aside by the court or withdrawn, or ceases to have effect, or would, if valid, cease to have effect, then if it is made to appear to the court —
>
> (*a*) that the notice was not given in good faith; or
>
> (*b*) that the person giving the notice attempted in any material respect to support it by misrepresentation or the concealment of material facts;
>
> the court may, on the application of the landlord, order that person to pay M
> to the landlord such sum as appears sufficient for compensation for damage or loss sustained by the landlord as the result of the giving of the notice.

Subsection (6) provides:

> In any case where under subsection (5) above the court has power, on the application of the landlord, to order a person to make a payment to the landlord, the court (whether or not it makes an order under that subsection) may, on the application of the landlord, order that any further notice given by that person under this Part of this Act of his desire to have the freehold . . . of the same house or any part of it, with or without any other property, shall be void if given within the five years beginning with the date of the order.

Against this statutory background, I turn to the history of this

A case, reciting the facts either as they are common ground or as they have been found by the judge. In 1968 Mrs Dymond's then husband and she purchased jointly a property known as Hillsborough Cottage, Atlantic Way, Westward Ho!, Devon, which is about 4 miles from Bideford. She and her husband brought up their three sons in this cottage until, unfortunately, their marriage broke down. The husband then moved out and the property became vested in Mrs Dymond's sole name. In early 1984 she was living there with her three sons, Lee, Simon and Timothy, then aged respectively 17, 15 and 12. The judge described the cottage as having three good bedrooms, substantial accommodation and a garden.

B For some time before 1984 Mrs Dymond had been coming frequently to Bideford. Her parents lived there and her mother was not well. In 1984 her father who, as the judge described, is a large property owner in Bideford, suggested to her that she should buy the lease of the house. It comprised a shop, which was let, and above it a flat or maisonette in which she could live. It was held under a long tenancy created on March 1 1896, whereby it was devised for a term of 99 years from December 25 1895. In due course she entered into this transaction and the lease was assigned to her by an assignment of October 26 1984 made in her favour by one of her father's companies for a monetary payment.

The judge found as follows:

C I have no doubt that the lease was assigned with the specific purpose of obtaining the ownership of the freehold under the Leasehold Reform Act. I have no doubt at all that the eyes of all the persons concerned were wide open, and that it was done with the intention of having the lease enfranchised at the earliest possible moment.

If Mrs Dymond was to acquire the right to have the lease enfranchised at the earliest possible moment, she was no doubt aware, or was made aware, that she had to occupy it as her only or main residence for the next three years. Her evidence was that she moved into the house on October 4 1984, that is to say about three weeks before the assignment was executed, and occupied it as her home at least until February 18 1988, which, as I shall explain hereafter, is the relevant date.

D The judge described the accommodation in the flat as smaller than that of Hillsborough Cottage and somewhat cramped. It does appear to have been somewhat cramped. In her oral evidence Mrs Dymond described it as having two bedrooms, a boxroom, a bathroom through one bedroom, and a kitchen/living room. When she moved in, Lee and Timothy went with her but, at least for the time being, Simon stayed on at the cottage. According to her evidence, thereafter her home was at the house. However, she did not sell the cottage. She continued to pay the mortgage interest on it. At some times the cottage was let and at some times her sons lived there, both when below school-leaving age and after they had left school. She looked after the garden at times and there was a room set aside for her flower decoration. She kept furniture in a E shed.

On November 2 1987 she served a notice under the 1967 Act stating her desire to acquire the freehold of the house. It is common ground that that notice was invalid, but it appears that at that stage she did not realise its invalidity. The judge found that while she treated the house as her main residence up to October 1987, thereafter she spent less and less time there and was spending her time at Hillsborough Cottage and at a house to which she referred as the "Manor House". The Manor House belongs to a Mr Wilson, an antiques dealer. The judge said that it appeared that she slept there two nights a week, but not on other nights.

On February 18 1988 Mrs Dymond served a second notice of her F desire to acquire the freehold. This, she asserts, was a valid notice. It is the relevant notice for our purposes. February 18 1988 is correspondingly the relevant date for the purpose of considering her claim.

The notice of her claim contained a schedule in common form which required the completion of a number of particulars following the pattern of para 6(1)(*d*) of Schedule 3 of Part II to the 1967 Act. Para 6 of the schedule to the notice required her to state:

Periods for which in the last ten years and since acquiring the tenancy the claimant

(i) has not occupied the house as his residence
(ii) has occupied the house as his residence

In the case of periods stated under (ii)

(a) state what parts (if any) of the house have not been in his own occupation and for what periods G
(b) state what other residence (if any) he has had and for what periods, and which was his main residence.

In answer to these items Mrs Dymond made no entry in response to (i). She answered (ii) by saying:

The claimant has occupied the maisonette as her only residence since the date of acquisition (and before, ie, by the 4th October 1984).

She answered (a) by saying:

The lock-up shop since the date of acquisition.

Perhaps most materially for present purposes, she answered (b) by saying "None".

On April 26 1988 the respondents served a notice by way of reply in H which they did not admit Mrs Dymond's claim, on the grounds, among others, that:

(b) You have not occupied the house as your only or main residence for a period of three years
(c) You were not in occupation of the premises as your only or main residence on February 18 1988:

This was followed by the issue of an originating application by Mrs Dymond, applying for enfranchisement of the property. In para 6 of the application she stated:

On the date when I served the notice referred to in paragraph 4 hereof

— that is the notice of February 18 1988 —

I have been a tenant of the house and premises under a long tenancy at a low J rent and I have been occupying it as my sole residence for the previous three years . . .

On October 20 1988 the respondents served an answer, in which para 6 of the originating application was denied. In particular it was denied that:

. . . the Applicant was at the date of the notice given by her residing in the premises and further it is denied that she had been occupying the premises as her sole residence for the previous three years.

On April 13 1989 Mrs Dymond served an amended application in which para 6 was amended to read:

On the date when I served the Notice referred to in paragraph 4 hereof, I have been a tenant of the house and premises under a long tenancy at a low rent and K I have been occupying it as my sole or main residence for the previous three years or for periods amounting to three years in the last ten years.

Not surprisingly, perhaps, this amended application provoked a request for further and better particulars. Later in April of 1989 further and better particulars were delivered of para 6 of the amended application. The first request read:

Please state, precisely, all the dates on which the Applicant contends that she was resident in the premises.

To that an answer was given:

The applicant does not keep a record or diary and cannot accurately recall or say where she was on each day between October 26 1984 and February 18 1988. However, the Applicant contends she was resident in the premises in the L sense that it was her permanent home at all times throughout that period. The Applicant has spent occasional nights at the home of her parents in Bideford, with friends in Plymouth, with friends in St Albans and with friends in Monkleigh. She has also been on foreign holidays as to which see below.

The second request read as follows:

Please specify all other places where, in the period October 26 1984 to February 18 1988, the Applicant has resided for an aggregate of more than two weeks, setting out, in addition the dates of such periods of residence.

Mrs Dymond's reply reads as follows:

The Applicant has had two foreign holidays in the period mentioned and each of those holidays were of about three weeks duration but the Applicant cannot recall the precise dates. The Applicant has spent occasional nights away from the premises as set out in answer to request 1 above. The Applicant M cannot say whether the total number of nights so spent away from home exceeds 40 in aggregate.

I pause to observe that neither of those answers makes any reference at all to Hillsborough Cottage.

On June 10 1988, however, Mrs Dymond's solicitors wrote a letter to the respondents' solicitors in which mention of the cottage was included. We have been told that this letter was not before the judge, but it has been admitted in evidence before this court by agreement of the parties. The final paragraph of that letter, which makes reference to Hillsborough Cottage, reads as follows:

Whilst it is of course for our client to establish her right under the Act, we think it only fair and reasonable that our client should know the facts upon

A which the denial has been made bearing in mind that the property at 17 Bridgeland Street is her only residence, although she does own one other property in Westward Ho! which is let as flats.

At the start of the hearing before the judge, as appears from his notes of the hearing, the respondents' counsel applied for, and obtained, leave to amend their answer by alleging material misrepresentation in the notice of February 18 1988. Their application for leave to amend was granted and, although no copy of an amended answer is available, it seems clear that thenceforward the question of misrepresentation became an issue in the proceedings.

The judge heard oral evidence from a number of witnesses, including Mrs Dymond herself, Mr Wilson and some three other witnesses on her side and three witnesses for the respondents. Apart
B from the question of misrepresentation, it would appear that there were four main issues argued before the judge which Mr Reynolds, Mrs Dymond's counsel in this court, has helpfully identified for our assistance as follows:

(a) Whether at the date when Mrs Dymond served her notice on February 18 1988, she had been occupying the house as her "only or main residence" within the meaning of section 1(2) of the 1967 Act for the requisite period

(b) Whether her notice failed to comply with the requirement of paragraph 6(1)(d)(ii) of Schedule 3 to the 1967 Act, in that it failed to state that she had during the relevant period another residence, namely Hillsborough Cottage.

(The effective issue there, of course, was whether the cottage was or
C was not a "residence" within the meaning of the Act.)

(c) If so, whether failure to comply with this requirement invalidated the notice or whether it was a mere "inaccuracy in the particulars" within the meaning of paragraph 6(3) of the Schedule, which did not invalidate the notice.

Then, after he had delivered his judgment, the judge, on the respondents' application, had to decide a further issue, namely:

(d) Whether Mrs Dymond should be barred under section 20(6) of the 1967 Act from serving a further notice of a claim for enfranchisement for a period of five years from the date of the judgment.

The judge decided issue (a) in favour of Mrs Dymond. As to that point, he said this (at pp 5 and 6 of his judgment):

> The respondents say that albeit the Applicant was residing in the flat until
> D October 1987, she was not living there until well after the assignment and that she effectively vacated the flat as her real residence after she assumed that the period under the Act was completed.
> I am not satisfied that she did not occupy the flat from the outset. On any logical point of view she would move in as soon as possible. I think she did so and that she treated it as her main residence to October 1987.
> Thereafter I think she spent less and less time there and was spending her time at the Manor House

— that is, Mr Wilson's house —

> and the cottage. Albeit that the burden is on the Applicant to show that she treated the flat as her main residence up to and on the date of the Notice, I think there is no satisfactory evidence to prove the contrary. The electricity and the telephone bills show she spent more and more time away, but I think it
> E is very difficult to say that the flat ceased to be her main residence. If I was left only with that decision, I would find in her favour and that the flat was her main residence at the material time.

The judge's decision on this issue (a) has been challenged by the respondents in a respondents' notice. His decision on issues (b), (c) and (d), however, are the subject of Mrs Dymond's appeal and I shall deal with them first in turn.

As to issue (b), the judge having quoted the requirements for a notice set out in the 1967 Act in Schedule 3 Part II, para 6(1)(*d*)(ii), continued as follows (at p 3 *et seq*):

> Mr Maurice

— counsel for the respondents —

> says that whether or not the applicant was residing in the cottage for any
> F length of time, the fact is that it was her residence. There was a room set aside for her flower decorations. She tended the garden at times. She paid the mortgage. Her children lived there — both when below school-leaving age and after they left school. She kept furniture in a shed. This was a residence available to her and should have been disclosed.
> Mr Edmunds for the applicant says that a residence means somewhere where you reside. If you do not reside there, it is not required to be disclosed. The particulars to be disclosed refer to a main residence and an "other residence". The "other residence" need not be disclosed unless it is an alternative residence.
> This is a difficult question, but in my view Mr Maurice's contention is the correct one. This was a residence available to live in, even if she made only occasional visits, and it was required to be disclosed by the Act.

In the schedule to her notice of February 18 1988, Mrs Dymond had specifically stated that since October 4 1984 she had occupied the G
house as her only residence and had had no other residence. On her behalf, Mr Reynolds has submitted that at no time during the period from October 4 1984 to the date of the notice had Hillsborough Cottage been one of her "residences" in the relevant sense and that accordingly she had completed the schedule to her notice entirely accurately. As he rightly pointed out, this submission involves consideration of two questions: (1) what is the meaning of the word "residence" as used in para 6(1)(*d*)(ii) of Part II of Schedule 3 to the 1967 Act; and (2) whether, on the evidence, Hillsborough Cottage at any time during the period from October 4 1984 to February 18 1988 was a "residence" within that meaning. For brevity I will henceforth refer to that para 6 simply as "para 6".

The wording of that paragraph itself shows that a dwelling does H
not have to be a "main" residence to qualify as an "other residence" within its meaning. On the other hand, I think that Mr Reynolds is right in submitting that the mere ownership of another dwelling which is capable of being used for residential purposes will not by itself necessarily mean that the owner also has an "other residence" within the meaning of the paragraph. He submitted that the owner of a house cannot be said to have had an "other residence" within that meaning, during a stated period, unless for significant times during that period he occupied the other dwelling as his residence.

For present purposes I am prepared to assume in favour of Mrs Dymond that it is not enough that the other house in question has been owned by the owner and has been available for his occupation, J
but that there must also have been a degree of actual occupation if a house is to qualify as a residence. Nevertheless, it seems to me that in every case it must be a matter of fact and degree whether a house constitutes some "other residence" within the meaning of para 6.

In this context I derive helpful guidance from a passage in the judgment of Eveleigh LJ in the case of *Poland* v *Earl Cadogan* [1980] 3 All ER 544, at p 550, where he said:

> I am of the opinion that a person may occupy premises as his residence, even though he is absent for a very long period. It may well be, however, that a long absence will necessitate the acquisition of another residence which will prevent the tenant from claiming that the premises were his "only or main residence", which is the requirement of section 1(2) of the Act. It seems to me
> that, if a person takes the tenancy of a house and furnishes it and keeps it ready K
> for his own habitation "whenever he pleases to go into it", he is, as Lush J said, an occupier though he may not reside in it one day in a year, and I would further say that he occupies it as a residence. Moreover, I would say that he occupies it as his residence provided that it is understood that the expression "his residence" is not used in a context which implies his main residence. It means he keeps it for himself to live in. If, however, he sublets the property, he ceases to occupy it as his residence because, as Moore LJ explained, the occupation is in the subtenant. It may not be inaccurate, however, still to refer to the premises as his residence. That in isolation would be a permissible description of the property, but it does not mean that he is in occupation.

In the present case the evidence showed that there were from time to time lettings of parts of Hillsborough Cottage up to September
1987. Though Mrs Dymond's solicitors' letter of June 10 1988 had L
referred to lettings, the evidence did not actually establish, so far as I am aware, any lettings after September 1987. On the other hand, the evidence and findings of the judge did establish a number of other matters, which showed that this was not merely a case of Mrs Dymond owning the cottage and having it available for her use. During the relevant period, from October 4 1984 to February 18 1988, it was furnished and, so far as it was not let, was actually used by her from time to time. She sometimes spent nights there and, according to the judge's findings, spent still more time there after October 1987. Her children were living there from time to time both before and after they left school; she kept furniture in a shed; she did gardening work there from time to time and there was a room set
aside for her flower decoration. In my judgment, on these findings it M
was fully open to the judge, as a matter of fact and degree, to find that the cottage constituted an "other residence" of hers within the meaning of para 6(1)(*d*)(ii), and I think he was right so to hold.

In these circumstances, subject to the point raised by issue (c), to which I shall shortly turn, in my opinion the judge was right to hold that the notice of February 18 1988 failed to contain the particulars required by para 6(1)(*d*)(ii) because it did not mention that Mrs Dymond had an "other residence" and, subject to that point, was invalid.

I now turn to issue (c). As to that issue, Mrs Dymond's case was, and is, that her failure to mention Hillsborough Cottage in her notice was an "inaccuracy in the particulars required" by para 6, within the

A meaning of para 6(3), and therefore did not invalidate the notice.

The judge dealt with this matter as follows (at pp 4 and 5 of his judgment). After referring to para 6(3), he continued:

> A mere inaccuracy in the notice can be corrected as a matter of right. I have been referred to the case of *Cresswell* v *The Duke of Westminster* (1985) 275 EG 461*, a decision of the Court of Appeal. I think it necessary before deciding whether this was a mere inaccuracy or whether it goes further to say something about the evidence I have heard.
>
> I am bound to say that I found the evidence of the applicant totally unsatisfactory and in some respects I did not believe her. If she thought an answer might be wrong or would not help her case she would say she could not remember, and on any material facts I tended to disbelieve her.
>
> She called Mr Wilson, who owns a large manor house and says he is an
> B antiques dealer. It appears he and she sleep together two nights a week but not on other nights. According to her she spends her nights at the flat. I cannot avoid saying that I found Mr Wilson's evidence unsatisfactory.
>
> There is some objective evidence to support the applicant's case — there are electricity and telephone bills to show she was living at the flat at various periods.
>
> I think the position is this on the evidence. Up to the time of the first notice, Mrs Dymond was treating the flat as her main residence although she had the Cottage as another residence available to her from time to time.
>
> She did not include it in the notice as she wished to avoid mentioning it if she could do so. It was not disclosed on the notice. This is a case where the notice clearly did not, on a fair view, relate to the true facts and it is therefore invalid.

The only authority to which we have been referred in this context is the decision in *Cresswell* v *Duke of Westminster* which was
C mentioned by the learned judge. In that case a bold argument was advanced on behalf of the tenant to the effect that anything that was incorrect in the notice, whether it was due to fraud, misrepresentation or negligence, was all just an "inaccuracy" in the relevant sense, which entitled the tenant to avail himself of para 6(3). The Court of Appeal unhesitatingly rejected that submission. Lord Donaldson MR said at p 152 of [1985] 2 EGLR 151 that he was not prepared to accept that for one moment. On the particular facts of that case, however, he went on to say that there was no question of fraud, bad faith or anything of that sort. Against that background, he observed:

> In the end I suspect that the court has to ask itself: "Looking at the facts as
> D they were and what was stated in the notice, can this fairly be said to be an inaccuracy, or is it simply a notice which does not on a fair view relate to the facts?" Where we draw the line I do not know, and I doubt whether it is in anybody's interests that I should attempt to draw that line.

The drawing of the line may present particularly difficult problems in a case such as *Cresswell*, where there is no question of fraud, bad faith or anything of that sort at all. In the present case, however, at least on the respondents' submissions, and on the judge's findings, we have a rather different situation. The judge found as a fact that the notice was "supported by concealment or misrepresentation of material facts". In making this finding he plainly had in mind the second limb of section 20(5) of the 1967 Act (para (*b*)), which I have read. If that finding of fact was justified, I think it clear that the
E omission to refer to Hillsborough Cottage in the notice cannot properly be described as an "inaccuracy" within the meaning of para 6(3). The purpose of the legislature in requiring a tenant to supply the information set out in para 6(1) is plainly to assist the landlord in deciding what course to adopt in response to the tenant's notice. If an omission by the tenant in his notice is deliberately calculated to mislead the landlord, in my judgment it cannot qualify as an "inaccuracy" for the purpose of para 6(3).

In these circumstances, Mr Reynolds naturally sought — and he did so strenuously and skilfully — to attack the judge's finding of concealment or misrepresentation of material facts. He pointed out that the allegation of deliberate misrepresentation was raised for the first time at the trial. He submitted that there was no sufficient
F evidence to support the judge's finding in this context.

Though it does appear that this allegation was raised for the first time at the trial, it also appears that there was no application by counsel then appearing for Mrs Dymond for an adjournment and no lack of a proper opportunity for her to deal with the allegation.

In the context of this allegation, the form of her pleadings seems to me to be very relevant. In her originating application, at para 6, she had stated that she had been occupying the house as her sole residence. In her amended originating application she had stated in para 6 that she had been occupying the house "as her sole or main
G residence". When asked for further and better particulars of that paragraph, she gave particulars in the form which I have read. I am bound to say that the particulars given seem to me to have been, at least on the face of them, extremely disingenuous. Though they referred in terms to "occasional nights" spent by her at the home of her parents in Bideford, with friends in Plymouth, with friends in St Albans and with friends in Monkleigh, no mention was made of any nights spent at Hillsborough Cottage; indeed, the particulars contained no mention of the cottage at all.

The judge's notes of the evidence record that at the end of Mrs Dymond's cross-examination she was asked about the omission in her pleadings of any reference to the cottage. Mr Reynolds, on the basis of rather fuller notes of her cross-examination taken by counsel
H who then appeared for her at the trial, has told us that, in fact, her further and better particulars were put to her and she made an answer to the effect that, when she gave the necessary instructions for those particulars, she did not remember spending nights at the cottage, and it did not seem important to her. The judge, having had the great advantage of hearing and seeing her give evidence, unfortunately formed the conclusion that her evidence was totally unsatisfactory and that in some respects he did not believe her. I have already quoted the passage from his judgment in which he said that if she thought an answer might be wrong, or that it would not help her case, she would say that she could not remember. He formed the view (stated at p 5 of his judgment) that she did not include any reference to the cottage in
J the notice because she wished to avoid mentioning it if she could do so.

In my judgment, there was ample material to justify his finding that her notice was supported by concealment or misrepresentation of the material facts. Though Mr Reynolds rightly pointed out that her solicitors' letter of June 10 1988 had mentioned her ownership of the house to the respondents' solicitors, this did not, in my opinion, negative the concealment or misrepresentation, because it did not reveal that at times during the relevant period Hillsborough Cottage had *not* been let, but had been used and slept in by Mrs Dymond.

From what I said earlier, it follows that, in my opinion, the judge was right to reject the argument that the omission of any mention of Hillsborough Cottage in the notice constituted an "inaccuracy"
K within the meaning of para 6(3) and correspondingly was right to decide issue (c) against Mrs Dymond.

As to issue (d), the judge simply said:

> This is a further application by the respondents under section 20(5) and (6) of the Act that no further application should be allowed for a period of five years. I note that the application for costs on an indemnity basis is not pursued. Mr Edmunds said that this application was not an attempt to cheat the Trustees of the Charity out of their property. I think this is a case where the applicant and her family sought to take advantage of the Act in a way which may be perfectly legal, but which was backed up by the minimal possible evidence. I order that any further notice given by the Applicant of her desire to have the freehold or an extended lease of the premises shall be void if given in the next five years.

L In my view, Mr Reynolds rightly and inevitably conceded that, unless this court were prepared to upset the judge's finding of concealment or misrepresentation of material facts, the judge's conclusion in regard to issue (d) must follow his conclusion in regard to issue (c); it could not be upset by this court.

For these reasons, in my judgment, having regard to my views on issues (b), (c) and (d) this appeal must fail. After we had intimated to Mr Monty, the respondents' counsel, that this was our view, we gave him the opportunity to address argument on the respondents' notice if he so desired, but for obvious reasons he did not seek to avail himself of this opportunity.

The judge's conclusions on issues (b), (c) and (d) were in the end all conclusions on matters of fact. Mr Reynolds, in an admirable
M argument, has said everything that could be said in support of this appeal but, for the reasons which I have given, in my opinion the judge's findings of fact on these three issues cannot be disturbed and I would accordingly dismiss the appeal.

Agreeing, **Ralph Gibson LJ** said: I have very little to add. The result for the applicant is disastrous, but that result is, in my judgment, fully in accordance with the intention of Parliament as set out in the Act of 1967. What I have to add is on issue (b). Mr Reynolds, among other submissions, contended that any occupation by the applicant of the cottage could not be held to have been occupation of an "other residence" within the meaning of para 6(1)(*d*)(ii) of Schedule 3 unless there was occupation of it for

---

*Editor's note: Also reported at [1985] 2 EGLR 151.

A significant periods of time. He pointed to section 1(2), which uses the words "occupying it as his only or main residence" and relied upon the contention that the applicant's occupation of the cottage would not have satisfied the test of "occupies the dwelling-house as his residence" under section 2 of the Rent Act 1977. Finally, he submitted that the learned judge concentrated on the availability to the applicant of the cottage as contrasted with its actual or intended use.

During the course of argument Slade LJ drew attention to the case of *Poland* v *Earl Cadogan*, to which he has referred in his judgment. That case was concerned with the primary question whether the tenant had proved the necessary period of occupation "as his only or main residence" under section 1 of the Act. The tenant in that case B had gone abroad from 1975 to January 1977, leaving the house to be sublet by agents. During the absence of the applicant and his wife, possession was taken by mortgagees. The decision of the county court judge was that the tenant had been in occupation as his only or main residence during that time. His decision was reversed by this court on the ground that, although absence — even prolonged absence — from the house did not by itself bring qualifying occupation to an end, it was necessary, if qualifying occupation was to continue during a period when the tenant was not actually living in the house, for the tenant to have an intention to reside there in the sense of keeping it ready for habitation by himself whenever he was pleased to go into it. The reasoning of their lordships supports, in my C judgment, the proposition that if a resident tenant of a house, the subject-matter of an application for the tenant to acquire the freehold under the Act, has, during the relevant period, a second house which is furnished and available for occupation as his residence which the tenant keeps ready for habitation by himself whenever he is pleased to go into it, and which, from time to time, the tenant uses for habitation by himself, that tenant has an "other residence" within the meaning of para 6.

It is not necessary, on the facts of this case, to decide whether the second house would qualify as an "other residence" if it had been kept available only, without the tenant's going into it at any time D during the relevant period. I incline to the view that it could so qualify. I would add, however, that if in such a case an applicant omitted reference to it in his notice it is probable, to say the least, that the omission would be held to be an inaccuracy within para 6(3) on the ground that the omission would have been unlikely to be calculated or intended to mislead the landlord.

I do not, however, accept that if an applicant has another residence within the test that I have set out, it does not qualify as an "other residence" for the purposes of para 6(1)(*d*) unless the applicant had occupied it for periods of time which, according to Mr Reynolds' submission, could be described as "significant".

It is clear that an "other residence" which is required to be E mentioned in the notice is not the "only or main residence" of the applicant, which must be the house the subject-matter of the application. The legislative purpose of the requirement must be, as Slade LJ has already said, to cause the applicant who has more than one residence to inform the landlord of the other residence — other, that is, than the house the subject-matter of the application — so that if the landlord wishes to do so, the landlord may investigate the facts so as to be able to test, and it may be to demonstrate, the inaccuracy of the assertion that the house the subject-matter of the application has indeed been occupied as the main residence of the applicant during the relevant period. The obligation to give particulars of the other residence is not qualified in the statute by reference to any necessary significance, or frequency of habitation of that other residence. F

In my judgment, this case was a good example of circumstances in which the landlords had real need to be told of the applicant's other residence in the cottage, so that they could investigate which was the applicant's main residence.

I would therefore uphold the judge's finding on this point and on the other matters for the reasons which Slade LJ has given.

The appeal was dismissed with costs.

LANDLORD AND TENANT

RENT ACTS

A

Court of Appeal

June 8 1990

(Before Lord Justice STOCKER and Lord Justice STAUGHTON)

SPRINGFIELD INVESTMENTS LTD v BELL

Estates Gazette January 19 1991

[1991] 02 EG 157

Rent Act 1977, Schedule 15, Case 16 — Cottage required for agricultural worker — No issue as to any of the conditions of B Case 16 except 16(*b*) — Whether "not later than the relevant date, the tenant was given notice in writing that possession might be recovered under this Case" — No question as to relevant date but an issue as to whether the tenant was given "notice" that possession might be recovered — Whether a certificate of fair rent constituted notice — Appeal from decision of county court judge, who held that the plaintiff landlords were entitled to possession of the cottage

Both before the county court judge and the Court of Appeal the matter was reduced to the single issue of the notice given under Case 16(*b*) — The tenant had never been employed by the landlords and the court was satisfied that the cottage in C question was required for occupation by a person to be employed by the landlords — The specific question which had to be determined was whether a certificate of fair rent issued by a rent officer could be regarded as constituting "notice" for the purpose of Case 16(*b*) — At first sight it seems unlikely that it could qualify for that purpose — A certificate of fair rent is, after all, a document obtainable by an owner to enable him to discover in advance what rent he will be able to recover legitimately under the Rent Act on letting a dwelling — It is the first stage in a process which has to be completed by the actual registration of a rent in pursuance of the certificate — How D could that be notice to a tenant that possession might be recovered under Case 16? — Nevertheless, both the county court judge and the Court of Appeal held that it did constitute such notice

It appears that a copy of the certificate of fair rent, which is a landlord's document, was given to the tenant at the time when she received the keys to the cottage — The document certified that £12 per week or £624 pa would be a fair rent for the cottage under a regulated tenancy — The certificate identified and described the dwelling, referred to the application form for particulars of furniture, mentioned that E there were no services provided and, under "Other terms of the proposed tenancy", set out the following information:

It is proposed to let on a regulated tenancy basis, subject to the provisions of Case 16, 15th Sch Rent Act 1977 (whereby repossession could be obtained to house an agricultural worker, employed, or to be employed by the Landlord), Standard repair etc terms, Landlord responsible for outside, structure and outside painting; tenant for inside [decorations] & garden and payment of rates direct.

The county court judge pointed out that no specific form was prescribed for the notice under Case 16(*b*) and held that the giving of the copy certificate, which referred to the Case and stated that repossession could be obtained under its F provisions, did all that was required — It was not necessary for the landlords to prove that particular attention had been drawn to the reference to the recovery of possession

On appeal objections were raised by the appellant tenant that a certificate for the benefit of the landlords was not a notice to the tenant; that the contents of the certificate were not adequate as a notice; that the notice required by Case 16 was intended to state specifically that possession might be recovered under the 1977 Act; that the document should make it abundantly clear that it would produce a compulsory situation so far as the tenant was concerned; and that the plain G purpose of the notice was to enable the tenant to appreciate that she would not be obtaining security of tenure under the letting — The authorities of *Fowler* v *Minchin* and *Bradshaw* v *Baldwin-Wiseman* were cited — The Court of Appeal rejected these objections — Although the document was headed "Certificate", it did draw attention specifically to Case 16 and the possibility of the recovery of possession — There was no prescribed form of notice and it should be observed that Case 16(*b*) itself mentioned "notice" and not "a notice" — The court considered that the giving of the copy certificate to the tenant satisfied the statutory conditions — Appeal dismissed

H

The following cases are referred to in this report.

Bradshaw v *Baldwin-Wiseman* (1985) 49 P&CR 382; [1985] 1 EGLR 123; 274 EG 285; 17 HLR 260, CA

Fowler v *Mirchin* [1987] 1 EGLR 108; (1987) 282 EG 1534; 19 HLR 224, CA

Pittalis v *Grant* [1989] 3 WLR 139; [1989] 2 All ER 622; [1989] 2 EGLR 90; [1989] 28 EG 126

Spurling (J) Ltd v *Bradshaw* [1956] 1 WLR 461; [1956] 2 All ER 121; [1956] 1 Lloyd's Rep 392, CA

This was an appeal by the tenant, Caroline Jane Bell, from a decision of Judge McNaught, at Gloucester County Court, granting the plaintiff landlords, Springfield Investments Ltd (the respondents to the appeal), an order for the possession of Shadyside Cottage, Hyde Lane, Whitminster, Gloucestershire. J

Charles J D Auld (instructed by A E Smith & Son, of Stroud) appeared on behalf of the appellant; Martin Rodger (instructed by Hyde Mahon Bridges) represented the respondents.

Giving judgment, **STOCKER LJ** said: This is a defendant's appeal from a decision of His Honour Judge McNaught made on November 20 1989, whereby he ordered that the plaintiffs recover possession of premises known as Shadyside Cottage, Hyde Lane, Whitminster in the County of Gloucester, on or before February 20 1990. The defendant seeks an order that the claim for possession should be dismissed.

The facts, shortly stated, are that the cottage in question was let to K the defendant in about September or October 1981 at a rental of £12 a week. Prior to this letting the defendant had occupied the cottage next door, which was called Sunnyside Cottage, rent-free as an employee in agriculture of the plaintiffs' predecessor in title. When this employment came to an end, upon the whole farm of which those two cottages form part being sold to the plaintiffs, the defendant moved in to Shadyside and a fair rent was fixed by the rent officer.

The claim for possession is brought pursuant to an amended particulars of claim, grounds for possession being Case 16 of Schedule 15 to the Rent Act 1977. There had earlier been abortive proceedings for possession in the county court, brought under the

A particulars of claim in their original unamended form, in which possession was claimed upon different grounds and in respect of which the notice to quit was defective. No more need be said of those proceedings.

The present claim, as has been stated, is based upon Case 16, Schedule 15 to and section 98(2) of the Rent Act 1977, that section rendering mandatory an order for possession if the circumstances of the case fall within Case 16. An effective notice to quit has in fact been served so far as these proceedings are concerned.

Case 16 of Schedule 15 to the 1977 Act reads:

> Where the dwelling-house was at any time occupied by a person under the terms of his employment as a person employed in agriculture, and
> (*a*) the tenant neither is nor at any time was so employed by the landlord and
> B is not the widow of a person who was so employed, and
> (*b*) not later than the relevant date, the tenant was given notice in writing that possession might be recovered under this Case, and
> (*c*) the court is satisfied that the dwelling-house is required for occupation by a person employed, or to be employed, by the landlord in agriculture.
> For the purposes of this Case "employed", "employment" and "agriculture" have the same meanings as in the Agricultural Wages Act 1948.

It was conceded at the trial that the opening sentence of the Case applied to the premises, which had previously been occupied by a person under the terms of his employment as a person employed in agriculture. Para (*a*) was admitted on the pleadings and the learned judge found as a fact that the provisions of para (*c*) were satisfied. There is no appeal against that finding. The issue before the county
C court and before this court therefore relates to para (*b*).

As I have already stated, the rent was fixed by the rent officer and his certificate dated September 7 was sent to the plaintiffs. It is therefore necessary to consider the terms of that certificate. It appears on pp 59 and 61 of the bundle, p 61 being in fact the reverse of p 59, and the relevant part of the certificate, which is headed: "Rent Act 1977, Rent Officer Service" and with the main heading "Certificate of Fair Rent", reads:

> This is to certify that £12 a week [or] £624 per annum (exclusive of rates) would be a fair rent for the dwelling-house referred to in Part 1 of the Schedule overleaf under a regulated tenancy on the other terms indicated in Part III of the said Schedule.

D Then overleaf on p 61 of the court bundle, there is the heading: "Schedule". The premises are stated as being Shadyside, Hyde Lane, Whitminster, Gloucester, and Part I gives the address of the premises and various other matters related to the accommodation contained within it. It also states that there is a "Small garden front, small lawn at back. Car space in farm-yard at side via farm gate off Lane", and other matters. "Part II: Furniture to be provided as specified in the application, Cupboards in Kitchen, Landlords fixtures". Services provided: none. "Part III" — this is the crucial part of this document — "Other terms of proposed tenancy". It is then stated:

> It is proposed to let on a regulated tenancy basis, subject to the provisions of Case 16, 15th Sch Rent Act 1977 (whereby repossession could be obtained to house an agricultural worker, employed, or to be employed by the Landlord),
> E Standard repair etc terms, Landlord responsible for outside, structure and outside painting; Tenant for inside [decorations] & garden and payment of rates direct.

Those were the relevant terms of the document in which it is said there was a notice which complied with Case 16.

There was a conflict of evidence at the trial regarding whether and in what circumstances that document was served by the plaintiffs upon the defendant. The judge held that the document was handed to the defendant at the same time that she received from the plaintiffs the keys, and that was on or before September 14. In that context reference was made to certain letters which appear at pp 55 and 57 of the bundle.

At p 55 there is a letter dated September 2 1981 by Mr Tiley, which
F reads:

> Dear Miss Bell,
> Further to our conversation wherein you said you are prepared to move into Shadyside at a rent to be agreed by the Fair Rents Officer. I have spoken to the Fair Rents Officer and he will be visiting the property on Thursday next at about 3.30pm and he has promised to let me have his Certificate within a few days.
> As soon as I receive this, I will call and see you with it and perhaps we can arrange a date when you will be moving.

At p 57 there is a letter dated September 14 1981, also by Mr Tiley. It is addressed to the defendant and reads:

> Dear Miss Bell,
> We confirm giving you the keys of Shadyside, also a copy of the Fair Rents Officers report by Mr W H Tiley. G

That means of course the Mr Tiley who gave the keys and the report.

> We would suggest that the day for taking occupation is the 3rd October.

September 14 was in fact a Monday.

By para 2(*d*) of Part III of Schedule 15 to the Act, "the relevant date" means the date of commencement of the tenancy. So that on the judge's findings the document, if it was a notice in writing that possession might be recovered under the Case for the purpose of para (*b*), complied with the time requirement of that paragraph. The issue, therefore, is whether the rent officer's certificate, by reason of Part III of the schedule to it, the terms of which have been read, was a notice for the purposes of that paragraph.

The learned judge found facts relating to the delivery of the H
document. He said:

> The question is: was this sufficient notice?
> The rules do not require a specific form and it is not suggested by the defendant that the form of words in this particular case is ineffective. I think that Miss Bell has forgotten about this particular document. I think that she has been absolutely honest in her evidence but she has forgotten. There is no particular form required, it is simply sufficient to show that the tenant was given notice in writing. I find that in this particular case the tenant was given notice in writing, and that the notice that was given was sufficient to satisfy Case 16. It is not necessary for the landlord to prove that he brought the tenant's attention to the specific terms of that notice. I, therefore, conclude that the landlord has established his ground for possession and I, therefore, order possession. J

There are two grounds of appeal. First, that the judge erred in law in holding that the plaintiff had given to the defendant a notice in writing that possession might be recovered under Case 16 in Schedule 15 to the Rent Act 1977. It is contended that the learned judge should have held that such documents as the plaintiff had given to the defendant did not constitute a notice required by Case 16. Second, further or alternatively there was no or no sufficient evidence upon which the learned judge could find that the defendant had been given notice in writing as required by the said Case 16.

At the hearing before us, Mr Auld has made it clear that his second ground of appeal is not a challenge to the learned judge's finding that in the physical sense the document was given to the defendant, but is really an alternative formulation of his first ground and really relates K
to the adequacy of the contents of the notice rather than its physical delivery.

There was an issue raised by the respondents in their skeleton argument as to the jurisdiction of the court, on the grounds that the construction of the document in question had never been argued before the learned judge and accordingly raised a point of law which this court would have had no jurisdiction to consider. That aspect of the matter was not pursued, in the light of the decision in *Pittalis* v *Grant* [1989] 3 WLR 139*, and it was conceded by Mr Rodger that the matter became one of discretion of the court. Accordingly, in the exercise of discretion we heard the appeal on the merits.

Mr Auld's submission is that the certificate to which reference has L
been made does not constitute notice for the purposes of Case 16 and that the judge's finding to the contrary is wrong, and not in accordance with authority, which supports his submissions. In that connection he cited before us the case of *Fowler* v *Minchin* (1987) 19 HLR 224†. That was a case concerned specifically with Case 16 of Schedule 15 to the Act. At p 231 Stephen Brown LJ (as he then was) said:

> It is quite true that the notice does not have to follow any particular form, but in my judgment it must state quite specifically that possession might be recovered under the provisions of the Act: that is to say, to make it clear that a situation would obtain which would be a compulsory situation so far as the tenant was concerned. In my judgment the alleged term of the so called agreement that the defendant would vacate on 28 days' notice if the plaintiff required it for a farm worker is no more than a voluntary undertaking that he M
> would do so.

In fact the factual matrix upon which *Fowler* v *Minchin* then was concerned is set out on pp 228 and 229 of the report and is relevant, in my view, to the ratio of the decision. Stephen Brown LJ said:

> The position at the trial was this. Until the actual hearing commenced or was about to commence the defendant's counsel had had no notice that reliance was to be placed upon a specific written document which it would be alleged complied with the requirement of sub-paragraph (b) of Case 16. During the course of the case, therefore, additional further and better particulars of the

*Editor's note: Also reported at [1989] 2 EGLR 90.
†Editor's note: Also reported at [1987] 1 EGLR 108.

A amended particulars of claim were given by the plaintiff. They are handwritten and we have a photostat copy of them:

1. *Date of written notice*

During the week prior to July 29 1978 *ie* in the week before four weeks prior to the date shown in the Rent Book.

2. *When and where it was given to the Defendant*

— and that is answered —

3. — this is the crucial part of the matter —

Exactly what it said

That the defendant was to keep the premises in a good state of repair. That the defendant would vacate on 28 days' notice if the plaintiff required it for a farm worker,

B — and lastly —

That the rent was to be paid weekly and was to be kept on a level with Tirley council house rents.

Thus the allegation, as I understand it, was that there was compliance with Case 16 by reason of the agreement that the defendant would vacate on 28 days' notice if the plaintiff required it for a farm worker. It is clearly noticeable that in that case it was not even contended that there had been any express reference to Case 16 of Schedule 15 to the Act. I find a little bit of difficulty in understanding how it could even have been arguable on those facts that there was such a compliance, but in so far as it manifestly was argued, then the passage of the judgment that I have read clearly indicates the ratio of the decision.

C The second case upon which reliance is placed and which has been cited to us is the case of *Bradshaw* v *Baldwin-Wiseman* (1985) 49 P&CR 382*. That was a case which concerned Case 11 of Schedule 15, which imported some element of discretion, in that possession could be ordered if it was just and equitable so to do, and it was in that context that Griffiths LJ said at p 385:

It is, however, a condition of such a right to recover the property that the landlord gives a written notice to the tenant at the time of the letting that he may seek to recover possession when he returns. The purpose of giving the written notice is obvious and important. It is of the utmost importance to a tenant that he should appreciate when he takes rented property whether or not he is obtaining a secure tenure. I can think of nothing likely to have a greater effect on the way people order their lives than the knowledge one way or the D other whether or not they have a secure home.

Mr Auld submits that the following propositions are to be deduced from those authorities, to which reference has been made: first, that Case 16 notice does not have to follow any particular form; second, it must state specifically that possession might be recovered under the provisions of the Rent Act 1977, and must be such as to make it clear that a compulsory situation would obtain so far as the tenant was concerned; and third, the purpose of the notice is to enable the tenant to appreciate that she is not obtaining security of tenure under the letting.

He therefore submits, in the light of those principles, that the judge was wrong in law in certain respects. First he says that the notice upon E which the landlord relies is not a notice at all. It is in fact, he submits, a certificate and indeed is so headed. In that context it seems to me relevant to observe that para (*b*) of Case 16 does not require "*a*" notice in writing. It requires notice in writing that possession might be recovered. Second, it is submitted by Mr Auld that the wording itself in the third part of the schedule to the document on p 59 commences with the words, "It is proposed", and he says that the wording does not in any event therefore constitute notice in any sense because it is preceded by those words "It is proposed" and that it represents therefore, at the highest, the landlords' proposals as to the terms upon which the letting would be made. However, it does seem to me that since the notice has to be given on or before the commencement of the tenancy, such a notice in the first instance will often, if not F always, be initially a proposal and, second, in my view, one has to look at the whole of the circumstances in which the document came into existence and its form.

If one refers to the facing page of the document on p 59, one sees that at the very outset, under what might be called the certificate itself, there is reference to the fact that it is to be a regulated tenancy on the other terms indicated in Part III of Schedule 15, and moreover the dwelling-house is the dwelling-house referred to in Part I of the schedule overleaf. It seems to be inevitable, therefore, that anybody who read that document at all with any form of understanding would necessarily look at the schedule to see the identity and full incidents of the property that was being let to them. For example, it includes the G fact that there is a car space in the farm-yard at the side and the entrance to it through the farm gate off the lane. That is a specific incident of the premises which were being let, and without reference to Part I of the schedule there would, or at any rate might be, a conflict as to the precise nature and extent of the premises let.

Part II describes the furniture specified and includes cupboards in kitchen and landlord's fixtures. The services provided are stated to be nil and when one comes to Part III there is what is contended to be a notice in accordance with the terms of Case 16. It is of course the question at issue as to whether it is or not, but I would refer again to the wording that it is "let on a regulated tenancy basis, subject to the provisions of Case 16, 15th Sch Rent Act 1977", so that there is an express and specific reference to that Case. But it goes further, H because there is an explanation in brackets of the substance of the effect of Case 16. For the words appear, "whereby repossession could be obtained to house an agricultural worker, employed, or to be employed by the Landlord".

Furthermore, there was no formal tenancy agreement. The certificate, the contents of which, so far as was relevant, have already been read, was handed over at the same time that the keys were handed over, and the evidence of Mr Tiley, which appears on p 39 of the notes of evidence, says:

I didn't know what to charge her so we got the rent officer, and he fixed it at £12 per week. He sent [the documents] through. Sent them to my company. Then we agreed date. I gave her a copy of the [document] when I gave her the J keys. I [looked] at [the] two copy letters, [of] 2nd Sept 1981 and 14th Sept 1981,

— which have been read —

posted to Miss Bell at Sunnyside. I did call and give her a copy of the certificate. I photocopied both sides of certificate. On back is reference to Case 16. Rent officer put it in because we might need it in the future.

So that the circumstances in which that document was handed over were concurrent with the handing over of the keys and the agreement of the date of commencement of the tenancy. The tenant went into possession and continued to pay the rent and to occupy the premises, and was of course still in occupation when the possession order was made. It seems to me that she must have been occupying as tenant on the terms of the document, albeit it is headed "certificate". K

The second way in which Mr Auld puts his case is that although there may be circumstances in which a landlord can rely upon Case 16 notice although it is contained in some other document, such circumstances will be rare and such cases will comply with the requirements of Case 16 only if the notice contained in that other document reasonably brings to the attention of the tenant that the notice relied upon indicates that possession might be required under Case 16, and he argues that this was not the case here, (1) because it is headed "certificate", and it may well be that the recipient, the defendant in this case, would not read any further than the first few lines to see what the rent had been assessed at; (2) that it was only a proposal — that really repeats the argument already considered — L and that it is contained at the end of a document, and that accordingly sufficient attention would not have been drawn to the tenant which might reasonably indicate to her that possession might be required under Case 16.

In my view, those arguments cannot prevail either. First, by the acceptance of the keys on entry into possession the tenant did so on the terms of that certificate, and those terms of course include conditions both for the benefit of the tenant and for her detriment. It was brought to her notice since it did contain those beneficial and detrimental conditions. Second, there is no form required for notice under Case 16. As I have already observed, notice is not necessarily "*a*" notice. The Act and the schedule could have specified a standard or statutory form, but it did not do so. This document, though M headed "certificate", did by its format draw attention specifically to Case 16 in Schedule 15. It seems doubtful if indeed these questions were argued below, but the learned judge found that the notice was sufficient to satisfy Case 16 and with that finding I agree.

Accordingly, for these reasons I would dismiss this appeal.

Agreeing, **Staughton LJ** said: Two points have been argued by Mr Auld for the tenant. First he submitted that the certificate spoke of terms which might be included, or were expected to be included, in some future tenancy agreement, and not terms actually included in an agreement. In point of form, he is correct. The certificate had to state terms which might be included in a future agreement, because

*Editor's note: Also reported at [1985] 1 EGLR 123.

they might have a bearing on what was a fair rent. But on the facts of this case, those became the actual terms of the tenancy agreement when a copy of the certificate, with the keys, was handed by the landlord to the tenant. At that moment the proposed terms became actual terms and the tenant should reasonably have appreciated that.

The second point has given me rather more difficulty. The document in question, on its face, purports to be a certificate of what is a fair rent and not notice of anything. Mr Auld submits that to qualify under Case 16 a notice must be such, in point of form, as is reasonably sufficient to bring to the attention of the tenant that the landlord will by law be entitled to recover possession for an agricultural worker.

There must be some such requirement as Mr Auld suggests. A notice would be bad if it were in such small print as to be legible only by those with the sharpest eyes or the strongest spectacles. It would be bad if it were buried in the middle of some document on a quite different topic, or written in a foreign language which the tenant did not understand, or enclosed in an envelope which had the appearance of containing junk mail. One is reminded of what Denning LJ said in *J Spurling Ltd* v *Bradshaw* [1956] 1 WLR 461 at p 466 in relation to exemption clauses:

> Some clauses which I have seen would need to be printed in red ink on the face of the document with a red hand pointing to it before the notice could be held to be sufficient.

Or one can compare the Consumer Credit Act and the regulations made under it, which provide that some particulars of a consumer credit agreement have to be inserted in a box with a certain degree of prominence.

Parliament has not followed that in the Rent Act 1977. Nor has Parliament said, as it sometimes does, that notice under Case 16 must be in a prescribed form, or as Parliament has on occasion said, must be in ordinary language which the tenant can understand. I am inclined to think that the notice must be given reasonable prominence, such that it is fairly drawn to the tenant's attention. There may be difficulties when the tenant is blind or illiterate or understands only a foreign language, but I need not say anything about those cases.

Was the notice fairly drawn to the tenant's attention in this case? I have had some doubts on that point. It is contained in a document which ostensibly deals on p 1 with what is a fair rent. On the other hand it is evidently a legal document and one which has to do with the tenant's occupation of the property. I would have wished to have had the county court judge's view on this point. He, after all, knows the sort of people who take tenancies of cottages which might be required for the agricultural population of Gloucestershire. If the judge had reached a conclusion one way or the other as to whether this notice was fairly drawn to the tenant's attention, I should have been very slow to differ from it. But he did not reach a conclusion because the point was not argued in that form before him. Fortunately, as Stocker LJ has said, we are now free from the old rule that points of law could not be argued at all in the Court of Appeal on appeals from the county court unless they had been raised in the court below. See *Pittalis* v *Grant* [1989] 3 WLR 139. There is still a discretion whether to allow a point of law to be argued if it has not been argued below, and this court does not normally allow a new point of law to be argued if some further finding of fact is necessary. It is, however, not so much a further finding of fact which would have been welcome in this case as the judge's view as to whether the information in Part III of the certificate fairly drew to the tenant's attention the landlord's possible right to recover possession under Case 16. On balance, I agree that it did fairly draw that to the tenant's attention.

Accordingly, I, too, would dismiss this appeal.

The appeal was dismissed with costs; the order for costs not to be enforced without further order; legal aid taxation for the defendant; possession to take place in 28 days, unless time extended by consent.

Court of Appeal

November 7 1990

(Before Lord Justice PURCHAS, Lord Justice McCOWAN and Sir Denys BUCKLEY)

HEDGEDALE LTD v HARDS

Estates Gazette April 20 1991
[1991] 15 EG 107

Rent Act 1977, Schedule 1, para 3 — Statutory tenancy by succession — Meaning of "residing with" in contrast with "living at" — Position where period of physical separation during the six months immediately before original tenant's death — Original tenant temporarily absent from the subject premises while recovering from effects of accident — Appeal allowed from decision of county court judge holding that the qualifying residence condition of para 3 of Schedule 1 had not been satisfied — Reasoning in judgment of Russell LJ in *Foreman* v *Beagley* followed

One of the difficulties in this case, as mentioned by Sir Denys Buckley, was that the county court judge had not made precise findings as to certain material dates — Thus it was not clear exactly when the appellant moved into his grandmother's flat — Subject thereto the factual background appears to have been as follows — The appellant was a young man born in 1969 who did not get on well with his stepfather, his mother's second husband, but who had a good relationship with his grandmother — She had been the tenant of the subject flat since 1942 and was at the material times the statutory tenant of it — The appellant had left his stepfather's house and moved into his grandmother's flat at a date not precisely known but apparently shortly after the grandmother had suffered an accident and broken an arm rather badly — The appellant was able to help her in a number of ways and was pleased to get away from his stepfather — However, the task of looking after an elderly lady of over 80 with a broken arm was too much for a young man of 19 and the grandmother left her flat to stay for a time with her daughter, the appellant's mother, and the stepfather — This stay lasted for about four months — In the meantime the appellant lived in the flat and it was his hope and intention to continue to live there when his grandmother returned, to look after her and to form with her a family unit — The grandmother's hope and intention was the same — She returned to the flat at about Christmas 1987 and lived in the flat with her grandson as family — Unfortunately it was not to be for long; she died on May 1 1988 — If the qualifying period of six months for the purpose of para 3 of Schedule 1 was to be computed from Christmas 1987, it had not been completed in time

The appellant's application for a declaration that he was a statutory tenant by succession within para 3 of Schedule 1 failed before the county court judge — The judge considered that the four months which the appellant had spent on his own in the flat, while his grandmother was living with his mother, were a fatal disqualification — The judge rejected the argument that the appellant should be considered as residing with his grandmother during that period because he intended to look after her when she returned and with her agreement form a family unit — The judge said "I reject the argument that future intention can be elevated to present residence with her"

The Court of Appeal disagreed with this conclusion — In their view the quality of the appellant's residence during the four months when his grandmother was temporarily absent was no different from its quality during the last four months when she was physically present in the flat — From the time when the appellant left his previous home with his mother and stepfather and took up his residence on a permanent basis in

his grandmother's flat he became resident there within the requirements of para 3 of Schedule 1 — This view was supported by the reasoning in *Foreman* v *Beagley*, although the actual result there was different — In the present case the appellant had been (1) a member of the original tenant's family, (2) a person residing with the tenant at the time of her death and (3) so residing for a period of six months immediately before her death, despite her temporary absence for part of the time — Appeal allowed and declaration granted

The following case is referred to in this report.

Foreman v *Beagley* [1969] 1 WLR 1387; [1969] 3 All ER 838, CA

This was an appeal by Derek Hards against an order made by Judge Batterbury, at Bromley County Court, granting Hedgedale Ltd, the landlords, possession of a flat at 79 Como Road, Forest Hill, London SE23. The appellant sought a declaration that he was entitled to occupy the flat as a statutory tenant by succession.

David Platt (instructed by Armstrong & Co) appeared on behalf of the appellant; Charles Salter (instructed by Wilson Barca) represented the respondents.

Giving judgment, **Purchas LJ** said: This is an appeal by Derek Hards (to whom I shall for simplicity's sake refer as "the grandson") against an order for possession of premises made by His Honour Judge Batterbury in favour of Hedgedale Ltd (to whom I shall refer as "the landlord") in the Bromley County Court on February 22 1990. The premises concerned were 79 Como Road, Forest Hill, London SE23. The grandson seeks a declaration that he is a statutory tenant by succession within the provisions of para 3 of Schedule 1 to the Rent Act 1977. I pause to remark that the amendments introduced to that Act by the Housing Act 1988 do not apply to the circumstances in this appeal.

The appeal raises a short but important point of law. It is convenient at the outset to refer to the two statutory provisions involved.

Section 2 of the Rent Act 1977 provides that Part I of Schedule 1 shall have effect for determining what person, if any, is a statutory tenant of a dwelling-house at any time after the death of a person who immediately before his death was either a protected tenant of the dwelling-house or a statutory tenant by virtue of the preceding subsection.

Schedule 1, para 3, omitting irrelevant matters provides that:

> . . . a person who was a member of the original tenant's family was residing with him at the time of and for the period of 6 months immediately before his death then, after his death, that person . . . shall be the statutory tenant if and so long as he occupies the dwelling-house as his residence.

The original tenancy was created in 1942, the tenant being a Mrs Phillips (to whom I shall refer as "the grandmother"). In about 1986, she, previously being a protected tenant, became a statutory tenant under the provisions of the Act.

The history can be shortly stated for the purposes of this appeal, partly on common ground and partly on findings by the judge:

1 On September 2 1987 the grandmother had an accident — not the first of such accidents I remark — which resulted in a serious fracture of her arm.

2 She then moved temporarily to live with her daughter, that is the appellant's mother, to recuperate.

The circumstances in the home of the mother (who was then living with her second husband, the grandson's stepfather) were that the grandson was not happy because, not unusually in these matters, he was at odds with his stepfather.

Although there has been some argument about the order of the next events, for the purposes of the moment I shall record that the grandson moved to the premises, with which this appeal is concerned, which were and remain the grandmother's home. Again there has been a dispute about this, but the judge found that certainly his intention was permanent, in the sense that he intended making his home with his grandmother when she returned to her home. They would live together there.

3 The grandmother returned at about Christmas 1987 and the grandson and grandmother lived together at the premises in circumstances which the judge found qualified within the provisions of para 3 of Schedule 1 saving only that the six months' period was not completed because, on May 1 1988, unhappily, the grandmother died.

The hearing took place over a period of nine months between May 23 1989 and February 22 1990. There were hearings on October 11 and December 4 as well as the two other dates. It is an unhappy example of the inconvenience caused by the listing difficulties in the county courts under the present arrangements. It could not have assisted the judge in his assessment of the evidence, which was by no means consistent even on the primary facts.

The grandson called one witness on the first occasion to support his case, a Mr White. His case as pleaded failed to mention the fact that the grandmother had returned to her daughter's home for a period of four months between early September (the date is not certain) and Christmas. Nor when giving his evidence did the grandson mention this. The grandson's case was that she was living at the premises "continuously", but there was a question as to whether he left his mother and stepfather before or soon after the grandmother left the premises.

It was not until the final hearing, that is the fourth occasion, when the mother gave evidence that the court was told that the grandmother had returned temporarily to stay with the daughter while the grandson stayed on at the premises. Her evidence is important, since this omission to disclose the case on the part of the grandson has been relied upon by Mr Salter, for the respondents, in his submissions before us.

The part of the note of her evidence-in-chief which is relevant to this matter reads as follows:

> Derek always lived with me. He moved out in August 1987, mainly because that would leave mum on her own if he did not.
> He did not get on with my second husband, Charles, who is West Indian. The move was to be permanent; I had lived with Charles since Derek was about 10 years old.

In cross-examination she said that she telephoned Mrs Parker, who was the lady who arranged the home helps for the grandmother, and continued:

> I said that mother had broken her arm badly and that she will move in with me because Derek won't be able to wash her.
> Mother was not at home because she was living with me from September 2 1987 until after Christmas 1987.

In re-examination she said:

> When living with me from September 2 1987 to past Christmas 1987, my mother left her dogs behind. She intended to go back to Como Road.

The judge carefully rehearsed this evidence and the other testimony dealing with these events. He referred to the evidence which I have just recited and said this:

> . . . It is implicit that Derek was there already in a caring role but a broken arm was too much to cope with. She then went on to say that her mother had moved to live with her. This lasted between September 1987 and Christmas 1987; in re-examination Mrs Williams said that Mrs Phillips' dogs had stayed in the flat and I am certain that if there were someone there they would have to feed them — and quite probably every day.
> It was a long time ago when the respondent gave his evidence. He said that he had lived at 79 Como Road since August 1987; he is now 19, having been born on December 31 1969. His gran died at the age of 83 — and he lived with her in the two-bedroomed flat. "I moved in because she had shortly before broken her arm and she could not manage by herself." To which arm break was he referring? The evidence speaks of two breaks — one in September 1987 and the other before Mrs Hooper moved in 1986 — If there are others the dates are not ascribed. He did not tell me that from September 2 1987 for over four months his gran had moved out to his mother's. He then went on to tell me that she had angina and the shakes; he sorted out her pills and looked after her and ran messages for her. She could not do everyday chores; he described how he helped her — shopping, tidying up, walking dogs, getting her pension. Then he said "My great aunt lived there about one year before I did. She left after the divorce went through; gran lived in the flat until one week before her death. I last worked just before she died. Mother paid for me at weekends". . .
> In cross-examination his mother said that he had "moved in". He did not mind as he did not get on with his stepfather. He moved in with his clothes and for the duration of his gran's lifetime. The home help came in but would not help me, he said.

Then the judge referred to Mr White. He did not accept his evidence that the grandson moved in in July or August 1978. He went on to consider the options which he had to decide; first, that the grandson moved in before the break to Mrs Phillips' arm; second, that he moved in after she broke her arm; third, that he moved in after her arm mended; and, fourth, he did not move in at all until it was too late to qualify. He then referred to the difficulties the witnesses had in their recollection and assessed the value of Mr White's evidence. It is significant that, although he specifically mentioned the omission on the part of the grandson to refer to the grandmother's absence for four months, the judge did not in any way include in his judgment criticism of the grandson as a witness or question his reliability as a witness. Later in his judgment, to which I shall come in a moment, the judge explained the position of the grandson in saying that the grandmother had resided throughout at the premises.

Mr Salter has pressed on us that the judge should have taken an adverse view of the grandson and that that would rebut any criticisms of his findings which are the subject of this appeal. I must confess that, notwithstanding the forceful way in which Mr Salter made his submissions, I was not assisted by that particular support that he sought to give to the judgment in line with his respondent's notice. I am satisfied that the judge assessed the evidence, was conscious of the inconsistencies and came to his own conclusion as to where the truth as to primary facts lay. It would be quite wrong, in my judgment, for this court to go behind such findings.

I now turn to the part of the judgment where those findings are summarised. At the relevant page the judge says:

> The findings of fact involve accepting some evidence and rejecting other. My evaluation of witnesses falls to suit. In my judgment I am quite satisfied that the date on which the tenant broke her arm if not September 2 was vey shortly before that and certainly within one cycle of home helps. Mrs Phillips did not allow the home help to come to an empty house. I accept Mrs Parker's evidence that Alice called there until September 2 1987, depending on how often she came.
>
> Before any of this evidence was gone into I ought to add that back in last May the respondent told me "I moved in because she had shortly before broken her arm" — in other words afterwards. That I believe is the truth of the position. It was an ingenuous answer right at the forefront of the evidence and was not calculated. He is a young man and I think you do remember if you are given the sudden and daunting task of looking after an old lady. Mrs Hooper's evidence, or the parts I accept, does not conflict — but walks around the point. Mrs Williams' evidence does conflict with it as she says that Derek moved out in August 1987 as he would have left her mother on her own. Mrs Hooper is wrong in saying that after she left that she returned almost every weekend — I accept Mrs Williams' evidence that the tenant was not there. Now the respondent continued in evidence to explain what he did for the old lady, that has already been rehearsed. I see no reason to doubt and find as a fact that after Christmas the old lady went back to the flat; it was not an ideal solution but she was an indomitable person who would not be put off. The good thing was that the respondent was in a position to give her a hand, he having moved in shortly after she broke her arm. That being the case the essential facts are these:
>
> Mrs Hooper departs; tenant breaks arm; respondent moves in; tenant moves back; then sadly on May 1 1988 the tenant dies — . . .
>
> The issue to be decided now is with what quality is the respondent's residence after she broke her arm to be invested?

I pause there to state that, in my judgment, the judge posed the correct test at that point of his judgment.

> [The grandson's] contention is that he can only have been "residing with" the tenant on my finding that he was acting as a caretaker looking after the animals, keeping out burglars — which is undoubtedly a useful thing. As to his motivation generally it wasn't just that he was willing to help gran, and immediately with the dogs, it was also that it gave him a great opportunity of getting away from home. He had lived with Mrs Williams' second husband since the respondent had been 9 or 10. This was almost manna from heaven, therefore. He may have been a bit daunted by looking after the lady and cooking meals when his habit seems to have been staying in bed. My answer is a firm "yes" — he was well-motivated when he moved out.
>
> As from the date when the old lady came back he resided with her within the meaning of para 3 to Schedule 1 of the Rent Act. But what of this period when living there on his own for four months? His case was "I was residing with her as I intended to look after her when she came back and I was willing to caretake and feed and water the dogs." He was not therefore on his own. In my judgment such an argument is artificial upon the facts: the persons residing with the tenant were Mrs Williams and her second husband. They were residing with each other. I reject the argument that future intention can be elevated to present residence with her.

That is the last quotation I seek to make from the judgment and it is at that point in the judgment that Mr Platt, for the appellant, has submitted that the judge fell into error in two respects. First, in finding that the tenant (the grandmother) was residing with her daughter and stepson and, second, in rejecting the argument that future intention can be elevated to present residence.

Mr Platt further submitted that as the judge had not made a specific finding as to whether the grandson's arrival at the premises preceded her departure or not, this court should draw an inference upon the facts found that it did. I would hesitate about this. There are certainly passages in the evidence, some of which I have recited in this judgment, that would support such an inference. There are also passages in the judgment upon which reliance could be made to indicate the way the judge's mind was working. However, in this instance I am content that it is not necessary for me to determine the validity of Mr Platt's second submission for the resolution of this appeal, which, in my judgment, can be dealt with under the main and central issue, namely what was the quality of the appellant's residence during the four months before the grandmother's return.

We have been referred most helpfully to one authority among others but this is the important one, the case of *Foreman* v *Beagley* [1969] 1 WLR 1387. In that case a widow who was a statutory tenant was in hospital for the last three years of her life. Her son first came to her flat to air the premises but eventually appeared to take up residence there for the last year of her life. After she died he claimed to be entitled to remain in possession of the flat as a statutory tenant under the provisions of the preceding Act to that with which we are concerned. The judge made an order for possession against him on the ground that he was not "residing with" his mother at the time of and for the period of six months immediately before her death. The defendant appealed and that appeal was dismissed.

One of the relevant facts was a statement that if and when the mother returned to the premises the son, unless there was any good reason not to do so, would have departed. It is, I think, necessary for me to refer only to one extended passage from the leading judgment, which was delivered by Russell LJ at p 1390 in which he said:

> It would seem that in concluding that the defendant was not "residing with" his mother, he considered that this could not be so "when she lived in hospital". Thus shortly stated this seems to be a misdirection, and was not supported before us as a proper ground for the conclusion. It is wrong to say that a tenant away at hospital cannot for that reason be "resided with". If a tenant spends the last two months of his life in hospital his widow at home would be residing with him at his death. The judge also, according to the note, said:
>
> > "The whole purpose of this section is to protect the person who has made their home with an elderly relative when that person has died. I think it goes against the whole tenor of the Act to give the same rights to someone who takes advantage of an elderly relative."
>
> In so far as this may have been a ground for the decision this was not supported before us.
>
> For the defendant it was initially contended that if, for the purposes of the statute, the mother was residing at the flat, and the defendant was actually residing at the flat, it necessarily followed that for the purposes of the Act the defendant was "residing with" his mother notwithstanding that they had never been physically together in the flat. This contention that "at" plus "at" *must* mean "with" did not stand up and was not pursued. Suppose a dwelling which could not possibly accommodate more than one person, it would be clearly be wrong to say that a relation who moved in while the tenant was in hospital could be described as residing with the tenant so as to become successor and statutory tenant in respect of the dwelling when the tenant died in hospital six months later. Or suppose that on a tenant going to hospital a member of her family simply moved in without reference to or permission of the tenant and resided there for the relevant period, he could not be reasonably described as residing with her.
>
> For the plaintiff it was initially contended that "residing with" was an impossibility when the two had never been in the dwelling-house at the same time. This contention was not pursued, I think rightly, on consideration of a hypothetical case in which, the daughter of a tenant mother having returned from Australia, the mother being in hospital, it was arranged between mother and daughter that they should share their lives in the mother's dwelling-house, the daughter to move in at once and the mother to join her there as soon as she came out of hospital: in such case if the mother died still in hospital six months later there would have been a "residing with".
>
> In the light of those considerations, can it be said that the defendant was "residing with" his mother for the purposes of para 7 of the Schedule? It is never very wise in these cases to generalise; but at the least it seems to me that, in the phrase in this context, the alleged second successor must be able to point to his situation as being a member of the tenant's household. If at one time he actually lived with the tenant not for example as subtenant of part but as part of the tenant's household, his continued residence during her absence in hospital would normally retain the quality of "residing with" the tenant. His continued residence would take its colour from the past situation. If there were no previous physical coincidence of actual residence there might on the facts nevertheless be a "residing with" as part of the tenant's household, as in the example of the daughter from Australia and the tenant mother, in which case the colour of "residing with" as a member of the mother's household would come from the understanding and arrangement as to what the "set-up" was designed to be.

I gratefully and respectfully adopt those remarks by Russell LJ and for my part consider that they accurately reflect the law.

Mr Salter submitted that "residing at" in law, and in this submission he was referring particularly to the earlier part of the extract which I have just cited, on the part of the grandmother during the four months' absence and "residing at" physically by the grandson at the premises could not amount to "residing with" the grandmother on the part of the grandson. He submitted that there had to be physical joint residence at the beginning of the qualifying period in para 3 of Schedule 1. Although he accepted that had there been such a joint residence with the proper intention, then a supervening event leading to their physical separation, eg admission to hospital after an accident, would not destroy the quality of the residence enjoyed by the grandson.

I cannot accept this distinction upon which Mr Salter's submissions were based either in common sense or in logic. It does not accord with the example given by Russell LJ of the daughter returning from Australia while her mother was in hospital. In my judgment, the matter can be resolved without resort to any artificial, strained or illogical distinctions such as those suggested by Mr Salter. As I have already said, the judge posed the correct test but, unhappily, failed to apply it.

In stating the proposition, in the extract from his judgment which I have cited, that the grandmother was residing with her daughter and stepson, that is at her daughter's home and not at the premises, the judge fell into error if, by making this statement, he was indicating thereby that this was a reason for saying that the grandmother could not be residing at the premises for the purposes of residing with the grandson who was physically there. The test where the grandmother was residing depends upon her intentions to return to what was always her home and remained her home until the final few weeks of her life, and that the grandson in her intentions and plans would be there residing with her as part of her family. With respect to the learned judge, I have come to the conclusion that his attention was distracted from the real test which he had himself provided earlier when he discussed the position in the daughter's home away from the premises, which on any version of the facts was a temporary one brought about by the grandmother's physical incapacities resulting from the accident. Second, where the learned judge said that he was unable to accept the submission that "future intention can be elevated to present residence" he also fell into error. The error was in excluding the intentions of the grandmother and of the grandson, either expressly or from the circumstances and the conduct of each of them as to the future use of the premises. This must always be an element in determining the quality of the residence of either or both of the relevant parties, namely the tenant grandmother and the putative statutory tenant by succession, the grandson, in the event of the grandmother's demise. For example, if their intention was that the grandson should merely be a caretaker while the grandmother was temporarily absent, so that he could look after the dogs, ward off the burglars, keep the house warm, then there would be no residence established on the grandson's part that would qualify under para 3 of Schedule 1. That was why the appellant failed in *Foreman* v *Beagley*, because there was clear evidence that, unless there had been some emergency, when the tenant returned the appellant was going to leave. There was no intention in *Foreman* v *Beagley* that the appellant would become part of the household of the statutory tenant. But if, as the judge found in this case, in relation to the post-Christmas four months, the intention was that the grandson should become a permanent member of the grandmother's household, then the fact that the grandmother was temporarily absent would not detract from the quality of the residence at the beginning of the qualifying period any more than it would do if it had occurred as an intervening event during the latter four months when the judge had found, on the clearest possible evidence, that the residence did qualify within para 3 of Schedule 1. The critical consideration is that he was there physically in the premises during the first four months and intended to remain as a member of the household when his grandmother returned and, moreover, it is quite clear from the evidence that the grandmother's intentions were the same, namely that she and her grandson would form a family unit occupying the premises for an indeterminate length of time. She was an elderly and frail lady, and it is not an over-extension that probably the intention was, as was said in the judgment, "for the rest of her life".

In my judgment, there is no reason to think that the quality of the residence was different during the first four months from the quality of the residence in the last four months, notwithstanding the absence of the grandmother at the beginning of the first period. I have already said that I cannot follow the logic or basis of the submission that, because the grandmother was not there at the beginning of the qualifying period, therefore in some way the ensuing quality of the residence was affected. I can see no ground for that basis. It follows, therefore, that, in my judgment, from the time that the grandson transferred himself from his previous home with his mother and stepfather and took up his residence on a permanent basis as the judge found at the grandmother's home, he became resident there within the requirements of para 3 of Schedule 1 to the Act. There is no dispute that he was a member of the original tenant's family, that he was there at the time of the death of the grandmother and, in my judgment, was also there for a period of six months immediately before the relevant death. I would, therefore, allow this appeal and grant the declaration sought in the notice of appeal.

McCowan LJ agreed and did not add anything.

Also agreeing, **Sir Denys Buckley** said: I just want to add an observation on a quite subsidiary matter.

It was, in my view, unfortunate that the judge did not find precisely the date of the grandmother's accident, the date on which the grandson moved into the grandmother's flat and the date on which the grandmother moved to the mother's home.

Evidence on these matters was in several respects inconsistent and incapable of reconciliation. But the dates in question were important to the conclusion of the issues which the learned judge had to decide. I feel sympathy with his difficulty in making satisfactory findings. But he should, I consider, have endeavoured to find these facts. His failure to do so has made the determination of this appeal more difficult than it would or might otherwise have been. I agree that this appeal should be allowed.

The appeal was allowed, with costs of appeal and below; leave to appeal to the House of Lords was refused.

Court of Appeal

November 23 1990

(Before Lord Justice DILLON and Lord Justice BINGHAM)

PALMER v McNAMARA

Estates Gazette May 4 1991

[1991] 17 EG 88

Rent Act 1977, section 12 — Resident landlord exception — Whether landlord qualified as a "resident landlord" for the purpose of the Rent Act despite the facts that the room retained by him did not contain a cooker and that he did not sleep there — Appeal from decision of county court judge holding that the plaintiff was a resident landlord and entitled to an order for possession of the flat occupied by the defendant — Judge's decision upheld by the Court of Appeal

The flat occupied by the defendant tenant formed part of a building which was not a purpose-built block of flats — So far there was no problem, but the questions which arose under section 12 related to the plaintiff landlord's occupation of other accommodation in the building — Was his occupation such as to confer on him the status of a resident landlord and thereby deny the benefits of protection under the Rent Act to the defendant tenant? The two specific questions were (a) was the landlord at the time when he granted the tenancy occupying another dwelling-house which also formed part of the building, and (b) did the landlord at all times since that tenancy was granted occupy that other dwelling-house as his residence?

The facts were that the tenant occupied a ground-floor room at the front of the house, while the landlord retained a ground-floor room at the back — He was in occupation of this back room at the date when he granted the tenancy of the front room and he thereafter continued to occupy the back room — He kept his books and typewriter there, as well as his television

A **set and video, a refrigerator, cutlery, electric kettle and a fold-up bed on which he could rest — He spent his days in this room — Practically all his possessions were there, but there was no cooker — The landlord had never learnt to cook — He either bought food, which he did not need to cook, or he bought ready-cooked take-away meals — For reasons connected with his health he required help to dress and undress and in consequence he did not sleep in his room but in the spare room of a friend who lived some distance away**

It was clear that a single room could be a dwelling-house, but it was suggested on behalf of the tenant that a room
B **without a cooker could not be described as a dwelling-house — It was also suggested that, as the landlord did not sleep in his room, he could not properly be said to be occupying it as a residence — In these two respects, it was submitted, the requirements of section 12 were not satisfied — Both of these submissions were rejected by the Court of Appeal — Using common sense, the room occupied by the landlord contained everything required to be his dwelling-house except a cooker, which he did not want — The judge below rightly dismissed the suggestion that because he did not want a cooker the room he occupied could not rank as a dwelling-house — As regards the other suggestion that the landlord could not be said to**
C **occupy the room as his residence, the court again agreed with the judge below — There was authority for interpreting "occupy as his residence" as equivalent to "occupy as his home" — Applying common sense again to what was a matter of fact and degree, the judge was entitled to take the view that the landlord occupied this back room as his home — As a result the tenant's appeal had to be dismissed**

The following cases are referred to in this report.

Beck v *Scholz* [1953] 1 QB 570; [1953] 2 WLR 651; [1953] 1 All ER 814, CA
Hampstead Way Investments Ltd v *Lewis-Weare* [1985] 1 WLR 164; [1985] 1 All ER 564; [1985] 1 EGLR 120; (1985) 274 EG 281, HL
D *Regalian Securities Ltd* v *Scheuer* [1982] EGD 591; (1982) 263 EG 973; 5 HLR 48, CA
Skinner v *Geary* [1931] 2 KB 546, CA

This was an appeal by the defendant, Kevin McNamara, against a possession order made by Judge Sumner, at Wandsworth County Court, in favour of the plaintiff, Oliver Rees Palmer, in respect of a room occupied by the defendant at 84 Drakefield Road, Tooting, London SW17.

Mark Rowland (instructed by Jenkins O'Dowd & Barth) appeared on behalf of the appellant; Michael Michell (instructed by Collingwood & Co) represented the respondent.

Giving judgment, **DILLON LJ** said: This is an appeal by the
E defendant in the action, Mr Kevin McNamara, who is by occupation a background artist for television animation, against a possession order made against him at the trial of this action in the Wandsworth County Court by Judge Sumner on February 12 1990. By that order it was ordered that the defendant give the plaintiff, Mr Oliver Palmer, possession of the ground-floor flat at 84 Drakefield Road, Tooting, London SW17, which the defendant has occupied as tenant of the plaintiff. The question is whether the defendant is entitled in respect of that flat to the protection of the Rent Act 1977 and that depends on section 12 of that Act, the side note of which is "Resident landlords". The question is whether the plaintiff qualifies, for the purposes of that section, as a resident landlord.

The section provides by subsection (1) that, subject to certain
F matters which are not material:

> . . . a tenancy of a dwelling-house granted on or after 14th August 1974 shall not be a protected tenancy at any time if —
> (*a*) the dwelling-house forms part only of a building and . . . the building is not a purpose-built block of flats; and
> (*b*) the tenancy was granted by a person who, at the time when he granted it, occupied as his residence another dwelling-house which —
> . . .
> (ii) . . . also forms part of the building . . .; and
> (*c*) subject to paragraph 1 of Schedule 2 to this Act, at all times since the tenancy was granted the interest of the landlord under the tenancy has belonged to a person who, at the time he owned that interest, occupied as his residence another dwelling-house which . . . also formed part of the building.

There is no difficulty over condition (*a*) in section 12(1). The building G
is 84 Drakefield Road, the flat occupied by the defendant forms part of it and the building is not a purpose-built block of flats. There has been no change in the landlord since the tenancy was granted; the plaintiff has been the landlord throughout and the questions that arise are, first, whether under condition (*b*) it can be said that when the tenancy was granted to the defendant in December 1987 the plaintiff who granted it occupied at that time another dwelling-house which also formed part of the building. There is further the question whether both at that time and for the purposes of condition (*c*) at all times since the tenancy was granted the plaintiff has occupied that other dwelling-house as his residence within the meaning of the section.

The factual position is that on the ground floor of 84 Drakefield H
Road there is a front room which is let by a tenancy to the defendant. It is a bedsit with a kitchen-diner at the rear of it. Beside that, leading towards the rear there is a hall and an inner hall. Off the inner hall there is a bathroom and wc and the ultimate arrangement between the plaintiff and the defendant was that that was to be shared. That, at any rate, was the case until the defendant took it upon himself to put a padlock on the door. There is a door from the bathroom out to the garden and the defendant was to have the right to use the garden, and at the back there was a room retained by the plaintiff. The question is whether, in the circumstances, the plaintiff's room is a dwelling-house and whether it is occupied and has throughout the tenancy been occupied by the plaintiff as his residence. J

It appears that the plaintiff occupied the whole of the ground floor of the house from about 1982, at which stage he separated from his wife. In around 1984 he granted a tenancy of the front room to a Miss Murray, who was an aunt of the defendant. Under the terms of that tenancy the plaintiff had shared use, if he wanted it, with Miss Murray not only of the bathroom and wc but also of the kitchen at the rear of the bed-sitting room let to Miss Murray. Miss Murray left towards the end of 1987 and the tenancy was then granted to the defendant with exclusive use of the kitchen at the back of the front room let to him, which indeed includes the kitchen-diner. The plaintiff continued to occupy, and I will come in a moment to his use of it, the back room on the ground floor, which he still occupies.

The plaintiff is a gentleman who is now 59 years old. He used to be K
engaged in advertising, but he retired from that several years ago. However, he does some part-time writing and it seems he has had a book published and has an interest in poetry.

In the rear room on the ground floor he keeps practically all his possessions. There are his books and his typewriter, a table and chairs, television, video, a refrigerator, a kettle, a sink, a fold-up bed which he used to use, cutlery and saucepans and obviously, because of the fridge and kettle, electric points. But there is no cooker. There is no cooker because the plaintiff does not cook and has never learnt to cook. When he wants meals in his room at the back, he either buys food which he can eat without cooking or buys hot take-away meals which he brings home and eats in his room. It is said that the room by L
itself does not constitute a dwelling-house because it has no cooker. It has of course a point, as I have said, and it would be very easy to install a small portable cooker or hotplate if the plaintiff wanted one, but he does not.

The test suggested for whether the room constitutes a dwelling-house is whether it had such qualities and facilities as permitted use for the ordinary purposes of living. It has been submitted to us that it is essential, for it to be a dwelling-house, that it must have facilities for cooking, eating and sleeping. It is not enough if it merely provides facilities for any two of those three, without the third.

We have been referred by Mr Rowland, for the appellant defendant, to various authorities, but they all have to be taken as decisions on the particular facts with which the court was dealing or M
as containing comments directed to the particular situation with which the court was dealing. It seems to me quite clear that in considering whether a room or a series of rooms is a dwelling-house one has to look not only at the physical surroundings but at the purpose. Thus in section 1 of the 1977 Act it is stated that "a tenancy under which a dwelling-house . . . is let as a separate dwelling is a protected tenancy for the purposes of this Act". If a flat is let unfurnished, at the moment of letting it will probably not have a bed in it and, if the tenant is going to provide his own portable cooker, it will not have a cooker in it, but I cannot believe that the criterion of the Act is so stringent on the need for a dwelling-house to contain a bed and a cooker that unfurnished lettings are not protected. Using

A common sense, as is permissible, the room occupied by the plaintiff contains everything required to be his dwelling-house, except a cooker, which he does not want. There could be a cooker there and it would then be a complete dwelling-house. I cannot take the view that just because he does not want a cooker, the room he occupies cannot rank as a dwelling-house. It is, of course, clear that a single room may be a dwelling-house. That is commonly the case in tenancies and it can apply equally well to the landlord's occupation when section 12 has to be considered. Therefore, I take the view that Judge Sumner was correct in holding that what the plaintiff retained was indeed a dwelling-house.

Does he, then, occupy it as his residence? He used to beyond any
B question. Indeed, in this field of law the terms used, which have now been part of the jurisprudence on the subject for 70 years, are very well understood — "residence", "dwelling-house", "occupied as his residence" — and so the cases that come before the courts tend to be cases where the facts are unusual. To that generality the present case is no exception. Apart from his inability to cook, which has had the result that he does not have a cooker in his room, the plaintiff suffers from ailments as a result of which he has not slept in the room since before the tenancy was granted to the defendant. The plaintiff's evidence is that he has diabetes, but, more seriously, he has a 3-inch shortening of his left leg and osteo-arthritis in his left hip which has caused serious deterioration. That is supported by a doctor's certificate, which refers to it as "long-standing hip joint disease".
C The effect of that, particularly if the climate is damp, is that he needs help to dress and undress himself. Therefore, since 1987 he has spent his nights in the spare room of the house of a very old friend, a Miss Ducker, who lives about a mile or mile-and-a-half away from Drakefield Road. The plaintiff has a car and can travel between the two houses. He uses her spare room and keeps a change of clothes there. He probably now has baths there, because of the difficulty in dressing and undressing, and sometimes has his evening meal or Sunday lunch with Miss Ducker at her home. He may keep the odd book there, but he does not keep his other possessions there. The furniture in the spare room is Miss Ducker's furniture.

It is said that because he sleeps at Miss Ducker's home he does not use his room at 84 Drakefield Road as his residence. He does not use
D Miss Ducker's spare room as a residence either, and he is therefore in the position, which is not impossible but perhaps unusual, of a man with accommodation but no residence. He spends his days at 84 Drakefield Road. He does his writing and reading there. He watches television and video and has meals that he brings in. But it is said that because he does not sleep there he does not occupy it as his residence.

The question whether a dwelling-house is occupied as a residence is commonly paraphrased as "does he occupy it as his home?": see the observations of Lord Brandon in *Hampstead Way Investments Ltd* v *Lewis-Weare* [1985] 1 All ER 564 at p 568A.* He said:

> My Lords, in order to determine this appeal, it is necessary to examine the more important cases decided between 1920 and 1968 on what is meant by the
> E occupation of a dwelling-house by a person as his residence, or, as it is put in many of the cases (without, in my view, any difference of meaning), the occupation of a dwelling-house by a person as his home.

The word "home" was used by Scrutton LJ in the leading case of *Skinner* v *Geary* [1931] 2 KB 546. It was also used by Lord Evershed MR in *Beck* v *Scholz* [1953] 1 QB 570 at p 575 and by May LJ in *Regalian Securities* v *Scheuer* [1982] 5 HLR 48.

Applying that test, which raises a question of fact and degree to be determined by applying ordinary common sense, I have no doubt that the judge was entitled, and indeed right, to take the view that the plaintiff occupied his back room as his home. That is enough to determine this appeal.

We were referred to authorities which are concerned with the
F question whether, if a person has ceased, for whatever reason, to occupy a dwelling-house as his home he can none the less save the position for the purposes of the Rent Act if he has an intention or desire to return to the dwelling-house, and we had an argument on whether it was relevant that at one stage the plaintiff had had it in mind to sell 84 Drakefield Road. But this does not, to my mind, arise as a question for decision if I am right in the view, which the judge also took, that he is currently occupying the room as his residence or home. I would dismiss this appeal.

Agreeing, **Bingham LJ** said: The learned county court judge set out the facts of this case in considerable detail and reviewed the relevant law with great care. Counsel has not persuaded me that he G
was guilty of any legal misdirection and there is no challenge to his factual conclusions. This makes the appellant's task very difficult because the field is one in which the facts and circumstances of individual cases are usually determinative. The appellant's task is even more difficult because the facts of this case are, in the literal sense, extraordinary. No closely comparable facts appear to have occurred in any reported case and they are perhaps unlikely to recur.

The first question for decision, as Dillon LJ has indicated, is whether the premises have the necessary characteristics of a dwelling-house for the purposes of the Act. Mr Rowland, for the appellant, submitted that they lacked an essential characteristic of a dwelling-house because there were no cooking facilities and therefore they H
could not be used for one of the more fundamental functions of living. To this, Mr Michell, for the landlord, pointed out that any unfurnished dwelling-house might not contain facilities for living, such as a bed or a table, but since these could readily be provided the premises were none the less a dwelling-house. By parity of reasoning he pointed out that the plaintiff could easily have provided rudimentary means for cooking and the only reason why he did not do so was because he either bought food out and brought it back to eat or he ate out. The learned judge accepted Mr Michell's submission, as, in my judgment, he was fully entitled to do. But for the plaintiff's unusual habits, the argument would not have been open and those habits do not, in my judgment, prevent these premises being a dwelling-house. J

The second question was whether the plaintiff occupied the premises as his residence. A long line of authority, to which Dillon LJ has referred, shows that the correct approach is to ask the simple and intelligible question: was the plaintiff in occupation of these premises as his home? The judge concluded that he was, despite the fact that for medical reasons he slept elsewhere. The learned judge's conclusion was, in my opinion, fully justified by the evidence and was correct. It is enough, however, for the purposes of this appeal, to say that there are no grounds on which this court could possibly say that he was wrong. I would also therefore dismiss this appeal.

The appeal was dismissed with costs, not to be enforced without leave of the court. Legal aid taxation on both sides ordered. K

Court of Appeal

December 4 1990

(Before Lord Justice NICHOLLS and Lord Justice MANN)

TELEVANTOS v McCULLOCH

Estates Gazette May 18 1991

[1991] 19 EG 118 L

Rent Act 1977, Schedule 15, Part I, Case 1 — Rent in arrear — Statutory tenancy — Tenant's counterclaim for damages for breach of covenant implied by section 11 of Landlord and Tenant Act 1985 to keep the matters mentioned in the section in repair and working order — Judge's award in respect of counterclaim greater than arrears of rent — Whether order for possession justified — Equitable set-off — Whether tenant came to court with clean hands — Tenant's appeal allowed

**The plaintiff in the county court proceedings claimed an order for possession on the ground of arrears of rent payable by the defendant tenant — It was common ground that the M
defendant had deducted and withheld from her rent payments a total of £2,274 — The defendant counterclaimed for damages for breach of the covenant to repair and keep in working order implied by section 11 of the Landlord and Tenant Act 1985, and the judge awarded the tenant £2,700 in respect of the breach of this covenant — Despite a balance of £426 in favour of the tenant, the judge made an order for possession in the landlord's favour — The defendant appealed — She submitted that the judge was wrong to make such an order as the fact was that in the result there was no unpaid rent**

*Editor's note: Also reported at [1985] 1 EGLR 120 at p 122.

The judge's reasoning appears to have been that at the time of the issue and commencement of proceedings there were arrears of rent because at that time the quantum of the defendant's counterclaim had not been ascertained — It was accepted that this reasoning was not as such sustainable — The subject of the defendant's counterclaim related to a period which, save possibly for a few days, was wholly before the start of proceedings — This meant that, even before the start, the tenant had an equitable set-off to the rent claim, subject, however, to one obstacle raised by the plaintiff

The obstacle was the application of the well-recognised equity principle, applicable alike to claims and defences, that "he who comes to equity must come with clean hands" — It was suggested by the plaintiff that the defendant tenant's hands had not been clean, but the only evidence of that before the judge was that the tenant had not been co-operative with the builder in arranging for the works to be done, causing delay and loss to the plaintiff and, on one occasion, behaving in an "unladylike fashion" when the glass in a door was broken — None of these went anywhere near precluding the tenant from relying on an equitable set-off — In any case the point was not taken in the court below and so was not open in the Court of Appeal

The judge had decided that there was rent in arrear, despite his findings on the tenant's counterclaim — He had therefore, in accordance with section 98(1) of the Rent Act 1977, to decide whether it was reasonable to make a possession order — Here again his conclusions and reasoning met with criticism by the court — He held that it was reasonable to make the order, but, in striking a balance in favour of the landlord, he alluded to personal and family considerations and failed to give weight to the essential points that the tenant's complaint about the inconvenience she was suffering as a result of the landlord's breach was well founded and that she was entitled to a sum for damages in excess of the rent withheld

The final matter was the plaintiff landlord's cross-appeal alleging that the award of £2,700 damages was excessive, in particular the £1,000 for general damages — The Court of Appeal rejected this and regarded the award as "unimpeachable" — Appeal allowed and cross-appeal dismissed, with costs

The following case is referred to in this report.

British Anzani (Felixstowe) Ltd v *International Marine Management (UK) Ltd* [1980] QB 637; [1979] 3 WLR 451; [1979] 2 All ER 1063; (1978) 39 P&CR 189; [1979] EGD 414; 250 EG 1183

This was an appeal by the defendant tenant, Patricia Williams McCulloch, from the decision of Judge Hill-Smith, at Westminster County Court, granting the landlord, Maris Vassos Televantos, possession of the ground-floor flat at 46 Rosebery Road, Muswell Hill, London N10.

Philip Engelman (instructed by Whittington Tilling & Knight) appeared on behalf of the appellant; Nicholas Hopkins (instructed by Chalker Smalley) represented the respondent.

Giving judgment, NICHOLLS LJ said: This is an appeal by the defendant from an order made by His Honour Judge Hill-Smith, sitting in the Westminster County Court, on April 11 1990. In the proceedings the plaintiff claimed possession from the defendant of a ground-floor flat at 46 Rosebery Road, Muswell Hill, London N10. The tenancy is a statutory tenancy and the basis on which the claim for possession was made was Case 1 of Schedule 15 to the Rent Act 1977. The particular ground relied upon was arrears of rent.

At the trial it was common ground that the defendant had deducted and withheld from her rent over a period of months sums totalling £2,274. The defendant counterclaimed for damages for breach of the lessor's covenant, which, admittedly, was implied in this lease by section 11 of the Landlord and Tenant Act 1985 — that covenant obliging the lessor to keep the matters mentioned in that section in repair and working order. The defendant claimed in the proceedings, although the precise words "set off" did not appear in her defence or counterclaim, to set off those damages against her liability in respect of unpaid rent.

The judge awarded the defendant £2,700 in respect of her counterclaim. Nevertheless, although that sum exceeded the agreed amount of unpaid rent, the judge made an order for the defendant to give up possession in five months' time. He ordered the defendant to pay the plaintiff's costs on the claim and the plaintiff to pay the defendant's costs on the counterclaim, those costs orders to be set off against each other.

I should further mention, because this was not recorded in the order as drawn up, that the judge also ordered that the defendant was to pay the agreed sum of rent arrears of £2,274. No doubt that sum was to be set against the plaintiff's liability to pay £2,700 and thereby be wholly extinguished, thus leaving the plaintiff obliged to pay the defendant the balance of £426.

The defendant's appeal is primarily based on the simple proposition that the judge erred in making an order for possession because, on his own findings, there was no unpaid rent.

For the purposes of this appeal the law on this point is clear enough. In his judgment the judge referred to the following summary set out in volume 1 of *Woodfall on Landlord and Tenant*, para 1/1464:

> Where the landlord expressly agrees to do repairs, there is no implied condition that if such repairs be not done the tenant may quit or is not liable for rent; but it was held in an old case that the tenant may do the repairs, and deduct the expense from the rent, and such would still seem to be the law. The tenant can always counterclaim for the expense if sued for the rent. He cannot recover more than the reasonable cost of complying with the covenant, and loses that right if he has denied the landlord the opportunity of complying with the covenant himself. A tenant, upon being sued for unpaid rent, may be able to rely upon his own counterclaim against the landlord for breach of a landlord's repairing covenant as according a defence by way of equitable set off to the claim for rent.

Having quoted the substance of that paragraph, the judge said:

> . . . it is clear, accordingly, that the law is that a tenant in such circumstances is not entitled unilaterally to reduce the rent by a figure that arbitrarily is selected by her or him or the tenant as representing a reasonable reduction for the rent. His or her remedy is to counterclaim in proceedings. In this case I find that there was no agreement in relation to any reduction in rent: and it follows from that that at the time that I have to consider, namely the issue of the proceedings, there were arrears of rent.

Thus the heart of the judge's reasoning seems to be that, because the quantum of the defendant's counterclaim had not been ascertained when the plaintiff started the proceedings, the rent was then in arrears. For my part, I have some difficulty with that conclusion. It is not a point, however, on which further exposition or consideration is called for in this case because, as I understand the case that has been presented to us on behalf of the landlord, subject to one point which I will mention in a moment it is accepted that that passage in the judge's reasoning is not supportable.

The defendant's counterclaim relates to damage, by way of inconvenience and otherwise, suffered by her while the flat was in a state of disrepair and while extensive works of repair were being carried out. Those works began in about July 1987 and ended the following May. The plaintiff issued his proceedings early in that month. Thus the subject of the counterclaim relates to a period which, save possibly for a few days, was wholly before the proceedings started. Thus, even when the proceedings were issued, the defendant had available to her a defence of equitable set-off to the rent claim. In due course, when the dispute over the defendant's claim was determined by the court, what then became revealed was that, indeed, the defendant had a defence to the whole of the claim for unpaid rent.

What is said on this point on behalf of the plaintiff is this. The principle set out in *Woodfall* is culled from the decision in the case of *British Anzani (Felixstowe) Ltd* v *International Marine Management (UK) Ltd* [1980] QB 637. There may be a defence by way of equitable set-off in this type of case, but the defence of equitable set-off depends upon an application of principles of equity. As an equitable defence the principle is not applicable in this case because the tenant did not come to the court with clean hands.

In support of that submission it is said that the judge preferred the evidence of the landlord to that of the tenant where there was conflict. He found that the tenant was not co-operative with the builder in arranging for the works of repair to be done and this caused delay and damage and loss to the plaintiff, and that on one occasion the tenant behaved in, to quote the judge's expression, an unladylike

A fashion when the glass in a door was broken. This point was not taken in the court below, and in my view it is not open in this court. This is just the sort of point which, if taken at the trial, can be expected to be the subject of evidence, including cross-examination. In any event, in my view, the point is misconceived. None of the matters relied upon goes anywhere near furnishing grounds for saying that, although the damages payable by the landlord to the tenant exceed the amount of unpaid rent, the tenant should be precluded from relying upon those damages by way of equitable set-off.

The judge, having reached the conclusion that there was rent in arrear, despite his findings on the counterclaim, then went on to consider, as he was bound to do, whether it was reasonable that there B should be an order for possession. What he said on that was this:

> I must balance one factor against the other. This property is a rent controlled property. It was purchased by the plaintiff's father some time ago and had it been of a rateable value highly in excess of this figure, of course the plaintiff would be entitled to possession. His common law right has been whittled down by statute. It is his bricks and mortar.
>
> The second aspect is that he has family and at one stage there were earlier proceedings, apparently improperly constituted, to obtain possession so that other members of his family might occupy: and there was again the possibility that the plaintiff himself, Basil's son, might move to London in pursuit of his work. That is one side. On the other side, the defendant is a single person, that is to say, she is living alone (in fact she is a married person). What has happened to her husband is not at all clear and her partner, Mr Cross, has disappeared from the scene. She is to complete her course of study in three C months' time. It is clear that until or after the period when rent was arbitrarily withheld by her, she complied with her obligation to pay rent. But I am satisfied from the whole evidence given in this case and the way it was given and the demeanour of the parties, that it is a case in which on balance I ought to accede to the landlord's claim for possession.

Had this point become material, I must say, with all respect to the judge, that I am wholly unable to accept that he reached a correct decision on the question of reasonableness. The factor, above all others, on the judge's findings which needed to be taken into account was that the sums withheld from the payments of rent were withheld because of a claim by the tenant that she was entitled to damages for the inconvenience that she was suffering as a result of the landlord's breach of covenant. In the event, the judge found that her complaint D was well founded and that she was entitled indeed to a sum in excess of the amount which she had withheld. In my view, when one takes that into account, I have the greatest difficulty in seeing how it could be regarded as reasonable in the circumstances of this case to say that an order for possession should be made. On that the judge was plainly wrong.

The third issue which has been put before us concerns a cross-appeal by the plaintiff against the judge's award of damages in the sum of £2,700. The background facts can be stated shortly. In May 1987 the London Borough of Haringey served a notice under section 9(1)(*a*) of the Housing Act 1957, requiring extensive works to be carried out within 120 days to bring the house up to a reasonable standard. That followed a complaint to the environmental health E officer by the tenants of the upstairs flat. The judge found that the complaint was made with, as he put it, the support, active or inactive, of the defendant. The judge was satisfied that defects existed, that considerable works of repair were necessary and that those works involved disturbance to the defendant.

The works took a substantial period. The initial estimate of the small builder who was engaged was 10 weeks, but in the event the works took 10 months. The judge said the builder had limited experience and was clearly without any real equipment or expertise. The defendant was irritated by the slowness and irregularity of the works, but she herself was not co-operative. There were difficulties in arranging dates for some of the work and responsibility for that fell F equally on both sides. The judge said:

> It is clear that she suffered a degree of inconvenience, spasmodically, greater and sometimes less, during the period when the quite extensive repairs G involving structural alterations and the repositioning of water tanks and the likes of that were taking place. It is clear on that basis that arrangements ought to have been made by the landlords either to let her use the flat above or otherwise to mitigate the inconvenience.
>
> I am satisfied that there was inconvenience, and on that basis a breach of the section 11 implied covenant.

The judge awarded the defendant the sum of £1,700 by way of general damages for breach of that covenant. There was also a claim for special damages for items such as loss of the contents of a freezer when the electricity was disconnected and the cleaning of carpets and curtains etc. Under that head the defendant claimed £1,753 and he awarded her £1,000, which, he said, amounted to generous H compensation.

On the cross-appeal the plaintiff contends that the award of £1,700 was excessive. It was submitted, in short, that a sum of the order of £1,000 pa, depending on the circumstances, is the typical figure to be awarded for distress, inconvenience and hardship. It is submitted that the period of inconvenience in the present case should not be equivalent to more than one year for the purpose of damages when one takes into account, first, the inevitable disruption which will follow from a landlord's complying with his repairing obligations and carrying out works; second, the obstruction of the defendant in this case; third, the fact that the defendant had and occupied alternative accommodation in Paris, at least intermittently, during the time when the main works were done; and, fourth, that there had J been no increase in the rent since the lease was granted in 1985.

It is submitted that a time equivalent of five or six months' inconvenience is in point in the present case and that the sum awarded for that period should be of the order of £500 to cover inconvenience while the works were being carried out. The principal inconvenience arose when the works were being carried out, and, in respect of the earlier period, the damages should not exceed a sum of the order of, say, £150. It is submitted that the difference between these figures and the amount awarded by the judge is so large as to demonstrate that the judge must have fallen into error when making his award.

I am unable to accept these submissions. Each case must depend on its own facts. Here the damages awarded were in respect of the whole K period of the tenancy from April 1985 onwards. No doubt the principal problems were in the 10 months while the work was being carried out, but the interference with the tenant's enjoyment was not confined to that time. In my view, the sum awarded cannot be regarded as plainly excessive.

The judge saw the witnesses and was able to form an opinion on the state of the premises, on the degree of inconvenience suffered from time to time by the defendant and on the plaintiff's culpability. The judge had in mind, because he mentioned these matters in his judgment, that the tenant was not co-operative and that it was only during the time the works were being carried out that the principal problems existed.

I can see no reason for thinking that the judge failed to take into L account and give due weight to the matters which were material and I can see no basis for saying that his award was plainly wrong. In my view, the award is unimpeachable. For these reasons, it must follow that, in my view, this appeal should be allowed and the cross-appeal dismissed.

Agreeing, **Mann LJ** said: There is one question upon which I would wish to reserve my position. It is this. Can it be said that, as at the commencement of proceedings in circumstances such as here obtained and were found to have obtained, that any rent lawfully due from the tenant had not been paid? The answer to that question is devoid of consequence in these appeals. It does not affect the result. M

The appeal was allowed and cross-appeal dismissed with costs.

LANDLORD AND TENANT
RENT REVIEW

A
Court of Appeal

July 25 1990

(Before Lord Justice NEILL, Lord Justice NICHOLLS and Lord Justice STAUGHTON)

PRUDENTIAL ASSURANCE CO LTD v TRAFALGAR HOUSE GROUP ESTATES LTD

Estates Gazette January 12 1991

[1991] 01 EG 103

B **Landlord and tenant — Rent review clause in lease — Arbitration on true construction of a provision — Application for leave to appeal from arbitrator's award — Arbitration Act 1979, section 1(3)(*b*) — Guidelines for grant of leave — Harman J in the present case felt real doubt about the correctness of the arbitrator's ruling and, applying the guidelines then given by the Vice-Chancellor in *Lucas Industries plc* v *Welsh Development Agency,* granted leave to appeal — However, on hearing that these guidelines were shortly to be considered by the Court of Appeal in *Ipswich Borough Council* v *Fisons plc*, Harman J also gave leave to appeal against his own decision to the Court of Appeal — That** C **court, applying the new guidelines in the *Ipswich* case, allowed the appeal**

The question in the present case which gave rise to the dispute was whether, on the true construction of the lease, certain suspended ceilings, light fittings and carpets were to be taken into account on the rent review — The answer made a difference to the reviewed rent, £850,000 pa if (as the landlords contended) they should be taken into account, £800,000 if (as the tenants contended) they should not — The arbitrator ruled that the equipment in question should not be taken into account — The landlords applied for leave to D **appeal under section 1(3) of the 1979 Act — As mentioned above, Harman J felt real doubt as to the ruling and granted leave — The lease, it should be said, included a provision that the items in question were to be treated as tenant's fixtures and fittings even though they were not removable unless replaced and even though they were to become the property of the landlord at the end of the lease**

In *Ipswich Borough Council* v *Fisons plc* Lord Donaldson of Lymington MR had explained that it might well be that in most rent review disputes the "standard terms" approach would be justified, so that a strong *prima facie* case of error by the arbitrator should be shown before leave to appeal is given E **— It was true that "strong" was an imprecise word and the degree of strength called for was left to the exercise of judicial discretion in the circumstances of the particular case — The bottom line was, however, that the judge concluded that there was a more or less strong, but still "strong," *prima facie* case that the arbitrator had erred in law — It was now law that this stricter test should be applied instead of the *Lucas* test that leave to appeal should be given if the judge was "left in real doubt" as to whether the arbitrator was right in law — Harman J had directed himself correctly in accordance with the law as it then stood, but since then the law had moved on —** F **Although in the present case cogent criticisms could be made of the reasoning of the arbitrator, the court was not satisfied that a strong *prima facie* case had been made out that he had erred in law — The appeal from Harman J's decision must accordingly be allowed**

The following cases are referred to in this report.

Antaios Compania Naviera SA v *Salen Rederierna AB* [1985] AC 191; [1984] 3 WLR 592; [1984] 3 All ER 229; [1984] 2 Lloyd's Rep 235, HL

Ipswich Borough Council v *Fisons plc* [1990] 2 WLR 108; [1990] 1 All ER 730; [1990] 1 EGLR 17; [1990] 04 EG 127, CA

Pioneer Shipping v *BTP Tioxide Ltd ("The Nema")* [1982] AC 724; [1981] 3 G WLR 292; [1981] 2 All ER 1030; [1981] 2 Lloyd's Rep 239, HL

Retla Steamship Co v *Gryphon Shipping SA* [1982] 1 Lloyd's Rep 55

In this appeal the Court of Appeal had to determine whether the judge, Harman J, was right to grant leave to appeal to the High Court under section 1(3) of the Arbitration Act 1979 from the award of the arbitrator in regard to the rent review provisions in the lease of a property known as Cementation House situated in the main road between Rickmansworth and Denham.

Kirk Reynolds (instructed by the solicitors' department of the Prudential Assurance Co Ltd) appeared on behalf of the landlords; Patrick Talbot (instructed by Ashurst Morris Crisp) represented the tenants, Trafalgar House Group Estates Ltd (sued as Trafalgar H House Group Premises Ltd).

Giving the first judgment at the invitation of Neill LJ, STAUGHTON LJ said: In March 1977 there was a lease of Cementation House which lies on the main road between Rickmansworth and Denham. The landlords were Prudential Assurance Co Ltd and the tenants were Trafalgar House Group Estates Ltd, sued as Trafalgar House Group Premises Ltd. The term was for 35 years and the initial rent was £380,000. There were to be rent reviews every five years. This case concerns the second review as at March 25 1987.

The third schedule in the lease provided that, in the absence of agreement, the revised rent should be determined at the option of the landlord either by an arbitrator or by an expert valuer acting as an J expert and not as an arbitrator, "such arbitrator or valuer to be a Fellow of Royal Institution of Chartered Surveyors and to be nominated by the President for the time being of that Institution on the application of the Landlord".

The parties were unable to agree on a revised rent, and Mr P L Banks FRICS was appointed arbitrator. The nature of the dispute then became clear, as was set out subsequently in the award:

> During the course of the reference to Mr Banks the parties agreed that the amount of the revised rent depended upon the correct answer to a question of law arising out of the terms of the lease. They also agreed alternative amounts for the revised rent, dependent upon whether the answer to the question was favourable to the Claimant or the Respondent. K
> The question and the alternative amounts were formulated as follows:
>
> > "Whether upon a true construction of the Lease . . . the annual rent payable from 25th March 1987 (the review date) should be £800,000 (as contended by the Respondent) or £850,000 (as contended by the Claimant) dependent on whether the suspended ceilings light fittings and carpets are to be taken into account on review."

It was then agreed that Mr Banks would retire from his post as arbitrator and be replaced by Mr Derek Wood QC.

By his award made on May 9 1989 he ruled that the equipment in question should not be taken into account. Consequently the revised rent was £800,000 pa and not £850,000. The landlords applied for leave to appeal under section 1(3) of the Arbitration Act 1979. That application came before Harman J on October 27 1989.

It might have been argued that the first point to be decided was whether the question of law could substantially affect the rights of the parties in terms of section 1(4) of the Act. After all, the amount in dispute was only $\frac{1}{17}$ of the revised annual rent. However, it was conceded that the rights of the parties could be substantially affected by the answer to the question of law, so I need say no more about that. In the case of *Retla Steamship Co* v *Gryphon Shipping SA* [1982] 1 Lloyd's Rep 55 a judge at first instance held that a point of law which affected only one-eighth of the total amount in the dispute did not substantially affect the rights of the parties. However, that was in the early days of the 1979 Arbitration Act. Furthermore, the decision was given on the afternoon of October 1 and may possibly need to be reconsidered one day.

The issue before the judge turned on the degree of conviction that he had to hold as to the correctness of the award before he could grant leave to appeal. At that time two decisions of the Vice-Chancellor had laid down that the *Nema* and *Antaios* guidelines, while applicable to disputes in commerce and trade, did not apply to rent review disputes. In the latter class of case the judge had only to feel real doubt as to whether the arbitrator was right before he could be justified in granting leave.

Harman J did feel real doubt and so he did grant leave. However, he was told that the second of the Vice-Chancellor's cases, *Ipswich Borough Council* v *Fisons plc,* was shortly to be considered by the Court of Appeal. So he took the unusual step of granting leave to appeal to this court, under section 1(6A), from his decision to grant leave to appeal from the arbitrator to the High Court. In my view, if I may say so, the judge was wholly right to take that course in the circumstances.

So we now have to determine whether or not the judge was right to grant leave to appeal to the High Court on the law as it now stands.

This court decided *Ipswich Borough Council* v *Fisons plc* [1990] 2 WLR 108* on November 24 1989. The only reasoned judgment was that of Lord Donaldson of Lymington MR. Nothing would be gained by repeating his analysis of the *Nema* and *Antaios* decisions, so I can turn at once to what he said about their application to rent review cases at p 114:

> The House of Lords' guidelines with regard to one-off contracts, that an obvious case of error should be shown, I think assumed that the effect would also be one-off. This might well be true of the last rent review, particularly if the remainder of the lease were relatively short. It would not, however, be true of earlier rent reviews. There the long-term effect on subsequent reviews would be analogous to the effect of decisions on standard terms, the only difference being that the same, rather than different, parties might be affected.
>
> Accordingly it may well be that in most rent review disputes the "standard terms" approach will be justified: a strong *prima facie* case of error should be shown. But "strong" is an imprecise term and I do not think that the House of Lords intended that the same degree of strength should be called for in every case. This is, after all, a matter which Parliament has left to the exercise of judicial discretion. Consistency of approach is important, but it must not negative a discretion based upon the facts of individual cases.
>
> So how strong is strong? No meter can be applied or indeed devised. It is a matter of relative values. If the chosen arbitrator is a lawyer and the problem is purely one of construction, the parties must be assumed to have had good reason for relying upon his expertise and the presumption in favour of finality or, to put it the other way round, the strength needed to rebut it will be greater. So too if the dispute really centres upon an issue calling for non-legal expertise, albeit with some underlying question of law, and the chosen arbitrator has that expertise. But if the chosen arbitrator is not a lawyer and the whole dispute centres upon a difficult question of law, less strength may be required.
>
> Similarly, the degree of strength will be affected by whether the clause in question is one of a class commonly encountered, so that others would benefit from an authoritative decision on its meaning or application, and I see no reason why some account should not be taken of the seriousness of the consequences to the parties of the arbitrator's error, if error there be. But the bottom line must always, I think, be that the judge concludes that there is a more or less strong, but still "strong", *prima facie* case that the arbitrator has erred in law. To adopt any other approach would be to fly in the face of the legislative preference for finality.

Harman J, through no fault of his own, did not apply that test. He instead applied the test laid down by the Vice-Chancellor in two cases as to whether he had real doubt that the award was right. Since that is no longer the right test in view of the decision of this court after Harman J's judgment, it falls to us to look at the matter afresh.

*Editor's note: Also reported at [1990] 1 EGLR 17.

I then turn to look at the dispute in this case and see if the *Ipswich Borough Council* test is satisfied. That test is whether the court forms a strong *prima facie* view that the arbitrator was wrong, although "strong" may be a relative term in the circumstances. I do not form such a view.

Lord Diplock in the case of *Antaios Compania Naviera SA* v *Salen Rederierna AB* [1985] AC 191 said that it was in general unwise of judges at first instance to express views on the merits either when granting leave to appeal or when refusing it. If I may say so, I can fully understand that ruling. If the judge refuses leave to appeal but makes observations, it may be thought he is providing a precedent, after what is supposed to be a fairly summary hearing. If he grants leave to appeal and expresses a view on the merits, he may embarrass or inhibit the judge who has to hear the substantive appeal.

It seems to me that these observations may apply equally to this court, although it is to be observed that in the *Ipswich Borough Council* case itself the Master of the Rolls did express some views on the substantive merits.

For my part I am content to say in this case, applying the test laid down by the Master of the Rolls in the *Ipswich Borough Council* case, that I would not grant leave to appeal to the High Court. Accordingly this appeal should be allowed.

Agreeing, **NICHOLLS LJ** said: Mr Reynolds has directed some powerful and, I think, cogent criticisms at part of the reasoning of the arbitrator. Despite this, he has not, in my view, made out a strong *prima facie* case that overall the arbitrator reached the wrong conclusion on the wording of this particular lease, which includes a provision that the items in question were to be treated as tenant's fixtures and fittings even though they were not removable unless replaced and even though they were to be become the property of the landlord at the end of the lease.

Accordingly, I, too, would allow this appeal. The judge directed himself correctly in accordance with the law as it then stood, but since then the law has moved on. Applying the test enunciated by this court in the *Ipswich* case, I also would refuse leave to appeal to the High Court under section 1(3) of the Arbitration Act 1979.

Also agreeing, **NEILL LJ** said: I agree with the opinion expressed by Nicholls LJ that cogent criticisms can be made of the reasoning of the arbitrator. Nevertheless I am not satisfied that a strong *prima facie* case has been made out that the arbitrator was wrong in law.

Accordingly, I, too, would agree that this appeal should be allowed for the reasons given by Staughton LJ.

The appeal was allowed with costs.

Court of Appeal

November 1 1990

(Before Lord Justice SLADE, Lord Justice NICHOLLS and Lord Justice FARQUHARSON)

HENNIKER-MAJOR AND OTHERS v DANIEL SMITH (A FIRM) AND OTHERS

Estates Gazette March 30 1991

[1991] 12 EG 58

Landlord and tenant — Rent review clause — Construction — Preliminary question — Trigger notice not served by landlords within time-limit laid down in clause — Time not of the essence but question as to date from which increase of rent specified in the belated notice became payable — Whether payable from the date from which it would have been payable if the landlords' notice had been served in time or whether payable from a year later — Position of surveyors who served the late notice and solicitors who served a second notice — Plaintiff landlords, who considered that they had lost a year's increase, commenced an action against surveyors claiming damages for alleged negligence — Surveyors contended that on the true construction of the lease there had, in fact, been no loss of a year's increase and plaintiffs complained that, if this construction were correct, the solicitors had been at fault in

too readily accepting the contrary view — A trial of the question of construction was ordered and Potts J decided against the surveyors — They brought the present appeal, which was allowed by the Court of Appeal

The plaintiff trustees owned a shop in Kilburn, London NW2, which had been let for a term of 35 years from March 25 1975 at an original rent of £7,500 a year — The rent was reviewable at five-year intervals — The review clause provided for a trigger notice by the landlords at any time before the expiration of the fifth, 10th, 15th, 20th, 25th and 30th years of the term — There were provisions for the tenant to serve notice of acceptance of the rent specified in the landlords' notice, for the parties to try to reach agreement if the tenant did not accept the specified increase, and for arbitration failing agreement — What happened was that the surveyors instructed by the plaintiffs failed to serve a notice by March 25 1980 and when they did serve one later it was defective because it failed to specify the amount of the increased rent — The plaintiffs then instructed solicitors, who served a proper notice in December 1980 — The surveyors, however, contended, in an action brought against them for negligence, that there had been no loss of a year's increase of rent because the notice served by the solicitors gave rise to an increase retrospectively payable from March 25 1980 — This prompted the plaintiffs to add the solicitors as defendants, complaining that they had too readily accepted the tenant's contention that a year's increase of rent had been lost — Clearly, there was a need for a preliminary determination of the true construction of the lease — This was carried out by Potts J, who held that the surveyors' construction was incorrect — The surveyors appealed by leave on this preliminary point

Clause 3 of the lease on the face of it provided for the lessors to serve a rent notice during the last six months of each succeeding year of the term and made no provision for a late notice — However, as a matter of law, in accordance with the principles established by *United Scientific Holdings Ltd* v *Burnley Borough Council*, time was not of the essence in regard to the service of a rent notice by the landlords in the present case — The result was that the belated notice served by the solicitors was effective to trigger an increase of rent — The question then arose as to the date from which the specified increase began to operate — Was it from March 25 1980 or March 25 1981? — The solution, as Slade LJ explained, was to read clause 3 and the lease as a whole and then consider what the parties might reasonably be taken to have intended to happen in the circumstances which actually occurred — The gap had to be filled by necessary implication — The underlying purpose of the rent review clause was to update rent every five years — As the time-table set out in clause 3 was not of the essence, it could not have been intended that the landlords would lose their right to have the rent increased merely because they had served their notice after, rather than before, March 25 1980 — The intention of the parties must have been that they would obtain, although belatedly, the clause 3 increase — It was true that they would be to some extent out of pocket by being paid later than they otherwise would have been, but after all they had been tardy in serving the notice

There was one more point — As at what date should the amount of the rent be assessed? — It would not be right for the landlords to enjoy the bonus of a larger increase, in a period of rising rents, by having served their notice out of time — The rent should therefore be calculated as at the latest date on which the landlords could have served a notice in strict accordance with clause 3, namely, March 24 1980 (or, *per* Nicholls LJ, March 24 or 25)

The appeal by the surveyors on the preliminary issue was therefore allowed and a declaration made that any increase of rent payable in accordance with a rent notice served between March 25 1980 and March 25 1981 was payable from the earlier date

The following cases are referred to in this report.

Glofield Properties Ltd v *Morley (No 2)* [1989] 2 EGLR 118; [1989] 32 EG 49
Liverpool City Council v *Irwin* [1977] AC 239; [1976] 2 WLR 562; [1976] 2 All ER 39, HL
United Scientific Holdings Ltd v *Burnley Borough Council* [1978] AC 904; [1977] 2 WLR 806; [1977] 2 All ER 62; (1977) 33 P&CR 220; [1977] EGD 195; 243 EG 43 & 127, HL
Weller v *Akehurst* [1981] 3 All ER 411; (1980) 42 P&CR 320; 257 EG 1259

This was an appeal by surveyors, Daniel Smith (a firm), from the decision of Potts J (reported at [1990] 1 EGLR 180; [1990] 20 EG 74) on the preliminary question of the true construction of a lease of shop premises at 108 High Road, Kilburn, London NW2. The action in which this preliminary issue was ordered to be tried was between the plaintiff trustees who were the freehold owners of the shop and the defendant surveyors, Daniel Smith, and the defendant solicitors, the partners in the firm of AF & RW Tweedie in 1980. Potts J decided against the surveyors on the construction of the lease and they brought the present appeal by leave of the judge.

Paul Collins (instructed by Brice Droogleever & Co) appeared on behalf of the appellants; Kenneth Farrow (instructed by Sharpe Pritchard) represented the respondent trustees; James Thom (instructed by Reynolds Porter Chamberlain) represented the respondent solicitors.

Giving the first judgment at the invitation of Slade LJ, NICHOLLS LJ said: This appeal concerns a rent review clause. It raises the question of how the provisions of such a clause are to be interpreted and applied when, in respect of dates specified in the clause, time is not of the essence and a rent review notice was served late and the clause made no express provisions for that eventuality.

The three plaintiffs are trustees. They own the freehold of shop premises at 108 High Road, Kilburn, London NW2. On March 19 1975 they granted a lease of the premises to Swears & Wells Ltd for 35 years from March 25 1975. The rent was reviewable at five-yearly intervals at the instigation of the lessors. The first defendants ("the surveyors") are a firm of chartered surveyors, known in 1980 as Daniel Smith, Briant & Done. In 1978 the plaintiffs instructed the surveyors to attend to the forthcoming rent review. Unfortunately, the surveyors failed to serve a rent notice under the rent review provisions by Lady Day 1980. They did give a notice on June 5 1980, but that was defective because it failed to specify the amount of the increased rent. The plaintiffs turned to their solicitors for assistance. They instructed the second to ninth defendants ("the solicitors") who, in 1980, practised in partnership under the name of AF & RW Tweedie. On December 15 1980 the solicitors served on behalf of the plaintiffs a second, and effectual, rent review notice stating that as from March 25 1980 the annual rent payable under the lease would be £30,000. In due course the increased rent was fixed at £22,500 pa. However, the tenants took the point that, having regard to the wording of the lease, the revised rent payable pursuant to that belated notice was payable only from the end of the then current year of the term, namely Lady Day 1981, which was the sixth anniversary of the commencement of the term. The plaintiffs, no doubt reluctantly, accepted that contention.

That meant that, by reason of the late service of the notice, the plaintiffs lost the amount of one year's rent increase, plus interest. So they commenced this action against the surveyors, claiming damages for professional negligence. The writ was not issued until February 7 1986, when the six-year limitation period had nearly run its course. Thus, events are now somewhat stale.

In the action the surveyors denied negligence. They took the point that even though the effective rent notice was not served until December 1980 nevertheless, on a proper interpretation of the lease, the rent increase to which it gave rise was payable retrospectively from the end of the fifth year. Therefore, so the surveyors contended, the plaintiffs had not lost one year's increased rent at all; the plaintiffs should not have accepted the tenant's contention on this point.

This prompted the plaintiffs to add the solicitors as additional defendants, claiming that if the surveyors were correct in their interpretation of the lease, then the solicitors were professionally

A negligent in accepting the tenant's contention "without demur", as the plaintiffs put it in their pleading. For their part also, the solicitors have denied any negligence.

On July 26 1989 Master Creightmore directed the trial of a preliminary question on the true construction of the lease. The question was whether any increase in rent payable in consequence of a rent notice served in the sixth year of the term — that is to say, between March 25 1980 and March 25 1981 — was payable as from the end of the fifth year, namely March 25 1980, or as from the end of the sixth year, namely March 25 1981. On October 17 1989, Potts J held in favour of the latter date. The surveyors have now appealed to this court by leave of the judge.

B I turn to the lease. The reddendum provides for the payment of an annual rent of £7,500 for the first five years of the term and that

> for the remainder of the said term such rent as shall be ascertained pursuant to the provisions of clause 3 hereof.

The rent was payable quarterly in advance on the usual quarter days.

Omitting immaterial words, clause 3 reads:

> At any time during the six months before the expiration of the fifth, tenth, fifteenth, twentieth, twenty-fifth and thirtieth year of the term hereby created the lessors may serve on the lessee a notice in writing (hereinafter called "the rent notice") providing for the increase of the rent payable hereunder as from the expiration of the year of the term then current to an amount specified in the rent notice and thereupon the following provisions shall have effect:
>
> (1) the lessee within one month after receipt of the rent notice may serve on the C lessors a written acceptance of the amount specified in the rent notice
>
> (2) If the lessee shall not serve such written acceptance as aforesaid, then the parties hereto shall forthwith consult together and use their best endeavours to reach agreement as to the amount of the increased rent to be paid hereunder as from the expiration of the said year but failing agreement within two months after service of the rent notice (or within such extended period as the parties hereto shall mutually agree) the question of whether any, and if so what, increase ought to be made in the rent payable hereunder as from the expiration of the said year shall be referred to the arbitration of a single arbitrator . . .
>
> (3) The arbitrator shall determine the question so referred to him by ascertaining the rent at which the demised premises might reasonably be expected to be let in the open market as between a willing lessor and willing lessee as at the date of the rent notice . . .
>
> D (4) The lessee hereby covenants with the lessors that if the rent shall be increased by agreement or arbitration in pursuance of this clause, the lessee will, as from the expiration of the relevant year pay the increased rent at the times and in the manner aforesaid and the rent as so increased shall remain payable until the expiration of the term hereby granted or until further increased under the terms of this lease, as the case may be.

Read on its own terms, clause 3 gives rise to no difficulty. The lessors may serve a rent notice during the last six months of each succeeding fifth year of the term. No other notice is contemplated. That notice must specify the amount of the rent payable "as from the expiration of the year of the term then current". If such a notice is served, the lessee can accept the amount specified within a month. If E he does not do so, the parties are to consult and seek to agree on, once more, and as one would expect, "the amount of the increased rent to be paid hereunder as from the expiration of the said year". If the parties have not reached agreement within two months after service of the notice, the question of what increase in rent shall be payable "as from the expiration of the said year" is to be referred to arbitration. When the rent increase has been fixed by agreement or by arbitration, the lessee is to pay the increased rent "as from the expiration of the relevant year". In each of those four instances the phrase beginning with the words "as from" is referred to the same point in time, namely the end of the year of the term within the last six months of which, as envisaged by the opening words of clause 3, the rent notice is to be given. For his part, the arbitrator is to determine F the open market rental "as at the date of the rent notice". This also fits in reasonably neatly, because clause 3 envisages that the notice will be served not more than six months before the date from which the rent increase will take effect.

Thus far there is no problem. As already noted, the lease envisages a rent notice being given only within the prescribed six-month period. On that footing the pieces fit together comfortably. But it is important to note, first, that the premise on which clause 3, spelling out the consequences of a rent notice, is drafted is service of the notice within the six months preceding the end of successive fifth years of the term and, second, that the end consequence of such a notice is a rent increase from the successive fifth anniversaries of the start of the term.

The difficulty in the present case is that, as a matter of law, and G conformably with the principles enunciated in *United Scientific Holdings Ltd* v *Burnley Borough Council* [1978] AC 904, time is not of the essence of the provisions in clause 3 regarding service of a rent notice. That is common ground between the parties. A rent notice served late, by which I mean a rent notice served after the expiry of the six-month period specified in clause 3, is still effectual to trigger a rent increase pursuant to clause 3. But, and this is of crucial importance in this case, clause 3 does not cater at all for what is to happen in that event. The draftsman has made no express provision for that occurrence. As a matter of language, the lease provides for a rent notice being given only on such a date as will mean that it is given in the latter half of one of the successive fifth years of the term, and it provides for the increased rent to run from the end of those years, H namely the successive fifth anniversaries of the beginning of the term. It is to be noted that the rent increase will run from those anniversary dates even though the agreement or arbitration process, in respect of a notice served before the anniversary, may not yield finality on the amount of the increased rent until after that anniversary date has passed.

How, then, is the clause to operate when a notice is served late? Since the lease does not cater for what is to happen when a notice is served late, it cannot be right to attempt to find the answer to that question solely by a literal application to a notice served late of provisions which were geared to a notice served in time. That would be to apply provisions which were intended to operate only in one J circumstance to another, materially different, circumstance, to which they were not directed at all. That would be to misuse the language in clause 3.

In this situation, the only sensible way in which to proceed is to read clause 3 and the lease as a whole and consider what the parties must reasonably be taken to have intended to happen if the event to which their review machinery is geared did not happen, namely a notice served in time, but instead the machinery is triggered by another event with which they did not expressly deal, namely a notice served late. In this case that necessitates some adaptation of the express language of clause 3. This is not to propound a novel or heterodox approach. It is simply the necessary consequence of having to apply the clause 3 provisions in a circumstance which the K parties did not have in mind and with which those provisions do not deal expressly. Clause 3 applies to a notice served late only by reason of giving to the clause an effect different from that which a literal construction of the language would produce. That being so, consequential adaptation of the strict language of the clause must follow, so far as necessary, to enable such a notice still to achieve the substance of the underlying purpose of the clause.

As I see it, therefore, the exercise is not simply one of construing the express language of clause 3. Rather, the exercise is one of implication, for the court is having to apply the clause 3 code in a circumstance for which it made no provision. There is a gap which it is *necessary* to fill. In filling that gap, the court is concerned to seek to L apply the provisions in clause 3 as nearly as can reasonably and fairly be done in the circumstances. The court is searching for what must be implied, the parties not having stated what was to happen if the notice was served late: see in this regard the familiar observations of Lord Wilberforce in *Liverpool City Council* v *Irwin* [1977] AC 239 at p 254. In parenthesis, I observe that the existence of such a gap in the contractual provisions, in the event of a notice being served late, may afford ground for supposing or contending that time was intended to be of the essence. If that were right, and time were of the essence, there would be no gap to fill. But once it is held that on the proper construction of the lease time is not of the essence, then the court has no alternative but to fill the drafting gap which is thus revealed.

If one applies this approach in the instant case, there is no room for M doubt. Had the rent notice been served in accordance with the clause 3 timetable, the rent increase would have been payable from Lady Day 1980. The underlying purpose of the clause was to update the rent every five years. Admittedly the timetable specified in clause 3 was not of the essence. Writ large, that means that admittedly the parties did not intend, or are not to be taken to have intended, that the lessors would lose their right to have the rent increased under clause 3 from Lady Day 1980 merely because they served their rent notice after, rather than before, that date. The parties must be taken to have intended that if the lessors were late with their notice, they would still get, albeit belatedly, the clause 3 increase. They would be out of pocket to the extent that, through their own fault, or through the

A
fault of their agents, they would be paid later than they otherwise would have been. That would be a loss which they would suffer through being late with their notice.

In my view, the arguments for the solicitors and the plaintiffs addressed to us fail to carry through and give full effect to their acceptance that time was not of the essence of clause 3. The arguments accept that the notice served in December 1980 triggered a rent increase under clause 3, but the arguments stop short of giving to that rent increase substantially the same consequences as those which clause 3 alone contemplates, namely that the increase will take effect from the relevant fifth anniversary date. In my view, the parties are reasonably to be taken to have intended that in the case of a notice
B
served after the relevant Lady Day, namely outside the clause 3 timetable, the ensuing rent increase under clause 3 would take effect from that Lady Day just as much as would a rent increase following upon a notice served in the six months leading up to the relevant Lady Day.

That is not quite the end of the matter. Clause 3(3) provides that failing agreement an arbitrator is to determine:

> the rent at which the demised premises might reasonably be expected to be let in the open market as between a willing lessor and a willing lessee as at the date of the rent notice . . .

Those words, "as at the date of the rent notice", are clear and unambiguous. It is no part of the function of the court to distort their plain meaning. The function of the court in this case, as I have
C
already sought to explain, is to apply the provisions of clause 3, which were geared to a notice served in time, to a different circumstance at which the clause was not directed. The parties cannot sensibly be presumed to have intended that, with market rents for the most part rising rapidly, a tenant should have to pay a higher rent for the ensuing five-year period if the lessors served their notice months out of time than it would pay if the rent notice had been served in time. The lessors are not to be better off than they would be if they had served their notice in accordance with the clause 3 timetable. In my view, the answer here lies in the arbitrator's determining the rent as at the latest date on which the lessors could have served a notice in accordance with the clause 3 timetable; that is to say, on March 24 or
D
25 1980. Such an implication is not inconsistent with the express provisions of subclause (3), because the provision in the subclause is concerned only with the assessment of market rent as at the date of a rent notice which has been served in time.

Furthermore, this application of that assessment provision in this case accords with the general observations of Nourse LJ in *Glofield Properties Ltd* v *Morley (No 2)* [1989] 32 EG 49 at p 50*:

> It is contrary to the whole purpose of a rent review provision that the reviewed rent should be fixed by reference to values prevailing at a date significantly later than the start of the period in which it is payable.

For my part, therefore, for these reasons I would allow this appeal and declare that any increase in rent payable in consequence of a rent notice served between March 25 1980 and March 25 1981 was payable
E
from the former date.

I mention one last point. It has not been suggested that this is a case in which the tenant was prejudiced by late service of the notice. That was a point which was touched upon in the House of Lords in the *United Scientific* case, especially in the speech of Lord Salmon. The matter might have stood very differently had that been the case, but, as I have said, that is not so here.

FARQUHARSON LJ agreed and did not add anything.

Also agreeing, **SLADE LJ** said: I will add something of my own since we are differing from the learned judge.

The House of Lords decision in *United Scientific Holdings Ltd* v
F
Burnley Borough Council [1978] AC 904 has given rise to a presumption that strict adherence to the timetable specified in rent review clauses is not of the essence of the contract between landlord and tenant. It was common ground in the court below, as in this court, that the lease now under consideration contains nothing sufficient to rebut this presumption.

Nevertheless, this case does well illustrate the frequent difficulties of applying the presumption in a case in which the draftsman of a lease plainly contemplated that the parties would in practice adhere to the timetable provided for by him as part of the machinery for ascertaining the reviewed rent and did not stipulate what was to happen if they did not. It is, of course, the intentions of the parties to G
the lease rather than the intentions of the draftsman which matter. Nevertheless, the intentions of the parties can be ascertained only from the words that they use through him.

The draftsman of this lease, in drafting clause 3, plainly did not direct his mind to what was to happen if the lessors for any reason failed to serve a rent notice on the lessees during the six months before the expiration of the fifth year of the term. We have none the less to presume that the parties did not intend that strict adherence to the timetable set out in clause 3 would be obligatory. Applying this presumption, we have to do our best to collect from the words used in the lease what they intended, or would have intended, to happen in the event of the lessors' serving the rent notice after the expiration of that six-month period — that is to say, during the course of the next H
five years.

Conducting this somewhat artificial exercise and, I hope, approaching the matter not too simplistically, I am reasonably confident of three points, which seem to me to highlight the correctness of the construction placed on the lease by Nicholls LJ.

First, I think that the parties clearly intended that the original rent of £7,500 reserved by clause 1 should be payable only for the first five years of the term. Second, they intended that thereafter — that is to say as from the expiration of the fifth year of the term — there should be payable a rent ascertained according to the formula provided for by clause 3. As to these two points I refer to the wording of clause 1 of the lease, which clearly and specifically provides that there shall be J
paid "for the first five years the clear rent of £7,500 per annum and for the remainder of the said term such rent as *shall* be ascertained pursuant to the provisions of clause 3 hereof" (the emphasis is mine). There is, I think, no such gap as there was in the case of *Weller* v *Akehurst* [1981] 3 All ER 411, which compels a construction that the old rent should continue to be payable after the expiration of the first five years of the term, in the event of a rent notice not being served during that period. Nor is there in principle any reason why, in a situation such as this, the new rent when ascertained should not be payable retrospectively: see the *United Scientific* case at p 935, *per* Lord Diplock.

On behalf of the respondents much emphasis was placed on the phrase at the beginning of clause 3 which states that a rent notice K
should provide for the increase of the rent payable thereunder "as from the expiration of the year of the term then current". It was suggested that this phrase could only mean "as from the expiration of the year of the term in which the notice was served". I think that this point is not well founded. In its context, in my judgment, the phrase merely bears the sense "as from the expiration of the fifth, tenth, fifteenth, twentieth, twenty-fifth or thirtieth year of the term as the case may be". It was merely a convenient means of expressing this concept with brevity. As I have already said, the draftsman was not contemplating the situation which would arise if a notice was served out of time.

The third point which seems to me to be clear on a reading of this L
lease as a whole is that the parties intended, or would have intended, that in the event of the late service of a rent notice after the first five years of the term, any increased rent, ascertained accordingly to the formula provided for by clause 3, would be ascertained as at the last date on which the notice could properly have been served; that is to say, as at March 24 1980. True it is that subclause (3) specified that the increased rent should be ascertained as at the date of the rent notice. However, the clause as a whole contemplated that the lessors should serve the notice before, rather than after, the expiration of the fifth year of the term. In this sense, therefore, the clause, as Nicholls LJ has pointed out, leaves a drafting gap. In my judgment, the proper inference must be that in the events which actually happened the parties would have intended that the rent should be ascertained as if M
the lessors had complied with the timetable provided for by the clause and served the notice before March 25 1980. The intervention of equitable principles by virtue of the presumption to which I have referred means that the lessors do not altogether lose their right to obtain an increased rent for the second five years of the term merely because they have been late in serving their notice. It cannot, however, mean that they are entitled in any way to benefit from their tardiness.

In concluding that the construction of the lease which is placed upon it by Nicholls LJ is the right one, I am supported by the consideration that it does no violence to the language of the lease. It does no more than fill some gaps which have to be filled and it

*Editor's note: Also reported at [1989] 2 EGLR 118 at p 119M.

A produces a result which seems to me by no means unfair either to lessors or to lessees. The lessors, despite their tardiness in initiating the rent review procedure, will have largely preserved their rights under clause 3. However, the delay in the determination of the new rent will have worked to the economic benefit of the lessees, since until the higher rent has been determined they will have had the use of the money which they would have to pay by way of rent if the lessors had acted more promptly: see and compare the *United Scientific* case at p 935 E-F *per* Lord Diplock.

For these reasons, and for the other reasons given by Nicholls LJ, I too would allow this appeal and concur in the form of order which he has proposed.

B *The appeal was allowed; plaintiffs and second to ninth defendants to pay costs of first defendants in Court of Appeal and below; counsel to prepare and agree draft order to be handed in. An application for leave to appeal to House of Lords was refused.*

Court of Appeal

October 8 1990

(Before Lord Justice DILLON, Lord Justice RALPH GIBSON and Lord Justice LEGGATT)

C F L C MANAGEMENT LTD v GANTON HOUSE INVESTMENTS LTD

Estates Gazette April 13 1991

[1991] 14 EG 107

Landlord and tenant — Rent review clause in underlease — Construction — Appeal from decision of Mr Michael Wheeler QC, sitting as a deputy judge of the Chancery Division — Whether reference in underlease to the "said term" was a reference to a term of the underlease or of the headlease — If D the latter, a trigger notice served by the underlessors was out of time, time being of the essence in this case — In support of the view that the notice was out of time the underlessees submitted that it was the intention of the parties that the provisions of the underlease should follow those of the headlease, so that the reviews of the rent should take place in the same years — The Court of Appeal agreed with the decision of Mr Michael Wheeler QC that there was no justification for this construction

The appellants were underlessees holding a term of 21 years from March 1 1974 for the residue of the term of 21 years (less E the last three days) from August 31 1972, ie until August 28 1993 — The respondents held the headlease of the premises for a term of 21 years from August 31 1972 — The underlease did not extend to the whole of the premises comprised in the headlease — The underlease contained provisions for rent review at the expiration of the 7th and 14th years of "the said term" — The review notice had to be given before the beginning of a clear period of two quarters commencing on one of the usual quarter days immediately preceding the relevant review date — The underlessors served a review notice on May 18 1987 to operate the rent review machinery of the underlease as from March 1 1988 — The appellant F underlessees argued that, on their true construction, the words "the said term" in the underlease were intended to refer to the headlease, the object being to follow the review provisions in the headlease — On that construction the underlessors' review notice was out of time and, because time was of the essence, invalid — Mr Michael Wheeler QC, sitting as a deputy High Court judge, rejected this argument — The wording of the underlease did not support it — The "said term" in the underlease plainly referred to the underlease itself, not the headlease — It would have been possible to draft the underlease so as to dovetail in the way suggested with the headlease, but this had not been done.

The Court of Appeal went through the relevant clauses of the underlease in detail and reached the same conclusion — The review notice had been given by the underlessors in time and was effective — Appeal dismissed G

No cases are referred to in this report.

This was an appeal by Ganton House Investments Ltd from the decision of Mr Michael Wheeler QC granting F L C Management Ltd (formerly Finger Lickin' Chicken Ltd) a declaration that a trigger notice served by them on the defendants (present respondents) to initiate a rent review of premises at 91/93 Gloucester Road, Bristol, was valid and effective.

Miss Claire Andrews (instructed by Millikin & Co) appeared on behalf of the appellants; Martin Farber (instructed by Jeffrey Green Russell) represented the respondents. H

Giving judgment, **DILLON LJ** said: This is an appeal by the defendant, Ganton House Investments Ltd, against a decision of June 28 1989 given by Mr Michael Wheeler QC, sitting as a deputy High Court judge in the Chancery Division. The question is as to the meaning of the words "the said term" in an underlease and, in particular, in the rent review provisions of the underlease. On this depends whether a review notice given by the plaintiff landlords on May 18 1987 was valid.

The review notice has to be given before the beginning of a clear period of two quarters commencing on one of the usual quarter days immediately preceding the relevant review date. Time is of the essence, so that if the notice is given too late the landlord loses his review. The review date is the expiration of the seventh or 14th year of the said term as the context requires. J

On the facts we are concerned with the expiration of the 14th year. To find the date of expiration of the seventh or 14th year you have to know when the said term began. Was it August 31 1972, the date of the headlease and of the commencement of the term under the headlease granted by the head landlord to the mesne landlord (the landlord by the underlease, the plaintiffs in these proceedings), which is what the defendants contend for, or was it, as the judge found and the plaintiffs contend, March 1 1974, the commencement of the term of the underlease between these parties which was an underlease of a part only of the premises comprised in the headlease? That is a question which fortunately has not required, as most cases on rent review clauses apparently are thought to by the Bar, a citation of endless other cases on general principles of rent review clauses or decisions of other judges on particular clauses. K

It is sufficient, in my view, to go to this particular underlease, which is dated May 1 1974. It is made between the plaintiff — or at least it is the plaintiff by a previous name; there has been a change of name — as lessor, and the definition of "the lessor" is to include "the reversioner for the time being immediately expectant on the term hereby created of the one part", and the defendants, who are called "the Tenant" of the other part. The term "the Tenant" is to include "successors in title and assigns of the Tenant". There is then a demise with exceptions and reservations, which do not matter. The habendum is as follows: L

To hold the demised premises unto the Tenant from the First day of March [1974] for the residue of the term of twenty one years (less the last three days thereof) from the thirty first day of August [1972] that is to say until the twenty-eighth day of August [1993] subject nevertheless to the proviso for determination hereinafter contained Paying Therefor during the said term yearly and proportionately for any fraction of a year the rent hereunder set out:

(a) For the first seven years of the said term the yearly rent of One thousand four hundred pounds: M

(b) For the next seven years of the said term either the rent reserved in Sub-Clause (a) hereof or the open market rental value of the demised premises at the review date whichever is the higher and in any case the same shall remain constant during the whole period referred to in this sub-clause:

(c) For the remainder of the said term either the yearly rent or the market rental value of the demised premises at the review date whichever is the higher and in any case the same remain constant during the whole period referred to in this sub-clause:

There is then a provision that —

the said rents shall in all cases be paid by equal quarterly payments in advance on the usual quarter days in every year without any deduction whatsoever except as authorised notwithstanding any agreement between the parties by any statutory enactment relating to income tax for the time being in force the

A first payment apportioned in respect of the period from the first day of March [1974] to the quarter day next hereafter to be paid on the execution hereof:

That, of course, was necessary on any view, because the term granted or, as it was put in the definition of "the Lessor", "the term hereby created" was to run from a date prior to the date of the underlease and as the rent was to be paid by quarterly payments in advance on the usual quarter days there had to be a payment for the broken period on the execution of the lease.

There is then the definition of "open market rental" and of "review date" — I need not go through those — and a provision for the giving by the landlord of a review notice, which I have already summarised.

B There were provisions in the review mechanism for reference to an independent surveyor. Those have not so far been implemented because of the defendants' contention that the review notice given by reference to the commencement of the term of the underlease was out of time because it should have been given by reference to the commencement of the term of the headlease which, as I have mentioned, started from a considerably earlier date. For that reason there has been little point in the independent surveyor proceeding to calculate what the rent ought to be for this period until it has been established that the review notice was valid.

After the review provisions there is a provision for an insurance rent to be paid from time to time throughout the said term. There are then in clause 2 "Tenants covenants". The opening words are as C follows:

The Tenant hereby covenants with the Lessor to the intent that the obligations may continue throughout the said term as follows:

(1) To pay during the said term the said reserved rents at the times and in manner aforesaid without any deduction except as aforesaid:

In (3) the covenant is:

To pay to the suppliers thereof all charges for gas and electricity (including meter rents if any) consumed in the demised premises during the said term:

In (5) the covenant is:

In the fourth ninth and fourteenth years of the said term and also in the last year thereof (howsoever determined) to paint in a proper and workmanlike manner all the inside wood iron and other parts heretofore or usually painted D of the demised premises [with all the details that follow in a lengthy painting clause.]

In (7) there is a covenant:

At all times during the said term to do and execute or cause to be done and executed all such works and to do all such things as under or by virtue of any Act or Acts of Parliament now or hereafter to be passed and byelaws rules and regulations thereunder are or shall be directed or necessary to be done or executed upon or in respect of the Tenant's user thereof by the owner lessee Tenant or occupier thereof . . .

There are other references. For instance, in clause (18)(a) there is a covenant:

Not to assign demise underlet or otherwise part with possession of the whole E of the demised premises for all or any part of the said term without the licence in writing of the Lessor and the Superior Lessor which shall not be unreasonably withheld . . .

One can then pass on to subclause (20), which is a covenant:

At all times during the said term to comply in all respects with the provisions and requirements of the Town and Country Planning Act 1962 and all regulations or orders made thereunder . . .

After clause 2 there is clause 3 containing the lessor's covenants and these include the usual covenant for quiet enjoyment:

That the Tenant paying the rent hereby reserved and observing and performing the several covenants and stipulations herein on the part of the Tenant contained shall peaceably hold and enjoy the demised premises throughout the said term without any interruption by the Lessor or any person F rightfully claiming under or in trust for the Lessor:

There is then the proviso for re-entry — if and whenever during the said term the rents fall into arrear or there is breach or non-observance of the covenants, or the tenant becomes bankrupt or goes into liquidation, and so forth. There are other references to using the words "the said term".

It is argued in particular by Miss Andrews, for the appellants (defendants), that the parties both knew that this was an underlease. They knew that the headlease contained review provisions to operate at the expiration of the 7th and 14th years of the term granted by the headlease, that is to say the term from August 31 1972, and so they must have intended that the provisions of the underlease should follow those of the headlease and the review should take place in the same years as the review of the rent in the headlease. The calculation, G of course, is not the same because the premises, the rent for which is to be reviewed, are only a part of the premises in the headlease. It could be that the underlease could have been drawn to achieve that result, but it seems to me quite plain that the "said term", as copiously used throughout this document and in particular in clauses 2 and 3, must mean the term "created by the underlease" and not the term created by the headlease.

Miss Andrews referred to a decision of the Court of Session on the construction of a Scottish trust-disposition and settlement* (that is to say a will) where the question was whether the said John Shepherd in a particular disposition meant John Shepherd the brother of the testatrix who had been appointed an executor or John Shepherd the nephew of the testatrix, the son of another brother, who was a H legatee. The court, holding that it was the nephew, referred to the aid to construction that a word is to be referred back to the proximate antecedent; so, says Miss Andrews, the proximate antecedent to the use of the words "pay during the said term" in the reddendum is a reference to the residue of the term of 21 years from August 31 1972. But I think the crux of the matter is that this was a demise from March 1 1974, and the important term for all these references is the term created by the underlease and granted thereby to the tenant during which all the obligations on each side were to apply. Therefore I have no hesitation in reaching the conclusion which the judge reached that the review notice given by the landlords was given in time and was valid and effective. Accordingly, I would dismiss this appeal. J

Ralph Gibson and **Leggatt LJJ** agreed and did not add anything.

The appeal was dismissed with costs.

*Editor's note: *Shepherd's Trustees* v *Shepherd* [1974] SC 60.

Chancery Division

November 15 1990 K

(Before Mr Justice VINELOTT)

FIRST LEISURE TRADING LTD v DORITA PROPERTIES LTD

Estates Gazette June 15 1991

[1991] 23 EG 116

Landlord and tenant — Rent review provisions in lease — Construction and effect of review provisions — Difficulties as to the assumptions which should be made by the arbitrator in determining the rent under review — Questions submitted for L preliminary determination by the court, by agreement of the parties and consent of the arbitrator, under section 2(1) of the Arbitration Act 1979 — Importance both of distinguishing between the hypothetical tenant and the actual lessee and at the same time ensuring that the hypothesis bears as close a resemblance to reality as possible — Complications in present case owing to the manner in which interests in adjoining premises had become interlinked

The questions brought before the court for preliminary determination in this case arose from the situation and association of two adjoining properties, the Greyhound Hotel, Croydon, and shop premises on the ground floor, at 46 M and 48 St George's Walk, over which part of the hotel extended — There had been a number of dispositions of interests, but at the material time the plaintiffs, First Leisure Trading Ltd, had come to hold both a long lease of the hotel premises and a long lease of the adjoining properties (referred to in the case as "the extension") — The defendants, Dorita Properties Ltd, held the reversion on the extension — The reversion on the hotel was held by a group not concerned in the present proceedings — The rent of the hotel was fixed, but the rent of the extension, although in the first years fixed but at a rising rate, became subject to review from the expiry of the

A **21st year of the lease (December 25 1989) — The reviewed rent was to be an open market rent excluding, as is common, the matters mentioned in section 34(1)(*a*) to (*d*) of the Landlord and Tenant Act 1954 — The lease contained a provision that the use of the premises was to be only as an extension to the Greyhound Hotel**

The originating summons raised five questions which are set out in full in the judgment — The answers to the first three were agreed by the parties — The remaining questions, 4 and 5, were as follows (question 4 as reformulated by agreement):

4 Can the arbitrator find that the plaintiffs are a possible hypothetical
B **tenant?**

5 Is the arbitrator required to assume that the possible hypothetical tenants bidding in the open market are restricted to those persons who are prepared to take the risk of seeking and obtaining the consent of the owner and occupier of the immediately adjoining premises formerly known as the Greyhound Hotel to use the demised premises as an extension to those said adjoining premises?

After considering the plaintiffs' submission that the answer to question 4 should be "No", and examining in detail the decision at first instance and on appeal in *F R Evans (Leeds) Ltd* v *English Electric Co Ltd*, Vinelott J decided that the answer to question 4 was that strictly the arbitrator was not entitled to assume that First Leisure either was or was not a
C **possible hypothetical tenant — He was, however, entitled to find as a fact that a lessee of the Greyhound Hotel, whoever he might be, would be a potential tenant — It was for the arbitrator to say what effect that fact, if found, would have as between a hypothetical willing lessor and a hypothetical willing lessee negotiating the terms of a hypothetical lease of the extension — It followed that the answer to question 5 must be "No"**

The following cases are referred to in this report.

Cornwall Coast Country Club v *Cardgrange Ltd* [1987] 1 EGLR 146; (1987) 282 EG 1664
D *Evans (FR) (Leeds) Ltd* v *English Electric Co Ltd* (1977) 36 P&CR 185; [1978] EGD 67; 245 EG 657
Inland Revenue Commissioners v *Clay* [1914] 3 KB 466; (1914) 83 LJKB 1425; 111 LT 484; 30 TLR 573
Norwich Union Life Insurance Society v *Trustee Savings Banks Central Board* [1986] 1 EGLR 136; (1986) 278 EG 162
Raja Vyricherla Narayana Gajapatiraju v *Revenue Divisional Officer, Vizagapatam* [1939] AC 302

This was an originating summons taken out by the plaintiffs, First Leisure Trading Ltd, for the purpose of seeking the determination by the court under section 2(1) of the Arbitration Act 1979 of a number of questions as to the assumptions that should be made by the arbitrator in determining the rent under the lease of the shop premises
E at 46 and 48 St George's Walk, Croydon, of which the plaintiffs were the lessees and the defendants, Dorita Properties Ltd, were the lessors.

Christopher Lockhart-Mummery QC and John Male (instructed by Paisner & Co) appeared on behalf of the plaintiffs; Kim Lewison (instructed by Brecher & Co) represented the defendants.

Giving judgment, **VINELOTT J** said: This originating summons raises a number of questions as to the construction and effect of a rent review clause.

The background is shortly as follows. By a lease dated May 16 1962 Croydon Centre Development Ltd demised premises, comprising basement, ground, first and mezzanine floors of a building, part of
F the Croydon Centre, and known as the Greyhound Hotel, to Courage Barclay & Simmonds Ltd for a term of 130 years from March 25 1962 at a fixed rent which is not subject to review. The lease contains covenants by the lessee, among others: (a) not to alter the main structure or the external appearance of the demised premises without obtaining the prior approval of the lessors, such approval not to be unreasonably withheld; and (b) not to permit the demised premises or any of them to be used other than as a high-class public house, or for such other purposes as the lessors shall from time to time in writing agree, such agreement not to be unreasonably withheld. The lease also contains a covenant in usual form not to assign or underlet without the lessors' licence in writing, such licence not to be unreasonably withheld.

The premises demised by the 1962 lease (to which I shall refer as G
"the Greyhound Hotel") extend at first-floor level over shop premises, part of the Croydon Centre development, which comprise the ground floor of 46 and 48 St George's Walk. By a lease dated October 7 1969 and made between the same parties, the lessor demised the two shops to the lessee for a term of $123\frac{1}{4}$ years from December 25 1968. The two terms will thus both expire on March 25 in the same year. The rent is a fixed but rising rent for each of successive periods of seven years and from the expiry of the 21st year (December 25 1989) a rent ascertained in accordance with the provisions contained in clause 2. Under clause 2(1), for the first 14 years from the end of the 21st year and for every successive period of 14 years until the end of the term, the rent is to be the market rent as therein defined or the yearly rent reserved for the immediately H
preceding period, whichever is the greater. The market rent is defined as:

The best rent that would be expected to be obtained in the open market for the demised premises excluding all matters referred to in subclauses (a) to (d) of section 34 of the Landlord and Tenant Act 1954 if let with vacant possession at the end of the said 21 years or the 35 years or of each successive 14 years on a lease for a term of 14 years and on similar terms.

The clause goes on to prescribe the procedure to be followed in ascertaining the market rent. The lessor is to give notice specifying what it considers to be the market rent and if the lessee gives notice in writing objecting to it and if the parties cannot subsequently agree, it is to be determined by arbitration. The clause does not in terms J
require the market rent to be ascertained as between willing lessee and willing lessor, but it is common ground that that requirement is implicit in the requirement that it be the best rent that could be expected to be obtained in the open market.

The 1969 lease provides that:

Except as to the premises demised the term of years granted and the rent reserved this demise is made on the same terms and subject to the same reservations and to the same lessees' covenants and conditions provisos agreements and declarations as are expressed and contained in [the lease of the Greyhound Hotel].

In clause 4 the lessee also covenants, among other things:

To use the demised premises only as an extension to The Greyhound Hotel.

At some time before March 7 1989 the reversion to the Greyhound K
Hotel and the reversion to the premises comprised in the 1969 lease (which I will call "the extension") was severed. The reversion to the Greyhound Hotel became vested in the Nestlé Co Ltd and the reversion to the extension became vested in the defendant, Dorita Properties Ltd ("Dorita"). By a licence dated March 7 1989 and made between Dorita, Courage Barclay & Simmonds Ltd (which had changed its name to Courage Group Ltd) and the plaintiff, First Leisure Trading Ltd ("First Leisure"), which was then known as Entam Leisure Ltd, Dorita gave its licence to the assignment of the lease of the extension by Courage Group Ltd to First Leisure and gave its consent to the carrying out of alterations to the extension by First Leisure in accordance with annexed plans. The alterations included L
the construction of a substantial staircase leading from the centre of the extension to the first floor above. Then by a deed dated March 8 1989 and made between the Nestlé Co, the Courage Group Ltd and First Leisure, the Nestlé Co gave Courage Group licence to assign the benefit of the lease of the Greyhound Hotel to First Leisure. The Nestlé Group also gave consent: (a) to the use of the Greyhound Hotel, as to the basement, as a high-class discotheque incorporating a high-class restaurant and bars and/or beer cellar; as to the ground floor as a high-class café and public bar and reception area; as to the first floor, as a high-class night-club with bars and restaurants and/or for private functions, banquets, dinner dances, conferences and exhibitions and/or offices and staff accommodation ancillary to the business conducted by First Leisure; and as to the second floor as M
offices and staff accommodation ancillary to that business; and (b) for First Leisure to carry out alterations to the Greyhound Hotel. First Leisure also entered into certain further covenants with the Nestlé Group restrictive of the use of the Greyhound Hotel, which it is unnecessary to set out.

The position that has obtained since the alterations were carried out is that the extension, which is not separated laterally from the Greyhound Hotel, is used as a reception area where there is a small bar with a commodious staircase leading from the centre to the night-club and ancillary rooms on the first floor.

The new rent for the extension was not agreed and an arbitrator was appointed. Questions having arisen as to the assumptions that

A should be made by the arbitrator in determining the rent, the parties agreed, with the consent of the arbitrator, that these questions should be submitted for determination by the court pursuant to section 2(1) of the Arbitration Act 1979.

Five questions are asked in the originating notice of motion. The parties are agreed as to the way in which the first three fall to be answered, but for completeness I will set them out in the way in which it is agreed they should be answered.

> Question 1: Is the arbitrator required to assume that the use of the demised premises under the hypothetical lease is restricted for use as a high-class public house or for such other purposes as the lessor shall from time to time in writing agree, such agreement not to be unreasonably withheld?

B It is agreed that this question falls to be answered in the affirmative. The 1969 lease incorporates the covenants in the 1962 lease, including a covenant not to use the extension for any purpose except that of a high-class public house or for such other purpose as the lessor may agree. For this purpose, the lessor is the person for the time being entitled to the reversion to the extension. First Leisure is in fact in breach of this covenant. The licence of March 7 1989 (unlike the deed of March 8 1989) did not include consent by the lessor to a change of use of the extension. However, it is common ground that, as the licence and the deed together with the assignment of the benefit of the two leases to First Leisure were part of a single transaction, Dorita as lessor of the extension could not refuse consent for this change of use.

C

> Question 2: Is the arbitrator required to assume that the demised premises must be used only as an extension to the immediately adjoining premises formerly known as the Greyhound Hotel?

It is agreed that the answer to this question is also in the affirmative. I should perhaps add that the covenant in the 1969 lease is negative in effect; that is, it is a covenant not to use the premises otherwise than as an extension to the Greyhound Hotel and not a positive covenant so to use it. Otherwise there would be a conflict between this covenant and the covenant as to user incorporated by reference from the 1962 lease if the lessor of the Greyhound Hotel permitted a use which the lessor of the extension was not prepared to permit and to which he could reasonably refuse his consent.

D

> Question 3: Is the arbitrator required to disregard the actual or former occupation of the demised premises by the plaintiff?

Dorita accepts that the answer to this question is also in the affirmative, with the qualification that the "disregard" is limited to the actual or former occupation of the extension and not of the Greyhound Hotel itself.

> Question 4: Is the arbitrator required to assume that the plaintiff was a possible hypothetical tenant in the market for the demised premises on the rent review date?
>
> Question 5: Is the arbitrator required to assume that the possible hypothetical tenants bidding in the open market are restricted to those persons who are prepared to take the risk of seeking and obtaining the consent of the owner and occupier of the immediately adjoining premises formerly E known as the Greyhound Hotel to use the demised premises as an extension to those said adjoining premises?

It is agreed between counsel for Dorita and First Leisure that question 4 is inaptly framed. The question should be whether the arbitrator can find that the plaintiff is a possible hypothetical tenant.

Mr Lockhart-Mummery submitted that the answer to question 4 as reformulated is "No". At the same time the arbitrator must have regard to the fact that in the real world First Leisure occupies the Greyhound Hotel under a long lease. It follows that the answer to question 5 is "Yes". The hypothetical new lessee would be one who is willing to take a lease of the extension on the footing that he could (subject to obtaining the lessor's consent) either sublet it to First Leisure or use it for some purpose ancillary to the use of the F Greyhound Hotel — for instance as a flower shop or selling newspapers or magazines or something of that sort. The market would thus be very restricted. That is reflected in the rental value advanced by the expert engaged by First Leisure. He has advised that the fair rent of the extension ascertained in accordance with the rent review clause is only £11,000 pa. The expert engaged by Dorita has advised that the fair rent is £60,000 pa.

The construction contended for by Mr Lockhart-Mummery has what seems at first sight a very surprising consequence. The *prima facie* purpose of a rent review clause is to enable the landlord to recover an increased rent which will reflect changes in the value of money and changes in the value of comparable premises in the neighbourhood. Any increases which stem from the past occupation of the premises by the lessee or from improvements he has made, or G from goodwill associated in his business, are to be disregarded. The purpose of providing that the rent is to be assessed on the footing that the premises are offered with vacant possession and without regard to the matters set out in paras (*a*) to (*d*) of section 34(1) of the Landlord and Tenant Act 1954, is to ensure that the parties so far as practicable are notionally put in the same position as they were when the rent was initially settled. The rent is to be the best rent that would be expected to be obtained in the open market; the effect of so providing is that special features which might make Dorita or First Leisure an eager or a reluctant lessor or lessee are also disregarded. But as Hoffmann J observed in *Norwich Union Life Insurance Society* v *Trustee Savings Banks Central Board* [1986] 1 EGLR 136* at p 137, there is "a presumption that the hypothesis upon which the H rent should be fixed upon a review should bear as close a resemblance to reality as possible". If Mr Lockhart-Mummery is right, a fact which would have been a material factor in ascertaining the initial rent, given that the lease contains a qualified covenant not to use the extension otherwise than as a high-class public house and a covenant to use the extension only as an extension to the Greyhound Hotel, has to be left out of account on reviewing the rent.

However, Mr Lockhart-Mummery submitted that the question is concluded in his favour by the unreported decision of the Court of Appeal, which was given on November 11 1977 upholding the decision of Donaldson J (as he then was) in *F R Evans (Leeds) Ltd* v *English Electric Co Ltd* (1977) 36 P&CR 185*. It is important in J understanding the judgment of Donaldson J and the judgment of Browne LJ, who gave a leading judgment in the Court of Appeal, to have in mind the precise question that fell to be decided and the context in which it arose. The lease in question was a lease of an area of some 60 acres which included factory premises of nearly 1m sq ft of floorspace. The lease contained a rent review clause which provided that the rent should be the rent at which the premises were worth to be let on the open market, as between willing lessor and willing lessee, but disregarding the matters there set out, which are in substance the matters set out in paras (*a*) to (*c*) of section 34(1). The case came before the court on a special case stated by the arbitrator after he had heard evidence and argument. He had found facts which are set out in the special case and of which the following are the relevant findings: K

> (1) There is no other property on the market that provides the accommodation and facilities provided by the subject property.
>
> ...
>
> (4) Apart from the present tenants, it is very unlikely on the available evidence and on the state of the market that there would have been a potential tenant in the market for the subject premises on the terms of the lease offered.
>
> (5) In October 1976, if the present tenants had not been in occupation of the subject premises, they would not have made any bid for the lease being offered.
>
> (6) The present tenants in October 1976 could have split their operation into parts and could have moved each of those parts into smaller alternative premises in the United Kingdom.

The rival contentions are summarised in the judgment of L Donaldson J:

> The tenants submit that the arbitrator should assume: (i) that in the hypothetical market there is at least one hypothetical tenant bidding for the lease, and (if it be found that there would be no other bidder than the English Electric Co Ltd) that that hypothetical tenant is the English Electric Co Ltd, and (ii) that the rent must be determined at an amount that the hypothetical tenant would agree with the claimants in the light of all the bargaining advantages and disadvantages that would have existed at October 1 1976, and in particular on the assumption that, if the parties failed to reach agreement on the rent, the hypothetical tenant was free to occupy two or more smaller properties as an alternative to taking a lease of the subject property.
>
> The landlords submit that the arbitrator should assume: (i) that there is a person (who may or may not be the English Electric Co Ltd) who is willing to take the subject property as a whole on the terms of the lease offered (other M than rent) and to enter into meaningful negotiations to arrive at a fair rent, that is to say, a rent that is fair in all the circumstances of the case including rental values prevailing in the area, and (ii) that the rent must be determined at an amount that the claimants would agree with such person in all the circumstances, including in particular the circumstances (if such be found to exist) (a) that there is no other person willing to take a lease of the subject property on the terms of the lease being offered, and (b) that there is no other property on the market that provides the accommodation and facilities provided by the subject property, but disregarding the possibility that the parties would fail to reach agreement on a rent for the subject property.

Donaldson J, commenting on the arbitrator's findings of fact, said:

*Editor's note: Also reported at (1977) 245 EG 657.

A (a) the fact that there is no other property on the market which provides the accommodation and facilities provided by the Walton Works is certainly relevant. Its effect on the negotiations has, however, to be balanced by two [*sic*] other factors. First, whilst the hypothetical tenant is a willing lessee, he is not an importunate one. He wishes to take a lease of the premises, but he is operating in a commercial field and in deciding what to offer by way of rent will take account, covertly or overtly, of the alternative of taking a lease of two or more premises. That is not to say that he would prefer that solution. That will depend on the level of rent which is under consideration. He is a willing lessee, and is quite content to take the Walton Works at the right price. It is just that he is not considering the proposition or negotiating in a vacuum.

(b) The fact that the property could be subdivided and let in parts or sublet and the advantages and disadvantages of that approach are irrelevant. The clause contemplates a letting as a whole.

B (c) The fact that it is very likely that the English Electric Co Ltd would have been the only potential lessee is relevant, but its relevance is indirect. It does not matter whether the only potential lessee was this company or the XYZ Co Ltd. What matters is that in the state of the market there was not likely to be more than one willing lessee. The effect of this fact is not, however, decisive because this single potential lessee is to be assumed to be a willing lessee — neither reluctant nor importunate, but willing. Just as the hypothetical lessor cannot rely overmuch on the fact that no property similar to the Walton Works is available on the market, so the hypothetical lessee cannot rely too much upon the fact that he has no competitors — he is, and is known to be, a willing lessee. Furthermore, it is known that he will remain a willing lessee so long as the willing lessor does not press his demand for rent beyond the point at which he is ceasing to act as a willing lessor and at which a willing lessee would cease to be such.

C If authority be required for the proposition that monopoly positions on either or both sides do not render hypothetical agreements impossible, it is to be found in *Tomlinson* v *Plymouth Argyle Football Co Ltd* [1960] 6 RRC 173, a rating case, and *Inland Revenue Commissioners* v *Clay* [1914] 3 KB 466, a taxation case.

(d) The fact that if the tenants had not been in occupation in October 1976 they would not have made any bid for the lease being offered is wholly irrelevant. I am quite prepared to accept that the tenants are most unwilling lessees. This does not matter. The arbitrator's concern is with the attitude of the hypothetical *willing* lessee, who is not in occupation of the premises. The clause assumes that there is such a person, and it is nothing to the point to prove that there was not.

(e) The fact that the tenants in October 1976 could have split their operation into parts and could have moved those parts into smaller alternative premises
D in the United Kingdom is also quite irrelevant. It is irrelevant because it confuses the hypothetical lessee with the actual tenants. Certainly the hypothetical lessee would consider this alternative, and it would affect the amount of the rent which he would be prepared to offer, but he would remain a willing lessee. The extent of the influence exerted by this alternative would depend on the extent to which more than one lease and divided premises might be expected to increase his costs. And it is not the English Electric Co Ltd's costs which come into the equation, but average and assumed costs of a hypothetical potential lessee.

Then, commenting on the tenant's contentions, Donaldson J accepted that it was to be assumed that there was at least one hypothetical tenant bidding for the lease, and that:

. . . the arbitrator would be justified in proceeding on the basis that, although
E there was one person who was willing to become lessee of the Walton Works, it was unlikely that there would be more than one.

But he rejected the proposition that:

. . . the potential lessee either is, or necessarily has any of the characteristics of, the English Electric Co Ltd. He is a complete abstraction, and, like the mule, has neither pride of ancestry nor hope of posterity. He is someone whose needs are such that, in relation to the Walton Works, he is a willing lessee.

Turning to the landlords' contentions, he said:

I accept that there is a person who is willing to take the Walton Works as a whole on the terms of the lease offered (other than rent), but I do not accept that that person may be the English Electric Co Ltd. For the purposes of fixing a new rent, he is not.

F Then a little later:

I accept that the rent has to be agreed in the light of all the circumstances which in fact affect the property and, in theory, affect the hypothetical lessor and lessee. Any circumstance which affects the actual landlord and the actual tenant but which would not affect the hypothetical lessor and lessee is irrelevant. I agree that these circumstances include, but stress that they are not limited to, the fact that it is unlikely that there will be more than one willing lessee and that in October 1976, which was the relevant date, there was no other property on the market which provided the accommodation and facilities provided by the Walton Works.

I do not think that the decision of Donaldson J properly considered assists Mr Lockhart-Mummery. The unusual feature of the case before him was that it was unlikely that there would be any bidder
G except the English Electric Co Ltd and that the English Electric Co Ltd would have been a reluctant bidder. English Electric tried to build on those foundations the claim that the market rent should be taken to be the rent that it would have been prepared to pay. The fallacy in that argument is that it assumed (contrary to the required hypothesis) that there would be no willing lessee. It was not suggested on behalf of the landlord or the tenant that the English Electric Co Ltd should be notionally excluded from the market altogether. I have already set out the rival contentions but I will repeat the salient points.

It was contended on the one hand that the arbitrator should have assumed that in the hypothetical market there is at least one hypothetical tenant bidding for the lease and (if it be found that there
H would be no other bidder than the English Electric Co Ltd) that the hypothetical tenant is the English Electric Co Ltd; and on the other that the arbitrator should assume that there is a person (who may or may not be the English Electric Co Ltd) who is willing to take the subject property as a whole.

The decision of Donaldson J, as I understand it, does not rest on any principle of law that the English Electric Co Ltd had to be treated as excluded from the market. English Electric in effect excluded themselves. The arbitrator had found as a fact, on the assumptions required by the terms of the rent review clause, that if English Electric had not been in occupation at the review date they would not have been potential lessees for the whole of the demised premises.

J That this is the correct analysis of the decision of Donaldson J is, I think, confirmed by the judgment of Browne LJ in the Court of Appeal. The main contentions advanced in the tenants' notice of appeal were:

[Ground 2] The learned judge erred in law in holding that for the purposes of clause 4(4) of the said lease and on a proper construction of that clause and having regard to the facts found by the arbitrator the willing lessee there referred to was an abstraction or a hypothetical person as opposed to being the appellants.

[Ground 3] The learned judge erred in law in holding that the fact that if the appellants had not been in occupation in October 1976 they would not have made any bid for the lease being offered was wholly irrelevant to the determination of the reviewed rent.

K In opening the appeal, Mr Glidewell, who appeared for English Electric, described ground 2 as "the crux of the case". Browne LJ observed that ground 3 was of no practical importance because there were no special factors that would have affected the landlords. He then summarised the argument of Mr Glidewell in the following passage:

Mr Glidewell said in opening that the crux of the case was ground 2 of the grounds of appeal and the tenants' contention (a)(i) set out in the second schedule to the award although, as I have said, in reply he said that ground 6 was the crux. Mr Glidewell submits that on the facts as found the hypothetical tenant must be taken to be, or to have the characteristics of, English Electric. If so, the tenants' bargaining position would have been much stronger in view of the findings that if English Electric had not been in occupation they would
L not have made any bid for the tenancy and that they could have split their operations and gone elsewhere into other premises. He says Donaldson J wrongly deprived the tenants of these bargaining weapons. As I have said, I agree with Donaldson J's reasons (which I have already read) for rejecting Mr Glidewell's main submission, and my view is reinforced by Mr Bernstein's respondent landlords' counsel's submissions, which I accept, namely (a) that the fact that the lease contemplates assignment by either party or both parties suggests that the parties did not contemplate that the rent to be fixed on review should be related to the particular circumstances of the landlord or the tenant who happened to be there at the date of review, especially as there was to be no review for 9 years after 1976; (b) that the logical result of Mr Glidewell's argument is that the premises would not have been let at all in October 1976, which is contrary to the assumptions required by clause 4(4) and makes nonsense of the clause; (c) — to some small extent — Mr Bernstein's point
M about the contrast between "*the* lessee" and "*a* lessee" in the various subclauses of clause 4; and (d) clause 4(4) (ii), which I think certainly tends to support the view that the assessment under the rent review clause is concerned with hypothetical people and not with the real tenant.

Then a little later he expanded his reasons by saying:

In my view, as a matter of general principle, special factors affecting the actual occupier cannot be attributed to the hypothetical tenant even if there is only one potential tenant. There must be many factories of which the only likely tenant is the actual occupier, and I cannot accept that in such a case it would be right to attribute to the hypothetical tenant, for example, the financial difficulties or the bad labour relations of the actual occupier. It is not altogether clear whether Donaldson J was in fact attributing to the hypothetical tenant the possibility of splitting his operations or whether he was merely saying that that was something which was inherent in the nature of

A the hereditament and which any hypothetical tenant would have in mind. If he was attributing to the hypothetical tenant the characteristic of English Electric in that respect, I think he may have been too favourable to the tenant; but I am prepared to assume for the purposes of this case that it is right to attribute to the hypothetical tenant that he could split his operations and go elsewhere. I agree with Donaldson J that the finding that if English Electric had not been in occupation in October 1976 they would not have made any bid for the tenancy is wholly irrelevant, because clause 4(4) requires it to be assumed that there was then a willing lessee and of course the tenants' argument produces the absurdity to which I have already referred.

Ground 6 on which Mr Glidewell relied in his reply was that:

> The learned judge erred in law in holding that the possibility of the parties failing to reach agreement was to be disregarded.

B As to that, Browne LJ said:

> As to ground 6 — that is the possibility of not reaching agreement — Mr Glidewell accepts that the arbitrator must assume that agreement on some rent will be reached, but he says that the parties must not be assumed to have known this while they were negotiating and that this would strengthen the bargaining powers of the tenant. It seems to me that this is really the same point as the points about the findings that English Electric would not have bid at all if they had not been in occupation and that they could have moved out.
>
> As I have said, under clause 4(4) one must assume vacant possession. There is, therefore, I think, really no distinction between this point about failure to reach agreement and the point about moving out. It is clearly wrong to talk about the possibility that English Electric or the hypothetical tenant would consider moving out — he is never in, on the assumptions. That is really the
> C same point as saying that English Electric would not have bid at all, and both these points are really contrary to the fundamental assumptions. But the possibility of the hypothetical tenant's breaking off negotiations and going elsewhere was obviously a point which would affect the negotiations and which it was for the arbitrator to weigh and decide how much effect it would have on the rent. It seems to me that on both these points — failure to reach agreement and the possibility of going elsewhere — what Donaldson J was saying was exactly that, that they were relevant factors but that the weight to be attached to them was a matter for the arbitrator.

Mr Lockhart-Mummery placed considerable reliance on the observations of Browne LJ that "as a matter of general principle, special factors affecting the actual occupier cannot be attributed to the hypothetical tenant even if there is only one potential tenant". I
D do not think that this passage will bear the weight he, Mr Lockhart-Mummery, seeks to put on it. In general, special features peculiar to the lessor or the lessee which, for instance, may make either (in Donaldson J's words) "reluctant or importunate" must be disregarded. So the arbitrator must disregard, for instance, any cash-flow problem which would make the lessor anxious to obtain a letting or any desire to expand which would make the lessee anxious to obtain further premises. It does not follow that the feature which makes the extension of greater value to any lessee of the Greyhound Hotel must be disregarded merely because First Leisure is the lessee of the Greyhound Hotel.

Mr Lockhart-Mummery also relied on the decision of Scott J in
E *Cornwall Coast Country Club* v *Cardgrange Ltd* [1987] 1 EGLR 146. The facts of that case are complex and unusual. In 1978 Daejan Investments Ltd ("Daejan") granted a lease for 26 years from December 8 1977 of 30 Curzon Street to Ladbroke Rentals Ltd ("Rentals"). Shortly after, Rentals, a member of the Ladbroke Group, sublet the premises to Ladup Ltd, another member of the Ladbroke Group, for the same term less three days. The lease contained a rent review clause similar to the rent review clause in the instant case except that, first, the "disregards" incorporated in the instant case by reference to section 34(1) are spelled out (subpara (a) required that any effect on rent of the fact that the tenant or his predecessor in title or some associate of the tenant had been in occupation had to be disregarded) and, second, that the clause
F required that there be disregarded:

> (d) Any addition to the value of the demised premises attributable to the gaming or justices licence which may be held in respect of such premises if it appears that having regard to the terms of the current tenancy and any other relevant circumstances the benefit of the licence belongs to the tenant or some associate of the tenant.

Later the headlease was assigned to the plaintiff and the sublease was assigned to the defendant, a member of the Lonrho Group. The premises had then been converted for use as a gaming club or casino and were occupied by Crockford's Ltd, another member of the Lonrho Group. Crockford's held a gaming licence in respect of the premises which was renewable annually. The peculiar difficulties that arose in that case all stem from the fact that Crockford's held a gaming licence. The factual setting is summarised by Scott J at G
p 148K:

> There were, I have been told, in the south Westminster area, on December 8 1983, 19 gaming establishments including 30 Curzon Street. There were 21 current gaming licences held in respect of premises in this area, including Crockford's 30 Curzon Street licence. The holders of two of these licences were, I have been told, seeking transfer of their licences from the premises in respect of which the licences had been granted to new premises. This explains the discrepancy between 19 gaming establishments and 21 licences. I have been told, also, that the policy of the Gaming Board and of the justices with jurisdiction in south Westminster was that the number of gaming establishments in this prime area should be strictly limited. I have been told that these authorities were not prepared to allow any increase in the number of licensed gaming establishments. These details may or may not be established
> by evidence before the arbitrator. They are by no means all common ground. H
> But if the situation has been correctly described, the position of Crockford's and of Crockford's current gaming licence becomes highly significant.

One of the questions asked was whether the arbitrator was "entitled to assume and/or find that the respondent (Cardgrange) was a possible hypothetical tenant in the market for the property on the rent review date." It is on the answer given to that question by Scott J that Mr Lockhart-Mummery particularly relies. Scott J said [at p 152 G-K]:

> Crockford's is a real company. Its stature in the gaming world is well known. The "vacant possession" hypothesis requires it to be assumed that Crockford's right of occupation and actual occupation of 30 Curzon Street
> have ended on December 8 1983. So is Crockford's to be assumed to be a J
> potential bidder in the open market for the hypothetical lease of 30 Curzon Street? If it is, then the hypothetical lessee must outbid Crockford's. The presence of Crockford's in the market without any right of occupation of 30 Curzon Street and without any other gaming premises would tend to push up the rent the hypothetical lessee would have to pay. Not surprisingly, Mr Barnes [counsel for the plaintiffs] submitted that I should answer the question in the affirmative.
>
> In my judgment, however, the right answer to this question is "No". The open market must be, so far as is possible, a market which corresponds with reality. The "vacant possession" hypothesis necessarily introduces an element of unreality. But as I have already tried to emphasise, the various hypotheses must, in my view, be taken no further than their terms make strictly necessary. It is not necessary for the purpose of giving effect to the
> "vacant possession" hypothesis that Crockford's should be treated as a K
> possible hypothetical tenant. On the other hand, so to treat Crockford's introduces hypothesis upon hypothesis. It requires Crockford's to be invested with hypothetical qualities and surrounded by hypothetical circumstances that do not correspond with reality. Thus the proposition that Crockford's is a possible hypothetical tenant requires the hypothesis that it would be in the market for 30 Curzon Street if 30 Curzon Street were vacant. That latter hypothesis is not a necessary consequence of the vacant possession hypothesis. It is no more than an arguable one. The proposition that, if it had to leave 30 Curzon Street, Crockford's would try to negotiate a transfer of its gaming licence to the new lessee of 30 Curzon Street is another arguable hypothesis. So is the hypothesis that Crockford's would have established itself in other premises and obtained a transfer of its licence to those premises. None of these hypotheses is the fact. None is demanded by the terms of the
> rent review provisions. Each is no more than an arguable consequence of the L
> "vacant possession" hypothesis. I reject them all. It is impossible to treat Crockford's as a possible hypothetical tenant without inventing for it hypothetical circumstances. So, in my view, Crockford's cannot be assumed to be a possible hypothetical tenant.

It is, I think, clear from the penultimate sentence that Crockford's had to be excluded from the market value not because there is any general principle that the tenant must be excluded but because of the special circumstance that Crockford's actually held a current gaming licence and under "disregard" (d) any addition to the value of the lease attributable to the gaming licence held by Crockford's had to be disregarded. The introduction of Crockford's as a potential bidder
required the hypothesis that Crockford's did not have any other M
premises to which its licence had been or could be transferred. So to treat Crockford's as a possible tenant "introduces hypothesis upon hypothesis. It requires Crockford's to be invested with hypothetical qualities and surrounded by hypothetical circumstances that do not correspond with reality".

In my judgment, therefore, there is no principle of law which requires the arbitrator, while having regard to the fact that First Leisure is the lessee of the Greyhound, to assume that First Leisure is not a potential lessee of the extension. First Leisure cannot be identified as the hypothetical lessee; adopting Donaldson J's words, "His profile may or may not fit that of First Leisure". But in ascertaining the best rent that would be obtainable in the open

A market, the arbitrator must take into account the potential value of the extension to a lessee of the Greyhound Hotel.

I was referred by Mr Lewison to two other decisions: the decision of the Court of Appeal in *Inland Revenue Commissioners* v *Clay* [1914] 3 KB 466, and the decision of the Privy Council in *Raja Vyricherla Narayana Gajapatiraju* v *Revenue Divisional Officer, Vizagapatam* [1939] AC 302. In *Inland Revenue Commissioners* v *Clay*, the value of a house had to be ascertained for the purposes of section 26 of the Finance (1909-10) Act 1910, which required the value to be ascertained on the footing that it had been sold on the valuation date in the open market by a willing seller. Its market value as a dwelling-house was £750; the trustees of an adjoining property
B run as a nursing home were willing to pay at least £1,000. The case for the Crown was that the special needs of the trustees had to be disregarded or alternatively that it should be assumed that the trustees needed only marginally more than £750 to secure the property. Both arguments were rejected by the Court of Appeal. Sir Herbert Cozens-Hardy (as he then was) said at p 472:

> An "open market" sale of property "in its then condition" presupposes a knowledge of its situation with all surrounding circumstances. To say that a small farm in the middle of a wealthy landowner's estate is to be valued without reference to the fact that he will probably be willing to pay a large price, but solely with reference to its ordinary agricultural value, seems to me absurd. If the landowner does not at the moment buy, land brokers or speculators will give more than its purely agricultural value with a view to reselling it at a profit to the landowner.

C In the Privy Council case land was compulsorily acquired by harbour authorities because it contained a spring yielding good drinking water which could be made available to oil companies and people engaged in harbour works and which had the particular attraction to the acquiring authority that water could be used to supply the local population whose wells, a source of malaria-carrying mosquitos, could then be closed. The fact that malaria was endemic in the area might otherwise have made the development of the harbour uneconomic. Lord Romer stated the principle applicable in cases of compulsory acquisition in a well-known passage:

> The compensation must be determined, therefore, by reference to the price which a willing vendor might reasonably expect to obtain from a willing
> D purchaser. The disinclination of the vendor to part with his land and the urgent necessity of the purchaser to buy must alike be disregarded. Neither must be considered as acting under compulsion. This is implied in the common saying that the value of the land is not to be estimated at its value to the purchaser. But this does not mean that the fact that some particular purchaser might desire the land more than others is to be disregarded. The wish of a particular purchaser, though not his compulsion, may always be taken into consideration for what it is worth.

Inland Revenue Commissioners v *Clay* was referred to by Donaldson J in *F R Evans (Leeds) Ltd* v *English Electric Co Ltd* (which I have cited) as authority for the proposition that "monopoly positions on either or both sides do not render hypothetical agreements impossible". These cases are also authority for the
E proposition that the potentiality of land which falls to be valued as between willing vendor and willing purchaser — its value for a particular purpose for a particular purchaser — must be taken into account. Mr Lockhart-Mummery stressed the danger of taking cases decided in the field of compulsory purchase or the ascertainment of the value of freehold land as a guide in the very different field of rent reviews. There there is no antecedent history to be disregarded and no continuing relationship as there is between lessor and lessee. But these cases seem to me a guide in determining the way in which the potential value of land for a particular purpose has to be brought into account in the hypothetical higgling in the hypothetical market place.

In my judgment, therefore, the answer to question 4 is that strictly the arbitrator is not required to assume either that First Leisure is or
F that it is not a possible hypothetical tenant. But he is entitled to find as a fact that a lessee of the Greyhound Hotel, whoever he might be, would be a potential tenant. It is for him to say what effect that fact, if found, would have as between a hypothetical willing lessor and a hypothetical willing lessee negotiating the terms of a hypothetical lease of the extension.

It follows that the answer to question 5 must be "No".

Court of Appeal G

February 14 1991

(Before Lord Justice PURCHAS and Mr Justice THORPE)

SOTERI AND ANOTHER v PSYLIDES AND ANOTHER

Estates Gazette June 22 1991

[1991] 24 EG 161

Landlord and tenant — Lease of business premises — Rent review clause — Collateral agreement in the form of undertaking by landlords that if lease were executed they H would install in the premises suitable lavatory and washing accommodation — Undertaking not fulfilled — Rent not paid — Claim for forfeiture of lease — Counterclaim for breach of collateral agreement — Judge's order giving judgment in favour of landlords on claim for possession with relief to tenants on terms — Appeal by landlords seeking implementation of rent review provisions — Appeal allowed with qualifications

The property concerned in this case was a ground floor and basement used by the tenants for the sale or hire of video cassettes — It was clear before execution of the lease that the premises were wholly lacking in lavatory and washing J accommodation — The landlords as a condition of execution entered into the collateral agreement mentioned above — Apart from being a necessity for the convenience of the tenants the installation was required to conform, *inter alia*, with the Health and Safety at Work etc Act 1974 — Unfortunately, the landlords' undertaking was not fulfilled and the tenants ceased to pay rent — Litigation became inevitable and the landlords claimed in a High Court action the forfeiture of the lease and the tenants counterclaimed for damages for breach of the collateral agreement — The trial judge's order was that the landlords were entitled to K possession and recovery of arrears of rent and mesne profits, abated by one-third on account of the landlords' continuing failure to provide the lavatory and washing accommodation — It was, however, part of the order that, on the landlords' fulfilling their obligations, the tenants should be relieved against forfeiture on payment of the abated rent and mesne profits — The landlords appealed and sought an order to provide for the implementation of the terms of the lease dealing with rent review — In their appeal the landlords complained that the one-third abatement ordered by the judge did not represent a proper valuation but was merely taking L figures "out of the air"

The lease, which was for a 10-year term, contained a rent review clause which became operative for the second five years — It provided, briefly, for an open market rent to be agreed or, failing agreement, to be determined by an independent surveyor — The writ in the present case had been issued and served well before the end of the first five years — The trial judge had rejected a submission that the judge should consider and deal in his order with the question of rent review — The Court of Appeal held that the judge was wrong in this ruling and that he should have attached some condition to the grant of relief against forfeiture which would preserve the M landlords' rights under the rent review provisions of the lease — The judge had not exercised his discretion at all in this respect — The court recalled that it could, and in suitable cases would, exercise its own discretion in substitution for that of the trial judge — The present was a suitable case — All the necessary materials were before the court — The main circumstance was the issue of the writ well before the termination of the first five-year period and well before the time when the landlords should have taken action under the provisions of the lease to obtain a review for the second period of five years — The court accordingly allowed the appeal on

A **this point and included in the order a provision preserving the rights of the landlords to bring into operation the rent review procedures**

The court rejected the complaint that, in arriving at an abatement of one-third of the rent, on account of the landlords' continuing breach of undertaking, the judge had merely taken a figure out of the air — On the contrary, he had carefully considered the effect on the tenants of the continuing breach, including the inconvenience caused to the tenants' staff — It was not necessary for the judge to go into the profits and losses of the business — There was no justification for the
B **court to interfere with the exercise of the judge's discretion in this respect**

The appeal was allowed to the extent set out in the judgment of Purchas LJ

No cases are referred to in this report.

This was an appeal by the landlords, Nicholas Soteri and Andrew Koumis, from an order made by Judge Harris QC at Westminster County Court. The appeal arose out of a lease and collateral agreement between the landlords and Kyriacos Konstantinou Psylides and Charalambos Manoras, the tenants, relating to a ground-floor and basement property at 21 Camberwell Church
C Street, London SE5.

Simon Wheatley (instructed by Armstrong & Co) appeared on behalf of the appellants; Brian Riley (instructed by John Constant & Co) represented the respondents.

Giving judgment, **Purchas LJ** said: This is an appeal by Nicholas Soteri and Andrew Koumis (to whom I shall hereafter refer jointly as "the landlords") from an order of His Honour Judge Harris QC made on October 12 1989 in the Westminster County Court. The appeal arises out of a lease and collateral agreement made between the landlords and Kyriacos Konstantinou Psylides and Charalambos Manoras (to whom I shall hereafter refer as "the tenants"). The property concerned was the ground floor and basement at 21
D Camberwell Church Street, London SE5. The premises were business premises and were used by the tenants for the sale or hire of Greek video cassettes. The judge made findings of fact which are not in dispute and for convenience I will read from the note of his judgment:

> Now the facts, as I find them are these, that these premises were purchased by the plaintiffs in about 1980 for a sum I am told in the region of £9,000. The premises had to have some work done to the roof but it was a shop premises on the ground floor and basement, which was operated prior to the lease, it seems, by the second plaintiff, Mr Koumis as a hirer out and/or vendor of Greek video films. When the lease was to be executed it was appreciated by the defendants that there was no lavatory or washing accommodation there and the evidence is quite clear and is not seriously challenged by the plaintiffs, that
> E Mr Soteri, on behalf of both of the plaintiffs, gave a promise or assurance, which formed a collateral agreement, in my judgment, that if the lease was to be executed the plaintiffs would undertake to put in these premises, as I say, suitable and lawful lavatory and washing accommodation so that the staff of the shop could use the same and that the owners of the shop premises being used as a shop could comply with the necessary statutory regulations and orders. I am satisfied that it was on that basis that the defendants agreed to take this lease, that is the sort of collateral contract that one often finds in this sort of case where undertakings are given by a vendor or a prospective lessor in respect of doing various works to the property. It is not necessary, in my judgment, for the lease to be amended, no doubt at great expense and parties often do this collateral agreement to deal with matters which are thought to be of no great expense and great moment. I am satisfied further that Mr Soteri told Mr Manoras, the second defendant, that about February was thought to be the date when in fact these works were to be done. I do not think Mr Soteri
> F knew when the works were going to be done because he left these matters of detail in respect of getting a grant from the local authority to his co-owner, the second plaintiff, Mr Koumis. However, I am satisfied on the evidence that February was the date when Mr Soteri gave this undertaking that the lavatory and washing facilities would be installed. Rent was paid for the first quarter on the execution of the lease, the defendants having gone into occupation in December, the lease not being executed until February. However, by the time of the next quarter day March 25 1983, no work had been done, no endeavours to get the lavatory installed had been taken by the plaintiffs, and the matter was left as it was at the date the defendants went in. Accordingly the defendants, as in my judgment was sensible for them, did not pay the rent, they no doubt were thinking that their failure to pay the rent might in fact crystallise the matter and bring things to a head.

In April of 1983 the landlords instructed solicitors, who demanded the rent, and the tenants replied that the landlords were in breach of G
the collateral agreement. That was the first exchange following the tenants' ceasing to pay the rent until something had been done about the collateral agreement.

In June and July 1983 the environmental health department of the local authority started to take enforcement action by way of serving a notice on the tenants including a schedule of works required to comply with the Health and Safety at Work etc Act. The tenants referred the matter to landlords. Things drifted on until the end of 1983. There was another gentleman, also called Mr Soteri, instructed by the landlords to try to effect some resolution of the impasse which had arisen between the parties. The judge commented on the attitude and conduct of the parties, but I do not feel it is necessary to review those matters as they have no impact on the issues raised in this H
appeal.

This long history of dispute dragged on. There were no new events of significance until 1986. In December 1986 the landlords issued a writ in the High Court claiming forfeiture of the lease for non-payment of rent and the tenants counterclaimed for damages for breach of the collateral agreement.

It is convenient here to refer again to the judgment, where the learned judge made further findings of fact:

> Accordingly, I am satisfied, as I have said, that there was a collateral agreement. I am satisfied also that it was not complied with. What the plaintiff, Mr Soteri, now says is that the work can be done to put in a proper lavatory and wash-basin within a very short time. Mr Soteri has in the interim J
> bought out his co-owner Mr Koumis, the property now being worth in the region of £80,000 and Mr Koumis has settled for his half, so he is no longer really concerned, he is now having no legal interest in the land at all. So what one has to try and consider is what damages the defendants are entitled to in respect of the failure of the plaintiffs to comply with that contract of a collateral nature. In my judgment although the agreement between Mr Soteri dealt with the matter up to December 1983, it was then envisaged by the parties that in fact the lavatory would be put in very shortly thereafter and the balance would be paid. It was not and there was a continuing breach and the longer it went on I am satisfied from Mr Manoras' evidence the more difficult it was to keep their staff on an equable level. They had to go down the road to the restaurant if they wanted to go to the lavatory, that seems to me to be an intolerable situation.

I interpose here to say that the judge is referring to the fact that, by K
chance, the tenant or a relative of his had another establishment in the same road and in that way they were able to relieve the difficulties arising out of the failure to provide proper lavatory accommodation.

> Accordingly, in my judgment, for the period when the contract was broken that the rent should be abated by one-third. Accordingly, that is the amount of the counterclaim which I have not, as at the present, calculated. However, no doubt the parties will be able to calculate and agree an abatement.

The order was drawn up by the judge and signed by counsel. Having adjudged that the plaintiffs were entitled to recover possession of the land and that they were entitled to recover a sum of some £6,635.63 for arrears of rent as abated, together with mesne profits, the order by para 3 provided as follows: L

> AND upon the plaintiff within 8 weeks of today installing a water-closet connected to the main drains and a water wash basin to comply with statutory regulations and orders and to the satisfaction of the local authority and the defendants' surveyor, THAT if when such works have been so completed the defendants do have relief from forfeiture provided they pay into the office of this court the sum due in respect of rent as abated, namely £6,633.63 together with £83.75 in respect of the mesne profits as abated and £3,610.67 in respect of interest together with the costs of the action as taxed or agreed within a period of 42 days of the date of such completion, and unless such sum is paid by such date the defendants shall thereupon give possession of the land to the plaintiffs.
>
> 4 AND THAT there be judgment for the defendants on their counterclaim for relief against forfeiture and for an abatement of the rent and mesne profits due in the amount of one third from the 25th of December 1983 until such time as the said works be completed and as provided for above, and that the M
> plaintiffs do pay half the defendants costs of the counterclaim on scale 3 to be taxed if not agreed.

The intent and purpose of that order was, therefore, first of all, to give judgment in favour of the plaintiffs on the claim for possession but to grant relief on terms set out in the order. The matter then rested at that point.

The landlords appeal against the order and seek an order to provide for the implementation of terms of the lease dealing with rent review.

I must here refer to the second ground of the appeal before coming back to the terms of the lease. Mr Wheatley, who appears for the appellant landlords, submits that the judge's assessment of an

A abatement of one-third of the rent for the period during which the landlords were in continuing breach of the collateral agreement was a figure reached, as he has submitted, "out of the air and upon no rational computation". He submits that the judge should have taken into account the market value, the loss of profits (if any) to the video cassette business, and so on.

I should also record one event during the currency of this long and tiresome history, namely that at one point the tenants made an effort to install a lavatory and washing facilities. Regretfully they did not do it very well and the situation remained that the premises were still in breach of the Health and Safety at Work etc Act. Mr Wheatley submits that that should have been in mitigation of damages, and I will return to that in a moment.

B I now turn to record the relevant terms of the lease. It is necessary only to recite from a proviso to the agreement to pay rent, on the second page of the document. The lease, as was stated in the judgment, was created on February 3 1983 and ran for 10 years. The proviso read as follows:

> PROVIDED ALWAYS AND IT IS HEREBY AGREED that the yearly rent payable by the Lessees hereunder during the last five years of the said term . . . time being computed from the date from which the term hereby expressed to have commenced shall be the amount of rent which was payable immediately before the Lessors have given notice of their intention to review the rent or such a sum (whichever shall be the higher) as shall be assessed as current market rent for the demised premises in the open market at that time . . .
>
> (a) such rent not being less than the rent hereby reserved as shall be agreed C in writing between the parties three months before the commencement of the second period

— That is after five years —

> (b) in the event of the parties hereto failing to reach such agreement on or before the date appointed . . . the current market value rent . . . for the second period shall be fixed or assessed by an independent Surveyor appointed by agreement by the parties hereto or (failing agreement as to such appointment within two months before the commencement of the second period time to be of the essence) by an independent Surveyor appointed for the parties by the President for the time being of The Royal Institution of Chartered Surveyors (upon the application of the Lessors) . . .
>
> (c) providing that if the Lessors have failed through inadvertence to exercise their right hereunder to require the review of the rent at the time D hereinbefore appointed they shall be entitled to require such review of rent on any succeeding rent day (hereinafter called "the late revision date") by giving the Lessees three months prior written notice and the yearly rent payable from such late revision date shall be the amount of rent which was payable immediately before the Lessors gave notice of their intention to review the rent . . .

The writ of course had been issued well before the end of the first period of five years. Mr Wheatley submits that that having been done and being an unequivocal repudiation of the lease, subject to the obtaining of a court order for possession on the grounds relied upon of failure to pay rent, it was not open to the landlords to take any action under the rent review clauses to which I have just referred. If they did, it would have been treated as a waiver of their seeking E possession and forfeiture of the lease. But Mr Wheatley submits that, when affording relief against forfeiture on terms which should arise from a consideration of all the circumstances by him, the judge should have applied his mind to the question of rent review.

It is fair to say that until we arrived at court this morning a crucial event was not disclosed within the court's papers. We are not concerned with whose fault that was, it does not matter, but it is of crucial importance. It is true that in the notice of appeal, had one read it with some degree of prescience and perhaps a little more, the point is fairly taken:

> 1. The learned judge was wrong and/or erred in law by:
>
> . . .
>
> F b. refusing to hear argument in relation to the rent review clause.

That arose in these circumstances, which are now disclosed in a copy of a letter passing between Mr Wheatley (who did not appear in the court below but, as I have said, now appears before us) and Mr Riley (who had the opportunity of appearing both in the court below and here). I need not read the contents of the letter. It does record total agreement between counsel as to what in fact happened. The tenant's counsel, Mr Riley, had already concluded his submissions on the main issues in the case as pleaded. The pleadings did not disclose any request for directions as to rent review from the court. At the end of his submissions Mr Hoffman (who has since unfortunately died) invited the judge to deal with the question of rent review. The judge commented that there was nothing before him and, furthermore, that he had not been referred to the appropriate clause G in the lease. In whatever language he used, he firmly declined to deal with the question of attaching a condition to the relief from forfeiture which reinstated the rights which would otherwise have been available under the terms of the lease (which I have cited) had the possession proceedings not been initiated as and when they were. In his judgment the learned judge in fact refers to the rent review clause in these terms:

> There was a rent review clause which has not been dealt with, there were various repairing covenants and there was the usual clause for forfeiture if the tenants did not pay their rent within 21 days. The claim by the plaintiffs is for forfeiture and possession on the ground, they claim in their writ, of £39,606.

That is a wholly exaggerated figure, and the judge comments on it. H

Mr Wheatley submits that the judge should have accepted Mr Hoffman's invitation to consider the matter and make some provision in his order so as to preserve the right otherwise arising in favour of the landlords from the terms of the proviso to the agreement for the payment of rent. He further submits that there was no requirement to plead this matter, nor was there any need to place before the court any evidence as to the details of the exercise of the right preserved by the terms of the lease, since the judge was not being asked to assess what the new rent ought to be or from what date it ought to run. The judge was merely being asked to deal with the strict provisions of the provisos of the lease with particular regard to two matters, namely that the notice which the landlords were obliged to give, ie two months before the commencement of the second period J of five years (and I quote from the lease) "time to be of the essence", failing an agreement between the parties three months before the commencement of the second period; and the relief clause, subclause (c), providing that the landlords ". . . if . . . through inadvertence to exercise their right . . . to require the review of the rent" could repair their omission within the terms of that proviso. Mr Wheatley rightly submits that in this case it could never be argued that the landlords had been inadvertent.

Mr Riley has submitted that that is not so and that the landlords could have safeguarded their position either by making some extra-curial offer or by applying to the court itself. With respect to his able argument on that matter, I am not persuaded by it. In my judgment, Mr Wheatley is right that the judge should have accepted the K invitation from Mr Hoffman, at least to consider the position which had then arisen and that he should have attached some condition to the relief from forfeiture which would preserve the landlords' rights under the provisions of the lease to which I have referred.

It is quite clear that the judge did not exercise his discretion at all in these circumstances. That being so, this court will in suitable cases intervene, because the failure to exercise discretion is the exercise of no discretion at all and that is plainly wrong if an exercise of discretion is required. And, moreover, in appropriate cases, this court will exercise its own discretion in substitution for the discretion of the trial judge. In my judgment, this is such a case. The matters to which the judge's attention should have been drawn are all before this L court. They are extremely short and simple. They rest upon the terms of the lease itself and the accepted history of the matter, the main incident being the issue of the writ well before the termination of the first period and well before the time at which the landlords should have taken action under the provisions of the lease to achieve a review of the rent for the second period of five years.

On this point, therefore, I would allow the appeal and would include in the order made by the judge a provision which would entitle the landlords as from the date of this order to take such steps as they saw fit to bring into operation the rent review procedures under the lease.

However, in that context, I must now turn to the second of Mr Wheatley's submissions. With respect to those, I am not persuaded M that the judge merely took a figure out of the air. As I read his judgment the judge carefully considered the effect upon the tenants of what he rightly described as a "continuing breach" of the collateral agreement. In my judgment, the incident to which I have referred of a somewhat ineffective attempt to mitigate the position by the tenants should not be held against them and certainly was not a feature which would have obliged the judge to make a different approach to the assessment of the damages from that which he adopted. In my judgment, he rightly looked at the situation on the ground, including the inconvenience caused to the tenants' staff. It was not necessary for him to go into the financial effects on the profits or losses of the business. The judge was correct to consider the

A effect of having to operate both by themselves and their staff without proper facilities and being exposed to the threat of proceedings under the Health and Safety at Work etc Act. The judge was right to ask himself the question, as he clearly did: what would be the appropriate rent to charge as a result of this failure to put the premises in proper order? He adopted that approach and came to a round figure for compensation for damages which he related to one-third of the rent. He was also correct to provide that it should run not from the beginning of the lease, because it was not until after the rent period for the first three months that the agreement envisaged that the work would be done. Thereafter until today, because I understand the work has still not been done, there is a continuing figure of damages running.

B For my part, I see no reason for this court to interfere with what was essentially a question of discretion, that discretion including the method of approach to computation of the overall general figure for damages. I appreciate there are other methods, but there is no reason to say that one method is more appropriate than another in the circumstances of this case. I would therefore not disturb the assessment of damages by the judge on the counterclaim or that part of his order which dealt with an abatement by one-third of the rent with effect from the end of the first rent period, that is March 25 1983.

With those qualifications and in that respect I would allow this appeal.

C **Thorpe J** agreed and did not add anything.

Chancery Division

March 1 1991

(Before Mr Justice HOFFMANN)

BUFFALO ENTERPRISES INC v GOLDEN WONDER LTD

Estates Gazette June 22 1991

[1991] 24 EG 171

D **Landlord and tenant — Rent review clauses in lease — Construction — Complications caused by reservation of "additional rent" — Whether the hypothetical letting should take into account the existence of this additional obligation — Question similar to that which arose in *Guys 'n' Dolls Ltd* v *Sade Brothers Catering Ltd* — Tenant's proposed construction would result in eliminating the additional rent at the first review, a paradoxical situation — Solution was that the additional rent should be disregarded in the reviews of the basic rent**

E **The lease in question in this case was of a factory in Wales and was granted for a term of 25 years from August 29 1978 — The lease reserved a basic rent of £43,400 pa for the first five years subject thereafter to review and also reserved an "additional rent" — The additional rent was 15% of a capital sum of £60,000 which had been spent on the demised premises — A clause in the lease provided that this 15%, ie £9,000 pa, was to be increased pro rata with the increase in the basic yearly rent — Thus the basic rental value of the premises as determined on each rent review date was to serve as an inflation index for the £9,000 fixed as the additional rent — The review formula for the basic rent provided for an F assessment of "the fair yearly market rental of the premises as it shall then be let on the terms and conditions (except as to yearly rent) herein stipulated for the residue of the said term of years" — The issue in the present case turned on the effect of the words "let on the terms and conditions (except as to yearly rent) herein stipulated"**

It was contended on behalf of the tenant that the hypothetical letting should be assumed to include the obligation to pay additional rent as it stood at the commencement of the lease — Hoffmann J pointed out that this construction would mean that at the first review date the additional rent would in practice disappear, just as in *Guys 'n' Dolls Ltd* v *Sade Brothers Catering Ltd* a similar construction would have resulted in the "fair rack rental market value" in that case being determined without the addition of the fixed figure of £7,500 — This is hardly likely to have been what the parties intended — The proper and sensible construction, which did justice to the wording, was to disregard the additional rent altogether for the purpose of reviewing the basic rent — Declarations to that effect were accordingly made G

The following case is referred to in this report.

Guys 'n' Dolls Ltd v *Sade Brothers Catering Ltd* (1983) EG 129, CA H

In this originating summons, in which the landlord, Buffalo Enterprises Inc, was the plaintiff, and the tenant, Golden Wonder Ltd, was the defendant, the question raised was as to the true construction of the rent review clauses of the lease of a factory in Wales.

Kim Lewison (instructed by Rayner De Wolfe) appeared on behalf of the plaintiffs; John Furber (instructed by Farrer & Co) represented the defendant.

Giving judgment, **Hoffmann J** said: This is an originating summons which raises a question of construction on a rent review clause. The lease was granted on January 12 1979. The demised premises were a factory in Wales, and the term was for 25 years from J August 29 1978. The lease reserved first what was described as a basic yearly rent of £43,400 for the first five years, subject thereafter to review in accordance with the provisions of clause 5(ii). Second, it reserved what was called an additional rent.

Before explaining how the additional rent was calculated I must refer to some other provisions of the lease. By clause 4(2) the tenant covenanted to use its best endeavours within nine months to carry out at its own cost to the reasonable satisfaction of the lessor certain works specified in a schedule. Those works were all for the improvement of the premises and its adaptation for the needs of the tenant. The obligation to pay for those improvements was, therefore, *prima facie* imposed upon the tenant, but by clause 4(i) the lessor covenanted to contribute £60,000 to the tenant's expenditure. By K clause 2(c) the tenant covenanted to contribute £13,500 to the landlord's expenditure under clause 4(i). One might well ask why the lease did not simply provide that the landlord was to make a contribution of £46,500 to the improvements. It may be that the form of the covenants had something to do with the provisions for obtaining industrial grants.

Those provisions are the background to the reservation of an additional rent which is said to be a sum equal to 15% on the capital defined by clause 4(i), that is to say the sum of £60,000 expended on the demised premises. Another way of saying the same thing would have been to say £9,000, and to make that assurance doubly sure the lease goes on to say that for the avoidance of doubt the maximum L additional rent during the first period of the lease is to be £9,000. As from the rent review date, that additional rent was to be increased in accordance with clause 5(iii). So one has a rent in two parts, a basic rent subject to review under clause 5(ii) and an additional rent subject to review under clause 5(iii).

Clause 5(ii) provides that the basic rent on review is to become:

> the fair yearly market rental of the premises as it shall then be let on the terms and conditions (except as to yearly rent) herein stipulated for the residue of the said term of years but assuming the premises to be fairly maintained and repaired in accordance with the lessee's covenants in this lease contained, there being disregarded any goodwill attached to the premises and any effect on the rent of any improvement carried out by any tenant with the landlord's permission. M

It is accepted on behalf of the landlord that the effect of the concluding words is to require the valuer to disregard the value of the improvements made under clause 4(ii). That makes perfectly good sense because the landlord's return for his expenditure on those improvements is already covered by the additional rent.

Clause 5(iii), dealing with the additional rent, says that it shall be increased at the rent review dates pro rata with the increase in the basic yearly rent. Thus the rental value of the premises as determined on each rent review date under clause 5(ii) is to serve, so to speak, as an inflation index for the sum of £9,000 fixed as the additional rent.

The issue in this case turns on the effect of the words "let on the terms and conditions (except as to yearly rent) herein stipulated", and in particular whether it requires the valuer to take into account the fact that the lease contains the obligations in clause 4 to make and to pay for the improvements, the obligation in clauses 2(c) for the tenant to repay to the landlord part of its contribution to those improvements and the obligation to pay additional rent.

It has been established by a number of cases in recent years that the guiding principle of construction in the case of rent review clauses is to achieve, unless the words positively drive one to a different result, a meaning which gives effect to the underlying commercial purpose of such a clause. That principle ordinarily requires one to assume that the hypothetical letting at the rent review date is of the premises as they exist in the circumstances which then exist and that it is a letting of what the landlord then actually has to offer, namely the residue of the term.

As at the date of this rent review clause the obligations under clauses 2(c) and 4 had been performed. They were not obligations which had any relevance to a letting of the residue of the term as it then stood. The principle of reality therefore suggests that those past obligations should be disregarded as being incapable of having any effect upon the hypothetical letting. Mr Furber, in a most attractively presented argument, said that they ought nevertheless to be taken into account because they expressly played a part in the calculation of the additional rent. While it is perfectly true that this is what the lease says, the reality was that once the obligations had been performed and the lease was running they no longer had any relevance to the rental obligation. They might as a matter of history explain why there was an additional rent, but all that would matter from the point of view of the hypothetical tenant was that he would have to pay a premium or addition of £9,000 a year, adjusted in accordance with clause 5(iii).

That brings one to the central point in the case which is whether the valuer in performing the valuation under clause 5(ii) should take into account the existence of the additional obligation under clause 5(iii). There is an obvious difficulty at once about taking into account that obligation as it would actually exist at the moment of the hypothetical letting. The tenant taking a letting of the residue at the rent review date would have to pay by way of additional rent not the original £9,000 but £9,000 increased in accordance with clause 5(iii). The extent of that increase, however, depends upon the valuer's answer to the question posed by clause 5(ii). One therefore has a circularity which the parties are highly unlikely to have intended. Mr Furber therefore contended that the hypothetical letting should be assumed to include the obligation to pay additional rent as it stood at the commencement of the lease, and he frankly acknowledged that this was the only escape from the circularity to which I have referred. The difficulty is, however, as he equally acknowledged, that it runs up against the presumption of reality. I would be willing to give the words that effect only if the language clearly required it or if the other alternative construction was so wayward that the parties could not be supposed to have intended it.

Mr Furber's submission was that the alternative construction did lead to a wayward result. It meant that the landlord would be getting an ever-increasing sum by way of additional rent, a sum which the effect of inflation could well carry to an amount exceeding his original capital contribution and that this amount could hardly be described, as it was in the lease, as a return for the landlord on his capital expenditure. For my part, I do not find that result so surprising that the parties must be taken not to have intended it. The evidence does not go into the background as to grants and so forth which may have led to this provision, but there is nothing improbable in a landlord's stipulating not only for a premium over the market rent but for that premium being linked to some index, which in this case was provided by the value of the premises themselves.

There is the additional difficulty that the construction for which Mr Furber contends, certainly if shorn of the provisions of clause 4 and clause 2(c), would mean that at the first rent review date the additional rent in practice disappeared. It was the improbability that the parties could have intended such a result which led the court in *Guys 'n' Dolls Ltd* v *Sade Brothers Catering Ltd* (1983) 269 EG 129 to hold that the parties must by necessary implication have intended a provision for additional rent to be disregarded when the market value was being determined. Mr Furber distinguishes that case on the grounds that the premium, the additional rent there reserved, was to be a constant amount throughout the period of the lease. In this case there is the provision in clause 5(iii) for its increase.

Mr Furber is able to rely upon the remarks of Robert Goff LJ where he expressly says that if the premium was going to be doubled at the next rent review date it would have been appropriate for the valuer to take that fact into account. I am bound to say, with all respect to Robert Goff LJ, that I find it difficult on the facts of that case to see why that should be so. But be that as it may. I am concerned only with the construction of this document, and given in particular the fact that the measure of the future increase of the additional rent is dependent upon the increase in the market rent under clause 5(ii), I am really left in no doubt that the proper construction is to disregard the additional rent altogether for the purposes of reviewing the basic rent.

Accordingly I will make declarations in terms of para 1 of the summons, as to which there is no issue before me, and also in terms of para 2*.

The plaintiffs were awarded the costs of the summons.

---

*Editor's note: Paras (1) and (2) of the summons read:

(1) A declaration that upon the true construction of a Lease made the 12th January 1979 between Hensher (Furniture Trades) Ltd (1), the Defendant (2), and Imperial Food Ltd (3) the obligations contained in clauses 2(c) and 4(i) and (ii) of the said Lease have been discharged by performance.

(2) A declaration that upon the true construction of Clause 5(ii) of the said Lease and in the events which have happened it is to be assumed, for the purposes of the rent review, that the premises comprised in the said Lease are to be assumed to be let upon the terms and conditions contained in the said Lease with the exception of

(i) the obligations contained in clause 2(c)
(ii) the obligations contained in clause 4(i) and (ii)
(iii) the liability to pay the additional rent referred to in clause 1 and clause 5(iii) of the said Lease.

NEGLIGENCE

Chancery Division

May 10 1990

(Before Sir Nicolas BROWNE-WILKINSON, Vice-Chancellor)

KNIGHT AND ANOTHER v LAWRENCE

Estates Gazette January 12 1991

[1991] 01 EG 105

Negligence — Action against receiver appointed by mortgagees — Borrowers had mortgaged tenanted properties, in many cases subject to rent reviews — Failure by receiver to serve trigger notices in time — Sale by mortgagees — Sale prices at auction depressed because rental income not increased in absence of review — Receiver held to have been negligent — Difficulties in quantification of damages

Plaintiffs in this case, a husband and wife, had mortgaged certain tenanted properties owned by them as security for debts incurred by the husband and a business partner — Some tenanted properties vested in the partnership were also mortgaged — The mortgagees were an insurance company — Many of the tenancies were subject to rent review provisions — The mortgagees appointed a receiver, the defendant in the present action — The plaintiffs' complaint against the receiver was that he negligently failed to serve trigger notices within the time-limits laid down under the appropriate rent review clauses, thereby missing the opportunity of obtaining increases of rent — As a result, when the mortgagees came to exercise their power of sale the prices realised were lower than they would have been if the rents had been increased through a timely exercise of the review machinery — It should be mentioned that neither the plaintiff husband's former partner who had become bankrupt nor the latter's trustee in bankruptcy were parties to the present action, an "oddity" which led to certain complications in the assessment of damages

The plaintiffs' claim was that the defendant as receiver was under a duty of care to them and that he was in breach by failing to serve rent review notices in time — The Vice-Chancellor found that the defendant was under such a duty and was in breach of it — There was a direct causal link between the failure to give the notice in due time and any diminution in the prices realised at the sale — The Vice-Chancellor rejected a submission that the defendant as receiver had no power to serve a rent review notice — He had such a power either under section 109 of the Law of Property Act 1925 alone or by that section as amplified by provisions of the legal charge — A suggestion that there had been contributory negligence on the part of the plaintiffs was dismissed

The quantification of damages was a matter of quite extraordinary difficulty, as the second half of the Vice-Chancellor's judgment shows — After a most detailed examination of items making up the total there were questions calling for further investigation which caused him to direct an inquiry as to damages — He was, however, able to arrive at a minimum sum of damages due to the plaintiffs of £18,200 and he made an order for interim payment of that sum — He directed that his order be not drawn up for 28 days, allowing time for the trustee in bankruptcy of the husband's former partner to make an application if so advised — He also declared that if, on inquiry, it was determined that Williams & Glyn's Bank or the City of Westminster Assurance Co Ltd were entitled to further payments, the amount would be payable by way of extra damages

The following cases are referred to in this report.

American Express International Banking Corporation v *Hurley* [1985] 3 All ER 564, DC

Pittortou Re, ex parte Trustee of the property of the bankrupt [1985] 1 WLR 58; [1985] 1 All ER 285

Standard Chartered Bank Ltd v *Walker* [1982] 1 WLR 1410; [1982] 3 All ER 938; (1982) 264 EG 345, CA

This was an action by Mr and Mrs J W Knight claiming against the defendant, Mr Harvey Lawrence, damages for alleged negligence when acting as a receiver appointed by the City of Westminster Assurance Co Ltd in respect of properties charged by the plaintiffs to the company for moneys borrowed. The negligence alleged was the defendant's failure to serve rent review notices on the tenants of the mortgaged properties within the time-limits laid down.

Charles Purle QC and Anthony de Garr Robinson (instructed by Ingham Clegg and Crowther & Laytons, of Preston) appeared on behalf of the plaintiffs; Glen Tyrell (instructed by Kennedys) represented the defendant.

Giving judgment, **Sir Nicolas Browne-Wilkinson V-C** said: In this case Mr and Mrs Knight, the plaintiffs, were borrowers under a legal charge. They claim damages for negligence against the defendant, Mr Lawrence, who was appointed as receiver of the mortgaged property by the lenders. The title of the action contains also the name of a firm of solicitors as third party, but for reasons which have not been explained to me the third party proceedings have not come before me.

The properties in question were all let to tenants, many of the leases containing rent review clauses. The time for serving certain trigger notices necessary to put in motion those rent reviews expired during the period when Mr Lawrence was the receiver of those properties. He did not serve the necessary notices. As a result, it is said, the rents were not increased as they would have been had the trigger notices been served. The mortgage properties having been sold by the mortgagees under their power of sale, it is said that because the rental income had not been increased the prices realised on the sale were depressed.

The background of the case is this. For very many years Mr Knight has been engaged in the business of developing tyre depots where people can take their motor cars to be equipped with new tyres and, I think, possibly certain other motoring accessories. In the distant past Mr Knight and Mrs Knight were engaged in that process on premises belonging to Mr and Mrs Knight jointly. Having developed the tyre depot the process was carried through by the tyre depot being sold off together with the benefit of a lease. As a result Mr and Mrs Knight were left with properties which were either freehold or long leasehold reversions subject to tenancies granted to those who had taken over the tyre depot.

In 1971 Mr Knight went into business with a Mr Sutton. The nature of the business was the same. They, in turn, came to own certain properties as partnership assets. Having been developed, the tyre

A depots were sold off with the benefit of leases granted to the purchasers. The partnership between Mr Knight and Mr Sutton borrowed heavily from the lenders, the City of Westminster Assurance Co Ltd. There were borrowings and charges on the properties prior to the legal charge with which I am concerned.

That legal charge was executed on May 13 1983. The borrowers in that legal charge were expressed to be not only the two partners, Mr Knight and Mr Sutton, but also Mrs Knight. The properties subject to the charge were not only the partnership properties belonging to Mr Knight and Mr Sutton but also the properties vested in the name of Mr Knight and Mrs Knight in the way that I have already mentioned.

The moneys borrowed amounted to £200,000. They carried interest at 16% per annum, increased to 18% per annum in the event
B of prompt payment not being forthcoming.

I must now shortly mention the properties in question, of which there are seven.

The partnership properties belonging to Mr Knight and Mr Sutton were a property at Skelmersdale, a property at Skipton, a property at Alverstone, a property at Fleetwood and a property at Preston. The remaining two properties, at Retford and Doncaster, belonged to Mr and Mrs Knight. All seven of those properties were charged under the legal charge to the City of Westminster Assurance Co Ltd, which I will call the City of Westminster.

The moneys borrowed were borrowed for the purposes of the partnership between Mr Knight and Mr Sutton, not for the purposes
C of Mrs Knight.

The leases in question are not all before me, but it is common ground that the rent review clauses in each case were for five-yearly reviews and they required the service of a notice not more than 12 and not less than six months before the review dates.

In relation to the Skelmersdale, Alverstone, Fleetwood and part of the Skipton properties, the first date on which a trigger notice (requiring a review on September 1 1984) could have been served was August 31 1983. On the same date the landlord became entitled to serve a trigger notice in relation to the first part of the Skipton premises — those premises being subject to five separate tenancies. On September 30 1983 the landlord was entitled to serve a rent review notice in relation to a second part of Skipton.

D The receiver was appointed by the City of Westminster on December 8 1983, at a time when there was still a considerable period which might elapse before it was necessary to serve the trigger notices. The time for service of the trigger notices expired in relation to the Skelmersdale, Alverstone, Fleetwood and the first part of the Skipton premises on February 28 1984. The right to serve a rent trigger notice on the second part of the Skipton premises expired on March 31 1984. At that time no rent review notices had been served, either by the receiver or by anyone else.

It has not been discussed before me whether time was of the essence for the service of such rent review notices or whether notices could be served at a later date. What has emerged is that when notices were
E eventually served there was no apparent protest by any of the tenants at the lateness of the notice.

The time for the service of the review notices having passed, the matter proceeded in this way. On May 14 1984, the City of Westminster decided to call in the loan. Up to that date Mr Lawrence's sole function had been as receiver of the rents and profits, but on May 14 he was instructed to consider the possibility of the sale of the property. Immediately thereafter he instructed Mr Roger Spyer [FRICS], a valuer and agent (who also acted for the City of Westminster), to act in relation to the sale. Mr Spyer proceeded to put in hand preparations for an auction sale, which eventually took place on December 11 1984.

F On June 6 1984, Titmuss Sainer & Webb, solicitors, called in the loans owed by Mr and Mrs Knight and Mr Sutton. On July 10 Titmuss Sainer & Webb gave notice to certain of the tenants in this form:

> We, Titmuss Sainer and Webb, as agents for Harvey Lawrence the receiver appointed by the City of Westminster Assurance Company Limited, pursuant to a legal charge dated the 13th May 1983

— and the parties are given —

> hereby give you notice that C Sutton and J W Knight require the rent reserved by the said lease to be reviewed on and as from the 1st September 1984.

So, at that stage Titmuss Sainer & Webb regarded it as proper to give the notices as agent for the receiver.

On the same day Titmuss Sainer & Webb sent copies of those notices to the plaintiffs' solicitors asking them whether Mr Knight or G
Mr Sutton had themselves previously served any notices under the rent review clause. On September 3 the reply was given that no notices had been previously served.

On November 14 the surveyor acting for the plaintiffs at that stage, Mr Clive Eckersley, forwarded to Mr Lawrence an offer that had been received of some £293,500 for all the properties. That offer, in the end, did not materialise because, I was told, the would-be purchaser got wind of the auction and declined to go on, knowing that there was an imminent auction.

The non-service of the necessary trigger notices having come to the attention of Mr Knight, he, through his solicitors, complained of the position. The matter came to light when the auction brochure was
produced showing only the old rents as being the recoverable rents H
and referring to the next review dates as being not 1984 or 1985 but 1989.

Mr Knight pressed the City of Westminster to postpone the auction until the increased rents had been agreed. There was resistance to that by the City of Westminster. They agreed to put off the auction provided there was a signed contract with another purchaser shortly before the date of the auction but, as I have said, that did not occur. Under continuing pressure, the City of Westminster began to feel some concern about the position.

Mr Spyer, the agent, had advised the City of Westminster that they should have comparatively low reserve prices at the auction sale.
Because of the pressure from Mr Knight's solicitors indicating that he J
would be inclined to hold the City of Westminster, or the receiver, liable for the failure to serve the rent review notices in time and a failure to agree them before the date of the auction, there were instructions given by the City of Westminster, acting through Mr Robert Browning, to Mr Spyer (their agent and one of the joint auctioneers) to agree with Mr Eckersley, agent for Mr Knight, what the reserve figures should be at the forthcoming auction. It is clear from the correspondence before me that Mr Spyer was acting on instructions which indicated that the reserves should be such as to prevent any subsequent complaint by Mr Knight against the City of Westminster.

The auction took place on December 11 1984. Mr Knight attended
that sale. He told me — and I accept — that in relation to those K
properties where the rent review had been missed, the auctioneer made a public announcement to that effect before the lot was put up.

The result of the auction was as follows. The Preston property was sold at its reserve price of £25,000, the Alverstone property was sold for £20,500, being £500 above the reserve, the Doncaster property was sold for £29,000, the Retford property for £26,050. Therefore, both the properties belonging to Mr and Mrs Knight were sold at above or at their reserve figures. However, the Skelmersdale, Skipton and Fleetwood properties did not sell, having failed to reach their reserves.

It is not clear to me how far there was any genuine bidding for any
of those premises. Mr Knight told me that there was a bid up to L
£109,900 for the Skipton property, being £100 below its reserve. At one stage in the correspondence Mr Spyer said that a bid of approximately that amount, I think he said £110,000, was received. At another stage in the correspondence Mr Spyer says that no bid was received. I do not find myself able to get any guidance as to the value of the properties deriving from any bids alleged to have been made at that auction. It is not clear the extent to which the auctioneer was taking bids off the wall, as opposed to genuine bids.

Those three properties having failed to sell, each of them being properties where the rent review had been missed, the next event that happened of direct importance was that Mr Sutton had a bankruptcy petition presented against him. A receiving order was made on that
petition on April 9 1985. At that stage, if not before, the partnership M
between Mr Knight and Mr Sutton came to an end.

As to the properties, steps were taken by the City of Westminster to renegotiate rents as to part of the Skipton property.

The Skipton property had five tenants. In a transaction which is complicated, and which I do not think it is necessary to expound, Motorway Tyres & Accessories Ltd, which had occupied the part of the premises which consisted of a modern warehouse and part of adjoining premises, gave up their tenancy of the part of the adjoining premises, which was taken over by the other occupier of those adjoining premises, ABC Engineering Ltd.

It is apparently common ground in this case that the old rent of what was previously leased to Motorway Tyres & Accessories of

£5,100 was, as a matter of negotiation, increased to £6,800 at that stage.

As to another part of the Skipton premises occupied by Retail Bakers (Yorkshire) Ltd, there had been a fire before the first auction on September 7. As a result of that the tenants were refusing to pay their rent. The property, so it is said, was underinsured. There is no evidence that any revised rent was ever agreed as to any part of the premises occupied by Retail Bakers.

In June 1985 the Skelmersdale property, which at the first auction had an agreed reserve of £40,000, was sold by private treaty for £27,000. It is not clear to me whether any agreed increase in rent pursuant to the rent review had been implemented or agreement reached prior to that sale by private treaty.

The Fleetwood premises had an old rent of £6,000 per annum. In the course of 1985 the surveyors acting for the tenants and for City of Westminster agreed that as a result of the rent review no increase should take place. The rent, therefore, continued at £6,000 a year.

The Skipton and Fleetwood properties were sold at public auction on November 28 1985. Skipton, which had had an agreed reserve of £110,000 at the first auction, was sold for £72,000. Fleetwood, which had a reserve at the first auction of £48,000, was sold for £41,000.

The proceeds of sale were paid over to the City of Westminster Assurance. The account, which is produced to me, from the City of Westminster Assurance shows that having continued to run the account and debit it with interest and with receipts, both of rent from Mr Lawrence, as receiver, and of capital moneys as proceeds of sale, the position was reached on March 25 1986 where there was still a debit balance on the account of £12,103.50. The account records as the last item on it "Interest written off £12,103.50". Therefore, according to the City of Westminster's records, there was a debt still owing from Mr and Mrs Knight and Mr Sutton of £12,103, but, as a matter of internal record-keeping, that debt has been written off.

There is one other matter that I should mention at this stage, since it enters into the computation of damages. There was a second mortgage on all the premises to secure a borrowing from Williams & Glyn's, now Royal Bank of Scotland. I will come to the significance of that later.

It is an oddity of this case, which is one of the facts that gives rise to difficulty, that Mr Sutton and his trustee in bankruptcy are not parties to the case at all. There are, therefore, only two out of the three owners of the property and borrowers subject to the legal charge making the claim against Mr Lawrence. Their claim in outline was that Mr Lawrence was negligent in failing to serve the trigger notices relating to each of the rent review clauses in due time, as a result of which new and higher rents were not fixed prior to the date of the first auction. That, in turn, it is said, led to a diminution in the purchase price paid and received for the properties subject to those tenancies. They allege that Mr Lawrence, as receiver, was under a duty of care to them and failed to discharge that duty. Mr Lawrence says that in the circumstances of this case, namely that he was a receiver appointed under section 109 of the Law of Property Act 1925, being a receiver only of the rents and profits, he was under no duty of care to the plaintiffs to ensure the service of the trigger notices. Alternatively, if he be wrong on that, he says that though he might be under a duty of care to Mr Knight as being one of the two owners of the properties in relation to which there was a failure to serve the necessary trigger notices, he could owe no such duty of care to Mrs Knight, none of whose properties were the subject-matter of the failure to serve the trigger notices.

Alternatively, Mr Lawrence submits that, if otherwise there would be a duty of care owed by him either to Mr or to Mrs Knight, he had no power under his appointment to serve rent review notices. He further contends that, since the period for the service of notices triggering the rent reviews was running prior to his appointment and no warning was given to him of the need for such notices, Mr Knight was guilty of contributory negligence.

Finally, if those submissions are wrong there is an extremely complex dispute as to the quantification of damages.

I must now fill in some of the details as to the nature of Mr Lawrence's appointment.

Under section 109 of the Law of Property Act 1925 mortgagees are entitled, in circumstances which existed in this case, to appoint a receiver. Under section 109(2) a receiver so appointed is deemed to be the agent of the mortgagor. Under subsection (3) the receiver has power to demand and recover all the income of which he is appointed receiver, and to give receipts therefor, and to exercise any powers which may have been delegated to him by the mortgagee pursuant to this Act. Then there are other provisions to which I need not refer.

Those powers are substantially widened by the legal charge in the present case. Clause 12 of the legal charge, so far as relevant, reads as follows:

> At any time after the principal monies hereby secured shall have become immediately payable, the lender may from time to time appoint by writing any person to be a receiver of the property or any part thereof, and may from time to time in writing remove any receiver so appointed from his receivership of the whole or any part of the said property and may appoint another in his stead.

Clause 12(2) provides that the receiver should be the agent of the borrowers. Clause 12(3) confers on the receiver power (a) to enter upon and take possession of any property in respect of which he is appointed or any part thereof; (b) to carry on or manage or concur in carrying on or managing the business of the borrowers; (c) to sell or concur in selling, to let or concur in letting and to accept or concur in accepting surrenders of leases of any part of the property. There is no express authorisation to serve rent review notices or to increase the rent in any other way. Under (d) he is given power to make any arrangement or compromise which he or the lender may think expedient and under (g) he is given power to do all such other acts and things as may be considered incidental or conducive to any of the matters or powers aforesaid which the receiver lawfully may or can do as agent for the borrowers. It is expressly provided that those powers are in addition to the statutory powers.

By the document appointing him he was appointed to be "the receiver of the rents and income of all the property comprised in the legal charges, with all the powers conferred on a receiver by the said Act, and by the said legal charge". Mr Lawrence gave notification to Mr and Mrs Knight and Mr Sutton that he had been appointed as receiver, but he indicated that he was appointed as receiver of the property, not simply of the rents and profits.

I will consider first whether Mr Lawrence was under any duty of care to Mr and Mrs Knight. I have no doubt whatsoever that he was under such a duty. As I have said, he was appointed the agent of the borrowers, Mr and Mrs Knight. That was his status in acting as receiver. It would be clear to anybody that in carrying out his duties, if they did include the duty or power to maintain the rent at its true market level by triggering the rent reviews, the failure to do so would adversely affect not only the value of the security to City of Westminster but also the value of the property both in terms of future income and capital value to the borrowers. In the case of investment properties of this kind, the rent receivable is a basic factor in the value of the asset.

By whatever standard one should now judge whether a duty of care arises, in my judgment that standard is fully met in the present case. In terms of neighbourhood it is plainly foreseeable that the owner of the property will be prejudiced if a rent review is missed. If the more recent proximity test is the appropriate yardstick to judge it by, the closeness of the relationship between principal and agent is a classic case where such proximity exists. If one has to judge it by reference to the existence of an established relationship giving rise to a duty of care, that, too, is satisfied, since such duty of care has long been established in relation to a situation of an agent.

In *American Express International Banking Corporation* v *Hurley* [1985] 3 All ER 564 it was held that a company receiver owed a duty of care to the guarantor of the company's debt. In *Standard Chartered Bank Ltd* v *Walker* [1982] 1 WLR 1410 a receiver appointed to realise the assets of a company under a debenture was held to owe a duty of care both to the borrower and to the guarantor of the loan. Accordingly, the duty of care arising from the position of a receiver is owed not only to his principal but also to those directly and immediately affected by his actions.

So far as Mrs Knight is concerned, the position is one stage further away. It is pointed out that the failure to serve rent review notices had no direct application to the Retford and Doncaster properties, which were the only properties in which Mr and Mrs Knight had an interest. No question of rent review arose in relation to those properties. Therefore, it is said that any duty which Mr Lawrence owed could apply only to those properties where there was a requirement to serve rent review notices. In my judgment, this is not a material distinction. In the legal charge Mrs Knight was expressed to be a borrower and was subject to the liability to pay the debts. Her properties, as well as Mr Knight's and Mr Sutton's properties, were charged to the bank. Where there are joint debtors owning different properties subject to a charge, the failure to act carefully in relation to any one of those properties is manifestly capable of having an adverse impact on other properties subject to the same charge, since any diminution in the value of one of the properties will increase the liability on the others. In my judgment, the duty of care was owed as much to Mrs Knight as to Mr Knight.

All that I have said so far is on the assumption that Mr Lawrence has the necessary power to serve the rent review notice. Plainly there could be no duty of care on him to do a thing which he had no power to do. Mr Tyrell, for Mr Lawrence, submits that there is no such power. He points out that in section 109 itself there is no express power to increase the rents in any way, and the power, if any, would have to be by implication. I do not find it necessary to express any view on whether in the absence of additional powers conferred by the legal charge there is or is not power in a receiver appointed under section 109 to serve a rent review notice, because in the present case the powers of the receiver are amplified in a major way.

I bear in mind a factor which I think is an important one, namely that although under clause 12 the City of Westminster could have appointed Mr Lawrence to be the receiver of the property, ie not simply of its income, and though Mr Lawrence held himself out as having been so appointed, the actual appointment was restricted to being a receiver of the rents and income of the property.

The question is, given such limited appointment, are his powers extended by clause 12 to enable him to do those things which are necessary to maintain the income and the rents and profits of the property? In my judgment they are. He is expressly given power to enter upon and take possession of any property in respect of which he is appointed. That will enable him to take possession of the rents and profits of the property. That being so, quite apart from his power under (3)(d) to make any arrangement or compromise which he may think expedient, he has power under (g) "to do all such other acts and things as may be considered to be incidental or conducive to any of the matters or powers aforesaid, and which the receiver lawfully may or can do as agent for the borrowers". It is hard to imagine, from the point of view both of the lender and of the borrower, anything more necessary or conducive to the receipt of the rents and profits of premises than the service of those notices necessary to keep up the rents and profits to their maximum.

I therefore hold that the receiver had power under the legal charge to serve the notices, a view in which I am fortified by the fact that when Titmuss Sainer & Webb came to serve the trigger notices out of time they, too, served them on his behalf and not on anybody else's behalf.

The next question is whether there has been a failure to discharge that duty of care. Mr Lawrence's own evidence seemed to me to demonstrate that there had been such a breach of duty. In my judgment, Mr Lawrence had a total misapprehension about the functions of a receiver. He regarded himself as being there to do what he was told by his appointor, the City of Westminster; provided he discharged what they told him to do he had discharged his functions. He was, in his own eyes, nothing but a rent collector.

That, to my mind, is an unhappy misapprehension of the functions of a receiver. Though he may be appointed by one party, his function is to look after the property of which he is receiver for the benefit of all those interested in it. He is not, even in commercial terms, the mere agent of the appointor; he is there to safeguard the property for all who have interest in it. Simply to regard himself as the collector of rents without any further function was to misapprehend the nature of his appointment. The position was made overwhelmingly clear by the expert evidence that I had from a very experienced receiver, Mr Christopher Morris, in which he deposed to the fact that even under section 109 (let alone a wider power) it is one of the first functions of a receiver in a case like this to get solicitors or others to review the position of the rent review clauses and to take such steps as are necessary to ensure that the reviews take place. I have no doubt that there was in this case a failure to come up to the standard of care required of the receiver.

I find as a fact that if the notices had been served in due time, that is to say shortly after Mr Lawrence became the receiver, by the date of the auction on December 11 1984 the rents would have been agreed.

I next come to the issue of contributory negligence. It is said that Mr Knight was at fault because he did not, during the period when he could have served trigger notices, do so. Moreover, Mr Tyrell further drew attention to a matter which arose at an earlier stage. Under earlier charges securing borrowing between Mr Knight and Mr Sutton on the one hand and the City of Westminster on the other, Mr Lawrence had been appointed as receiver on two earlier occasions. One of those occasions was from February 18 to May 13 1983. There are two letters in the correspondence before me showing that on February 28 1983 and on March 2 1983 there was an exchange of letters between Parker & Co, agents and surveyors acting on behalf of Mr Knight, and the tenant of one of the Retford properties, confirming their agreement to an increase in rent under the rent review provisions. It is said that that shows that in the earlier receivership Mr Knight was regarding it as still the function of his agent to deal with rent reviews under the leases. Moreover, it is said that Mr Knight should have alerted the receiver to the need to serve the trigger notices.

I cannot find any contributory negligence in this case. There could have been no obligation in common prudence to serve the rent review notices at the first occasion that they could be served. When the receiver was appointed in December 1983 there was still approximately three months left in which review notices could have been served. It was not a question of a last-minute failure to serve a notice.

As to the position in the earlier receivership and the fact that Mr Roger Parker [ARICS] of Parker & Co was writing during the period of that receivership, in my judgment that is in no sense decisive, since the letters in question were written very shortly after Mr Lawrence's appointment as receiver, and it is not clear at what stage the agreement which that letter confirms was reached.

Finally, as to the failure to alert the receiver to his function, it is, in my judgment, the common understanding that, when you are told (as were the borrowers in this case) that a receiver has been appointed, the receiver has taken over the management of that property and it is not for the true owner to intervene at that stage. It does not seem to me that it can be said that a failure at that stage for the person who has been displaced by the receiver to take steps to alert the receiver can amount to contributory negligence.

I therefore find that Mr Lawrence is liable for breach of duty. The question then arises of what damages are payable.

This has proved to be a very complicated problem which has absorbed a large proportion of the time of this over-prolonged hearing.

There was a preliminary point taken by Mr Tyrell to the effect that there was no evidence that the failure to serve the notices triggering the rent review clauses had led to any failure to agree rents subsequently. There was no evidence of any tenant refusing to agree an increased rent or seeking to negotiate a lower rent because of late service of the notices. That is correct. But the damage in this case flows not from the failure to agree the higher rent in due course but from the failure at the time of the sale to have agreed the rents. These properties were put up for auction on the basis that their review dates had been missed and, therefore, would-be purchasers were faced with a property which at the time of the auction did not have an agreed rent and, on the basis of the announcement, would not have a higher rent until 1989.

There is a letter from Mr Spyer of December 12 1984 reporting the result of the first auction in which he expressly says that the sale prices or bids would have been depressed by the fact that the reviews had been missed, and it is noteworthy that all the properties in relation to which there was no doubt as to the rent reviews did sell at that first auction at the agreed reserve prices.

I therefore find that there is a direct causal link between the failure to give the notices in due time and any diminution there may be in the prices obtained on the sale.

The plaintiffs' claim has been variously formulated. Basically what is said is this. If the rents had been increased, all the properties would have sold at the first auction and at prices higher than those eventually realised. As a result, there would have been a surplus of the proceeds of sale over and above the sum necessary to satisfy the City of Westminster's claim. It was alleged that, if all had been in order, the properties would have sold for some £334,750, whereas the debt due to the City of Westminster was in the region then of £220,000. It is said that the surplus would have accrued to the plaintiffs, not to Mr Sutton, on the following grounds. Since the Retford and Doncaster properties — which I will call the second properties — were not partnership properties but belonged to Mr and Mrs Knight separately, as between Mr and Mrs Knight on the one hand and Mr Sutton on the other, Mr and Mrs Knight would be entitled to be exonerated from the debt to the extent that the other partnership

A properties — which I will call the first properties — were capable of discharging the debt to City of Westminster. Therefore the first tranche of any surplus which might have been realised if the properties had sold at their proper price would have gone to recouping to Mr and Mrs Knight, as the owners of the second properties, the debt which had been discharged out of them amounting to £55,050.

Then, it is the plaintiffs' case, any surplus above that figure would, under the terms of the partnership between Mr Sutton and Mr Knight, be applicable in paying the partnership debts. The first properties were partnership property. Since there was no such surplus the debts were not so paid. Mr Knight alleges that he has paid the debts of the partnership amounting to some £57,000. Therefore, B he claims, he has suffered the loss of that amount through the failure to have available the increased proceeds of sale which could have been used to pay the debts.

Finally, the plaintiffs say that any ultimate surplus would be divisible 50/50 between Mr Knight and Mr Sutton's trustee in bankruptcy: on that ground his claim in negligence is limited to 50% of the ultimate surplus.

The starting point is to try to ascertain what would have been the sale price of the properties if on December 11 1984 the rent reviews had properly taken place. That involves two stages. First, to find out at what the reviewed rents would properly have been agreed. Second, to find out, taking into account the higher rent payable as a result of C such review, what increased price would have been obtainable for each of the properties.

To that end I have had a mass of expert evidence, some before me, some in the correspondence. For the plaintiffs, Mr John Royle [ARICS] gave evidence that earlier in 1983 he had conducted a detailed valuation for mortgage purposes of all these properties. In the process of that he had seen the properties close to the relevant time and had made his valuations at that time. He struck me as being a careful witness. However, he did not have the benefit, when he made that valuation in early 1983, of the experience which the other valuers have had of the prices actually realised at the sale in December 1984. Moreover, I formed the impression that he was not always fully familiar with the local conditions of each of the tyre depots. For D example, in relation to Skelmersdale he produced an extremely high rent, quite out of line with the estimates put forward by other valuers. I think it may well be that he was not aware in that instance — and possibly in others — of the detailed local conditions applicable.

The other expert called by the plaintiffs was Mr Eckersley, who in the far past had acted for Mr Knight in the acquisition of these properties. He had been brought back into the matter late in 1984. In the meantime, Mr Parker, to whom I have referred, had been doing the surveying and agency work on behalf of Mr Knight. Mr Eckersley again struck me as being a careful witness, but he suffered from two disadvantages. He had not recently seen the premises and he worked entirely on the basis of estimates of rent put forward by Mr Parker in E a letter written in late 1984. Mr Parker himself was not called and I have no basis on which to judge the rightness of Mr Parker's estimate of what rents could be achieved.

The second disadvantage in Mr Eckersley's approach was that in finding the multiplier he applied to the uplifted rent (ie the years' purchase) he fixed it by reference to the years' purchase reflected in those reserved prices agreed between himself and Mr Spyer before the first auction, producing an average years' purchase of 8.95. This is substantially above the years' purchase rate adopted by other valuers and, in my judgment, it is a defective basis for taking the years' purchase. Unknown to Mr Eckersley, in the circumstances I have mentioned Mr Spyer was under some pressure to agree high reserves with a view to ensuring that the City of Westminster was not F subjected to later litigation. I do not find an agreement between valuers under those circumstances as in any sense necessarily reflective of what was the open market value.

I find a much more reliable guideline to be the years' purchase obtained on those properties which did sell at the first auction, when there was no trouble in relation to rent reviews, that is to say the Retford, Doncaster and Preston properties. The average years' purchase on those properties varied between 7.14 and 8, giving an average of 7.4. That seems to me closer to a realistic and contemporaneous assessment of what was the kind of years' purchase that a person buying a tyre depot for investment purposes was prepared to pay.

The defendant's expert evidence consisted of Mr Norman Green [FRICS], who gave evidence as to the values of all the properties save G the Skipton property. He struck me again as being a careful witness. My doubts about him are limited to this: I am not sure how far he was taking account of the specialist nature of tyre depots. The evidence, which I accept, is that for some reason tyre depots appear to be quite a popular type of investment property, not necessarily directly referable to the value of similar warehouses.

All the three witnesses I have mentioned I found to be reliable. But valuation is not an exact science; everybody knows that there is a range within which both rents and prices can be fixed. I formed the view, which is not an uncommon one in these cases, that those valuers called on behalf of the plaintiffs were giving me figures towards the top end of the range and those called for the defendant were giving me figures towards the bottom end of the range. I think I have to take H that into account in reaching any view.

There remains a further valuer, Mr John Paget [FRICS], called on behalf of the defendant to value the Skipton property. I am afraid I did not find his evidence wholly satisfactory. He had entered into the forensic battle with considerable gusto. He seemed to me to be going outside the realms of the bottom and top ends of the range. His attitude was totally dogmatic; even when it was demonstrated to him that he had approached the valuation under a misapprehension as to which part of the premises was occupied by Motorway Tyres & Accessories Ltd, he was quite unable to accept that he was wrong. I attach considerably less weight to his views than I do to those of the other surveyors. J

I now turn to the individual properties. First, Skelmersdale. This was a specialist tyre depot in a new town. The more important factors applicable to it were that at that date — though I believe not now — Skelmersdale was not a favoured area for industrial development; the new town corporation was itself constructing further warehouses which were on the market and unlet. Mr Purle, for the plaintiffs, adamantly objected to my taking evidence of comparables from Mr Green. He referred me to certain authority which says that though he can give his expert view he cannot tell me what actually happened on comparables. I find that authority surprising, but I ignore specific facts which Mr Green gave me. The fact remains that I can certainly entertain his view that there was in Skelmersdale at the time a surplus of warehouse-type accommodation. K

Mr Green said that the rent would have been agreed at £5,200, giving rise to a capital value of £34,500. Mr Royle, in my judgment, went way over the top and did not take account of the local conditions in Skelmersdale in reaching a figure of £7,675 for rent and a capital value of £58,500. Mr Parker, in his letter, put a figure on the rent of £6,020, on the basis of which Mr Eckersley, applying seven years' purchase, reached a capital price of £42,000.

Doing my best to put those factors together, it seems to me right to take an annual rent which would have been agreed at £6,000 on a seven years' purchase, giving rise to a price which would have been realised of £42,000.

Next, as to the Fleetwood property, the evidence of Mr Green was L that the rent was a full one at £6,000 per annum before any review. He pointed out, as appears to have been the case, that when, after the first auction, steps were taken to agree an increased price surveyors on both sides agreed that no increase was appropriate. The property actually sold for £41,000 on that rent at the second auction. Mr Royle suggested that the appropriate rent was £6,975, giving at eight years' purchase a capital value of £51,750.

In my judgment, the fact that no increase was, in fact, agreed when the parties got around to discussing it is a major endorsement of Mr Green's view. I do not think the failure to trigger that rent review clause gave rise to any loss in rent and, accordingly, did not give rise to any diminution in the capital value realised. I therefore put its value at the sale price actually achieved of £41,000. M

As to Alverstone, there is not a major difference between the plaintiffs' valuation and the defendant's. The plaintiffs claim that its value is £23,650; the defendant £22,730. There being such a little difference, and the factors being so minor, it seems to me appropriate to take a figure of £23,000 as being its capital value.

As to Preston, that sold at auction. There was no problem about the rent review clause, and, therefore, its value is £25,000.

Up to now I have been dealing with comparatively simple cases where the properties in question have been tyre depots on their own.

I now come to Skipton, which is much more complex. As I have indicated shortly, Skipton was a composite property. Unhappily, it is the one over which there is the widest difference of view. The plaintiffs claim that its real price would have been £136,920, whereas the defendant's case is that it would have been £67,500, a remarkable figure, considering that it actually sold for £72,000.

There were three parts to the Skipton property. There was a purpose-built, modern warehouse, used as a tyre depot, tenanted by Motorway Tyres & Accessories Ltd, a subsidiary of a public limited company and, therefore, a good covenant.

Next door to that purpose-built warehouse was a two-storey building, the ground floor of which was initially occupied by Motorway Tyres & Accessories Ltd under its lease. The upper floor was occupied by ABC Engineering Ltd, not what, in surveying terms, is called a first-class covenant.

The third part was a property not favourably described by any of the surveyors, an old Victorian warehouse in a considerable state of dereliction occupied, in three separate parts, by Carpet Consignment Ltd, Ellesmere Press Ltd and Retail Bakers. The rent reviews were missed in relation to that part of the premises occupied by Motorway Tyres and that part occupied by Retail Bakers. The fire that I have mentioned was in the part occupied by Retail Bakers.

The evidence is that when you get multiple occupancy of a mixed kind like this it is inclined to depress values. In this case is it not possible to say of all the premises that they are the subject-matter of a good tenant's covenant.

The rent as fixed by Mr Paget I reject. In fact, the figure which he advised would have been the reviewed rent was actually less than that being received at the time of the second auction, even though no rent review had been agreed in relation to Retail Bakers' part of the premises at the time of the second auction.

I even more affirmatively reject his application of 4.4 years' purchase to the premises. I accept his evidence that given the mixed nature of the property, and having regard to the fire, it attracts a much lower years' purchase than do other single-entity tyre depots. But to apply a 4.4 years' purchase — which he did by reference to the years' purchase actually obtained at the second auction — seems to me to overlook the fact that at the second auction there was still no rent agreed in relation to the Retail Bakers' part of the premises.

Mr Royle put a rent on all the parts of it together at £16,875, and Mr Parker £16,375. It seems to me that, given that by the time of the auction it was apparently attracting a rent of £16,350, these figures are not out of line, and I would therefore put a rent of £16,500 per annum on the Skipton premises.

Turning now to its capital value, as I say, I do not accept Mr Paget's multiplier of 4.4. Mr Eckersley, who achieved a value of £136,920, was applying a years' purchase of 8.4. In my judgment, that is much too high. As I have said, the actual prices achieved at the December 1984 auction averaged about 7.4. The Skipton premises are less attractive as an investment than were the other ones because of the multiple occupancy and because of the fire.

Mr Royle I found impressive on this aspect of the case. He had gone through a very careful valuation of the rents, dividing up the Skipton premises into two parts; unit A, being the part leased to Motorway Tyres and I & B Electrical, and the rest of the properties, which he regarded as unit B. He paid attention to the fact that it was a poorer section of the building with poorer covenants, and in relation to unit B he took a yield on existing rents of 20% and 25% on revised rents. That is a proper reflection of their lesser value. In relation to unit A, he adopted a yield of between 10 and 12%. Having gone through that operation, he produced for unit A a total capital value of £70,704, and for unit B £37,908, being a figure of £108,000, but he then took account of the other adverse factors, including, as I understand it, the fact that one was buying a single composite unit where one had to take the bad with the good, and rounded the figure down to £100,000. That seems to me by far the most reliable evidence that I have of the value of Skipton, and I therefore adopt it.

That gives a total value to the first properties of £231,000. In addition, there were the second properties, which realised £55,050, giving rise to a value which would, in my judgment, have been realised at the sale on December 11 1984 if all the properties had been the subject-matter of proper rent reviews by that date.

They would have been subject to the liability to pay auctioneers' and solicitors' costs. My mathematics is notoriously unreliable. I make that at 3.25%, plus VAT £10,689, leaving net proceeds of sale of £275,361, say £275,300.

The actual sum realised for all the properties was £240,550. Therefore, in round figures, if sold with rents properly reviewed, the properties would have achieved £35,000 more than they did.

But that is not the measure of loss recoverable by the plaintiffs. The damages have to be assessed on the basis that they are to be put back into the position they would have been in had the sale taken place on that basis, that is to say if there had been a total realisation of £275,300. It is at this stage that the real complications are capable of arising.

Out of that surplus there would have had to be paid not only the debt of City of Westminster but also a sum for premiums unpaid on policies. I have not to date mentioned this further wrinkle. There was a second charge executed on May 13 1983, being a charge by Mr Knight alone on a life policy taken out by him. The evidence shows that there were arrears on payments payable under that policy as at the date of completion for the December 1984 auction, namely January 16 1985, amounting to £14,816.04. In addition to those unpaid premiums, there was the second mortgage to Williams & Glyn's amounting to £22,928.41 as at January 16 1985. There is a dispute as to how, in computing damages, I should have regard to those two factors.

First, the unpaid premiums. Mr Purle says that they should not be deducted. He puts his argument in this way, that under the mortgage of the policy there is a covenant that "the policy holder will, during the continuance of the security, punctually pay all premiums". That is clause 3(4). Clause 3(5) provides that if he does not do so, it "should be lawful for the lender", that is the City of Westminster, "to pay the same and to recover the same from the policy holder". Clause 3(6) provides that "any sums paid or expended by the lender in effecting a new policy or in respect of premiums are to be added to the principal monies" and I quote "for the time being on this security". So, says Mr Purle, unless there is evidence that City of Westminster, in fact, applied money in paying the premiums on the policy, they would have no right to recover them from Mr Knight.

In my judgment, that is to overlook the terms of the mortgage itself, the principal charge, which describes the property as "Being charged with the payment in accordance with the covenants herein contained of all principal monies, interest or other monies hereby covenanted to be paid by the borrowers". Clause 5 of the principal mortgage contains a covenant "to produce the payment to the lender of the premiums due under the policy of assurance referred to in the schedule to a collateral mortgage of even date herewith", ie the policy mortgage. Therefore, there is a covenant by all the borrowers under the principal charge to procure the payment of the premiums due under the policy of assurance. In my judgment, that includes those premiums within the moneys charged on the properties by the principal mortgage. I am unable to accept Mr Purle's distinction between moneys covenanted to be paid and moneys which the borrowers under the principal charge procure to have paid. Plainly, it seems to me, the purpose of the interaction between the two policies was to secure the payment of the premiums under the collateral mortgage. Therefore, in my judgment, the unpaid premium policies would have been deductible from any surplus recovered and cannot feature in the damages recoverable by Mr and Mrs Knight.

If there had been any evidence as to what would have been the surrender value of the policy that would have been secured if the premiums had been paid, that might have been a recoverable head of damage, but there was no such evidence.

It is common ground, on the basis that the unpaid premiums would have been recovered, that the sum which City of Westminster would have deducted from the net proceeds of sale on January 16 1985 would have amounted to £235.69, leaving a surplus, after discharging all liabilities to City of Westminster, of £39,631. But that is not the end of the story.

If there had been a surplus of that amount in hand on January 16 1985, it would have had to be paid over to Williams & Glyn's as second mortgagee. The amount outstanding under the mortgage was £22,928, leaving a net surplus of £16,703. Mr Purle says that in computing the damages payable I should not deduct from the amount of the damages the sum which would have had to be paid to Williams & Glyn's on January 16 1985. There is evidence before me — and a witness was called who was delicately examined by both sides — to the effect that for internal purposes only Williams & Glyn's had, in fact, written off that debt.

Mr Purle says it would be quite wrong in those circumstances to deduct the amount of the debt since, though it is written off internally, the fact remains that Williams & Glyn's may be able and keen to recover that debt now from Mr Knight if he is put in funds.

There is some support for that — though I noted that nobody asked the question direct of the representative of Williams & Glyn's — in a letter sent by Williams & Glyn's to Mr Lawrence's solicitors this month, just before the trial started, which ends: "We trust you will keep us fully advised of the outcome of the proceedings", which sounds as though they have not abandoned all interest in Mr Knight.

The position seems to me extremely complex, since the quantum of the loss suffered by Mr and Mrs Knight must depend on whether they are still liable to Williams & Glyn's for that amount; not only liable, but likely to have to find it.

A similar point, though arising in a slightly different way, is the net balance, £12,103.50, written off by City of Westminster. That is written off in their internal accounts. There is no evidence that they will not now turn against Mr Knight and seek to recover it. None of that £12,103 would have been due if the properties had sold on December 11 1984 and the City of Westminster, on the basis of the computation I have already made, had been paid off fully by January 16 1985. Therefore, that £12,000 odd, if demanded, would be an additional head of damage.

Now, very understandably, both sides have invited me to try to reach a conclusion as to whether those debts will be demanded by Williams & Glyn's and City of Westminster respectively. I have every sympathy with the desire to produce finality in this case, but there really are limits to what a judge can find as a fact. I am quite unable to form a conclusion as to what will happen in relation to a matter which is entirely within the volition of a party who is not before me. It is a novel position to me.

I find it quite impossible to find the necessary facts to make a concluded finding at this point. The position seems to me to be this. If the properties had realised their full value, and the City of Westminster had been paid off on January 16 1985, the sums due to Williams & Glyn's would have been deducted. As a result, there could have been no further liability on Mr Knight either to the City of Westminster or to Williams & Glyn's. But Mr Knight would never have had this sum of money in hand as cash. So, the position is that the sums would have been paid off, the Williams & Glyn's debt would have been paid off, and Mr Knight would have been free from liability. To allow the deduction of the Williams & Glyn's debt as a once-and-for-all deduction, but not to give Mr and Mrs Knight damages which will represent the sum they have to pay to Williams & Glyn's, if Williams & Glyn's now demand payment, would not be to put them back into the position they would have been apart from the negligence of Mr Lawrence.

What I propose to do is to direct an inquiry as to damages, including in particular an inquiry as to whether any, and if so what, sum is due and owing by the plaintiffs to City of Westminster or Williams & Glyn's and whether such sum is presently being demanded.

In the meantime, I propose to make an order under Ord 29, r 11 for interim payment in respect of damages, since I am entitled, in a case where there is a judgment against a respondent for damages to be assessed, to make an order for interim payment of the damages.

I am quite satisfied, and will, if required, make a final declaration, that in any event the plaintiffs are entitled to the surplus proceeds of sale, namely £39,631, less the sum which would have been payable to Williams & Glyn's, namely £22,928; the balance, the net surplus, being £16,703.

I will add to that an approximation of the lost rental income between the date on which the rent should have been agreed and the date of sale, which I will put at £1,500, making a rounded figure of a minimum sum of damages of
£18,200.

I will also declare that if as a result of the inquiry it proves that Williams & Glyn's or City of Westminster are, in fact, demanding payment of any sum, and are entitled so to demand, the final assessment of damages will include the additional sum so demanded.

How, then, are we to deal with the £16,703? It is argued that the plaintiffs have not shown that they are entitled to the whole of that sum because Mr Sutton was also an owner of the premises. The position seems to me to be clear. Although as between City of Westminster on the one hand and Mr and Mrs Knight and Mr Sutton on the other, all three of them were liable for the debt, as between Mr and Mrs Knight on the one hand and Mr Sutton on the other the properties primarily liable for the payment of the debt are the partnership properties for the benefit of which the moneys were borrowed. The second properties (the properties belonging to Mr and Mrs Knight) were brought in as additional security and they are almost in a position comparable to that enjoyed by a surety. It would seem to me to follow as a matter of ordinary principle that, as between the three borrowers, the debt should be primarily payable out of the first properties, that is to say the partnership properties, in exoneration of the second properties. That, as I say, seems to be a matter of general equitable principle, whether it be called marshalling or the equity of exoneration. The case of *Pittortou Re, ex parte Trustee of the Property of the Bankrupt* [1985] 1 WLR 58 seems to me just one example of the general principle in application.

Therefore, if the surplus had been realised Mr and Mrs Knight would have been entitled to insist that the debt to the City of Westminster, and to Williams & Glyn's, should be paid primarily out of the first properties, that is to say any surplus would have come to them as owners of the second properties up to a figure of £55,050.

Therefore, it is shown that in their individual claims for negligence they have suffered the loss up to any figure of surplus up to £55,050. Accordingly, the figure of £18,200 which I have declared is the quantum of the damage does belong to them exclusively and is recoverable by them to the exclusion of Mr Sutton.

Mr Tyrell, for Mr Lawrence, objects that I cannot decide that point in the absence of Mr Sutton because under Ord 15, r 4(2) it is provided that "where the plaintiff in any action claims any relief to which any other person is entitled jointly with him, all persons so entitled must, subject to the provisions of any Act, and unless the court gives leave to the contrary, be parties to the action".

In my judgment, that does not apply to the present case. I do not think the claim for breach of duty is a joint claim. Each of the persons, Mr Knight, Mrs Knight and Mr Sutton, has a separate claim, though the quantification of their claims involves the rights of third parties.

However, I am deeply sympathetic to Mr Tyrell's submission that, however clear the position seems to me in law, in the absence of Mr Sutton as a party I cannot finally determine the matter as against him.

There is, therefore, the theoretical possibility that the trustee in bankruptcy of Mr Sutton, having witnessed Mr Knight's success in this action, might himself start separate proceedings alleging the same breach of duty against Mr Lawrence and then contend that my decision as to the person to whom the damages are payable is erroneous. That would, obviously, be most undesirable and would be a result that I would be reluctant to reach in the extreme.

What I propose to do to meet that position is to direct that my order be not drawn up for 28 days, that Mr Lawrence's solicitors, or the plaintiffs' solicitors, communicate with the trustee in bankruptcy telling him that he has 28 days in which to apply to me if he wishes to contend that what I have decided is wrong, in which case I will entertain the application.

Finally, there was considerable argument as to what the position would be if the surplus had exceeded £55,050, that is to say the amount required to exonerate the second properties. As, in my judgment, that position can never arise, it is unnecessary for me to determine the point. I record, in case — contrary to my hopes — this case goes on appeal, that the evidence given by Mr Knight as to the amount of the partnership debts paid by him was plainly not satisfactory. I stopped counsel cross-examining him on it because I felt that this case, which has already, through nobody's fault, lasted far longer than the sum at stake demands, would never have come to an end if, in the presence of the court, we had sought to take the partnership account in open court through cross-examination. If the point had been relevant I would have directed an inquiry as to the amount of the partnership debts.

I therefore determine finally that the minimum damages suffered by the plaintiffs amount to £18,200 and I make an order for interim payment of that sum. I also declare that if, on inquiry, it is determined that Williams & Glyn's or City of Westminster are demanding further payments, the amount of such payment is payable by way of extra damages.

Official Referees' Business

November 5 1990

(Before His Honour Judge PETER BOWSHER QC)

WATTS AND ANOTHER v MORROW

Estates Gazette April 13 1991 and April 20 1991

[1991] 14 EG 111 and [1991] 15 EG 113

Negligence — Claim against building surveyor — Husband and wife wished to purchase a trouble-free country property — Reliance on surveyor's report that, although there were defects, the house was overall in sound condition and that defects could be dealt with by regular maintenance — In fact there were defects calling for repairs involving substantial expenditure — Finding of negligence — Measure of damages — Cost of repairs or difference in value — Whether additional damages for distress and inconvenience appropriate — Authorities considered — Damages on cost of repair basis plus damages for distress and inconvenience awarded — Breakdown of marriage not allowed under latter head

The plaintiffs, who were married at the date of the purchase but were subsequently divorced, were a professional couple, high earners with a higher potential, but likely to be overstretched by the cost of the proposed purchase — They were anxious to ensure that the purchase of the property, a substantial and attractive house near Blandford, Dorset, with three-quarters of an acre of garden and three acres of paddock, would not involve them in further expense beyond normal maintenance — The price offered was £177,500 — The defendant, a chartered building surveyor, was instructed by the plaintiffs to carry out a structural survey — The plaintiffs asked for an "in-depth report" and for comments on value — The fee for the report was £400 plus disbursements and VAT — The defendant produced a lengthy report which, although it drew attention to a number of defects, was on the whole reassuring — He advised that the defects to which attention had been drawn could be treated on an on-going basis and he was "satisfied that a comfortable and largely trouble-free dwelling of considerable charm can be attained" — However, a few months after the plaintiffs took possession a builder, employed partly to remedy certain of the defects mentioned by the defendant, discovered, among other matters, that the roof was in a seriously defective condition which would require expensive building work to put right — The plaintiffs in alarm instructed other surveyors to carry out a full structural survey, not mentioning that one had already been carried out — The new survey report showed a very different picture from that presented by the defendant — There were serious defects calling for substantial expenditure, including the cost of a major overhaul of the roof and work to chimneys, main walls, windows and floor timbers — It became clear that the defendant's report had been unduly optimistic and reassuring, leading the plaintiffs to the reasonable but incorrect conclusion that no major repairs would be required in the immediate future

After hearing evidence from experts on both sides and considering authorities, the judge concluded that the defendant had been not only wrong but negligent — Granted that a difference of opinion does not necessarily imply negligence, the defendant had been negligent by falling below the standard of reasonable care and skill required of the ordinary professional man exercising the same functions as himself — Further, although a departure from the recommendations contained in the RICS practice note on "Structural Surveys of Residential Property" was not in itself an indication of negligence, there had been one departure by the defendant which was of some materiality — He had dictated his survey report (not notes for it) directly into a dictating machine as he walked round the property — Incidentally this meant that he had no notes to disclose on discovery of documents — This practice led to a report which was "strong on immediate detail but excessively, and I regret to have to say negligently, weak on reflective thought" — The judge rejected a suggestion of contributory negligence on the part of the plaintiffs — He found that the reasonable cost of repairing the relevant defects was £33,961.35

The next question was the correct measure of damages — The plaintiffs claimed damages based on the cost of repairs plus general damages for inconvenience and distress, with interest under both heads — Alternatively, they claimed damages calculated on the "difference of value" basis — On this basis the damages were the difference between the value of the property at the date of purchase, namely £162,500 (according to unchallenged evidence at the trial) and the price paid by the plaintiffs in reliance on the defendant's report, £177,500, a difference of £15,000 — The judge at the request of the parties set out the sums under each alternative; £33,961.35 on the repairs basis, £15,000 on the difference of value basis, plus appropriate interest at 15% in each case — He also held that on either basis the plaintiffs were entitled in addition to damages for distress and inconvenience, because a prospective house-buyer expects to obtain, *inter alia,* "peace of mind and freedom from distress" — He assessed this additional amount at £8,000 (£4,000 in respect of each plaintiff) — As a matter of policy, however, he disregarded the breakdown of the plaintiffs' marriage as an allowable head of damage

In the end the judge, following largely his own judgment in the case of *Syrett* v *Carr & Neave,* held that the correct basis in the circumstances of the present case was the cost of repairs — The result was that the plaintiffs were awarded £33,961.35 plus £8,000, a total of £41,961.35, plus interest at 15% (from the appropriate dates) on the £33,961.35 and interest at 2% on the £8,000 from the date of issue of the writ

The following cases are referred to in this report.

Bigg v *Howard Son & Gooch* [1990] 1 EGLR 173; [1990] 12 EG 111
Clark v *MacLennan* [1983] 1 All ER 416
Hayes v *James & Charles Dodd* [1990] 2 All ER 815
Hipkins v *Jack Cotton Partnership* [1989] 2 EGLR 157; [1989] 45 EG 163
Maynard v *West Midlands Regional Health Authority* [1984] 1 WLR 634; [1985] 1 All ER 635, HL
Midland Bank Trust Co Ltd v *Hett, Stubbs & Kemp* [1979] Ch 384; [1978] 3 WLR 167; [1978] 3 All ER 571
Philips v *Ward* [1956] 1 WLR 471; [1956] 1 All ER 874, CA
Perry v *Sidney Phillips & Son* [1982] 1 WLR 1297; [1982] 3 All ER 705; [1982] EGD 412; (1982) 263 EG 888, CA
Steward v *Rapley* [1989] 1 EGLR 159; [1989] 15 EG 198
Stewart v *HA Brechin & Co* 1959 SC 306 (Outer House)
Strover v *Harrington* [1988] Ch 390; [1988] 2 WLR 572; [1988] 1 All ER 769; [1988] 1 EGLR 173; [1988] 9 EG 61
Syrett v *Carr & Neave* [1990] 2 EGLR 161; [1990] 48 EG 118
Wilsher v *Essex Area Health Authority* [1988] AC 1074; [1988] 2 WLR 557; [1988] 1 ALL ER 871, HL

The plaintiffs in this case, Ian Roscoe Watts and Lesley Mary Samuel Watts, claimed damages from the defendant, Ralph Morrow FRICS, a building surveyor, alleging negligence by the defendant in carrying out a structural survey of a property purchased by the plaintiffs in Dorset, near Blandford and the River Stour.

Jonathan Acton-Davis (instructed by Goodman Derrick & Co) appeared on behalf of the plaintiffs; Iain Hughes (instructed by Pinsent & Co) represented the defendant.

Giving judgment, JUDGE PETER BOWSHER QC said: The plaintiffs claim damages for negligence from the defendant in his capacity as a building surveyor.

The first plaintiff has degrees in economics from the universities of Warwick and Oxford. He is employed as a stockbroker by Barclays de Zoete Wedd UK Equities Ltd. The second plaintiff has a law degree from Cambridge University and is a solicitor employed as a director of Kleinwort Benson Ltd since 1986. She took her articles with Slaughter & May and later qualified to practise as a lawyer in

A New York State. Both plaintiffs are extremely intelligent and well used to reading reports.

The plaintiffs were divorced in April 1990, having separated early in 1989.

In the summer of 1986, when the plaintiffs were still happily married, they were living in a house in Cloudesley Road, Islington, London N1, where the second plaintiff still lives. They both had extremely demanding jobs in the City, and they decided to look for a country home for use at the weekends and during holidays. They had no children and intended to have no children. They concentrated their search upon Dorset.

Since they both had little leisure time, they both wanted a house which would be, so far as possible, trouble-free. They wanted to buy B a house into which they could move without undertaking any substantial works of repair. Although each had a substantial income, their incomes were already stretched and they felt that though they could buy a new house within a budget of £170,000, they could not afford to buy a house which required the carrying out of any expensive repairs, nor did they have the time or the surplus energy to get involved in doing up a house which required extensive repairs or improvements.

Mrs Watts had in mind a small country cottage with roses round the door, but soon accepted that there were few such cottages in Dorset. It seems that Mrs Watts was at the time even more occupied with business than her husband, and as a result he was more involved C in looking for properties than she was. At the time the plaintiffs were considering buying a country house, Mrs Watts became involved in several major transactions which absorbed all her attention at weekends and also took her abroad. It was Mr Watts who found and fell in love with Nutford Farmhouse, and Mrs Watts visited it only once before they made an offer. Her view was that it was larger than she had hoped for, "but it was very beautiful, a house with a heart, and difficult to resist".

The house is L-shaped and probably built in the 18th and 19th centuries, though described in the selling agents' particulars as 17th century. It is a substantial and attractive house in attractive surroundings, having three quarters of an acre of garden and a further three acres of paddock. It is set in the countryside about three D quarters of a mile from Blandford and not far from the River Stour and the vendor was willing to let the purchaser have fishing rights in the river. The main living accommodation is on the ground and first floors, but there are four rooms in the attic lit by dormer windows. There was no bathroom or lavatory in the attic and the roof was not insulated, so the attic rooms were really useful only for overflow accommodation in the summer. The plaintiffs wanted at least two bedrooms so that they could invite friends and business clients to stay, and on the first floor there were three bedrooms, though only one bathroom.

The asking price for the house was £175,000, so it was outside the budget on which the plaintiffs had decided. The selling agents told E Mr Watts that there was another prospective purchaser who was interested in buying at the asking price of £175,000 and they advised him to make his "best offer". The two plaintiffs decided that the most they could spend would be £177,500. An offer in that sum was acceptable to the vendors, but that was certainly more than Mrs Watts thought they should spend on a second home, and the plaintiffs agreed together that before proceeding they would get a full and detailed survey to ensure that there were going to be no unexpected costs. Although they were high earners, this purchase would overstretch them. Accordingly, Mr Watts asked the vendors' agents to recommend a local surveyor. They recommended three, and Mr Watts chose the defendant at random from among the three.

The defendant, Mr Morrow [FRICS], is in practice on his own, and F when Mr Watts rang his office on August 19 1986 he spoke to the defendant's wife. I accept Mr Watts's unchallenged evidence of his conversation, which was corroborated by a note taken by Mrs Morrow, who did not give evidence. Mr Watts said that he wanted a full structural survey of the property and he wanted to be sure that the price offered of £177,500 was the right one in the current market. Mr Watts explained that he and his wife lived in London and would not be around to carry out any investigations into the condition of the house, that they had looked at the house from the point of view of its attractiveness and amenity value but had paid no regard to its state of repair. Mrs Morrow's note was written on a printed form entitled "Structural Survey Request". In a box headed "Special Instructions", Mrs Morrow wrote:

> In-depth report required! G
> Paying £177,500, and would appreciate your comments on value.

The defendant said in evidence that he is a chartered building surveyor, not a valuer, and that as a matter of ordinary practice his wife would explain that he was not a valuer, but that was not put to Mr Watts, and there is no evidence that either the defendant or his wife ever said any such thing to Mr Watts.

The defendant did not confirm the instructions in writing, but on August 26 he wrote to Mr Watts enclosing a structural report on the property. In the covering letter he invited Mr Watts to contact him if there were any matters he wanted to discuss or if he could be of any help "in the orchestration of works of renovation and repair recommended in my report". The fee charged was £400 plus disbursements and VAT. H

The report was detailed and lengthy, running to 27 pages. It contained no comment about valuation. The report mentioned defects and made recommendations for repairs, but both Mr and Mrs Watts found it reassuring. On the basis of their experience of the cost of repairs to their London home, and as amateur householders, they estimated that the cost of the recommended repairs together with the cost of converting a part of the house into a wine store would be about £3,000. Mr Watts telephoned Mr Morrow and asked him to recommend a builder who would do the work on the wine store and also remedy the defects identified in Mr Morrow's report. Mr Morrow recommended a Mr Cyril White, who traded under the name Okeford Crafts. Mr Watts asked the defendant whether the price J they had offered was fair and whether the repairs he had recommended would result in substantial expenditure. The defendant replied that the valuation was fair and that no repairs would be substantial in terms of cost.

A short time later, the plaintiffs would both receive greater cash rewards for their work, but at that time, though by no means poor, they were over-extended and in Mr Watts's words, were "strapped" for cash. It was very important indeed to them that they would not find themselves involved in unexpected expenditure, and they relied on the defendant's advice to avoid paying more than the property was worth and to guard themselves against repair bills which they could not afford.

Relying on the defendant's written and oral reports, the plaintiffs K proceeded with the purchase of the house. As a result of the vendors' requests, contracts were not exchanged until October 15 1986 and completion was delayed until April 10 1987. So, although they made an offer, which was not binding in law, in August 1986, the plaintiffs did not get possession until eight months later. Meanwhile, they received an offer of a mortgage from Kleinwort Benson on ordinary commercial terms (not on terms showing special favour to an employee). That offer of a mortgage, which was accepted, was accompanied by a mortgage valuation report by Fox & Sons dated September 15 1986, valuing the property at £172,000. The plaintiffs quite reasonably paid little attention to this report because the defendant had made a full structural survey for them and in their view L the Fox & Sons report was limited to protecting Kleinwort Benson to the extent of their loan. Indeed, the report itself contained the following note:

> IMPORTANT NOTICE TO THE APPLICANT
>
> 1 This report has been obtained solely to enable Kleinwort Benson Ltd to consider whether the property is suitable security having regard to the amount of the proposed loan. It is NOT a market valuation or structural survey and you are strongly advised to engage an independent survey to protect your own interests.
>
> 2 The valuer has made this report without any liability to you. Kleinwort Benson Ltd, its agents and the valuer are not responsible to you for the accuracy or opinions in this report.

At the trial, the evidence (unchallenged in this respect) of Mr Peter M Wadey FRICS of Poole, Dorset, a surveyor and valuer called on behalf of the plaintiffs, was that in October 1986 the value of the house in its true condition was £162,500 and its value in the condition as described by the defendant was £172,500. Mr Wadey considered that there was no significant difference in the valuation between August and October 1986. On that basis, in reliance on the defendant's advice, the plaintiffs paid for the property £5,500 more than the property was worth on the defendant's assessment of its condition and £15,000 more than it was worth on Mr Wadey's assessment of its condition.

Mr Wadey became involved in this way. After taking possession in April 1987, the plaintiffs started working on the garden and soon found that one of the boundary walls was in a bad state. Through his solicitors, he asked the defendant to give further advice. That matter was dealt with and forms no part of this action. In May 1987, Mr Watts contacted Mr White in accordance with the defendant's recommendation for a quotation for work on the wine store and for remedying certain defects identified by the defendant, namely the roof flashings and defective windows. In June, Mr White went to the house and then telephoned Mr Watts in London. Mr White said that the first-floor windows were "useless" and should be replaced but, more important, commented on the roof (which he had not been asked to look at) by saying that: "One decent fall of snow and the whole roof will come down". By lifting a few tiles, Mr White had discovered that some laths had been so worn away over the years that they no longer existed in many places and that the wooden pegs placed through holes in the handmade tiles to suspend them from the laths had in many instances rotted away. Mr Watts was very disturbed indeed by those comments but was sufficiently canny not to take the unsupported word of a builder that a large amount of expensive building work was required on his house, so, with advice from his solicitors, Goodman Derrick & Co, he decided to ask another independent surveyor to prepare another full structural survey on the premises, without disclosing the fact that this was a check on a previous surveyor's report. Mr Watts went to Bernard Thorpe & Partners, who instructed Mr Wadey who acted as an independent consultant to their firm. Mr Wadey is a FRICS and has been working as a building surveyor and valuer in Dorset for the last 22 years. Towards the end of July 1987 Mr Wadey was asked by Bernard Thorpe & Partners to undertake a full structural survey of Nutford Farmhouse for the plaintiffs. Mr Wadey was not aware that the plaintiffs had already bought the property, nor was he aware that a previous survey had been carried out on their behalf by the defendant. When he later discovered the true position, he was rather annoyed.

There was an important difference between the conditions of Mr Wadey's survey and the conditions of the defendant's survey. One of the important features, as it turns out, was the floor of the landing on the first floor. When the defendant inspected the floor of the first-floor landing, it was covered by a fitted carpet. The defendant lifted a corner of the carpet, but Mr Wadey had the advantage of seeing under the floor, because he was asked to lift carpets and floorboards, particularly on the landing on the first floor, and in the bathroom, and did so with the help of a handyman employed by the plaintiffs. If I were to accept Mr White's evidence in its totality, I would also find that Mr Wadey had the benefit of Mr White drawing his attention to certain defects. But I accept Mr Wadey's evidence of his inspection that Mr White was not on site at the time. Mr White must have confused another visit to the site in his memory. Another important difference was that, unlike the defendant, Mr Wadey felt that the survey was sufficiently difficult to require him to return on a second day to complete his inspection, which took place on July 29 and 30 1989.

The reports of Mr Wadey and Mr Morrow were very different both in form and content. Mr Wadey reported that there were a number of defects which would involve current expenditure or future liability. In particular, he said, the roof was due for a major overhaul of the coverings including stripping tiles, repairing/replacing the roof structure as required, rebuilding the top courses of brickwork, specialist woodworm treatment, felting, rebattening and retiling. The chimneys and facing bricks to the main walls required repointing in places, and some bricks in the chimneys required cutting out and renewing, and lead flashings needed to be installed. The window casements and frames generally needed to be upgraded and replaced. The first-floor timbers needed specialist woodworm treatment and refixing of firring pieces and extensive renewal of floor boardings. Mr Wadey also referred to other matters not the subject of this action.

The defendant in his report pointed out many defects and made many recommendations for repair. It is accepted on behalf of the plaintiffs that many of those recommendations were sound. But the general criticism made is that, as a recurring theme in the report, the defendant pointed out a defect and then gave a reassurance that the defect could be dealt with as part of ordinary ongoing maintenance and repair. While criticisms have been made of the adequacy of the last two paragraphs headed "Conclusion", the conclusion does go far to summarise the tone of the report, though not all its detail. The conclusion in full is as follows:

> Despite earlier minor settlement I found the overall dwelling-house to be sound, stable and in good condition. Attention is particularly required to the sometime eradication of wood borer in roofing timbers, and to minor works of eradicating continued dampness in ground-floor walls. Ideally some insulation would be introduced to upper ceiling areas, but in general, the defects referred to within this report can normally be identified as being associated with regular maintenance required of a building of this age and type.
>
> The foul drainage arrangements, while apparently adequate, and [*sic*] not ideal, and some further rationalisation and improvement of this may be required in the sometime future, but with attention given to the various aspects referred to within this report, which could well be attended to on an ongoing basis, I am satisfied that a comfortable and largely trouble-free dwelling of considerable charm can be attained.

In citing that conclusion, I do not forget that the defendant has more than once stated, quite properly, that his report should be read as a whole. Indeed it should, and it must be stressed that the defendant has taken great care to mention many matters, about many of which there is no complaint. But nowhere in the report is there any warning that it would be necessary or desirable to undertake an immediate and heavy expenditure on the property. The defendant contends that his report is not only not negligent but is also correct in its statements of fact and opinion, and that Mr Wadey (supported by Mr White) was wrong to recommend the substantial immediate works which have been undertaken. It is also pleaded on his behalf that if the plaintiffs suffered the damage claimed, their damage was caused or contributed to by their own negligence in failing to read or pay sufficient attention to the body of the defendant's report.

Having received the defendant's report, both plaintiffs read it. I have no hesitation in accepting their evidence that they read the whole of the report and read it with care. They very reasonably concluded that there was nothing to suggest that any major repairs would be required in the foreseeable future and that nothing was required which could not wait until the two plaintiffs between them had the funds to spare for such work. I also accept the evidence of the plaintiffs that it was in reliance on the report of the defendants that they bought the house. I accept the evidence of Mrs Watts that if she had known of the defects as set out in Mr Wadey's report she would not have gone ahead with the purchase, and I accept the evidence of Mr Watts that if he had known of the defects as set out in Mr Wadey's report he would not have gone ahead with the purchase or would have done so only if he had been able to negotiate a substantial reduction in the price. Mr Watts said that if he had received Mr Wadey's report before the purchase, he would have got quotations for the work recommended on the roof and to the floors and windows and would have sought to negotiate a reduction in price, and if unsuccessful in that attempt he "would have walked away".

I have no hesitation whatsoever in rejecting the plea of contributory negligence made against the plaintiffs.

It is necessary to consider whether the defendant was not only wrong in giving his report but also negligent. Everyone makes mistakes, and before making a finding that a professional man has not only made a mistake but also been negligent in making it, one looks with care at the evidence.

On behalf of the plaintiffs there were called two experts. Mr Wadey had previously given evidence as an expert valuer, but he does not confine his expertise to valuation. Unlike the defendant, he is not a chartered building surveyor, but he has sat and passed those papers in the RICS examination which are concerned with building surveying and he has considerable experience in practice of surveying buildings. He had the considerable advantage of being asked to make an ordinary structural survey of the property before it was put into a different state.

Also called on behalf of the plaintiffs was Mr John Lloyd, a FRICS who is now a partner in Hurley Lloyd Thorpe, a partnership formed as a result of the decision of his former firm, Bernard Thorpe & Partners, to cease handling residential property. Mr Lloyd does not himself do surveys, but he has long experience in advising purchasers and vendors on the sale and purchase of properties on the basis of surveys undertaken by employees responsible to him. He did not see the building in its unaltered state. He was not trained as a building surveyor, but as a part of his training as a chartered surveyor he had to take papers in building construction. He in fact qualified as a surveyor in land agency and agriculture and for many years acted as adviser to many large estates including Sandringham. But Mr Lloyd does have the expertise required to assess reports of surveys of

A buildings. He naturally and properly accepted that it is possible for two careful and competent surveyors to reach different opinions. But, after reading the two reports of Mr Wadey and Mr Morrow, he was firmly of the opinion (which I accept) that it was not possible for two careful and competent surveyors to reach opinions so far apart as Mr Wadey and the defendant on the subject of the roof, the bulging of the walls, the first-floor flooring, and the windows. Although he did not say so, and it would have been inadmissible for him to have done so, it follows from Mr Lloyd's views that whichever of Mr Wadey and Mr Morrow was wrong was also negligent. Mr Lloyd preferred Mr Wadey's report because of the amount of money that was spent on the property and because Mr Wadey was supported by the evidence of Mr White. The fact that much money was spent on the
B house dealing with the matters to which I have referred is not of itself evidence that Mr Wadey was right: it was submitted on behalf of the defendant that the reason for the money being spent was that Mr White made excessive criticisms of the building in order to get work for himself, and that for reasons of self interest he was unwilling to undertake small works of repair like replacing one or two tiles, instead insisting that the whole roof had to be replaced. I do not accept that criticism of Mr White. He was, after all, recommended to the plaintiffs by the defendant. Moreover, although he tended to express himself both in and out of court in language more vivid than would be chosen by a professional man, he seemed to me to be an honest and careful builder who could be trusted to do good work and
C not to make work for himself by giving excessively pessimistic views. Over a fairly long period of time, he gained and kept the trust and confidence of the plaintiffs, who are not people to be easily taken in. The value of the work — the subject of the complaint in this action — is only about 20% of the total value of the work which Mr White has done on the property. Even if he had been inclined to deceive the plaintiffs with bad advice (and I do not believe that he was so inclined) it was not in his interest to do so. Contrary to the expectations of the plaintiffs, their financial situation improved much more quickly than they had expected, and quite apart from the matters the subject of the complaint in this action, about £150,000 has been spent on repairs and improvements to the property. If Mr White had lost the confidence of the plaintiffs he would have lost a
D great deal of work. I should add that Mr Watts still owns the property and uses it as a second home. Quite apart from the evidence of Mr White, I wish to stress that the evidence of Mr Wadey was given in restrained and wholly convincing terms.

Called on behalf of the defendant was Mr Simon Gregory, who is a Bachelor of Science in Building Surveying and is also a FRICS, building surveying division. He is a chartered building surveyor who does building surveys continually. He volunteered certain criticisms of the defendant's report and accepted further criticisms in cross-examination.

Since I am asked to take the serious step of condemning a professional man as having been negligent in his profession, it is
E important that I should remind myself of the legal issues involved.

The standard of care required of the defendant is the same as that required of any other surveyor or other professional man. The standard required is that of the ordinary skilled man exercising the same skill as himself. He must "display and apply reasonable care and a reasonable standard of professional competence in doing the work [he was] employed to do": *Stewart* v *H A Brechin & Co* 1959 SC 306 *per* Lord Cameron. Moreover, a difference of view between the defendant and other professional men does not of itself and without more ado establish negligence on the part of the defendant.

> The standard of reasonable care and skill allows for a margin of differing opinion and even a degree of error. It is not enough to prove that the surveyor
> F gave the wrong advice or that where there is room for two views he took the wrong view: *Jackson and Powell, Professional Negligence*, 2nd ed, p 136.

Lord Scarman expressed a similar view in a medical negligence case, *Maynard* v *West Midlands Regional Health Authority* [1984] 1 WLR 634 at p 638:

> Differences of opinion and practice exist, and will always exist, in the medical as in other professions. There is seldom any one answer exclusive of all others to problems of professional judgment. A court may prefer one body of opinion to another: but that is no basis for a conclusion of negligence:

Counsel for the defendant has criticised the nature of the evidence called on behalf of the plaintiffs to prove negligence. It is necessary to consider the nature of the evidence required in cases such as this. Counsel for the plaintiffs cited a passage from the judgment of Oliver J in *Midland Bank Trust Co Ltd* v *Hett, Stubbs & Kemp* [1979] Ch G
384 at p 402:

> This new plea does raise an issue of law not apparent upon the original pleadings, namely, what is the scope of a solicitor's duty when he is consulted about a particular aspect of a problem — is he entitled to confine himself to the particular matters for which he is retained to advise or must he consider all the circumstances affecting the underlying data including hypothetical circumstances or risks to which his attention is not directed and upon which his advice is not specifically sought?
>
> As to this I have heard the evidence of a number of practising solicitors . . . I must say that I doubt the value, or even the admissibility, of this sort of evidence, which seems to be becoming customary in cases of this type. The extent of the legal duty in any given situation must, I think, be a question of law for the court. Clearly, if there is some practice in a particular profession, H
> some accepted standard of conduct which is laid down by a professional institute or sanctioned by common usage, evidence of that can and ought to be received. But evidence which really amounts to no more than an expression of opinion by a particular practitioner of what he thinks that he would have done had he been placed, hypothetically and without the benefit of hindsight, in the position of the defendants, is of little assistance to the court; whilst evidence of the witnesses' view of what, as a matter of law, the solicitor's duty was in the particular circumstances of the case is, I should have thought, inadmissible, for that is the very question which it is the court's function to decide.

It is to be noted that Lord Oliver was considering the scope of the duty owed by a solicitor in a particular circumstance; he was not considering the standard of care to be adopted with regard to matters within that duty. Even so, he plainly considered it proper to adduce evidence of accepted practice within the profession or of standards J
laid down by a professional body and there are some cases where the absence of such evidence will be fatal to the plaintiff's claim. One such case was *Strover* v *Harrington* [1988] Ch 390*. In that case negligence was alleged against a surveyor who had reported that a house was served by main drainage through another property when in fact the drains went into a cesspool. The surveyor made that report relying on what he was told by the vendors. There was no evidence that it was negligent for a surveyor to rely on what he was told by the vendors about the drains, it being necessary to make a special test to check the information, and it was alleged that the surveyor was negligent in stating without qualification that there was main drainage. Sir Nicolas Browne-Wilkinson V-C expressed the view that K
it would be imposing too high a duty on a surveyor to hold him liable for not qualifying his report in this respect and added:

> That would be my view of the matter if it were for me to decide; but in my judgment it is not for me to apply my yardstick of what is or is not negligent. As a professional man, Mr Butler can only be held liable for negligence in the conduct of his profession if he acted in a way which no surveyor of ordinary skill would be guilty of, if acting with ordinary care. No expert evidence was led as to the practice of careful surveyors, and the burden is on the purchasers [the plaintiffs] to prove negligence. Therefore, in my judgment, there is no evidence on which I could hold Mr Butler to have been negligent, even if, contrary to my own inclinations, I thought he had been.

I do not find that statement of the Vice-Chancellor to be inconsistent with the earlier statement of Lord Oliver, since Lord Oliver was L
considering the legal question of the existence of a duty and the Vice-Chancellor was considering the standard of care once the duty is established, and both regarded the leading of evidence as to professional practice as relevant. In the latter case, where the standard of care was under consideration, such evidence was not only relevant but also necessary.

So far as professional practice is concerned, the plaintiffs led evidence of a practice note of the Royal Institution of Chartered Surveyors headed "Structural Surveys of Residential Property". The defendant at first said that he was unaware of that note and then qualified his evidence by saying that he had read it not as a separate document but in a book. The defendant certainly departed from that practice note in many respects. Despite those departures, he could M
have come to the right answer in the end, but his departures from (and, I have to say, ignorance of) the guidance given by his professional body made it more difficult for him to come to the right answer and contributed to his failure to come to the right answer.

The most extraordinary practice adopted by the defendant was not specifically disapproved by the institution, but the absence of disapproval was, I suspect, due to their failure to realise that some members might adopt the course taken by the defendant. It is the practice of the defendant to dictate his survey as he walks around the property during his inspection. He does not dictate notes into a

---

*Editor's note: Also reported at [1988] 1 EGLR 173.

dictating machine, he dictates his survey report into a dictating machine on site. When he returns to the office, he gives the tapes to his secretary, who types them up and the report is then amended and sent to the client. That was the practice adopted by the defendant on this occasion, so that he had no notes to disclose on discovery of documents. It also led to his report being lengthy and diffuse and to its conclusion being inadequate. That practice is inconsistent with the spirit of the practice note, which contains the following recommendation:

Site Notes
It is recommended that all details of the property be recorded in writing, together with the date of inspection, the names of individuals present at the time, weather conditions affecting the inspection, sources of information and any other relevant matters. These notes, both typed and handwritten, should be retained for as long as practicable . . .

Apart from what was in the report, the defendant had no notes either written or typed. Moreover, he was unable to give evidence of his first impressions because his first drafts were not retained and were destroyed. While he did not say so in as many words, Mr Gregory, giving evidence for the defendant, clearly did not approve of this practice.

Many of the departures from the recommended practice adopted by the defendant may not have produced important errors, but this departure, in my judgment, was serious. It led to a report which was strong on immediate detail but excessively, and I regretfully have to say negligently, weak on reflective thought.

As to the effect of a departure from general practice, counsel for the plaintiffs relied on the decision of Peter Pain J in *Clark* v *MacLennan* [1983] 1 All ER 416. Counsel expressly disclaimed any reliance on the suggestion in that decision that the burden of proof shifts where a departure from normal practice has been shown, but none the less relied on other parts of the decision. In the light of the disapproval of that decision by the House of Lords in *Wilsher* v *Essex Area Health Authority* [1988] AC 1074, I think it would be unsafe to rely on any part of that decision.

As regards the roof, the defendant devoted four pages to the roof covering and three pages to the roof construction. Any summary of those pages is likely to lead to the criticism that I have failed to give adequate attention to the text as a whole. But I have to say that my overall impression is one of reassurance being given to the plaintiffs. For example, it was said in relation to the northern pitch of the roof that there were some failures of the battens supporting the tiles, but that observation was qualified with the comment (which I find to be unjustified in its reassurance) that "while this is not seen to be particularly important at present, it all combines to indicate that a degree of ongoing maintenance and repair is required to maintain this roof in sound and serviceable condition". Turning to other parts of the roof, the defendant used such terms as "sound and satisfactory" and "satisfactory for the time being". In relation to the roof as a whole he said: "a degree of ongoing maintenance and repair will be required to replace occasional tiles becoming damaged, particularly by frost". Two pages later on he stated: "Clay tiles are affixed to very slender battens, which as stated elsewhere have failed in minor pockets, and occasional repair will be required as part of ongoing maintenance, but in general the overall covering of the roof is in my opinion sound, and should remain adequately so for a period of, say, 25 years before major reroofing may have to be contemplated." Two pages later, when referring to the construction of the roof, the defendant referred again to the 25-year life of the roof. He wrote:

Thereafter, no direct inspection of rafters was possible, except by lifting occasional tiles on the roof covering, and this indicates that timbers throughout the roof are subject to widespread attack by wood borer, to a point where a fair degree of brittleness exists, and where tiling battens are also suffering. *It is therefore recommended* that, as a minimum, access be gained to the apex of the roof void, and that treatment be applied as far as is possible to arrest such attack. I would accept that the likelihood of further significant damage being caused within the suggested twenty-five year life of this roof would be minimal, and as such therefore treatment could be deferred if so required until the roof covering has been removed and full access is available to timbers throughout . . . For the same reasons that no access is available to the roof void, there is no insulation available across ceilings to the upper floor but, given the fact that these areas are apparently only used as occasional overspill areas, this may not be particularly important, with the main space acting as insulation to the ground and first floor areas of accommodation. However, if it is intended that the attic floors be brought into full use, serious consideration should perhaps be given to undertaking all required repairs and re-roofing of the main covering, in order to provide opportunity to both treat timbers and to introduce insulation as necessary. As stated however, it would be my view that all of this could well be deferred.

The plaintiffs had no intention of bringing the attic floors into use (apart from occasional use in the summer) and so they saw no reason at all why any extensive work (other than annual replacement of a few tiles) would be required on the roof for 25 years. On the evidence before me, I am in no doubt that the impression so given to the plaintiffs in that regard was quite wrong and negligent. On the other hand, having later been persuaded by Mr White and Mr Wadey that the roof had to be generally overhauled, the plaintiffs have put in hand all those works which went with the general overhaul of the roof which would go to make the attic a normally habitable part of the house. It emerged during the trial that the first plaintiff has gone even further. Prompted by an application for listed building planning consent put in by a neighbour, Mr Watts has applied for consent for the addition of further dormer windows to the house. He does not have any intention at present to put in those extra dormer windows, but he thinks that this is a good moment to make the application in case he should in the future decide that he wants to enlarge and improve the usable area of the attic.

The outward bulging of the walls was another problem with the property. By careful and detailed reference to parts of his report, the defendant was able to show that he was aware of the problem, but nowhere did he meet it head-on and make plain and unambiguous recommendations in respect of the problem. Instead, he made one of his inappropriately reassuring comments: "The overall wall is sound, stable and in good condition" — a comment which he repeated more than once in differing terms.

The defendant paid attention to the floors both in the attic and on the first floor. Attention at the trial was concentrated mainly on the floor of the first-floor level. The defendant referred to certain defects but again gave reassurance. It is again difficult to select quotations from long passages, but the flavour of the report is sufficiently given by the following:

There is again evidence that the floor deck has been repaired (though not renewed) in relatively recent times, which would again be seen to be an integral part of providing access to apply treatment to timbers to eradicate wood borer. There is a fair degree of creaking in floorboards, where fixings are obviously a little loose, both as part of those repairs and overhaul and as part of the installation of services. Detailed examination of timbers throughout the floor construction was of course impossible by virtue of fully fitted coverings and furnishings, but all evidence is that the floor has been comprehensively repaired and treated in recent times to provide a sound and secure arrangement. As before, any guarantee in respect of treatment should be identified. In the absence of any guarantees being available, it should perhaps be accepted that further treatment of timbers may prove to be required in the future but I would not expect this to be necessary within the foreseeable future.

Here, as elsewhere, the defendant made plain the limitation on his inspection because of the fitted carpets and furnishings. But instead of saying that he was unable to express a firm view, he gave reassurance. He should not have given even a qualified assurance, because even without lifting carpets there was evidence which should have made it plain that there was need for further investigation. There were loose floorboards and defective floorboards and the floor was on a slope. The defendant qualified his reassurance by questioning whether a guarantee was in existence in relation to the floor and the treatment of the timbers but, having raised the query, he then gave further reassurance by saying that, in effect, even if there was no guarantee, further treatment of the timbers "may prove to be required in the future but I would not expect this to be necessary within the foreseeable future". There are many more words to this section of the report than I have quoted, but I have to say that my impression from the whole of the section is that the general structure of the floor was reported as sound and satisfactory (despite the limited nature of the inspection), and that whether there is a guarantee in existence as to the treatment of the floor, treatment of the timbers was not expected to be necessary within the foreseeable future. To express such a view was, in the light of the limited nature of the inspection, negligent. If the defendant had said that because of the limited nature of the inspection he could not express any view about the floor, he could not have been criticised, though he might have been asked to go back and make a more detailed inspection. On the inspection he did make, he could not reasonably express the views which he did express.

The other subject of the report criticised by the plaintiffs at the trial was the report on the windows. There was a substantial amount of rot

A and degeneration in the windows. Some of this the defendant pointed out, though without using the word "rot". Although some defects were pointed out, repeated reassurances were given: "adequately sound for the time being . . . at present existing windows are in my opinion acceptable for a property of this type . . . generally adequate . . . adequately sound for the foreseeable future . . .". What was recommended was "ongoing maintenance and repair" and "maintenance and making good is required in the short term, with longer term repairs becoming necessary". Again, that was far too optimistic. Many windows were no longer weathertight, and some sections were decayed by wet rot.

For all these reasons, I find that the defendant is liable to the plaintiffs in the respects advanced at the trial (which were not all of
B the criticisms advanced in the statement of claim in its original form).

Although the reasonableness of the cost of repairs has been contested, I find that the reasonable cost of the repairs of the defects in respect of which this claim has been limited was £33,961.35.

The plaintiffs claim damages in the amount of those repairs plus interest plus general damages for inconvenience and distress and interest at 2% from the date of the issue of the writ on the latter amount of damages. In the alternative to the cost of repairs, the plaintiffs claim damages calculated by reference to the "loss of value".

I have been asked to assess the damages for each of the alternative bases and, since an appeal is not unlikely whatever may be my
C decision, I gladly do so.

If the appropriate measure of damages is the cost of repairs, the damages are £33,961.35. The plaintiffs ask for 15% interest on those damages, being the Judgment Act Rate. The defendant submits that the appropriate rate should be the Special Account Rate, which at all material times was less than 15%. In the circumstances of this case it seems to me that the plaintiffs should be awarded 15% on the damages for repairs as from each date on which money was paid for repairs.

If the appropriate measure of damages is the "difference in value", I assess the damages at the difference between the value of the property at the date of contract, £162,500, and the sum which the plaintiffs paid in reliance on the defendant's statement as to its value,
D namely £177,500, a difference of £15,000. On that basis the plaintiffs should have interest at the rate of 15% from the date of the payment of the price.

The defendant's counsel quite properly submits that the plaintiffs have never pleaded any retainer to carry out a valuation. But the evidence has been given of the request for an opinion as to the price to be paid and of the fact that the defendant approved that price. That evidence is, in my view, relevant to the amount of the damages claimed.

In either event the plaintiffs are further entitled to damages for distress and inconvenience, and they claim under this head the cost of a holiday. On behalf of the defendant it was submitted that damages
E for distress and inconvenience are not recoverable at all in this class of case. Reliance was based upon the decision of the Court of Appeal in *Hayes* v *James & Charles Dodd* [1990] 2 All ER 815. In that case Hirst J awarded to each of two plaintiffs the sum of £1,500 for anguish and vexation. The action was a solicitors' negligence action in which it was found that a solicitor had been negligent in wrongly advising that there was an important right of way over some land leading to property which was bought for a motor repair business and maisonette. The Court of Appeal held that no damages were to be awarded for anguish and vexation. In the light of *Perry* v *Sidney Phillips & Son* [1982] 1 WLR 1297* and a long line of cases decided by official referees, the decision of *Hayes* v *Dodd* cannot have been intended to rule that in no case of a negligent house survey can
F damages be awarded for distress and inconvenience. For the reasons given by the Court of Appeal in *Hayes* v *Dodd* there was inadequate discussion in that case of the authorities on this point. But while damages for inconvenience and distress can plainly be awarded in appropriate cases against negligent surveyors instructed to survey residential houses, *Hayes* v *Dodd* helpfully illustrates features for which damages of that sort should not be awarded in the present state of the law. To give an example, it is plain that a surveyor instructed to advise on a purely commercial point such as the condition and value of a shipment of new Egyptian potatoes would not be liable for damages for distress and inconvenience if he gave negligent advice

*Editor's note: Also reported at (1982) 263 EG 888.

even if he added to the effects of that breach of duty by perversely G
denying liability and thereby putting the plaintiffs into a lengthy and physically debilitating fear of impending financial failure.

Staughton LJ said, at p 823G:

In *Perry* v *Sidney Phillips & Son* . . . damages were awarded for the distress, worry, inconvenience and trouble which the plaintiff had suffered while living in the house he bought, owing to the defects which his surveyor had overlooked. Lord Denning MR considered these consequences were reasonably foreseeable. Kerr LJ stated a narrower test:

> "So far as the question of damages for vexation and inconvenience is concerned, it should be noted that the deputy judge awarded these not for the tension or frustration of a person who is involved in a legal dispute in which the other party refuses to meet its liabilities. If he had done so, it would have been wrong, because such aggravation is experienced by H
> almost all litigants. He awarded these damages because of the physical consequences of the breach, which were all foreseeable at the time. The fact that in such cases damages under this head may be recoverable, if they have been suffered but not otherwise, is supported by the decision of this court in *Hutchinson* v *Harris* (1978) 10 Build LR 19."

I would emphasise the reference to physical consequences of the breach. I am not convinced that it is enough to ask whether mental distress was reasonably foreseeable as a consequence, or even whether it should reasonably have been contemplated as not unlikely to result from a breach of contract. It seems to me that damages for mental distress in contract are, as a matter of policy, limited to certain classes of case. I would broadly follow the classification provided by Dillon LJ in *Bliss* v *South East Thames Regional Health Authority* [1987] ICR 700 at 718:

". . . where the contract which has been broken was itself a contract to J
provide peace of mind or freedom from distress . . ."

It may be that the class is somewhat wider than that. But it should not, in my judgment, include any case where the object of the contract was not comfort or pleasure, or the relief of discomfort, but simply carrying on a commercial activity with a view to profit. So I would disallow the item of damages for anguish and vexation.

Sir George Waller (at p 818) and Purchas LJ agreed with Staughton LJ and Purchas LJ added at p 826:

I would approach this special and restricted head of damage rather in the same way as the courts approach the question of pecuniary loss dissociated from physical damage caused by negligence. Damages of that nature are recoverable only when the special relationship between the parties involved demonstrates that the one has in mind liability to pay pecuniary loss and the K
other relies on that assumption of responsibility. It is a very far cry from a submission that damages can be recovered by a litigant who is involved in the frustration and hassle inevitably arising out of a breach of contract or tort of this sort, including that involved in seeking his remedies at court.

Very limited argument was directed at the trial to this point beyond the citation of *Hayes* v *Dodd* and *Perry* v *Sidney Phillips*. It seems to me on the authorities to be clear that in appropriate circumstances a negligent surveyor of a residential property which he has undertaken to survey for a prospective purchaser of the property who intends to live there may be liable to his client in damages to compensate him or her for inconvenience and distress arising out of living in the property, not out of litigation about it. That seems to me to be the case whether the position be rationalised by reference to "special L
relationship" or "a contract to provide peace of mind or freedom from distress". A prospective buyer of a house goes to a surveyor not just to be advised on the financial advisability of one of the most important transactions of his life but also to receive reassurance that when he buys the house he will have "peace of mind and freedom from distress". In fact, most prospective purchasers have the higher aspirations of positive happiness and enjoyment rather than the mere absence of their opposites.

As to the amount of damages to be awarded under this head, I have been reminded of my decision in *Syrett* v *Carr & Neave*, July 31, 1990, unreported*. In that case I said:

The plaintiff claims damages for inconvenience and distress. She plainly is entitled to such damages on a scale which is not excessive but modest: see *per* M
Lord Denning MR in *Perry* v *Sidney Phillips* [1982] 1 WLR 129 at pp 1302, 1303. A most helpful article by Miss Kim Franklin in (1988) *Construction Law Journal* at p 264 makes interesting comparisons between the misery caused by people being in an unsatisfactory home with the distress caused by ruined holidays and minor personal injuries. The plaintiff has been restrained in describing her distress and inconvenience, but it has been real, substantial, and prolonged, lasting from about October 1988 until early next year. It has been much more serious than a spoilt fortnight in Mallorca though not as bad as the experiences of some householders. I assess those damages on a moderate scale in the sum of £8,000.

In the present case, the plaintiffs have both described the distress

*Editor's note: Reported at [1990] 2 EGLR 161.

A and inconvenience which they suffered in the house, albeit limited to weekends. The statement of Mr Watts, supported by his former wife, was in the following terms:

> The quality of our life was very seriously affected. We had bought the house on the basis of the assurances of the defendant that apart from regular maintenance, the property would not require attention. In the event, for eight months we spent almost every weekend staying in a building site. The conditions were quite deplorable. At various times there was scaffolding around the entirety of the house, the first-floor floorboards and the first-floor windows were removed in sequence. There was constant dirt and dust permeating the air. Moving our possessions from room to room in accordance with the builder's requirements was a recurrent nightmare and our possessions and furniture became dirty. From time to time there was no internal sanitation and we could not bathe. Lesley and I were forced to spend B extended periods out of our first-floor bedroom, living and sleeping in the unheated, undecorated attic rooms. We were most reluctant to entertain our friends and were completely unable to entertain clients.
>
> Throughout the period of building works Lesley and I continued to spend weekends at the house. As there was no supervising architect or surveyor employed the only supervision was that provided by me. I believed it to be essential that I spent as much time as possible with the builders to provide them with an incentive to get on with it, to ensure that the work was done to the requisite standard and to provide guidance with the numerous minor problems which arose. I met the builder at the property virtually every weekend.
>
> Our summer holiday in 1988, which we had planned to spend in the house, was spent in a hotel in the Western Isles at an unplanned cost of £880. Both C Lesley and I were employed in extremely stressful occupations in the City of London. Weekend relaxation was essential to us both in terms of our physical well-being and our capacity to perform our jobs to the requisite standard. The denial of that relaxation by virtue of the building works caused us both great distress. I believe that it was, at the very least, a contributory cause to the unfortunate breakdown of our marriage, which culminated in our separation late in 1989.

The plaintiffs were criticised during the trial for not having appointed an architect or surveyor to supervise the works. It may be that such a professional might have negotiated some reduction in the building costs (though I am not persuaded by any evidence that that would have been so), but it is certain that the costs of such a professional would have added to the amount of the claim against the D defendant, and the plaintiffs would still have had the anxiety of supervising the professional man.

The cost of the holiday was, I thought, abandoned by the plaintiffs during the opening of the case, but it was brought back in later. As one would expect, there was a degree of imprecision in the evidence of the plaintiffs about their intentions to spend their annual holidays in the country home. I conjecture that they would probably, if all had gone well, have spent their holidays in early years in Dorset and in some, but not all, later years elsewhere. I do not think it right to award damages for the cost of the holiday in 1988 as special damages, but I do think that the holiday should be taken into account in assessing general damages for distress and inconvenience. On the evidence, they probably could have so organised the work that E nothing was being done at the house during their holiday period, but again, at that time they really needed to get away and try to forget the unhappiness caused to them by the works at Nutford Farmhouse.

The point has been made that the plaintiffs have voluntarily undertaken works going far beyond the works the subject of this action and therefore their alleged distress and inconvenience should be discounted. The answer of the first plaintiff seems to me to be sensible and well founded. Once it became necessary to do certain works, other works naturally followed. For example, once the floor had to come up, it was only natural and sensible that the wiring and underfloor plumbing also should be dealt with, even though those matters would not normally have been considered for some years.

The evidence about the effect on the plaintiffs' marriage raises a F question of important principle which was not argued at all in this case. I have no difficulty in accepting the evidence of both plaintiffs that their unhappiness arising out of the problems in relation to Nutford Farmhouse contributed to the breakdown of their marriage. But quite apart from more general questions of policy, there are two specific difficulties about assessing damages in relation to such a matter. First, some marriages are strengthened by a shared coping with adversity and others are weakened. Second, if it were to be accepted that damages should be assessed for the contribution made to the breakdown of a marriage caused by a negligent report by a surveyor it would be necessary to conduct something like an old-style "fault" divorce action within a professional negligence action to determine the proportion of blame to be attached to the negligent report as opposed to other causes of stress between the parties. G Happily, counsel have not sought to embark on any such inquiry. However, in the light of the authorities to which I have already referred, it seems to me that as a matter of policy this is a head of damage which should be disregarded and I do disregard the breakdown of the plaintiffs' marriage in assessing damages.

As to the amount of distress, it seems to me that this case is very similar to the case of *Syrett* v *Carr & Neave.* The totality of the distress and inconvenience seems to me to be assessable on a modest scale in a totality of £8,000 for the plaintiffs jointly, or £4,000 each.

It remains for me to decide whether the main head of the plaintiffs' damages should be assessed as the cost of the repairs or the difference in value.

On this topic also I have been reminded of my judgment in *Syrett* v H *Carr & Neave,* and in particular of the following passage:

> The major item of the plaintiffs' costs is the cost of repairs. I have to decide whether those costs should be recoverable as a simple head of damages or whether I should award damages on the basis of diminution in value, which in this case is heavily influenced by the cost of repairs.
>
> In a solicitor's negligence action, *County Personnel (Employment Agency) Ltd* v *Alan R Pulver & Co* [1987] 1 WLR 916 at p 925, Bingham LJ stated eight principles to be applied in assessing damages. The first four of those are of particular relevance in this case and are as follows:
>
> "(1) The overriding rule was stated by Lord Blackburn in *Livingstone* v *Rawyards Coal Co* (1880) 5 App Cas 25 at p 39, and has been repeated on countless occasions since: the measure of damages is 'that sum of money which will put the party who has been injured, or who has suffered, in the J same position as he would have been if he had not sustained the wrong for which he is now getting his compensation or reparation'.
>
> "As Megaw LJ added in *Dodd Properties (Kent) Ltd* v *Canterbury City Council* [1980] 1 WLR 433 at p 451: 'In any case of doubt, it is desirable that the judge, having decided provisionally as to the amount of damages, should, before finally deciding, consider whether the amount conforms with the requirement of Lord Blackburn's fundamental principle. If it appears not to conform, the judge should examine the question again to see whether the particular case falls within one of the exceptions of which Lord Blackburn gave examples, or whether he is obliged by some binding authority to arrive at a result which is inconsistent with fundamental principle.'
>
> "(2) On the authorities as they stand the diminution in value rule appears almost always, if not always, to be appropriate where property is acquired following negligent advice by surveyors. Such cases as *Philips* v *Ward* [1956] 1 K WLR 471; *Pilkington* v *Wood* [1953] Ch 770; *Ford* v *White & Co* [1964] 1 WLR 885 and *Perry* v *Sidney Phillips & Son* [1982] 1 WLR 1297, lay down that rule and illustrate its application in cases involving both surveyors and solicitors.
>
> "(3) That is not, however, an invariable approach, at least in claims against solicitors, and should not be mechanistically applied in cases where it may appear inappropriate. In *Simple Simon Catering Ltd* v *J E Binstock Miller & Co* (1973) 228 EG 527 the Court of Appeal favoured a more general assessment taking account of the 'general expectation of loss'. In other cases the cost of repair or reinstatement may provide the appropriate measure: the *Dodd Properties* case [1980] 1 WLR 433, 456, *per* Donaldson LJ. In other cases the measure of damage may properly include the cost of making good the error of a negligent adviser: examples are found in *Braid* v *W L Highway & Sons* (1964) 191 EG 433, and *G & K Ladenbau (UK) Ltd* v *Crawley & de Reya* L [1978] 1 WLR 266.
>
> "(4) While the general rule undoubtedly is that damages for tort or breach of contract are assessed as at the date of breach (see for example, *Miliangos* v *George Frank (Textiles) Ltd* [1976] AC 443, 468, *per* Lord Wilberforce), this rule also should not be mechanistically applied in circumstances where assessment at another date may more accurately reflect the overriding compensatory rule. The *Dodd Properties* case both affirms this principle and illustrates its application."
>
> Following the guideline set out by Megaw LJ in *Dodd Properties (Kent) Ltd* v *Canterbury City Council* [1980] 1 WLR 433 it seems to me that the appropriate measure of damages to put the plaintiff in the same position as she would have been in if she had not sustained the wrong is the cost of repairs plus the costs ancillary to those repairs plus compensation for the blight on her happiness since October 1988. If she had received a careful rather than a M negligent report, she would not have bought this property and she would have bought a property which would have been worth what she paid for it and would not have required large expenditure in repairs and would have given her much enjoyment already. It does not seem to me that this approach falls within one of the exceptions mentioned by Lord Blackburn and I therefore turn to consider whether there is some binding authority which obliges me to arrive at some different result.
>
> The two principal binding authorities which I have to consider are *Philips* v *Ward* [1956] 1 WLR 471 and *Perry* v *Sidney Phillips & Son* [1982] 1 WLR 1297. In *Philips* v *Ward,* the Court of Appeal held that *on the facts of that case* (see Morris LJ at p 476, line 5) the appropriate measure of damages was the difference in value between the value of the property as it was described in the defendants' report and the value as it should have been described. In that case, the difference in value was £4,000 and the cost of repairs at the time of

A purchase was £7,000. The work of repair had not been done by the date of the trial, and the Court of Appeal held on the facts of that case that it was not reasonable to have delayed doing the repair (*per* Morris LJ at p 475). Both of those facts differ from the facts of the present case. The date when the infestation by death-watch beetle was discovered in *Philips* v *Ward* is not clear from the report. In the statement of facts on p 471 it is stated that the plaintiff bought in June, 1952 and moved in and "it was then found that the timbers of the house were badly affected by death-watch beetle". The only reason for postponing the doing of the works put forward by the plaintiff appears to have been that the plaintiff wanted the property to be available for inspection at the date of the hearing. That reason was, not surprisingly, rejected by Morris LJ (p 475 first paragraph). I infer that the purchaser in *Philips* v *Ward* discovered the death-watch beetle soon after moving into the property and he had a choice between selling the property again at its true value, making a loss
B which was less than the then cost of repairs, or doing the repairs. In those circumstances, to award as damages the cost of doing the repairs would have given him a benefit to which he was not entitled in the absence of a warranty from the surveyor as to the state of the property. That is not the case here where instant resale of the property would not have left the purchaser with a loss less than the cost of repairs and, more important, the purchaser had no reason to make an instant sale of the property because she did not know of the defect until two years after moving in and by then she was so heavily involved with the property that it was reasonable for her to keep it and repair it. It seems to me to be wholly artificial to tell a plaintiff that her damage is to be measured in relation to a loss which she could have taken by making a sale two years before she learned that there was a loss to be realised. That does not seem to me to be what the Court of Appeal was saying in *Philips* v *Ward* on the facts of that case.

C In *Perry* v *Sidney Phillips* also the defects unreported in the negligent report seem likely to have been discovered very soon after the plaintiff moved into the property. The defects were a leaking roof and a septic tank giving off an offensive smell (see p 1300H). Again, the plaintiff had a choice of either cutting his loss by reselling or undertaking the necessary repairs. He chose not to resell until after the trial, and he did not have the means to undertake repairs. Oliver LJ in that case (at p 1304G) rejected a submission based on *Dodd Properties* that the loss should "be assessed on the basis of some hypothetical value at the date of trial because the plaintiff has chosen — as he did in this case — to retain the property and not to cut his loss by reselling it". In the same case, at p 1302, Lord Denning said:

"The general rule of law is that you assess the damages at the date of the breach so as to put the plaintiff in the same position as he would have been in if the contract had been properly performed. Even if the claim be laid in tort
D against the surveyor, the damages should be on the same basis."

If the plaintiff is unaware of the breach until two years after the breach took place there may be many reasons for not applying that general rule and, in my judgment, there are cogent reasons for not applying that general rule in this case. Those cogent reasons are that the plaintiff did not have an opportunity of cutting her losses at or about the date of purchase and that she has acted reasonably throughout and that all the expenses which I find she has suffered or is likely to suffer are natural and probable results of the negligent survey. The plaintiff did have an opportunity of cutting her losses two years after purchase of the property in October 1988, but in her circumstances at the time it seems to me to have been reasonable for her to decide not to do so, and in any event there is no evidence before me as to what would have been the financial results of her selling at that time.

Accordingly, to return to the words of Megaw LJ in *Dodd Properties,* I do
E not feel that I am "obliged by some binding authority to arrive at a result which is inconsistent with the fundamental principle".

I therefore hold that the plaintiff is entitled to damages comprising the cost of repairs together with the ancillary costs associated with those repairs and damages for distress and inconvenience together with return of the fee paid for the report.

I should add that in coming to that decision my attention was invited to the decision of Scott Baker J in *Hipkins* v *Jack Cotton Partnership* [1989] 2 EGLR 157.

In the present case, I have also been asked to consider the decision of the Court of Appeal in *Steward* v *Rapley* [1989] 15 EG 198* and the decision of Judge Hicks in *Bigg* v *Howard Son & Gooch* [1990] 12 EG 111†. I regret to say that in the circumstances of this case I do not
F derive any assistance from those two cases. As Purchas LJ said in the former case, it raised no new point of law.

I would only add, further to that citation from my previous decision, a reference to a passage from *Jackson and Powell on Professional Negligence,* at p 176, to the effect that *Philips* v *Ward* [1956] 1 WLR 471 and *Perry* v *Sidney Phillips*

are only binding authorities for the proposition that the plaintiff is not to be put in the position which he would have occupied if the negligent surveyor's report were true. He is only to be put in the position which he would have occupied if he had been given a competent report.

G The circumstances of the present case are remarkably similar to the facts of the case in *Syrett* v *Carr & Neave.* The plaintiffs received the report of the defendant in late August 1986. Through no fault of the plaintiffs, exchange of contracts did not take place until October 1986, and completion did not take place until April 1987. The plaintiffs did not delay after that time, but it was not until August 3 1987 that Mr Wadey gave his report on the property. If the plaintiffs had then sought to cut their losses by reselling, they would have incurred very considerable costs in reselling (in the shape of agents' and solicitors' fees and stamp duty) and in finding a new property if that had been their chosen course of action.

In addition, they might have incurred a very substantial loss on the resale in very different market conditions and when they were selling
H what would have become a suspect house, and they would have had to devote much time to the sale of the house. In the very difficult position in which they found themselves, I find that they acted entirely reasonably in deciding to repair the premises rather than resell.

In the circumstances, I take the view that for the reasons given in *Syrett* v *Carr & Neave* the plaintiffs should be awarded damages assessed as the cost of repair of the premises in the relevant respects, namely £33,961.35.

I therefore find that the plaintiffs are entitled to damages from the defendants in the sums of £33,961.35 and £8,000, totalling £41,961.35. In addition the plaintiffs should receive interest at the rate of 15% on the various sums making up the total of £33,961.35 J
calculated from the dates on which those sums were paid by the plaintiffs, together with interest at the rate of 2% on £8,000 from the date of the issue of the writ.

House of Lords

April 18 1991

(Before Lord KEITH OF KINKEL, Lord BRIGHTMAN, Lord GRIFFITHS, Lord OLIVER OF AYLMERTON and Lord LOWRY) K

SWINGCASTLE LTD v ALASTAIR GIBSON (a firm)

Estates Gazette May 4 1991

[1991] 17 EG 83

Negligence — Mortgage valuation — Damages — Accrued mortgage interest — Important decision of House of Lords, reversing decision of Court of Appeal, on measure of damages — Review of authorities — *Baxter* v *F W Gapp & Co Ltd* "not an attractive precedent" — Lenders entitled to be put in the same position as if the negligence had not occurred but not L entitled to receive compensation for lost interest at the rate which the borrowers had contracted to pay — The fallacy of the lenders' case was that they had been trying to obtain from the defendant valuer compensation for the borrowers' failure and not the proper damage for the valuer's negligence — Judgment to be entered for the defendant valuer in the county court without the need for any fresh assessment of damages

Lord Lowry, with whose speech all the other lords agreed, stated at the outset the main issue of the appeal — It was whether a person who lends money at a high rate of interest in reliance on a negligent valuation, and who claims to have suffered loss as a result, can recover from the negligent valuer, M as part of his damages, interest at the contractual rate which the lender was entitled to be paid by the borrower and which remained outstanding at the termination of the transaction between the lender and the borrower — The answer given by the House of Lords was "no", but this result was reached only after considerable litigation and analysis of the authorities

The action was by mortgagees against a chartered surveyor who had made a valuation of a house for the information of the mortgagees, a finance company lending on properties at the high risk or "non-status" end of the market — The valuation, of £18,000, as on a forced sale with vacant

*Editor's note: Also reported at [1989] 1 EGLR 159.
†Editor's note: Also reported at [1990] 1 EGLR 173.

possession, was admittedly made negligently, the house in question not being worth anything like so much — It was also admitted that but for the valuer's negligence the lenders would not have made any loan to the borrowers on the property — On the basis of this valuation the lenders advanced the sum of £10,000 on the security of a mortgage of the house — The interest rates under the mortgage were high, 36.51% while the borrowers continued to make payments in accordance with the agreement but increased to 45.619% in the event of default — The borrowers in fact soon defaulted and after possession proceedings the mortgagees took over the property — They eventually sold it for £12,000 — In the county court action the mortgagees claimed damages under the following heads: the principal sum advanced of £10,000; general disbursements and interest on them; legal costs arising out of the possession proceedings; estate agents' commission on the sale of the property; and (the controversial item) the interest accrued under the mortgage contract — On the other side of the account was the £12,000 recovered from the sale of the house and the small amount of repayment of the loan which the borrowers had been able to make — The county court judge gave judgment in favour of the mortgagees of the balance of £7,136.41 and the valuer appealed

In the Court of Appeal, as indeed throughout the proceedings, the appellant claimed that the mortgagees' calculation of their loss was made on the wrong basis — The correct basis, on the appellant's submission, was that the mortgagees were entitled to be placed in the position that they would have been in if they had received a correct and competent report from the valuer and had consequently made no loan to the borrowers — This would have meant in essence a claim for the recovery of the £10,000 lent plus an amount calculated on the footing that the mortgagees had for two years (or whatever was the correct period) been deprived of the use of the £10,000, plus, of course, solicitors' and estate agents' charges, credit being given for the £12,000 recovered on the sale of the house and the small amount of repayment made by the borrowers — The Court of Appeal recognised the force of the valuer's submission but considered that they were bound by the authority of *Baxter* v *F W Gapp & Co Ltd,* supported by the later decision in *London & South of England Building Society* v *Stone,* to decide otherwise — Although cogent criticisms could be made of *Baxter* v *Gapp,* it was a decision of the Court of Appeal which could not be said to have been made *per incuriam* — The *ratio decidendi* covered the question of interest, even if the rate of interest was not specifically considered — However, the judgments in the Court of Appeal, despite unanimity in dismissing the valuer's appeal, recognised that questions of principle raised by the case had not been resolved

The valuer's appeal to the House of Lords provided the opportunity to reconsider the matter from first principles — These had been laid down in classic, but necessarily general, statements by Parke B in *Robinson* v *Harman* and Lord Blackburn in *Livingstone* v *Rawyards Coal Co* — The gist of these was that when a party sustains a loss, whether by breach of contract or tort, he should be placed, so far as money can compensate him, in the same position as if he had not suffered the wrongful act for which he is entitled to compensation

After a detailed examination of *Baxter* v *Gapp,* both before Goddard LJ (sitting as an additional High Court judge) and before the Court of Appeal, Lord Lowry said in his speech that it was "not an attractive precedent" — Apart from other criticisms, its approach seemed contrary to principle, the principle being that "the aggrieved party was entitled to be placed in the same position as if the wrong had not occurred, and not to receive from the wrongdoer compensation for lost interest at the rate which the borrower had contracted to observe" — What the lenders had lost in the present case, in addition to their other damages, was the use of the £10,000 while it was perforce locked up in the loan — The fallacy of the lenders' case was that they had been trying to obtain from the valuer compensation for the borrowers' failure and not the proper damage from the valuer's negligence

When it came to actual figures there was a lack of evidence as to how the lenders financed the loan or as to how the money, if not lent, could have been profitably employed — Taking the figures which had been mentioned of £10,000 for the advance and £401.35 and £983.25 respectively for estate agents' and solicitors' charges on the one side and £12,000 for the sale and £1,734 for repayments on the other, the lenders were £2,349.40 in credit — If, however, 12% pa interest were assumed for the use of the £10,000 locked up in the loan, the amount for two years would be £2,400, a figure not far removed from the £2,349.40 (Lord Lowry said that 12% would correspond to the 9% allowed by Ralph Gibson J in *Corisand Investments Ltd* v *Druce & Co*) — The lenders did not appear and did not even put in a written case on the hearing of the present appeal — In view of this and of the picture deducible from the above figures it would be neither fair nor sensible to remit the action to the county court for a fresh assessment of damages — Accordingly, the appeal would be allowed, the order of the Court of Appeal set aside and judgment directed to be entered for the defendant valuer in the county court — The valuer was awarded the costs of the appeal to the House and of the proceedings in the courts below

The following cases are referred to in this report.

Baxter v *F W Gapp & Co Ltd* [1938] 4 All ER 457
Chekiang, The [1926] AC 637
Livingstone v *Rawyards Coal Co* (1880) 5 App Cas 25, HL
London & South of England Building Society v *Stone* [1983] 1 WLR 1242; [1983] 3 All ER 105; [1983] EGD 921; (1983) 267 EG 69
Lowenburg, Harris & Co v *Wolley* (1895) 25 SCR 51
Perry v *Sidney Phillips & Son* [1982] 1 WLR 1297; [1982] 3 All ER 705; [1982] EGD 412; (1982) 263 EG 888, CA
Radford v *De Froberville* [1977] 1 WLR 1262; [1978] 1 All ER 33; (1977) 35 P&CR 316
Robinson v *Harman* (1848) 1 Exch 850; [1843-60] All ER Rep 383; 18 LJ Ex 202; 13 LTOS 141
Scholes v *Brook* (1891) 63 LT 837; 64 LT 674
Susquehanna, The [1926] AC 655

This was an appeal by the valuer, Alastair Gibson, a firm (Mr Alastair Gibson FRICS) from the decision of the Court of Appeal (reported at [1990] 2 EGLR 149; [1990] 34 EG 49) dismissing an appeal from the decision of Judge Harris QC, at Westminster County Court, in which he awarded damages to the plaintiff company, the present respondents, Swingcastle Ltd, in respect of the valuation for mortgage of a house at 36 North Road, Audenshaw, Manchester.

Roger Toulson QC and Roger Stewart (instructed by Reynolds Porter Chamberlain) appeared on behalf of the appellant valuer; the respondents did not appear and were not represented.

In his speech LORD LOWRY said: This is an appeal by leave of your Lordships' House from an order of the Court of Appeal (Neill and Farquharson LJJ and Sir John Megaw) affirming a judgment for £7,136.41, together with interest and costs, given by His Honour Judge Harris DSC QC in an action which had been brought by the respondents, a finance company known as Swingcastle Ltd ("the lenders"), against the appellant, who is a chartered surveyor, and remitted to the county court for trial.

The main issue in the appeal is whether a person who lends money at a high rate of interest in reliance on a negligent valuation, and who claims to have suffered loss as a result, can recover from the negligent valuer, as part of his damages, interest at the contractual rate which the lender was entitled to be paid by the borrower and which remained outstanding at the termination of the transaction between the lender and the borrower.

The facts are that at the beginning of 1985 Mr John Clarke and Mrs Kathleen Ann Clarke, his wife, who lived at 36 North Road, Audenshaw, Manchester, were in financial difficulties. Their home ("the property") was mortgaged and they owed £1,700 to the Leeds Permanent Building Society and £6,300 to Cedar Holdings Ltd.

Being anxious to pay off these debts, they went to a firm called Richard Murtagh & Co ("the brokers") whose business included organising finance for borrowers by putting them in touch with lenders. The brokers, having found out what Mr and Mrs Clarke ("the borrowers") required and what security they were offering, instructed the appellant (whom I shall call "the valuer") to survey the property and give a report, for the benefit of whoever might become the lending principals of the brokers, on its condition and its value on a forced sale with vacant possession.

The valuer placed that value at £18,000 and, relying on his valuation, the lenders on February 11 1985 lent to the borrowers £10,000 secured by a first charge over the property. The conditions of the loan reflected the fact that the borrowers were high risk (or "non-status") borrowers, since (a) the entire loan was used to pay the debts already mentioned of £1,700 and £6,300 and the brokers' fee which amounted to no less than £2,000; (b) the loan was repayable by 120 monthly instalments of £275, which meant that the annual percentage rate of interest was to be 36.51 per cent calculated in accordance with regulations made under the Consumer Credit Act 1974, with an alternative rate of 45.619 per cent calculated in accordance with the First Schedule to the Moneylenders Act 1927; (c) in the event that the borrowers went into arrears with the payment of interest, the lenders were entitled to add the outstanding interest to the principal sum then outstanding and charge interest on the whole at the rate of 45.619 per cent; and (d) an extra charge of six months' interest was payable in the event of early redemption of the charge by the borrowers.

The borrowers quickly fell into arrears and, after possession proceedings, they surrendered possession of the property to the lenders on June 30 1986. On February 9 1987 the lenders exchanged contracts for the sale of the property to a third party for £12,000 and completion took place on February 27 1987.

After an exchange of correspondence the lenders issued on September 4 1987 a writ endorsed with a statement of claim against the valuer claiming damages for loss allegedly sustained by the lenders in reliance upon the valuer's negligent valuation of the property. The lenders pleaded their loss as follows:

PARTICULARS OF LOSS AND DAMAGE

Amount required to redeem the Loan as at February 27 1987	£19,912.96
Payment to Messrs Cordingleys	401.35
Legal costs and charges of Messrs Brand Montague & Co. in connection with the possession proceedings in the Ashton-under-Lyne County Court and the subsequent sale of the property on behalf of the Plaintiffs as mortgagees in possession	983.25
	£21,297.56
Less sale proceeds of property	£12,000.00
	£9,297.56

13. By reason of the matters aforesaid the Plaintiffs suffered a shortfall in respect of the amount due to them under their loan to the said Mr and Mrs Clarke and under the security of their Legal Charge and they thereby suffered loss and damage.

14. Further the Plaintiffs are entitled to and claim interest pursuant to Section 35A of the Supreme Court Act 1981 at such rate and for such period as the Court thinks fit.

(Cordingleys were estate agents and Brand Montague & Co were the lenders' solicitors.)

The defence denied negligence, averred that the measure of damages alleged in para 13 of the statement of claim was incorrect and set out as an algebraic formula what was alleged to be the correct measure of damages.

Liability was admitted at the trial in the county court and it was further admitted that but for the valuer's negligence the lenders would not have made any loan to the borrowers. Thus the sole issue was the measure and the amount, if any, of the lenders' damages against the valuer on the basis that, if he had made a competent valuation, the lenders would not have made the loan.

About some of the figures and the way in which they should be taken into account there was no dispute. The lenders' outgoings included the loan of £10,000, the estate agents' charges of £401.35 in connection with the sale of the property and the costs of the solicitors amounting to £983.25 for the work described in the particulars. On the other side of the account the lenders had received a total sum of £1,734 from the borrowers between February 1985 and April 1986 and £12,000 from the purchasers of the property.

The disputed ingredient of the lenders' claim was the sum of £19,912.96, which gave credit for the receipt of £1,734, as mentioned above, and was described in the particulars as "Amount required to redeem the loan as at February 27 1987", the date of the sale of the property when the transaction between the lenders and the borrowers was deemed to be concluded. The lenders arrived at this figure, which was in itself accurate, by calculating the amount of principal and interest outstanding, having regard to the initial rate of interest of 36.51 per cent and also the penal rate of 45.619 per cent which they became entitled to charge when the borrowers defaulted, as I have already described. They also claimed £344.99 for "disbursements" and interest thereon of £129.31. To the total thus arrived at they added an extra charge of six months' interest which, according to the loan agreement, was to be payable by the borrowers in the event of early redemption.

The judge disallowed the last element in the claim on the ground that it was not a sum which was owing as interest on February 27 1987. I might point out that according to condition 8 of the conditions annexed to the agreement the extra charge of interest was claimable only if the borrowers repaid the whole of the loan. In all other respects, including the charge for interest on disbursements, he allowed the claim and gave judgment for £7,136.41 with interest and costs.

The valuer's case before action brought and throughout the proceedings has consistently been that the lenders' calculation of their loss was made on the wrong basis and that the damages should have been assessed on the basis that the lenders were entitled to be placed in the position that they would have been in if they had received a competent report from the valuer and had consequently made no loan to the borrowers; more explicitly, the lenders' damages are to be calculated on the footing that they should: (1) take credit for the payment of £10,000 to the borrowers, £401.35 to their estate agents and £983.25 to their solicitors; and (2) give credit for £1,734 paid by the borrowers and £12,000 received on the sale of the property and also on the footing that they have for two years (or such shorter period as might be thought right to allow) been deprived of the use of the £10,000 which they would not have lent but for the negligent valuation.

In the Court of Appeal [1990] 1 WLR 1223*, however, this submission fared no better than it had done in the county court. Neill LJ reviewed a number of authorities, including *Baxter* v *F W Gapp & Co Ltd* [1938] 4 All ER 457, affirmed in the Court of Appeal [1939] 2 KB 271, and *London & South of England Building Society* v *Stone* [1983] 1 WLR 1242†, both of which had been successfully relied on in the county court, and, having set out the defendant's argument at p 1229C-G, said "I see the force of these criticisms", but, having then concluded that *Baxter* v *Gapp* was a binding authority, was in favour of dismissing the appeal "on the short ground that this court is bound by authority and that the judge was right to assess the damages in the way that he did."

The lord justice then considered the case as if it were free from authority (p 1230B), commencing with the classic statement of principle by Lord Blackburn in *Livingstone* v *Rawyards Coal Co* (1880) 5 App Cas 25 at p 39:

> where any injury is to be compensated by damages, in settling the sum of money to be given for reparation of damages you should as nearly as possible get at that sum of money which will put the party who has been injured, or who has suffered, in the same position as he would have been in if he had not sustained the wrong for which he is now getting his compensation or reparation.

He divided lenders' actions arising out of negligent valuations into two types, the first being where the valuer's client had lent more than he would have done if competently advised, and the second where, if competently advised, he would (as in the present case) not have lent at all. In his consideration of the second type he said at p 1231G:

> A number of approaches are possible, including the following. (*a*) The lender could be awarded the unpaid interest owed by the borrower at the date when the security was realised. This was the method adopted in *Baxter* v *F W Gapp & Co Ltd* [1939] 2 KB 271. But to award damages on this basis is in effect to treat the valuer as the guarantor of the contract of loan. In the absence of authority I would for my part reject this solution. (*b*) The lender could be awarded a sum equivalent to the amount he would have earned by way of

*Editor's note: Also reported at [1990] 2 EGLR 149.
†Editor's note: Also reported at (1983) 267 EG 69.

interest on another loan if he had had the money available for this purpose. In my view, however, such an award should not be made in the absence of evidence that the money lent would have been used for another transaction. This evidence would have to be directed to proving an unsatisfied demand for loans and I anticipate that such evidence might seldom be forthcoming. Moreover, even if evidence of a lost transaction were available, I see no reason why the interest should be at the default rate rather than at the ordinary rate provided for in a standard contract for this type of business. (*c*) The lender could be awarded a sum equivalent to the interest which would have been earned if the sum had been placed on deposit. (*d*) The lender could be awarded a sum to represent the loss of the opportunity to invest the money elsewhere. This was the solution adopted by the Supreme Court of British Columbia in *Seeway Mortgage Investment Corporation* v *First Citizens Financial Corporation* (1983) 45 BCLR 87, where it was said, at p 101: "What the plaintiff lost then was the opportunity to invest its $50,000 in a security which had the same risks except that the appraisal would be accurate."

I do not propose to express any concluded view about these methods of assessment. I do not consider that any one of the last three methods of assessment would necessarily be right to suit all cases. It would depend on the evidence.

The lord justice, although he does not in principle appear to have been in favour of method (*a*) above, concluded his judgment by saying (p 1232D):

In the present case, however, for the reasons which I have stated, I consider that the court is bound by the decision in *Baxter* v *F W Gapp & Co Ltd* [1939] 2 KB 271.

I would dismiss the appeal.

Farquharson LJ neatly summed up the valuer's case and his own reaction to it at p 1233E:

The plaintiff's claim, he argues, in effect puts it in the same position as if the defendant had warranted or guaranteed the performance by the Clarkes of their contractual obligations, which he had not. Neither had he warranted the correctness of his valuation. Counsel concedes that the plaintiff could properly recover as damages any loss it sustained by reason of losing the use of the £10,000 advanced under the mortgage, but that there would have to be evidence, which there was not, that the plaintiff could have applied the £10,000 in another mortgage investment or at the very least to have earned interest by placing it on deposit.

As a matter of principle, I would, for my part, find these arguments hard to resist, but it is submitted by Mr Wulwik on behalf of the plaintiff that the weight of authority shows that the judge's decision as to the measure of damages is correct.

He noted the textbook writers' distinction between the two categories of case mentioned by Neill LJ and continued (p 1233H):

In the case of a mortgagee who would not have made any advance but for the negligent valuation the measure of damages will be the difference between the sum advanced and the sum recovered on sale of the property plus any consequential loss and expenses: see *Clerk and Lindsell on Torts,* 16th ed (1989), p 28, para 11-45. This definition does not, of course, answer the present problem and it is necessary to look at the authorities upon which it is based.

The lord justice then referred to *Baxter* v *Gapp* and *London & South of England Building Society* v *Stone* and concluded (p 1235B):

In the result, although the authorities are not entirely satisfactory in resolving the question whether a mortgagee in these circumstances is entitled to recover loss of contractual interest under the mortgage from the negligent valuer, they are binding on this court.

I would accordingly dismiss the appeal.

Sir John Megaw, as well as holding that *Baxter* v *Gapp* was a binding authority, rejected in principle the measure of damages put forward by the valuer and drew an analogy with the cases, such as *Perry* v *Sidney Phillips & Son* [1982] 1 WLR 1297,* where a house is purchased above its true value as a result of a negligent valuation. He also stated that he could see no practical basis for any distinction in the assessment of damages according to whether the lender, if he had not been misled as to the value of the security, would have lent a smaller amount or would not have lent at all. The judgment concluded as follows (p 1236D):

In the absence of authority to the contrary, I should regard the correct principle of damages as being that the claim for damages arises when the loss is incurred. The loss is incurred if and when the lender, validly under the loan contract, realises the security, and the amount realised is less than the amount due under the loan contract, including expenses properly incurred, and allowing for all payments made by the borrower. There is one important qualification. The valuer should not be liable for a greater amount than the amount of his original overestimate of the value compared with the true market value as at the date of the valuation. Any shortfall in the proceeds of the realisation above that amount should not be regarded as being caused by the negligent valuation.

*Editor's note: Also reported at (1982) 263 EG 888.

Subject, possibly, to that qualification, which did not arise in *Baxter* v *F W Gapp & Co Ltd* [1939] 2 KB 271, it appears to me that the principle which I have suggested above is wholly in accord with that case.

My Lords, I approach the assessment of damages in the first place by reference to principle, including the statement of Lord Blackburn which was cited by Neill LJ. The relevant part of the paragraph from *Clerk and Lindsell* (11-45) (incorporating the footnotes in the text) reads as follows:

Where money is advanced by a mortgagee on the basis of a negligent valuation the measure of damages will vary dependent on whether, had a proper valuation been made, the mortgagee would have made no advance at all, or advanced a lesser sum. Where the mortgagee can establish he would have made no advance, the measure of damages will be the difference between the sum advanced and the sum recovered on sale of the property, plus any consequential losses and expenses: *Baxter* v *F W Gapp & Co Ltd* [1938] 4 All ER 457. In *Corisand Investments Ltd* v *Druce & Co* (1978) 248 EG 315 Gibson J found that on a proper valuation a lesser sum would have been advanced and awarded the plaintiff the difference between the sum advanced and that sum. The defendants in such an action by a mortgagee can no longer generally set off against the prima facie measure of damages the amount which the mortgagee could recover by enforcing personal covenants by the mortgagor to repay the sums borrowed: *London & South of England BS* v *Stone* [1983] 1 WLR 1242.

I would also refer to *McGregor on Damages* (15th ed 1988). Para 1212 states:

Claims against surveyors and valuers by clients may be by those who have bought property on the strength of a negligent survey or valuation or by those who have lent money on mortgage in reliance on a negligent survey or valuation of the property mortgaged. The measure of damages is substantially different in the two cases so that they call for separate treatment.

The learned author then considers purchasers of property, and at para 1217 comes to mortgagees of property negligently surveyed or valued:

Claims by lenders to intending purchasers or to owners on the strength of a negligent survey or valuation of the property to be purchased or already owned have produced a different pattern of damages. For one thing, the emphasis shifts from the purchase price of the property to the amount loaned on the security of the property . . .

The text proceeds at para 1218:

While the starting point of the plaintiff's loss is still generally the amount of money that he has paid away, though here not by way of price to his seller but by way of loan to his mortgagor, the amount which falls to be deducted from the amount paid away is not, as it is in the case of the paying purchaser, the real value of the property; rather is it the amount that the plaintiff would have lent had the defendant produced a proper valuation.

Baxter v *Gapp* is then cited as authority for the proposition that the difference between the amount lent and the amount that would have been lent on foot of a proper valuation:

. . . may not always be the true measure of the lender's actual loss: the loss of the money advanced may be increased by expenses and reduced by receipts.

The further comments in the same paragraph on *Baxter* v *Gapp* do not help to elucidate the problem with which your Lordships have been presented, while the discussion of *London & South of England Building Society* v *Stone* shows that the ultimate effect on the damages of the lenders' rights against the borrowers depends on the facts of each case.

My Lords, it is now time for me to take a closer look at *Baxter* v *Gapp,* which has been regarded by all three members of the Court of Appeal as the authority for accepting the lenders' approach to the claim for interest.

At first instance most of the judgment of Goddard LJ was devoted to the question of liability. Certain facts, however, mentioned at [1938] 4 All ER at p 461 are relevant to the now disputed question of interest. The property had been negligently valued at £1,800 and was ultimately sold for £850. The plaintiff had advanced £1,200 on first mortgage. Goddard LJ said at p 461B:

A further £150 was advanced at a much higher rate of interest — 9 per cent, reducible to 8 per cent if punctually paid. I ought to have said, perhaps, that the client had interest, not at $4\frac{1}{2}$ per cent, the amount of interest contemplated on the first mortgage, but at 5 per cent. I dare say that he did that because he thought, after hearing what Mr Humfrey had said, that there was some risk in the matter.

The trial judge dealt with damages at pp 465-6. Starting with a basic figure of £500 (£1,350 advanced less the sale price of £850), he concluded at p 466F:

If he has done all that he can be required to do, so far as realising the property is concerned, you get at once the measure of the loss which he has suffered.

A The loss which he has suffered is not challenged in this case as a figure. Taking all the various items, and giving loss of interest, which lies in the ordinary way, and insurance premiums and credits allowed, the expenses of the sale, and one thing and another, and giving credit for the purchase price, the plaintiff says that he has suffered a loss of £742 16s 7d. That, I think, is the measure of the damage which he is entitled, and fairly entitled, to recover from the defendants in this case. I accordingly give judgment for that amount, and, if it is asked for, I think it must be with interest at the rate of $4\frac{1}{2}$ per cent from June 3.

In the All England Reports ([1939] 2 All ER 752) the appeal is reported in full, but it is interesting to note that, while the £1,200 advanced on first mortgage and the £150 advanced on second mortgage are mentioned in the headnote, the second mortgage is not B referred to in any of the judgments. In referring to the damages at p 757B MacKinnon LJ said:

As a result of lending this money upon this mortgage (the plaintiff) has sustained a certain pecuniary loss — namely, the loss of his interest during 2 years and the loss of his capital, so far as it has been repaid by the price paid by Mrs Raposo (the purchaser). That resulting figure — amounting to £720 (*sic*), I think — the judge has awarded the plaintiff as damages. In principle, apart from any authority, it seems to me that that is the right basis on which to assess the damages.

In the Law Reports ([1939] 2 KB 271) the case is reported only on the question of damages. The amount of the first and second advances, made in August and October 1935 respectively, and their rates of interest of 5 and 8 per cent, as well as the amount of the C judgment for £742 16s 7d, are correctly set out in the law reporter's summary of the facts, and among the heads of damage making up the total is included "the amount of the interest which the mortgagor had failed to pay". The argument of counsel for the appellant was, it appears, confined to the proposition that the proper measure of damages was the difference between the value actually arrived at by the surveyor and the value at which he ought to have arrived. Not surprisingly the question of the rate of interest was not referred to. The respondent's counsel were not called on. In his judgment MacKinnon LJ made the observations on damages to which I have already referred above (although he is reported as mentioning a total of £742 16s 7d). The cases cited in the judgment, namely, *Scholes* v *Brook* (1891) 63 LT 837; 64 LT 674 and *Lowenburg, Harris & Co* v D *Wolley* (1895) 25 SCR 51 are really of no assistance.

My Lords, I have gone into this amount of detail for two purposes: (1) to show that the trial judge calculated the claim for interest by reference, so far as I can make out, to what the mortgagor had failed to pay under her contract; but also (2) to indicate that the appropriate rate of interest (as to which a very small pecuniary difference must have been involved) was never in issue between the parties or in the mind of the court either at first instance or on appeal. The real issue, which has occasioned no difficulty in the present case, was whether, in addition to the difference between the capital amounts, the plaintiff could recover as damages his consequential loss. Having regard, however, to point (1) above, the case is, almost by accident I E would suggest, an authority for the proposition by which Neill and Farquharson LJJ found themselves, I think reluctantly, bound.

Your Lordships will recall two observations in particular about *Baxter* v *Gapp* which were made by Neill LJ. At p 1230A he said:

In the Court of Appeal detailed argument was directed to the measure of damages, and, even if no reference was made to the rate of interest, it seems to me to be impossible to exclude this one element of the award from the general approval given to the judge's method of assessment.

I have already quoted the passage in the lord justice's judgment at p 1231G where he set out four approaches and concerning method (a), "the method adopted in *Baxter* v *Gapp",* said:

But to award damages on this basis is in effect to treat the valuer as the guarantor of the contract of loan.

F Farquharson LJ commented on *Baxter* v *Gapp* as follows (p 1234B):

The most important case is *Baxter* v *Gapp & Co Ltd.* Here again the mortgagee would not have advanced money on a house but for a negligent survey. Goddard LJ [1938] 4 All ER 457, sitting as an additional judge of the King's Bench Division, held that the measure of damage was the whole loss sustained by the mortgagee and included the expenses of abortive sales, insurance premiums, a builder's account for upkeep of the property, the mortgagee's expenses and disbursements and the selling agent's commission upon the ultimate sale of the property, in addition to the principal advanced and the interest unpaid by the mortgagor. Goddard LJ's decision was also upheld by this court [1939] 2 KB 271. Mr Toulson seeks to discount the weight of that authority by pointing to the fact that Goddard LJ said at p 466F that "the loss suffered is not challenged as a figure," and says this issue was not argued. However, I think Mr Wulwik is right when he submits that the reference is to the mathematics of calculating the figure rather than to the G measure of damages which was clearly in dispute. While this case is authority for Mr Wulwik's submission in the present case, once again the principle involved does not seem to have been argued; ie, whether the interest should be the contractual amount in the mortgage deed or what represents the loss of use of the capital sum.

The lord justice continued at p 1234E:

Corisand Investments Ltd v *Druce & Co* (1978) 248 EG 315 was a case where the mortgagees lent a greater sum than they would have done but for the negligent valuation. Ralph Gibson J rejected the argument that the mortgagees were entitled to the 30 per cent interest they would have got if the borrowers had performed their contractual obligations.
However, in *London & South of England Building Society* v *Stone* [1983] 1 WLR 1242 O'Connor LJ in referring to that decision, said: "For some reason which is not clear to me Ralph Gibson J rejected the plaintiffs' claim for H 'interest.'" In *Stone's* case the issue was whether the plaintiff mortgagee was obliged to pursue its rights against the mortgagors and take any amount recovered into account when suing the negligent valuer, so the decision is not in point. Nevertheless, O'Connor LJ held, at p 1255, that *Baxter* v *Gapp & Co Ltd* [1939] 2 KB 271 was binding on the Court of Appeal. Furthermore Stephenson LJ said, at p 1261:
"Though the identity of the borrowers was of no interest or relevance to the valuer or his duty to the lenders, the amount of the advance and the length of time for repayment were. Though the mortgage deed by way of legal charge gave the lenders a right to foreclose for other breaches of covenant, the object of securing the loan by charging the property was to secure repayment of the sum lent with interest under the covenant to repay. The borrowers' obligation to repay was therefore not so collateral or remote as to be disregarded altogether in measuring the lenders' loss, and although the valuer was not a J party to the mortgage, and the mortgage was literally res inter alios acta, it was a transaction with which the valuer's report and valuation and breach of duty were closely connected, as now demonstrated by the decision of Park J in *Yianni* v *Edwin Evans & Sons* [1982] QB 438."

(The rate of interest actually awarded by Ralph Gibson J was 9 per cent, the Short Term Investment Account Rate for most of the relevant period.)

My Lords, *Baxter* v *Gapp* is not an attractive precedent. For one thing, it does not clearly exemplify the proposition contended for by the lenders, even if that proposition can be teased out of it; second, the dispute was about all the plaintiff's consequential damage and the pecuniary effect of the difference in interest rates (say, $\frac{1}{2}$ per cent in relation to the initial advance of £1,200 and $3\frac{1}{2}$ per cent in relation K to the much smaller second advance of £150) was relatively insignificant; and, third and most important, the approach, if carefully scrutinised, seems contrary to principle: the aggrieved party was entitled to be placed in the same position as if the wrong had not occurred, and not to receive from the wrongdoer compensation for lost interest at the rate which the borrower had contracted to observe.

The approach of the valuer in this case and the analysis of Neill LJ, which I have reproduced above, seem to me to be correct. What the lenders lost, in addition to their other damages was the use of the £10,000 while it was perforce locked up in the loan. I say "perforce" because I do not overlook the duty of the injured party to mitigate his loss or the fact that, once the borrowers had well and truly defaulted, the lenders had access to their remedy and thereby to their money. L

There is, as Neill LJ perceived, no cut and dried solution to calculating the amount of damages in cases of this kind. It depends on the evidence. Mr Toulson, who presented the appellant's case to your Lordships both clearly and persuasively, illustrated that point by reference to certain observations of Oliver J, as my noble and learned friend Lord Oliver of Aylmerton then was, in *Radford* v *De Froberville* [1977] 1 WLR 1262. That was a different kind of case, but the points made in the following passages are most apposite.

(1) at p 1268G:

As to principle, I take my starting point from what, I think, is the universal starting point in any inquiry of this nature — that is to say, the well known statement of Parke B in *Robinson* v *Harman* (1848) 1 Exch 850, 855 which is in M these terms:
"The rule of common law is, that where a party sustains a loss by reason of a breach of contract, he is, so far as money can do it, to be placed in the same situation, with respect to damages, as if the contract had been performed."

(2) at p1269G:

One of the difficulties about any question of damages is that there are so few general principles and that such as there are have, at times, been expressed in ambiguous and even contradictory terms. The matter was well expressed by Viscount Haldane LC in *British Westinghouse Electric and Manufacturing Co Ltd* v *Underground Electric Railways Co of London Ltd* [1912] AC 673, 688, where he says:
"In some of the cases there are expressions as to the principles governing the measure of general damages which at first sight seem difficult to harmonize.

A The apparent discrepancies are, however, mainly due to the varying nature of the particular questions submitted for decision. The quantum of damage is a question of fact, and the only guidance the law can give is to lay down general principles which afford at times but scanty assistance in dealing with particular cases. The judges who give guidance to juries in these cases have necessarily to look at their special character, and to mould, for the purposes of different kinds of claim, the expression of the general principles which apply to them, and this is apt to give rise to an appearance of ambiguity.

Subject to these observations, I think that there are certain broad principles which are quite well settled. The first is that, as far as possible, he who has proved a breach of a bargain to supply what he contracted to get is to be placed, as far as money can do it, in as good a situation as if the contract had been performed.

The fundamental basis is thus compensation for pecuniary loss naturally
B flowing from the breach; but this first principle is qualified by a second, which imposes on a plaintiff the duty of taking all reasonable steps to mitigate the loss consequent on the breach, and debars him from claiming any part of the damage which is due to his neglect to take such steps."

(3) at p 1270F:

There is, I think, a danger in elevating into general principles what are in truth mere applications to particular facts or situations of the overriding general principle as enunciated by Parke B; and on more than one occasion attention has been drawn to the undesirability of the application in this area of rigid rules or practices: see, for instance, the speech of Viscount Dunedin in *The Susquehanna* [1926] AC 655, 661 and that of Lord Sumner in *The Chekiang* [1926] AC 637, 643, 644.

(4) at p 1271A:

C As it was put by Denning J in *Duke of Westminster* v *Swinton* [1948] 1 KB 524, 534: "The real question in each case is: what damage has the plaintiff really suffered from the breach?"

Radford v *De Froberville* was a contract case, whereas the present action is founded on tort, but the principles to which I have just drawn attention apply equally (see *The Susquehanna* and *The Chekiang*) and the statement of Parke B in *Robinson* v *Harman* is entirely consistent with that of Lord Blackburn in *Livingstone* v *Rawyards* and leads to the same result. Of course, a different standard of remoteness may apply (though not necessarily with different results) in contract and tort, but I need not develop the point here.

My Lords, it is clear that the lenders ought to have presented their
D claim on the basis that, if the valuer had advised properly, they would not have lent the money. Where they went wrong was to claim not only correctly that they had to spend all the money which they did but incorrectly that the valuer by his negligence deprived them of the interest which they would have received from the borrowers if the borrowers had paid up. The security for the loan was the property, but the lenders did not have a further security consisting of a guarantee by the valuer that the borrowers would pay everything, or indeed anything, that was due from them to the lenders at the date, whenever it occurred, on which the loan transaction terminated. The fallacy of the lenders' case is that they have been trying to obtain from the valuer compensation for the borrowers' failure and not the
E proper damages for the valuer's negligence.

My Lords, having considered the principle, I come back to the figures. In his notice of appeal the valuer submitted that the sum due to the lenders was £66.22, but there has been no indication of how that sum was arrived at. Taking the figures I have mentioned of £10,000, £401.35 and £983.25 on one side and £12,000 and £1,734 on the other, the lenders are £2,349.40 in credit. Clause 9 of the mortgage agreement provided that the lenders could debit the mortgage account with "all costs expenses and disbursements incurred directly or indirectly in relation thereto". This no doubt explains the claim for £344.99, though not the claim for interest thereon of £129.31. But I find it impossible to say whether the whole or part of the £344.99 should be taken into account against the valuer
F or simply be treated as administrative expenses of the mortgage chargeable by agreement against the borrowers.

In the absence of any evidence as to how the lenders financed the loan or evidence showing how the money, if not lent to the borrowers, could have been profitably employed, I consider that 12 per cent interest, which would correspond to the 9 per cent allowed by Ralph Gibson J in *Corisand,* is the proper rate at which to recompense the lenders for being deprived of their £10,000. The actual time was two years, which would yield a result of £2,400, but one may ask whether it was reasonable for the tortfeasor to bear the liability up to the date of sale in February 1987, possession of the property having been surrendered on June 30 1986. Moreover, it is not clear how a calculation of damages would be affected by the incidence of tax or whether this is a case in which it would have been reasonable for the G
court to contemplate partial recovery by the lenders against the borrowers: see *London & South of England Building Society* v *Stone* (*supra*).

It was for the lenders to furnish the evidence by which to prove their case on the correct basis. Admittedly, the correct basis in this case has only now been established. But the lenders did not even put in a written case or appear on the hearing of this appeal, although they knew what the issue was. Having regard to that fact and to the picture which is deducible from the figures already available, I consider that it would be neither fair nor sensible to remit the action to the county court for a new assessment of the damages.

Accordingly, I would allow the appeal, set aside the order of the Court of Appeal and direct judgment to be entered for the defendant H
in the county court. I would also order the lenders to pay the valuer's costs of the appeal and of the proceedings in the courts below.

LORDS KEITH OF KINKEL, BRIGHTMAN, GRIFFITHS and OLIVER OF AYLMERTON agreed with the speech of Lord Lowry and the reasons given in it and did not add any observations of their own.

Court of Appeal

January 21 1991

(Before Lord Justice STUART-SMITH and Lord Justice MANN) J

PFEIFFER AND ANOTHER v E & E INSTALLATIONS

Estates Gazette June 8 1991

[1991] 22 EG 122

Negligence — Action against heating engineers — Complaint about inspection of a heating plant — Defendants instructed as a specialist firm to test heating system for plaintiffs about to purchase house — Defendants' report indicated nothing wrong, but soon after plaintiffs moved in the system began to be unsatisfactory — The Gas Board advised examination by K
other specialists and in the meantime warned plaintiffs not to operate the system — Subsequently it was condemned and plaintiffs commenced proceedings against defendants — County court judge decided that defendants had followed a standard practice and were not liable — Plaintiffs appealed

Plaintiffs were advised by the surveyors who surveyed the house before purchase to obtain specialist advice as to the state of the gas-fired heating system — The plaintiffs asked the surveyors to select and instruct a specialist firm and in due course the defendants were instructed — The defendants' report stated that a full working test had been carried out and that the whole unit was operating satisfactorily — L
Unfortunately, as mentioned above, this was not so and the second firm of specialists appointed on the advice of the Gas Board found a crack in the heat exchanger and further inspection revealed a more serious crack — Plaintiffs had to install a new unit as the age of the old unit meant that replacement parts could not be found — Plaintiffs commenced proceedings alleging negligence on the part of the defendants — The claim was pleaded in contract and tort — The case was actually presented to and dealt with by the county court judge in tort, although there clearly was a contract between the plaintiffs and the defendants — The duty M
of care, however, was the same on either basis

Certain facts were established — Cracks in a heat exchanger were potentially dangerous, were more likely to occur towards the end of a system's life, and were then a matter for condemnation of the unit — The system in this case was about 24 years old, its expected life was 25 years, but the defendants were found to have reasonably believed that it was 15 years old — It was a fact that there were cracks in the heat exchanger at the date of the defendants' inspection — One of the cracks could be seen by the use of a torch and a mirror, but the more serious crack could not be seen without removing the burner

What the defendants actually did at the inspection was to see that the gas was properly connected, to fire the unit and run it for forty minutes — They inspected the flame in the burner of the heat exchanger and saw that this appeared to be working correctly — They concluded that the heat exchanger was functioning satisfactorily — They did not carry out an inspection of the heat exchanger by the torch and mirror method — The plaintiffs' experts said that there should always as a matter of course be an inspection by the torch and mirror method — The defendants' experts said that the standard practice for a survey, as distinct from a servicing, of the system would not be to search for cracks in the heat exchanger unless there was a suspicion that the burners were not operating correctly or there was evidence of carbon monoxide in the atmosphere — In evidence, however, the defendants' expert qualified this statement by reference to the known age of the unit — At 20 years and even at 15 years (which defendants believed to be the age) the defendants should have been put on notice of the possibility of cracks

The judge came to the conclusion that as the defendants were following the recognised practice for a survey, as they reasonably believed the age of the unit to be not more than 15 years, and as nothing had occurred during the test to raise suspicions or put them on inquiry, they should not be held liable for failing to make a search of the inside of the heat exchanger — The judge accordingly dismissed the action — The Court of Appeal disagreed — In their view, in the case of a system which was reaching a suspect age (even taking the age as 15, although in fact it was 24) it was not sufficient simply to rely on the standard combustion test — The safety of the occupants of the house could have been at risk — The plaintiffs were entitled to expect that a full inspection had been carried out, as the report had undertaken to do — Appeal allowed; damages for plaintiffs agreed at £1,500, to include costs

No cases are referred to in this report.

This was an appeal by the plaintiffs, Mr and Mrs Pfeiffer, from the decision of Judge Goodman, at Croydon County Court, dismissing their claim for damages against the defendants, E & E Installations, for breach of contract and/or negligence in the inspection of the heating system at 63 Woodcrest Road, Purley.

Michael Harris (instructed by Percy Holt & Co, of Purley) appeared on behalf of the appellants; Alexander Hill-Smith (instructed by B K J Lewis) represented the respondents.

Giving judgment, **Stuart-Smith LJ** said: This is an appeal by the plaintiffs from the judgment of His Honour Judge Goodman, sitting in the Croydon County Court on January 9 1990, whereby he dismissed their claim for damages against the defendants for breach of contract and/or negligence in the execution of their duty as heating engineers when instructed to carry out an inspection of the heating plant at 63 Woodcrest Road, Purley, which the plaintiffs intended to purchase.

The plaintiffs had instructed Pollard Machin, a firm of surveyors, to carry out a survey of the house before their purchase, but Pollard Machin indicated that so far as the heating system was concerned they were not in fact competent to give an opinion about it and advised that a specialist firm should be instructed. The plaintiffs left it to Pollard Machin to instruct such a specialist firm, and in due course the defendants were instructed.

Instructions were given to them on or about November 9 1987, and on November 13 1987 the two Mr Luckhursts carried out an inspection — Mr Edward Luckhurst being a heating engineer.

On November 14 the defendants reported to the plaintiffs via Pollard Machin in these terms:

> Re: Central heating report for 63 Woodcrest Road, Purley, Surrey.
> A full working test and inspection of the central heating system was carried out at the above property on November 13 1987.

The system is then described, being a gas-fired, warm-air unit. The unit in fact was wrongly described as a "Lennox GH9-52" whereas in fact it was "Lennox GH6-52".

The report continues:

> The system was installed around 1972-1973 and we understand from the present owners that it is covered by a Service Contract with SEGAS.

The controls are then described and the report continues:

> The unit, controls and closeable grill were all operating satisfactorily at the time of our test.

There was no indication in that report that there was anything wrong with the unit.

The plaintiffs agreed to purchase the house; completion took place and they moved in on December 18 1987. That was in the middle of a cold snap.

Mr Pfeiffer turned on the central heating, which appeared to work well at first, but when it was turned down low the pilot light went out. That happened on a number of occasions. The Gas Board were called in, and they recommended the services of another firm of specialist engineers, Nordec Services, and warned that in the meantime the system should not be operated.

Owing to the Christmas holidays it was not until December 30 that Mr Mills, an engineer employed by Nordec Services, inspected the system and found a crack in the heat exchanger. Mr Mills in fact prepared a report, but was not called as a witness at the hearing before the judge because he had left the employ of Nordec Services, but on January 6 1988 Mr Mills showed the crack to the plaintiffs, and Mr Pfeiffer gave evidence to the effect that he had seen the crack; the defendants were also shown the crack by Mr Mills. But Mr Edward Luckhurst who attended at the house on that day said he could not see a crack and thought that it was merely rust.

Further inspections were carried out by Mr Garland, who was employed by Nordec Services, and a Mr Friend, an expert instructed on behalf of the defendants. In the course of further examination, it transpired that there was not only the crack on the left-hand side of the unit, described as a hair crack being the crack Mr Mills discovered, but there was a more serious crack on the right-hand side of the unit. The system was then condemned; it was common ground that if there were any cracks it was not satisfactory to be used. No parts could be replaced because the system was too old. Therefore the plaintiffs had a new heating unit installed which was in being and working by January 11 1988.

The plaintiffs then commenced proceedings, it being their case that the defendants should have discovered and reported upon the crack which was found by Mr Mills. The claim was pleaded both in contract and in tort, and in my view there was a contract between the plaintiffs and the defendants effected through the agency of Pollard Machin, but, for reasons which are not entirely clear, the judge appears to have dealt with the matter entirely on the basis of a duty in tort — and indeed that appears to have been the way in which the plaintiffs' case was presented to him in the court below.

In my judgment, it matters not whether the duty arose in contract or in tort; it was the same duty to carry out a test and inspection and report with reasonable care and skill of a competent heating engineer and to report carefully and accurately to the plaintiffs the result of such test inspection.

The plaintiffs called two expert witnesses, Mr Garland and Mr Avery, who were both employed by Nordec Services. Their evidence was to the effect that there should have been an inspection of the heat exchanger by using a torch and a mirror and that if that had been done the hair-line crack which was discovered by Mr Mills and seen by those who subsequently inspected it (with the exception of Mr Luckhurst) would have led to the system being condemned. The defendants did not carry out such an inspection at the time of their examination. What they in fact did was to see that the gas was properly connected; they fired the unit and ran it for some forty minutes. It appeared to be working satisfactorily — that is to say they inspected the flame of the burner in the heat exchanger which appeared to be functioning correctly — and they concluded that the heat exchanger was satisfactory.

In addition to Mr Edward Luckhurst, the defendants called Mr Friend, a heating engineer. He said in his evidence that it was not usual practice, when carrying out a survey, to make a visual inspection of the heat exchanger unless the engineer was put on notice by the burner test that something was wrong. The judge accepted that evidence. However, it is to be noted that Mr Friend qualified that evidence in his cross-examination in relation to the age of the system and I shall refer to that qualification in due course.

As to the facts as found by the judge, it seems to me that the following matters were established:

(1) If there is a crack in the heat exchanger, that is a potentially dangerous situation because carbon monoxide gases can escape from the burner chamber into the airduct and airvents and will be circulated around the living area of the house. Quite plainly, depending on the concentration and the time over which it continues, that is a potentially dangerous situation.

(2) Cracks in a heat exchanger of this kind are caused by expansion and contraction over a period of time; it is a function of metal fatigue and such cracks are more likely to occur in an old system. They develop, first of all, from a hair-line crack into a more serious crack.

(3) In fact this heating unit was some 24 years old and, as such, had reached the end of its useful life — or, indeed, expected life of some 25 years. However, the defendants did not know that. They assumed, correctly, that the system had been installed at the time when the house was built, but apparently they were informed that the house was built in 1972/73 and consequently that the system was 15 years old. That was in fact wrong; the house was built some 10 years earlier.

The judge held, however, that the defendants' view was reasonable that the system was about 15 years old.

(4) The crack on the left-hand side of the unit could be seen if a torch and mirror were used without removing the burner in the heat exchanger chamber. Indeed, as I have indicated, that crack was seen by everybody except Mr Luckhurst, who considered it to be a rust mark. The other more serious crack could not be seen without removing the burner.

That, in my judgment, is the effect of what the judge said at p 46 of his judgment:

> If of course the cracks were obvious and could have been easily seen by any heating contractor, they would certainly have been in breach of their duty as heating engineers if they failed to notice them or failed to report on them when they sent their report to Pollard Machin by whom they had been retained. But these cracks were not obvious, they were inside the heat exchanger and could only have been seen at the time of the Luckhursts' visits on November 13 by the use of a torch and mirror on which to reflect the cracks high up on the clams inside the heat exchanger. Indeed, I doubt whether there would have been any possibility of them seeing the one on the right as all the witnesses really agree in view of the presence of the burners.

(5) It is clear from the judge's exposition there that the cracks were present at the time of the defendants' inspection.

(6) If there is a crack of any description in a burner of this age, the system had to be condemned.

The judge, as I have indicated, accepted the evidence of Mr Friend. Mr Friend drew a distinction between a "service" and a "survey". When dealing with Mr Friend's evidence on this point, the judge said:

> He (Mr Friend) said that he would not expect an engineer carrying out a survey of this kind to search for cracks in the heat exchanger unless he suspected the burners were not operating correctly or that there was evidence of carbon monoxide in the atmosphere.

On the other hand, Mr Friend had said that if a service was being carried out, then the burners would have to be removed and a proper inspection made.

In my view, the judge does not grapple with the problem in terms of the contract because, as I have indicated, what the defendants said they were doing was to make a "full working test and inspection", and it is plain, in my judgment, that the word "full" qualifies both "test" and "inspection".

There was no evidence as to precisely what the defendants were instructed to do but, in my judgment, the proper conclusion is that they were instructed to do what they said they were purporting to do as I have already indicated and it seems to me the plaintiffs were entitled to expect that a full working test and inspection had been carried out. The defendants' report was not in any way qualified to the extent that a full test and inspection had not been carried out or was not able to be carried out.

The judge appears to have thought that the defendants were instructed to survey the system and in those circumstances he accepted Mr Friend's evidence that a survey would not require a full inspection of the heat exchanger other than to look at the burners in operation. But, as I have indicated, his evidence was qualified in cross-examination, and I refer to p 54 of the judge's judgment in relation to that matter:

> Does the practice which I have described

— that is to say simply relying on a visual examination of the system in action and not a visual examination of the heat exchanger itself —

> mean that there is no need to search for cracks or to warn of the possible existence of cracks however old the heating system is? The answer to this, I think, is clearly no when an appliance is clearly nearing the end of its useful life and when wear and tear will be taking their full toll. This was accepted by Mr Friend. He did not accept in cross-examination that this unit was 24 years old. At that time its actual age was unconfirmed but he agreed that, as a heating engineer who knew that that was its age, he would recommend a thorough examination of the heat exchanger which would include taking out the burners to look for cracks with a torch and a mirror. He also agreed that if the heat exchanger was, say, 18 to 20 years old, as he himself had thought it might have been, that could raise the possibility that there was considerable wear and tear including possible cracks. He also said that even at 15 years the defendants could have been on notice or that at such an age there would have been an indication of possible cracks or problems.

The judge then records that he did not have a note of that particular answer in cross-examination. It is unfortunate that that was so because, in my judgment, it was a crucial answer and clearly qualified what Mr Friend was saying was the proper practice in relation to such a heat exchanger.

The judge's note of Mr Friend's evidence is as follows:

> This age would alert to the possibilities of unit having considerable wear and tear including possible cracks. Cracks can develop over years or not be there one week and there the next. Most cracks start from metal fatigue due to heating and cooling.

Unfortunately in that passage it is not clear what age Mr Friend is referring to; it appears that he was referring to the age of the system as he believed it to be, namely 15 years — but that may not be so.

Later on in Mr Friend's evidence the judge's note is as follows:

> I'd remove burners if I suspected heat exchanger cracked. I would not suspect without physical sign, not just age . . . If I knew appliance was 24 years old I would recommend customer to have a thorough examination of heat exchanger and that would include taking out burners and looking for cracks etc with torch and mirror.

Some further light can be shed upon this question because it is plain that in the course of argument counsel for the plaintiffs was relying on this part of Mr Friend's cross-examination. The judge's note is as follows:

> Counsel queried whether my note, which I have marked with (an) asterisk in this typescript, was a full record. Plaintiffs' solicitors' note has after "possible cracks" the words "Yes even 12-14 years defendants on notice". Defendants' note says "age given by defendant in report would also (have given an indication)".

It seems to me that the judge's record of Mr Friend's evidence in the last sentence of the passage to which I have referred makes it plain that where the system is getting to an age — such as 15 years or so — it is reaching a time when cracks may be expected. At such an age, there would have been an indication of possible cracks or problems. The judge then addressed himself to the question whether, in those circumstances, it was sufficient for the defendants not to inspect other than the manner in which they did. The judge said:

> I have not found this aspect of the case an easy one and there is some force in the plaintiffs' counsel's submission that in the case of a 15-year-old machine the unit was beginning to become suspect. However, in the light of the whole of the evidence here I have come to the conclusion that, as the defendants were following a recognised practice in carrying out their task and inspection on a unit which they reasonably thought was only 14 to 15 years old and not 23 to 24 years old as it clearly was, and as nothing occurred during their test and inspection to put them upon enquiry, they should not be held liable for failing to search the inside of the heat exchanger for possible cracks merely because this unit had reached the age of 15 years.

As I have indicated, it seems to me that the effect of the expert evidence was this, that those called on behalf of the plaintiffs were saying that there should always be an inspection of the heat exchanger to see if there are cracks, regardless of the age of the unit.

By the time Mr Friend had completed his cross-examination, he had, in my view, very substantially qualified his evidence-in-chief that no such examination as the plaintiffs' experts recommended was necessary and that where the age of the unit was, as the plaintiffs' solicitors' note records, 12 to 14 years — or, as the judge appears to have accepted, 15 years — it was reaching a suspect age and there was an indication of possible cracks and problems. It was not sufficient simply to rely on the combustion test.

Moreover, it seems to me that in the passage to which I have just referred in the judge's judgment there is, with respect to him, something of a *non sequitur*. It is quite plain that a crack will not necessarily be revealed by the burner test — that is to say simply by lighting it up and looking at it and seeing whether the flame looks satisfactory. If it were so, then it would be impossible to account for the cracks in this case. Assuming that the defendants correctly observed satisfactory combustion, then it is plain, and I think was

A accepted by all witnesses, that that test would not of itself necessarily reveal cracks in the heat-exchanger unit.

Nevertheless, the judge said that there was no need to inspect the heat-exchanger unit because nothing occurred during the burner test and during the rest of the inspection to put the defendants upon inquiry. With great respect to him, that seems to me to miss the gravamen of what Mr Friend was saying and the gravamen of the plaintiffs' case, namely that where a unit such as this is reaching the end of its life, then notwithstanding that the burner test may be satisfactory, a competent heating engineer is on notice that there may be cracks in the chamber and that, in those circumstances, it is necessary to look for them. It may be that the more serious crack on the right-hand side of the unit was not visible without removing the B burner, but the crack on the left-hand side was visible and I see no reason to think that if the inspection recommended by Mr Garland and Mr Avery had been carried out it would not have been discovered. But in any event, if the defendants had approached their task as it would seem Mr Friend was recommending, then perhaps they should have asked the vendors if they could remove the burners, and if their request was refused, then they should have qualified their report by pointing out to the plaintiffs that they had not been able to carry out a full examination of the system.

This is a case in which the safety of the occupants of the house may be at risk; no doubt the system had not yet reached that stage — one does not know — but prudence requires, as it seems to me, that when a unit has reached the stage where deterioration and cracks may C occur, it is not enough to rely simply upon a test which, it is common ground, will not necessarily reveal cracks. Had the defendants carried out their inspection by looking with a torch and mirror and seen nothing, the position might have been different, but they did not look in that way and, in my judgment, the plaintiffs were entitled to expect that a full inspection was carried out, as the report indicates that it was. There was no qualification of any kind in the report and, in my judgment, the judge unhappily failed to give effect to the qualification which, in my view, was made in the evidence of Mr Friend.

It is plain that his evidence did not make an impact on the judge as perhaps it should have done because he did not have a full note, but D nevertheless, in my judgment, it was a crucial qualification to the standard practice, and it was not enough simply to say that the defendants here followed the standard practice in the case of a system which they believed to be 15 years old but which was in fact 24 years old.

In those circumstances I would allow this appeal.

Mann LJ agreed and did not add anything.

The appeal was allowed. By agreement judgment for the plaintiffs for £1,500, to include damages and costs. Costs in the county court on Scale 2.

A Court of Appeal

July 5 1990

(Before Lord DONALDSON OF LYMINGTON MR, Lord Justice RALPH GIBSON and Lord Justice TAYLOR)

ELLIOTT v LONDON BOROUGH OF ISLINGTON

Estates Gazette March 16 1991

[1991] 10 EG 145

Nuisance — Tree invading adjoining owner's property — B Tree belonging to local authority and not subject to a tree preservation order — Prolonged negotiations without agreement — Authority "passionately" desirous of keeping tree — The adjoining owner succeeded in obtaining a mandatory injunction requiring the defendant council to remove the tree — The defendants appealed but the grant of the injunction was affirmed by the High Court — The defendants then appealed to the Court of Appeal — Their appeal was dismissed — Principles discussed

The tree was an ancient horsechestnut tree growing in St Paul's Shrubbery — The spot where the tree was growing abutted on a high wall which was on the boundary of the C plaintiff's property — The trouble was that the tree was encroaching, pushing the wall further into the plaintiff's territory and extending its branches over the wall — There was no doubt that this was a nuisance in the legal sense and gave rise to a cause of action — The local authority's view was that the tree should be left where it was but that the plaintiff should be offered some assistance in regard to his wall — Unfortunately, the plaintiff did not find the offers, one of which would have created "the arborial equivalent of a cat flap for the tree", attractive and he issued a writ claiming a mandatory injunction requiring the defendant authority to D remove the tree — This was granted in the absence of any defence by the authority, but further litigation followed — The authority applied to have the judgment set aside, this application was refused and, on an appeal by the authority, the matter came before the Court of Appeal

The judge below had taken the view that the plaintiff was entitled to the full enjoyment of his rights and that an award of damages would not have been adequate protection — The judgment was subjected to some relatively minor criticisms but the main argument centred round the application of the test is for damages in lieu of injunction set out by A L Smith LJ in *Shelfer* v *City of London Electric Lighting Co* and followed E by subsequent authorities — The rule was that a plaintiff was entitled to an injunction to restrain violation of an established common law right unless there were special circumstances — It was a good working rule that damages in lieu of an injunction may be given if the injury to the plaintiff's legal rights was small, that it was capable of being estimated in money; that it could be adequately compensated by a small money payment and that the case was one where it would be oppressive to the defendant to grant an injunction

The appellants submitted that the above-mentioned tests F were satisfied in the present case, but the Court of Appeal disagreed — In their view the authorities showed that it was not the function of the courts to license breaches of the rights of citizens — It was true that there were exceptional circumstances in which it would be right to refuse an injunction and award damages — However, in the present case the *Shelfer* tests were not, in the court's view, satisfied — The injury to the plaintiff's rights was not small — Although it was perhaps capable of being estimated in terms of money, it was not correct that the breach could be compensated by a small money payment — As regards oppression, the court G could not see that an order to remove the tree would be in the least oppressive — The appellant authority could buy a mature tree and dig a hole for it further away from the respondent's wall — Appeal dismissed

The following case is referred to in this report.

Shelfer v *City of London Electric Lighting Co Ltd* [1895] 1 ChD 287; [1891-4] All ER Rep 838, CA

This was an appeal by the London Borough of Islington from an order of Judge Rice, sitting as a High Court judge, rejecting an application to set aside a decision of Kennedy J granting the plaintiff, Peter Albert Elliott an injunction, in default of defence, requiring the H defendants to remove a tree which was invading the plaintiff's property. The tree, which was owned by the defendant authority, was growing in St Paul's Shrubbery at a spot abutting a high wall on the boundary of the plaintiff's (the present respondent's) property at St Paul's Place, Canonbury.

Jeremy Sullivan QC and Adrian Cooper (instructed by Barlow Lyde & Gilbert) appeared on behalf of the appellants; Michael Harris (instructed by Lynch Hall & Hornby, of Harrow, Middlesex) represented the respondent.

Giving judgment LORD DONALDSON OF LYMINGTON said: Mr Peter Elliott is the owner of a house at St Paul's Place, Canonbury, London N1. The defendants, the Islington London Borough J Council, are the local authority for the area and are the owners of an ancient horse chestnut tree growing in St Paul's Shrubbery. St Paul's Shrubbery, and in particular the spot where the tree is growing, abuts a high wall on the boundary of the plaintiff's property.

The tree is said to be of considerable interest and amenity value, although it is not subject to a tree preservation order. Mr Sullivan QC says that that is not to be taken as reflecting in any way upon the character of the tree, it being the practice of many local authorities not to place tree preservation orders on trees which are in their own ownership.

What matters is that the tree and the plaintiff's wall have manifestly been on a collision course for a number of years. That collision has now taken place and the wall has been deflected into the K plaintiff's property, both at its base and, to a greater extent, further up the wall. Furthermore, the branches of the tree apparently extend over the wall into the plaintiff's property.

Prolonged negotiations have taken place between Mr Elliott and his advisers and the local authority and their advisers because the authority passionately desire to keep this tree and appear to be prepared to go to almost any lengths to do so, including visiting us in this court. There is no suggestion that this tree is not a nuisance in the technical sense of the word or that it is not invading the plaintiff's property. There is an issue as to how serious the matter is, but that there is a cause of action in nuisance is not disputed.

The local authority say that the proper way to deal with this matter is to leave the tree where it is and to give Mr Elliott assistance in the matter of his wall. They have offered to do two things in the alternative. They are quite prepared to rebuild his wall in such a way as to provide an alcove for the greater happiness and further growth of the tree. The alcove would, of course, necessarily be built entirely on the plaintiff's land. Alternatively, they say, if the plaintiff does not like that, they are prepared to put a precast concrete bar slightly above ground level and rebuild the wall on top of that with pillars at each end. In effect, this would create what I suppose would be the arborial equivalent of a cat flap for the tree. Whether cats would also use it is dismissed by Mr Sullivan as a detail which can easily be discussed! At any rate, those are the offers with, in addition, an offer to put right any further defects in the wall which might be caused by the tree's future growth.

The plaintiff was not impressed by these offers; nor, indeed, by the authority's attitude generally, but it is not relevant to go into that. What matters is that he decided that he would assert what he conceived to be his legal rights by issuing a writ claiming a mandatory injunction requiring the defendants to remove the tree. For reasons which are obscure and certainly do not matter, the defendants failed to put in a defence to this claim and, in consequence, Kennedy J gave judgment, in the absence of such a defence, granting the plaintiff an injunction.

Horrorstruck, the authority applied to the court to have that judgment set aside, and that application was heard by His Honour Judge Rice, sitting as a High Court judge, who, after considerable argument, decided to affirm the injunction.

I think I can really give the main thrust of his judgment if I refer to two pages of his reasoning:

> The main ground for refusing an injunction, in my view, is the question of public interest: in other words, balancing the rights of the public against those of the plaintiff, the owner of the land in respect of which the nuisance is being exercised. The courts have considered those two conflicting interests, first of all in *Miller* v *Jackson* [1977] QB 966 where the court decided that although there was a nuisance being caused to an adjoining occupier nevertheless the playing of cricket is of such public importance and perhaps importance to the nation that accordingly the owner of the land in respect of which the nuisance was being permitted should be restricted to damages.
>
> That decision was impliedly criticised in the case of *Kenaway* v *Thompson* [1981] QB 88 and that was a case involving noise nuisance with the noise being caused by power boats and motor boats on a lake near the plaintiff's premises. In the course of his judgment Lawton LJ, who gave the one judgment in the case, referred to the earlier authority which has been the basis of all subsequent decisions, namely *Shelfer* v *City of London Electricity Light Co* [1895] 1 ChD 287. He referred to an earlier decision, part of the judgment of Lindley LJ in that case:
>
> > . . . ever since Lord Cairns's Act was passed, the Court of Chancery has repudiated the position that the Legislature intended to turn that Court into a tribunal for legalising wrongful acts; or in other words, the Court has always protested against the notion that it ought to allow a wrong to continue simply because the wrongdoer is able and willing to pay for the injury he may inflict. Neither has the circumstances that the wrongdoer is in some sense a public benefactor (eg a gas or a water company or a sewer authority) ever been considered a sufficient reason for refusing to protect by injunction an individual whose rights are being persistently infringed . . .
>
> I am quite satisifed in this case and indeed there is no doubt — and Mr Cooper [for the council], if I may say so, very realistically at the onset of his cogent submissions accepted — that there is a nuisance here which is interfering substantially with the rights of the plaintiff. In the circumstances, even though it may well be that this tree doubtless has a certain aesthetic appeal — it is placed at the entrance, I am told, or near a public path, which doubtless the residents of Islington are wont to go — and although, therefore, its removal may upset people and, to some extent, reduce the amenities that are available in this area, nevertheless in my view the plaintiff is entitled to have his rights protected and damages will not adequately protect them. There is no way in which his wall can remain undamaged while the tree remains there. There is bound to be, even if the wall were rebuilt completely by the local authority, with this tree, given its size and position, further damage. In the circumstances, therefore, in my view, the plaintiff is entitled to relief which Kennedy J gave him . . .

Mr Sullivan makes two points on that. The first is that he says that the learned judge was mistaken in thinking (if indeed the judge did think) that Mr Cooper was conceding more than that there was here a nuisance in law. He makes the point that Mr Cooper never conceded that the nuisance was substantial.

Reading the judgment as a whole, for my part, I am quite satisfied that the judge was under no such misapprehension. Mr Cooper could not sensibly have been conceding that the damage was substantial if he was relying upon the test in *Shelfer's* case to say that this was a case in which damages would be an adequate remedy.

Slightly more cogent is the criticism of the judge's statement that there is no way in which the plaintiff's wall can remain undamaged while the tree remains there, "There is bound to be . . . further damage." Mr Sullivan says, I think rightly, that on the evidence the highest that one could put it was that there was a possibility of further damage.

Against that background Mr Sullivan says that the learned judge's judgment really was not founded upon a detailed examination of the facts, and that in a case such as this it is necessary that there should be a detailed examination of the facts and a full deployment of evidence such as is really only possible on a full trial. He accepts, of course, that the reasons why there was not a full trial was because his clients, the local authority, failed to put in an appearance, so that the form of these proceedings in the end was an application to set aside the judgment. That is his first submission.

For my part, I do not think that there is any necessity for a further or fuller trial. It seems to me that the facts here are reasonably clear, subject to the point that damage is not inevitable but there is a possibility of further damage.

Mr Sullivan then applies himself to the tests which were adumbrated by A L Smith LJ in the *Shelfer* case when he said:

> In my opinion, it may be stated as a good working rule that —
> (1) If the injury to the plaintiff's legal rights is small,
> (2) And is one which is capable of being estimated in money,
> (3) And is one which can be adequately compensated by a small money payment,
> (4) And the case is one in which it would be oppressive to the defendant to grant an injunction: —
> then damages in subsitution for an injunction may be given.

For my part, I would emphasise the word "may" because it seems to me that I would only be echoing the authorities, and in particular the decision in *Shelfer's* case, if I say that it is not the function of the courts to license breaches of the rights of citizens. Compulsory purchase under statute is a well-known concept and is subject to well-known protections, but it is not for courts to add to that burden on the citizen a system whereby, as in this case, they will grant, for a fee payable to the plaintiff, a compulsory lease of land to accommodate the roots of this tree in addition to putting up with the nuisance which it creates.

That there are exceptional circumstances in which it would be right to refuse an injunction and to grant damages is not to be gainsaid but, if one looks at these tests, I am far from satisfied that the injury to the plaintiff's legal rights is small. I think it is very considerable. In inches, no doubt, his wall has only been moved a relatively small distance, but nobody who has sat in these courts and has any experience whatever of these boundary disputes would be under the slightest misapprehension as to the strong feelings of emotion and, indeed, irrationality which flow from any incursion over the boundary of somebody's property.

I am very far from accepting that the injury to the plaintiff's legal rights is small in having this tree pushing his wall in the way in which it is common ground that it does. Mr Sullivan says that the judge may have equated "small" in this context with "minute" and was merely saying that the damage was not minute. That arises out of a reference to "minute" earlier in his judgment. I very much doubt whether this was in the judge's mind and it seems to me that he was merely saying, as I have been saying, that the injury was not small.

It is said by Mr Sullivan that it is one which is capable of being estimated in money. That, no doubt, is true, though there might be some slight difficulty in putting a money value on the loss of amenity to the plaintiff, not only in having the tree there but also the effect of its branches, which may affect light and air in his garden. But, if one comes on to the third point that it is one which can be adequately compensated by a small money payment, I would not regard that qualification as being met. Mr Sullivan says that the monetary payment here takes the form of rebuilding the wall, and that is a relatively minor matter. I would not accept that it is a relatively minor matter and, in any event, that is not the end of it because, as the authority conceded, one has to look at the future. They say that this aspect is met by their undertaking to go on repairing the wall if necessary but, if one looks at that in the round, I do not regard that as the equivalent of a small monetary payment.

Finally, we get to the point as to whether it would be oppressive to the defendants to grant an injunction. I think it is at that point that

A the enthusiasm of Mr Sullivan and his clients for the case really kindles because he would like to turn this case into a major authority on the extent to which the interests of the public can be taken into account: in this case, as he submits, the removal of the public's tree would be oppressive in the extreme.

I doubt whether this is a case in which to decide any such thing. It would appear from *Spry on Equitable Remedy* (1984 ed) at p 385 that

> . . . account is taken by the court of any hardship or inconvenience that the making of an order may cause either to particular persons who are not parties or to the public generally. Specific relief has not often been refused on this ground, for it is generally found in such cases that any hardship or inconvenience to third persons or to the general public is relatively indirect and small as against the inconvenience or hardship that would be caused to the
> B plaintiff if he were denied specific relief. In particular it is not generally appropriate that specific private rights should be denied in order to give rise to an indefinite advantage to the general public.

I would fully accept that statement as to the improbability of the situation arising in which the interests of the public were decisive, and draw particular attention to the last sentence that I have cited to the C
effect that it is not generally appropriate that specific private rights should be denied in order to give rise to indefinite advantages to the general public. Were it otherwise, the court would, as I see it, be legislating to deprive people of their rights.

Third parties in this sort of context who might well be taken into account, I would have thought, were adjacent owners who, although not parties to the action, might well have a considerable interest. I cannot see that an order to remove the tree would be in the least oppressive in this case. The defendants, if they want to have a tree here, can go out and buy themselves a mature tree. They can dig a hole in which to put it rather further away from the plaintiff's wall. It is his wall, it is his garden and there is no reason why he should not have full D
enjoyment of it.

Accordingly, I would affirm the learned judge's judgment.

RALPH GIBSON and **TAYLOR LJJ** agreed and did not add anything.

Appeal dismissed with costs.

RATING

A

Court of Appeal

March 12 1990

(Before Lord Justice NEILL, Lord Justice NICHOLLS and Lord Justice BUTLER-SLOSS)

R v NORTHAMPTONSHIRE LOCAL VALUATION COURT AND OTHERS, EX PARTE ANGLIAN WATER AUTHORITY

Estates Gazette February 9 1991

[1991] 05 EG 138

B

Rating — General Rate Act 1967, section 31 and Schedule 4 — Rating of statutory water undertakings on cumulo-value basis — Appeal arising out of error in completion of proposal form — "Potentially a very costly error, for about £3-4m hang on the outcome" — Error in water authority's office consisted in naming as the subject of the proposal a sewage treatment works which had ceased to exist instead of an existing sewage works about half a mile away — The fact that the valuation officer correctly understood what hereditament was intended was insufficient to validate proposal — Successful appeal by rating authority reversing decision of Popplewell J and C **upholding conclusion of local valuation court**

Assessment on the cumulo-value basis was advantageous to statutory water undertakings and the present disputed proposal was one of a number of such proposals intended to preserve this advantage for sewage treatment works by taking action before March 10 1988 — By so doing it was possible for undertakings to claim cumulo-value assessment for such works despite the provisions of the Local Government (Finance) Act 1988 — That Act had negated the effect of the decision in *Cakebread (VO)* v *Severn Trent Water Authority* in which the Court of Appeal had held that the Water Act 1973 D **had widened the scope of "statutory water undertakers" in section 31 of the 1967 Act so that sewage works were covered — The 1988 Act did not, however, apply to proposals made before March 10 1988 — Thus water authorities which had made such proposals for the deletion of their non-water supply hereditaments from the valuation lists were able to enjoy the rating advantages of this for the time being**

The disputed proposal form submitted by the applicants referred by mistake to the "Rushden Sewage Treatment Works", which had ceased to function in 1981, instead of the Broadholme Sewage Treatment Works, which appeared in the valuation list as a hereditament with a rateable value of E **£84,150 — The subsequent events were curious — No query was raised by the valuation officer about a proposal concerning a property no longer in existence and no longer shown in the valuation list — However, in the space at the foot of the form, headed "For use in the Valuation Office Only", a box which formed no part of the proposal, a member of the valuation officer's staff had entered particulars which clearly identified the subject of the proposal as the Broadholme Works — Moreover, subsequent correspondence, including the sending of a copy of the disputed proposal to the rating authority, the notification of an objection, and a proposal for** F **an increase in the rateable value from £84,150 to £85,400, all assumed that the disputed proposal concerned the Broadholme Works — In due course the local valuation court held that the disputed proposal was invalid, a clerical error having led to the wrong identification of the subject hereditament — In judicial review proceedings, however, Popplewell J quashed this decision and held the disputed proposal valid — The rating authority appealed**

The only issue on the appeal was whether the disputed proposal form did sufficiently identify the Broadholme Works as the subject of the proposal — The appellant rating G **authority's case was that the disputed proposal form did not in fact propose an alteration in respect of the Broadholme Works and the treatment of it in the valuation office did not change the position — The respondent water authority argued that the test to be applied in construing the proposal was the objective one of what would reasonably be understood from reading the form, but subject to the subjective gloss that if the proposer had made a mistake, and the valuation officer had correctly understood the true intention, then the proposal as so understood would be effectual although others reading the form would not have so understood it — The court rejected** H **this argument and also dismissed any application of the principle of estoppel (although no argument had been submitted on that ground)— Rating authority's appeal allowed**

The following cases are referred to in this report.

Cakebread (VO) v *Severn Trent Water Authority* [1987] RA 290, CA

R v *Winchester Area Assessment Committee, ex parte Wright* [1948] 2 KB 455; [1948] 2 All ER 552; [1948] 41 R&IT 348, CA

This was an appeal by the rating authority, East Northamptonshire District Council, from the decision of the Queen's Bench Divisional Court, Popplewell J, who had quashed a J decision of the Northamptonshire Local Valuation Court given in favour of the rating authority and held valid a disputed proposal by the respondent Anglian Water Authority for the deletion from the valuation list of a particular sewage treatment works. The decision of Popplewell J is reported at [1989] 2 EGLR 168.

Jeremy Sullivan QC and Peter Village (instructed by Sharpe Pritchard, agents for solicitors' department, East Northamptonshire District Council) appeared on behalf of the appellants; William Glover QC and Richard Glover (instructed by Beechcroft Stanleys) represented the respondents.

Giving the first judgment at the invitation of Neill LJ, **NICHOLLS** K **LJ** said: This appeal arises out of an error made in filling up a form. It is potentially a very costly error, for about £3 to £4m hang on the outcome.

The Anglian Water Authority is the regional water authority for an area which includes the area for which the East Northamptonshire District Council (to which I shall refer as "the rating authority") is the rating authority. In 1985 the valuation officer at Kettering was the officer in charge of the valuation list for the East Northamptonshire rating area. The Anglian Water Authority is a statutory water undertaking. There are special statutory provisions affecting the rating of the property of such undertakings. Under section 31 of the

A General Rate Act 1967, as it stood in 1985, the rating values of hereditaments occupied for the purposes of statutory water undertakings otherwise than as dwellings were ascertained according to a special formula. An amount, described in Schedule 4 as "the cumulo-value", was ascertained for the undertaking as a whole. This amount was apportioned among rating districts in which water hereditaments of the undertaking were situated. The amount apportioned to any rating district was to be the rating value of those hereditaments for the purposes of the rating lists.

By March 1985 the water industry was aware that the effect of certain provisions in the Water Act 1973 might be that the reference to statutory water undertakers in section 31 was no longer limited to water supply. If that were so, the special statutory provisions, B whereby the water undertakings' cumulo-value was apportioned over their water hereditaments, would cover also hereditaments such as sewage treatment works. This would be advantageous to water authorities in so far as, so it seems, the cumulo-value was calculated and fixed by reference only to property used for water purposes.

The Anglian Water Authority (which I will call "the water authority") operates many sewage treatment works. So, quite properly, the water authority decided to protect its position in case in due course the courts held that this was the effect of the legislation. In March 1985 the water authority served on the Kettering valuation officer proposals, on the prescribed forms, for the assessments C relating to some 40 sewage works to be expunged from the valuation list, on the ground that the water authority's occupation was deemed to be within the cumulo-value of the water undertaking.

It was important that these proposals should be served before the end of March, because a proposal which leads to an alteration in the valuation list has effect in relation to the rate current at the date when the proposal was served on the valuation officer, as from the commencement of the then current rate period. This is the effect of section 79(1). The rate period is the period of a year starting on April 1. So if the proposals were served by the end of March 1985, and were successful, they took effect from April 1 1984.

If I may look ahead for a moment. In December 1987, in the case of D *Cakebread (VO)* v *Severn Trent Water Authority* [1987] RA 290, this court held that the Water Act 1973 did have the effect of widening the scope of the expression "statutory water undertakers" in section 31 of the 1967 Act. The effect of that decision was promptly negatived by section 122 of the Local Government (Finance) Act 1988, by inserting into the 1967 Act a new section 31. But that amendment did not apply to any proposal made before March 10 1988. Thus water authorities which had made proposals at the relevant time for the deletion of their non-water supply hereditaments from the valuation lists were able to enjoy the rating advantages flowing from this until the new legislation took effect.

I return to the history of the present case. Among the proposals for alteration of the valuation list submitted by the water authority in E March 1985 was one dated March 28 1985. This stated, at the top of the form, as the address of the property to which the proposal related:

Rushden Sewage Treatment Works
Northampton Road
Rushden

The rating area was stated to be "East Northants DC" and the rating district "Rushden". That is the proposal which has given rise to all the difficulties. I shall refer to it as "the disputed proposal".

There were no sewage treatment works in Northampton Road, Rushden, in March 1985. There were no sewage treatment works known as the "Rushden Sewage Treatment Works". Nor was there any such hereditament on the current valuation list. There had been F such works, but they had ceased to function in 1981. The works had been dismantled and the site grassed over. The hereditament had been removed from the valuation list in July 1982.

However, about half a mile away, to the north-west of the site of the former Rushden Sewage Works, across the River Nene and beyond some old gravel pits, the water authority did have another, quite separate, sewage works. They were the Broadholme Sewage Treatment Works and their address was and is Ditchford Lane, Irthlingborough. The Broadholme works had been constructed in 1975 and extended at various times up to 1984. They appeared as a hereditament in the valuation list with a rateable value of £84,150. They were in the same rating area as the former Rushden works, but in a different rating district.

The water authority served no proposal naming the Broadholme G works as the subject property. Clearly an error was made within the water authority's offices. By mistake, among the 40 proposals prepared and served, no proposal was served naming the Broadholme works as the subject property; but, also by mistake, a proposal was served naming the old Rushden works, no longer in existence and no longer on the valuation list.

Precisely what happened within the offices of the valuation officer when the disputed proposal was received and processed is not clear. Given that the Rushden works hereditament no longer existed, one obvious possibility is that a member of the valuation officer's staff would have raised a query over the telephone with the water authority. But there is no evidence of any communication at this stage. All we know is what the documents record. These show certain H particulars inserted at the foot of the disputed proposal form, in a space which is headed

FOR USE IN THE VALUATION OFFICE ONLY.

I shall refer to that space as "the valuation office box". Completed in the valuation office box, on the disputed proposal form, was this: against the word "Description" was written "Sewage Disposal Works", and against the word "Address" was written "Broadholme Sewage Treatment Works Ditchford Lane (Anglian Water Authority) Irthlingborough". Presumably these particulars were completed by a member of the valuation officer's staff on or near the date, March 28 1985, stated in the box as the date on which the proposal was received and acknowledged. Also stated in the box was J the amount of the rateable value assessed in respect of the Broadholme Sewage Treatment Works in the valuation list, namely £84,150.

A valuation officer is required by section 70 of the Act to send to the relevant rating authority a copy of every proposal for the alteration of the valuation list served on him. On April 2 Mr A G Frost, who was then the valuation officer for the Kettering area, sent to the rating authority a copy of the disputed proposal form. The copy which was sent included, in the valuation office box, the insertions made therein by the valuation officer. Thereafter, on May 31 the valuation officer sent to the water authority a formal notification that he objected to the disputed proposal. In that notification he stated, as the address of the subject property, the K Broadholme Sewage Treatment Works. The valuation officer then became obliged under the Act to send to the clerk of the local valuation panel a copy of the disputed proposal and a note that he, the valuation officer, objected to the proposal. The valuation officer did that on June 14.

Under section 74(3), that operated as an appeal by the water authority against the valuation officer's objection. Both the water authority and the rating authority were notified by the valuation officer that he had taken this step, in letters which identified the Broadholme works as the subject property.

The appeal did not come on for hearing for some considerable time. Meanwhile, in March 1986, the valuation officer proposed that L the rateable value of the Broadholme works should be increased to £85,400 to take account of structural alterations which had been made. The water authority objected, and that led to a further appeal to the same valuation court. Both appeals came on before the court in March 1988. By consent the appeal relating to the valuation officer's proposal was adjourned. Regarding the other appeal, on March 31 1988 the valuation court decided that the disputed proposal was invalid and that the Broadholme works could not in consequence be expunged from the valuation list pursuant to that purported proposal. There was, they held, a duty on the maker of a proposal to identify the hereditament which was the subject of the proposal. Here there was no opportunity to confuse the two sewage works. A clerical error had led to the wrong identification of the subject M hereditament.

It is that decision which is the subject of these judicial review proceedings brought by the water authority against the Northamptonshire local valuation officer, Mr Handley, who now has responsibility for the relevant valuation list, and the rating authority. On March 14 1989 Popplewell J quashed the decision of the valuation court and declared that the disputed proposal was good and valid. Before us now is an appeal by the rating authority against that decision.

I turn next to the material statutory provisions. Section 69 of the 1967 Act provides for the making of proposals for the alteration of current valuation lists. So far as material, section 69(1) provides:

> A (1) . . . any person (including a rating authority) who is aggrieved —
> (*a*) by the inclusion of any hereditament in the valuation list; . . .
> may at any time make a proposal for the alteration of the list so far as it relates to that hereditament.

Subsection (2) provides for valuation officers to make proposals for alterations of valuation lists; and subsection (3) likewise provides for a rating authority to make proposals.

Section 69(5) provides as follows:

> Every proposal under this section must —
> (*a*) be made in writing; and
> (*b*) specify the grounds on which the proposed alteration is supported; and
> (*c*) comply with any requirements of any regulations made by the Minister with respect to the form of such proposals and otherwise with respect to
> B the making thereof.
> and every such proposal made otherwise than by the valuation officer must be served on the valuation officer.

It is not disputed that these requirements are mandatory.

Regulations have been made pursuant to that provision. The Valuation Lists (Proposals for Alteration) Regulations 1974 provide, so far as material:

> . . . every proposal shall be in the form set out in Schedule 1 to these regulations or in a form substantially to the like effect.

The form in Schedule 1 is substantially the same as the form on which the disputed proposal was made, save that the prescribed form does not include the valuation office box. Thus, as one would expect, the
C legislation contemplates that the proposal will contain a description of the hereditament sufficient to enable those concerned to act in response to the proposal to identify the hereditament intended to be the subject of the proposal. They will include those on whom the proposal, or a copy of it, is required to be served. They will also include, depending on the circumstances, the valuation officer, the occupier of the hereditament to which the proposal relates and the rating authority.

We were referred to the decision of the Court of Appeal in *R* v *Winchester Area Assessment Committee, ex parte Wright* [1948] 2 KB 455. The court there was concerned with an earlier rating Act, the Rating and Valuation Act 1925, but nothing turns on any differences between that Act and the current Act for present purposes. At p 459
D Scott LJ made some observations which are equally applicable to the 1967 Act:

> It has been judicially recognised that, in passing the Rating and Valuation Act, 1925, it was the intention of Parliament that an aggrieved person should be able to act upon many of its provisions, including section 37, without legal assistance. Greer LJ said in *The King* v *West Norfolk Assessment Committee* 94 JP 201, at page 203:
> "It is not intended by the statute that his difficulties should be such that he would have to go to an expert or to a lawyer to draw up his proposal."
> The language of proposals, by whomsoever made, should therefore be read without too much legal strictness; none the less the requirements of subsections (1) and (2) must be substantially satisfied, if the "proposal" is to be effective and valid.

E Scott LJ then considered the authorities concerning the statutory requirement to specify the grounds on which a proposed alteration is supported. At the end of his judgment, at p 463, he expressed his conclusion in these words:

> I find it difficult to harmonise all these decisions but state my conclusions thus: the proposal must give sufficient information to enable the assessment committee to know (i) whether an increase or decrease is asked for (ii) to which of the existing valuations in the rate book the proposal refers, and (iii) what is the ground of complaint as to the existing valuation on that head, and it is enough to state "incorrect" or "unfair" unless there is some unusual ground, in which case it ought to be specified.

Thus, and this is common ground in the present case, the question — and it is the only question which arises on this appeal — is whether
F the disputed proposal form did sufficiently identify the Broadholme works as the subject of the proposal.

Before turning to consider that question, there are some preliminary points to be noted. First, the valuation office box is not part of the proposal. It does not appear in the prescribed form. Its heading and position make clear that it was not intended to be completed by the proposer as part of his proposal, and this box was not completed by the water authority in this case. Second, there is no statutory provision enabling the valuation officer to amend a proposal. If the valuation officer wishes to make a different proposal, he puts forward his own proposal under section 69(2). Third, following from this, whatever entries the valuation officer may choose to make in the valuation office box, or indeed on any other part of the form after it has been served on him, cannot be G
regarded as thereby becoming part of the proposal. Fourth, depending on the circumstances, it will or may be otherwise if the valuation officer is authorised by the proposer to make changes or additions on his behalf. Such a change or addition may then take effect as though the proposer had attended upon the valuation officer and himself amended his proposal. No such authority was given in this case. Nor is it contended that the valuation officer here purported to amend the proposal on behalf of the proposer. Fifth, leaving aside a case in which an authorised change is subsequently made, a proposal is either a valid and effectual proposal regarding a particular hereditament when served, as then containing all the necessary ingredients of a proposal, or it is not. The valuation officer's understanding of the proposer's intention may be useful H
evidence of whether the necessary description of a hereditament is or is not sufficient to an informed local reader of the form. But it is no more than that. A misreading or misunderstanding of a proposal, duly recorded by the valuation officer in the valuation office box, could not prejudice the validity of a proposal which was in fact regular in all respects.

Mr Glover, for the water authority, advanced a submission to the following effect. He submitted that the test to be applied in construing the proposal is the objective one of what would reasonably be understood from reading the form, but subject to this subjective gloss, namely that if the proposer has made a mistake and in fact the valuation officer correctly understands the intended J
hereditament, then the proposal form is effectual even though others reading the form would not have so understood it. Mr Glover placed reliance on the language used by Scott LJ in the *Winchester* case in the second of the quotations I have made from his judgment where he referred to the need for the proposal to give sufficient information "to enable the assessment committee to know" various matters. The functions of an assessment committee in making and maintaining valuation lists have now been taken over by valuation officers.

I cannot accept this submission. I do not find in Scott LJ's judgment any support for such a qualified objective test. Moreover, and more fundamentally, I do not think that Scott LJ was addressing himself to distinctions between a so-called objective test and a so-called subjective test in the construction of proposal forms. There the K
question before the court was whether the grounds relied on had been sufficiently stated, when what was stated was only "revised assessments". Scott LJ was not considering what would be the position if, because of special knowledge acquired from extrinsic sources, a valuation officer would read, or was prepared to treat, the proposal in a different way from that in which it would be understood by a reader, who had available to him no more than the information generally available to any informed reader of a proposal: in particular, the contents of the proposal itself, knowledge of the locality and knowledge of the contents of the valuation lists.

In my view, the true position is that the adequacy of identification of the subject hereditament on a proposal, whether made by a L
ratepayer or a valuation officer or the rating authority, falls to be determined according to the ordinary principles of construction: how would the proposal reasonably be understood by those on whom the proposal or a copy is to be served? In answering this question, as with all questions of construction, common sense is to be applied plentifully. But special knowledge peculiar to the valuation officer, such as what he or a member of his staff may have been told orally or in a covering letter by the proposer, is not normally material. The proposal form is itself intended to be adequately definitive of the property affected. It is a public document, available for inspection by ratepayers under section 108.

I return now to the question which I have identified as the question of construction arising on this appeal. The valuation court correctly M
addressed itself to that question and held that the disputed proposal did not give sufficient identification of the Broadholme works. The judge, likewise, addressed himself to the question, but he reached the opposite conclusion. He said this, at p 11 of the transcript of his judgment*:

> . . . What I have to ask myself, applying what Scott LJ said, is: "Does the proposal give sufficient information to enable the valuation officer to know to what hereditament this proposal relates?" The answer is clear. The valuation officer did know, from reading the proposal, to what hereditament this proposal related. The valuation officer used some common sense and realised that the proposal related to Broadholme and not to Rushden. He

*Editor's note: Also reported at [1989] 2 EGLR 168 at p 171E.

> realised from the nature of the proposal that it was in relation to a sewage treatment works. He realised that Rushden was no longer in existence as a sewage treatment works, and he had sufficient information from that proposal to know that it related to Broadholme. I therefore do not have to consider what some other valuation officer might have been led to believe if he had not come to that conclusion. Nor do I have to consider other matters which were canvassed in argument as to what conclusion a person might draw from a particular proposal. In the end the question is essentially one of fact. The fact is that this valuation officer knew, from the proposal form, that it related to Broadholme. It was not in any way due to the Anglian Water Authority that he was enabled to come to that conclusion, because the proposal related to a different hereditament on the face of it. But the information from the proposal led the valuation officer to have the information which was required. It therefore does not seem to me possible to say that this proposal is invalid.

I confess to having some difficulty with parts of the learned judge's reasoning. In so far as what he said amounted to an acceptance of the submission made to us by Mr Glover, I must respectfully part company from him. In this case a valuation officer or other person looking at the disputed proposal would quickly have appreciated that something had gone wrong in the completion of the form.

There was no longer a sewage works at Rushden. The application sought to expunge from the valuation list a hereditament which was not to be found thereon.

But where does this lead? When considering whether the disputed proposal sufficiently identified the Broadholme works as the property intended to be the subject of the proposal, it is helpful to identify what was the nature of the mistake made in completing the form. As to that, the scanty material available seems to me, if anything, to point in the direction that the error was not one of misdescription in the sense that the writer directed his mind at a particular place and then, inadvertently, wrote down an inaccurate description of that place. In the present case, the right inference seems to me to be, if anything, that when the member of staff within the water authority's office filled in the disputed proposal and wrote in the name and address of the Rushden works and stated the rating area and rating district appropriate for that address, the place he was intending to identify was the former sewage works of the water authority of that name and at that address. He indeed was in error in referring to those works. He should not have made a proposal in respect of those works. It may be that he was working from an old, out-of-date, list. But when he wrote down "Rushden", he meant the place which bore that name and address and no other. He ought to have made the proposal not in respect of that place, the old Rushden works, but in respect of the sewage works bearing a different name and at a different address. He slipped up, as we all do from time to time.

If that be the right inference to draw, it would not be in the least surprising if the address given on the disputed proposal form were not sufficient to identify the Broadholme works as the hereditament to which it related. If the person who filled in the form was incorrectly addressing himself to the Rushden works, it would indeed be remarkable if, by good chance, he had succeeded in identifying not those works but some other works as the subject of the proposal, being the very works which he would have made the subject of his proposal if he had not misdirected himself.

But even if that is not the right inference to draw regarding the nature of the error, it seems to me none the less clear that the name and address inserted on this form were, self-evidently, insufficient to identify the Broadholme works, a quite distinct and separate sewage works, as the intended subject-matter. I do not think that the response within the valuation office takes the point any further in this case. The valuation officer's apparent readiness at that time — I say "at that time" because his stance changed later — to treat the proposal as relating to the Broadholme works does not overcome the difficulties I have mentioned.

The fact that the valuation officer treated the disputed proposal as relating to Broadholme, and acted on that footing, at once makes one think in terms of principles of estoppel. But no case has been advanced before us on this ground and, in my view, rightly so. Likewise, in respect of the rating authority and its failure to take action to query the position, even though the disputed proposal, with the valuation office box completed, was served on the rating authority in 1985.

The end result is that, in my view, this appeal succeeds. The disputed proposal form did not propose an alteration to the valuation list in respect of the Broadholme works. That was the position on March 28 1985. Nothing that happened thereafter changed that position. This result is unsatisfactory, in the way that often happens when a mistake has been made. The unfortunate consequences were compounded by the delay in the matter coming before the valuation court. So far as we know, it was only when the appeals were being prepared for hearing in February 1988 that the rating authority noticed the discrepancy on the disputed proposal form. Thus it came about that it was so long before the mistake was noticed. I would allow this appeal.

Neill and **Butler-Sloss LJJ** agreed and did not add anything.

The appeal was allowed, with costs of appeal and both sides to have half costs below; leave to appeal to House of Lords was refused.

TOWN AND COUNTRY PLANNING

A
Court of Appeal

March 16 1990

(Before Lord Justice SLADE, Lord Justice PARKER and Lord Justice GLIDEWELL)

NEWBURY DISTRICT COUNCIL v SECRETARY OF STATE FOR THE ENVIRONMENT AND ANOTHER

Estates Gazette February 16 1991

[1991] 06 EG 164

B **Town and Country Planning Act 1971 — Enforcement notice — Appeal from decision of deputy Queen's Bench judge — Mobile home and caravans stationed on small area of appeal site, the rest being used for agriculture — Inspector held to have failed to consider alternative methods of achieving the correct result, taking account of the parties' representations — Appeal from deputy judge allowed and owner-occupier's case against enforcement notice remitted to Secretary of State to redetermine in the light of court's opinion**

The appeal site consisted of a small area (OS 9744) of 0.4 of an acre, on which stood a stable block, together with an area C used for grazing horses of about five acres — Planning permission had been granted for the stable block — The present litigation arose from an enforcement notice served by the planning authority on the owner-occupier, Mrs Gore, alleging a breach of planning control consisting of a change of use to the mixed uses of agriculture, the stationing of a mobile home for residential purposes and the stationing of two touring-type caravans (the home and the caravans are for convenience referred to as "the caravans") — The notice required the cessation of the use and the removal from the land of the caravans — Mrs Gore, appealed to the Secretary of D State, seeking planning permission to keep the caravans on OS 9744 and representing that the five acres which had been used for agriculture should not have been included in the enforcement notice — The course taken by the inspector was to allow the appeal, quash the enforcement notice and grant planning permission in respect of the whole appeal site for the continuation of the mixed use of agriculture and stationing of the caravans — The permission was subject to conditions, including a scheme to be approved by the planning authority for the siting of the caravans — The inspector's intention was that the caravans should remain on OS 9744 and that there should be no caravans on the five acres — The planning E authority appealed to the High Court under section 246 of the 1971 Act — Mr Malcolm Spence QC, sitting as a deputy judge of the Queen's Bench Division, dismissed the appeal — The planning authority appealed to the Court of Appeal

After hearing submissions on behalf of the planning authority and the Secretary of State (Mrs Gore was not represented), the Court of Appeal concluded that, once the inspector had decided that it was appropriate to grant permission for the retention of the caravans but to restrict them both in number and in location to OS 9744, there were three possible courses which he could have taken — These F were:

(1) to vary the enforcement notice by deleting OS 9744 from the site to which it applied, to grant planning permission for the retention of the three caravans on OS 9744 (with landscaping conditions) and to uphold the enforcement notice in relation to the five acres;

(2) to grant planning permission as before for the retention of the caravans on a site limited to OS 9744 (with landscaping conditions) and to allow the appeal by quashing the enforcement notice;

(3) to take the course which he in fact took, by allowing the G appeal generally and granting planning permission for the mixed use of the whole appeal site subject to conditions such as he imposed

The first alternative was available to the inspector to adopt as a matter of law — It might have been unfair to Mrs Gore, who wanted only planning permission for OS 9744 and who represented that there had been no breach of planning control in respect of the five acres; this argument was, however, one for the Secretary of State rather than the court — The second alternative, which would have left the existing use of the five acres as its lawful use, would effectively have achieved what H both Mrs Gore and the planning authority had wanted — It did not appear that the inspector had considered adopting either the first or the second alternative, but it was his duty to do so — As he intended to grant planning permission for the retention of the caravans on OS 9744, he was under a duty as a matter of law to consider the best way of achieving this object and to do so in relation to the representations made to him

The appeal against the decision of Mr Malcolm Spence QC was accordingly allowed — The case of Mrs Gore against the enforcement notice was remitted to the Secretary of State to redetermine in accordance with the judgment of Glidewell LJ, J whether by a rehearing or other procedure as the Secretary of State decided

The following cases are referred to in this report.

Bath City Council v *Secretary of State for the Environment* (1984) 47 P&CR 663

Hammersmith London Borough Council v *Secretary of State for the Environment* (1975) 73 LGR 288; 30 P&CR 19, DC

This was an appeal by the planning authority, Newbury District Council, from a decision of Mr Michael Spence QC, sitting as a deputy judge of the Queen's Bench Division, who had dismissed the K council's appeal from the inspector's decision in relation to an enforcement notice served by the appellants on the owner-occupier of the appeal site. The enforcement notice was concerned with the stationing of a mobile home and two caravans on a small part of the appeal site, which was at Church Lane, Burghfield, Berkshire.

John Steel (instructed by Sharpe Pritchard) appeared on behalf of the appellant authority; Robert Joy (instructed by the Treasury Solicitor) represented the Secretary of State for the Environment; Mrs Gore, the second respondent, did not appear and was not represented.

Giving the first judgment at the invitation of Slade LJ, **GLIDEWELL LJ** said: The land to which this appeal relates is an area of nearly five acres ("the appeal site") at Church Lane, Burghfield, Berkshire. It is owned and occupied by the second respondent, Mrs Gore. On the northern part of the appeal site, OS 9744, an area 0.4 of an acre in extent, there stands a stable block for which planning permission was granted after its erection in 1986. The remainder of the appeal site is used for grazing horses. I shall call this "the five acres".

On May 4 1988 the appellants, Newbury District Council ("the council"), served upon Mrs Gore an enforcement notice which related to the whole of the appeal site. This alleged that there had been a breach of planning control, namely a change of use of the land to the mixed uses of:

(i) agriculture;
(ii) stationing a mobile home for residential purposes;
(iii) stationing two touring-type caravans.

The notice required Mrs Gore to cease to use the land for the stationing of the mobile home and the two touring-type caravans and to remove them from the land.

On May 20 1988 Mrs Gore appealed to the first respondent, the Secretary of State for the Environment, against the enforcement notice on one sole ground, namely that planning permission ought to be granted for the development to which the enforcement notice related. The appeal was dealt with by way of written representations. The representations submitted on behalf of Mrs Gore said that the appeal site "includes land in the order where no breach of planning control has occurred". This was accompanied by a plan showing housing development in Burghfield, which delineated most of OS 9744 by a dashed red line with the notations: "Where planning permission ought to be granted for mobile home and caravan" and "The only land to which the Enforcement Notice is relevant". In my view, it is clear that Mrs Gore's agent was submitting on her behalf:

(i) that she was seeking planning permission to retain the mobile home (which I shall call a caravan) and the two touring-type caravans on OS 9744, and

(ii) that the five acres were used only for agriculture, that there had been no development on this land and thus that it ought not to have been included in the ambit of the enforcement notice.

It is only right to say that Mr Malcolm Spence QC (sitting as a deputy judge of the Queen's Bench Division), from whose decision this is an appeal, did not read these representations as conveying so clear a message.

The appeal was decided by an inspector of the Department of the Environment. He dealt with the issue about the five acres first. At para 4 of his decision letter, dated December 21 1988, he said:

> You say that the notice is at fault because it includes land on which you contend that no breach of planning control has occurred. However I think it right that the notice should relate to the appellant's land as a whole since the allegation in the notice refers, *inter alia*, to items, namely caravans, which are physically capable of being stationed anywhere on the appeal site. I take the view too that the allegation in the notice reasonably describes the use being made of the enforcement notice site as a whole at the date the notice was issued, and that that mixed used was in breach of planning control.

The inspector then went on to consider whether it would be right to grant planning permission for the retention of the mobile home and the two caravans on OS 9744. He concluded that if no more caravans were brought on to the land, if they were sited only on OS 9744 and if some further landscaping were undertaken

> the continuing presence of the caravans would not cause demonstrable harm to the appearance of the area.

He said that, in his view, his requirements could be secured by conditions on the planning permission.

Having so concluded, the course adopted by the inspector was to allow the appeal, to direct that the enforcement notice should be quashed and to grant planning permission in respect of the whole appeal site for:

> the continued use of land . . . for a mixed use of agriculture, the stationing of a mobile home for residential purposes, and the stationing of two touring-type caravans, subject to the following conditions:
>
> 1. Not more than one mobile home shall be stationed on the land for residential purposes at any time, and not more than two touring-type caravans shall be stored on the land at any time;
>
> 2. Within three months of the date of this letter a scheme for the siting of the mobile home and two touring-type caravans on the land should be submitted to the Local Planning Authority for approval. Any mobile home or touring-type caravans stationed on the land shall be sited in accordance with the scheme within three months of its approval.

There was a third condition requiring a scheme for landscaping to be submitted to and approved by the council.

It is quite clear that what the inspector intended to achieve was that the three caravans should remain on OS 9744 and that there should be no caravans on the five acres. He envisaged that this would be achieved by the council only being willing to approve a scheme for the siting of the caravans which put them all on OS 9744.

The council appealed to the High Court under section 246 of the Town and Country Planning Act 1971 against the inspector's decision. On March 2 1989 Mr Malcolm Spence QC dismissed the appeal. With leave granted by him, the council now appeal to this court.

An appeal to the Secretary of State against an enforcement notice lies under section 88 of the 1971 Act. The ground of appeal relied on by Mrs Gore is contained in section 88(2)(*a*).

Section 88A, so far as is material, provides:

> (1) On the determination of an appeal under section 88 of this Act, the Secretary of State shall give directions for giving effect to the determination, including, where appropriate, directions for quashing the enforcement notice or for varying its terms.
>
> (2) On such an appeal the Secretary of State may . . . give directions for varying its terms, if he is satisfied that the . . . variation can be made without injustice to the appellant or to the local planning authority.

By section 88B:

> (1) On the determination of an appeal under section 88 of this Act, the Secretary of State may —
>
> (*a*) grant planning permission for the development to which the enforcement notice relates or for part of that development or for the development of part of the land to which the enforcement notice relates;
>
> (*b*) . . .
>
> (*c*) determine any purpose for which the land may, in the circumstances obtaining at the time of the determination, be lawfully used having regard to any past use of it and to any planning permission relating to it.
>
> (2) In considering whether to grant planning permission under subsection (1) of this section, the Secretary of State shall have regard to the provisions of the development plan . . . and to any other material considerations; . . .
>
> (3) Where an appeal against an enforcement notice is brought under section 88 of this Act, the appellant shall be deemed to have made an application for planning permission for the development to which the notice relates and, in relation to any exercise by the Secretary of State of his powers under subsection (1) of this section —
>
> (*a*) any planning permission granted under that subsection shall be treated as granted on that application; . . .

Section 92 of the 1971 Act deals with the effect of planning permission on an enforcement notice. Section 92(1) provides:

> If, after the service of a copy of an enforcement notice, planning permission is granted for . . . the continuance of a use of land, to which the enforcement notice relates, the enforcement notice shall cease to have effect in so far as it requires steps to be taken for . . . the discontinuance of that use . . .

An appeal to the High Court against a decision of the Secretary of State or an inspector appointed on his behalf under section 246 may only be made on a point of law.

Mr Steel, in an able argument on behalf of the council, submits that the course adopted by the inspector was undesirable from the public point of view, in that, under the planning permission granted by the inspector, though the council could control the actual siting of the caravans, they could not control the use of the five acres for purposes ancillary to the occupation of the caravans, eg for the parking of motor vehicles in connection with the residential caravan. If the planning permission had been limited to OS 9744, he submits, it would have been a breach of planning control for the five acres to be used for any purpose other than agriculture. Moreover, if the enforcement notice had been upheld in relation to the five acres, an attempt to use the five acres for any purpose other than agriculture would have been a breach of the enforcement notice for which the council could prosecute. In other words they would not be required to serve a fresh enforcement notice in order to ensure that the five acres were not used for any unauthorised purpose.

Mr Steel submits that, as a matter of law, where a course can be adopted which upholds an enforcement notice without imposing any disadvantage on the landowner, this course should be adopted. In support of his submission he refers us to two decisions of the Divisional Court.

In *Hammersmith London Borough Council* v *Secretary of State for the Environment* (1975) 30 P&CR 19, a house was being used to accommodate large numbers of young people who came to London from overseas on holiday. The council served an enforcement notice

A on the owner alleging a material change in the use of the house to use for the purposes of a guest house. On appeal, the Secretary of State held that the phrase "guest house" was not an appropriate description of the actual use of the house and that some such phrase as "holiday hostel" would be more appropriate. He took the view that this was a material error in the enforcement notice and therefore allowed the appeal and quashed the notice. On an appeal by the council, the divisional court held that there would have been no injustice to the owner of the house if the Secretary of State had amended the enforcement notice to relate to whatever he thought was an appropriate description of the use to which it was being put, and then upheld the notice. At p 21 Lord Widgery CJ, in a judgment with
B which the other members of the court agreed, said:

> It seems to me that the Secretary of State is perfectly entitled, and indeed bound in an appropriate case, to choose his own label to describe the activity which is complained of if he thinks that the language employed by the draftsman of the notice is unsuitable. It is, I think, the Secretary of State's duty to try to get the enforcement notice in order, and if one of the defects in the notice is that an unsuitable label has been used I am entirely in favour of the proposition that the Secretary of State can take steps to substitute a different and more suitable label.

This decision was followed by Woolf J in *Bath City Council* v *Secretary of State for the Environment* (1984) 47 P&CR 663, where the enforcement notice related to a failure to obtain listed building consent.

C Mr Jay, for the Secretary of State, argues that these authorities do not apply to the situation in the present case, since the inspector has not upheld the enforcement notice nor sought to do so. On the contrary, he has allowed the appeal by granting planning permission for the retention of the three caravans on OS 9744. In reply, Mr Steel makes two alternative submissions. First, he submits that the principle in the authorities which I have just quoted applies equally where the Secretary of State decides that the proper course is to grant planning permission in relation only to part of the land subject to the enforcement notice and to uphold the notice in relation to the rest of the site. In effect, he submits, this is what the inspector was trying to achieve in the present case. Alternatively, since Mrs Gore's agent in his written representations on her behalf had made it entirely clear
D that what she was seeking was planning permission to retain the three caravans on OS 9744, it was wrong of the inspector to grant a planning permission which was wider than the permission Mrs Gore was seeking.

Once the inspector had decided that it was appropriate to grant planning permission for the retention of the three caravans, but to restrict them both in number and to OS 9744, there were, in my view, three possible ways in which he could achieve this objective. They were:

(i) To vary the terms of the enforcement notice by deleting OS 9744 from the site to which it applied, to grant planning permission for the retention of the three caravans on OS 9744 with conditions as to
E landscaping, and to uphold the enforcement notice in relation to the five acres.

(ii) To grant planning permission as before for the retention of the three caravans on a site limited to OS 9744 with conditions as to landscaping, and to allow the appeal by quashing the enforcement notice.

(iii) To follow the course he adopted by allowing the appeal generally and granting the planning permission for the mixed use of the whole appeal site, subject to conditions such as those he imposed.

Alternative (i) was, in my view, available for the inspector to adopt as a matter of law. He was empowered to vary the enforcement notice by deleting OS 9744 from its ambit by section 88A(1). He had already decided that it was appropriate that the enforcement notice should
F apply not merely to OS 9744 but to the whole appeal site. Thus, he could properly have upheld the enforcement notice prohibiting a mixed use of the truncated appeal site, the five acres.

Mr Jay, however, points out that the adoption of this course would have been unfair to Mrs Gore. Her agents had made it clear that she only wanted planning permission for OS 9744. It would be unfair to uphold the enforcement notice in relation to the five acres when it was not alleged that there had been any breach of planning control on this land. This argument, however, is one for the Secretary of State rather than for this court to consider.

If the inspector had considered the first alternative, but decided not to adopt it because he considered it would be unfair, the second alternative was open to him. It would have given Mrs Gore what apparently she was asking for and at the same time would not have G
allowed any ancillary use of the five acres. In other words, the lawful use of the five acres would remain its existing use. Indeed, the inspector could have gone further and made use of his power under section 88B(1)(*c*) to determine that the five acres might only lawfully be used for the purposes of agriculture. This would effectively have achieved what both Mrs Gore and the council wished. Of course, if Mrs Gore had thereafter sought to use the five acres for any other purpose another enforcement notice would have been necessary.

It is not clear from the inspector's decision letter whether he ever considered taking either the first or the second course. Since what he intended to do was to grant planning permission for the retention of the three caravans on OS 9744, he was, in my view, under a duty as a H
matter of law to consider the best way of achieving this, and to do so in relation to the representations made on behalf of Mrs Gore. As I have already said, he could as a matter of law have upheld the enforcement notice by adopting the first course. Alternatively, if he had adopted the second of the courses I have outlined above, he would simultaneously have given effect to the representations made on behalf of the appellant to him, Mrs Gore, and achieved what the council wanted at least as their second alternative, that is to say the restriction of any development on OS 9744.

I think it can properly be said that the principle enunciated by Lord Widgery CJ in the *Hammersmith* case applies to the unusual situation with which the inspector was confronted in the present case. I thus think that as a matter of law he was under a duty to consider the J
first and second alternatives and that he failed to do so or at least gave no indication that he had done so.

While I am reluctant to disagree with the views of somebody so experienced in this field of law as Mr Spence, for the reasons I have given I would allow this appeal. If my brethren agree, the order I propose is that Mrs Gore's appeal against the enforcement notice should be remitted to the Secretary of State for him to redetermine in accordance with the opinion I have expressed above. Whether this will require a rehearing, or what procedure he should adopt in order to effect a redetermination, will be a matter for the Secretary of State himself to decide.

PARKER LJ and SLADE LJ agreed, and did not add anything. K

Appeal allowed with costs.

Queen's Bench Division

January 11 1991

(Before Mr David WIDDICOMBE QC, sitting as a deputy judge of the division)

CAMBRIDGE CITY COUNCIL v SECRETARY OF STATE FOR THE ENVIRONMENT AND ANOTHER L

Estates Gazette March 9 1991

[1991] 09 EG 119

Town and Country Planning Act 1971, section 246 — Enforcement notices — Appeal to High Court by council against decision of inspector who had upheld owners' appeal in respect of enforcement notices — Question as to whether the demolition of two houses constituted development — "A question which like a ghost has haunted planning law for many years" — Authorities considered — Held, laying the ghost to rest (subject to possible appeal), that the demolition M of the houses in the present case did constitute development — Appeal from inspector allowed

The history of events leading up to this appeal was briefly as follows — The owners of the two houses which were the subject of the proceedings were developers of an adjoining office block and had purchased the houses in order to provide additional car parking and with a view to enhancing the visual aspect by landscaping and to improving road safety — After temporary planning permission to use the houses as site offices for the development had run out, the owners began to demolish them — The houses were not listed buildings and

A **were not in a conservation area — The city council (the present appellants) obtained an interlocutory injunction restraining the owners, after they had declined a request to stop, from carrying out any further demolition — Enforcement notices, complaining of a breach of planning control, were issued by the council, referring to the removal of slates, tiles and other roof materials — The owners appealed against the notices to the Secretary of State and an inspector appointed by him allowed the appeal and quashed the notices — The inspector decided that the operations so far were the first stages of a projected total demolition which was not part of preparatory**
B **works for the erection of a building and was not such as to constitute an engineering operation — He concluded that the operations did not constitute developments requiring planning permission — The city council appealed to the High Court under section 246 against the inspector's decision (an appeal only allowed on points of law)**

In his judgment Mr David Widdicombe QC reviewed in detail the inspector's decision letter and a number of authorities which had been cited and discussed — The judge held that the inspector had erred in law in misunderstanding the effect of the decision in *London County Council* v *Marks & Spencer Ltd*, but that the case was of little relevance to the
C **present and the error was not material — The inspector also erred in law in his understanding of the decision in *Iddenden* v *Secretary of State for the Environment*; and in that case the error was material — The inspector seemed to have thought that the *Iddenden* case laid it down that in no circumstances could total demolition constitute development, whereas it merely said that the demolition of unsubstantial structures, like Nissen huts, was not development — Because of his misunderstanding of the *Iddenden* decision the inspector had failed to pose the key question, namely, whether in the present case the demolition was a building operation — The judge**
D **took the view that the demolition of these two houses was a building operation, ie one of the "operations normally undertaken by a person carrying on business as a builder" — He said "the pulling down of ordinary dwelling-houses like these seems to me unquestionably within the scope of a builder's business" — Thus the demolition of the two houses constituted development — The council's appeal was allowed and the case was sent back to the Secretary of State — Leave to appeal was granted**

Unless Mr Widdicombe's judgment is disturbed on appeal he has laid this ghost to rest

E The following cases are referred to in this report.

Coleshill & District Investment Co Ltd v *Minister of Housing and Local Government* [1969] 1 WLR 746; [1969] 2 All ER 525; (1969) 68 LGR 334; 20 P&CR 679; [1969] EGD 736; 211 EG 727, HL

Iddenden v *Secretary of State for the Environment* [1972] 1 WLR 1433; [1972] 3 All ER 883; (1972) 26 P&CR 553, CA

London County Council v *Marks & Spencer Ltd* [1953] AC 535; [1953] 2 WLR 932; [1953] 1 All ER 1095; (1953) 3 P&CR 409; 51 LGR 329, HL

This was an appeal by Cambridge City Council under section 246 of the Town and Country Act 1971 against the decision of an inspector appointed by the Secretary of State for the Environment whereby he allowed appeals by Milton Park Investment Ltd against
F two enforcement notices served by the city council in respect of alleged breaches of planning control at 21 and 23 Milton Road, Cambridge.

Christopher Lewsley (instructed by Nabarro Nathanson) appeared on behalf of the appellants; Michael Kent (instructed by the Treasury Solicitor) represented the first respondent, the Secretary of State; Robert McCracken (instructed by Ginn & Co, of Cambridge) represented the second respondent, Milton Park Investment Ltd.

Giving judgment, **Mr David Widdicombe QC** said: this is an appeal under section 246 of the Town and Country Planning Act 1971 (now section 289 of the Town and Country Planning Act 1990) against the decision of an inspector, acting on behalf of the Secretary of State for the Environment, whereby he allowed appeals against G
two enforcement notices served by the Cambridge City Council ("the council") in respect of alleged breaches of planning control at 21 and 23 Milton Road, Cambridge.

Section 246 and the other provisions of the 1971 Act have now been repealed and re-enacted in the consolidating legislation of 1990; but as the proceedings up to this point have been conducted under the provisions of the 1971 Act, to keep things simple I will continue to refer to those provisions. An appeal lies to this court only on a point of law.

The enforcement notices were issued on October 3 1989. The breach of planning control alleged in each notice was:

> . . . the demolition of part of the premises involving the removal of all slates H
> or roof tiles, battens and other materials from the main roofs of the premises, the removal of all slates or roof tiles and roof timbers from the roof of the single-storey rear extension to the premises, and the making of holes in the ceilings of some of the first-floor rooms in the premises without having previously obtained the benefit of planning permission.

The requirements of each notice were:

> 1. to cease the alleged breach of planning control;
> 2. to replace the slates or roof tiles on the main roof of the premises with materials to match those currently in situ on the roof of the front two-storey bay-window extension of the premises;
> 3. to replace the slates or roof tiles, battens, roof timbers and other materials on the roof of the single-storey rear extension to the premises to match those currently in situ on the roof of the front two-storey bay-window
> extension of the premises; J
> 4. to make good and repair the ceilings in the first-floor rooms of the premises with materials to match the existing ceilings.

The history of the matter is as follows. The second respondents (whom I will call "the owners"), a subsidiary of the Sir Robert McAlpine Group, were the developers of an office block of about 140,000 sq ft in Milton Road, Cambridge, known as the Westbrook Centre. With a view to providing additional car parking for the development, enhancing its visual aspects by landscaping and improving highway safety for vehicles using the approach road, the owners purchased two semi-detached houses, 21 and 23 Milton Road, which adjoined the office block. These houses were used, with temporary planning consent, as site offices for the development. K
When the temporary planning consent ran out, and renewal was refused, the owners began the demolition of the houses. The houses are not listed buildings, nor are they in a conservation area.

The council asked them to stop and, when they did not do so, obtained an *ex parte* injunction in the High Court restraining them from carrying on any further demolition of the properties. The work then stopped. Building preservation notices under section 58 of the Town and Country Planning Act 1971 were issued, but these were challenged by the owners and, as I understand it, were not confirmed by the Secretary of State, presumably because the buildings were not of special architectural or historic interest. Instead, the enforcement notices, the subject of these proceedings, were issued, alleging a breach of planning control and requiring the reinstatement of the L
houses.

The owners appealed against the enforcement notices, and the appeal was dealt with by way of written representations. In their notices of appeal the owners claimed that there was no breach of planning control because demolition did not require planning permission; and they referred to the decision of the Court of Appeal in *Iddenden* v *Secretary of State for the Environment* [1972] 1 WLR 1433 in support of that proposition.

The council, in their written representations, said the case was one of partial demolition and drew attention to the fact that in a letter to the Secretary of State, the owners had justified the removal of the roof and the other work which had been done as necessary in order to M
prevent the occupation of the houses by squatters. They claimed that the case was covered by the decision of the House of Lords in *Coleshill & District Investment Co Ltd* v *Minister of Housing and Local Government* [1969] 1 WLR 746. On the merits of the matter, they drew attention to the shortage of housing in Cambridge and the size of their "waiting list".

In their response to the council's representations, the owners stated that the works carried out were the initial stages of the proposed total demolition of the properties. The intention was to demolish them utterly and level them to the ground.

The inspector issued his decision by letter dated May 15 1990. After referring to the proceedings, he went on:

3. The appeal sites are a pair of semi-detached two-storey houses on the western side of Milton Road, immediately south of an access road which leads to a complex of recently-completed office buildings. At the time of my inspection the condition of the houses was substantially as set out in the notices although the timbers of the main roofs, which remain in place, had been temporarily re-covered in what appeared to be heavy duty polythene. The properties are virtually identical, and as the two notices and appeals are founded on the same facts and raise the same issues I will deal with them together.

. . .

5. In support of ground (*b*) you submit, in the light of the case of *Iddenden and Others* v *Secretary of State for the Environment and Another* [1972] 1 WLR 1433, that demolition does not require planning permission. Although the houses have so far been demolished only in part, your clients intended to and began to demolish them entirely; that action was interrupted when the Council obtained a High Court injunction restraining your clients from continuing with works of demolition, and subsequently served a Building Preservation Notice and issued the present enforcement notices.

6. In the Council's submission the decision of the House of Lords in the case of *Coleshill and District Investment Co Ltd* v *Minister of Housing and Local Government* [1969] 2 All ER 525 shows that partial demolition, as in this instance, can amount to building operations constituting development requiring planning permission. They distinguish the decision of the Court of Appeal in *Iddenden* on the grounds that that case was concerned with the total demolition of buildings of a temporary nature, whereas the *Coleshill* case dealt with the partial demolition of permanent buildings. In the Council's submission the latter judgment is therefore directly applicable to the present circumstances.

7. The letter dated 15th May 1989 from your clients' parent company to the Secretary of State, a copy of which was put in by the Council, clearly indicates that the appeal properties were acquired by your clients with a view to providing additional car parking for the adjoining office development, enhancing its visual aspect by landscaping, and improving highway safety for vehicles using the access road. The total demolition of the properties is implicit in those proposals, and I am satisfied that that was your clients' intention.

8. In the light of the *Coleshill* case I accept that demolition — whether total or partial — is capable of being development. The Courts had previously decided that total demolition and clearance of the site may constitute works for the erection of a building and therefore development (*London County Council* v *Marks & Spencer Ltd* (1953) 3 P&CR 409, HL); however, your clients propose to use the sites of the appeal properties for car parking, landscaping and highway improvements rather than for the erection of any building, and it therefore seems to me that that judgment does not lend support to the Council's case. The facts in *Coleshill*, on which the authority seek to rely, were highly unusual in that what was to be demolished, namely blast walls and embankments erected as external parts of an explosives store, were not buildings in the ordinary sense of the word; moreover, as the Council emphasise, that case was concerned with partial demolition only.

9. In Milton Road the operations undertaken so far have been the first stages in the projected total demolition of the houses, not their partial demolition with a view to alteration, and in my view the facts are comparable to those in *Iddenden* rather than in *Coleshill*. The projected total demolition was not to form part of preparatory works for the erection of a building, and nor has it been or would it be of such scale or complexity as to constitute an engineering operation in its own right. In the light of the foregoing, and having taken into account all the points made in the representations, I conclude that the operations carried out do not constitute development requiring planning permission. The appeals will therefore succeed on ground (*b*), and no other aspect of the matter falls to be considered.

He allowed the appeals and quashed the enforcement notices.

Mr Lewsley, for the council, accepted that the inspector had found as a fact in para 7 of the decision letter that total, not partial, demolition of the houses was intended. He identified two errors of law in the decision. The inspector had erred (1) in para 8 in his interpretation of *London County Council* v *Marks & Spencer Ltd* [1953] AC 535 and (2) in para 9 in his statement that the facts of this case were "comparable to those in *Iddenden* rather than in *Coleshill*".

As to (1), Mr Lewsley said that the inspector was wrong to reject the *Marks & Spencer* case. There was no relevant distinction between "works for the erection of a building", the phrase considered in the *Marks & Spencer* case, and "works for the development of the site for car parking, landscape and highway improvements", which was this case. The parties had not referred to the *Marks & Spencer* case in their written representations, the inspector had introduced it himself, without any reference back to the parties.

As to (2), *Iddenden* was distinguishable from the present case because of the nature of the structures there — Nissen huts and a lean-to workshop. The inspector had misunderstood the case.

Generally, Mr Lewsley's proposition was that where demolition, as here, formed part of the works necessary for the redevelopment of the site, or for a material change of use of the site, it was part of the development and required planning permission.

I can say straight away that I think Mr Lewsley is right on his first point. The *Marks & Spencer* case, which I will consider in more detail later in this judgment, was concerned with the phrase "works for the erection of a building", but no special significance attached to the fact that the development consisted of a building, apart from the fact that that was the phrase in the statute.

The Court of Appeal, who were upheld by the House of Lords, held that the phrase covered demolition of the old building on the site as well as the erection of the new one. The same reasoning would support the inclusion of the demolition of the old building if the development proposed had been something other than a building — if it had been the development proposed for the site in the present case, for example.

Mr Kent, for the Secretary of State, and Mr McCracken, for the company, did not seriously argue otherwise. Mr Kent said that the case was anyway of only marginal relevance and Mr McCracken drew attention to the passages in the judgments which said that demolition by itself was not development.

So I think there is an error of law in the decision here. Whether the error was a material one is something I must consider in a moment.

As I have had occasion to say before, I think it is unwise for an inspector in his decision letter to introduce new points of substance without reference back to the parties. In some cases questions of natural justice may arise, and where, as here, it is a matter of law, it renders an appeal to this court all too likely. I also find it surprising that, in a case which raises such a difficult and important question of law, the Secretary of State did not undertake the decision himself rather than delegate it to an inspector.

Before considering Mr Lewsley's second point, about the *Iddenden* case, I must look at the decision of the House of Lords in *Coleshill*. That was the case of the wartime ammunition depot. It consisted of six buildings, each surrounded by a blast wall about 9 ft high, and a grass-covered earth embankment sloping down from the top of the wall to the ground about 8 ft or 10 ft away. In order to put the buildings to a storage use appropriate for peacetime, the owners started to remove the embankments and walls. An enforcement notice was served in respect of the removal of the embankments, and the owners made an application for determination whether removal of the blast walls required planning permission.

The minister upheld the enforcement notice and on the application determined that planning permission was required. He held that the embankments and walls were integral parts of each building and that their removal constituted development requiring planning permission. The removal of the embankment was an "engineering operation", and the removal of the walls was a "building operation", namely the structural alteration of a building, within the definition of "development" in what are now sections 22(1) and 290(1) of the 1971 Act.

The House of Lords unanimously upheld this decision. In doing so, all the learned law lords rejected the suggestion that the issue in the case was whether demolition by itself was "development". The right approach was (1) to ascertain the operations involved and (2) to ask whether they came within the definition of "development" in the Act. This approach is clearly stated in the speech of Lord Morris of Borth-y-Gest, who gave the first opinion. He said at p 752:

> My Lords, one question that was persistently raised in this appeal was formulated as being whether demolition constitutes development for the purposes of the Town and Country Planning Act 1962. Neat and arresting as the question so expressed may seem to be it is not in fact the direct question which calls for our decision. If someone propounded a question of comparable generality such as whether modernisation constitutes development someone else might ask for a ruling as to whether renovation constitutes development. No one of these inquiries has precision. If development needs permission, which in most cases it does, and if development is defined, as in the Act it is, the true path of inquiry first involves ascertaining exactly what it is that it is desired to do, or exactly what it is that has been done, and then to see whether that comes within the statutory definition of development. Once some completed or projected work or operation is fully and clearly described then the words of definition can be applied. It is unnecessary and may be misleading to give the work of operation some single labelling word and then try to apply the definition to that word. We are here concerned with actual operations and not with possible operations or with those which can for the future be imagined. Why, then, introduce and interpose some general word of description when precise words of description are at hand? Why gaze into the crystal when one can read the book?

A The other law lords adopted the same approach. I need not examine all the speeches, but I can cite a passage from Lord Pearson at pp 770 to 771, which is to the same effect:

> The argument for the appellants was mainly based on the proposition that demolition, or demolition in or by itself, or demolition per se, does not constitute development. I think there is in this proposition some truth but only a limited amount of truth. On the one hand, there is in section 12 and in the relevant definitions in section 221 no mention of or reference to demolition or removal or any such operation. Therefore, an operation does not qualify as development by virtue of being a demolition or removal operation. It is not right to say "This is a demolition or removal operation: therefore, it is development." On the other hand, there are not in section 12 or in the relevant definitions in section 221 any words excluding operations from being B development if they are demolition or removal operations. An operation is not disqualified for being development because it is a demolition or removal operation. It is not right to say "This is a demolition or removal operation: therefore, it cannot be development." Notwithstanding that an operation is a demolition or removal operation, one still has to see whether it comes within the scope of development as defined in section 12 assisted by section 221. It may be within the definition of "building operations", eg, because it constitutes a structural alteration of a building or because it is such an operation as to be normally undertaken by a person carrying on business as a builder. It may be an engineering operation. Whether it is or not any of those things depends on the facts of the particular case.

The House of Lords held that the minister's finding that the embankments and walls formed an integral part of each building was a finding of fact which contained no error of law; that the removal of C the embankments was an "engineering operation"; and that the removal of the blast walls was a "building operation" that is a "structural alteration" of the buildings which materially affected their external appearance. The works were therefore "development" and required planning permission.

The first question in the present case is, therefore, what are the operations which fall to be considered? I think that this question was answered by the inspector in para 7 of the decision letter, which I will read again:

> The letter dated 15th May 1989 from your clients' parent company to the Secretary of State, a copy of which was put in by the Council, clearly indicates that the appeal properties were acquired by your clients with a view to D providing additional car parking for the adjoining office development, enhancing its visual aspect by landscaping, and improving highway safety for vehicles using the access road. The total demolition of the properties is implicit in those proposals, and I am satisfied that that was your clients' intention.

The inspector is here rejecting the council's argument that it was a case of partial demolition, but I think that in doing so he finds as a fact that the demolition of the houses is linked with the redevelopment of the site for car parking, landscaping and highway improvements in connection with the adjoining office development. The total demolition of the houses is "implicit", he says, in the redevelopment proposals. It is not suggested that there is any error of law in para 7; there was ample evidence before the inspector to E support this conclusion.

Do these operations constitute "development" within the meaning of the Act? That expression is defined in section 22(1) of the Act to mean "the carrying out of building, engineering, mining or other operations in, on, over or under land, or the making of any material change in the use of any buildings or other land". Section 22(2) lists certain exceptions to that definition.

In section 290(1) there are the following further definitions:

> "building" includes any structure or erection, and any part of a building, as so defined, but does not include plant or machinery comprised in a building;
> "building operations" includes rebuilding operations, structural alterations of or additions to buildings, and other operations normally undertaken by a person carrying on business as a builder;
> F "engineering operations" includes the formation or laying out of means of access to highways;
> "land" means any corporeal hereditament, including a building . . .;
> "means of access" includes any means of access, whether private or public, for vehicles or for foot passengers, and includes a street.

It is common ground in this case that the proposals for car parking, landscaping and highway improvements constitute "development" and require planning permission, though Mr McCracken said that the proposals might only be "development" because they constituted a material change of use. No details of the work necessary for the proposals was in evidence, but it seems unlikely to me that only a change of use is involved. Be that as it may, the disputed element is the demolition of the houses.

G Mr Lewsley says that demolition of the houses is development because it is part of the totality of the proposals for the site. He says that it is not necessary to consider whether it would be development in the absence of proposals for the use of the site afterwards. Whether this proposition is correct is something I must consider in a moment.

It was in connection with this question that the inspector introduced the *Marks & Spencer* case. That case concerned a site in Marylebone which had planning permission for redevelopment under the Town and Country Planning Act 1932. The old buildings on the site were demolished in 1939, but the erection of the new buildings was prevented by the war. After the war the project was resumed, and the question arose whether there was deemed planning permission for the operations under the Town and Country Planning H Act 1947. This depended on section 78(1) of the Act, which granted deemed planning permission where "any works for the erection of a building have been begun but not completed before the appointed day".

The Court of Appeal [[1952] Ch 549], by a majority, held that the "works for the erection of a building" included the demolition of the old buildings and that the project had deemed planning permission under the Act of 1947. This was upheld by the House of Lords.

For the reasons I gave earlier, I think that the inspector was wrong to dismiss this case because it concerned a building. The same reasoning would apply whatever the proposed development of the site. However, I agree with Mr Kent for the Secretary of State that the J phrase "works for the erection of a building" is not one which calls for consideration in the present case, so that the case has little if any relevance. It decided that demolition was part of the works for the redevelopment of the site, but it does not supply the answer to the question whether demolition as part of the proposals for the site is within the definition of "development" in the Act of 1971.

In my judgment, therefore, the inspector's error of law on this point was not a material one; the case is of little relevance, though not for the reason he gave.

Mr McCracken, for the owners, also took the view that the case was concerned with a different question and he drew attention to two passages in the judgments in which it was said that demolition by itself is not development. In the Court of Appeal, at p 563, Jenkins K LJ said:

> Returning to the phrase against that background, it is "works for the erection or alteration of a building". That phrase seems to me to be a phrase of wide import, and the inference is that it was adopted so as to cover a wide field of work. If the legislature had intended to confine the application of section 78(1) to cases where buildings had been begun but had not been completed, inevitably the section would have run: "Where the erection or alteration of a building has been begun but not completed". Here we have "works for the erection or alteration of a building", so as to include, in terms, operations which are not in themselves building operations. For my part I find irresistible the conclusion reached by the judge as to the meaning of these words when he said: "It is clear that the erection of a building need not have been begun, because otherwise no meaning would have been given to the words 'works L for'." It is, therefore, in my view, not necessary, in order to bring a case within the subsection, that one should be able to point to some work of construction on the site and say: "That is part of the new building the erection of which 'has been begun'." It is enough if, on the facts, one can conclude that on the site in question operations have been carried out which are part of the totality of operations necessary on that site for the purpose of carrying to completion a particular building project. Where it is shown by the evidence that a building owner had in view a particular building project to be carried out on a site already built upon, and that his intention to carry out that project had never been abandoned, then work such as the demolition of the buildings already on the site, as a necessary preliminary to the carrying out of the building project, would, in my judgment, be "works for the erection or alteration of a building" within the meaning of the subsection.

In the House of Lords, Lord Normand, with whose speech the M other law lords agreed, said at p 541:

> As regards the other reason which influenced the Master of the Rolls, I have not found it necessary to consider in what circumstances permission may have been required to demolition works as such under the Act of 1932, for, as I have said, I agree with the opinions of Jenkins and Morris LJJ, who assumed that demolition works did not require a planning permission.

I think Lord Normand must have said that the point was "assumed", because there is nothing in the reports, as far as I can see, to suggest that it was an issue in the case.

I return to the inspector's decision letter. He next deals with the *Coleshill* case. He says [in para 8]:

> The facts in *Coleshill*, on which the authority seek to rely, were highly unusual in that what was to be demolished, namely blast walls and embankments

A erected as external parts of an explosives store, were not buildings in the ordinary sense of the word; moreover, as the Council emphasise, that case was concerned with partial demolition only.

The fact that the ammunition stores in that case were not "buildings in the ordinary sense of the word" seems to me irrelevant; the point is that they were buildings within the definition in the Act. He is, however, correct to point out that *Coleshill* was a case of partial demolition and (in his para 9) that the demolition of the blast walls was held to be "development" because it was a structural alteration of a building. The demolition of the embankments was development because it was an engineering operation. I think, therefore, that the inspector was right to conclude that the present case is not comparable on its facts to *Coleshill*.

B In his para 9, the inspector repeats the finding that the present case was one of total demolition and says: "in my view the facts are comparable to those in *Iddenden* rather than in *Coleshill*". This brings me to Mr Lewsley's second point of law, that the inspector misunderstood the *Iddenden* case.

Iddenden v *Secretary of State for the Environment* [1972] 1 WLR 1433 was an enforcement notice appeal about land at Alton in Hampshire. In the Divisional Court, Willis J stated the facts as follows, at p 1435:

> The premises in respect of which the enforcement notice was served was a complex of buildings and land put to various relevant uses near Alton which originally had consisted, so far as the buildings are concerned, of three nissen C huts and a lean-to. The site had formerly been used as an agricultural small holding and depot for a contracting and repair business, but when the landowners bought the premises in 1967, they were minded to carry on a manufacturing use on the premises and applied for planning permission to pull down the existing buildings and to re-erect new buildings more suitable for their purposes. This application was refused in March 1968, and in the result they chose to erect three buildings without planning permission. They included the only building relevant to this appeal, which is described in the schedule to the enforcement notice as "an Arcon type prefabricated building measuring about 150 feet by 20 feet near the front of the premises." What was asserted in the enforcement notice was that the buildings had been erected within four years from February 1970 without the grant of planning permission, and the landowners were required to demolish and remove all unauthorised buildings.

D The owners challenged the enforcement notice claiming that it was invalid because it only required the removal of the new building and did not also require the reinstatement of the Nissen huts and lean-to workshop. That argument received short shrift from the Divisional Court, which held that the local planning authority had a discretion as to what steps to require. In the Court of Appeal the Divisional Court's ruling on that was upheld, but a second point arose, namely whether demolition of the huts and lean-to was development within the Act. On that, Lord Denning MR said, at p 1439:

> The words of the section are "to remedy the breach." That means the breach of planning control. What was the breach in this case? If it was the pulling down of the old nissen huts or the old workshop, then it may be that in order to remedy the breach they would have to be restored. But I do not think their E demolition was a breach of planning control. Whilst some demolition operations may be development — see *Coleshill and District Investment Co Ltd* v *Minister of Housing and Local Government* [1968] 1 WLR 600, [1969] 1 WLR 746 — the demolition of buildings such as these was not development. Mr Iddenden did not need planning permission to pull them down. No doubt the pulling down of the old and the erection of the new was all one combined operation by the workmen. But in planning law they are different operations. Mr Iddenden only required planning permission for his new buildings and their user. The only breach by him of planning control was the unauthorised erection of the new buildings, the concrete base, and the user of the factory. The enforcement notice required him to remedy that breach. It told him the steps he had to take. It was perfectly good.

Mr Lewsley says that the explanation of this passage lies in Lord F Denning's use of the phrase "buildings such as these". He was dealing with Nissen huts and a lean-to workshop and saying that, whatever might be the case with more substantial buildings, the removal of these items was not development. The inspector seemed to think that Lord Denning was drawing a distinction between partial demolition and total demolition, and he wrongly relied on that.

Mr Kent for the Secretary of State said that the inspector had not tied himself to the *Iddenden* case — he merely said it was more comparable than *Coleshill*. Mr McCracken drew attention to Lord Denning's statement that "No doubt the pulling down of the old and the erection of the new was all one combined operation by the workmen. But in planning law they are different operations".

The problem with *Iddenden* is that this second point dealt with by the Court of Appeal is treated in a fairly summary manner, no doubt G because it was not essential to the decision, the enforcement notice being valid for the reasons given by the Divisional Court and upheld by the Court of Appeal. But I think it is clear what was decided.

First, I have no difficulty in accepting Mr Lewsley's point about the phrase "buildings such as these". I think the Master of the Rolls was emphasising the insubstantial nature of the structures, saying in effect that it does not require an engineering or building operation to remove them. Second, Lord Denning's point that in planning law demolition is a different operation, to be considered separately from any other works with which it is combined, I understand to be saying that the nature of the operations is not changed because of their association with other works. The question whether demolition is development has to be considered whether or not the demolition is H part of a larger scheme.

It seems to me that this must be so — in the present case, even if the demolition of the houses is considered as part of the overall scheme for the site, the question still arises whether the demolition is an operation within the definition of "development" in the Act. So I am not convinced by Mr Lewsley's argument that demolition is development merely because it is part of a larger scheme for the site. I cannot see how that changes the nature of the operation or alters the question whether such operation is within the definition. I think *Iddenden* is against him on that.

Nevertheless, I think the inspector did err in his understanding of the *Iddenden* case. He seems to have thought that it laid down that in J no circumstances can total demolition constitute development. In my view, it does not say that. It merely says that the demolition of insubstantial structures like Nissen huts is not development. The question of the demolition of substantial buildings like dwellinghouses is left open. So I think that the inspector did misunderstand the *Iddenden* case and thought that it went further than it did.

If my analysis of the cases is correct, the ultimate question in this appeal is, therefore, whether demolition of these houses is development within the definition in the Act. This is a question which like a ghost has haunted planning law for many years. The time has now come when the ghost must be laid to rest — I see no escape from that task. K

There has been no decision that total demolition of a building cannot be development. It has been discussed in the cases and differing views have been expressed, but there appears to be nothing binding on me. I accept the inspector's view that the demolition here is not an "engineering operation" — there is no reason to doubt that. But he does not pose the key question, whether it is a "building operation", because he thought it was not open to him to consider that in view of *Iddenden*.

But that question is open, for the reasons I have given, and it has to be answered. I take the view that the demolition of these houses is a "building operation", that is an "other operation normally undertaken by a person carrying on business as a builder". There are L no doubt many demolitions which are not normally the business of a builder — a nuclear power station was mentioned in argument, and there will be other examples, less extreme than that; and there will be debatable cases where evidence may be desirable — but the pulling down of ordinary dwellinghouses like these seems to me unquestionably within the scope of a builder's business, whether he goes on to develop the site afterwards or not. The Nissen huts and lean-to workshop in *Iddenden* are distinguishable.

Lord Pearson in the *Coleshill* case, at p 771B, in the passage I have cited, thought that this was a question which might have to be considered, and he referred specifically to the words on which I have relied.

The alternative question is whether the demolition of these houses M is an "other operation in, on, over or under land". I prefer not to express a view on that. It is a difficult question on which there is little or no authority. I only note that in *Coleshill* the majority of their lordships all thought that some limitation must be imposed on the words "other operations" by reason of their proximity to or association with "building, engineering or mining operations".

Mr Kent said that it would be anomalous if demolition were development where there was evidence of an intention to redevelop the site but not where there was no evidence of such intention. On the view I have taken, that anomaly does not arise.

It would also be anomalous, in my view, if demolition of part of a house, perhaps the major part, is development and requires planning permission (because it is a structural alteration) but demolition of the whole house is not development and can be carried out without permission. There is no logic in that, and that anomaly also, on the view I have taken, does not arise.

I can summarise my conclusions as follows: (1) The inspector erred in law in his understanding of the *Marks & Spencer* case, but the error is not material because the *Marks & Spencer* case is of little relevance in any event. (2) The inspector erred in law in his understanding of the *Iddenden* case, and the error is material because he failed to consider whether the demolition of the houses was a building operation. (3) The demolition of these houses is not "development" merely because it is part of larger proposals for the development of the site. (4) The demolition of these houses is "development", namely "a building operation", being an "other operation normally undertaken by a person carrying on business as a builder".

Finally, I should mention two points made by Mr McCracken. First, he points out that the council's case before the inspector was based on the proposition that the demolition was only partial. Mr McCracken said that the inspector could hardly be blamed, therefore, if he treated the case as one in which the issue was between partial and total demolition. It is true that the council's case was based on partial demolition, and the inspector rejected that. But I do not think that excuses a decision which is wrong in law. The fact that the council put the case the way that they did did not deprive the inspector of any factual material. Once he had the material, it was his responsibility to get the law right.

Second, Mr McCracken drew attention to the wording of the enforcement notices. The breach of planning control alleged is "the demolition of part of the premises" and the work done is specified and alleged to be without the benefit of planning permission. But this is factually correct — that is as far as the demolition had got. I cannot see that any different wording would be called for to allow for the fact that total demolition was intended.

The appeal will therefore be allowed and the case remitted to the Secretary of State with the opinion of the court for rehearing and determination by him.

Appeal allowed with costs. Leave to appeal granted.

Court of Appeal

November 20 1990

(Before Lord Justice FOX and Lord Justice BUTLER-SLOSS)

EAST HAMPSHIRE DISTRICT COUNCIL v DAVIES

Estates Gazette March 16 1991

[1991] 10 EG 149

Town and Country Planning Act 1971 — Enforcement notices — Site used as a winter stop-over for mobile homes, caravans and equipment of travelling showman — Appeal by local authority from refusal by deputy High Court judge of interlocutory injunction — Principles in leading case of *American Cyanamid Co* v *Ethicon Ltd* — Balance of convenience — General rule that a local authority should try the effect of criminal proceedings available to them before seeking the assistance of the civil courts: *Stoke-on-Trent City Council* v *B&Q (Retail) Ltd* — Local authority's appeal from refusal of injunction dismissed

The local authority in this case (the present appellants) were concerned about breaches of planning control consisting of the unauthorised use of a site for the purpose of a mobile showman's winter quarters — The authority's concern gave rise to a multiplicity of proceedings — The present defendant had purchased land including the relevant site from the previous owner when the site was already the subject of an enforcement notice requiring the discontinuance of its use for a mobile home and the removal of the offending structure — The defendant disregarded warnings and was found to have stationed on the site some five residential caravans or mobile homes and 11 other vehicles and trailers — He intended to use the site as a convenient stop-over for his caravans and equipment — The early months of 1990 witnessed a considerable amount of litigious or procedural activity — The local authority issued three summonses against the defendant for failure to comply with the enforcement notice served on his predecessor — The authority also issued a writ claiming an injunction to restrain the defendant from residing in a mobile home, caravan or otherwise on the site and from storing on it mobile homes, caravans, trailers and showground or funfair equipment — The defendant applied for planning permission to use the site as a showman's winter quarters — The authority's writ was served on the defendant and the summonses were adjourned — The application for an interlocutory injunction came before the High Court (Mr Patrick Bennett QC, sitting as a deputy judge) and was dismissed; hence the present appeal by the local authority — The defendant's application for planning permission was refused and the authority served a further three enforcement notices on him, alleging breaches of planning control in consequence of the present user of the site

The deputy judge's dismissal of the authority's application for an interlocutory injunction was based on his view that the language of the 1985 enforcement notice was not wide enough to cover the placing of a mobile home on the site by the defendant — The Court of Appeal reached the same result by a different route — As this was an application for an interlocutory injunction, it had to be determined in accordance with the *American Cyanamid* principles — There was a serious question to be tried and the test to be applied was the balance of convenience — Relevance factors were whether damages would be an adequate remedy (and here the answer was clearly "no") and whether the authority's undertaking in damages (if an injunction were granted but the authority lost at the trial) would be an adequate remedy — In the present case, however, there was superimposed on the question of the balance of convenience a further question as to the appropriateness of granting injunctions at all to a local authority in cases, such as the present, where Parliament has made special provisions for penalties for breaches of the relevant statutes: *per* Lord Templeman in *Stoke-on-Trent City Council* v *B&Q (Retail) Ltd* — In the present case the effect of the criminal law in relation to the enforcement notices had not yet been tested, as the proceedings had been adjourned — Other points to be taken into account were that there had been no permanent or long-term damage to the site; that decisions on the planning and enforcement notice appeals were due to take place soon; that the defendant lived in a mobile home on the site and had nowhere else to store his vehicles and equipment; and that the showman's winter season was now approaching when it would be difficult for showmen to find winter quarters — Also, to turn the defendant out when the planning appeals were still pending could, if permission were granted, inflict damage on the defendant disproportionate to the temporary protection of any public interest — The balance was clearly in favour of refusing an injunction — Appeal dismissed

The following cases are referred to in this report.

American Cyanamid Co v *Ethicon Ltd* [1975] AC 396; [1975] 2 WLR 316; [1975] 1 All ER 504, HL

Stoke-on-Trent City Council v *B&Q (Retail) Ltd* [1984] AC 754; [1984] 2 WLR 929; [1984] 2 All ER 332, HL

This was an appeal by East Hampshire District Council from a decision by Mr Patrick Bennett QC, sitting as a deputy judge of the Queen's Bench Division, dismissing an application by the council for an interlocutory injunction restraining the defendant, Mr William Davies, by himself, his servants or agents (a) from occupying or residing in a mobile home or residential caravan or otherwise on the site at Bowleswood Farm, Headley, Bordon, Hampshire, and (b) from using the site for the storage of mobile homes, residential caravans, vehicle trailers and showground or funfair equipment.

A James Findlay (instructed by Sharpe Pritchard, agents for the solicitor to East Hampshire District Council) appeared on behalf of the appellant council; Sebastian Head (instructed by Fox & Kent, of Chichester) represented the respondent.

Giving judgment, **FOX LJ** said: This is an appeal by East Hampshire District Council ("the council") from an order of Mr Patrick Bennett QC, sitting as a deputy judge of the Queen's Bench Division, dismissing the council's application for an interlocutory injunction.

The case concerns land at Bowleswood Farm, Headley, Bordon, Hampshire. I will refer to it as "the site". It is within the council's district. The council brings the action under the powers conferred by B section 222 of the Local Government Act 1972. The council is a local authority within the provisions of section 272 of that Act.

The site was formerly part of a larger agricultural estate and more recently has been divided into sub-plots of two or three acres each. The site is situate on the northern side of the B3002. Directly opposite the land, and on the south side of the B3002, are Bramshott and Oakshott Commons, both of which are owned by the National Trust and which constitute a designated site of special scientific interest covering about 370 ha and including extensive tracts of heathland. The site itself has been included in Hampshire County Council's Bramshott and Ludshott countryside heritage area.

That designation is at present on a non-statutory basis. The public C has access to the National Trust land.

The site is not obscured from the public road except by a small hedge. There is also a 2-m high steel mesh fence bordering the site but it does not prevent persons passing on the public road from seeing the whole extent of the site.

It is the council's case that part of the site is being used by the defendant unlawfully and in breach of the planning legislation as an area for a showman's winter quarters. That involved the placing of vehicles, including mobile homes, on the site and, say the council, constitutes a material change of use. That change of use was unauthorised and a breach of the planning laws. There are upon the land, according to the evidence, five residential caravans or mobile homes, together with 11 assorted vehicles or trailers. The council's D view is that these vehicles are a particularly offensive sight in this area of highly attractive countryside.

A planning authority (the council in the present case) may issue an enforcement notice in respect of a breach of planning control requiring the breach to be remedied. A person having an interest in the land may appeal against the notice and, if he does, the notice has no effect pending the final determination or withdrawal of the appeal (see section 88(10) of the Town and Country Planning Act 1971 [now section 175(4) of the Town and Country Planning Act 1990]).

An enforcement notice has effect against the owner of the land for the time being but, since appeals must be within four weeks from the notice, the owner for the time being may be out of time for appealing.

E If an effective enforcement notice is not complied with, the person in default is liable to a fine.

On October 5 1981 temporary planning permission was granted to a Mr Swoish for the placing of a mobile home on an area including the site. On October 5 1984 the temporary planning permission expired.

On January 3 1985 an enforcement notice ("the 1985 enforcement notice") was served, taking effect on February 14 1985. This required that there be permanent discontinuance "of the use of the land as a site for a mobile home" and the removal of the mobile home from the land. Mr Swoish appealed against that enforcement notice. The appeal was dismissed on October 8 1985.

On September 22 1989 the defendant purchased from Mr Swoish an area of land which included the site. Mr Swoish retained about 13 F acres.

On November 7 1989 Mr Curtis, a development inspector in the planning department of the council, visited the site and met the defendant, who told him that he intended to use the land as a winter stop-over for his caravans and equipment. Mr Curtis gave the defendant a copy of the 1985 enforcement notice. On November 24 1989 Mr Curtis visited the site again and found on it five residential caravans or mobile homes and 11 other vehicles or trailers.

On January 8 1990 the council issued three summonses against the defendant in respect of failure to comply with the 1985 enforcement notice.

On January 10 1990 the writ in this action was issued. The relief sought was (*inter alia*) an injunction restraining the defendant by himself, his servants or agents (a) from occupying or residing in a G mobile home or residential caravan or otherwise howsoever on the site, and (b) from using the site for the storage of mobile homes, residential caravans, vehicle trailers and showground or funfair equipment.

On January 22 1990 the defendant applied for planning permission to use the site as a showman's winter quarters (that is to say from October 31 to March 1 each year). On January 23 1990 the writ was served. On the same day, January 23, the three summonses came before the Alton magistrates and were adjourned for the defendant to obtain legal advice.

On February 6 1990 the present application for an injunction came before Mr Patrick Bennett QC and was dismissed. H

On February 20 1990 the three summonses again came before the Alton Bench and by consent were adjourned *sine die* pending the determination of the present appeal.

On March 6 1990 the council gave notice of appeal against the decision of Mr Bennett.

On March 7 1990 the defendant's application for planning permission was refused.

On March 23 1990 the council served three further enforcement notices upon the defendant in respect of alleged breaches of planning control in consequence of the present user of the site.

The decision of the deputy judge turned largely on the effect of the 1985 enforcement notice. The background to that notice was the planning permission granted to Mr Swoish in October 1981. That J provided that for three years from the date of the permission or until the dwelling ceased to be occupied by the appellant (Mr Swoish), whichever should first happen, "the use hereby permitted" shall cease. The permitted use was the retention of a mobile home on the land. Schedule 1 to the enforcement notice states the premises to which the notice applies; those premises include the site. Schedule 2 states the alleged breach of conditions, ie that the use of the land as a site for a mobile home has not ceased after the expiration of the three-year period. Schedule 3 states the steps required to be taken to remedy the breach, ie "permanently discontinue the use of the land as a site for a mobile home and remove the mobile home from the land".

The deputy judge took the view that, as a matter of construction, K the language of the 1985 enforcement notice was not wide enough to cover the placing of a mobile home on the land by the defendant. He therefore held that the notice does not necessarily have an application wide enough to include the admittedly unlawful activities (of the defendant) on the land. In these circumstances he refused an injunction.

Accepting that there may be questions of construction as to the ambit of the 1985 enforcement notice, I do not think it is appropriate to attempt to resolve them on an interlocutory application or that their resolution in favour of the defendant is sufficient to determine the present application. The 1985 notice and the three notices served this year are a basis for criminal proceedings (assuming them to be valid and enforceable according to their terms). As I have mentioned, L the 1990 notices are under appeal and, therefore, of no effect pending the appeal. The 1985 notice, whatever its construction, adds little to the council's contention that the defendant's present activities in any event constitute unlawful development on the site and should be stopped until judgment in the action or further order.

This is an application for an interlocutory injunction and must be determined in accordance with the *American Cyanamid* principles. As to those, there is, in my view, a serious issue to be tried in the action. It is not, I think, in dispute that the defendant's activities are in breach of the planning law. The issue to be tried is whether it is a proper case for the grant of a permanent injunction. The next question is whether damages at law would be an adequate remedy for M the council. They would not. Damages are a wholly inappropriate and inadequate remedy for a local authority in these circumstances. Next, would the council's undertaking in damages, if an injunction were granted but the council lost at the trial, be an adequate remedy for the defendant? This gives rise to questions having regard to the position of travelling showmen. But, in any event, the matter under the *American Cyanamid* principles becomes a question of the balance of convenience (see *American Cyanamid Co* v *Ethicon Ltd* [1975] AC 396 at p 408E). I will return to that, but there is superimposed upon the question of the balance of convenience a further question as to the appropriateness of granting injunctions at all to a local authority in cases, such as the present, where Parliament

A has made special provisions for penalties for breach of the relevant statutes.

In *Stoke-on-Trent City Council* v *B&Q (Retail) Ltd* [1984] AC 754 (a case concerned with breaches of the Sunday trading legislation) Lord Templeman, whose speech was concurred in by the other members of the House, said at p 776:

> The right to invoke the assistance of the civil court in aid of the criminal law is a comparatively modern development. Where Parliament imposes a penalty for an offence, Parliament must consider the penalty is adequate and Parliament can increase the penalty if it proves to be inadequate. It follows that a local authority should be reluctant to seek and the court should be reluctant to grant an injunction which if disobeyed may involve the infringer in sanctions far more onerous than the penalty imposed for the offence. In B *Gouriet* v *Union of Post Office Workers* [1978] AC 435 Lord Wilberforce said at p 481, that the right to invoke the assistance of civil courts in aid of the criminal law is "an exceptional power confined, in practice, to cases where an offence is frequently repeated in disregard of a, usually, inadequate penalty . . . or to cases of emergency . . ." In my view there must certainly be something more than infringement before the assistance of civil proceedings can be invoked and accorded for the protection or promotion of the interests of the inhabitants of the area.

Lord Templeman went on to say that as a general rule a local authority should try the effect of criminal proceedings before seeking the assistance of the civil courts but, in the *Stoke-on-Trent* case, the council were entitled to take the view that the traders would not be deterred by the maximum fine, which was substantially less than the C profit which could be made from illegal Sunday trading.

The position in the present case is as follows. First, the effect of the enforcement of the criminal law has not been tested at all. No action can be taken by way of prosecution in respect of the 1990 enforcement notices because their operation is stayed by the statute pending the appeals against them. So far as the 1985 enforcement notice is concerned, proceedings before the magistrates have been instituted by the council but have been adjourned, by consent of both sides, pending the outcome of the present appeal.

Second, it is a case in which the defendant has not done permanent or long-term damage to the land and, pending the outcome of the planning and enforcement notice appeals, he offers an undertaking not to do such damage (though the precise terms of any such undertaking may require consideration).

Third, the planning and enforcement notice appeals will be heard D in a few weeks, ie on December 4. The decision will not follow immediately but will be given within a period estimated at between one and three months. I should mention that though the matter was heard before the deputy judge in February it did not come before the Court of Appeal until the Michaelmas Term. Apparently, there was some misunderstanding as to the intention of the council to pursue the appeal and it was struck out of the list.

Fourth, the defendant's evidence is that he lives in a mobile home on the site and has nowhere else where he can store his vehicles and equipment. The legislation, I should observe, shows a degree of protectiveness in relation to a resident in a mobile home in that section 90 of the Town and Country Planning Act 1971 [section 183 E of the 1990 Act] provides that a stop notice shall not prohibit the use of land as a site for a caravan occupied by any person as his only or main residence.

Moreover, the showman's winter season is now upon us and it is difficult for showmen to find winter quarters. To turn the defendant out now when the planning appeals are still pending could, if planning permission were granted on the appeals, cause damage to the defendant which might be disproportionate to the temporary protection of any public interest.

Looking at the whole matter, I think that the balance of convenience is in favour of refusing an injunction. I am not criticising the council for starting the proceedings; they are properly concerned F for the public interest in the amenities of the countryside. Nor am I saying that the defendant has a licence to remain indefinitely. I am saying only that, having regard to the onset of the winter season, the difficulties likely to face the defendant in finding quarters for the remainder of the season and the fact that the planning appeals are only a few weeks away tilt the balance of convenience, in a case where the criminal penalties have not been applied, in favour of the defendant at this stage. I would, therefore, dismiss the appeal.

Butler-Sloss LJ agreed and did not add anything.

The appeal was dismissed with costs in the Court of Appeal; the costs' order below to stand.

For further cases on this subject see p 283

VENDOR AND PURCHASER

A
Chancery Division

November 27 1990
(Before Mr Justice HOFFMANN)

SPIRO v GLENCROWN PROPERTIES LTD AND ANOTHER

Estates Gazette January 19 1991
[1991] 02 EG 167

**Law of Property (Miscellaneous Provisions) Act 1989 —
B Vendor and purchaser — Option to purchase — Notice given by purchaser exercising option within the stipulated time — Purchaser, however, failed to complete — Whether contract on which vendor relied complied with the provisions of section 2 of 1989 Act — Whether the "contract for the sale . . . of an interest in land" for the purposes of section 2 was the agreement by which the option was granted or the letter by which the option was exercised — If by the latter it did not comply with section 2 — Held that the agreement by which the option was granted was the contract within section 2 and complied with the statutory requirements**

**Section 2 of the Law of Property (Miscellaneous
C Provisions) Act 1989 provides that a contract for the sale or other disposition of an interest in land can be made only in writing and only by incorporating all the terms which the parties have expressly agreed in one document, or, where contracts are exchanged, in each — The terms may be incorporated in a document either by being set out in it or by reference to some other document — The document incorporating the terms or, where contracts are exchanged, one of the documents incorporating them (but not necessarily the same one) must be signed by or on behalf of each party to the contract**

D In the present case, if the relevant document for the purposes of section 2 was the agreement by which the option was *granted*, there was no difficulty — The agreement was executed in two exchanged parts, each of which incorporated all the agreed terms and had been signed by the vendor and purchaser respectively — If, however, the letter *exercising* the option was to be regarded as constituting the contract, it was signed only by the purchaser and so did not comply with section 2

**The way in which the problem came before the court was as follows — On the defendant purchaser's failure to complete, the plaintiff vendor, after serving notice to complete and
E issuing a writ for specific performance, rescinded the contract — The plaintiff obtained judgment in default of defence against the purchaser for damages to be assessed — The second defendant in the action was the guarantor of the defendant purchaser's obligations — The matters now before Hoffmann J were a summons by the plaintiff for judgment under Ord 14 against the guarantor and a summons by the defendant purchaser to set aside the judgment against it — The parties agreed to treat the present hearing as the trial of the action**

**It seemed to Hoffmann J to be plain, apart from authority, F
that section 2 was intended to apply to the agreement creating the option and not to the notice exercising it — There was some authority supporting this view, for example the speech of Lord Macnaghten in *Helby* v *Matthews* and the case of *Re Mulholland's Will Trusts* — It was, however, argued by the defendant purchaser that an option to purchase was merely an irrevocable offer, which did not become a contract for the sale of land until it had been accepted by the notice exercising the option — There was some authority tending to support this view, although in different contexts, but, as Diplock LJ pointed out in *Varty* v *British South Africa Co*, to speak of an G
enforceable option as an "irrevocable offer" was juristically a contradiction in terms and could only be justified by way of metaphor or analogy — The truth was, Hoffmann J said, that an option to purchase was a relationship *sui generis* — In some ways it resembled an offer and in some ways a conditional contract, but it did not conform precisely with either — Hoffmann J's conclusion was that the construction which most closely accorded with Parliament's intention in section 2 of the 1989 Act was to regard the agreement granting the option as the contract for sale for the purposes of that section — The most helpful authority in support of this view was perhaps *Re Mulholland's Will Trusts* — Judgment for H
plaintiff against both defendants**

The following cases are referred to in this report.

Beesley v *Hallwood Estates Ltd* [1961] Ch 105; [1961] 2 WLR 36; [1961] 1 All ER 90; [1961] EGD 366; (1960) 177 EG 71, CA
Griffith v *Pelton* [1958] Ch 205; [1957] 3 WLR 522; [1957] 3 All ER 57, CA
Helby v *Matthews* [1895] AC 471
K (Enduring Powers of Attorney), In re [1988] Ch 310; [1988] 2 WLR 781; [1988] 1 All ER 358
Kennewell v *Dye* [1949] Ch 517; [1949] 1 All ER 881
Laybutt v *Amoco Australia Pty Ltd* (1974) 48 ALJR 492
London & South Western Railway Co v *Gomm* (1882) 20 ChD 562
Mulholland's Will Trusts, Re [1949] 1 All ER 460 J
Richards v *Creighton Griffiths (Investments) Ltd* [1973] EGD 205; (1972) 225 EG 2104
Sainsbury (J) plc v *O'Connor (Inspector of Taxes)* [1990] STC 516
United Scientific Holdings Ltd v *Burnley Borough Council* [1978] AC 904; [1977] 2 WLR 806; [1977] 2 All ER 62; (1977) 33 P&CR 220; [1977] EGD 195; 243 EG 43 & 127, HL
Varty v *British South Africa Co* [1965] Ch 508; [1964] 3 WLR 698; [1964] 2 All ER 975, CA

The plaintiff in the action was the vendor, Trevor David Spiro. The property subject to the option to purchase was in Finchley, North London, and the price was £745,000. The first defendant was the purchaser, Glencrown Properties Ltd, and the second defendant was the first defendant's guarantor, Bennoy Berry. K

Beverly-Ann Rogers (instructed by Denton Hall Burgin & Warrens) appeared on behalf of the plaintiff; Michael Douglas (instructed by Paul Shrank & Co) represented the defendants.

Giving judgment, **HOFFMANN J** said: This is an action for damages for breach of a contract to buy land. On November 14 1989 the plaintiff granted an option to the first defendant ("the purchaser") to buy a property in Finchley for £745,000. The option was exercisable by notice in writing delivered to the vendor or his solicitors by 5 pm on the same day. The purchaser gave a notice exercising the option within the stipulated time. He failed to complete and the vendor, after serving a notice to complete and issuing a writ for specific performance, rescinded the contract. The second defendant, Mr Berry, is guarantor of the purchaser's obligations. On May 1 1990 the vendor obtained judgment in default of defence against the purchaser for damages to be assessed. There are now before me a summons by the vendor for judgment under RSC Ord 14 against Mr Berry as guarantor and a summons by the purchaser to set aside the judgment against it. Since both summonses raise the same short point of law and there are no other issues in the case, the parties have agreed to treat this hearing as the trial of the action.

The only question for decision is whether the contract upon which the vendor relies complied with the provisions of section 2 of the Law of Property (Miscellaneous Provisions) Act 1989, which came into force on September 27 1989, some seven weeks before the grant and exercise of the option. It is a question which has produced a lively debate in conveyancing journals. The relevant provisions are as follows:

> 2(1) A contract for the sale or other disposition of an interest in land can only be made in writing and only by incorporating all the terms which the parties have expressly agreed in one document, or, where contracts are exchanged, in each.
>
> (2) The terms may be incorporated in a document either by being set out in it or by reference to some other document.
>
> (3) The document incorporating the terms or, where contracts are exchanged, one of the documents incorporating them (but not necessarily the same one) must be signed by or on behalf of each party to the contract.

If the "contract for the sale. . .of an interest in land" was for the purposes of section 2(1), the agreement by which the option was *granted*, there is no difficulty. The agreement was executed in two exchanged parts, each of which incorporated all the terms which had been agreed and had been signed by or on behalf of the vendor and purchaser respectively. But the letter which *exercised* the option was of course signed only on behalf of the purchaser. If the contract was made by this document, it did not comply with section 2.

Apart from authority, it seems to me plain enough that section 2 was intended to apply to the agreement which created the option and not to the notice by which it was exercised. Section 2, which replaced section 40 of the Law of Property Act 1925, was intended to prevent disputes over whether the parties had entered into a binding agreement or over what terms they had agreed. It prescribes the formalities for recording their mutual consent. But only the grant of the option depends upon consent. The exercise of the option is a unilateral act. It would destroy the very purpose of the option if the purchaser had to obtain the vendor's countersignature to the notice by which it was exercised. The only way in which the concept of an option to buy land could survive section 2 would be if the purchaser ensured that the vendor not only signed the agreement by which the option was granted but also at the same time provided him with a countersigned form to use if he decided to exercise it. There seems no conceivable reason why the legislature should have required this additional formality.

The language of section 2 places no obstacle in the way of construing the grant of the option as the relevant contract. An option to buy land can properly be described as a contract for the sale of that land conditional upon the exercise of the option. A number of eminent judges have so described it. In *Helby* v *Matthews* [1895] AC 471, which concerned the sale of a piano on hire-purchase, Lord Macnaghten said:

> The contract, as it seems to me, on the part of the dealer was a contract of hiring coupled with a conditional contract or undertaking to sell. On the part of the customer it was a contract of hiring only until the time came for making the last payment.

In *Griffith* v *Pelton* [1958] Ch 205, which raised the question of whether the benefit of an option was assignable, Jenkins LJ said at p 225:

> An option in gross for the purchase of land is a conditional contract of such purchase by the grantee of the option from the grantor, which the grantee is entitled to convert into a concluded contract of purchase, and to have carried to completion by the grantor, upon giving the prescribed notice and otherwise complying with the conditions upon which the option is made exercisable in any particular case.

In the context of section 2, it makes obvious sense to characterise it in this way. So far, therefore, the case seems to me to be clear.

The purchaser, however, submits that I am constrained by authority to characterise an option as an irrevocable offer which does not become a contract for the sale of land until it has been accepted by the notice which exercises the option. It follows that the "contract for the sale . . . of an interest in land" within the meaning of section 2 can have been made only by the letter.

The first case upon which the purchaser relies is *Helby* v *Matthews*; the very case in which, as I have said, Lord Macnaghten characterised an option to purchase as a conditional contract to sell. Lord Herschell LC and Lord Watson, however, expressed themselves rather differently. Lord Herschell said:

> . . . when a person has, for valuable consideration bound himself to sell to another on certain terms, he may, in popular language, be said to have agreed to sell, though an agreement to sell in this sense, which is in truth merely an offer which cannot be withdrawn, certainly does not connote an agreement to buy, and it is only in this sense that there can be said to have been an agreement to sell in the present case.

Lord Watson said:

> In order to constitute an agreement for sale and purchase, there must be two parties who are mutually bound by it. From a legal point of view the appellant was in exactly the same position as if he had made an offer to sell on certain terms, and had undertaken to keep it open for a definite period.

It is, however, important to read these statements in the context in which they were made. The question in *Helby* v *Matthews* was whether, during the currency of the hire, the hirer had "agreed to buy" the goods within the meaning of the Factors Act 1889. The purpose of the Factors Act was to give a buyer, who was in possession but had not yet acquired title, the power to confer title upon a third party who took in good faith. It is not surprising that the House of Lords decided that a person who had at the relevant time no obligation to acquire or pay for the goods had not "agreed to buy" them within the meaning of the Act. The language and purpose of the statute requires one to look at the arrangement from the buyer's point of view. And the essence of an option is that while the seller may be said to be conditionally bound, the buyer is free. In the *Helby* v *Matthews* context it was therefore true to say that, pending exercise of the option, the position of the buyer was *as if* he had been made an offer which the seller could not withdraw.

But the concept of an offer is of course normally used as part of the technique for ascertaining whether the parties have reached that mutual consent which is a necessary element in the formation of a contract. In this primary sense, it is of the essence of an offer that by itself it gives rise to no legal obligations. It was for this reason that Diplock LJ said in *Varty* v *British South Africa Co* [1965] Ch 508 at p 523:

> To speak of an enforceable option as an "irrevocable offer" is juristically a contradiction in terms, for the adjective, "irrevocable" connotes the existence of an obligation on the part of the offeror, while the noun "offer" connotes the absence of any obligation until the offer has been accepted.

This does not mean that, in Lord Diplock's opinion, Lord Herschell and Lord Watson were speaking nonsense. They were not using "offer" in its primary sense but, as often happens in legal reasoning, by way of metaphor or analogy. Such metaphors can be vivid and illuminating but prove a trap for the unwary if pressed beyond their original context. As I said recently in another connection, K (*Enduring Powers of Attorney*), *In re* [1988] Ch 310 at p 314:

> . . . there are dangers in reasoning from the metaphor as if it expressed a literal truth rather than from the underlying principle which the metaphor encapsulates.

Here the underlying principles are clear enough. The granting of the option imposes no obligation upon the purchaser and an obligation upon the vendor which is contingent upon the exercise of the option. When the option is exercised, vendor and purchaser come under obligations to perform as if they had concluded an ordinary contract of sale. And the analogy of an irrevocable offer is, as I have said, a useful way of describing the position of the purchaser between the grant and exercise of the option. Thus in the recent case of *J Sainsbury plc* v *O'Connor (Inspector of Taxes)* [1990] STC 516 Millett J used it to explain why the grantee of an option to buy shares did not become the beneficial owner until he had exercised the option.

But the irrevocable offer metaphor has much less explanatory power in relation to the position of the vendor. The effect of the "offer" which the vendor has made is, from his point of view, so different from that of an offer in its primary sense that the metaphor is of little assistance. Thus in the famous passage in *London & South Western Railway Co* v *Gomm* (1882) 20 ChD 562 at p 582 Sir George

A Jessel MR had no use for it in explaining why the grant of an option to buy land confers an interest in the land upon the grantee:

> The right to call for a conveyance of the land is an equitable interest or an equitable estate. In the ordinary case of a contract for purchase there is no doubt about this, and an option for repurchase is not different in its nature. A person exercising an option has to do things, he has to give notice of his intention to purchase, and to pay the purchase money; but as far as the man who is liable to convey is concerned, his estate or interest is taken away from him without his consent, and the right to take it away being vested in another, the covenant giving the option must give that other an interest in the land.

The fact that the option binds the vendor contingently to convey was the reason why an option agreement was held to fall within section 40 of the Law of Property Act 1925: see *Richards* v *Creighton* B *Griffiths (Investments) Ltd* (1972) 225 EG 2104, where Plowman J rejected a submission that it was merely a contract not to withdraw an offer. Similarly in *Weeding* v *Weeding* (1861) 1 J&H 424) Page-Wood V-C held that the grant of an option to buy land was sufficient to deem that land converted into personalty for the purposes of the grantor's will, even though the option had not yet been exercised when he died. The Vice-Chancellor said:

> I cannot agree with the argument that there is no contract. It is as much a conditional contract as if it depended on any other contingency than the exercise of an option by a third person, such as, for example, the failure of issue of a particular person.

Thus, in explaining the vendor's position, the analogy to which the C courts usually appeal is that of a conditional contract. This analogy might also be said to be imperfect, because one generally thinks of a conditional contract as one in which the contingency does not lie within the sole power of one of the parties to the contract. But this difference from the standard case of a conditional contract does not destroy the value of the analogy in explaining the *vendor's* position. So far as he is concerned, it makes no difference whether the contingency is within the sole power of the purchaser. The important point is that "his estate or interest is taken away from him without his consent".

Griffith v *Pelton*, to which I have already referred, was another case in which the irrevocable offer analogy was unhelpful. The rule is that an offer in the primary sense can be accepted only by the person D to whom it is made. The offeree cannot assign to someone else the right to accept an unaccepted offer. In explaining why the benefit of an option *could* be assigned, so that someone other than the grantee could accept it, Jenkins LJ (in the passage which I have already cited) characterised the option as a conditional contract and said:

> The conditional contract constituted by the grant of the option is a chose in action the benefit of which can (if the terms of the contract are such as to show that it is not merely personal to the grantor) be assigned by the grantee to anyone he chooses, subject to any restriction imposed by the contract as to the persons in whose favour assignment is permissible.

Similarly in the Australian case of *Laybutt* v *Amoco Australia Pty Ltd* (1974) 48 ALJR 492, following *Kennewell* v *Dye* [1949] Ch 517, E the conditional contract analogy was used by Gibbs J to explain why an option could be enforced against the estate of the grantor, notwithstanding the rule that an ordinary offer lapses upon the death of the offeror.

The purchaser's argument requires me to say that "irrevocable offer" and "conditional contract" are mutually inconsistent concepts and that I must range myself under one or the other banner and declare the other to be heretical. I hope that I have demonstrated this to be a misconception about the nature of legal reasoning. An option is not strictly speaking either an offer or a conditional contract. It does not have *all* the incidents of the standard form of either of these concepts. To that extent it is a relationship *sui generis*. But there are ways in which it resembles each of them. Each analogy F is in the proper context a valid way of characterising the situation created by an option. The question in this case is not whether one analogy is true and the other false, but which is appropriate to be used in the construction of section 2 of the Law of Property (Miscellaneous Provisions) Act 1989.

There is only one case in which, as it seems to me, the adoption of the irrevocable offer metaphor was allowed to dictate the result without regard to the context. This was *Beesly* v *Hallwood Estates Ltd* [1960] 1 WLR 549 in which Buckley J decided that an option was not a "contract . . . to convey or create a legal estate" within the meaning of that part of the definition of an estate contract in section 10(1) of the Land Charges Act 1925. He arrived at this conclusion on the ground that the option was not a contract to convey but only an irrevocable offer. It seems to me, with respect to Buckley J, that this G was a misuse of the irrevocable offer metaphor. The purpose of including estate contracts in the Land Charges Act 1925 was to enable a purchaser to obtain notice of contracts which created interests binding upon the land. For this purpose, as Sir George Jessel MR pointed out in *Gomm's* case, there is no difference between an option and an ordinary contract of sale. In both cases the land is bound by an agreement which entitles a third party, either conditionally or unconditionally, to demand a conveyance. A purposive construction of section 10(1) therefore requires that one characterise the option from the point of view of its effect on the land in the hands of the grantor. For this purpose, it is more appropriate to regard it as a conditional contract than an irrevocable offer. In fact, Buckley J was able to give effect to the manifest intention of Parliament because the H definition of an estate contract went on to say "including a contract conferring . . . a valid option of purchase". He was therefore able to hold that an option was an estate contract, but only by treating the word "including" as extending the meaning of the previous words. In my judgment this was unnecessary. The option would have been an estate contract even without the additional words.

Mr Douglas, for the purchaser, relied strongly upon the decision of the House of Lords in *United Scientific Holdings Ltd* v *Burnley Borough Council* [1978] AC 904 as authority for the universal application of the irrevocable offer characterisation. That case concerned the rule that the conditions for the exercise of an option, including any time stipulations, must be strictly complied with. The J rule had been developed by analogy with the rule that an ordinary offer can be accepted only by strict compliance with the condition which it lays down. The Court of Appeal had extended the analogy to a notice under a rent review clause in a lease. The House of Lords held that this further extension was unjustifiable. Their lordships distinguished the ordinary option to purchase or to extend the lease from the activation of a rent review clause. In the former case, said Lord Simon of Glaisdale:

> . . . the parties, on the exercise of the option, are brought into a new legal relationship. It was argued . . . that the rent review clauses were also such unilateral terms. I cannot agree. The operation of the rent review clauses does not at all change the relationship of the parties, which remains that of landlord and tenant throughout the currency of the lease whether or not the machinery K of the rent review clauses is operated.

The reasoning of the other Law Lords was similar. It was rightly submitted on behalf of the purchaser in this case that the House of Lords thereby endorsed the irrevocable offer analogy for ordinary options. But the endorsement was in the context of the rule that there must be strict compliance with the conditions of acceptance. The case is no authority for extending its application in a different context.

Perhaps the most helpful case for present purposes is *Mulholland's Will Trusts, Re* [1949] 1 All ER 460. A testator had let premises to the Westminster Bank on a lease which included an option to purchase. He appointed the bank his executor and trustee and after his death the bank exercised the option. It was argued for his widow and children L that the bank was precluded from exercising the option by the rule that a trustee cannot contract with himself. Wynn-Parry J was pressed with the irrevocable offer metaphor, which, it was said, led inexorably to the conclusion that when the bank exercised the option it was indeed entering into a contract with itself. But the learned judge held that if one considered the purpose of the self-dealing rule, which was to prevent a trustee from being subjected to a conflict of interest and duty, the only relevant contract was the grant of the option. The rule could only sensibly be applied to a consensual transaction. While for some purposes it might be true to say that the exercise of the option brought the contract into existence, there could be no rational ground for applying the self-dealing rule to the unilateral exercise of M a right granted before the trusteeship came into existence. The learned judge quoted the passage I have cited from Sir George Jessel MR in *Gomm's* case and said:

> As I understand that passage, it amounts to this, that, as regards this option, there was between the parties only one contract, namely the contract created by the provisions in the lease which I have read creating the option. The notice exercising the option did not lead, in my opinion, to the creation of any fresh contractual relationship between the parties, making them for the first time vendors and purchasers, nor did it bring into existence any right in addition to the right conferred by the option.

The contrast between this passage and my citation from Lord Simon of Glaisdale in *United Scientific Holdings* is a striking illustration of how in different contexts the law can accommodate

A analogies which appear to lead to diametrically opposing conclusions.

In my judgment there is nothing in the authorities which prevents me from giving section 2 the meaning which I consider to have been the clear intention of the legislature. On the contrary, the purposive approach taken in cases like *Mulholland* encourages me to adopt a similar approach to section 2. And the plain purpose of section 2 was, as I have said, to prescribe the formalities for recording the consent of the parties. It follows that, in my view, the grant of the option was the only "contract for the sale or other disposition of an interest in land" within the meaning of the section and the contract duly complied with the statutory requirements. There must be judgment for the plaintiff
B against both defendants with costs.

Court of Appeal

November 5 1990

(Before Lord DONALDSON OF LYMINGTON MR, Lord Justice NOURSE and Lord Justice RUSSELL)

C & G HOMES LTD v SECRETARY OF STATE FOR HEALTH

Estates Gazette February 16 1991
C **[1991] 06 EG 170**

**Vendor and purchaser — Restrictive covenants — Alleged breaches by Secretary of State for Health — Houses purchased by Secretary of State and intended to be used in pursuance of the government's "Care in the Community" programme, whereby persons who have suffered mental disability are being returned from hospitals to live in the community — Questions as to whether this use was in breach of restrictive covenants contained in the transfer of the properties to the Secretary of State — Whether the use
D intended was use as a private dwelling-house only — Whether use constituted a "detriment" within a covenant prohibiting a nuisance, annoyance, danger or detriment — Appeal from decision of Ferris J, who held that the use proposed would be a breach of both covenants — Court of Appeal agreed in regard to the private dwelling-house covenant, but not in regard to the suggested detriment**

**The plaintiff company, which was the housebuilding arm of the Cheltenham & Gloucester Building Society, claimed against the defendant Secretary of State that the proposed use by the latter of two houses purchased by him was in breach of
E the restrictive covenants contained in the transfers — One of these covenants prohibited the use of the properties for any purpose other than as a private dwelling-house — The other relevant covenant prohibited any nuisance, annoyance, danger or detriment to the transferor or the owners or occupiers of other parts of the estate — The purpose for which the Secretary of State purchased the houses was to implement the "Care in the Community" programme — This involved using each of the houses as a home for people who had suffered in the past from mental disability, but were now being returned from hospitals to be integrated into the community — The occupants, four in each house, would have the services
F of a trained nurse and other support staff, including night service, supplied by the National Health Service — The Secretary of State appealed against the decision of Ferris J that the use of the properties constituted a breach of both restrictive covenants**

In considering the alleged breach of the covenant restricting use to that of a private dwelling-house the court examined a number of authorities — It was clear that in several respects the present case met the requirements for use as a private dwelling-house — The number of residents in each house was small, not more than four; the relationship between them was that of a group of friends or at any rate associates; and the occupants made no payment for their board or lodging — G
Although the owner did not reside there himself, he did everything in his power to make each house the home of the residents, giving them responsibility for domestic duties and decisions — However, there was a vital difference from a *private* dwelling-house — The occupants remained within the continuing care for which the Secretary of State was responsible — They occupied these particular houses because they were former in-patients of a nearby hospital — They were being treated and rehabilitated, with a view to a full return to the community, by the National Health Service under the responsibility of a minister of the Crown — Praiseworthy as H
these arrangements were, the court agreed with Ferris J that they did not constitute the normal use of a private dwelling-house

The other covenant alleged to have been breached was that which prohibited acts which might constitute a nuisance, annoyance, danger or detriment — It was not suggested that there was any question of nuisance, annoyance or danger — The submission was that there had been a "detriment" to the owners or occupiers of other parts of the estate by way of "impaired marketability" of their properties — The only evidence of this appeared to be that in a buyers' market an astute purchaser of a property had been able to make use of the J
general nature of the programme to obtain a small reduction in the purchase price of a property — The conclusive objection to the claim under this head was, however, that the only detriment contemplated by the covenant was one which affected the enjoyment of part or parts of the estate — The financial loss suggested, even if it could be proved, was not within this covenant — The court therefore differed from Ferris J in his finding that there had been a breach of this covenant — The result was that the appeal would be allowed in part

K

The following cases are referred to in this report.

German v *Chapman* (1877) 7 ChD 271; 47 LJCh 250; 37 LT 685; 42 JP 358; 26 WR 149, CA
Heglibiston Establishment v *Heyman* (1977) 36 P&CR 351; 246 EG 567, CA
Rolls v *Miller* (1884) 27 ChD 71
Segal Securities Ltd v *Thoseby* [1963] 1 QB 887; [1963] 2 WLR 403; [1963] 1 All ER 500
Tendler v *Sproule* [1947] 1 All ER 193, CA
Thorn v *Madden* [1925] Ch 847

This was an appeal from the decision of Ferris J (reported at [1990] 1 WLR 1272) whereby he held that the use proposed to be made by the Secretary of State for Health of two properties on the Charlcombe L
Park Estate, Dean Hill Lane, Weston, Bath, would constitute breaches of restrictive covenants by which the Secretary of State was bound. The respondents (plaintiffs below) were C & G Homes Ltd, a wholly-owned subsidiary of, and the housebuilding arm of, the Cheltenham & Gloucester Building Society.

John Macdonald QC and David Van Hee (instructed by Things & Long, of Bath) appeared on behalf of the appellant; David Oliver QC and J Stephen Whitaker (instructed by Laytons, of Bristol) represented the respondents.

Giving the first judgment at the invitation of Lord Donaldson MR, **Nourse LJ** said: Covenants against use for the purposes of any trade or business and for any purpose other than as a private dwelling- M
house are often found in conveyances and leases of residential property. The effect of such covenants in regard to well-recognised uses, for example as a lodging-house, is no longer in doubt, although the question being one of fact and degree, the circumstances of individual cases from time to time cause difficulty. Here we are confronted with a novelty: the use by the Secretary of State for Health, through the agency of a district health authority, of each of two adjoining houses on a new development as a permanent home for four persons in his care, being people who have suffered in the past from mental disability and are now to be returned to the community under the current policy of ceasing to maintain in hospitals those who no longer need to be there.

A The background to the case and the material facts as they stood at that time are set out in the judgment of Ferris J at [1990] 1 WLR 1272 and need not be repeated. The judge held that the use which the Secretary of State was making of the houses was in breach of two of the covenants contained in the transfers by which they were acquired. He made a declaration accordingly and ordered an enquiry as to damages. The Secretary of State appeals to this court. In the six months or so which have elapsed since the trial he has continued to use the houses as before. It is not suggested that that was not a proper course for him to take, the assumption no doubt being that if it had been a case in which an injunction could have been issued a stay would have been granted pending the appeal.

B Like the judge, I will start with the covenant in para 24.2 of the third schedule to each of the transfers:

> 24. Not at any time within ten years of the date hereof: . . .
> 24.2 to carry on or from the Property or any part or parts thereof any trade business or manufacture whatsoever (with the exception of the profession of a Solicitor Doctor Dentist Private Teacher Accountant Architect or other professional person provided that any necessary local authority approval is obtained for such use) and not to use the said dwelling-house nor to use the said garage for any purpose or purposes other than those incidental to the enjoyment of a private dwelling-house.

At [1990] 1 WLR 1272 at pp 1275E to 1276F, in describing the use to which the houses were to be put, the judge quoted at some length from an affidavit of Mrs Maureen Teresa Wykes, who is employed by
C the Bath District Health Authority as locality manager for the Bath Mental Health Unit, her principal concern being to plan and provide mental health services for the City of Bath. In another part of her affidavit Mrs Wykes described how the residents had been fully involved in the selection of furnishings for their own rooms and in the choice of colour schemes, also in the selection and purchase of the smaller items for their homes. In a second affidavit before the judge she told of the ways in which the residents had been able to develop their participation in normal domestic duties and decisions. Now, in a third affidavit which has been admitted without objection in this court, Mrs Wykes has brought the supervisory and supportive arrangements up to date. They can be summarised by saying that from the middle of September the numbers of staff in attendance
D have been reduced, there being either one trained nurse and one untrained person or two untrained persons to provide domestic support during the day and one untrained to provide domestic support at night and another untrained to sleep in. It is anticipated that later in the year a single untrained person will cover all duties at night. Mrs Wykes has also described the medical and other services of which the residents have the benefit. She states that, apart from those of the patient services manager, all the services are available to persons living in their own homes who have need of them. She adds that there have been no incidents or friction with other occupiers of houses on the estate and that the residents have settled in well.

The question whether the use which is being made of the two
E houses is in breach of the covenant in para 24.2 depends upon the true construction of the covenant. At the outset of his argument Mr Macdonald QC, for the Secretary of State, emphasised that it was entered into in 1989, at a time when the Town and Country Planning (Use Classes) Order 1987 had introduced a new class C3 for dwelling-houses in these terms:

> Use as a dwelling-house (whether or not as a sole or main residence) —
> (a) by a single person or by people living together as a family, or
> (b) by not more than 6 residents living together as a single household (including a household where care is provided for residents).

Mr Macdonald accordingly submitted that since these houses are being used as dwelling-houses in the eyes of the planning legislation, the covenant should be construed to the like effect. On the
F assumption, which is no doubt correct, that the planning view of the matter was part of the factual background known to both covenantor and covenantee, it is certainly a factor to be taken into account in construing the covenant. No less to be taken into account is that parties to a conveyancing transaction, having entered into a covenant in a long-established and familiar form, must have intended that it should have the effect which earlier authorities have said that it has.

Consistently with the fundamental rule that due effect must be given to all of the words in which the parties have expressed themselves, I regard this covenant as having two limbs: (1) not "to carry on or from the property . . . any trade business or manufacture . . ."; and (2) "not to use the said dwelling-house nor to use the said garage for any purpose or purposes other than those incidental to the enjoyment of a private dwelling-house". That was the initial view of G
Ferris J. But on consideration he thought that he ought to approach the covenant as a whole, looking at both limbs together as opposite sides of the same coin. In making that approach he was influenced by what was said by Lindley LJ in *Rolls* v *Miller* (1884) 27 Ch D 71 at p 87, a case where the covenant was only against the carrying on of any trade or business, from which it was implied that the property was not to be used otherwise than as a dwelling-house. Although in the end it may make no difference, I think that, if the judge had been referred, as we were, to *German* v *Chapman* (1877) 7 Ch D 271, he would have thought that his initial view was correct. I should add that the second limb, although expressed at greater length, was treated by both sides as the equivalent of "not to use the said dwelling-house H
. . . for any purpose or purposes other than those of a private dwelling-house".

I will consider the five authorities to which we were referred in chronological order. In *German* v *Chapman* a conveyance of freehold property contained a covenant by the purchaser not to erect or build on the land more than four dwelling-houses and also that no

> dwelling-house, or other building so to be erected or built upon the same land shall at any time hereafter be used or occupied otherwise than as and for a private residence only, and not for any purpose of trade.

It was held by this court that the proposed erection, by the trustees of a charitable institution for the daughters of missionaries, of a large building capable of holding 100 girls who were to be boarded, lodged and educated there was a breach of the covenant not to use or occupy J
the land otherwise than as and for a private residence. It was acknowledged that no trade was involved in the project, but it was held that the covenant had two separate limbs which must be separately considered. At p 277 James LJ said:

> . . . it is impossible to say that in the ordinary use of English language the using of a place for a large institution (although a very charitable and beneficial institution) for the reception of one hundred girls to be lodged, boarded, and taught at that place, is using it as a private dwelling-house. To my mind such an institution is no more a private residence only than a club of gentlemen or a working men's club, or any other establishment of that kind, where a great number of persons are brought together. Whose private residence can it be said to be?

In *Rolls* v *Miller* the lease of a house contained a covenant by the K
tenant that he should not:

> use, exercise, or carry on, . . . in or upon the premises hereby demised any trade or business of any description whatsoever . . .

It was held by this court that the proposed use of the house as a "Home for Working Girls", where working girls and young women between the ages of 15 and 25 who were without situations and temporarily unable to support themselves would be provided with board and lodging, would be a breach of the covenant, notwithstanding that no payment was required or made. At p 86, having held that no trade was being carried on, but that it was not essential that there should be payment in order to constitute a business, Cotton LJ, in a passage quoted by Ferris J [1990] 1 WLR L
1272 at p 1278, concluded that what was being done must be considered as carrying on the business of a lodging-house. At p 88 Lindley LJ came to the same conclusion. The judgments emphasised that a superintendent and staff were employed to receive and look after the girls on the owners' behalf and that the owners themselves did not live there. Ferris J found *Rolls* v *Miller* the most helpful of the authorities which were cited to him. Having referred to the conclusions of Cotton and Lindley LJJ, he said at p 1278:

> In my judgment, the same has to be said here. I am asked by Mr Macdonald to say that this covenant must be construed according to the standards of 1989, not those of 1884 when *Rolls* v *Miller* was decided, but although that may be right as a matter of principle, I cannot see what has changed in the meaning of language, or in the way that things are done, which enables me to distinguish M
> this case from that case.

In *Tendler* v *Sproule* [1947] 1 All ER 193 a tenancy agreement of a house contained an agreement by the tenant:

> . . . not to use the said premises or any part thereof for any trade or business but keep the same as a private dwelling-house only . . .

After the contractual tenancy had expired the tenant held over as a statutory tenant and later took in "two paying lodgers". It was held by this court that that was a breach both of the covenant against business use and also of the covenant to keep the premises as a private dwelling-house only. In referring with approval to the decision of Tomlin J in *Thorn* v *Madden* [1925] Ch 847, Morton LJ said, at p 194:

A I think that the real gist of the decision is that the taking in of paying guests is a business and that a house which, or part of which, is used to take in paying guests is not a house which is being kept as a private dwelling-house only.

In *Segal Securities Ltd* v *Thoseby* [1963] 1 QB 887 a lease of a maisonette contained a covenant by the tenant:

to use the demised premises for the purpose of a private residence in the occupation of one household only.

Residing with the tenant was a woman friend of hers, who shared the accommodation, meals and expenses on terms that yielded no profit to the tenant, and another woman, who had come in answer to an advertisement and who lived an independent life, there being no real sharing of meals and social life in the sense that she lived as part of a
B family. Having referred to *Thorn* v *Madden* and *Tendler* v *Sproule*, Sachs J, at p 894, said:

To my mind, in the way of life of 1962, the mere taking in of a single paying guest who shares the family life so far as practicable would not, save in exceptional circumstances, be regarded by a reasonable man as a breach of a covenant to use the house as a private residence only; nor would I myself willingly hold that to be such a breach; nor do the authorities on analysis bind a court so to hold. It is in each case a question of fact and of degree whether the taking in of paying guests is of an order that, having regard to all the circumstances, constitutes a breach of the covenant in question.

The judge thought that had the friend been the only person there with the tenant, no breach would have been established. However, he held that the circumstances of the other woman's coming and being there
C caused there to be a breach, albeit by no great margin.

In *Heglibiston Establishment* v *Heyman* (1977) 36 P&CR 351 a lease of a flat contained a covenant that the tenant would not:

. . . use or permit the said flat . . . to be used for the purpose of any art trade or business whatsoever . . . but shall and will use the said flat as and for a private residence only and in particular will not take in boarders or lodgers.

The flat was occupied for some of the time by the tenant and for all of the time by his son and the woman with whom he was living as man and wife and by an old school friend of his as well. The three permanent occupants paid no rent, but they shared the household expenses, including the gas and electricity bills and the food. The county court judge found that the woman was treated as if she was the son's wife and that the son's friend was treated as his brother. He
D therefore held that they were not lodgers or boarders and that there was no breach of the covenant. His decision was affirmed by this court. In delivering the leading judgment, Browne LJ, at p 360, said that there was all the difference between a person who made fixed periodical payments and a person who merely agreed to share part of the expenses of a single household. He approved the approach of Sachs J in *Segal Securities Ltd* v *Thoseby*.

These authorities show that the question of fact and degree which has to be answered in each case will involve a consideration of all or some of the following matters: the number of occupants; the degree of permanence of their occupancy; the relationship between them; whether payment is made or not and, if so, whether it is only a
E contribution to expenses or something more; whether the owner or lessee resides there himself and, if not, whether he has people there to supervise and support those who do.

In *German* v *Chapman* and *Rolls* v *Miller* charitable boarding institutions which made no charge to the occupants were held in the one case not to be a private residence and in the other to be a lodging-house. In both cases there were many occupants, whose occupation was either temporary or not more than short term. In *Rolls* v *Miller* importance was attached to the fact that the owners employed a superintendent and staff and did not reside there themselves. *Tendler* v *Sproule* and *Segal Securities Ltd* v *Thoseby* establish that the presence of no more than two (or even one) paying guests or lodgers can cause there to be a breach of both limbs of the covenant.
F *Heglibiston Establishment* v *Heyman*, on the other hand, shows that the presence of three other persons who are, or are treated as being, members of the tenant's family, who pay no rent and who merely share the household expenses, will not bring about a breach.

What are the essential circumstances of the present case? Taking, wherever there may be a doubt, the view more favourable to the Secretary of State, I would say that they are these. The number of residents of each house is small, no more than four. The period of their residence is permanent or indefinite. The relationship between them is that of a group of friends or at any rate of associates. They make no payment for their board or lodging. The owner does not and cannot reside there himself. He does everything within his power to make each house the home of the residents, giving them, so far as
G practicable, responsibility for domestic duties and decisions and, so we were told by Mr Macdonald, the right to determine the composition of their group. But the owner remains under a statutory duty to care for them. He does that by means of the attendance of two staff during the day and one at night and the provision of the services of the patient services manager and regular medical services.

Although I do not find it necessary to express a concluded view on the first limb of the covenant, I doubt whether the Secretary of State is carrying on a business on or from either of these houses. The small number of residents, the permanence of their residence, the relationship between them and the absence of any payment for their board and lodging together suggest to my mind that it is not the business of a lodging-house. And, while I accept that "business" is a
H word of wide application, my impression — and when you have assembled all the features of a case it is really no more than a matter of impression — is that it is not a business of any other kind.

Is, then, the Secretary of State using these houses for any purpose or purposes other than those of a private dwelling-house? I think that he is. The basic proposition inherent in Mr Macdonald's argument was that because each house is the private dwelling-house of those who reside there, therefore it is being used by the Secretary of State as a private dwelling-house. I think that it might well be possible to say that he is using the houses as dwelling-houses. But I cannot agree that he is using them as "private" dwelling-houses.

We were not referred to any judicial definition of a private
J dwelling- house. It seems that judges, no doubt wisely, have been content to say whether, in any given set of circumstances, the description is or is not satisfied. The definition of a private house given in the *Shorter Oxford English Dictionary* (1933) is:

The dwelling-house of a private person, or of a person in his private capacity.

Where the owner himself is in occupation it can usually be said that he is using it as *his* private dwelling-house. But he can still use it as *a* private dwelling-house without occupying it himself, for example where he lets it to another individual for use as his private dwelling-house. Use as or for the purposes of a private dwelling-house seems to assume that there is at least one private individual who, whenever he chooses, can occupy the house as his own, even though he may not be
K in actual occupation, for example where he allows his children and some friends to live there.

In the present case no emphasis of the residents' ability to treat the house as their permanent home nor of the advanced degree of autonomy which is given them in the running of it can obscure the continuing responsibility of the Secretary of State for their care or the incidental powers which are vested in him. In particular he has the power to determine a resident's licence to occupy the house and in an emergency he would have both the power and the duty to do so without notice. More significant in practice is his responsibility to provide supervision and support for the residents, a responsibility which has not been diminished by the reduction in numbers of the staff who are needed to administer it. Moreover, the responsibility
L being one which belongs to a minister of the Crown, the houses are in that sense used for public and not for private purposes.

In a commendable search for a label, Mr Oliver QC, for the plaintiffs, referred us to the definition of "a residential care home" in section 1(1) of the Registered Homes Act 1984:

[an] establishment which provides or is intended to provide, whether for reward or not, residential accommodation with both board and personal care for persons in need of personal care by reason of old age, disablement, past or present dependence on alcohol or drugs, or past or present mental disorder.

That certainly shows that these houses might be residential care homes within the definition of the 1984 Act. It does not show conclusively that they are not private dwelling-houses within the
M contemplation of the covenant.

In summary, I would say that if a house cannot fairly be described as someone's private dwelling-house, it cannot be said to be being used as such. I therefore ask myself the question which was asked by James LJ in *German* v *Chapman*: Whose private dwelling-house can it be said to be? It is not the Secretary of State's. And I am unable to say that it is the residents'. In the end, although by somewhat different route, I have come to a clear conclusion that Ferris J's decision on the first question was correct. I would therefore affirm his declaration and his order for an inquiry as to damages, so far as they relate to the covenant contained in para 24.2

I can deal more briefly with the covenant contained in para 20 of the third schedule to each of the transfers:

A 20 Not to cause or permit or suffer to be done in or upon the Property any act or thing which may be or become a nuisance annoyance danger or detriment to the Transferor or owners or occupiers for the time being of other parts of the Estate.

Mr Oliver submitted that, by allowing the houses to be occupied as he has, the Secretary of State has caused or permitted or suffered to be done in or upon each of the houses some act or thing which has caused a detriment to the plaintiffs, within the terms of this covenant. It is not suggested that there has been a nuisance, annoyance or danger. Nor is it suggested that there has been a detriment to the owners or occupiers of the other parts of the estate.

The detriment alleged is a financial loss to the plaintiffs caused by what the judge described as "the impaired marketability" of other B houses on the estate. The three instances put forward and the evidence in support of the alleged loss, such as it is, are summarised by the judge at [1990] 1 WLR 1272 at pp 1279H to 1280E. In this court at any rate the detriment principally, if not exclusively, relied on was the reduction of £3,000 in the price and the other benefits achieved by Mrs Midgley on the purchase of plot 20.

Mr Macdonald's primary submission was that the covenant does not impose any restriction on the persons who may occupy the property. It restricts only the acts or things which the occupants, whoever they be, may do there. Although Ferris J thought that that was too narrow a view, I suspect that it accords with the construction which most conveyancers would put on a covenant in this form, again C a very familiar one.

Be that as it may, there is a further and conclusive objection to the plaintiffs' arguments on this second question. The only detriment within the purview of the covenant is one which affects the enjoyment of some part or parts of the estate. That inevitably follows from the linking of "detriment" with "nuisance, annoyance, danger", all of which affect the enjoyment of land, and the reference to "the Transferor or owners or occupiers for the time being of other parts of the Estate", which confirms that the detriment etc must be suffered by the plaintiffs as owners of the part or parts of the estate which for the time being they retain. A financial loss suffered in their exploitation of the retained land, but dissociated from their enjoyment of it, is something with which the covenant is not in the D least concerned.

Although that is enough to dispose of the second question in favour of the Secretary of State, I will add that I am very doubtful whether, on the evidence which was before the judge, it was established that the plaintiffs had suffered any relevant financial loss. To take only the case of Mrs Midgley, admittedly it was shown that she had succeeded in wielding the Secretary of State's intended user of the houses as a weapon to obtain a reduction of £3,000 in the purchase price and other benefits. It was not shown, at all events not satisfactorily shown, either that Mrs Midgley would not have purchased if the plaintiffs had stood their ground or that the reduced price, less the cost of providing the other benefits, fell short of what E the plaintiffs would have got for plot 20 elsewhere.

For these reasons I respectfully disagree with Ferris J's view that there has been a breach of the covenant contained in para 20. I would therefore discharge his declaration and his order for an inquiry as to damages, so far as they relate to that covenant. To that extent, and to that extent only, I would allow the Secretary of State's appeal.

Agreeing, **Russell LJ** said: For those who are residing in the two adjoining houses on the Charlcombe Park Estate, with which this case is concerned, the result of this appeal will be deeply disappointing. I have no doubt that they regard the properties as their home, just as much as any other house in the development is the home of the family residing there.

But, unhappily, this cannot be the approach which the lawyers F must adopt. The covenant in para 24 of the third schedule to each of the transfers prohibited the use of the properties "for any purpose . . . other than . . . a private dwelling-house". The emphasis must be on the word "private". The houses are owned but not occupied by the Secretary of State. The residents do not occupy as tenants. They occupy because, and only because, they happen to be former patients of a nearby hospital who are being treated and rehabilitated by the National Health Service pursuant to the "Care in the Community" policy. This is the obligation of the Secretary of State, and the former in-patients remain under the statutory care of the Secretary of State who provides, within the houses, staff to fulfil his statutory duty.

I do not believe that any objective description of two houses occupied by the patients and administered by the NHS could ever embrace the phrase "private dwelling-house". For my part, because G the policy of the Secretary of State is to take the patients out of the environment of the hospital I do not regard the houses as hospital annexes or indeed any other type of institution. While recognising the imperfections of any description in the new and unique policy of the Secretary of State, I would regard and describe the houses as small NHS homes for people who have suffered from mental disability. Whatever be an apt description, however, and in agreement with the Master of the Rolls and Nourse LJ, I am abundantly satisfied that the houses, in their current occupation and ownership, are not private houses, and that consequently the Secretary of State is in breach of covenant.

As to para 20 of the third schedule to each of the transfers, I too take the view that the "detriment to the transferor" is directed to the H enjoyment by them of other parts of the estate. Nothing of the kind was alleged or proved and, in my view, the financial loss alleged is not caught by para 20.

For these reasons I agree with the order proposed by Nourse LJ and, to the limited extent he indicates, would allow the appeal.

Also agreeing, **Lord Donaldson of Lymington MR** said: This appeal is *not* about the merits of the "Care in the Community" policy. No one has said a word against it. What this appeal is about is something quite different.

Every year thousands of houses change hands. In many, possibly most, of these cases the transfer is subject to covenants, that is to say conditions. Where what is being sold is a "second hand" house, there J may be little scope for altering the conditions which probably date from when it was built. In the case of a new house, the builder or developer has a much freer hand, but it is his hand and no one else's. It is also the case that where an estate is being developed, it is usual for all houses to be sold subject to the same covenants. Whether the house is new or second hand, the covenants have to be agreed between the seller and the buyer before a sale can take place. If the parties are unable to agree, the seller must find another buyer and the buyer must look elsewhere for a house.

In the present case C & G Homes Ltd ("C & G") and the Secretary of State did agree upon the covenants which would govern the transfer and no doubt each thought they had a common K understanding as to their meaning. It was only after completion that it emerged that C & G thought that they meant one thing and that the Secretary of State thought that they meant another.

It is not the first time, and will assuredly not be the last, that such a dispute has arisen between a willing seller of a house and his willing buyer. If the parties cannot compose their differences, the courts have to decide which (if either) is right and what the covenants mean. That, and that alone, is what this appeal is all about.

The conditions agreed by both C & G and the Secretary of State included the following, which appear in the third schedule to the transfer:

20 Not to cause or permit or suffer to be done in or upon the Property any L act or thing which may be or become a nuisance annoyance danger or detriment to the Transferor or owners or occupiers for the time being of other parts of the Estate.

. . .

24 Not at any time within ten years of the date hereof

. . .

24.2 To carry on or from the Property or any part or parts thereof any trade business or manufacture whatsoever (with the exception of the profession of a Solicitor Doctor Dentist Private Teacher Accountant Architect or other professional person provided that any necessary local authority approval is obtained for such use) and not to use the said dwellinghouse nor to use the said garage for any purpose or purposes other than those incidental to the enjoyment of a private dwellinghouse.

The Secretary of State contends that neither covenant prevents his M using the two houses for the purpose for which he bought them, namely implementing the "Care in the Community" programme. C & G contend that one or other or both covenants do have this effect. It may well be that it would be a serious blow to the Secretary of State's programme if C & G were right. Furthermore, it may well be that C & G attaches an exaggerated importance to these covenants. Neither consideration can influence us in reaching a decision. We have simply to decide what the covenants mean.

Mr John Macdonald QC, appearing for the Secretary of State, rightly submits that these two covenants have to be construed in the light of circumstances as they prevailed in 1989 when the transfers of the two houses were executed. Early decisions of the courts must be

A approached with caution, lest they be based upon circumstances which have since changed. Nevertheless I have no doubt that the general guidance contained in the judgment of Lindley LJ in *Rolls* v *Miller* (1884) 27 Ch D 71, at p 87 as to the approach to be adopted is still good law. He said:

> Now the first question to be considered is what is the object of this covenant. The covenant must be construed consistently with that object, and on the other hand, something may fall within the scope of the covenant which does not fall within the words. One must look, therefore, at both the words and the object.

In other words, if the covenant is to require or prevent a particular course of action, it must fall within both the object and words of the covenant.

B I have already set out the words. What then is their object? I think that it is plain. It is to ensure that the Charlcombe Park estate is developed as a normal residential estate. To this end words have been used in covenant 24.2 which are apt to prevent the houses being used for professional or business purposes, other than in connection with specified professions which might fit in with a residential estate. Thus the use of the houses as boarding or guest houses or as student hostels would be excluded. To this end words have also been used in covenant 20 to ban conduct which may be a nuisance, annoyance or danger to other owners or occupiers. All this is reasonably clear, but it does not solve the central problem.

In many respects the way in which the Secretary of State's houses C are being used is quite indistinguishable from the way in which the other houses on the estate are being or will be used. All are being used as dwelling-houses. Indeed it is central to the "Care in the Community" policy that this should be the case. The intention is that the occupants shall be weaned from their reliance upon institutionalised care and progressively assisted to take their full place in the community. Furthermore, although in the last century it would have been very unusual to have four occupants from different families sharing the same house, this today is common at least in the case of students and young people. It is therefore necessary to see whether there is anything else which distinguishes the purpose for which these houses are being used from the purposes for which private dwelling-houses are normally used.

D Private dwelling-houses are used for the accommodation of the owner or a tenant and his family. They are also used to accommodate small groups, such as students, who club together and take a joint tenancy of the house. There are other permutations, but a common feature is that at least one of the occupants has an interest in the house as owner or as tenant or that the occupiers have a family relationship with the owner, eg a case in which parents make a house available for occupation by children or grandchildren. That is not this case, for none of the occupants of these houses owns or has a tenancy and, of course, they cannot properly be described as members of the Secretary of State's family.

Describing the nature of the use being made of the houses, and E attaching a label, is not conclusive, but it may be illuminating. In so doing I am concerned with the position as it is, rather than as it may become when the process of accustoming these occupants to life in the community is further advanced. They remain patients in the care of the National Health Service to the extent at least that they are ineligible for social security benefits and rely upon the National Health Service for money to buy the requirements of the household, such as food. This does not matter of itself — other people live on allowances — but the source of the allowance and the relationship between the occupants and the hospital underlines the special nature of their occupation and thus the special nature of the use which is being made of the houses. They have a resident house leader who is a registered nurse and two support workers on a daily basis, all of F whom are employed by the National Health Service. If a label can be attached to such a use, it seems to me to be that of "hospital annexe" or "mental health hostel". Praiseworthy though this no doubt is, I have to say that, in my judgment, this is not a use which would be regarded as a normal use of a *private* dwelling-house. It follows that I have to conclude that the Secretary of State is in breach of covenant 24.2.

The position in relation to covenant 20 is quite different. Once again I have to look to the object and to the words. It is not directed to the use being made of the property, that being the subject-matter of covenant 24.2. It is directed instead at conduct in or upon the premises which causes or may cause nuisance, annoyance, danger or detriment to other owners (which incidentally would at present include C & G) or occupiers of other parts of the estate or to C & G. G No complaint whatsoever is being made in relation to the conduct of the occupants in or upon the premises. The evidence relied upon as constituting a detriment to C & G relates solely to the use of the premises made by the Secretary of State and amounts to no more than that in a buyers' market a particular purchaser was astute enough to use the general nature of that use as a lever to obtain a small reduction in the purchase price.

There has been no breach of covenant 20.

To that extent only I agree that the appeal should be allowed.

The appeal was allowed in part but the respondents were awarded the costs of the appeal.

H

Chancery Division

July 16 1990

(Before Mr Justice KNOX)

MILLERS WHARF PARTNERSHIP LTD v CORINTHIAN COLUMN LTD

Estates Gazette June 1 1991 and June 8 1991

[1991] 21 EG 115 and [1991] 22 EG 124

Vendor and purchaser — Contract for grant of lease — Conditional contract — Performance of conditions — Provision that if conditions were not satisfied by a specified date either party might "at any time thereafter" rescind the agreement — Not all conditions satisfied by specified date — Purchaser gave notice of recission some four months after specified date — Questions as to purchaser's entitlement to rescind — Dearth of English authorities on particular point — Australian cases cited — Conclusion that words "at any time thereafter" mean what they say J

The plaintiff company in this case had acquired a docklands site for redevelopment and had entered into an agreement with K **the defendant company to grant a lease to the latter of a particular flat, one of 22 flats into which an old warehouse was being converted — The premium payable for the leasehold flat was £465,000 and there was a deposit of 10% of that sum — The conversion scheme was subject to a large number of consents, approvals and conditions required by different authorities, including planning and listed building permissions, building regulations and approval of steps taken in response to conditions — The agreement between the plaintiff and defendant companies stipulated that the plaintiff (described in the agreement as the vendor) should use all** L **reasonable endeavours to satisfy all the various conditions and requirements and that, if they had not been satisfied before June 30 1988, either party might "at any time thereafter" serve written notice on the other rescinding the agreement — In fact not all the consents had been obtained or the conditions fulfilled by June 30 1988 and some were not implemented by the date when the defendant company gave formal notice of rescission — The defendant's claim to rescind was challenged by the plaintiff in the present proceedings**

On these facts a question of some general interest was raised as to the position in regard to a right of rescission where there is a condition precedent to the performance of a contract to M **grant a lease and the condition is required to be performed by a fixed date — If the right of rescission is expressed to be exercisable at any time after that fixed date, is it nevertheless no longer exercisable if the condition is satisfied after the fixed date but before the right to rescind is exercised? It appeared that no English authority gave the answer to this question and certain Australian decisions cited by counsel did not provide clear guidance — Accordingly, Knox J had to resolve the problem by reference to general principles**

The judge considered that if the right of rescission were to be regarded as not exercisable where the condition was

A **satisfied after June 30 but before October 27 1988, this could only be as a result of one of two alternative arguments — The first would be the apparent incompatibility between a right of rescission and a contract which had become unconditional — The second was the assumed need for an implied term qualifying the right to rescind "at any time thereafter" by the addition of words indicating that the time must be before the condition was satisfied — The judge was not persuaded by either argument — So far as the suggested incompatibility was concerned, there was no substance in the objection; the concept of a vested but defeasible right was one which was**
B **quite common in property law — He saw no need for an implied term and came to the conclusion that the words "at any time thereafter" should be given their plain and natural meaning — That conclusion was fatal to the plaintiff's case — He accordingly dismissed the plaintiff's action and gave judgment in favour of the defendant's counterclaim for a declaration and the return of the deposit of £46,500 with interest**

The following cases are referred to in this report.

Aberfoyle Plantations Ltd v *Cheng* [1960] AC 115; [1959] 3 WLR 1011; [1959] 3 All ER 910
C *Gange* v *Sullivan* (1966) 116 CLR 418
Mardorf Peach & Co Ltd v *Attica Sea Carriers Corporation of Liberia* [1977] AC 850; [1977] 2 WLR 286; [1977] 1 All ER 545; [1977] 1 Lloyd's Rep 315, HL
Perri v *Coolangatta Investments (Proprietary) Ltd* (1982) 149 CLR 537
Property & Bloodstock Ltd v *Emerton (No 2)* [1967] 2 All ER 839
Sandwell Park Colliery Co Ltd, Re [1929] 1 Ch 277
Suttor v *Gundowda Proprietary Ltd* (1950) 81 CLR 418; [1950] ALR 820

The plaintiff company, Millers Wharf Partnership Ltd, sought in these proceedings to enforce an agreement for the grant to the defendant company, Corinthian Column Ltd, of a lease of Flat 2A Millers Wharf, St Katharine's Way, London E1. The defendant company claimed that it had effectively rescinded the agreement and
D counterclaimed for a declaration to that effect and the return of the deposit.

Nicholas Patten QC and Timothy Harry (instructed by Hardwick & Co) appeared on behalf of the plaintiff; John V Martin (instructed by Wilson-Smith & Co) represented the defendant.

Giving judgment, **Knox J** said: In this action between Millers Wharf Partnership Ltd as plaintiff and Corinthian Column Ltd as defendant, the basic facts were not the subject of dispute. There was an agreement that the proof of evidence of a Mr Brian Ingwell should stand as evidence and I take the basic facts from that document.

The plaintiff in this action is a company that was acquired for the purpose of purchasing and carrying out a redevelopment of an
E existing warehouse, Millers Wharf, fronting the Thames near St Katharine Docks. The development consisted of converting Millers Wharf into 22 flats. The original drawings for the development were prepared by Terry Farrell Partnership, who were architects to the development from the outset until the end of 1987. At this point Pascal Watson of 146 Buckingham Palace Road became project architects. MWPL purchased the site in approximately June 1987. There then follows a description of the site by reference to a plan on which there are five different coloured pieces of land. The warehouse itself (which was the subject of the redevelopment) is a substantial site edged red, and that was acquired freehold. The second part of the site was edged green on the plan referred to and consisted of a jetty in front of the warehouse and fronting, obviously, on to the river. That
F was acquired leasehold from the Port of London Authority, which owned the freehold. To the right of the warehouse and of the jetty was first an area, reading from north to south, edged blue. This was a public highway falling under the jurisdiction of the London Borough of Tower Hamlets, which is the local highway authority. Next, below that to the south, comes an area edged brown on the plan called Alderman's Stairs, which is a listed structure. The part of the stairs up to the front wall of the warehouse also falls under the jurisdiction of the London Borough of Tower Hamlets as highway authority. That part of the stairs in front of the front wall of the warehouse falls under the jurisdiction of the Port of London Authority. Finally, to the south of Alderman's Stairs, is the final area of land, edged yellow on the plan, which is a causeway to which access is gained by going down the stairs. That, too, falls within the jurisdiction of the Port of G
London Authority.

When the plaintiff acquired the warehouse and jetty there was an existing planning permission. This had been granted in 1983 to Land Securities Management Ltd. As Millers Wharf itself is a listed building, Land Securities Management Ltd had also obtained listed building consent.

The development contemplated by Land Securities Management Ltd was, again, a redevelopment of the warehouse. However, when the plaintiff purchased the warehouse and jetty there were sufficient differences between the scheme contemplated by the plaintiff and that which was the subject of the existing planning permission and listed building consent that a fresh application for planning permission and listed building consent was necessary. That fresh H
application for planning permission and listed building consent was therefore submitted to the London Docklands Development Corporation, which was the relevant planning authority, on July 14 1987. Those applications were made on behalf of the plaintiff by Terry Farrell Partnership.

The redevelopment contemplated by the plaintiff not only involved conversion of the warehouse but also involved works to the Alderman's Stairs and the jetty. Essentially it was contemplated that the jetty would become a public footpath. The public would then be given a direct means of access to the jetty from the Alderman's Stairs. No direct means of access to the jetty from the stairs existed at the outset because such was the angle of the stairs at the point where they J
became level with the jetty that they were some 3 ft or 4 ft below the level of the jetty.

The scheme contemplated by the plaintiff, therefore, envisaged rebuilding the stairs to a sufficient level to create direct access and also creating a passageway through the corner of the warehouse.

At a planning meeting held at the London Docklands Development Corporation on September 14 1987, attended by Terry Farrell Partnership, the corporation determined that the proposed works to the stairs did not fall within the ambit of the existing planning application. A supplementary planning application was therefore made on September 25 seeking planning permission and listed building consent for the works to the stairs and jetty.

The London Docklands Development Corporation wrote on K
November 23 to those representing the plaintiff, reporting the decision of the corporation. The letter includes the following:

> Further to our recent telephone conversations, I can confirm that your applications for planning permission and listed building consent for the development of the above site were considered by the Corporation's planning committee on the 10th November 1987 and that the committee resolved as follows:
> 1 The Corporation enter into a supplemental agreement to revise the s52 agreement attached to application TH/83/118 for the provision of a public riverside walkway and link to St Katharine's Way.
> 2 The applicant having entered into the said agreement, the Corporation do thereupon consent to the development set out in the application for planning permission and listed building consent, subject to the following conditions. L

The first one was the standard one:

> The development hereby permitted must be commenced not later than the expiration of five years from the date of this permission.

The only other one which I need read in full is the one numbered 3, which reads:

> Full details, including colour and samples of the facing materials to be used on the building, should be submitted to and approved by the local planning authority before any work on site is commenced, such details to relate in particular to:
> (a) penthouse roof cladding and walls;
> (b) riverside walkway;
> (c) Alderman's Stairs. M

Pausing there for a moment, the evidence before me was that that condition was common form down to but not including the passage that reads: "such details to relate in particular to" (a), (b) and (c). So far as the latter is concerned, it obviously was specifically directed to this particular proposal, and the evidence was that the penthouse roof cladding and walls were to be a new structure otherwise than in brick, which was of what the original warehouse was built and which it was not, apart from amendments to openings such as windows and doors, proposed to alter.

By the time that resolution was passed, the agreement which is the subject of these proceedings had been made. It was dated October 7 1987 and was made between the plaintiff and the defendant. The

A defendant then had a different name, but nothing turns on that. It was recited that the vendor (which is the name given to the plaintiff) proposed to redevelop the property known as Millers Wharf, St Katharine's Way, London E1 (that is defined as "the estate"). Second, that a company named Millers Wharf Management Co Ltd ("the management company") had been formed for the purpose of managing the common parts of the estate. Third, that there was an intention in the plaintiff that shares in the management company should be issued to the purchasers of flats in the estate.

The operative part of the agreement, so far as material, reads as follows. Clause 1 is the critical clause in this case:

> 1(a) This agreement is conditional on:
>
> B i. The grant of a planning permission satisfactory to the vendor to permit the redevelopment of the estate in accordance with the plans prepared on behalf of the vendor and so as to enable the vendor to comply with its obligations herein.
>
> ii. The vendor obtaining all other permission, consents and approvals required for such redevelopment of the estate.
>
> The vendor shall use all reasonable endeavours to satisfy such conditions, but if they have not been satisfied on or before 30th June 1988, either party may at any time thereafter serve written notice on the other rescinding this agreement with effect from service of such notice, whereupon National Condition 10(2) shall apply.

Clause 2 contains the main obligations to grant the lease which is in issue. It reads in subpara (a):

> Subject to the terms of clause 1, the vendor shall grant as beneficial owner and C the purchaser shall accept a lease ("the lease") of the property ("the Property") known as flat 2A Millers Wharf, St Katharine's Way, London E1, the approximate position of which is shown by red edging on plan no 2 attached to the lease.

Then there was the usual provision for the lease to be in an attached draft lease form. Clause 3 contained the consideration, a premium to be paid by the purchaser for the grant of the lease of £465,000, and it provided that a deposit of 10% of that sum, in addition to any preliminary deposit paid to the vendor's agents, should be paid to the vendor or the vendor's solicitors as agents for the vendor on or prior to the exchange of the agreement.

There was then a proviso which allowed a variation in the amount D of the premium if the gross floor area of the property after completion of its construction varied by more than 5% from the area shown red on the plan mentioned above.

Clause 4 contained the vendor's obligations in relation to its building operations. It reads as follows:

> The vendor will, at its own expense, complete the construction of the property in accordance with the provisions of the schedule hereto.

The schedule contained three paragraphs. (a) reads:

> The property will be constructed:
>
> i in a good and workmanlike manner;
>
> ii in accordance with the provisions of the planning permission and other consents to be obtained by the vendor pursuant to clause 1 and/or relevant building regulations;
>
> E iii in accordance with the brief specification attached hereto.

The brief specification had obviously been provided by a third party, the Philip Pank Partnership, and it gave details of the standard to which the redevelopment work was to be conducted. It was prefaced by two paragraphs under the heading "Generally" which read as follows:

> All materials and workmanship will conform with the relevant requirements of the building regulations, London Building Acts and other by-laws, rules and regulations issued by the local authority;
>
> All internal and external structural works, including beams, columns, external walls, internal walls and partitions, fireproofing, means of escape, drainage and all services and the like will be carried out under the supervision of the district surveyor, environmental health officer, drainage inspectorate F and the fire officer;
>
> All works will comply with the requirements of the planning department and permissions received from the local council.

Going back to the schedule which was annexed to the agreement, para (b) reserved to the plaintiff the right to make variations. It reads:

> The vendor shall be entitled without obtaining the approval of the purchaser and without any right of compensation or rescission accruing to the purchaser to make such variations involving departure from the specifications and plans for the property as shall be required by the vendor or the planning authority provided that:
>
> i the quality of the fixtures and fittings to be supplied by the vendor are of no less quality; and
>
> ii the location of the property within the building of which it forms part remains substantially similar.

The third paragraph need not be read: it dealt with the car park G access and the pathways. The only other provision that needs to be read is the provision for completion, which by clause 5(a) was to take place 14 days after the purchaser or his solicitors were notified in writing by the vendor or its solicitors that the property had been completed pursuant to clause 4 thereof and vacant possession of the property should be given to the purchaser on completion of the grant of the lease.

On January 6 1988, plans for building regulation approval were deposited. In February 1988 the building work started. On March 9 1988 the London Borough of Tower Hamlets, which was the local authority to whom the plans for building regulation passing had been submitted, rejected them. They wrote a letter dated in March 1988 H which, so far as material, reads as follows:

> Notice of Rejection of Plans. Premises, Millers Wharf, St Katharine's Way, E1. Date of deposit of plans, 6th January 1988

— then there is the number of the plans and the numbers of the applicant's plans were also attached —

> Proposed Building Work or Material Change of Use: conversion of warehouse into flats with additional sixth storey.
>
> The plans deposited have been rejected under s16 of the Building Act 1984 for the following reasons:
>
> The plans are defective in that they do not contain sufficient information to determine whether the proposed work will comply with the requirements of regulation 4(a), (b), (c), (f), (h), (j) and (k) as in attached list.
>
> The plans show that the proposed work would contravene regulation 4(b) as specified in the attached list. J

The attached list was a very lengthy one, but it was common ground that the real thrust of that rejection was concerned with the fire precaution work which, the local authority took the view, was defective from the point of view of the building regulations. In particular, in the list of contraventions under the heading "Flat 2A" the following was found:

> If the front door is located at the head of the stairs, then the travel distance to the common stair is greater than 15 metres and not acceptable; or, if the front door is at the foot of the stair, then travel distance from the bedroom is greater than 7.5 metres. An alternative escape route is required.

And there were other requirements and shortcomings in relation to the plans specified by the local authority. K

On March 31 1988, there was a formal grant of the planning permission which was foreshadowed in the resolution that I read earlier passed by the London Docklands Development Corporation on November 10 of the preceding year. This grant reads:

> The London Docklands Development Corporation as the local planning authority within the meaning of the above-mentioned Acts and Order in the pursuance of its powers thereunder hereby permits the following development.
>
> Location: Millers Wharf, St Katharine's Way, E1.
>
> Description: Conversion and extension of existing warehouse to form 22 residential units involving partial demolition of a listed building and alterations to the listed Alderman's Stairs.

The permission was stated to be granted subject to specified L conditions which were in fact the same conditions as the one that had been mentioned in the resolution of November 10 of the previous year. In particular, it included condition number three, which I have already read once and need not read again.

At the same time, listed building consent was granted in very similar terms. It is not necessary to read that in any detail. All that needs to be mentioned is that it was made subject to conditions which repeated some but not all of the conditions attached to the planning permission. It did, in particular, repeat condition number three regarding the submission of full details, including colour and sample of facing materials, which I have read and do not repeat.

Third and last, there was made on the same day a section 52 M agreement between the plaintiff (described as "the Developer") of the first part, the London Docklands Development Corporation (which is called "the Corporation") of the second part and the London Borough of Tower Hamlets (called "the Council") of the third part. It recited that the developer owned the warehouse in fee simple and had the benefit of a lease of the jetty, that the corporation was the local planning authority and that the council was the highway authority. It then contained a series of definitions of which I need read only some. The site was defined as "the land shown edged red and edged green on the plan". The delineation and colouring of the various pieces of land are as I described them at the beginning of this judgment by reference to the map, which is in fact annexed to this

A section 52 agreement. The walkway was defined as "the walkway to be provided as part of the development on part of the site which is shown edged blue on the plan". The jetty was "the jetty edged green on the plan". The stairs were "Alderman's Stairs edged brown on the plan". The works were "the construction of the walkway and the works to the jetty" as thereinafter described. The causeway was "the causeway which is edged yellow on the plan".

The operative part of the agreement need be read only so far as clause 3 thereof is concerned. That contained this covenant by the developer (which is, of course, the plaintiff):

> The developer hereby covenant and undertake with the Corporation separately with the Council so as to bind the site as provided for in the beforementioned statutory provisions that upon implementation of the
> B development the developer will:
>
> (a) take all necessary steps and diligently lay out, construct and otherwise complete in a proper and workmanlike manner all necessary works for the provision to the reasonable satisfaction of the Corporation and the Council of the walkway across the site and so that the expression "all necessary steps" shall include *inter alia* application for all consents, approvals and licences required by statute or otherwise.
>
> (b) take all necessary steps and diligently lay out, construct and otherwise complete in a proper and workmanlike manner all necessary works for the alterations to the stairs so as to provide access from the walkway to the stairs to the reasonable satisfaction of the Corporation and the Council.

And the same definition of "all necessary steps" was added.

> (c) take all necessary steps and otherwise complete in a proper and
> C workmanlike manner all necessary work to refurbish, resurface and light the jetty and to connect the jetty to the walkway to be provided on the adjoining development on the land to the west of the site to the reasonable satisfaction of the Corporation and the Council and to refurbish and resurface the causeway.

Again, the same definition of "all necessary steps" was added. Then (d) I need not read, but (e) reads:

> Not to commence works of demolition, construction, development or rebuilding on the site or any part thereof in pursuance of the development until the plans and specifications of the works have been approved by the Council and the Corporation, such approvals not to be unreasonably withheld or delayed.

On May 9 the plaintiff reacted to the rejection of the plans for building regulation purposes by appealing against that rejection. On
D May 13 the plaintiff took an alternative measure to surmount the difficulty posed by the rejection of the plans submitted for building regulation passing by sending in revised plans for building regulation approval. They involved a modification of the access to the flat that was proposed to be sold to the defendant, and that in fact had the effect of reducing by an extent, which was not insignificant, the square area of the flat. The revised plans thus submitted were treated by the authority, Tower Hamlets, as having been deposited three days later, on May 16.

That was the state of affairs which obtained when June 30 came and went, June 30 being the date by which the conditions mentioned in clause 1(a)(i) and (ii) of the agreement had to be satisfied. At that
E stage planning permission and listed building consent had been obtained, but there was attached to each of them the condition three, which in turn required a consent which had not yet been obtained, ie the one relating to facing materials. Second, the building regulation plans had not yet been passed. The first edition had been rejected and an appeal had been lodged against that rejection. The later plans, with modifications, had been submitted but had not yet been passed. Third, the approvals that were required pursuant to the section 52 agreement in clause 3(a), (b), (c) and (e) had not yet been obtained.

The events after June 30 which are relevant start with July 8, when the London Borough of Tower Hamlets passed the modified plans for building regulation approval. That was done by a letter of that date which said, *inter alia*:

> F The plan has been passed under s16 of the Building Act 1984 subject to (certain further conditions).

On July 26 the London Borough of Tower Hamlets approved the works pursuant to the section 52 agreement that were proposed. The letter dated that date to the plaintiff's representatives said:

> I confirm that the details of the parapet wall as shown on your drawing are acceptable.

On August 9 the London Docklands Development Corporation similarly gave an approval pursuant to the section 52 agreement by a letter of that date to the plaintiff's agents, saying:

> Thank you for your letter of 4th August and accompanying drawings detailing the lighting proposals for the jetty and walkway at Millers Wharf. I can advise you that the proposals as set out (on certain drawings which are identified) are acceptable. This approval is given in pursuance to clause 3(f)(i) G
> of the s 52 agreement and does not negate the obligation of the developer to secure the various detailed approvals required by the conditions of the planning permission and listed building consent for this development.

On September 5 the plaintiffs were persuaded that it was necessary for them to apply for planning permission in respect of the revisions that had been brought about to comply with the original rejection of the plans submitted under the building regulations. In the event, nothing turns on that particular application, but it was in fact ultimately granted but later. The works involved were described as "repositioning of internal fire escape and elevation or alterations in the course of construction involving alterations to a listed building".

It was only on October 26 1988 that the defendant, through its solicitor, exercised what it claimed to be its right to rescind. The letter H
from the defendant's solicitors to the solicitors then acting for the plaintiff included the following:

> Under the terms of the agreement of the 7th October relating to the flat and to the car parking space, the same is conditional upon the grant of planning permission satisfactory to your clients and to your clients obtaining all other permissions, consents and approvals required. The contract thereafter provides that if these conditions have not been satisfied on or before the 30th June 1988 either party may at any time thereafter serve written notice on the other rescinding the contract.
>
> Your clients have not obtained all other permissions, consents and approvals required for the redevelopment in the terms of clause 1(a)(ii) of each contract. Our client is therefore exercising his right to rescind both agreements.

And there is a formal rescission of the two agreements which were J
involved. I have only so far referred to the agreement relating to the flat. There was a separate but similar one relating to the car parking space on which there are no separate considerations that were argued before me.

Finally, the letter asks for a return of the deposit, which is what condition 10(2) of the National Conditions of Sale, 20th ed, which was incorporated in the contract, provides for if the right to rescind is properly exercised.

After the right of rescission certain events occurred. The London Docklands Development Corporation gave planning permission for the fire escape provisions. An appeal against the rejection of the plans for building regulation purposes was successful on November K
17 1988. On October 3 1989 the London Docklands Development Corporation gave written consent to the works that were required by condition three attached to the planning permission and to the listed building consent.

There were three grounds in the end advanced and insisted on on behalf of the defendant in support of its claim to have successfully rescinded. The first was that planning permission and listed building consent was given only conditionally on March 31 1988 and that one of those conditions, number three, itself required an approval to be given. Either, it was submitted, this was an element in the requirement of condition 1(a)(i) as a planning permission or it was within the provisions of clause 1(a)(ii) as another consent required for redevelopment of the estate in accordance with the plans L
prepared on behalf of the vendor and so as to enable the vendor to comply with its obligations therein.

Second, building regulation plans were not passed until after June 30 1988. Third, the consents required by clause 3(a), (b), (c) and (e) of the section 52 agreement were not obtained until after June 30 1988.

As regards the first of those three grounds for rescission, since the consents were not given until after the notice to rescind was given, the only way in which the plaintiff can rebut this defence is by showing that, as a matter of construction, those consents are not within the provisions of clause 1(a)(i) or (ii).

So far as the consents attached to the section 52 agreement are concerned, it was accepted by the plaintiff (and I see no reason to doubt that the concession was a perfectly proper one) that they were M
indeed "other consents" within the meaning of clause 1(a)(ii) and the rebuttal to the defence on this score was therefore limited to the proposition that the conditions became satisfied before the rescission notice, which was therefore too late.

This latter point involves the proposition that where there is a condition which is not a promisory condition but which is precedent to the obligation of performance of a contract to grant a lease and is required to be performed by a fixed date, a right of rescission expressed to be exercisable at any time after that fixed date will no longer be exercisable if the condition is satisfied after the fixed date but before the right to rescind is exercised. It is submitted that the failure of the condition before the fixed date only rendered the

A contract voidable and that once the condition was satisfied it had ceased to be voidable and became not only unconditional but enforceable.

The researches of counsel on each side failed to discover any authority in English law which either supports or rejects that proposition. I was referred at the outset to the well-known Privy Council authority of *Aberfoyle Plantations Ltd* v *Cheng* [1960] AC 115, where the Privy Council set out the general principles in the opinion of the board which was delivered by Lord Jenkins. Those principles are quoted in the headnote, but I take them from the opinion at p 124, where Lord Jenkins said:

> But, subject to this over-riding consideration, their Lordships would adopt, as warranted by authority and manifestly reasonable in themselves, the
> B following general principles: (i) Where a conditional contract of sale fixes a date for the completion of the sale, then the condition must be fulfilled by that date; (ii) where a conditional contract of sale fixes no date for completion of the sale, then the condition must be fulfilled within a reasonable time; (iii) where a conditional contract of sale fixes (whether specifically or by reference to the date fixed for completion) the date by which the condition is to be fulfilled, then the date so fixed must be strictly adhered to, and the time allowed is not to be extended by reference to equitable principles.

And reference was made, notably, to *Re Sandwell Park Colliery Co Ltd* [1929] 1 Ch 277, a decision of Maugham J.

The *Aberfoyle* decision does not, in my judgment, govern any issue disputed in these proceedings. First of all, it is accepted on behalf of the plaintiff that time was of the essence in relation to the
C coming into operation of the right to rescind immediately after June 30 1988 and there is therefore no question in relation to that aspect of the matter. Second, the question whether what Lord Jenkins said regarding the foundation for the general principle was directed to there not being any contractual bonds at all between the parties or, on the other hand, to there being a contractual relationship but not one of sale and purchase, does not arise for decision in the present case. That makes it unnecessary for me to decide whether to follow the detailed analysis of Ungoed-Thomas J in *Property & Bloodstock Ltd* v *Emerton* [1967] 2 All ER 839, which was approved by Sachs LJ in the Court of Appeal [1968] Ch 94 at p 191, or, on the other hand, the doubts expressed by Danckwerts LJ in that latter court at p 116, with whose judgment Sellers LJ agreed.

D Whether the former or the latter provides the correct analysis of the jurisprudential basis for what the Privy Council held to be the general principles applicable in relation to conditional contracts is not a matter that arises for decision here. The reason for that is that it is common ground between the parties that in this case the plaintiff and the defendant were contractually bound to each other before June 30 1988, the date limited for the satisfaction of the conditions. That must clearly be right since mutual obligations were imposed by the agreement of October 7 1987. The defendant was obliged to pay and did, I understand, pay the deposit and, perhaps more important, the plaintiff was placed under a specific obligation to use all reasonable endeavours to satisfy the conditions and there is no doubt
E but that that had contractual force.

On the question of principle mentioned earlier, there are certain matters that are agreed between the parties. First of all, it is common ground that the contract became voidable and not automatically void by the non-satisfaction of the conditions on or before June 30 1988. Second, as I have already mentioned, it is accepted that time was of the essence as regards the arising of the right to rescind on June 30 1988 and that a satisfaction, even the next day, would have been inadequate if a notice to rescind had been served before the condition was satisfied. Third, the defendant accepted that on the happening of the condition — and I am now looking at the section 52 condition where approval was given by the two authorities on July 26 and August 9 respectively — the contract did become unconditional. But
F the fourth point which is in issue is whether there remained a right to rescind after the happening of those conditions connected with the section 52 agreement.

It should also be observed that there is no relevant claim that there was a waiver or affirmation of the contract after knowledge of the right to rescind was acquired by the defendant. Indeed, the evidence was that, so far as the section 52 agreement was concerned and the consents given pursuant to it, the defendant had no knowledge of such matters until discovery in these proceedings took place.

The question therefore is one of principle, but it was submitted on behalf of the plaintiff that although English authority is lacking there is authority in cases decided in the High Court of Australia. The first of these was *Suttor* v *Gundowda Proprietary Ltd* (1950) 81 CLR 418. That was a case concerned with a contract for sale by Suttor, the G
appellant, to Gundowda Proprietary Ltd, the plaintiff, of certain pastoral, that is to say, non-arable, land. The particular clause in issue that contained the condition that was in issue read (it is set out at p 421):

> In the event of the consent of the Treasurer not being obtained within two months of the date hereof or within such further period as may be mutually agreed upon by the parties hereto, this contract shall be deemed to be cancelled, and upon the vendor returning to the purchaser any deposit paid herein neither party shall be under any further liability to the other for any sum for damages, cost or otherwise.

The date of the agreement was October 20 1947 and the treasurer who was referred to appears to be a government official who was required to give his consent to certain transactions under legislation H
then in force in New South Wales.

The relevant defence and the answer to it are at p 423 in the report. The defence was that the consent of the treasurer was not obtained before the expiration of a two-month period. The facts in relation to that are stated at p 423:

> With respect to the defence that the contract came to an end under its own terms, it appeared that the consent of the Delegate of the Treasurer to the contract of 20th October was obtained on 5th January 1948. The two-month period referred to in cl 12 of that contract expired on 20th December 1947. The defendant therefore submitted that the contract was to be deemed to be cancelled upon the expiry of that period, and in effect then came to an end. There were three answers put forward on behalf of the plaintiff to that submission. It was said that the time fixed in cl 12 was not of the essence of the J
> contract, and that although the consent was not obtained within two months that did not bring the clause into operation at once, but still allowed a reasonable time to obtain the consent; further that there was an agreement between the parties within the terms of cl 12 itself to extend the time for a reasonable time beyond 20th December; and further that after 20th December the defendant by a clear recognition of the continued existence of the contract forbore to assert the provision of cl 12 against the plaintiff and waived the cancellation which that clause might have brought about.

The evidence on this aspect of the matter is stated at p 425:

> On 27th December 1947, Mr McManamey with a Mr Allworth went to Gundowda at the defendant's invitation. Allworth was interested in the plaintiff company and its finances. He was an accountant with some knowledge of pastoral accounts and was, apparently, the accountant to the K
> company or person who was providing the finance for the transaction. His evidence was taken before the suit commenced as he then had but a short time to live, and he died some little time after the evidence was taken. He and McManamey stayed at Gundowda until 31st December 1947, and there were a number of conflicts between them and the defendant as to what occurred during their stay. Their evidence showed that they and the defendant all treated the contract as still being on foot, and that the defendant had been informed that consent would be obtainable within a very short time, and that he led them to believe that the contract would be completed by him.

Then there is a statement of what the various matters that were the subject of the negotiations were, which are not particularly important for the purposes of this decision.

The case was heard at first instance by Roper CJ in equity and he L
decided this on the facts:

> His Honour accepted the evidence of McManamey and Allworth on the matters relating to the delay in obtaining the consent of the Delegate to the Treasurer, the extension of time therefor . . .

and the other matters that I did not mention specifically but which were the subject of the negotiations between McManamey and Allworth and the vendor.

So far as the law was concerned, his Honour said this:

> If time were of the essence in cl 12 of the contract it appeared to his Honour that it was no longer of the essence in the light of that conduct by the parties and that any cancellation under cl 12 was in fact waived by the defendant.

And then he proceeded to deal with the fairly numerous other M
defences.

When the matter came to appeal from the Chief Justice's grant of specific performance of the contract, the argument that was addressed by the appellant was as follows — it is at p 428.

> There was not any waiver before 20th December 1947. All the conversations and actions are consistent with the fact that the contract was then on foot. After 20th December there was not any waiver. There cannot be a waiver of a condition in a contract which has legally come to an end. There was not any finding of estoppel. Everything was consistent with negotiations for a new contract or a variation of the old one which would in fact be the making of a new contract. The onus is upon the respondent to prove that there was an extension of the old contract.

That is sufficient as a background for a consideration of what was

A said by the High Court of Australia. It will be noted that the pleaded issues were, first, that time was not of the essence, second, that there was an agreement to extend time and, third, that after time had run out there was a recognition of the continued existence of the contract and a waiver.

The evidence which was accepted was that the defendant, with knowledge of the fact that the date for avoidance or rescission had passed, went on negotiating terms on the footing that the contract was still going to be completed by him and at first instance there was a specific holding that there was a waiver. It was against that holding that the argument in the High Court of Australia was directed, namely to the proposition that there was no such waiver.

B Against that background I turn to the judgment of the court, which consisted of Latham CJ and Williams and Fullagar JJ, and I take it up at p 440, where the judgment of the court says this:

> So far as cl 12 is concerned, there are, in our opinion, two answers to the contention of the appellant.
> In the first place McManamey was at Gundowda from 29th November to 8th December 1947 and he gave evidence of conversations between himself and the defendant which his Honour accepted and which are quite sufficient to prove an oral agreement prior to 20th December that the time for the Treasurer's consent should be extended for a reasonable period after that date.

Then the court decides that the Statute of Frauds, because not pleaded, is irrelevant. Going on a little later in the judgment:

> C In the second place, although cl 12 in terms provides for an automatic avoidance of the contract on the occurrence of a specified event, that is (even if no agreement for an extension of time were made) by no means the end of the matter. The effect of contractual provisions of this character was discussed and explained in *New Zealand Shipping Co Ltd* v *Société des Ateliers et Chantiers de France* . . . .

And his Lordship continues with a quotation from Lord Atkinson's judgment in that case, which I need not read. It is to the effect, stated in the judgment of the High Court of Australia, picking it up at p 441:

> But we are of opinion that the *New Zealand Shipping* case requires the same construction to be given to the contract in both classes of case. The provision in question is to be construed as making the contract not void but voidable. The question of who may avoid it depends on what happens. If one party has
> D by his default brought about the happening of the event, the other party alone has the option of avoiding the contract. If the event has happened without default on either side, then either party may avoid the contract. But neither need do so, and, if one party having a right to avoid it does not clearly exercise that right the other party may enforce the contract against him. This is, we think, the view of Lord Shaw and Lord Wrenbury in the *New Zealand Shipping* case, and it is consistent with what was said by Lord Finlay L-C.

They then turn to what actually happened at Gundowda between December 27 and 31 1947 and in relation to that they say:

> The construction of cl 12 which we have adopted is that the clause did not automatically cancel the contract of 20th October 1947 but only brought it to an end if after that date one of the parties exercised its option to cancel it. On 6th January 1948 the defendant's solicitor wrote to the plaintiff's solicitors
> E that cl 12 of the contract spoke for itself and this may have been intended to be a notice that the defendant considered the contract to be cancelled. But there was no clear statement that the contract was considered by the defendant as cancelled until 15th January 1948 when the defendant's solicitor wrote to the plaintiff's solicitor that "the consent of the Treasurer was not obtained within the period of two months from the date of the contract and therefore the contract is no longer effective after 20th December 1947". But this letter was obviously written on the view, with which we do not agree, that cl 12 effected an automatic cancellation of the contract when the Treasurer had not consented by 20th December 1947. His Honour accepted the evidence of McManamey and Allworth of the events that occurred at Gundowda between 27th and 31st December 1947, and it is clear from this evidence that the defendant was treating the contract as still on foot although he was asking for certain variations to which McManamey agreed provided the directors of the
> F plaintiff approved. Before the defendant's solicitor purported to cancel the contract the consent in writing of the Treasurer to the transfer had been obtained on 5th January, 1948, and the cancellation was therefore too late.

Treating the contract as still on foot, in my judgment, refers to the waiver argument, and I do not read it as a decision that where there is an express right to rescind and no question of waiver the happening of the condition upon which the right to rescind depends destroys of itself the right of rescission. Accordingly, I do not find an authority which in any event would not be technically binding upon me but would be treated by me with the greatest respect that determines the issue that I have to decide as a matter of principle.

The other Australian authority which is arguably relevant to this issue is *Perri* v *Coolangatta Investments (Proprietary) Ltd* (1982) 149 CLR 537. The headnote states the facts conveniently shortly: G

> A contract for the sale of land was made on 7 April 1978. It fixed no time for completion. A special condition provided that the contract was subject to the purchasers' completing the sale of their property at Lilli Pilli. On 17 July 1978 the vendor gave a notice requiring the purchasers to complete by 8 August 1978. They did not do so, and on 10 August 1978 the vendor gave them notice rescinding the contract. On 29th September 1978 he sued them claiming a declaration that the contract had terminated on or about 10 August 1978. The purchasers did not complete the sale of the Lilli Pilli property until 13 June 1979. They cross-claimed for specific performance of the contract.
> Held (1) that the special condition made the completion of the sale of the purchaser's property a condition upon which their obligation to complete the purchase of the vendor's property depended . . . . (2) That the special condition obliged the purchasers to complete the sale of their property within a reasonable time. H

and *Aberfoyle* was applied for that purpose.

> (3) by Gibbs CJ and Stephen and Brennan JJ, that a reasonable time for the fulfilment of the special condition had expired by September 1978, and it had then been open to the vendor to avoid the contract without first giving any notice limiting a reasonable time for completion. It had done so by commencing proceedings.

Wilson J held that reasonable time had expired on July 17 1978 and therefore the contract was terminated on August 10 and, at the other extreme, Mason J took the opposite view, that a reasonable time had not expired by July 17 1978 and that the notice of August 10 did not rescind the contract, and accordingly he dissented from the judgment of the remainder of the court, that the contract had been avoided J
before the purchaser's claim was instituted for specific performance.

The headnote also contains a proposition attributed to Gibbs CJ and Stephen, Wilson and Brennan JJ. The proposition that is set out under their names reads:

> When the time has elapsed for performance of a condition which is not a promissory condition, but a condition precedent to the obligation to complete a contract of sale, either party, if not in default, can elect to treat the contract as at an end if the condition has not been fulfilled or waived. Notice does not have to be given calling on the party in default to complete the contract or fulfil the condition.

Gibbs CJ, at p 541, said:

> In my opinion special condition 6 made the completion of the sale of the property at Lilli Pilli a condition precedent to the performance of certain of K
> the obligations of the parties under the contract, including the obligation of the respondent to complete the sale.

He then makes observations about conditions precedent and the necessity for a point of time to which they are either precedent or, if they are subsequent, subsequent and continues:

> However, provided the effect of a condition is clearly understood, its classification may be merely a matter of words. The condition in the present case was not a condition precedent to the formation of a binding contract. It is clear that a binding contract came into existence immediately upon signature, and that the parties to it were from that moment subject to certain obligations. For example, the appellants became liable to pay a deposit. Further, there was implied a promise by the appellants that they would do all that was reasonable to find a buyer for the Lilli Pilli property and to complete a sale to him. L

Pausing there for a moment, those are obviously points of similarity with the present case, where it is common ground that there was a binding contract and it is equally common ground that the plaintiff was under an obligation to obtain the consents.

There then follows, in Gibbs CJ's judgment, a passage in which he discusses the question whether in the *Aberfoyle* decision there was thought to be a condition which prevented the formation of a binding contract at all. I need not go into that aspect of the matter. It will suffice for me to state the Chief Justice's conclusion, where he said at p 543:

> Nevertheless, it probably does not matter in the present case whether the condition is described as "precedent" or "subsequent" provided that it is M
> understood that its non-fulfilment did not prevent a binding contract from coming into existence but did have the effect that the respondent was under no obligation to complete the sale unless the condition was fulfilled or waived.

He then stated the question that he saw as being the one for decision.

> The question then arises whether the respondent was entitled to terminate the contract without first giving to the appellants a notice requiring them to fulfil the condition, or to perform the contract, within a reasonable time.

He then refers to *Aberfoyle* as authority that no notice is needed when the contract expressly states the time by which the condition is to be fulfilled, and discusses the question whether and to what extent that should apply in a case such as the one before him where there was

A no time fixed for the satisfaction of the condition.

After that discussion, he refers to an earlier case, to which I too was referred, *Gange* v *Sullivan* (1966) 116 CLR 418, which I have found of no assistance in deciding this particular problem. Gibbs CJ quoted from the judgments of some of the judges in that case. He then said this, which is the first of the passages particularly relied upon by the plaintiff:

> The authority principally relied on by the Court

meaning in that case the court in *Gange* v *Sullivan*

> was *Suttor* v *Gundowda Propriety Ltd*, a case of condition subsequent. It was not suggested in either of those cases that it was necessary to give any notice to complete The non-fulfilment of the condition gave the party not in default a
> B right to avoid the contract, but if that party did not exercise the right the other party might enforce the contract against him. In *Suttor* v *Gundowda Propriety Ltd* the attempt to cancel the contract was made too late, since the condition had been fulfilled in the meantime. It appears to have been assumed that no notice to complete need be given before the right to avoid the contract, was exercised. In both cases however a time for fulfilment of the condition was expressly fixed by the contract.

It is first to be observed that what Gibbs CJ was looking at was the question of necessity for a notice. Second, it is to be observed that in describing the effect of *Suttor* v *Gundowda* in terms that the condition had been fulfilled in the meantime and therefore the attempt to cancel the contract was made too late, it is entirely explicable on the footing that there was either, on the one hand, a
C waiver, or, on the other, an extension of the time for the satisfaction of the condition, which were the two grounds that were pleaded and, in my judgment, specifically upheld in the judgment of the High Court of Australia in the *Suttor* case.

His lordship in the *Perri* v *Coolangatta* case continues:

> *Suttor* v *Gundowda Propriety Ltd* and *Gange* v *Sullivan* are consistent with *Aberfoyle Plantations Ltd* v *Cheng* and support the view that where a conditional contract fixes the date by which the condition is to be fulfilled the contract may be terminated if the condition has not been fulfilled when that date arrives, and that it is unnecessary to give any prior notice to the other party. The question is whether a similar conclusion should be reached when no time for fulfilment of the condition is fixed by the contract, and it accordingly must be performed within a reasonable time. There is little
> D authority on the question, and it must be approached from the standpoint of principle.

There too, in my judgment, it is clear that what the learned Chief Justice was primarily concerned with was the necessity or lack of necessity for a notice calling for completion. Second, the reference to "when the date arrives" is clearly, as a matter of English, a reference to the date when the condition had to be fulfilled. His lordship was not in any way concerned with the effect of the satisfaction of the condition after the specific date that was mentioned in the contract.

Finally, at p 546, there is this passage when the Chief Justice has come to the conclusion that a notice is not necessary where there is no time specifically fixed for the satisfaction of the condition any more
E than there is when time is fixed. He said:

> For these reasons I consider that when the time has elapsed for performance of a condition which is not a promissory condition, but a condition precedent to the obligation to complete a contract of sale, either party, if not in default, can elect to treat the contract as at an end if the condition has not been fulfilled or waived, and that it is not necessary first to give a notice calling on the party in default to complete the contract or fulfil the condition.

That, of course, is the passage from which the quotation in the headnote is taken. The reference to non-fulfilment as well as waiver does not, in my judgment, amount to a statement of principle that satisfaction of a condition after a fixed date for its occurrence has passed of itself destroys an express right of rescission. The point was not in terms argued either in the *Suttor* case or in *Perri* v *Coolangatta*.
F What Gibbs CJ said (and other members of the court agreed with) was in any event clearly *obiter* in the *Perri* v *Coolangatta* case because there was no fixed time in that case. For all those reasons, it does seem to me that the plaintiff does not find such a clear statement of principle in these Australian cases as I think would be requisite to shed light on a matter unaffected by any English authority. Accordingly, I have to deal with the case on general principles rather than by reference to specific authority as a guide.

The argument for the defendant was primarily based on what was claimed to be a close analogy between the case with which I am concerned and the situation that arises where a party to a contract is in breach of an obligation going to the root of the contract to do an act by a certain time, whereupon the other party, the innocent party, has an election whether to treat the contract as at an end because of that G
repudiatory breach. The latter right is not lost on the authorities if before the repudiation is accepted the act which was not done in time is done, albeit late. I was referred to *Mardorf Peach & Co Ltd* v *Attica Sea Carriers Corporation of Liberia* [1977] AC 850. I need not refer to that authority because the proposition as such was not disputed. What was disputed was that the analogy was an exact one, or one which I should treat as binding or useful. I accept the plaintiff's argument on this score: that the analogy is less than perfect because the court does undoubtedly deal differently with cases where there has been a breach of contract on the one hand and cases where a condition which is not dependent on fault on either side is not fulfilled on the other.

I do not find that the analogy is sufficiently close, although H
undoubtedly there is considerable similarity, for me to treat it as a guide without further consideration of the principles involved. It seems to me that the most satisfactory way of resolving the puzzle is, first of all, to look at what the condition itself provides. I need not repeat clause 1(a)(i) and (ii) which sets out what the conditions are, but the remainder of the clause reads as follows:

> The vendor shall use all reasonable endeavours to satisfy such conditions but if they have not been satisfied on or before the 30th June 1988 either party may at any time thereafter serve written notice on the other rescinding this agreement.

And then the consequences are set out.

If the right of rescission is to be lost on the occurrence of the J
condition after June 30 1988, it seems to me that it can be on only one of two alternative arguments. The first is that there is a repugnance between a right of rescission and an unconditional contract so that the two are incapable of co-existing. The second is that it would be right, as a matter of construction, by implication, to qualify the right to rescind at any time thereafter by making it one to rescind at any time thereafter but before the condition is satisfied. There are many ways in which the necessary implication could, as a matter of grammar, be inserted into the sentence. The latter way of advancing the case was not, in fact, argued on behalf of the plaintiff and, in my judgment, entirely justifiably.

The first argument can perhaps usefully be tested by looking at K
what the position would be if, immediately after the words "on or before the 13th June 1988" there had been included the words "but before the 27th October 1988". In such circumstances, it would of course have been plainly unarguable that one could imply an earlier date. But that would only exclude that aspect of the matter. It would not necessarily exclude an argument based on a repugnance between a right of rescission on the one hand and the existence of an unconditional contract for sale and purchase or grant and receipt of a lease on the other. But, in my judgment, there is no such repugnance between those two. The concept of a vested but defeasible right is one which is quite common in property law. Examples which come to mind are class gifts, which are capable of expanding after members of
the class have obtained a vested interest, and perhaps more common L
interests which are vested under a trust subject to being defeated by the exercise of trustees' overriding powers of appointment.

The two conceptions of, on the one hand, a right of rescission so as to avoid a contract and on the other hand a contract which has become unconditional due to the happening of an event satisfying a condition precedent are quite separate as a matter of law. I am satisfied that they can perfectly well co-exist at one and the same time and upon that basis I do not see that there is any necessity for saying that this clause which says "at any time thereafter" should not be treated as meaning what, in terms, it says.

I have, of course, not dealt with the case which might well be
common if there was a significant amount of delay after the right to M
rescind arose or something that amounted either to waiver by, or an estoppel against, the party with the right to rescind. An affirmation of the contract as, in my judgment, happened in the *Suttor* case would be another example. None of that is in issue in the present case and it is only in relation to the mere happening of the condition with which I am concerned. On that score, I see no reason for not giving the provisions of the clause their natural meaning and that, in fact, is fatal to the plaintiff's case.

The other two grounds upon which the defendant relied were very fully and carefully argued before me, but in the circumstances it does not seem to me to be necessary or indeed useful for me to express my views upon them. I propose, subject to counsel's submissions as to

A the form of order, to dismiss the action and to give judgment for the defendant on the counterclaim for the refund of the deposit, but I will hear counsel on the form of order.

The plaintiff's action was dismissed and the defendant's B *counterclaim for the return of the deposit of £46,500 with interest was allowed; the plaintiff to pay costs on the standard basis.*

LANDS TRIBUNAL
CAPITAL TRANSFER TAX

A BAIRD'S EXORS v COMMISSIONERS OF INLAND REVENUE

LANDS TRIBUNAL FOR SCOTLAND
(LTS/CTT/1988/1)
June 14 1990
(Lord ELLIOTT (President), WM HALL FRICS and A R MACLEARY FRICS)

Estates Gazette March 9 1991 and March 16 1991
[1991] 09 EG 129 and [1991] 10 EG 153

B **Renunciation of agricultural tenancy — Mainly arable farm at Montrose, Scotland — Whether renunciation of tenancy a chargeable transfer for capital transfer tax purposes in terms of s20(2) of Finance Act 1975 — Valuation of tenancy interest transferred — Whether ratio in *Inland Revenue Commissioners* v *Crossman* applied to valuation of interest other than company shares — Whether transferor's interest in tenancy a whole or a half share — Appeal by transferor's executors against assessment to capital transfer tax at £138,000**

Tenancy held since 1935 on implied reletting ("tacit C relocation") under original lease granted in 1921 to transferor and his father — Original lease expressly excluded assignment without landlord's consent — In 1936 tenancy interest made over to transferor by his father and in 1959 landlord accepted transferor's son as joint tenant — Following death of his son, transferor in 1977, with landlord's agreement, renounced tenancy in favour of daughter-in-law and grandson

Valuation by district valuer based on open-market value in terms of s38(2) of 1975 Act, assuming hypothetical sale of tenancy interest not prevented by prohibition against assignment — Such assumption followed House of Lords' D decision in *Crossman* holding that restriction of transfer of company shares to be disregarded for purposes of hypothetical sale but that resultant valuation nevertheless to take account of some restriction continuing in purchasers' hands — So far as DV aware, tenants' interests in farms in Scotland were never sold, so that there was no available evidence of open-market sales — DV therefore had looked at nearest secondary evidence in four different circumstances: (1) 10 "golden handshake" cases, where landlords had paid tenants to give up their tenancies, (2) 11 sale-and-leaseback transactions, where farmers had sold to financial institutions below full vacant possession value and retained their tenancy, E (3) eight compulsory acquisition settlements, and (4) two open market lettings in the area — DV in no doubt that in hypothetical situation agricultural tenancies had substantial value, varying up to 50% of vacant possession value — Taking "robust view", DV valued at £138,000, being £125,000 for tenant's interest (25% of vacant possession value) and £13,000 for value of tenant's improvements

In taxpayers' valuer's opinion, value nil — It was almost invariable practice in Scottish agricultural leases to incorporate a prohibition against assignees and subtenants — Thus, in his view, non-assignable leases were in themselves of F no value, as tenants unable to sell in open market — Detailed criticisms of DV's approach and of supporting evidence — No alternative valuation evidence proffered

Taxpayers' main argument was that s20(2) of 1975 Act introduced entirely new concept of valuation related, for first time in taxation law, to diminution in value of a transferor's interest — While conceding that s38(1) introduced open-market criterion, taxpayers contended s20(2) remained ruling valuation provision — As tenancy interest transferred was unmarketable, owing to prohibition against assignment, value of transferor's estate not diminished G

Inland Revenue contended that what was deemed to be sold in hypothetical open market was right to be tenant under relevant lease — S38 did not therefore innovate in any way on similar test in s7(5) of now-repealed Finance Act 1894, hence decision in *Crossman* applicable not only to 1894 Act but also to 1975 Act — Renunciation of tenancy gave transferees right to make large profits from farm — Tenant's interest therefore a valuable asset and as such capable of being valued

In tribunal's opinion, s38 alone directed how a chargeable asset is actually to be valued, namely at price which property might reasonably be expected to fetch if sold in open market at time of transfer — S20(2) indicates only what is chargeable H transfer and not how to calculate any diminution in value — *Crossman* and subsequent case of *Lynall* v *IRC* (also concerning share transfer) read by tribunal as being of general principle in all taxation cases where open-market criterion introduced for valuation purposes — In absence of any alternative valuation proffered by taxpayers, tribunal accepted DV's valuation but reserved opinion as to whether, in future case, there might not be another valuation approach

Extent of transferor's interest at date of renunciation considered — Held that 1959 agreement, which provided for introduction into lease of transferor's son as joint tenant but J without specific words of survivorship, then ruled — Tribunal not persuaded that use of words "as joint tenant" could in themselves be taken to imply a *jus accrescendi* in survivor — Thus, transferor entitled to only one half share in tenancy at renunciation date — Value of relevant half share therefore determined at £69,000, being half Inland Revenue's determination

The following cases are referred to in this report.

Anderson v *Moray District Council* 1978 SLT 37 K
Ashcroft, In re [1927] 1 Ch 311
Attorney-General for Ireland v *Jameson* [1905] 2 IR 218
Bede Steamship Co Ltd, In re [1917] 1 Ch 123; 86 LJ Ch 65; 115 LT 580; 33 TLR 13, CA
Chandler v *Director of Public Prosecutions* [1964] AC 763; [1962] 3 WLR 694; [1962] 3 All ER 142, HL
Coats v *Logan* 1985 SLT 221
Corrie v *Macdermott* [1914] AC 1056
Director of Public Prosecutions v *Schildkamp* [1971] AC 1; [1970] 2 WLR 279; [1969] 3 All ER 1640, HL
IRC v *Crossman* [1937] AC 26
IRC v *Duke of Westminster Trustees* [1936] AC 1
Lowe v *Inland Revenue Commissioners* 1985 SLT 12

Lynall v *Inland Revenue Commissioners* [1972] AC 680; [1971] 3 WLR 759; [1971] 3 All ER 914, HL
Magistrates of Banff v *Ruthin Castle Ltd* 1944 SC 36
Minister of Transport v *Pettitt* (1968) 20 P&CR 344; 67 LGR 449; [1969] RVR 26, CA
O'Brien v *Benson's Hosiery (Holdings) Ltd* [1980] AC 562; [1979] 3 WLR 572; [1979] 3 All ER 652; (1979) 53 TC 241, HL
Odeon Associated Theatres Ltd v *Glasgow Corporation* (1973) 27 P&CR 271; 1974 SLT 109
Reid's Trustees v *Macpherson* 1975 SLT 101
Robertson's Trustees v *Robertson* 1982 SLT 22
Salvesen's Trustees v *Inland Revenue Commissioners* 1930 SLT 387
Smith v *Grayton Estates Ltd* 1960 SC 349; 1961 SLT 38
Stewart v *James Keiller Co Ltd* 4 F 657
Wakerley v *St Edmundsbury Borough Council* (1977) 33 P&CR 497; 251 EG 921
Watson v *Secretary of State for Air* [1954] 1 WLR 1477; [1954] 3 All ER 582; (1954) 5 P&CR 13; 164 EG 524, CA
Young v *Gerard* 6 D 737

K Mure QC (instructed by Biggart Baillie & Gifford WS, of Edinburgh) appeared for the appellant executors; B Gill QC and J McNeill (instructed by the Solicitor of Inland Revenue) represented the respondents.

Giving their opinion, **THE TRIBUNAL** said: The appellants are the executors of the late George Alexander Baird ("the transferor"), who died in 1985, having on December 22 1977 renounced his tenancy in the farm of Mains of Dun, Montrose. This renunciation, to which his landlord agreed, was in favour of his daughter-in-law, Mrs Charlotte M Baird, and his grandson, George G Baird — his son having, on September 24 1977, been killed in a car accident.

By notice of determination the Inland Revenue have now determined that this renunciation constituted a chargeable transfer for the purposes of capital transfer tax in terms of sections 19(1), 20(2) and (5) and 38(1) of the Finance Act 1975. This they assess in the sum of £138,000. The appellant taxpayers deny that this renunciation involved a transfer for value or could have any open market value, as the tenancy interest then renounced, in terms of the lease, could not have been transferred without the landlord's consent. In any event, they dispute the Inland Revenue's valuation, contending that, at the relevant date, the transferor's interest was only in a half share of the tenancy.

Section 20(2) of the 1975 Act provides:

> Subject to subsections (3) and (4) below, a transfer for value is any disposition made by a person ("the transferor") as a result of which the value of his estate immediately after the disposition is less than it would be but for the disposition; and the amount by which it is less is the value transferred by the transfer.

The appellants' case is that there was no such diminution in value of the transferor's estate caused by the said renunciation. The Inland Revenue in contrast rely principally on the terms of section 38(1):

> Except as otherwise provided by this Part of this Act, the value at any time of any property shall for the purposes of Capital Transfer Tax be the price that the property might reasonably be expected to fetch if sold in the open market.

Similar provisions to the above have now been re-enacted in section 3 of the Inheritance Tax Act 1984 (formerly the Capital Transfer Tax Act 1984).

Statement of facts

Mains of Dun was originally leased by a lease dated July 8, 12 and 20 1921 between Mrs Alice Foote or Kennedy Erskine as trustee of the late Augustus Erskine on the one part and Thomas Baird and his son, George Alexander Baird, on the other part, in the following terms:

> To the said Thomas Baird and George Alexander Baird (who are hereinafter referred to as "the tenants" under which expression are also included their successor, or successors, in the tenancy of the said subjects) and to the survivor of them, and the heirs of the survivor, but excluding heirs-portioners, and declaring that the eldest heir-female shall always succeed without division, and also excluding all assignees whether legal or voluntary, and all subtenants . . .

Assignees were therefore expressly excluded.

By subsequent minute of agreement dated August 1 and 17 1936 between the said Mrs Alice Foote or Kennedy Erskine, the said Thomas Baird and George Alexander Baird, the said Thomas Baird, with the express concurrence and consent of the said Mrs Alice Foote or Kennedy Erskine, assigned, conveyed and made over to the said George Alexander Baird and to his heirs:

> my whole right and interest in and to the foregoing Lease with power to the said George Alexander Baird to occupy and possess the subjects thereby let in the same way and manner as if the said Lease had originally been granted in favour of the said George Alexander Baird alone.

By minute of agreement dated December 14 and 15 1959 relative to the original 1921 lease of Mains of Dun and also the relative minute of agreement dated August 1 and 17 1936, David Blythe Bogle WS, as *curator bonis** for the then proprietor, Augustus John Kennedy Erskine of Dun, of the one part, and the said George Alexander Baird and his son George Donald Baird both residing at the said farm of Mains of Dun, of the second and third parts, Bogle accepted for the proprietor earlier in the sentence at the request of and with the consent of the latter accepted the said George Donald Baird as joint tenant with the said George Alexander Baird. They also thereby bound themselves "all jointly and severally regularly to pay the rent and to implement all the obligations, terms and conditions of the said Lease".

After George Donald Baird's death on September 24 1977 and by typewritten letter dated December 9 1977, George Alexander Baird made the following request to the then proprietor, Mrs Millicent Lovett:

> I, George Alexander Baird hereby request that the tenancy of Mains of Dun Farm, Montrose, be transferred to the names of Mrs Charlotte Mary Baird, Mains of Dun Farm and her son George Grant Baird. I hereby relinquish the said tenancy, Yours faithfully, George Baird.

Underneath there was written in holograph:

> I agree to the above arrangement, Millicent Lovett, 22/12/77.

The said letter with its holograph† subscription was not an agreed document but was nevertheless spoken to in evidence by Mr Thomas Gibson Smith, chartered surveyor, Savills. At that time he managed the estate of Mrs Millicent Lovett of Bromley House, which included Mains of Dun, as well as two other farms, and of which the Baird family had been tenants for many years. While Mr Smith was forgetful of the precise circumstances under which this letter was prepared he nevertheless clearly recollected the approach made to him in 1977 by the Baird family requesting that the tenancy of Mains of Dun be transferred to Mrs Charlotte Baird and her son George Grant Baird, who was then in minority. This was shortly after the death of the latter's father, George Donald Baird, in the car accident. He confirmed that when he raised with Mrs Millicent Lovett the Baird family's request for the transfer he was not surprised that she readily agreed as, in his words: "she saw it as the natural progression, in the circumstances". No formal document, however, appears to have been executed and the transfer merely rested on the terms of this informal letter. This letter dated December 9 1977 is the document of transfer on which the present levy to capital transfer tax is based by the Inland Revenue, taking, however, the date of Mrs Lovett's agreement thereto on December 22 1977 as being the operative date of the transfer.

Following the transfer of the lease, Mains of Dun was subsequently farmed by Charlotte Baird and her son George Grant Baird in terms of the said arrangement. They also, for this purpose, entered into a contract of co-partnership with the transferor dated January 17 1978, but with date of entry backdated to September 24 1977, being the date of death of George Donald Baird.

As at the date of transfer, the lease of Mains of Dun was continuing on tacit relocation‡ at a rent of £6,400 per annum (representing £15 per acre for the better quality land).

It is a predominantly arable farm extending to 638 acres on either side of the Montrose basin. It has a substantial two-storey, detached dwelling-house with garden enjoying a good outlook in an elevated position. As at the date of valuation, there were eight cottages on the farm each comprising three rooms, kitchen and bathroom. Three of these were occupied by farm workers and a fourth by a retired employee. The steading was in course of being rebuilt. Much of it has now been demolished to make way for proposed new cattle courts and general purpose sheds. The steading comprised a general purpose shed; grain drier and storage bins; open hay shed; dutch barn and cart sheds. The value of the tenant's improvements to the farm has been agreed between parties at £13,000.

**Curator bonis*: The person appointed by the court to manage the affairs of a minor or a person under a disability.
†*Holograph*: A document in the handwriting of the grantor or signatory which is valid without being witnessed.
‡*Tacit relocation*: Implied reletting. It applies where the express term of a lease has expired without notice given to terminate it.

Valuation evidence

Taxpayers' valuer

Valuation evidence on behalf of the appellant taxpayers was given by Mr James M P Galbraith ARICS, of Cluttons, who holds a diploma in rural estate management from the Royal Agricultural College, Cirencester. For several years he has been involved in the sales, valuations and management of agricultural estates and is the partner responsible for the management of Leck Agricultural Estates in Scotland extending to over 23,000 acres.

He was asked to advise whether a non-assignable lease had any value in the hands of a tenant and, consequently, whether the value of the estate of the transferor was diminished as a result of the transfer of his tenancy interest in Mains of Dun in 1977. He said that, in his experience, a prudent landlord of agricultural land would choose his tenant carefully and would not offer for lease an assignable tenancy, as he would be primarily concerned with his farm being maintained and operated in such a way that its profitability was assured. It was the personal ability of an individual tenant to fund, run a successful business and maintain or improve the farm which really mattered. For example, on offering the farm of Balmachie, Carnoustie, to let on the open market in 1978 his firm selected the 28th highest offer at a rent of £70 per acre from among 142 offers which ranged from £20 per acre to £100 per acre. The criteria used in the selection of the successful candidate were proven ability, adequacy of working capital and other personal factors. Accordingly, to allow a tenant under his lease a right of free assignation would be poor, if not negligent, estate management. For these reasons, it was almost the invariable practice in all Scottish agricultural leases to incorporate a prohibition against assignees and subtenants.

In Mr Galbraith's opinion, non-assignable leases were therefore in themselves of no value, as the tenant would be unable to sell his interest in the open market. There was, for this very reason, no evidence of non-assignable leases having been offered for sale in the open market. The only value that could properly be said to have been transferred by the transferor's renunciation in 1977 was the value of the improvements which the tenant would have been able to claim by way of statutory compensation or in terms of the lease.

Mr Galbraith, who was conversant with the district valuer's approach to valuation, then proceeded to criticise the latter's reliance on certain "golden handshake" cases (see *infra*). These were cases where landlords had been willing to make a payment in order to extinguish a burden on their land; but he said this was different to a third party seeking to acquire a right to occupy a farm under the obligations of the lease. Under a "golden handshake" transaction, the landlord was stepping into the shoes of his tenant and the benefit to him was entirely different. He was then able to obtain the alternative benefits of "marriage value" by selling the farm at an enhanced value with vacant possession; or reletting it at current market value; or retaining the farm for his own occupation. In his experience, a landlord's willingness to acquire vacant possession of a let farm depended very much on his own personal circumstances and his view of the estate. The cost of reletting the farm could outweigh any increase in rental value and, in any case, landlords frequently preferred the family of the sitting tenant to continue. The landlord's personal considerations rather than open market forces thus applied and there was no general rule governing the size of any payment which might be agreed. In the "golden handshake" cases cited by the district valuer sums varying between 18.5% and 50.7% of vacant possession value had been paid to tenants indicating that there was really no common basis. In the present case, moreover, it was his understanding that the proprietor had absolutely no interest in repossessing the present farm and the rent and other terms of the lease remain unchanged.

Commenting on the Inland Revenue's "sale and leaseback" cases, Mr Galbraith said that in these cases owners of land sold off their interest in the freehold (often to an institution) but retained an agricultural tenancy, which could give them a capital sum with which to effect improvements. But they usually realised less under the sale than they might have done by simply disposing of the farm with vacant possession. The district valuer's comparisons again indicated a very wide range of discounts on vacant possession values from 25% to 46%. In such "sales and leasebacks" the ex-owner and prospective tenant were in equal negotiating positions with the prospective landlord to agree lease terms tailored to meet the tenant's own requirements. The tenant indeed had a choice of prospective landlords and market forces therefore applied. The prospective tenant could select his landlord; but if a lease were offered for sale no such choice would be available.

Commenting on the Inland Revenue's list of settlements of compensation on compulsory purchase, Mr Galbraith observed that a sitting tenant was entitled to compensation under the Land Compensation (Scotland) Act 1973 in respect of his unexpired term. There was, however, no generally accepted approach by valuers of simply taking 25% of the vacant possession value of the land as the district valuer had done in the present case. This was really only his own "rule of thumb" approach. In Mr Galbraith's view, this evidence of settlements was anyway irrelevant, as it was not open market evidence and the sums were all relatively small owing to the small acreages involved. In such cases it was always in the interest of both parties to reach agreement rather than risk an appeal. Taking 25% of vacant possession value also took no account of the terms of the particular leases or the rents passing. It merely provided a convenient method of settling awkward small claims for compensation. Mr Galbraith offered no opinion as to what value was appropriate.

The evidence of open market rents, however, showed that there might be an element of profit rent until the next rent review. Capitalisation of the difference between the existing rent passing and the revisionary rent at the next review, however, would show only a *de minimis* value if applied to this tenancy interest.

There was indeed no accepted general rule upon which the value of an interest in an agricultural tenancy might be calculated and this militated against the view that such an interest could have any value. Taking 25% of the vacant possession value was, however, misguided. Even if a potential purchaser for such an interest could be found, his payment would also have to be shared between the landlord and the tenant. Ultimately, in Mr Galbraith's view, the landlord's power of veto would command at least the major share; and he said this was yet a further complication in the way of attempting to value the tenancy interest on its own. In his opinion, its value was therefore nil.

District valuer

Valuation evidence on behalf of the Inland Revenue was given by Mr James G Menzies ARICS, who has been district valuer, Tayside District, since 1980. He comes from a farming family and has been engaged in the valuation of agricultural property in East and Central Scotland for the past 24 years. He said he had been instructed that the renunciation by the transferor on December 9 and 22 1977 was a chargeable transfer for the purposes of capital transfer tax with an effective date of December 22 1977.

Mr Menzies said that his valuation had to be based on the open market value of the tenancy interest which was renounced in terms of section 38(2) of the Finance Act 1975. He had also been instructed that, following the decision of the House of Lords in *Inland Revenue Commissioners* v *Crossman* [1937] AC 26, he should assume, for the purposes of assessing the open market value, that the hypothetical sale of the tenancy interest in the open market would not be prevented by the prohibition against assignation contained in the lease.

Mr Menzies said that he had valued the tenant's interest in Mains of Dun disregarding the landlord's power of veto and assuming a hypothetical sale on the open market but with the hypothetical purchaser taking the tenancy subject to the other existing lease terms. As, however, in reality tenants' interests in farms are never sold as far as he was aware, there is no evidence available of such sales. He was therefore confronted with the problem of how to assess the value of the present interest without the benefit of directly comparable market evidence. Valuers, however, were frequently called upon so to value property interests in such situations. They had then simply to look at the nearest secondary evidence available and also to rely on their own experience and judgment in arriving at an opinion of value.

The district valuer had therefore found evidence of value in four different circumstances. These were, first, "golden handshake" cases, where landlords had paid tenants to give up their tenancies. In such cases landlords have paid substantial sums of money to induce agricultural tenants to renounce their tenancies. In Mr Menzies' experience the market value in 1977 of a let farm protected under the Agricultural Holdings Acts, but not large enough to be of interest to a financial institution, was unlikely to be greater than 50% of its vacant possession value. It was of benefit to a landlord wishing to sell his let farm first to buy out his tenant's interest if this could be done for less than the difference between vacant possession value and the

A value of the landlord's existing interest subject to tenancy. The evidence available to him indicated that sums varying between 18.5% and 50% of vacant possession value had been paid to tenants in such situations. What each landlord was in effect doing was buying the tenant's interest in the tenancy. The sums paid therefore varied, depending on the circumstances of each landlord and tenant. The transactions, however, went some way to indicating the open market value of each tenant's interest. Ten farms were listed, six of which were situated in Angus. In all these instances the market for the tenant's interest was restricted to the actual landlord. In a hypothetical sale in the open market he thought that the situation might be different, as the actual landlord (as potential purchaser) might also have to compete with outsider prospective tenants seeking
B to step into the shoes of the tenant as seller. Because of that competition, the market would be assured and prices would, in his opinion, be greater, since they would not depend on a single prospective purchaser.

Mr Menzies' second examples were "sale and leaseback" transactions set out in a list containing 11 such transactions in the south of Scotland. He said that in the late 1970s and early 1980s many farmers had sold their farms to financial institutions in return for a price less than their full vacant possession value on condition, however, that they obtained a tenancy of the farm. These farmers had the option to sell at full vacant possession value but instead chose to accept a reduced price but with a secured tenancy. In effect, they
C were buying the tenancy of the farm for the difference between the vacant possession value and the restricted price paid by the investing institution, sacrificing between 25% and 43% of the vacant possession value. Included in his list were also four instances of newcomers who bought farms on the open market with vacant possession and immediately resold them subject to a leaseback arrangement. These farmers were thus prepared to go to the trouble and expense of buying a farm, reselling it subject to a tenancy from a financial institution at a high rent and under fairly onerous repairing conditions.

In Mr Menzies' opinion, in the hypothetical situation he was now required to consider, farmers would have eagerly grasped the chance of buying into the existing tenancy of Mains of Dun at its then modest
D rent and on favourable repairing conditions.

Mr Menzies then turned to the question of compensation payable to tenants on their dispossession following the compulsory acquisition of tenanted farmland. His list contained eight compensation settlements negotiated by himself, all at 25% of vacant possession value and involving relatively small acreages compulsorily acquired for road widening up to a maximum of 16 acres. In these cases the acquiring authority was the Scottish Development Department. Additional payments were also made for severance, disturbance, and any tenant's improvements involved.

Mr Menzies finally relied, as an indication of value, on the level of open market rents as at December 1977. He said that few farms had
E then been offered for letting on the open market, which was not surprising when the capital value of a farm when let was normally no more than 50% of its vacant possession value. The only two open market lets of which he was aware in Angus around 1977 were of Leys of Dun, Montrose, which is adjacent to Mains of Dun and was let on a tenant-repairing basis in 1977. Over 69 offers were received, at least one of which was in excess of £75 per acre. An offer of £61 per acre was accepted at a time when the neighbouring tenant of Mains of Dun was paying a rent equivalent to £15 per acre for similar quality land on a landlord-repairing basis.

The other was the open market let in 1978 of Balmachie Farm, Carnoustie, let on the open market for £70 per acre. Many offers were understood to have been received some of which were in excess of £75
F per acre. It was thus clear that there was no shortage of prospective tenants prepared to offer very high rents in order to secure the tenancies in question. Mr Menzies, however, saw in these high rents the presence of high premiums unrelated to true agricultural value.

In Mr Menzies' view, all the foregoing evidence left him in no doubt, in the hypothetical situation he was required to assume for capital transfer tax purposes, that agricultural tenancies had a substantial value and that value could vary up to 50% of the vacant possession value depending upon circumstances.

Turning finally to his estimate of the vacant possession value of Mains of Dun itself as at December 22 1977, he remarked that the three prior years had been of remarkable prosperity for Scottish arable farms and land prices had risen rapidly throughout the period to record levels. Malting barley was achieving record prices well into G
1977 compared with earlier years. Exceptional profits together with continued optimism and falling interest rates therefore led to a considerable demand for good agricultural land to buy or lease. The **Estates Gazette** Review of Agricultural Land in 1977 gave details of the rapidly increasing value of agricultural land at that time, particularly in Scotland. The potential value of sitting tenants' interests was therefore obvious.

In the light of this favourable situation, and having regard to the adequate state of the buildings at the valuation date, Mr Menzies' opinion of the vacant possession value of Mains of Dun was as follows:

164 acres good arable land at £1,400 per acre	£229,600	H
145 acres second-quality arable at £1,100 per acre	159,500	
176 acres third-quality arable at £600 per acre	105,600	
81 acres permanent pasture at £200 per acre	16,200	
52 acres poor grazing at £50 acre	2,600	
18 acres road and buildings	—	
Total	513,500	
Less value of tenant's improvements	13,000	
ie vacant possession value less tenant's improvements	£500,500	

Having determined the vacant possession value, Mr Menzies then turned to the value of the tenant's interest renounced in December 1977. As there was no directly comparable evidence available, he looked to the secondary evidence, from which it was clear that where J
farmers or landowners were able to buy tenancies they were prepared to pay substantial sums. Looking, therefore, to all the factual evidence available to him and taking "a robust view", he was of the opinion that the open market value of the tenant's interest in Mains of Dun in December 1977, on a hypothetical sale on the open market, would have been between 25% and 50% of its vacant possession value. In view, however, of the uncertainty in the hypothetical situation, he proposed a value of only 25% of vacant possession value in order to eliminate any special considerations which might be present in his secondary evidence. His calculation of the £138,000 is as follows:

Vacant possession value *less* tenant's improvements	£500,500	K
Tenant's interest whereof at 25%	125,125	
Add value of tenant's improvements	13,000	
Total	£138,125	
But, say	£138,000	

Parties' submissions

Counsel for taxpayers

Mr Mure QC, on behalf of the appellant taxpayers, submitted that section 20(2) of the Finance Act 1975 introduced, for the first time, an entirely new principle for measuring a charge to tax on the transfer of capital. This charge was not on the actual value of the property passing but, in contrast, on the amount by which the transferor's own L
estate was diminished. Earlier Revenue cases under earlier legislation had therefore to be read in the light of this innovation. In particular, the estate duty charging provisions of sections 1 and 2 of the Finance Act 1894 dealt with charges to estate duty on property passing or deemed to pass on death. While section 7(5) of that Act might be regarded as similar in wording to section 38(1) of the Finance Act 1975, as also involving the open market criterion, nevertheless section 38(1) should be read as subordinate to the new section 20(2), which was now the ruling provision. One could, therefore, apply the open market criterion under section 38(1) only where there could be shown to be a measurable diminution in value of the transferor's own estate by the transaction in question. Here there was none, as the tenancy interest renounced in 1977 in favour of others contained an M
express prohibition against assignation. Even if it then had some value, this could not be ascertained satisfactorily. The earlier estate duty cases did not, therefore, provide guidance in the present case.

In re Ashcroft [1927] 1 Ch 311; *Inland Revenue Commissioners* v *Crossman* [1937] AC 26 and *Lynall* v *Inland Revenue Commissioners* [1972] AC 680 were distinguishable, as they applied only to the transfer of shares in companies. The "*Crossman* principle" had not been applied anywhere in the United Kingdom except to company shares and had certainly never been sought to be applied to an unassignable agricultural lease. While in *Crossman* a hypothetical sale of the company shares had been required to be assumed, it was inappropriate to insist upon such an assumption in the case of the

A renunciation of an unassignable lease. Assets such as shares had an ascertainable value in the open market. But there was no open market value for the tenancy interest transferred in the present case. It was not, therefore, a transfer for value in terms of section 20(2). Hence one had to distinguish between assets with a market value and those without; only the former being marketable like shares in a private company. Assets without a market value could not, therefore, involve a "transfer of value" in terms of section 20(2).

In the present case, the 1921 lease contained a discretionary veto in the landlord which might be exercised arbitrarily and entirely in his own interest. Shares in private companies might have varying types of restriction but nevertheless had to be exercised *bona fide* and not B arbitrarily: see *In re Bede Steamship Co Ltd* [1917] 1 Ch 123 and also *Stewart* v *James Keiller Co Ltd* 4 F 657, which indicated that directors in this regard must act not capriciously but in the interests of the company as a whole. In the present case, however, there was nothing to stop the landlords vexatiously refusing an assignation of the lease. Indeed, in the case of agricultural leases, there was nothing to preclude a payment being made not only to the tenant to induce him to leave but also to the landlord himself for his consent to an assignation. There was thus a distinction between readily transferable assets like company shares and those subject to close restriction like agricultural leases.

Mr Mure also emphasised that the renunciation in 1977 had not resulted in any cash passing or anything with a cash worth; and, C furthermore, that the rental and other conditions of the lease had remained unaltered.

Counsel then submitted that, in any event, no proper comparative evidence had been produced to put a value on this particular transaction. This again showed that there could be no market value in such a transaction. Before such a transaction could occur there had also to be some coincidence of interest between landlord and tenant, also perhaps involving a three-way agreement. Any price fixed in such a context could never be a proper open market transaction. This particular assignation was not, therefore, a "transfer of value" in terms of section 20(2), as it could never occur in the absence of the landlord's consent. As the value could not be satisfactorily D determined, the district valuer had attempted to do so on misplaced analogies. His calculations left entirely out of account the landlord's power of veto. His evidence regarding "golden handshakes" and "sales and leasebacks" simply did not reflect open market values. His evidence relating to statutory compensation on compulsory purchases related only to small acreages as opposed to entire farms. His evidence on open market rents and rentals was also of little assistance. Where a lease could not be assigned without the landlord's consent, it therefore had no value and the Revenue's argument failed.

Counsel finally submitted that even if the tribunal disagreed with his legal submissions, a positive value still had to be capable of E calculation. If incapable, a charge to capital transfer tax could not be levied. The tribunal should therefore return a nil valuation. In reply to a question from the tribunal, counsel agreed that, as no alternative valuation evidence had been proffered, should the tribunal be against the appellants on these submissions, they would have to support the district valuer's valuation.

On the separate question of whether the renunciation in 1977 was of a whole or a half share in the tenancy of Mains of Duns, Mr Mure pointed out that the original lease in 1921 to Thomas Baird and the transferor had a survivorship provision. It was for 14 years and since 1935 had continued on tacit relocation. By minute of agreement in 1936 Thomas Baird dropped out but remained guarantor and the transferor became sole tenant. Under the minute of agreement dated F December 14 1959, the proprietor, at the request and with the consent of the transferor, had accepted his son, Donald Baird, "as joint tenant with the said George Alexander Baird of the subjects of the said lease" as from Martinmas 1959. The joint tenants then bound themselves "and their respective heirs executors successors and representatives whomsoever all joint and severally" to pay the rent and fulfil all the other leasehold provisions. While there was a survivorship provision in the original lease, this had been effectively altered in 1936: and again in 1959 when they were accepted as joint tenants without reference to a survivorship provision. This all went to confirm the transferor's interest as being in a half share only which would pass under his will or on intestacy. There could be no accretion. Counsel referred to *Robertson's Trustees* v *Robertson* 1982 SLT 22; *Coats* v *Logan* 1985 SLT 221; *Young* v *Gerard* 6 D 737; G

Magistrates of Banff v *Ruthin Castle Ltd* 1944 SC 36 and *IRC* v *Duke of Westminster Trustees* [1936] AC 1.

Mr Mure also submitted that on Donald Baird's death in 1977 his widow, Mrs Charlotte Baird, succeeded to his half share in the tenancy under the residue clause of his will under which she was sole beneficiary. However, it was immaterial whether his interest passed under his will or on intestacy; the important matter was that it did not accresce, as the transferor did not have more than a one-half share.

Counsel for Inland Revenue

Mr Brian Gill QC, on behalf of the Inland Revenue, said that the case raised two major issues. First, whether the agreement dated H December 9 and 22 1977 constituted a transfer of value in terms of section 20 of the Finance Act 1975 so as to make it chargeable and, second, what was the value of the interest so transferred? A transfer of value in terms of section 20(2) was a disposition made by a person as a result of which the value of his estate immediately after the disposition was less than it would be but for the disposition. It was section 38(1), however, which laid down the method by which the property so transferred was to be valued for the purposes of capital transfer tax — namely its value on the open market. What was deemed to be sold in the hypothetical open market was the right to be tenant under the relevant lease. Section 38 did not, therefore, innovate in any way upon the similar test laid down in section 7(5) of the Finance Act 1894. Hence the decision in *Crossman* was applicable J not only to the 1894 Act but also to the Finance Act 1975.

In the present case, the appellants' expert witness, Mr Galbraith, had given evidence without reference to any of these statutory provisions and had merely directed his attention to whether the particular tenancy could in fact be marketed. Under the section 38(1) hypothesis, however, the prohibition against assignation simply flew off. Mr Galbraith's conclusion, however, as to nil value was simply based on the absence of an actual open market where such tenancies were traded. He saw the one as the automatic consequence of the other.

Mr Gill then questioned whether, in any event, this tenancy really was unassignable, as the landlord could easily lift the restriction. Every lease could in fact be assigned if the landlord consented. The K veto in this lease was not absolute because, even if the tenant were elderly, the landlord would still have no early prospects of possession unless his tenant was without heirs or legatees to whom he could bequeath or transfer his interest under section 20 of the Agricultural Holdings (Scotland) Act 1949 or section 16 of the Succession (Scotland) Act 1964. In this respect, Scottish agricultural legislation was different from the English and the Scottish agricultural tenant was broadly in the same position as that of the occupier.

What the tribunal were now being asked to accept by the appellants was that the transferor's renunciation of his tenancy in favour of his grandson and daughter-in-law was of no value when in reality it gave them the right to make large profits from the farm. The interest of the L tenant was therefore a valuable asset and as such should be capable of being valued. Each tenant acquired a right *in rem* in respect of the property able to be vindicated against all the world, including the landlord herself. He was also protected by security of tenure under section 3 of the 1949 Act, so that his right continued by tacit relocation beyond the contractual period.

The basic question was therefore whether the transferor's estate in 1977 had been depleted and, if so, by how much in value. The asset in question certainly had a value to the transferor and by parting with it his estate was indeed diminished by a quantifiable amount. That this might be difficult to value did not mean that it had no value at all.

Mr Gill then referred to Mr Menzies' four different indications that the property interest renounced in 1977 did have some value. M These were, first, that a person's right to be tenant enabled him to exact a payment from his landlord for the renunciation of that right. Second, that in 1977 extra sums were being paid as "key money" in open market rental offers considerably over the agricultural market value simply to acquire the right to be tenant. This was a problem which required to be remedied under the Agricultural Holdings (Scotland) (Amendment) Act 1983, which was designed to exclude this element in rental determinations by arbiters and the Scottish Land Court. Here indeed was statutory recognition of the extra value created by a scarcity of lets.

He added that there was further recognition of the value of tenant's interests (despite a prohibition against assignation without

A consent) in sections 98 and 99 and Part V of the Finance Act 1981 introducing amendments to capital transfer tax and in relation to Scottish agricultural leases neither of which amendments would have been necessary if an interest in a farm tenancy was not a chargeable transfer on death. Another statutory pointer in the same direction was section 177(1) of the Capital Transfer Tax Act 1984, which excluded the value in certain circumstances. There were also various cases in which agricultural tenancy interests had required to be valued on compulsory acquisition. Examples were *Minister of Transport* v *Pettitt* (1968) 67 LGR 449; *Watson* v *Secretary of State for Air* [1954] 1 WLR 1477 and (1956) 6 P&CR 51 and *Wakerley* v *St Edmundsbury Borough Council* (1977) 33 P&CR 497. Also this tribunal's own decision in *Anderson* v *Moray District Council* 1978
B SLT 37, where the tribunal itself had recognised that an agricultural tenant displaced on compulsory purchase of the land had a valuable interest which required to be compensated under section 44 of the Land Compensation (Scotland) Act 1973.

Mr Gill submitted that, if a farm tenancy had a value, the mere presence of a prohibition against assignation could not be used to negative that value, for in every case this would obviate section 38 of the 1975 Act. Section 38(1), however, like other taxation legislation, introduced the open market criterion, requiring it to be assumed that there was a sale in a hypothetical open market disregarding any contrary direction. The principal authority to this effect was *Crossman (supra)*, which was not confined only to shares in
C companies and was indeed binding on the tribunal. So was the similar decision in *Lynall (supra)*. The ratio in *Crossman* had also been applied by the tribunal to value an interest other than shares in *Lowe* v *Inland Revenue Commissioners* 1985 SLT 12. Only new legislation could therefore now alter the application of the "*Crossman* principle". All that section 20(2) really did was to identify what asset was to be valued and the measure of value was the amount by which the estate was depleted. Section 38 then showed how to put a figure on it. It was also the transferor's property interest which strictly required to be valued — not the recipient's.

The appellants' submissions as to the distinction between a prohibition on the transfer of company shares as compared with the assignation of a lease failed because the latter was in reality no more
D absolute than under company law in relation to shares. See, in particular, *Palmer on Company Law*, para 40-12. In this case the tenancy interest, despite the prohibition against assignation, had indeed been assigned on no less than three occasions. When it came to the stage of valuation, however, with a hypothetical sale in the open market assumed, then a hypothetical purchaser required to be assumed as holding the land subject to all the pre-existing leasehold conditions: see *Salvesen's Trustees* v *Inland Revenue Commissioners* 1930 SLT 387; *Corrie* v *Macdermott* [1914] AC 1056 and *Odeon Associated Theatres Ltd* v *Glasgow Corporation* 1974 SLT 109.

Mr Gill finally submitted on the question of valuation that, once there was an asset of undoubted value, it was never beyond the range
E of a valuer to place an appropriate value upon it as indeed was done in *Anderson* and *Wakerley (supra)*. In the present case, no valuation at all had been submitted for the appellants and therefore the district valuer's method was to be preferred. The latter had adopted four tests all indicating the presence of value in a tenancy interest even when subject to a prohibition against assignation.

On the question of the extent of the transferor's interest and whether this had been that of sole or joint tenant, Mr Gill submitted that his interest in 1977 was in the whole tenancy (1) owing to accretion or (2) because the interest of Donald Baird had not been bequeathed: see *Reid's Trustees* v *Macpherson* 1975 SLT 101 at p 109. It had certainly not been intimated under section 20 of the 1949 Act. Nor had any steps since been taken to confirm it under section 16
F of the 1964 Act. If, therefore, the transferor's interest had not vested in Donald Baird, no one held it as at December 22 1977 either as executor, legatee or acquirer. In any event, the transferor had been accepted as sole tenant by the landlord and that was assented to by the widow and grandson by their acceptance of the two half-shares in the whole tenancy. The whole transaction was therefore inexplicable on any other basis than that the transferor had been accepted on all sides in 1977 as sole tenant.

*Esto** the transferor transferred only a one half-share and some other party held the other half, it still did not follow that the transferor's half-share had no value. This was because he had a weapon against the other joint tenant because of his ability to bring the lease to an end. He also had the power to deliver vacant possession G
to the landlord on service of a notice to quit: see *Smith* v *Grayton Estates* 1960 SC 349. On that assumption, his half-share should be valued at half the district valuer's valuation, *viz* £69,000.

**Esto*: Granted that.

Decision

This case turns on the interpretation of certain sections of the Finance Act 1975. Section 19(1) introduced capital transfer tax, "charged on the value transferred by a chargeable transfer". Section 20 provided:

> (1) The following provisions of this section shall have effect, subject to the other provisions of this Part of the Act, for determining for the purposes of capital transfer tax what is a chargeable transfer and what value is transferred by a chargeable transfer. H
> (2) Subject to subsections (3) and (4) below, a transfer of value is any disposition made by a person ("the transferor") as a result of which the value of his estate immediately after the disposition is less than it would be but for the disposition; and the amount by which it is less is the value transferred by the transfer.

Section 23 defined a person's estate as "the aggregate of all the property to which he is beneficially entitled . . .".

Section 38 provided:

> (1) Except as otherwise provided by this Part of this Act, the value at any time of any property shall for the purposes of capital transfer tax be the price which the property might reasonably be expected to fetch if sold in the open market at that time . . .

The main argument on behalf of the appellant taxpayers is that J
section 20(2) of the Finance Act 1975 introduced an entirely new concept of valuation related, for the first time in taxation law, to the diminution in value of the transferor's estate. In counsel's submission, section 20(2) was not only the main charging provision but also served to indicate the basis of the valuation itself, namely the diminution in value to the transferor's estate. While the appellants concede that section 38(1) introduced the open market criterion, they nevertheless contend that section 20(2) remains the ruling valuation provision. As the tenancy interest transferred was unmarketable at the time of transfer (owing to the prohibition against assignation), the value of the transferor's estate was not diminished and hence the tribunal must find the value of the transfer to be nil.

In the tribunal's opinion, however, section 20(2) strictly indicates K
only what is a chargeable transfer and not how to calculate any diminution in value. For the latter one must turn to section 38(1), as the format of the statute seen as a whole helps to indicate. In *Director of Public Prosecutions* v *Schildkamp* [1971] AC 1 Lord Reid commented at p 10 on "whether and to what extent it is permissible to give weight to punctuation, cross-headings and side-notes to sections in the Act". He observed (at p 10) that these were realistically also part of the legislative process and said:

> If the authorities are equivocal and one is free to deal with the whole matter, I would not object to taking all these matters into account, provided that we realise that they cannot have equal weight with the words of the Act.

See also Lord Upjohn (at p 28) commenting on an earlier view of L
Lord Reid in *Chandler* v *Director of Public Prosecutions* [1964] AC 763 at p 789.

While the main weight has therefore to be given to the actual words of sections 20(2) and 38(1) of the 1975 Act, it is now legitimate, as an aid to their construction and to gauging their interrelation, to note that both sections appear in Part III of the Act, which deals with capital transfer tax. The cross-heading "Main charging provisions" appears above sections 19-36, thus indicating that section 20(2) is intended to be a main charging provision. The cross-heading "Rate of tax and valuation" appears only above sections 37 and 38 with the side-note "Rate of Tax" beside section 37 and "Valuation" beside section 38.

In the tribunal's opinion, it is therefore the valuation section, 38, M
which alone directs how a chargeable transfer is actually to be valued, namely at the price which the property might reasonably be expected to fetch if sold in the open market at the time of the transfer.

Under section 38(1) Parliament has also directed that the valuation of any property at any time shall be made *upon the hypothesis* that the property was then offered for sale in a hypothetical open market. This is a clear statutory direction — however difficult it may sometimes be to apply — and it is that direction itself which requires us to disregard any prohibition against sale so as to enable a sale in the hypothetical open market to take place. The basic flaw in the appellants' approach is therefore in their failure to give effect to the statutory hypothesis contained in section 38(1).

A Section 7(5) of the now-repealed Finance Act 1894 similarly introduced the open market criterion for determining, for the purposes of estate duty, the value of property passing or deemed to pass on death. In *Crossman,* the House of Lords was required to decide what was the proper basis of valuation under section 7(5) of shares in a limited company where the right of transfer was restricted by its articles of association. The House decided (by a majority of 3 to 2, overruling the Court of Appeal) that the restriction on transfer had simply to be disregarded for the purposes of assuming a hypothetical sale but that the resultant valuation should nevertheless still take account of the same restriction continuing in the hands of the purchasers. In so doing they approved *Attorney-General for Ireland* v *Jameson* (1905) 2 IR 218 and *Salvesen's Trustees* v *Commissioners*
B *of Inland Revenue* 1930 SLT 387.

In *Crossman,* Viscount Hailsham LC, in commenting on the contrary decision of the Court of Appeal, said (at p 41):

It seems to me that this construction involves treating the provisions of section 7(5) as if their true effect were to make the existence of an open market a condition of liability instead of merely to prescribe the open market price as the measure of value. The right to receive the price fixed by the articles in the event of a sale to existing shareholders under subclause 14(a) is only one of the elements which went to make up the values of the shares. In addition to that right, the ownership of the share gave a number of other valuable rights to the holder, including the right to receive the dividends which the company was declaring, the right to transmit the share in accordance with article 34, subclauses 1, 2 and 3, and the right to have the shares of other holders who
C wished to realize offered on the terms of article 34, subclause 14(a). All these various rights and privileges go to make up a share and form ingredients in its value. They are just as much part of the share as the restriction upon the sale. The construction placed upon the statute by the Court of Appeal seems to me to ignore all these elements in the value of the share and to treat as its value what, in truth, is only the value of one of the factors which go to make up that share. But the purpose of section 7, subsection (5) is not to define the property in respect of which estate duty is to be levied, but merely to afford a method of ascertaining its value. If the view entertained by the Court of Appeal were correct, it would follow that any property which could not be sold in the open market would escape estate duty altogether.

In his concurring speech, Lord Roche said (at p 73):

It seems to me to be important to remind oneself that section 7, subs. 5 of the
D Act is simply a provision for estimating by means of a hypothesis the value of property which has passed otherwise than by an actual sale and transfer and may be incapable of so passing. But the section has to be applied and can be applied to all such cases. As to property which by its inherent or essential nature cannot be actually sold or transferred the application of the section was explained as I should like to explain it by my noble and learned friend, then Russell J, in the case of *In re Cassel* [1927] 2 Ch 275, 281. The learned judge says: "Next, how is the principal value of that property to be ascertained? The machinery provided by the Act is section 7, subs. 5 supplemented to some extent by section 7, subs. 8. That machinery does not exactly fit the present case, because from the personal nature of the provision, the benefit of it could not be sold at all; an outside purchaser would not obtain delivery of the goods. Nevertheless the machinery must be made to fit, and I think it can be made to fit. It was made to fit in the Irish case of *Attorney-General for Ireland* v
E *Jameson* [1905] 2 IR 218. The value cannot be ascertained by reference to market price because there can be no market.

An attempt was made in *Lynall (supra)* to persuade the House of Lords to reverse their previous majority decision in *Crossman* in a case which also dealt with the valuation for estate duty purposes of shares in a private company. There was also a similar restriction on transfer. The House, however, unanimously approved *Crossman,* as also the preceding Irish decision of *Jameson (supra)* and the Scottish decision of *Salvesen's Trustees (supra).* Lord Pearson, in particular, approved the following passage from the judgment of Lord Fleming in *Salvesen's Trustees* at p 391:

If the articles of association be complied with, a sale in the open market in a reasonable sense seems to be impossible. The petitioners argued that the
F maximum price the shareholder can obtain for his shares in the open market is determined by the best price he can obtain in the closed market, namely, £1. But it appears to me that if this argument is well founded, it merely demonstrates that there cannot be a real sale in the open market under the articles. The Act of Parliament requires, however, that the assumed sale which is to guide the commissioners in estimating the value is to take place in the open market. Under these circumstances, I think that there is no escape from the conclusion that any restrictions which prevent the shares being sold in the open market must be disregarded so far as the assumed sale under section 7(5) of the Act of 1894 is concerned. But on the other hand, the terms of that subsection do not require or authorise the commissioners to disregard such restrictions in considering the nature and value of the subject which the hypothetical buyer acquires at the assumed sale. Though he is deemed to buy in an open and unrestricted market, he buys a share which, after it is transferred to him, is subject to all the conditions in the articles of association, G
including the restrictions on the right of transfer, and this circumstance may affect the price which he would be willing to offer.

The appellants in the present case have endeavoured to distinguish the decisions in *Crossman* and *Lynall* on the ground that they should be regarded as confined in their application to the valuation of shares in private companies the transfer of which was restricted. The tribunal, however, read those decisions as being ones of general principle in all taxation cases where the open market criterion is introduced for valuation purposes; and, further, that this is to be approached in the first place by disregarding any prohibition or restriction on the transfer of the property which would render impossible its sale in the open market. For, of course, section 7(5) of H
the 1894 Act governs all kinds of property passing or deemed to pass on death and not only company shares. Furthermore, if regard were paid to a prohibition against alienation not only would it render ineffective in most cases the open market criterion of value but such a power of veto could also be inserted in various circumstances where its actual use would be unlikely but would nevertheless serve to avoid liability to taxation. This could apply not only to capital transfer tax but also, for example, under section 150(1) of the Capital Gains Tax Act 1979 or as formerly under the Development Land Tax Act 1976, section 7(1). The House of Lords in *O'Brien* v *Benson's Hosiery (Holdings) Ltd* [1980] AC 562 has since held that the sum of £50,000 received by a company in return for the surrender of its rights under a service agreement constituted the disposal of an asset for the J
purposes of corporation tax notwithstanding that those rights had not been assignable and had had no market value.

Crossman and *Lynall* have since been followed by the Second Division in *Odeon Associated Theatres Ltd* v *Glasgow Corporation* 1974 SLT 109 in a case which was not concerned with company shares. This was an appeal from this tribunal and involved the assessment of compensation on the compulsory purchase of a disused cinema under rule (2) (section 12(2)) of the Land Compensation (Scotland) Act 1963), being: "The amount which the land if sold in the open market might be expected to realise." The disused cinema had been bought by a wholly-owned subsidiary of a holding company who then leased it back on a 55-year lease for an annual rent of £3,500. The dispute was as to the effect of this K
continuing lease on a hypothetical open market sale and whether the presence of the lease would increase the value of the interest acquired. The land element had otherwise been agreed at £1,000. The court followed *Crossman, Lynall* and *Salvesen's Trustees.* Lord Kissen said:

I return to rule (2). A hypothetical sale in the open market has to be assumed. While the personal restrictions may not enable Ranks to prevent a sale, the attitude of Ranks is of considerable importance . . . the correct view is, I think, that while restrictions are disregarded for the purposes of the hypothetical sale taking place, the hypothetical buyer takes his purchase with the restrictions still attached to them.

Lord Fraser (later Lord Fraser of Tullybelton), after referring to L
Crossman and *Lynall,* said (at p 117):

But although the estate duty cases are not directly in point, they do, in my opinion, indicate the principle which should be applied in this case. I have reached the opinion that the value of the *dominium utile** on the open market should be estimated subject to the existing lease to Ranks, but on the terms that the rent and the tenant's other obligations would be exigible only if the purchaser were a subsidiary of Ranks and that they would cease to be exigible if the purchaser were an outside third party.

On this basis the case was remitted to the tribunal to reconsider the question of value on that basis, but they found that, as a matter of valuation, the existence of the lease would not enhance the purchase price.

In *Lowe* v *Inland Revenue Commissioners* 1985 SLT 12 this M
tribunal, for the purposes of assessing development land tax, required to determine the market value of agricultural land at the date of disposal when this land was subject to an option in favour of Wimpey Ltd. Section 7(1) of the Development Land Tax Act 1976 provided, like other taxation statutes: "For the purposes of this Act, the market value at any time of an interest in land is the consideration which that interest might reasonably be expected to fetch on a sale at that time in the open market." The tribunal (see p 15) said: "We have to assume that there was a sale in the open market in March 1979

**Dominium utile:* The expression is a relic of feudal law, where the *dominium directum* was the right of the feudal superior and the *dominium utile* the right of the vassal. For most purposes today *dominium utile* can be equated with full beneficial ownership.

disregarding any right in Wimpey, if they chose to exercise it, to prevent such a sale." In so doing they followed *Crossman, Lynall* and *Odeon*.

The tribunal are not persuaded that the present transfer of the lease of Mains of Dun involved was of no value whatsoever; or that it was an interest so difficult to assess that a nil value should be returned. The appellants' submissions to this effect are really based upon the presence, rather than the disregard, of the landlord's right of veto. What the case of *Crossman*, followed by *Lynall*, however, decided was that any such restriction which prevents the application of the open market criterion must be disregarded for the purposes of the *hypothetical* open market sale; although, on such a sale, the property had thereafter to be valued in the hands of the hypothetical purchaser as subject to the same restriction. This is hardly an unfair test to apply, for, as already remarked, such a restriction could be deliberately inserted into articles of association or, as here, into a lease extinguishing its value for tax purposes, whereas in many cases it may in reality be readily transferable with the landlord's consent. In the present case, indeed, the lease in question has actually been so transferred on several occasions including the very renunciation which gives rise to the present determination and to which Mrs Lovett appears to have readily agreed as being a natural progression within the tenant's family.

Looking at the position from the viewpoint of the transferor's own estate, it appears that by renouncing his tenancy interest in favour of his grandson and daughter-in-law, he then deprived himself of the profits of a valuable arable farm in Angus during a buoyant period in local agriculture. He also deprived himself of the right to occupy the farmhouse and various other buildings. This was, of course, in return for the payment of a rent which, however, was substantially below open market rents then prevailing and, in particular, the rent of the adjoining farm. It will be seen that this involved profit rent — although the rent might have been raised at the next rent review. In renouncing his tenancy and transferring it (with the landlord's consent) to his daughter-in-law and grandson he also deprived himself of security of tenure even extending under Scots law to his successors. All these valuable rights also then came to be enjoyed by the transferees.

That a farm tenant's interest has a value, despite the exclusion of assignees in most Scottish agricultural leases, is established by the district valuer's list of "golden handshake" cases. Landlords were, in 1977, prepared to buy out their secure tenants at sums of up to one-half of the vacant possession values of the farms in question. This indicates that the leasehold interests concerned then had a value, particularly when premiums were being paid for lettings in the open market.

Also indicative of value are the district valuer's "sale and leaseback" transactions, which at least signify some value in the secure tenant's interests which thereby result. Another indication that a tenant's interest has an open market value, despite a clause prohibiting assignation, is to be found in the compulsory purchase cases. We have already referred to *Odeon (supra)*. Examples of the compulsory purchase of let farms and the consequent payment of compensation to the agricultural tenant as well as to the owners are *Wakerley* and *Anderson (supra)* which, however, adopted a profit rental approach to valuation under section 12(2) of the Land Compensation (Scotland) Act 1963 which, like section 38(1) of the 1975 Act, also adopts the open market criterion of value. If the appellants' present argument were correct, however, it would appear to be open to an acquiring authority to contend that the open market value of a tenant's interest which had been compulsorily acquired was nil in view of the landlord's prohibition against assignation. This, however, may be affected by the establishment of a loss of profits claim as was presented in the *Wakerley* case.

The tribunal also agree with Mr Gill that, if the appellants' present contentions were correct and a tenant's unassignable interest in an agricultural holding has no value, sections 98 and 99 of the Finance Act 1981 and section 177(1) of the Capital Transfer Tax 1984 (which consolidates and re-enacts part of the Finance Act 1975) would not have needed to be passed.

The district valuer has therefore adopted what he himself described as a "robust approach" and valued the whole tenancy interest at 25% of its vacant possession value. Total vacant possession value is agreed by joint minute between parties at £513,500 (including £13,000 for tenant's improvements). As no alternative valuation, other than a nil one, has been proffered for the appellants, the tribunal as at present advised therefore accept the district valuer's valuation of £138,000, being £125,000 for the tenant's interest and £13,000 for the value of the tenant's improvements. We would, however, reserve our opinion as to whether, in a future case, there may not be another approach to such a valuation.

Extent of transferor's interest

As already indicated, the original lease of Mains of Dun as an agricultural holding was entered into in July 1921 between (1) the then proprietor, Mrs Alice Foote or Kennedy Erskine, and (2) Thomas Baird and the transferor as her tenants. We repeat the relevant words of let, which were as follows:

> Lets to the said Thomas Baird, and George Alexander Baird (who are hereinafter referred to as "the Tenants" under which expression are also included their successor, or successors, in the tenancy of the said subjects) and to the survivor of them, and the heirs of the survivor, but excluding heirs-portioners . . .

The original lease in 1921 thus involved a destination over to, or a *jus accrescendi* in, the survivor of the two named tenants.

By minute of agreement, however, dated August 1 and 17 1936 between the same landlord and the same tenants, one of them, Thomas Baird, then assigned and made over to his co-tenant (the transferor) and to his heirs his whole right and interest in the lease with power to him

> to occupy and possess the subjects thereby let in the same way and manner as if the said Lease had originally been granted in favour of the said George Alexander Baird alone, and I hereby renounce my whole right and interest in and to the said Lease.

Under this 1936 minute of agreement the destination over was therefore removed and the transferor became sole tenant with no possibility on his death of accretion in favour of anyone else.

By further minute of agreement, however, dated December 14 and 15 1959, the *curator bonis* to the new proprietor agreed to a further fundamental alteration which had the effect of introducing the transferor's son into the lease as co-tenant along with his father, thereby altering the destination of the lease — although most of its terms and conditions remained the same.

The relevant words were as follows:

> The proprietor at their request and with the consent of George Alexander Baird accepts the said George Donald Baird as joint tenant with the said George Alexander Baird of the subjects of the said lease as from the term of Martinmas 1959 (declaring that the provisions of the said lease excluding certain successors shall apply to the said George Donald Baird as well as to the said George Alexander Baird) and binds himself and his foresaids to implement all the obligations, terms and conditions of the said lease as modified by the correspondence a copy of which is annexed and signed as relative hereto so far as incumbent on the proprietors and the said George Alexander Baird and George Donald Baird bind themselves and their respective heirs, executors, successors and representatives whomsoever all jointly and severally regularly to pay the rent and to implement all the obligations, terms and conditions of the said lease modified as aforesaid so far as incumbent on them or either of them as tenants thereunder and it is hereby agreed and declared that except as altered by these presents or already implemented, the said lease modified as aforesaid shall remain in force.

The new co-tenant, George Donald Baird, however, unexpectedly predeceased his father (the transferor) when he was killed in the car accident on September 24 1977. The question now raised is whether his interest thereupon accresced to his father or, being only in a separate one-half share, could have passed under the general residue clause of his will to his widow, Mrs Charlotte Mary Baird. If the latter, then the Revenue's present claim for capital transfer tax based on the transferor's renunciation of his tenancy interest as at December 22 1977 would be halved.

Mr Gill, for the Inland Revenue, has raised the question, if the transferor's interest in September 1977 was only in a one-half share, what thereafter happened to his son's other one-half share which should have passed by bequest to his mother? He commented that such bequest had never been notified to the landlord. Also that the renunciation agreement in December 1977 appeared to be based upon the transferor's own supposition that he then owned the whole tenancy interest; for in the letter dated December 9 1977, he wrote: "I hereby relinquish *the* tenancy."

The appellant taxpayers, however, are the transferor's executors and this tribunal are required to assess the market value of the transferor's tenancy interest that was renounced as at December 22 1977. This in turn depends on what had already happened to the

A tenancy at the date of the son's death on September 24 1977 and what was thereafter the transferor's interest in the continuing lease which he later renounced. We do not consider that the appellant taxpayers can be in any way barred by anyone's subsequent actings from now contending that in law there was no accretion of the son's interest to the transferor.

As at September 24 1977, it was the 1959 agreement which ruled. That agreement provided for the introduction of the transferor's son "as joint tenant with the said George Alexander Baird". For the Revenue it is contended that the survivorship clause contained in the 1921 lease (as varied in 1936 and 1959) still applied in the circumstances of the new joint tenant so as to bring about accretion in
B September 1977. On behalf of the appellant taxpayers it is contended that when the transferor's son was accepted in 1959, as joint tenant, excluding certain successors, this confirmed that the transferor's interest was confined to a one-half share.

The tribunal are not persuaded that the use of the words "as joint tenant" can in themselves be taken to imply a *jus accrescendi* in the survivor. In *Rankine on the Law of Leases in Scotland* (3rd ed, at p 83) the learned author states:

> A common or joint lease creates a common or joint right, entitling each lessee, unless it be otherwise arranged, to insist on joint possession and management and involves common or joint liability.

On pp 83 and 84 he distinguishes leases to two lessees and their heirs from a lease to two lessees and the survivor and their heirs:

> C (a) if the lease be taken to A and B and their heirs, each lessee has during life an equal interest both in kind and degree, and upon the death of either his right devolves, not to the survivor but to his own heir . . . (b) if the lease be granted to A and B and the longest liver and their heirs, the right is common during their lives, with accretion to the survivor. Each during his life is vested with a right so qualified, that were he to assign his portion the assignation would terminate with his life, when the survivor acquires the sole right.

It will be seen from the above passage that the existence of a *jus accrescendi* depends not on the use of the term "joint" as opposed to "common" but whether there is a survivorship destination.

In the case of *Coats* v *Logan* 1985 SLT 221, which was concerned with the transfer on death of a tenant's interest in an agricultural lease, Lord Ross (at p 225) referred with approval to the following
D distinction between joint and common property as stated in *Magistrates of Banff* v *Ruthin Castle Ltd* 1944 SC 36:

> As Lord Justice-Clerk Cooper (as he then was) explained in that case, one of the difficulties is a terminological one, in that the expression "joint property" is used in differing senses. Strictly, however, the term "joint" is used to describe the class of right typified by the ownership of co-trustees, and the term "common" to describe that typified by the ownership of two or more persons in whom the right to a single subject has come to be vested, and each of whom is entitled by his separate act to dispose of his separate share. As Lord Justice-Clerk Cooper said "So far as has been traced, there is no instance of a joint right in the strict sense having been held to exist except in persons who were inter-related by virtue of some trust, contractual or quasi-contractual bond — partnership or membership of an unincorporated association being
> E common examples — and it seems to me that such an independent relationship is the indispensable basis of every joint right. The distinctive feature of the right of such joint proprietors is the *jus accrescendi*, which excludes the possibility of separate shares in the several joint owners, and still more emphatically, excludes the possibility of severance of the tie, except, of course, by dissolution of the relationship on which the joint ownership rests".

Nevertheless, the term "joint tenancy" continues to be more F
loosely used. For instance, in *Smith* v *Grayton Estates Ltd* 1960 SC 349 the First Division referred to an agricultural lease in favour of two tenants as a "joint tenancy": see Lord President Clyde at p 354 and Lord Sorn at p 356. In *Gill's Law of Agricultural Holdings in Scotland* it is observed (at p 123) that a tenancy in common containing a survivorship destination is invariably referred to as a lease to joint tenants.

The effect of the 1959 minute of agreement was not to set up a joint tenancy in the strict sense indicated by Lord Cooper *(supra)* of an indivisible right of tenancy held by a body such as trustees. It was in favour of two named individuals described as "joint tenants", thereby holding *pro indiviso** under a tenancy in common whereby G
their respective shares, in the absence of other direction, would normally descend to their successors — unless of course there was a survivorship destination.

The crucial question, therefore, in the present case is not whether this was a "joint tenancy" in the stricter or looser sense of those words but whether the words used in the 1959 agreement "as joint tenants" are sufficient in themselves to import a survivorship destination, thus excluding the possibility of continuing separate shares in the event of the predecease of one of them. The tribunal consider that these words cannot of themselves import a survivorship destination; nor can the survivorship destination contained in the 1921 lease be regarded as still extant, when it was clearly abrogated in H
1936. On the authority of the passage quoted from *Rankine* and the inferences to be drawn from the cases to which we have referred, a survivorship destination must be clearly expressed. As appears from *Smith* v *Grayton (supra)*, the words "joint tenancy" or "joint tenants" are often loosely used and cannot therefore be taken to imply a *jus accrescendi* which, being a destination over, needs to be clearly expressed rather than left to implication. Unlike the original lease of 1921, the 1959 minute of agreement contained no express survivorship clause. On the contrary it contained the specific words: "(Declaring that the provisions in the said lease excluding certain successors shall apply to the said George Donald Baird as well as the said George Alexander Baird)". But this was a reference to the provisions of the 1921 lease *excluding* heirs-portioners, assignees, J
subtenants and trustees for behoof of creditors from which we are unable to deduce the positive inclusion or reinstatement in 1959 of a survivorship destination which had previously been eliminated under the agreement of 1936.

Whatever the transferor's own views on the matter may have been, the tribunal therefore conclude that the transferor was entitled to only a one-half share in the tenancy of Mains of Dun as at the date of his renunciation of the lease on December 22 1977.

Hence we determine the value of the relevant half share in the tenancy to be £69,000, being half the Inland Revenue's determination of £138,000.

As agreed between parties, we meanwhile reserve all question of K
expenses, regarding which, failing agreement, parties may now submit written submissions.

**Pro indiviso:* In undivided shares.

LANDS TRIBUNAL

COMPENSATION

A GREENS MOTOR HOLDINGS LTD v PRESELI PEMBROKESHIRE DISTRICT COUNCIL

(REF/30/1989)

March 27 1990

(T HOYES FRICS)

Estates Gazette January 26 1991

[1991] 03 EG 139

Freehold interest in Haverfordwest 50-year-old, single-storey, former motor-car showroom premises adapted to B supermarket use — Occupied by supermarket company on 25-year lease with 5-yearly rent reviews and 13½ years to run at valuation date in January 1989 — Situated on fringe of town's shopping centre — Claim for £570,000 — Acquiring authority's figure £280,000 — Five issues in dispute: (1) rental value at rent review date in September 1987; (2) rental value at valuation date; (3) capitalisation rate; (4) allowance for future roof repairs; (5) whether increased recoverable rent resulting from 1987 rent review was compensatable

***Rental values* — Rent review negotiations at stalemate prior to authority's entry, lessee contending no increase over C passing rent of £33,000 pa justified in view of imminent acquisition —Lessor had sought increase to £50,000 pa — Both parties' valuers proceeded on investment basis and assumption that rent review procedure concluded — Claimant's figure £55,789 (£4.75 psf), authority's £35,235 (£3 psf), in each case supported by wide range of comparables — Tribunal concludes parties in 1987 would have met in climate of established and continuing rental growth for variety of retail premises in area and that rent of £50,000 pa (about £4.20 psf) would have been agreed in 1987 — At January 1989 figure would have been £55,000 pa (about £4.70 psf)**

D ***Capitalisation* — Claimant's valuer applied 10% to both 1987 and 1989 rental values after making deduction to cover lessor's estimated liability for cost of roof repairs — Authority's valuer took 12.5% including 1% for roof repair liability —Comparables indicated yield broadly in range of 9% to 12%, with well-located premises of reasonably modern design and car parking showing 10% or below and those less well located or older above 10% — In tribunal's judgment, to reflect subject premises' advantages and disadvantages as an investment required 11% to be applied to both 1987 and 1989 rental values, with capital values adjusted for cost of future roof repairs**

E

***Roof repairs* — Both parties' experts accepted physical life of roofing was 20 to 30 years — Tribunal considered prudent investor-purchaser would anticipate need for early renewal and notionally make immediate provision for financing lessor's 50% share of estimated cost — Deduction of £30,000 determined**

***Reviewed rental recovery* — Claimant contended acquiring authority, by including subject premises in CPO, was responsible for 1987 rent review being undetermined at date of entry and that position protected by section 9 of Land F Compensation Act 1961 — But for scheme lessor and lessee would have agreed rent on review in 1987 or been able to operate lease's review mechanism — If scheme to be ignored under section 9, its consequences likewise to be ignored and lessor should not be denied 17 months' increased rent — Tribunal rejected as inconsistent with rule (2) authority's countersubmissions that if there were loss of rent it arose from claimant not pursuing review mechanism to finality and that any loss was not relevant matter for tribunal's jurisdiction — In absence of compulsory acquisition, claimant would, in tribunal's view, have concluded agreement in respect of G revised rent prior to any sale or have sold freehold interest with right in purchaser to conclude review and recover any additional unpaid rent from lessee — Arithmetically, additional rent not recovered was £24,000, but in open market freeholder would need to incur expense in finalising review or make allowance to purchaser to cover expected costs — Moreover, under rule (2) it is not actual loss which falls to be compensated but rather addition to market value, or expected sale price, which has to be quantified and expressed as compensation — Determination of £15,000 — Total award of £471,000 — Costs against authority**

H

No cases are referred to in this report.

Vernon Pugh QC (instructed by Greathead & Whitelock, of Pembroke) appeared for the claimants, Greens Motor Holdings Ltd; Phillip Rees (instructed by the solicitor to Preseli Pembrokeshire District Council) represented the acquiring authority.

Giving his decision, MR T HOYES said: This is a reference to determine the compensation payable on the compulsory acquisition of former motor-car showroom premises, used as a supermarket since 1977, at Cartlett Road, Haverfordwest, Dyfed, South Wales.

The claimant company seeks compensation in the sum of £570,000 for the freehold interest, which is subject to an occupation lease to J Kwiksave Ltd. The acquiring authority contend that compensation should be £280,000.

Counsel for the claimant called Mr M A Lawley BSc FRICS and counsel for the acquiring authority called Mr P B Kelly FRICS.

The respective valuations are attached hereto at Annexure 1.

ANNEXURE 1

Valuation by M A Lawley BSc FRICS for the claimant

(i) Rent from 2.9.87			
11,745 sq ft @ £4.75	£55,789		
less 5% for roof costs	2,789		
	53,000		K
YP 3½ years 10%	2.8284		
		£149,905	
(ii) Reversion at 2.9.92			
11,745 sq ft @ £5.25	£61,661		
less 5% for roof costs	3,083		
	58,578		
YP perp 10% def 3½ years	7.1717		
		420,103	
		570,008	
	Say	£570,000	

A **Primary valuations by P B Kelly FRICS for the acquiring authority**

(i)	(a)	Assumed current rent payable (as reviewed in 1987)			
		Gross internal area 11,745 sq ft @ £3.00 psf, pa		£35,235	
		YP 3½ years @ 12.5%		2.69	£94,782
	(b)	Current ERV			
		GIA 11,745 sq ft @ £3.25 psf, pa		38,171	
		YP perp 12.5% defd 3½ years		5.32	203,069
					£297,851
(ii)	(a)	Assumed current rent payable (as reviewed in 1987)		35,235	
		YP 3½ years at 12.5%		2.69	94,782
B	(b)	Reversion to current ERV as @ 2.9.92 as before		38,171	
		YP 10 years @ 12.5%	5.53		
		PV of £1.00, 3½ years @ 12½%	0.66	3.65	139,325
	(c)	Reversion to site value in 2002			
		One-third acre @ £300,000 per acre		100,000	
		PV of £1.00 for 13½ years @ 8%		0.35	35,000
					£269,107
(iii)		Present rent		33,000	
		YP perp @ 12%		8.33	
				274,890	£275,000
(iv)		Taking the near mean of (i), (ii) & (iii)			
C		Open market value of freehold interest			£280,000

Check valuation by P B Kelly FRICS

(a)	Assumed current rent payable (after 1987 review)		
	GIA 11,745 sq ft @ £3.00 psf, pa	£35,235	
	YP 3½ years @ 11.5%	2.77	£97,600
(b)	Current ERV		
	GIA 11,745 sq ft @ £3.25 psf, pa	38,171	
	YP perp 11.5% defd 3½ years	5.97	227,880
		Total	£325,480

D Haverfordwest is the former county town of Pembrokeshire and since the creation of the county of Dyfed is the principal town in the western part of that county. The current resident population is 14,000 and the shopping population is increased significantly by tourists in the summer months. The A40(T) passes through the town and there is a connection to the main London to Fishguard railway.

The subject premises, now demolished, were situated on the western side of Cartlett Road, a dual carriageway section of the A40 forming part of the inner ring road sited to the east of the town centre. The premises were bounded on the south and west by Jubilee Gardens, a local street, which provided the only vehicular and the main pedestrian access points into the supermarket building. Immediately across Jubilee Gardens was a surface car park with 300 E spaces controlled by the local authority: the main short-stay car park in the town. The western boundary of this car park was formed by the Western Cleddau River over which there was a footbridge giving access to the long-established principal shops in High Street, Bridge Street, Castle Square and Victoria Place. The subject premises were therefore located upon the eastern fringe of the town centre, with easy road access and car parking close by; they were thus separated, but not isolated, from the traditional shopping centre of the town.

On September 23 1988 the Secretary of State for Wales confirmed the Preseli Pembrokeshire District Council (Haverfordwest Town Centre Redevelopment) Compulsory Purchase Order 1987. Notice to treat was served on December 22 1988 and notice of entry on the same date, requiring possession on January 23 1989. The acquiring F authority duly entered on this date, which is agreed to be the valuation date.

The subject premises were included within the compulsory purchase order so as to promote the redevelopment of the site together with other land, including the adjacent local-authority-owned car park, for new shopping and revised car-parking arrangements. The preferred scheme, selected in November 1986 and now being implemented, uses the site of the subject premises for multi-level car parking. The planning proposals up to the mid-1980s left the subject premises undisturbed and available for retail use in accordance with a permanent planning permission dated March 24 1977. In the absence of the redevelopment scheme, in the "no scheme" world, the freeholders could therefore have reasonably anticipated the continued retail use of the buildings and site. G

At the date of entry the subject premises comprised the main supermarket within the town centre of Haverfordwest. The agreed gross internal floor area was 11,745 sq ft. The single-storey buildings were originally constructed in about 1938 in brick walls, rendered externally and roofed in corrugated asbestos sheeting on steel trusses. They were used up to 1977 as motor-car showrooms following which they were adapted to use as a supermarket by the predecessor of Kwiksave Ltd.

The buildings almost covered the total site of 1,336 m² save for a goods off-loading bay at the north-west corner connecting to Jubilee Gardens. The perimeter of the building was designed so as to display a pronounced curve to the south and west frontages.

At the date of valuation the freehold interest was subject to a lease H dated June 30 1977 held by Kwiksave Ltd for 25 years from September 2 1977 and subject to rent review at every fifth year of the term. The lease was on full repairing and insuring terms save for the roof, in respect of repairs to which a sharing formula operated between lessor and lessee. Rent was payable quarterly in arrears and the permitted user was retail except for the sale of motor vehicles, motor spare parts and accessories. The passing rent at the valuation date was £33,000 pa, agreed as from September 2 1982, but the rent review operative from September 2 1987, although set in motion, had not been determined.

The issues in dispute material to the amount of compensation payable for the freehold interest can be summarised as follows: J

(1) The rental value in accordance with the terms of the subsisting lease at the review date of September 2 1987.

(2) Likewise at the valuation date, January 23 1989.

(3) The capitalisation rate to be applied to the above rental values in order to calculate capital value.

(4) The method of allowing for the residual liability of the freeholder, if any, in respect of the future periodic repair and/or renewal of the roof of the subject premises.

(5) Whether the effect of the redevelopment proposals, the scheme, upon the ability of the lessor to agree and recover an increased rent above £33,000 pa from September 2 1987 to January 23 1989 consequent upon rent review in 1987 is compensatable by the acquiring authority and in what sum. K

(1) and (2) *Rental value at September 1987 and January 1989*

The relevant notice seeking a review of the rent was served by the lessor, but negotiations had reached a stalemate between landlord and tenant prior to entry by the acquiring authority. In effect the lessee had contended that owing to the imminent compulsory acquisition of the premises no increase over the passing rent of £33,000 was justified. The lessor had contended that this factor should be ignored and sought an increase to £55,000 pa. There the matter rested and it is agreed that, but for the acquisition, the procedure detailed in the lease would be available to the parties to resolve the dispute.

The valuations of the two expert witnesses proceeded on an L investment basis and that, in order to derive a capital value for the period January 23 1989 up to September 2 1992, the date of the next rent review, it was necessary to assume the rent review procedure had been concluded. There was, however, no agreement upon the rent to be assumed. Mr Lawley spoke to £55,789 (£4.75 per sq ft) and Mr Kelly £35,235 (£3 per sq ft). In evidence both valuers sought to support their respective rental values by reference to a wide range of comparables comprising supermarkets and large shops in other towns in west and central South Wales.

The first schedule to the lease requires that the rent on the review date shall be "a yearly rent of amount equal to the rack rental value of the demised premises". Rack-rental value is defined as "the best yearly rent at which the demised premises could be let in the open M market" having regard to (in summary):

(a) the unexpired residue of the term (15 years at September 2 1987);

(b) all other terms of the lease including future rent reviews;

(c) current rental values for similar properties in the neighbourhood (assuming vacant possession of the demised premises) but ignoring the effect on value of any goodwill or of improvements effected by the lessee.

There was no dispute between the experts as to the interpretation of these assumptions. It was accepted in evidence that the user clause in the lease did not materially affect rental value for general retail purposes and while improvements or alterations had or may have

A been effected by the lessee in 1977, the effect upon rental value, if any, was trivial in 1987.

The main difficulty faced by the witnesses was identifying contemporary (1987) rents for similar properties, supermarkets or large shops enjoying comparable trading locations and similar physical attributes in terms of the shape of the building, quality of the accommodation, goods off-loading facilities, etc.

I have inspected externally all the comparables cited in order to illuminate the evidence given. The lease directs the valuer to the rental value of comparable premises and this is also the most reliable method of arriving at the rent which the lessor and lessee might have been expected to agree in September 1987. I derive no assistance in this exercise from the comparison of rateable values and agreed rents B used by Mr Lawley as a general check upon his opinion derived from comparables. Indices of rental growth over time and actual rental increases for small shops in Bridge Street, Haverfordwest, were of little assistance, save to demonstrate generally a measure of buoyancy in the retail letting market during the 1980s.

The valuations of both experts have regard to rental growth at the subject premises between September 1982 and 1987, although that adopted by Mr Kelly was admitted to be minimal. His reason for the small increase was that basically he regarded the actual passing rent of £33,000, agreed in 1982, as "very full". I reject this assertion as not substantiated in evidence and lacking in commercial logic in the instant case, where both parties to the transaction were substantial C and were, most probably, although not positively established, advised professionally. On this general evidence I conclude that in 1987 the parties would have met in a climate of established and continuing rental growth for a variety of retail premises in South Wales generally and Haverfordwest in particular.

After viewing the comparables, considering the most relevant transactions, namely those which could have been known to the parties in September 1987 or within a reasonable period thereafter, and making adjustments for the different attributes of the comparables and the subject premises highlighted in evidence, I conclude that a rent of £50,000 pa (about £4.20 per sq ft) would have been agreed in September 1987.

By the valuation date both valuers accepted that any rent agreed in D 1987 would have been "reversionary", namely less than the full market rent upon the basis of the terms of the lease. Implicit in this assumption is a continuation of rental growth; Mr Lawley to £61,661 from £55,789 (about 10.5%) and Mr Kelly to £38,171 from £35,235 (about 8.3%). A consideration of the rents paid for the quoted comparables, at or close to January 1989, leads me to the conclusion that the rental value of the subject premises was £55,000 at the relevant date (about £4.70 per sq ft).

For the purpose of the assessment of compensation it is therefore determined that the rack-rental value of the subject premises in accordance with the lease terms was £50,000 on September 2 1987 and £55,000 on January 23 1989.

E

(3) *Capitalisation or investment yield*

The evidence of Mr Lawley was that the yield should be 10%, which he applied to both his 1987 and 1989 rental values after making a deduction to cover an estimated liability of the lessor in respect of the cost of repairs to the roof. Mr Kelly did not make an explicit deduction for future roof repairs, but had regard to this potential liability in selecting a yield of 12.5%. In evidence Mr Kelly indicated that in his selected yield of 12.5% he had attributed 1% to the roof repair liability and he spoke to an alternative valuation based upon a yield of 11.5% as a check upon his primary valuation. From this alternative valuation figure he then deducted a capital sum in respect F of the cost of future roof repairs. On a common basis, before deduction for any liability for roof repairs, the range of yield applied to their respective rental values by the witnesses was 10% to 11.5%.

From the evidence it is apparent that these two figures of yield were derived from comparable transactions but with emphasis in the expert view taken upon different investment criteria. In brief, Mr Lawley took as his principal criterion the quality of the covenants by the lessee to pay rent etc; Kwiksave Ltd is an established multiple trader of substantial quoted public company status. Mr Kelly, while acknowledging the quality of the covenant afforded by Kwiksave Ltd, also had regard to the detailed lease terms, which he considered contained provisions not universally acceptable to investors; for example, rent payable quarterly in arrear. He also paid considerable attention to the physical quality of the buildings upon the site, not G purpose-designed as a supermarket and some 50 years old.

Any decision by an investor as to the required yield on capital outlay will inevitably seek to balance a number of factors and be taken in the context of the prevailing market. In the instant case this market is illustrated by the comparables, the generality of which indicate a yield broadly in the range 9% to 12%. Well-located premises of reasonably modern design in either town centres or adjacent thereto with available car parking demonstrate yields of 10% or below and those less well located or older above 10%. To fit the subject premises into this pattern, so as to reflect its advantages and disadvantages as an investment, requires, in my judgment, a yield of 11%, this yield to be applied to both the rental values determined above and the resultant capital values to be subject to H specific adjustment in respect of the cost of future roof repairs as considered below.

The evidence relating to the value of sites for redevelopment in the locality was of limited assistance owing to the special circumstances surrounding several transactions at or close to the valuation date. There was some common ground between the parties that the site of the subject premises had potential for redevelopment for retail purposes in the absence of compulsory acquisition, although no detailed scheme was put in evidence. In any event, at the valuation date, the subsisting occupation lease had $13\frac{1}{2}$ years to run with the probability, therefore, that no redevelopment proposals could take place within that time. Neither party contended that the subject J property was so ripe for immediate redevelopment that the lessor and the lessee would, as a matter of commercial prudence in the market, jointly promote a scheme of redevelopment and share any resulting profits between them. To an extent, both valuers, Mr Kelly in particular, viewed the value inherent in the site as underpinning their investment valuations. It is realistic to assume that an investor would take a similar view at the valuation date.

(4) *Liability for roof repairs*

The lease at clause 2(5) provides that:

> the Lessee shall only be expected to keep the roof of the demised premises in a wind and water-tight condition during the continuance of the term and if it is subsequently agreed between the parties hereto that the said roof needs renewing either completely or substantially then the cost of such renewal shall K be borne equally by the Lessor and the Lessee with the further proviso that should a new roof prove necessary and be erected during the last five years of the term hereby granted then the entire cost thereof shall be borne by the Lessor.

In evidence both experts accepted that the physical life of corrugated asbestos-sheet roofing was of the order of 20 to 30 years. Neither was able to confirm the age of the existing roof in January 1989 or to speak in detail from site inspection notes taken at about that time as to its condition. The matter rested upon inference; Mr Kelly taking the view that specific provision would not have been made in the lease in 1977 if the parties were confident that the roof was then in good condition and with ample future life. Mr Lawley L relied upon the position that he had no knowledge of the lessee notifying the lessor that major repairs, by, for example, partial renewal, were required. His assumption was, therefore, that at all times up to January 1989 the roof was satisfactory and would continue to be so for some time into the future. Cost estimates for complete renewal of the roof were produced in evidence in the sums of £45,622 and £62,358 excluding VAT; the former assumes renewal in an identical material and the latter in a more efficient modern substitute.

In their valuations both experts anticipated the need for the renewal of the roof at some future date and effectively made deductions from capital value accordingly — Mr Lawley £30,000; Mr Kelly £45,480, the difference between his primary valuation at M £280,000 net and his secondary valuation at £325,480 before allowance.

As the lease had some $13\frac{1}{2}$ years unexpired at January 1989 and the roof was of a considerable age, a prudent investor-purchaser would anticipate the need for early renewal and notionally make immediate provision for financing the lessor's 50% share of the estimated cost using the modern substitute covering. In round figures this sum is £30,000, with any interest accruing to this reserve,if and when actually invested, being available to meet future rises in building costs.

Bringing these matters together produces a capital value for the freehold interest at January 1989 as detailed below:

Assumed passing rent		£50,000	
YP for 3.5 years @ 11%		2.773	£138,650
Current rental value		55,000	
YP in perpetuity after 3.5 years @ 11%		6.318	347,490
	Total		486,140
Deduct provision for roof repairs			30,000
	Capital value		456,140
		Say	£456,000

(5) *Rental recovery*

Upon rent review under the lease there is an express presumption that rack-rent is to be determined having regard to the unexpired residue of the term of the lease. This is the period for which the lessee may expect to continue to occupy the demised premises at the review date. Any material variation to that period, especially an event which will drastically reduce it, such as the imminent compulsory acquisition of the reversion coupled with the certain dispossession of the lessee, will have an impact in the open market upon rack-rent value; secure occupation becomes precarious and therefore of less value. This was the contention of the lessee in response to the lessor's notice of rent review to be operative from September 2 1987. For all practical purposes, there the matter has rested since November 1987, the rent review was outstanding at the date of entry and the lessee has paid to the lessor the then reserved rent of £33,000 pa up to the date of entry.

Mr Pugh, for the claimant, contended that the acquiring authority, by including the subject premises within the compulsory purchase order, is responsible for the rent review due in 1987 being undetermined at the date of entry. In cross-examination Mr Kelly accepted that it was known in about December 1986 that the subject premises were to be included within the redevelopment scheme. Mr Pugh submitted as a matter of law that the 1987 rent review under the lease would have to reflect the reality of the position at the review date; in effect a curtailed period of continued occupation consequent upon imminent compulsory purchase of the freehold and dispossession of the lessee.

Although he cited no authorities, he contended that the position of the claimant was protected by section 9 of the Land Compensation Act 1961, which requires the rules for the assessment of compensation in section 5 to be applied ignoring the scheme of the acquiring authority. Section 9 of the Act of 1961 reads:

> No account shall be taken of any depreciation of the value of the relevant interest which is attributable to the fact that (whether by way of designation, allocation or other particulars contained in the current development plan, or by any other means) an indication has been given that the relevant land is, or is likely, to be acquired by an authority possessing compulsory purchase powers.

In effect, Mr Pugh contends that had there been no scheme the lessor and lessee would have agreed the rent on review in 1987 or, in default in agreement, have been able to operate the mechanism for determination provided in the lease. If the scheme is to be ignored under section 9 of the Land Compensation Act 1961, it follows that all its consequences should likewise be ignored and the lessor should not be denied the increased rent receivable for the period September 2 1987 to date of entry on January 23 1989, some 17 months. The lessor is thus entitled to look to the acquiring authority for compensation in respect of this item of loss.

In response, Mr Rees, for the acquiring authority, submitted, first, that if there was any loss it arose from the claimant company not pursuing to finality the rent review mechanism contained within the lease. Second, this tribunal is required to determine the value of the freehold interest in the subject premises on January 23 1989, and any loss of rent during the preceding 17 months is not a relevant matter for their jurisdiction.

I reject the submissions of Mr Rees as inconsistent with rule (2) of section 5 of the Land Compensation Act 1961. In the absence of the compulsory acquisition of the freehold interest the claimant company would have either concluded an agreement in respect of a revised rent prior to any sale or sold the freehold interest with the right in the purchaser to conclude the review and to recover any additional unpaid rent from the lessee back to September 2 1987. Indeed, an identical situation would prevail upon a voluntary sale of the freehold reversion between the claimant company and the acquiring authority and any unpaid, but legally recoverable, rent would influence the agreed price. The provisions of the Land Compensation Acts in the instant case are intended to secure the same result with respect to price, expressed in terms of compensation, as upon a voluntary sale.

The primary valuations of both experts proceeded upon the basis that a rent, other than the £33,000 pa passing immediately prior to the date of entry, should be imputed for the first three and a half years even though the actual rent review relevant to that period of time is undetermined. The logic of this approach must rest upon the supposition that but for the compulsory acquisition a review would have taken place, or will in the future take place, so as to entitle the freeholder to recover rent not yet actually paid. Furthermore, in the real world, unclouded by compulsory acquisition, the relevant recovery period would extend back to the contractual review date in the lease.

A comparable was cited in evidence of the sale of a freehold interest with an outstanding rent review, but this affords no assistance as to the appropriate increase in the price which might be expected in the instant case. Arithmetically the amount of the additional rent not recovered can be calculated as follows:

Assumed reviewed rent at September 2 1987		£50,000
Less passing rent paid		33,000
	Increase	17,000 pa
£17,000 pa for 17 months equals		£24,083
	Say	£24,000

To secure this sum in the open market the freeholder would need to incur expense in finalising the rent review or make an allowance to the purchaser of the freehold to cover the expected costs if a sale is effected prior to the rent review being concluded. Moreover, under rule (2) of section 5 of the 1961 Act it is not the actual loss which falls to be compensated but rather the addition to market value, or expected sale price, which has to be quantified and expressed as compensation. In the absence of direct evidence on the point, but having regard to the sum of £24,000 above and seeking to read the mind of an investor-purchaser, the compensation for this item is determined in the sum of £15,000.

The total compensation awarded for the freehold interest subject to the subsisting lease is therefore in the sum of £471,000 to which will be added a surveyor's fee in accordance with Ryde's Scale (1984).

Having opened the sealed offer lodged by the acquiring authority, the amount thereof is less than the award. The acquiring authority shall pay the costs of the claimant in the reference, such costs unless agreed, to be taxed by the Registrar of the Lands Tribunal on the High Court Scale.

LONDON & PROVINCIAL POSTER GROUP LTD v OLDHAM METROPOLITAN BOROUGH COUNCIL

(REF/249/1987)

March 30 1990

(J C HILL TD FRICS)

Estates Gazette February 2 and 9 1991

[1991] 04 EG 148 and [1991] 05 EG 146

Advertising site, with three 48-sheet panels, in Oldham area — Claim for £73,846 for value of site, disturbance and severance and injurious affection — Council's primary figure £7,493 — Effect on value of council's dual policies being pursued at material date, a discontinuance policy on advertising sites and an environmental improvement policy

In early 1980s council developed, and pursued vigorously, policy of reducing proliferation of advertising sites along road where subject site situated — In August 1980 council resolved to hold discussions with owners of subject site to see if alternative site could be found and, if these were unsuccessful, to serve a discontinuance notice — That resolution unimplemented — From 1982 discontinuance notices served in batches, starting west of subject site and, by September 1985, affecting sites to east, leaving subject site well behind

A **and isolated — During 1982-84 council also operating policy of environmental improvement — Subject site included in general improvement area and, in March 1983, included in CPO — Agreed valuation date September 1985**

Claimants' valuer effectively assumed use of site for advertising would continue indefinitely — Market would have concluded 1980 resolution had ceased to be of any consequence by 1985 — Even if discontinuance notice served, there would have been reasonable chance of successful appeal — Council's valuer in effect assumed market would regard discontinuance action as both certain and immediate —
B **Following resolution to acquire subject site, service of discontinuance notice became unnecessary, but if site had been removed from CPO council would have sought to serve such notice**

Not in issue that s9 of Land Compensation Act 1961 applied and that any reduction in value of site resulting from compulsory acquisition had to be disregarded — Tribunal considered, however, that any depreciating effect of environmental improvement policy's existence must also be disregarded but did not agree that discontinuance policy came within s9's scope — In artificial "no scheme" world the market, in September 1985, could have concluded, in
C **tribunal's view, that although discontinuance action was still likely the situation was so different from that of 1980 that chances of successful appeal were reasonably high**

As starting point for its decision on valuation issues, tribunal took parties' valuations based on no threat of discontinuance action, namely claimants' valuer's sole valuation and council valuer's second alternative valuation of £24,114 — (The Tribunal rejected both council valuer's preferred valuation and his first alternative valuation) — In respect of rental value, tribunal determined £1,750 pa per 48-sheet panel to give total of £5,250 pa to which it applied the
D **claimants' valuer's YP of 8 to produce £42,000 site value — Adjustments to disturbance claim produced capital value on "no threat" basis of £61,000 — Severance and injurious affection claim (at 10% of current site value) rejected — While loss of subject site could perhaps have adversely affected value of nearby sites in claimants' ownership, any claim would have to be based on diminution in value of such other sites and could not be calculated as arbitrary percentage of value of subject site**

From stage of concluding that potential purchasers would be reasonably optimistic of successful appeal, tribunal had
E **virtually no assistance from valuation evidence and had "perforce to take a broad view" — Potential purchasers, however optimistic, would in tribunal's view, have required substantial discount on "no threat" value — Therefore, tribunal applied 50% to site value of £42,000 to give £21,000 — While "not entirely happy" in applying some arbitrary discount both to site value and to disturbance claim, tribunal considered that claimants, faced with same uncertainty as to continuance of their business as potential purchasers would be, would take much the same view and settle at risk factor of 50% of capital value of £61,000 — Award accordingly of £30,500**

F **If tribunal wrong in concluding that only council's environmental improvement policy, and not their discontinuance policy, fell to be disregarded under s9 of 1961 Act, total compensation on basis of disregarding both policies would be £61,000**

The following cases are referred to in this report.

London Borough of Hackney v *Macfarlane* (1970) 21 P&CR 342; 68 LGR 682, CA
Shaw v *London Borough of Hackney* (1974) 28 P&CR 477, LT
Trocette Property Co Ltd v *Greater London Council* (1974) 72 LGR 701; 28 P&CR 408
Watson v *Warrington Borough Council* (1979) 250 EG 977, LT

David Elvin (instructed by Blythe Liggins, of Warwick) appeared G
for the claimant; Andrew Jeffries, solicitor for Oldham Metropolitan Borough Council, represented the acquiring authority.

Giving his decision, MR J C HILL said: This is a reference to determine the amount of compensation payable to the claimant for the acquisition of its freehold interest in a site at Oldham Road, Failsworth, Oldham, Lancashire.

The matters at issue are the value of the site and the amount of compensation, if any, payable for disturbance, severance and injurious affection.

Mr David Elvin, counsel, appeared for the claimant and called Mr C C Manley ARICS and Mr A Stephenson, a regional director of Mills
& Allen Ltd, which is the parent company of the claimant company. H
Mr A Jeffries, solicitor for Oldham Borough Council, appeared for the compensating authority and called Mr J M Robinson BSc FRICS and Mr I S Fellows, both of Oldham Metropolitan Borough Council.

The site is situated on the south side of Oldham Road (A62), between the Lamb Inn public house to the east and 78 Oldham Road to the west. There is a small street immediately behind the site serving the rear of properties in Hobson Street, which is a residential street running parallel to Oldham Road. The site has a frontage to Oldham Road of about 80 ft and a depth of about 60 ft, the area being about 530 sq yds. At the material date the claimant maintained three advertising panels on the site, the panels being of the 48-sheet size, ie
20 ft long by 10 ft high. The centre panel was parallel to Oldham Road J
and the two outer panels were angled inwards towards the centre of the site frontage so that each was visible to traffic approaching from the direction in which the outer panels faced.

From the statement of agreed facts and the evidence the following information emerged: the freehold interest in the subject site was acquired by one of the claimant's predecessor companies in 1958. Planning permission was obtained in March 1960 for the erection of four advertising panels, although at some subsequent date the number of panels actually installed on the site was reduced to three. The original express consent had a limited life and, having expired, the display continued by deemed consent by virtue of regulation 13(1) of the Town and Country Planning (Control of Advertisements)
Regulations 1984. The effect of the transition from express to K
deemed consent was that a discontinuance notice could have been served by the local planning authority at any time requiring that the use of the site for advertising purposes should cease if it considered it expedient to do so in the interests of amenity or public safety. If a discontinuance notice had been served the claimant could not have claimed compensation for the removal of the advertising panels, as the use for such purpose had commenced subsequent to August 1 1948 (regulation 30(1) of the 1984 regulations). The regulations make provision for an appeal to the Secretary of State where a discontinuance notice is served.

In the early 1980s Oldham Borough Council developed and
pursued vigorously a policy of reducing the proliferation of L
advertising sites which then existed on the Oldham Road between Oldham town centre and the Manchester city boundary to the west. The policy was stated in minute 52/1982 of the council's development services committee, where it was recommended that the removal should be sought of all advertising hoardings:

(i) whose size and format were not compatible in terms of scale with their surroundings;
(ii) which did not perform the task of screening unsightly land or buildings whose appearance might prove difficult to improve;
(iii) where the hoarding owners could not reach agreement with the council over a scheme by which the owners achieved some environmental improvement to their site in return for the
hoarding being allowed to remain, perhaps in a modified M
form.

The minute also listed the first batch of 10 properties for which authorisation to instigate removal proceedings was sought. Between 1982 and 1988 discontinuance notices were authorised and served in batches in a progression from west to east.

Before the policy was formally adopted, however, the council had already resolved on August 13 1980 (minute 100/1980 of the Development Services Committee) to hold discussions with the owners of the subject site to see if an alternative site for the display of the advertising panels could be found and that if the discussions were unsuccessful a discontinuance notice should be served. It does not appear from the evidence that any serious discussions about a

A possible relocation took place until 1983. The first batch of notices which followed minute 52/1982 covered other advertisement sites in the section of Oldham Road in which the subject site is located. The second batch, in 1983, started about half a mile east of the subject site and the third batch, in 1984, started over a mile to the east.

During 1982–84 the council were also operating a policy of environmental improvement in general improvement and housing action areas within the borough. The subject site was included in the Hobson Street General Improvement Area, Failsworth, and on March 30 1983, before any action had been taken on the August 1980 minute, the council's housing committee resolved that the subject site should be included in the Albert Street West (Failsworth)
B Compulsory Purchase Order 1983 for the purpose of providing local amenity public open space. Following a public local inquiry, the order was confirmed by the Secretary of State on May 22 1985. The council made a general vesting declaration which took effect on September 30 1985, actual possession of the site being taken on December 2 1985. It is agreed that the valuation date and the date from which interest on the compensation moneys will run is September 30 1985.

Against this background it is perhaps not altogether surprising that the parties take widely differing views as to the effect of the council's policies on the value of the site. Mr Manley arrives at a total figure of £73,846; of the three valuations submitted by Mr Robinson his preferred valuation totals £7,493.

C For the claimant, Mr Manley argued that the existence of the council's unimplemented 1980 resolution dealing with possible discontinuance action would not have had a significant effect on the value of the subject site if it had been sold in the open market in September 1985 and his valuation, which is reproduced at appendix 1, effectively assumed that the use of the site for advertising purposes would continue indefinitely. His reasoning was that potential purchasers would most likely be the four or five main advertising companies who would be very familiar with the market and in particular would be aware of the council's policy towards advertising sites in the Oldham Road area. Such purchasers would know that the site had been used continuously for advertising purposes for the previous 25 years. They would be aware of the 1980 resolution and would see that sites subject to the subsequent discontinuance resolutions were situated progressively further away from the subject site so that by mid-1985 the sites affected by the last resolution (minute 32/1985 of the Development Services Committee) before the valuation date were at the Oldham town-centre end of Oldham Road, G ie at the opposite end to the subject site.

Potential purchasers would also have noted that although most of the sites had been dealt with in batches of about 10 sites, the council had shown by authorising discontinuance action in respect of a single site at the corner of Oldham Road and Fiscal Way in mid-1984 that they were prepared to deal with individual sites if it were considered expedient to do so. The council's policy as set out in minute 52/1982 was not intended to result in the removal of all advertising sites along Oldham Road and the size of the subject site, with its potential for improvement and landscaping, and the presence of nearby industrial development, suggested that even if the council had served a discontinuance notice in accordance with the 1980 resolution there would have been a reasonable chance of a successful appeal. H

Mr Manley's conclusion was that in the fiercely competitive market for advertising sites operators accepted that a much lower degree of security of tenure than that enjoyed by the occupier of normal business premises and the ever-present possibility of discontinuance action on sites not held with express planning consent were normal day-to-day risks. Such operators would take the view in 1985 that the council no longer intended to implement the 1980 resolution and that because of the lapse of time it was even possible that implementation was no longer authorised and they would not be particularly concerned about the possibility of discontinuance.

Mr Fellows, for the compensating authority, said that following the housing committee's resolution on March 30 1983 to take steps to J acquire the subject site, the serving of a discontinuance notice became unnecessary. He confirmed that the reasons for not serving a notice were correctly set out in a letter from the council to the claimant's solicitor on November 2 1988, the relevant sections of which read:

> The reasons for not serving the Discontinuance Notice between the authorisation in July 1980 and April 1983 is as follows. The 1980 report was submitted (with liaison with the Borough Planning Officer) by the Area Improvement Team, a multi-disciplinary officer team generally under the Director of Environmental Health involved in implementing policy in Housing General Improvement Areas. That team was disbanded in 1981 following cutbacks in public expenditure, and its planning functions for G.I.A.s reverted to the Borough Planning Officer. It was the Borough K Planning Officer who submitted a report on the Review to the Development Services Committee on the 2nd June 1982.
>
> It was owing to the loss of continuity when the team was disbanded that the Notice was not served.
>
> At the same time, the Borough Planning Officer was continuing with the environmental works side of the G.I.A. programme, on a reduced scale. He reported on consultations with residents to the Development Services Committee, and the result was the decision to make the Albert Street West, Failsworth CPO for open space purposes.
>
> The decision not to proceed with the Discontinuance Notice after June 1982 was based on two factors. It was felt to be duplication of effort, given the operation of the CPO. Secondly, there was some delay in getting the service of discontinuance notices under way, based on inexperience, and it was felt that the CPO procedure would be as quick. You will be aware that the L Discontinuance Notices in respect of 8, 18 and 185 Oldham Road, Failsworth (the sites at 1 Lindwood Street, 2 Newton Street and 4 Heywood Street referred to in the Development Services Committee Minutes of June 1982) were served in July and August 1983. As the procedure was implemented however, service of notices speeded up and decisions were found to be coming quickly from the Department of the Environment.

In his view, the subject site had not been forgotten and if for any reason it had been removed from the compulsory purchase order the council would have sought to serve a discontinuance notice. He was personally responsible for replying to local search inquiries. Question III in part II of the standard form of inquiry specifically asked if the council had resolved to serve a discontinuance notice; as long as the 1980 resolution had not been rescinded he had a duty to M disclose it and he considered that the disclosure would have effectively prevented the sale of the subject site for advertising purposes. By the end of 1984, 15 discontinuance notices had been served in respect of sites along Oldham Road; there had been 14 appeals of which 13 had been dismissed or withdrawn and one appeal had been allowed. The latter concerned a 16-sheet panel in a commercial area.

The display on the subject site was one of the largest in Oldham Road and although discussions had taken place with the claimant in 1983 and 1984 about a possible relocation on another site of the display on the subject site, or a possible reduction in the size of the display coupled with landscaping, the council had eventually decided

APPENDIX 1
Valuation by Mr C C Manley ARICS on behalf of claimant

Value of interest in site				
Freehold free from chief rent: approx 391m^2 — estimated site rental value @ Dec 2 1985 based on 3 × 48s				
D advertisement panels at RV £2,000/48s		£6,000pa		
Capitalised at YP		8		£48,000
Disturbance:				
1. *Loss on structure panels:* 3 × 48s panels at cost £4,500 depreciated 20%			£3,600	
2. *Loss of profits:* Gross income from 3 × 48s panels:				
1 × £400 month				
1 × £300 month				
1 × £296 month				
£996 × 12		£11,952pa		
Adjust for:	£ pa			
E (i) Site RV	6,000			
(ii) Rates	155			
(iii) Management Repair etc @ £148/16s sheet equivalent	1,332			
(iv) Interest on capital structure £3,600 @ 10%	360	7,847pa		
net adjusted profit		£4,105		
Capitalised at YP		4.25	£17,446	21,046
Severance and injurious affection to claimant's other sites held with subject site				
Estimated at 10% of current site value				4,800
F				£73,846

A that they had no similar land to offer as it was considered that 48-sheet hoardings were too large in that locality, as most of the nearby buildings were of residential scale. It was also thought that the claimant's proposals would not meet the council's objectives as the subject site was fundamentally unacceptable.

He agreed in cross-examination that the only way the council's objective of environmental improvement could be achieved was by compulsory purchase but contended that the removal of the display alone would be a benefit. He also conceded that if it had been decided to take discontinuance action on the 1980 resolution as late as 1985 it would probably have been advisable to have referred the matter back to the appropriate committee before implementation, although he saw no grounds for thinking that the 1980 resolution would have been B rescinded. He also accepted that the council would have been aware that the service of notice would have had a depressing effect on the value of the subject site, since the effect of such a notice had been specifically referred to in connection with another compulsory purchase order in minute H1 of the development services committee minutes of April 2 1982. He contended, however, that this was not a matter which was considered by the planning department.

The parties' valuations were argued at length during the five-day hearing and I will deal with them by examining in turn each of the major components, ie rental value, capitalisation rates, disturbance and severance and injurious affection.

C *Rental value*

For support for his rental value Mr Manley relied on nine transactions. Four of these related to sites owned by the City of Salford, which he said had a policy of seeking tenders for council-owned sites, and which showed rents varying from £1,050 pa to £2,160 pa per 48-sheet panel (where the panels were larger or smaller than 48 sheet he had adjusted the rents pro rata to provide a 48-sheet equivalent); one was a site licensed by the claimant from Trafford Borough Council at £1,538 pa per equivalent 48 sheet; one was licensed by the claimant from Bury Borough Council at £2,062.50 pa per 48 sheet. The remaining three, which were all in Oldham, were said by Mr Manley to have come to his knowledge only after he had prepared his valuation of the subject site and showed equivalent 48-D sheet rents of between £1,370 pa and £3,187.50 pa. The latter rent was obtained for a site adjoining the Guido public house in Oldham Road, about half a mile to the east of the subject site, and the rent had been fixed before a section 52 agreement had been entered into with the planning authority which resulted in the expenditure by the tenant of £6,100 for landscaping. From the evidence and his experience he had arrived at a rental of £2,000 pa for each of the 48-sheet displays on the subject site, which produced a total rental value of £6,000 pa.

APPENDIX 2

Valuations by Mr J M Robinson BSc FRICS on behalf of acquiring authority

E *Valuation one ("Preferred valuation")*

Assuming that the existence of the council's resolution to serve a discontinuance notice would have an effect on the valuation

(i) Present rental value		
3 × 48 sheet panels @ £1,000 pa each	£3,000pa	
*YP for 1 year @ 12%	0.89	£2,670
(ii) Reversion to site value after 1 year		
530 sq yds at capital value £10 per sq yd	£5,300	
YP in perpetuity deferred 1 year @ 10%	0.91	£4,823
Total compensation		£7,493

*1. It is considered that one year is a generous estimate of the life of an advertising site subject to discontinuance proceedings, in the normal course of events.

F

Valuation two ("First alternative valuation")

Assuming that the tribunal decides that the existence of the council resolution to serve a discontinuance notice should be ignored but assuming that the council's policy on advertising hoardings would have an effect on the valuation, in view of the risk of the council taking action against the subject site.

(i) Present rental value			
3 × 48 sheet panels @ £1,000 pa each		£3,000pa	
Capitalised at YP		2	£6,000
(ii) Loss of fixtures			
3 × 48 sheet panels @ £1,500 each			
depreciated by 50% for age	£750 each		
Assume total loss			2,250
(iii) Loss of profits			
Maximum gross income target			
2 panels @ £300 per month			
1 panel @ £400 per month			
Monthly total £1,000			
Annual maximum gross income target		£12,000	
Less			
reduction for voids, charitable lettings, client and trade discounts, say 50%		6,000	
Estimated true gross annual income		£6,000	
Less			
Outgoings			
rental value	£3,000		
rates	155		
management and services overheads			
9 × 16 sheet @ £148 per 16 sheet	£1,332		
Interest on capital			
Depreciated value of 3 × 48 sheet panels, say £2,250 @ 10%	225	4,712	
Adjusted net profit		1,288	
	YP	2	2,576
Total compensation			£10,826

NOTE: 1. It is not considered that any payment should be made in respect of severance/injurious affection because of the increase in value caused to the claimant's other existing sites which, by being reduced in number, will accrue scarcity value.

J

Valuation three ("Second alternative valuation")

Assuming the tribunal decides that neither the existence of the council resolution to serve a discontinuance notice nor the council's policy on advertising hoardings would have any effect on the valuation.

(i) Present rental value			
3 × 48 sheet panels @ £1,200 pa each		£3,600pa	
Capitalised at YP		5.5	£19,800
(ii) Loss of profits			
estimated true gross income		£6,000	
Less			
outgoings			
rental value	£3,600		
rates	155		
management etc	1,332		
interest on capital	225	5,312	
		£688	
	YP	3	2,064
(iii) Loss of fixtures			
as calculated in valuation 2			2,250
Total compensation			£24,114

NOTE: 1. As in valuation 2, no compensation payable in respect of severance/injurious affection.

L For the compensating authority, Mr Robinson took a much more pessimistic view. He said that a potential purchaser making a local search would be made aware of the 1980 resolution authorising possible discontinuance action. This knowledge would have had the effect of depressing the rental value, as the life of the site for advertising purposes would be limited and this would also reduce the usefulness of the site for long-term advertising campaigns. He produced details of four lettings of advertising sites effected by his council, the equivalent 48-sheet rentals ranging from £600 pa to £750 pa, and details of seven lettings effected by other local authorities in the Oldham area with equivalent 48-sheet rentals ranging from £400 pa to £1,300 pa. All the sites were let on licences and to reflect the difference between the limited security of tenure provided by these licences and the freehold interest enjoyed by the claimant he had M uplifted the equivalent 48-sheet rents by 25% in six cases and by 10% or 12½% in the other five cases where, for various reasons, he thought that there was a greater degree of security of tenure than in the cases to which he had applied the 25% uplift.

Mr Robinson accepted that the four sites owned by his council were inferior to the subject sites, but considered that the site in Heaton Lane, Stockport, with an equivalent 48-sheet rent of £1,300 pa (which he uplifted by 12½% to £1,462 pa), was a much better site than the subject site. From this evidence and taking into account his views concerning the restrictive effect of the existence of the 1980 resolution, for the purpose of his valuation one (which he regarded as the appropriate valuation in this case and which I shall refer to as his preferred valuation) he had arrived at a rental value of £1,000 pa for each of the 48-sheet panels on the subject site, ie a total of £3,000 pa. He had also used the same rental value in his valuation two (the first of his two alternative valuations which assumed that the 1980 resolution had to be ignored, but that the council's general policy on advertising hoardings would have had an adverse effect on value), but had increased his rental value to £1,200 pa per panel in his valuation three (the second of his alternative valuations), the latter being based on similar assumptions to those made by Mr Manley, ie effectively ignoring the existence of the 1980 resolution and the council's policy on advertising hoardings. Mr Robinson's three valuations are set out in appendix 2.

Capitalisation rates

Mr Manley referred to eight compulsory purchase settlements involving advertising sites which he had personally negotiated on behalf of the present claimant with Salford and Trafford councils. In six of the cases the capitalisation rate had been agreed at 9 years' purchase and $8\frac{1}{2}$ and 8 years' purchase in the other two cases. The sites were all freehold or very long leasehold. He considered the appropriate multiplier to be applied to his rental value in the present case to be 8 years' purchase, ie to show a return of $12\frac{1}{2}$%.

Mr Robinson relied on an acquisition of two related sites in Oldham Road, which had been dealt with by him personally and were situated almost opposite the subject site. His analysis of the transactions produced a years' purchase figure of $5\frac{1}{2}$ (ie a return of 18.18%) and he had adopted this YP for his second alternative valuation. For his preferred valuation, which assumed that a purchaser would not expect to have more than 12 months' occupation of the site, he had capitalised his rental value to show a return of 12%. For his first alternative valuation, which assumed that a purchaser would ignore the 1980 resolution but would still consider that there was a high risk of the council taking action in accordance with its known policy on advertising sites, he had used a years' purchase of 2.

Disturbance

In his preferred valuation Mr Robinson made no allowance for disturbance, arguing that the service of a discontinuance notice would preclude any claim for loss of profits or for the removal of the advertising panels. He did, however, make provision for disturbance in both of his alternative valuations. Mr Manley said that the cost of erecting a 48-sheet panel advertisement at the valuation date was about £1,500 for each panel, ie a total of £4,500. The latter figure was agreed by Mr Robinson and used by him in his alternative valuations. The valuers differed, however, in the amount they had allowed for age and condition. Mr Manley said that the panels on the subject site had been vandalised from time to time and the consequent need to repair or renew parts of the panels had resulted in the structure being well maintained, so that he considered that a 20% reduction for depreciation was adequate. Mr Robinson regarded 50% as the appropriate depreciation figure. The respective figures for loss of the panels were therefore £3,600 and £2,250.

Although the actual income from the site at the relevant date was discussed at length during the hearing and, despite the fact that the only evidence of income put forward by the claimant was a computer report which showed panel availability and rentals of the subject site and other sites only for the month of September 1985, both valuers had adopted an annual gross income figure of about £12,000. Both valuers made deductions for their respective views as to rental value; both valuers allowed for interest on capital at 10% of the depreciated value of the panels; both made identical deductions for rates and for management and repair. The major difference was that Mr Robinson made a deduction of 50% from the gross income for voids, charitable lettings and client and trade discounts, while Mr Manley made no deduction. To the resulting net profit figures Mr Manley applied a years' purchase figure of 4.25, producing a capital sum of £21,046. Mr Robinson applied a years' purchase figure of 2 in his first alternative valuation, giving a capital figure of £2,576, and a YP of 3 in his second alternative valuation, which produced a capital figure of £2,064 (the difference between the two alternative valuations being caused by the different rental values adopted).

Severance and injurious affection

Mr Manley included in his valuation an item for severance and injurious affection. He argued that an advertising company, such as the claimant's company, needed to have a portfolio of complementary sites in any particular area and that the loss of a site would reduce the effectiveness of the remaining sites by reducing the representation which could be offered to potential advertisers. He estimated the loss to the other sites held by the claimant in the Oldham Road area as being equivalent to 10% of the value of the claimant's interest in the subject site. He relied on four compensation settlements which he had agreed with Salford City Council, three of which included severance losses based on 10% of the value of the sites involved and one which was based on 20% of the site value. His capital figure under this head was £4,800. Mr Robinson made no allowance for severance in any of his valuations; he considered that any reduction in the value of other sites held by the claimant would be offset by the increased value of those sites resulting from the decreased numbers and the consequent scarcity value. Mr Jeffries argued that for a valid claim for severance to arise there had to be actual severance of the subject site from land held with that site and that even if the latter view were held to be wrong any claim for loss would have to be based on the diminution in value of the adjacent sites and not calculated as a percentage of the value of the subject site.

I was referred to the following cases: *London Borough of Hackney* v *Macfarlane* (1970) 21 P&CR 342 (CA); *Trocette Property Co Ltd* v *Greater London Council* (1974) 28 P&CR 408 (CA); *Shaw* v *Hackney London Borough Council* (1974) 28 P&CR 477 (LT); *Watson* v *Warrington Borough Council* (1979) 250 EG 977 (LT).

During the course of the hearing I visited the subject site and a number of other sites selected by the parties.

Decision

At the material date Oldham council were pursuing the two policies to which I have referred earlier, ie a discontinuance policy and an environmental improvement policy. It is not in issue that section 9 of the Land Compensation Act 1961, which says:

> No account shall be taken of any depreciation of the value of the relevant interest which is attributable to the fact that (whether by way of designation, allocation or other particulars contained in the current development plan, or by any other means) an indication has been given that the relevant land is, or is likely, to be acquired by an authority possessing compulsory purchase powers

applies to the valuation of the subject site and that any reduction in the value of the site resulting from the compulsory acquisition has to be disregarded. Mr Elvin argued that the council's environmental improvement policy also fell foul of section 9 and that any depreciating effect of the policy also had to be disregarded. In *Trocette Property Co Ltd* v *Greater London Council* (1974) 28 P&CR 408 at p 417 Megaw LJ said, when dealing with the interpretation of section 9:

> The phrase "attributable to" is, I think, deliberately used in section 9 in order to ensure greater scope for flexibility in its application than would have been achieved by other phrases, such as "caused by".

The subject site was clearly in the class of properties within the scope of the environmental improvement policy and likely to be compulsorily acquired. It seems to me, therefore, that any depreciation in the value of the subject site resulting from the existence of the policy would clearly be attributable to the policy, to secure the implementation of which the council both had and used compulsory powers, and must be disregarded.

Mr Elvin went further, however, and argued, albeit somewhat tentatively, that the council's discontinuance policy also came within the scope of section 9. I do not agree. The Town and Country Planning (Control of Advertisements) Regulations 1984 from which the council derived their powers to require discontinuance in appropriate cases make no provision for compulsory acquisition and the fact that the council possessed compulsory purchase powers under other statutes cannot, in my view, bring the discontinuance policy into the ambit of section 9.

It was presumably open to the council to have proceeded with the implementation of the 1980 resolution and, if they had been successful in securing the removal of the advertising hoardings, then to have compulsorily acquired the site to secure the required environmental improvements. The removal of the advertising panels would not have resulted in the payment of compensation by the council and the site would have subsequently been acquired at the appropriate open market value, which would have taken into account any potential uses of the site within the statutory definition of value.

The council did not take such two-stage action, however, nor did they rescind the 1980 resolution before exercising their compulsory powers. That the council knew the effect of a discontinuance notice followed by a compulsory purchase order was apparent from the Development Services Committee minute of April 2 1980, to which I have referred earlier, but Mr Fellows said that the consequences of not rescinding the 1980 resolution were never considered by his department when it was decided to seek a compulsory purchase order in respect of the subject site. It remains that the 1980 resolution was still on the record in September 1985 and, coupled with the effect of section 9, it appears to me that the result is a situation divorced from reality in that in September 1985 both the fact that the subject site was subject to a compulsory purchase order and the existence of the environmental improvement policy had in effect to be ignored, while the dormant 1980 resolution and the discontinuance policy could not, in my view, be disregarded. Such unreality is not, however, an uncommon consequence of the operation of section 9 and an earlier example of unreality created by the latter section moved Lawton LJ to remark in *Trocette* at p 420:

> The assessment of compensation in cases such as this is a most difficult task calling for the judicial use of fertile imagination.

Although he went on to say:

> It is important that this statutory world of make-believe should be kept as near as possible to reality. No assumption of any kind should be made unless provided for by statute or decided cases.

What, then, would potential purchasers have made of this mixture of reality and unreality in September 1985? They would be aware, from having made a local search, or earlier, of the council's 1980 resolution. In reality they would also be aware of the council's environmental improvement policy in general and of the Albert Street West (Failsworth) Compulsory Purchase Order 1983 in particular and inquirers would have been told that discontinuance action had been superseded by the compulsory purchase order. In the hypothetical world with which we are concerned here, however, only the existence of the 1980 resolution and the council's discontinuance policy would be known and the council would have had to find some other reason to explain why, five years after the 1980 resolution, no action had been taken to implement the resolution. Purchasers would know, from their market knowledge or from inquiry, that from 1982 onwards the council had vigorously pursued a policy (and pursued it so vigorously that Mr Manley described it as a "brutal" policy) of reducing the number of advertising sites in the Oldham Road area. They would be aware that the "rolling programme" of resolutions and discontinuance actions had started to the west of the subject property in 1982 and that by September 1985 the programme was affecting sites about a mile to the east, leaving the subject property well behind and isolated.

Although there was a time-lag between resolution and discontinuance action in respect of the sites covered by the resolutions between 1982 and 1986, which was at least partly caused by the high proportion of appeals, there was no precedent for a delay of five years without the service of a discontinuance notice (the 1980 resolution in fact differed from subsequent resolutions in that it proposed discussions about a possible relocation of the advertising panels with discontinuance action to follow if the discussions failed. In reality no discussions took place until after compulsory purchase proceedings had commenced in 1983, so that it can only be a matter of conjecture as to whether discussions would ever have taken place in the absence of such proceedings).

Mr Manley assumed that the market would have concluded that the 1980 resolution had ceased to be of any consequence and valued the subject site as if no problems existed. Mr Robinson, in his preferred valuation, in effect assumed that the market would regard discontinuance action as both certain and immediate. Mr Fellows said that there was no question of the subject site having been forgotten and that if for any reason the compulsory purchase had not proceeded he was certain that his council would have implemented the 1980 resolution. He conceded during cross-examination, however, that because of the lapse of time, if implementation had been considered in 1985 the matter would probably have first been referred back to the appropriate committee. He expressed confidence that the original resolution would have been confirmed. He may be right, but in the hypothetical situation it seems to me that confirmation of the 1980 resolution cannot be a foregone conclusion. The service of a discontinuance notice is subject to the provisions of regulation 16(1) of the Town and Country Planning (Control of Advertisements) Regulations 1984, which says:

> The local planning authority, if they consider it expedient to do so in the interests of amenity or public safety may serve a notice under this regulation (referred to in these Regulations as a "discontinuance notice") requiring the discontinuance of the display of an advertisement which is displayed with consent deemed to be granted under these Regulations, other than an advertisement of a description specified in regulation 9, or requiring the discontinuance of the use of a site for the display of such an advertisement . . .

(the remainder of regulation 16(1) is not relevant).

Clearly the council considered it expedient to contemplate discontinuance proceedings in 1980 (albeit with the alternative site discussion proviso) when the proliferation of advertising sites in the Oldham Road area was at its maximum. By 1985, however, the situation had changed considerably. The number of sites in the Oldham Road area had been reduced, and further reductions were in train, and the council were aware that although a discontinuance notice could result in the removal of the advertising panels from the subject site it would not achieve the council's primary environmental improvement requirement for the site, ie that it should be cleared and landscaped. Whether it was still expedient in 1985 that the advertising panels should be removed must be the subject of considerable doubt, but I think that, even in the unreal situation, potential purchasers could have expected that the council might still have instituted discontinuance action.

What is perhaps more relevant, however, is the view potential purchasers would have taken of the likelihood of a successful appeal against discontinuance action. The appeal rate was high — 14 appeals had been made against the 15 notices served in 1983 and 1984 — and as the subject site was one of the largest in Oldham Road I think it is safe to assume that if a discontinuance notice had been served in respect of the subject site an appeal would have followed. We now know that the failure rate on appeal was also high — Mr Fellows said that 13 of the 14 appeals against enforcement notices in Oldham Road in 1983 and 1984 were eventually dismissed or withdrawn. No doubt this is factually correct, but it appears from the documents that by September 1985 of nine appeals in the Oldham Road area which had been the subject of decision by the Secretary of State seven had been dismissed and two had been allowed, one of the latter involving a site which, although just off Oldham Road, had a relatively large display comprising two 48-sheet panels.

A feature of the decisions of the Secretary of State is that he consistently makes the point that appeals must be determined on their own merits and by reference to considerations of amenity and, where applicable, public safety. Mr Fellows described the Hobson Street General Improvement Area (of which the subject site forms part and which was wholly on the south side of Oldham Road) as being:

> of mixed use, mainly shops, some residential, but more residential in use and appearance than other sites in vicinity

by "appearance" I understood Mr Fellows to mean "scale"

As far as it goes, this description is correct, but I noted in the report of the inspector who held the public inquiry into the Albert Street West Compulsory Purchase Order that he described the order lands and the surrounding area as:

> mixed in character with predominantly industrial, commercial and residential uses . . . to the east there is the Lamb Inn, to the south and west terraced houses and to the north on the opposite side of Oldham Road there are a petrol filling station, a Territorial Army Centre and a mixture of commercial and industrial premises.

It seems to me that the Secretary of State is more likely to have considered any appeal in the context of the inspector's description of the area rather than in Mr Fellows' more localised description.

Since it was conceded by Mr Fellows that discontinuance action, although it would have resulted in the removal of the advertising panels, would not have achieved any environmental improvement in the site itself, it is debatable whether the view from the rear of the houses in Hobson Street would, as Mr Fellows said, have been improved by the replacement of the advertisement panels with the aspect of a derelict site. It may be, nevertheless, that discontinuance action would have been successful if it had been taken in the early stages of the rolling programme when it was generally accepted that there was a large mass of unsightly hoardings in the Oldham Road area, but by September 1985 and in the unreal situation then existing I think the council, unable to disclose what appears to have been the real reason for the delay, ie the compulsory purchase proceedings,

A would have found increasing difficulty in sustaining the original grounds for discontinuance action.

I think it is also relevant to note that the council granted planning consent in November 1983 for a large (64-sheet) advertising panel on a site adjoining the Guido public house in Oldham Road to the east of the subject site and under the terms of a section 52 agreement the lessee had to landscape part of the site. At the time the planning consent was granted the rolling programme included sites around and to the east of the Guido site and although it could be argued that the latter was one of the few sites for which the grant of planning consent was not inconsistent with the discontinuance policy, in the unreal "no scheme" world the absence of discontinuance action on B the subject site might have encouraged prospective purchasers to think that, particularly if they offered some landscaping treatment, they might at least avoid discontinuance action even if they did not achieve the security of occupation of the Guido site.

It is not, however, for me to attempt to anticipate the decision of the Secretary of State in the event of an appeal and I can only attempt to anticipate the conclusion at which the market for the subject site would have arrived in September 1985. Making the best I can of the evidence and arguments, I think that in the artificial "no scheme" world the market could have concluded that, although discontinuance action by the council was still likely, the situation was so different from that which existed in 1980 that the chances of a successful outcome to an appeal were reasonably high. A successful C appeal would have given the site a more or less indefinite life for advertising purposes and would have produced something approaching the maximum value for the site, perhaps likely to be exceeded only if there had been a specific planning consent for advertising purposes. I think the price paid would reflect the risk of not succeeding on appeal; I do not see that it could be related to any time-scale and it would allow for the risk either by an increase in the capitalisation percentage which would otherwise have been applied to the rental value or by an arbitrary reduction in the "no risk" capital value.

In practice, I think either alternative might have been adopted, but since both valuers have prepared valuations on the basis that there D was no threat of discontinuance action (Mr Manley's only valuation and Mr Robinson's second alternative valuation), these valuations at least provide some grounds for direct comparison and I propose to take these "no threat" valuations as the starting point for my decision on the valuation issues.

Before examining these valuations, I should perhaps at this point say that I reject both Mr Robinson's preferred valuation and his first alternative valuation.

In his preferred valuation he assumed the site to have a life for advertising purposes of only one year. I regard this period as wholly unrealistic and Mr Robinson himself had to concede during the hearing that his "generous" estimate of one year, even if discontinuance action had been certain, was inadequate in the light E of Mr Fellows' evidence that the minimum period for enforcement action by his council to have become effective was 18 months initially and later extended to over three years.

In his first alternative valuation Mr Robinson had used a figure of 2 YP to capitalise the rental value and, although he was unable to recall the actual period on which he had based the YP, he calculated during the hearing that if he adopted the same interest rate which he had used for his preferred valuation, ie 12%, the period would have been $2\frac{1}{2}$ years. Since Mr Robinson accepted that both the interest rate and the rental value in his first alternative valuation needed adjustment, I cannot place any reliance on the latter valuation and I would in any event regard $2\frac{1}{2}$ years as inadequate, bearing in mind F that his assumption in this valuation is that the council's 1980 resolution should be ignored and that the only threat was the council's discontinuance policy generally. If the 1980 resolution had not existed and the subject site had not been included in the rolling programme by 1985, by which time the programme was involving sites over a mile to the east, I think the market would have been justified in drawing the conclusion that it was not likely to be included in any future programme.

Rental value

Looking at the two "no threat" valuations, I have found some difficulty in reconciling the respective valuers' approach to the issue of rental value. Mr Robinson, as I have said earlier, produced four lettings by his council, all of which he had increased by an unsupported figure of 25% to allow for the short periods of security G involved. It emerged during his evidence that Oldham council generally negotiated rents with the tenants, unlike the adjoining Salford council, which had a policy of seeking tenders for sites in Salford ownership. Even allowing for differences in quality and position —and Mr Robinson accepted that the subject site was superior to all four of the Oldham sites — the difference between both his Oldham rents and his other rental evidence and the rentals of Mr Manley's comparables, all of which were in Trafford, Salford or Oldham, is so marked, despite Mr Robinson's arbitrary uplifts, that I cannot accept the latter's view of the rental value of the subject site. I am not satisfied, however, that Mr Manley's comparables entirely support his opinion of rental value. The rentals of six of his comparables range from £1,050 pa to £1,700 pa per 48-sheet panel, H two had rents of £2,062 pa and £2,160 pa and one produced £3,187.50 pa. The latter, the Guido site mentioned earlier, was a letting by tender of a site with planning permission and was one of the few sites in the Oldham Road area which was free from any kind of threat. In these circumstances it was probable that the site had a premium value which was reflected in the rental achieved and could not therefore be directly compared with the subject site. There was some discussion during the hearing about the consequences of the landscaping requirement arising after the rent had been fixed on the Guido site. Mr Manley argued that the cost of the landscaping increased the rental value, but it seems to me that although the tenants may in effect be paying more than they intended, the rental value is the bargain J struck in the conditions which were contemplated at the time of the tender. Making what I can of the evidence, as to the relative rents of the various sites, I have reached the conclusion that at the material date the rental value of the subject site was £1,750 pa per 48-sheet panel, giving a total rental value of £5,250 pa.

Mr Manley's eight capital value comparables produced YPs ranging between 8 and 9. Mr Robinson relied on a settlement in 1983 in respect of two related sites in Oldham Road which he analysed to show a YP of 5.5. As the latter settlement involved both the freeholder and the lessee and as the resultant YP, which is the equivalent of a yield of 18.18%, is wholly consistent with the 12% yield adopted by Mr Robinson for a term of only one year in his preferred valuation, I have little option but to accept the evidence of K Mr Manley and his YP of 8 to be applied to the rental value of the subject site.

Disturbance

By the end of the hearing the only remaining items of dispute in the claim for disturbance were the rate of depreciation to be applied to the structure on the site and the capitalisation of the adjusted net profit. Mr Manley contended for a depreciation rate of 20% and Mr Robinson for 50%. As the panels and the supports were regularly maintained or renewed I think a figure nearer to that of Mr Manley is justified and I have settled at 25%. Mr Manley's analysis of settlements he had made in connection with the compulsory purchase of six sites in the Salford area showed YPs of 4 in five cases and 4.25 in L the other case for the capitalisation of net profit. Mr Robinson's multiplier of 3 YP he derived from settlements in Sheffield and Chesterfield. I prefer the evidence of the local settlements, but I heard no reason why the YP should be higher than the 4 YP achieved in five of the six Salford cases.

Severance and injurious affection

This leaves only the claim for severance and injurious affection. While it may be that the loss of the subject site could have adversely affected the value of other nearby sites also in the ownership of the claimant, I agree with Mr Jeffries that even if such other sites could be regarded as being held with the subject site for the purpose of section 7 of the Compulsory Purchase Act 1965, any claim for injurious M affection or severance would have to be based on the diminution in the value of the other sites and could not be calculated as an arbitrary percentage of the value of the subject site, notwithstanding the agreements Mr Manley appears to have reached in respect of other settlements. Mr Manley conceded in cross-examination that Oldham's policies had had the effect of increasing the rental value of sites which were not subject to discontinuance action and Mr Robinson might well be right in saying that any diminution in value resulting from the loss of the subject site would be offset by betterment to the remaining sites. Since I cannot accept Mr Manley's approach, however, it follows that I must reject any claim for severance or injurious affection.

Bringing together the various elements, the "no threat" valuation, therefore, emerges as follows:

Rental value:				
3 × 48 sheets @ £1,750 pa per sheet				£5,250 pa
YP				8
Capital value				42,000
Disturbance:				
Loss of panels and supports: replacement costs		£4,500		
Less depreciation @ 25%		1,125	£3,375	
Loss of profits:				
Gross income: panel 1		4,800		
panel 2		3,600		
panel 3		3,552		
Total		11,952		
Less:				
Rental value	£5,250			
Rates	155			
Management, repairs etc	1,332			
Interest: 10% on £3,375	338			
Voidage on panel 2: 12% on £3,600	432			
Commission on panel 3: 15% on £3,552	533	8,040		
Adjusted net profit		3,912		
YP		4	15,648	19,023
				61,023
Capital value, say				£61,000

Even in the unreal "no scheme" world, however, and although potential purchasers are likely, as Mr Manley said, to be the major advertising companies who would be accustomed to working with considerably less security of tenure than normal commercial tenants, such purchasers would have to take a view as to the likelihood of successful discontinuance action. I have said earlier that my conclusion from the evidence is that they would have had grounds for being reasonably optimistic that if the matter went to appeal they would have achieved a successful outcome. Having reached this stage, however, I have virtually no assistance from the valuation evidence, as Mr Robinson's preferred valuation assumes virtually imminent discontinuance (I have already dealt with my reasons for rejecting the latter valuation and the first of his alternative valuations) and Mr Manley's valuation ignored the possibility of discontinuance in the foreseeable future, and I have, perforce, had to take a broad view.

However optimistic potential purchasers would have been, I think they would have still required a substantial discount on the "no threat" value and on balance I put the discount at 50%. Applied to the site value component of the "no threat" valuation, this would give a figure of £21,000. If there were no other considerations, I think I might have settled for a discount higher than 50%, but there is, however, the "fall back" value of the site if discontinuance action was subsequently taken and was successful. The only evidence before me as to the possible alternative use value of the site was that of Mr Robinson, who put the value for use as a car park at £5,300, and in the absence of any comment from Mr Manley I have no way of knowing whether the use as a public house car park would in fact have produced the optimum value for a site having a good frontage to Oldham Road and situated in a mixed-use area. Even the figure of £5,300 does, however, appear to provide a significant element of underwriting for the outlay of £21,000.

It is well established that a claim for disturbance represents the value to the claimant of the loss of his goodwill and is not necessarily what the market would pay for the goodwill. I am not entirely happy, therefore, in applying the same arbitrary discount both to the site value and to the disturbance claim, but in the circumstances of this case I think that the claimant, faced with the same uncertainty as to the continuance of his business as would be faced by potential purchasers of the site, would take much the same view as the latter and I would, accordingly, settle at a risk factor of 50% for the total of the "no threat" valuation.

I therefore award to the claimant total compensation in the sum of £30,500 together with surveyor's fees on Ryde's Scale 1984.

In the event, however, that I be wrong in my conclusion that only the council's environmental improvement policy, and not the discontinuance policy, falls to be disregarded as a consequence of the operation of section 9 of the Land Compensation Act 1961, it follows that if both policies were to be disregarded the total compensation would then be £61,000.

Having opened the sealed offer of compensation lodged by the acquiring authority I find that the amount thereof is less than the amount of my award; accordingly the acquiring authority will pay to the claimant its costs of this reference, such costs, unless agreed, to be taxed by the Registrar of the Lands Tribunal on the High Court Scale of Costs.

CRASKE AND OTHERS v NORFOLK COUNTY COUNCIL

(REF 121/1988)

November 30 1990

(HIS HONOUR JUDGE MARDER QC)

Estates Gazette March 30 1991

[1991] 12 EG 69

Powers of Lands Tribunal to amend interim award — Errors and oversights by parties in claims and evidence — Four shops in Norwich parade — Application of slip rule — Purely clerical errors or miscalculations distinguished from errors or oversights in production of evidence or argument — Whether interim award received by parties in unsigned and undated form required perfecting

Following protracted hearing, determinations made by tribunal of total compensation payable for freehold interest and on other issues — But, as result of decision to exclude certain land from acquisition, parties given opportunity to make further representations — Interim decision therefore issued in respect of matters determined — At further hearing to deal with reserved issue claimants raised number of issues concerning some errors and oversights brought to light since earlier hearing: (1) error in agreed floor areas in terms of zone A for one shop resulting from failure to take account of extension completed after figures agreed, so that award over £31,000 less than it should have been; (2) consequent adjustment in valuation of goodwill needed, otherwise overpayment of £3,716 would be made; (3) failure by claimants in one case to include element for loss of rents resulting in award being £2,367.36 less than it would have been; (4) omission to claim for employees' termination payments incurred on closure of some shop businesses

Claimants submitted that as tribunal's decision was interim award it was open to tribunal to take account of all these matters and to make necessary adjustments in formulating final award — Tribunal sat in judicial capacity with a court's jurisdiction, in control of its own practice, and entitled to recall an order before drawn and perfected in order to correct errors or oversights — Decision still capable of amendment, since it was delivered expressly as interim decision, was not dated or signed when received and therefore remained to be perfected

Tribunal, however, agreed with acquiring authority's submission that tribunal was entirely a creature of statute and its jurisdiction and power stemmed wholly from Lands Tribunal Act 1949, the rules made thereunder and those sections of Arbitration Act 1950 applied by r38 of 1975 Rules — It followed that tribunal may make interim award but, subject thereto, tribunal's award is final and binding on parties subject only to power under slip rule (1950 Act, s17) to correct in any award any clerical mistake or error arising from any accidental slip or omission — Tribunal's earlier decision was an interim award within s14 of 1950 Act — Final draft certainly signed by Member and absence of signature and date on parties' copies was no more than administration slip — Interim award complied with requirements of r54(1) of 1975 Rules and nothing more remained to be done to perfect it — It

A **was decisive of issues it dealt with and was final and binding on parties — Member *functus officio* in respect of those issues subject only to slip rule**

In tribunal's view, distinction to be drawn between purely clerical error or miscalculation and error or oversight in production of evidence or argument — Tribunal entitled to correct former but cannot remedy latter once decision given — Accordingly, no basis on which tribunal entitled to reopen issues (3) and (4) above — As to issue (1), however, error of measurement by claimants' valuer, and not demurred from by authority's valuer, was "clerical mistake or error arising from B **accidental slip or omission" within meaning of s17 of 1950 Act which tribunal had power to correct — Parties invited to resubmit recalculation of freehold interest award when precise floor areas and proportions agreed — Slip rule also applied to correct issue (2) and award accordingly reduced by £3,716 — Issue originally reserved for further consideration together with questions of interest and costs to be dealt with at later hearing**

The following cases are referred to in this report.

Fidelitas Shipping Co Ltd v *V/O Exportchleb* [1966] 1 QB 630; [1965] 2 WLR 1059; [1965] 2 All ER 4; [1965] 1 Lloyd's Rep 223, CA
Inchcape, Re [1942] 1 Ch 394
C *Merediths Ltd* v *London County Council* (1957) 9 P&CR 258
Mulholland v *Mitchell* [1971] AC 666; [1971] 2 WLR 93; [1971] 1 All ER 307, HL
Neigut v *Hanania* The Times, January 6 1983
Pittalis v *Sherefettin* [1986] 1 QB 868; [1986] 2 WLR 1003; [1986] 2 All ER 227; [1986] 1 EGLR 130; (1986) 278 EG 153, CA
R v *Cripps, ex parte Muldoon* [1984] QB 686; [1984] 3 WLR 53; [1984] 2 All ER 705; (1984) 82 LGR 439

John Holt (instructed by Daynes Hill & Perks, of Norwich) appeared for the claimants Basil Craske, Victoria Craske and Elizabeth Mutch; John Male (instructed by the solicitor to Norfolk County Council) for the acquiring authority.

Giving his decision, JUDGE MARDER said: This decision is issued D further to the interim decision which I made following a hearing in January 1990. I held a further hearing for the purpose of dealing with reserved matters on November 30 1990.

Background

The reference arose from a compulsory purchase order made by Norfolk County Council in 1986 whereby the county council were authorised to acquire all or most of the forecourts of a parade of shops [nos 363A, 363B, 365 and 367] in Aylsham Road, Norwich, for purposes of a road improvement scheme. In March 1988 the owners of the shops required the county council to purchase the whole of their interest in the properties by virtue of section 8 of the Compulsory Purchase Act 1965. The county council refused to E accept an obligation to purchase the whole interest until a late stage at the hearing in January 1990; in the light of the county council's belated acceptance of the obligation to purchase, the issue at the hearing was thereafter confined to determination of the compensation payable to the claimants in respect of a purchase of the whole of the order properties.

I set out in the earlier decision the issues which fell to be determined as follows:

(1) What is the value of the freehold interests to be acquired (it is agreed that they are to be valued as at the date of my award)?
(2) What sums are payable for injurious affection in respect of (a) the effect on trading profits of the disruption caused by the construction works in progress and (b) the loss of profits suffered F on completion of the scheme and directly attributable to the scheme?
(3) What sums are payable for disturbance calculated on the basis of extinguishment of the business as at the date of my award?

Upon the evidence presented at the protracted hearing, I determined those issues in the manner which appears from p 21 onwards of my earlier decision. To summarise my findings, I made an award of a total sum of £474,000 in respect of the freehold interests in the four subject shop properties. I made a further award in respect of loss of profits of three of the shop businesses caused by disruption during the period of construction of the road scheme; a further award was calculated by reference to the continued loss of profits caused by the adverse effect on trade of the completed road G scheme; and finally an award for disturbance in respect of one of the businesses only (the Boundary Fish Bar) based on extinguishment of that business.

It is to be noted that in valuing the freehold interests I stated (p 17): "they (the valuers for the acquiring authority and for the claimants) were agreed that the freehold interests in the subject properties fell to be valued by taking 10 years' purchase of the net annual rental income, in each case ignoring the effect of the road scheme. They have further agreed the areas of each of the properties and the areas in each to be valued in terms of zone A".

There remained an issue outstanding as to whether an area of land at the rear of no 363 was to be included in the conveyance to the council, and whether any adjustment to the compensation fell to be H made if that land were to be included or excluded. I determined that the land in question was excluded from the land to be acquired, and my decision ended with these words:

> Whether or not the exclusion of this land from the property to be acquired has an effect on valuation is a matter on which the parties must be given an opportunity to make further representations. I propose therefore to issue this as an interim decision and to invite any further submissions from the parties on this matter and as to interest on my award and as to the costs of the reference

Thus it was that a further hearing came to be held on November 30 1990.

Claimants' application

J On November 30 1990 Mr John Holt of counsel appeared for the claimants. He was prepared to deal with the issue which had been reserved as to the effect of excluding the backland and the parties had exchanged proofs of evidence on that matter. Mr Holt sought, however, to raise a number of other issues which had come to light since the earlier hearing. I can summarise those as follows:

(1) *no 365*
In regard to valuation of no 365 an error has been exposed in respect of what had been thought to be correct and agreed floor areas in terms of zone A. The error resulted from a failure to take account of an extension to the shop completed after the figures had been agreed; in consequence, the award in respect of no 365 was over £31,000 less than it should have been. K
(2) *no 363B — Hairdressers' Salon*
By an oversight the claimants had failed to include an element for loss of rents to the landlord in valuing the freehold interest. This resulted in an award which is £2,367.36 less than it would have been if such an element had been claimed and allowed.
(3) By an oversight the claimants had omitted to claim in respect of employees' termination payments incurred on closure of some of the businesses.
(4) *no 365 — Boundary Fish Bar*
As explained in a letter dated June 26 1990 by the claimants' solicitors, an adjustment in the valuation of the goodwill should follow from the corrected floor areas referred to in (1) above. Unless so adjusted the claimants would receive an overpayment in L respect of this item of £3,716.

Claimants' submissions

Mr Holt submitted that the tribunal's decision had been an interim award and therefore it was open to the tribunal to take account of all these matters and to make the necessary adjustments in formulating the final award. The Lands Tribunal was sitting in a judicial capacity and possessed the jurisdiction of a court, in control of its own practice, and entitled to recall an order before it was drawn and perfected for the purpose of correcting errors or oversights. Mr Holt referred to the note to Ord 35 r10 of the Rules of the Supreme Court, the reference being 35/10/5 in the *Supreme Court Practice* and to the case there cited, namely *Pittalis* v *Sherefettin* [1986] 1 QB 868*. M

I note that this was the case of the county court judge who, having delivered an oral judgment, almost at once decided that he had been wrong and reassembled his court to hear further argument, as a result of which he revised his order. The Court of Appeal held that course to be justified in an exceptional case (see p 879 F to G). There is in the report no suggestion of an error or oversight of the parties.

Mr Holt also referred to the case of *Mulholland* v *Mitchell* [1971] AC 666. However, that case, in my view, is concerned with the quite different issue as to the jurisdiction of the Court of Appeal as an appellate tribunal to admit fresh evidence arising after trial. I think

*Editor's note: Also reported at [1986] 1 EGLR 130.

this is also true of the passage which Mr Holt cited from *Phipson on Evidence* (14th ed) at p 295 in which the words of Lord Wilberforce in his speech in *Mulholland* v *Mitchell* at p 680 are quoted: "positively . . . it may be expected that the courts would allow fresh evidence when to refuse it would affront common sense or a sense of justice".

Mr Holt also referred me to the unreported case of *Neigut* v *Hanania* (*The Times* January 6 1983), but I found the reference too abbreviated and devoid of facts to be of value.

Mr Holt further submitted that the Lands Tribunal was entitled to act under the slip rule, available by virtue of section 17 of the Arbitration Act 1950 applied to the Lands Tribunal by rule 38 of the Lands Tribunal Rules 1975.

In any event, and apart from the slip rule, he pointed out that the tribunal decision was delivered expressly as an interim decision; that it was not dated or signed when received; and therefore remained to be perfected and thus was still capable of amendment.

However, if the matter fell to be dealt with under the slip rule, then he submitted that the power of amendment was a wide one, as the Court of Appeal explained in the course of the court's judgment in *R* v *Cripps, ex parte Muldoon* [1984] QB 686 at p 695D.

Acquiring authority's submissions

Mr John Male appeared for Norfolk County Council and opposed the applications made. Mr Male submitted that the Lands Tribunal was entirely the creature of statute and that, so far as relevant to this case, its jurisdiction and powers stemmed wholly from the Lands Tribunal Act 1949, the Lands Tribunal Rules 1975 and those sections of the Arbitration Act 1950 applied by rule 38.

Section 14 of the 1950 Act empowered the making of an interim award; the Lands Tribunal had made an interim award in respect of all matters save those specifically reserved for further consideration; that interim award was decisive of the issues determined, and the tribunal was now *functus officio* in respect of those issues. He cited the decision of the Court of Appeal in *Fidelitas Shipping Co Ltd* v *V/O Exportchleb* [1965] 1 Lloyd's Rep 223: see in particular Lord Denning MR at p 229 and Diplock LJ at pp 230 and 231.

The Lands Tribunal's own practice was in accord with the law as explained in the *Fidelitas* case: see *Merediths Ltd* v *LCC* (1957) 9 P&CR 258.

With regard to the slip rule derived from section 17 of the Arbitration Act 1950, an oversight by the claimants was not a ground for the application of the slip rule. Mr Male submitted that there was no case in which a court had applied the slip rule in order to do the claimants' work in relation to a substantive issue. Evidence or argument omitted by error is not a clerical mistake or error within the meaning of section 17 of the 1950 Act, nor is a mistake in calculation by one party. The case of *Pittalis* was that of the judge's error which the Court of Appeal held he was entitled to put right. The case of *Mulholland* related specifically to the power of the Court of Appeal to admit fresh evidence under Ord 59 r10(2). There was no comparable power in the Lands Tribunal, and no comparable function in the Lands Tribunal hearing a reference at first instance.

Mr Male did not seek to dispute the floor areas now put forward as corrected by the claimants, nor did he seek to deny that there may be an element of unfairness in retaining the error in the county council's favour. If, contrary to his submission, the tribunal had discretion to correct the award under the slip rule, then that discretion ought to be exercised only on terms as to costs incurred and thrown away by the acquiring authority.

Decision

In my judgment, Mr Male is correct in his basic submission. The jurisdiction and practice of the Lands Tribunal is derived only from statute and there is no equivalent to the "lacuna" provision in section 76 of the County Courts Act 1984.

The jurisdiction of the tribunal is to be found wholly within the confines of the Lands Tribunal Act 1949 and, so far as compensation is concerned, is limited to those matters set out in section 1(3) of the Act. The procedure of the tribunal is regulated by rules made under section 3(6); those rules may apply to the tribunal any of the provisions of the Arbitration Acts. The current rules are the Lands Tribunal Rules 1975 as amended. Rule 38 applies (*inter alia*) sections 14, 16 and 17 of the Arbitration Act 1950 to all proceedings of the tribunal. Rule 54 deals with the form of a decision of the tribunal.

It follows from these provisions, first, that the tribunal may, if it thinks fit, make an interim award (1950 Act, section 14); but, second, that subject thereto the award of the tribunal is final and binding on the parties (1950 Act, section 16) subject only to the power of the tribunal under the "slip rule" (1950 Act, section 17) to correct in any award any clerical mistake or error arising from any accidental slip or omission. As to the form of an award, rule 54(1) of the 1975 rules requires no more than that the decision of the tribunal shall be given in writing together with a statement of the tribunal's reasons for its decision.

The decision made by me and published to the parties in April 1990 under the heading of "interim decision" was an interim award. The fact that the parties may have received it in a form undated and unsigned is nothing to the point. I certainly signed the final draft, and the absence of a signature and date on the copies sent to the parties is no more than an administrative slip. The interim award complied with the requirements of rule 54(1) and nothing more remained to be done in order to "perfect it".

Accordingly, the interim award was decisive of the issues which it dealt with. It was final and binding on the parties as to those issues. I am *functus officio* in respect of those issues subject only to the provisions of the slip rule.

This view of the matter is consistent with the decision of the Court of Appeal in the *Fidelitas* case: see especially the judgment of Lord Denning MR at p 229 and of Diplock LJ (as he then was) at p 231. It accords also with the tribunal's own decision in the case of *Meredith Ltd* v *LCC* (1958) 9 P&CR 258.

The only question remaining, therefore, is whether any of the matters sought to be raised by the claimants can be dealt with by correction under the slip rule.

It is clear that the correcting power under the slip rule is not confined to errors of the court or tribunal; it may extend to error of the parties: *Re Inchcape* [1942] 1 Ch 394 and see *R* v *Cripps, ex parte Muldoon* [1984] QB 686 at p 695.

In my view, however, there is a distinction to be drawn between a purely clerical error or miscalculation on the one hand and, on the other, an error or oversight in the production of evidence or argument. The tribunal may be entitled to correct the former but cannot remedy the latter, once a decision has been given.

Accordingly, I can find no basis upon which the tribunal could be entitled to reopen those issues which Mr Holt has sought to reopen in respect of (a) the landlord's loss of profit as an element in the valuation of the hairdresser's shop (no 363B) or (b) the payments made to various employees on termination of their employment as an element of the disturbance claim.

However, I view the position differently in relation to the calculation of floor areas in the case of no 365. It was my task to determine the basis of valuation of the freehold interest, and I did that by consideration of the market evidence, and reached a conclusion as to the appropriate zone A values expressed in the last two paragraphs on p 21 of my interim decision. I have already noted that this was an agreed basis on which the valuation should proceed. What followed from that conclusion was a purely mathematical formulation, based on what were assumed to be agreed floor areas and agreed apportionments of value to different areas.

I consider that the error of measurement made by the claimants' valuer, and not demurred from by the county council's valuer, is a "clerical mistake or error arising from an accidental slip or omission" within the meaning of section 17 of the Arbitration Act 1950 which I have power to correct. In so far as the matter is one of discretion, I am in no doubt that this is a proper case to exercise that power. The mistake was that of both parties, or at any rate a mistake of the claimants acquiesced in by the county council. In the circumstances, to fail to correct the error would bring about an unjust loss of compensation to the claimants and the corresponding enrichment of the county council in a considerable sum.

However, as there appeared at the last hearing to remain an area of disagreement between the parties as to the precise degree of error. I propose to do no more at this stage than to invite the parties to submit a recalculation of the award in respect of the freehold interest in no 365 when the precise floor areas and proportions have been agreed.

I consider also that the slip rule should be applied to correct the further error relating to the valuation of the goodwill of the Boundary Fish Bar at no 365, which I understand the parties are agreed is consequential upon the measurement of floor areas. That has resulted, so I was informed, in an agreed overpayment to the claimants of £3,716 and that element of my award will be reduced accordingly.

There remain the issues which I originally reserved for further

A consideration, ie the effect, if any, on the valuation of excluding the back land at no 363 from the conveyance to the acquiring authority and the questions of interest and of costs. In the absence of agreement, these must be dealt with at a further hearing.

Since drafting this further decision. I have considered a letter dated December 7 1990 addressed to the tribunal by the claimants' solicitors. That letter sets out in detail what happened between the solicitors and tribunal staff concerning the receipt by the solicitors of my earlier interim decision in an undated and unsigned form, though bearing my typed signature.

I find nothing in this letter which would cause me to change the view I have expressed earlier in this further decision, namely that the earlier decision made by me and published to the parties in April 1990 B was an interim award within the meaning of section 14 of the Arbitration Act 1950 and an award which complied with rule 54(1) of the Lands Tribunal Rules 1975.

ABBEY HOMESTEADS (DEVELOPMENTS) LTD v NORTHAMPTONSHIRE COUNTY COUNCIL

(REF/45/1984)

October 11 1990

(V G WELLINGS QC, THE PRESIDENT)

C Estates Gazette April 13 1991 and April 20 1991

[1991] 14 EG 119 and [1991] 15 EG 121

Land at Towcester, part of area of residential development, compulsorily acquired for school — Planning agreement under s52 of Town and Country Planning Act 1971 — Restrictive covenant concerning subject land — *Pointe Gourde* principle — Whether restrictive covenant part of scheme — Whether s9 of Land Compensation Act 1961 applied

Larger area the subject of s52 agreement made in 1976 with district council — In 1973 letters had been exchanged with D county council concerning necessity for such agreement and draft agreement provided — Covenant in both draft and final agreement that 1.30ha of larger area be reserved for school purposes — At or about same time as agreement planning permission granted for residential development — Such development in due course carried out except on reserved area — In 1983 county council made CPO for 1.23ha of reserved land, taking possession in May 1984 — Certificate of alternative development for residential development of reserved land granted by district council — Planning permissions given in 1983 and 1984 for erection of school on that land

E Following decision of tribunal on preliminary point of law, Court of Appeal decided that relevant covenant in planning agreement was a restrictive covenant running with the land and was intended to be permanent — Court made declaration to effect that compensation be determined on basis that the land was affected by the restrictive covenant

Claim for £369,000 on basis of open market value with benefit of planning permission for residential development and as if there were no restrictive covenant — On same basis acquiring authority valued at £185,400 but primary contention was for £12,750 as agricultural land on basis that F restriction applied — Claimants' valuation supported by two comparables, a former cinema site in Towcester and a residential site in Oundle — Authority relied on further Towcester transaction — In tribunal's opinion, these transactions required so much adjustment as to undermine respective valuers' conclusions

Claimants' case put in four ways:

(1) Imposition of the restrictive covenant was itself part of the scheme underlying the acquisition and resulted from that scheme and any diminution in value was entirely caused by the scheme and to be ignored under *Pointe Gourde* principle

(2) Alternatively, an indication was given in 1973 that the land was or was likely to be compulsorily acquired and under G s9 of Land Compensation Act 1961 any depreciation attributable to that indication was to be ignored

(3) Alternatively, any diminution in value was caused not by the covenant but by the scheme operating on the covenant and inhibiting its removal — Accordingly, the effect of the scheme must be disregarded under *Pointe Gourde* principle as must any residual diminution attributable to the perceived cost of getting the covenant removed

(4) Alternatively, if any diminution in value was not to be disregarded, such diminution would operate to reduce the value by not more than 5%

After considering authorities on the definition of H "scheme", the tribunal found that:

A. Notwithstanding the 7-year interval between the making of the s52 agreement and the CPO, the agreement must have contemplated that a CPO would or might be required — Thus, the imposition of the restrictive covenant was part of the scheme underlying the acquisition

B. The execution by the claimants of the s52 agreement was a response, or reasonable response, within meaning given to word "indication" in s9 of 1961 Act in *Jelson Ltd* v *Blaby DC per* Stephenson LJ — Depreciation in value of claimants' interest was attributable to restrictive covenant in s52 J agreement — Alternatively, the agreement itself constituted a s9 "indication" — In further alternative, the proffering of the draft agreement or engrossment of it did so — Thus s9 applied

Tribunal agreed with claimants that Court of Appeal in its decision was dealing with real world, but tribunal in assessing compensation was not concerned with real world and was unfettered by the court's decision and declaration — Award of £300,000 — Alternative award of £255,000 on basis of claimants' submissions (3) and (4), the tribunal deducting 15% from open market value instead of 5%

The following cases are referred to in this report. K

Bird v *Wakefield Metropolitan District Council* (1978) 37 P&CR 478; [1978] EGD 533; 248 EG 499; [1979] JPL 25, CA

Birmingham City District Council v *Morris & Jacombs* (1976) 33 P&CR 27; [1976] EGD 513; 240 EG 539, CA

Fraser v *City of Fraserville* [1917] AC 187; 116 LT 258; 33 TLR 179, PC

Hertfordshire County Council v *Ozanne* [1989] 2 EGLR 19; [1989] 43 EG 182

Jelson Ltd v *Blaby District Council* [1977] 1 WLR 1020; [1978] 1 All ER 548; (1977) 75 LGR 624; 34 P&CR 77; [1977] EGD 309; 243 EG 47; [1977] JPL 579, CA

Lucas and Chesterfield Gas & Water Board, Re [1909] 1 KB 16

Myers v *Milton Keynes Development Corporation* [1974] 1 WLR 696; [1974] 2 All ER 1096; (1974) 72 LGR 420; 27 P&CR 518; [1974] EGD 405; 230 EG 1275, CA

Salop County Council v *Craddock* (1969) 213 EG 633, CA L

Wilson v *Liverpool Corporation* [1971] 1 WLR 302; [1971] 1 All ER 628; (1971) 22 P&CR 282; [1971] EGD 144; 217 EG 987, CA

Michael Burke-Gaffney QC and Mark West (instructed by Dickins Batty & Bidder) appeared for the claimants, Abbey Homesteads (Developments) Ltd; John Whittaker (instructed by the solicitor to Northamptonshire County Council) represented the acquiring authority.

Giving his decision, **THE PRESIDENT** said: This is a reference by Northamptonshire County Council (which I will call "the county council") for the determination of the amount of compensation to which the claimants, Abbey Homesteads (Developments) Ltd (which M I will call "the company") is entitled for compulsory acquisition by the county council of 1.23 ha of land belonging to the company and situate off Balmoral Close, Towcester, Northamptonshire.

The 1.23 ha (3.04 acres) was part of an area of 13.88 ha (34.3 acres) which the company purchased prior to 1974 from the former Northamptonshire County Council. In March 1976 or thereabouts the company made two outline applications to the district council for planning permission for residential development of the 13.88 ha. On June 9 1967 the district council and the company entered into an agreement under section 52 of the Town and Country Planning Act 1971 [now section 106 of the Town and Country Planning Act 1990].

A plan attached to the agreement shows the 13.88 ha outlined with

A a thick black line. The agreement contains six recitals, the fourth of which is:

The district council wishes to secure that in the event of planning permission being granted to the applications or one of them the development of the land shall in the interests of the proper planning of the area be restricted and regulated in accordance with the provisions of this agreement (including provisions of a financial character).

Clause 1 of the operative part provides, in effect, that pursuant to the provisions of section 52 of the 1971 Act the company agrees, declares and covenants with the district council that from the date on which planning permission for residential development of the 13.88 ha is granted the land shall be subject to the conditions restricting and regulating the development specified in the first schedule to the
B agreement in addition to any conditions or limitations subject to which planning permission may be granted as set out in the second schedule, together with any further conditions or limitations imposed on any appeal by the Secretary of State for the Environment. The relevant covenant is contained in para 5 of the first schedule to the agreement and is as follows:

An area of one point three (1.30) hectares adjacent to the playing field and amenity open space areas shall be reserved for school purposes.

Clauses 3 and 4 of the operative part of the agreement provide for payment of sums of money by the company to the district council and for the manner in which the sums are to be allocated by the district council, namely for the provision of the sewerage and sewage disposal facilities and a flood alleviation scheme. The first schedule
C to the agreement, in addition to the covenant to which I have referred, also contains provisions with respect to surface water drainage, the construction of roads and bridges and phasing of the development. The second schedule contains the conditions, which, according to the sixth recital, the district council proposed should be attached to any planning permission granted by it in respect of either of the applications therefor. Para 9 of that schedule deals with the provision of open space. Para 12 thereof contains a condition as follows:

An area 1.30 hectares in the approximate position indicated on the deposited plan 710/2 adjacent to the playing field and amenity open space areas shall be reserved for school purposes . . .

D At or about the date of the section 52 agreement, outline planning permission for the residential development of the 13.88 ha by the erection thereon of 343 units, incorporating associated open space, a site for a primary school and provision for local shopping, was granted by the district council. The planning permission was subject to the same conditions as those set out in the second schedule to the agreement. Approval of reserved matters was given by the district council as to 272 dwellings, garages, casual parking spaces, roads and ancillary drainage works on July 7 1977 and, as to the residue of the 343 units, on May 25 1978. Thereafter, the company carried out the permitted development on the 13.88 ha but omitted from the development a piece of land in the north-west corner of the 13.88 ha and that piece of land so omitted has an area of 1.30 ha.

E On July 7 1983 the county council, as education authority, made the Northamptonshire County Council (Towcester, New County Primary School) Compulsory Purchase Order 1983 under the Education Act 1944 and other enabling Acts. The purpose of the order was the acquisition of the 1.23 ha. The residue of 0.07 ha of the 1.30 ha to which the relevant restrictive covenant and planning condition apply was omitted from the order.

In the statement of reasons accompanying the compulsory purchase order, after referring to the planning permission and the section 52 agreement, it was stated:

The residential development has now been largely completed. The county council considers there is a need for a primary school site within the
F development to serve an area which includes the new housing since the capacity of the existing schools in Towcester is not sufficient to cope with the anticipated increase in the numbers of children of school age . . .

At the end of the statement of reasons it was stated that it had not been necessary to consider alternative sites, since the order land was reserved for school purposes under the section 52 agreement.

No objection to the compulsory purchase order was made by the company or at all and it was duly confirmed. Notice to treat was served on December 30 1983 and the county council took possession of the 1.23 ha on May 18 1984, which date accordingly is the valuation date.

Another company, Colroy Homes Ltd, owned approximately 20 ha of undeveloped land adjoining the 1.23 ha on its south-western
G side. In June 1976, that company was granted outline planning permission by the district council for the residential development of that land. The district council and Colroy Homes entered into a section 52 agreement dated June 4 1976. The agreement contained covenants on the part of Colroy Homes Ltd relating to phasing, density and open space including the provision of playing fields in the flood plain. No part of the land of Colroy Homes Ltd was subject to the compulsory purchase order but two small areas of that land immediately adjoining the 1.23 ha were acquired by the county council by private treaty from that company in order to square off what would otherwise have been an irregular boundary.

The company claims compensation for the land acquired in the sum of £369,000 together with appropriate interest and fees. The
H county council's case is that the effect of the restrictive covenant contained in para 5 of the first schedule to the section 52 agreement to which the company was party is to reduce the value of the land acquired to agricultural value, namely £12,750. As will be seen, the company puts its case in a variety of ways. The minimum sum for compensation for the interest acquired claimed by it on any basis is £350,550.

Earlier decisions

In my decision on preliminary points of law in this matter dated November 9 1984 I decided that para 5 of the first schedule to the section 52 agreement between the district council and the company dated June 4 1976 (a copy of which is contained in an appendix to that
J decision) was positive in sense rather than negative and, therefore, did not establish a restrictive covenant, that that paragraph applied to the whole of the 13.88 ha, that it was of limited duration, that is to say until the planning objective was achieved by the provision of a school site of 1.30 ha within the 13.88 ha, and that since that paragraph did not create a restrictive covenant, it did not as a matter of law preclude a valuation under rule (2) of section 5 of the Land Compensation Act 1961 as at the valuation date on the basis of the value of the 1.30 ha for residential purposes. I further held that the county council did not have the benefit of para 5 of the first schedule to the agreement. In that decision I dealt fully with the planning history and need not repeat it here.

K Upon a case stated by me to the Court of Appeal, that court [on March 19 1986] upheld my decision that the county council did not have the benefit of para 5 of the first schedule to the agreement but reversed my other decisions. The court held that the relevant paragraph was not void for uncertainty, that the identity of the 1.30 ha was intended to be ascertained and was ascertained in the course of construction by the developers themselves. The court further held that the agreement did create a restrictive covenant running with the land, which restrictive covenant was intended to be permanent. The court granted declarations to give effect to the decisions in the judgments, one of which declarations was in effect as follows: the amount of compensation to be paid by the county council for the compulsory acquisition from the company of 1.23 ha of land
L adjoining Balmoral Close, Towcester, be determined on the basis that such land is affected by the said restrictive covenant: see (1986) 53 P&CR 1 at p 13.*

On August 2 1983 planning permission for the erection on the 1.23 ha of a 245-place primary school was given. However, on June 6 1984 planning permission for the erection on that land of a 315-place primary school and nursery unit was given.

On August 10 1983 the district council granted to the company a certificate of alternative development in respect of the 1.23 ha. The certificate provided that if the land in question were not proposed to be acquired by any authority possessing compulsory purchase powers, planning permission would have been granted for residential
M development subject to conditions which, *inter alia,* were that before any development was commenced, detailed plans, drawings and particulars of the layout, number, siting, design and appearance of the proposed dwellings and the means of access thereto, together with landscaping and screen walls/fences should be submitted to and approved by the local planning authority and the development should be carried out in accordance therewith; and that no development should take place on that part of the site within the flood plain as indicated on the plan attached to the certificate. The reason given for the last-mentioned condition was that "the land is liable to flooding".

---

*Editor's note: Also reported at [1986] 1 EGLR 24.

A The company, through its counsel, Mr Michael Burke-Gaffney QC and Mr Mark West, accepts that the interest to be valued is the freehold of the 1.23 ha but subject to the relevant restrictive covenant. Counsel say that the Court of Appeal in its judgments and declarations granted by it was dealing with the real world but that the Lands Tribunal now, in assessing compensation, is not talking about the real world and is unfettered by those judgments and declarations.

Counsel for the company called Mr Paul Westoby BSc (Est Man) ARICS, principal of the firm of Habgood & Mammatt, chartered surveyors of Witney, Oxfordshire. Mr John Whittaker, for the county council, called Mr Anthony David Woods, assistant education officer of the county council; Mr Graham John Court, an
B architect and assistant director (design), major projects, in the department of land and buildings of the county council since January 1989 and previously thereto that council's principal architect since 1974; and Mr John Reginald Hodges ARICS, employed by the county council as assistant director (estates) in their land and buildings department.

Parties' valuations

Mr Westoby, in his valuation 1, assessed the open market value of the interest acquired in the 1.23 ha as at May 18 1984, with the benefit of a planning permission for residential development, and as if there were no restrictive covenant affecting that land, in the sum of £369,000. Mr Hodges' valuation, as at the same date and on the same
C basis, was £185,400.

It is obvious that the value of the 1.23 ha, as at the valuation date, was seriously depreciated by the effect of the restrictive covenant. Mr Westoby did not indicate what the amount of that depreciation might be. He did not need to do so, because the gravamen of his evidence was that that depreciation in value (whatever might be its amount) was to be ignored.

Neither sought to value the land expressly as a site for use for school purposes. Mr Hodges could find no demand for use for such purposes and valued the land, on the basis that the restriction applied, as agricultural land, in the sum of £12,750. Mr Westoby thought that that sum even on that basis was insufficient because the comparable on which it was based, unlike the 1.23 ha, did not have
D the benefit of a planning permission for residential use. However, Mr Westoby did not proffer an agricultural value of his own.

The amount of the depreciation attributable to the restrictive covenant must be, or must lie between:

(1) £369,000 and £12,750 or
(2) £185,400 and £12,750.

The first decision which I must make is as to the open market value of the land on the valuation date with the benefit of planning permission for residential development, but on the assumption that the land was not subject to the restrictive covenant.

Mr Westoby supported his valuation 1 by two comparable
E transactions. The first related to the site of a former cinema, that is to say, 1.25 acres of land adjacent to the A5 road on its western side in the centre of Towcester. Mr Westoby made a number of adjustments to the price of £80,000 paid.

First, he added to the price of £80,000 a percentage based on the difference in the national index of residential land values at July 1983 and May 1984. For that reason he added 20% to the price, but I agree with Mr Whittaker that, if this adjustment is made, the percentage addition should be 15.5 (the difference in the index between 135 for the second half of 1983 and 156 for the first half of 1984). Mr Westoby added to the price so adjusted the cost of demolishing the buildings, namely according to him £25,000, but I think that it is now agreed that the cost should be £18,000. Mr Westoby then added 10%
F because the cinema site did not have the benefit of an actual planning permission. Mr Hodges would allow 2.5% on that account, but I shall apply 10%. To the balance Mr Westoby added 20% because the 1.23 ha is a greenfield site, which in the open market would have a higher value than the former cinema site. Mr Hodges would have allowed 5% on that basis, but I accept Mr Westoby's figure of 20%. Accordingly, the analysis of the old cinema site transaction becomes:

£80,000 plus 15.5%	£92,400
Add demolition costs	18,000
	110,400
Add 10% for lack of planning permission	11,040
	121,440
Add 20% for greenfield site	24,288 G
	£145,728

£145,728 for 1.25 acres equals £116,582 per acre
£116,582 per acre × 2.47109 equals £288,085 per ha

Thus the price paid under the cinema site transaction, if adjusted in this way, would support a value for the 1.23 ha as at the valuation date of £354,345 (£288,085 × 1.23).

Mr Westoby's second comparable is a transaction at Oundle in respect of 0.95 acre with a residential planning permission for 10 units sold by public auction on November 29 1984 at the price of £91,000. Mr Westoby, as per his appendix 4-4c, adjusted that price to
£106,326 per acre. He expressed the opinion that residential land H
values in Towcester would be 15% higher than in Oundle:

£106,326 per acre + 15% = £122,275 per acre
A value of £122,275 per acre × 2.47109 = £302,152 per ha
That price multiplied by 1.23 ha = £371,647

Mr Hodges thought that there was no comparison to be made between residential areas in Oundle and those in Towcester. I see some force in that opinion, but do not disregard the Oundle transaction completely, at all events, if used as a check.

The main transaction relied upon by Mr Hodges was the sale in June 1984 of land in Watling Street, Towcester, for the price of £240,000 for 3.12 acres. From that price Mr Hodges deducted
£20,000 as the value of a retained office on the site and added J
demolition costs of £20,000 (agreed). He then added £24,000 (10% of £240,000) because the land had no planning permission; he deducted £48,000 (20% of £240,000) for improved location and added £12,000 (5%) for risk of abnormals. The adjusted price thus became £228,000 which for 3.12 acres represents £73,077 per acre.

I think that this transaction is useful, but I agree with Mr Westoby's criticism of it in that it is on the wrong side of the busy A5 road and is divided by a stream. I do not think that the deduction of 20% or any percentage for improved location should be made. It is not a greenfield site and for that reason is inferior to the 1.23 ha.

Mr Hodges made use of the value of £73,077 derived from his Watling Street transaction as follows:

2.5 acres at £73,077		£182,692.50 K
0.54 acres at £5,000 (flood plain)		£2,700.00
		£185,392.50
	Say	£185,400.00

I agree with Mr Westoby that land in the flood plain, though it could not be developed, could be used as playing fields or other open space and ought to be accorded residential value. For that reason, Mr Hodges' valuation on his own figures ought to be increased to £222,154.

In my opinion, the prices obtained in the transactions relied on by
Mr Westoby and Mr Hodges require so much adjustment as to L
undermine the conclusions which they seek to draw from them.

I propose to determine the value of the 1.23 ha as at the valuation date in the sum of £300,000. I note that if no adjustment at all were made to the price of £91,000 paid for the Oundle property, that transaction would support a valuation for residential purposes of an Oundle property as at November 29 1984 of £236,703 per ha, which would equate to £291,245 for 1.23 ha. It is improbable that the reference land of 1.23 ha could be worth less than that figure.

The four ways in which the company puts its case are as follows:

(1) The imposition of the restrictive covenant in the section 52 agreement was itself part of the scheme underlying the acquisition and resulted from that scheme, and any diminution in the value of the
relevant interest (the freehold subject to the covenant) was entirely M
caused by the scheme; and under the *Pointe Gourde* principle that diminution is to be ignored.

(2) Alternatively, an indication was given on September 10 1973 that the relevant land was or was likely to be acquired by an authority possessing compulsory purchase powers, and under section 9 of the Land Compensation Act 1961, any depreciation of the value of the relevant interest attributable to that indication is to be ignored.

(3) Alternatively, any diminution in the value of the relevant interest was caused not by the covenant itself but by the scheme operating upon the covenant and inhibiting its removal. Accordingly, the effect of the scheme in so inhibiting the removal of the covenant must be disregarded, as must any residual diminution

A attributable to the perceived cost of getting the covenant removed either voluntarily or by application to the Lands Tribunal, under the principle in *Pointe Gourde*.

(4) In the alternative, if, contrary to the company's contention, any diminution in the value of its interest attributable to the existence of the covenant or of the scheme or attributable to the indication is not to be disregarded, then such diminution will operate to reduce the value of the said interest by not more than 5%.

It would seem that the sum claimed by the company by reference to the first two of these four bases is £369,000 and that, by reference to each of the third and fourth bases, the sum claimed is £350,550 (that is to say, £369,000 less 5%). I am not sure that I have understood the
B difference between the third and fourth bases. Mr Whittaker said that the third basis is just another way of saying that the restrictive covenant is part of the scheme. Mr Burke-Gaffney said that the third basis assumes that the covenant is not part of the scheme, that it is an independent covenant which on its own has no meaning and will achieve nothing without the scheme; if the land is not required for school purposes, he said, the covenant is bare.

Mr Whittaker said that he did not deny the *Pointe Gourde* principle. He accepted that there was a scheme of acquisition but alleged that there was no diminution in value attributable to it. It was the restrictive covenant, he said, which caused the diminution but the restrictive covenant was not part of the scheme; it was imposed by the local planning authority for proper planning purposes. He accepted
C that the restrictive covenant caused a diminution in value of the 1.23 ha so that the land was worth less with the restriction than without it. The company said that it had to be ignored, but, he said, that was wrong.

The company's first basis

According to the company, the start of the scheme was a memorandum from the chief education officer of the pre-1974 Northamptonshire County Council to the county planning officer with reference to the Towcester Development Plan, stating that it would be sound policy if a site of 3 acres, adjacent to a new public recreation ground towards the south-western boundary of the development, were reserved for another primary school. In this
D connection I was referred to the contract relating to the purchase by the company of the 13.88 acres and to two of its conditions, as well as to a report made in 1973 by the company's architects. It would seem that in that year an application by the company to the former county council for outline planning permission had been made and had been accompanied by that report. On July 27 1973, the company's solicitors wrote to the clerk of the former county council stating:

I understand that before the outline planning permission can be granted a section 52 agreement must be entered into and I should be obliged if you could let me have a draft of this as soon as possible together with details of the proposed conditions that will be inserted in the planning permission.

On August 3 1973 the clerk of that council replied:

I confirm that a section 52 agreement will be necessary before the outline
E planning permission can be granted.

On September 10 1973 a draft of the proposed section 52 agreement was provided to the company. The parties mentioned in the draft were the former county council, Towcester Rural District Council and the company. The draft contained recitals, *inter alia,* as follows:

(A) The county council is the local planning authority for the administrative county of Northampton;

(B) The district council is the public health authority and exercises powers under the Land Drainage Acts 1930-1961 for the Rural District of Towcester.

(C) . . .

(D) . . .

F (E) The county council and district council jointly and separately wish to secure that in the event of planning permission being granted to the application the development of the land shall in the interests of the proper planning of the area be restricted and regulated in accordance with the provisions of this agreement . . .

(F) . . .

(G) . . .

Para 5 of the first schedule to the draft contained a proposed restrictive covenant in precisely the same words as those of para 5 of the first schedule to the agreement dated June 4 1976, to which agreement I have already referred. On the same date outline planning permission was granted to the company and the section 52 agreement between Colroy Homes and the county council was entered into.

Mr Whittaker submitted that there was no evidence of a proposal G to put a primary school on the 1.23 ha, as opposed to the Colroy Homes land, prior to 1976. The architects' 1973 report had been made on behalf of both Colroy Homes and the company and it was not clear from that report whether the proposed school was to be on the 1.23 ha or on the Colroy Homes land. Mr Whittaker said that no scheme could be identified prior to March 1977, that is to say, when the company applied for approval of reserved matters, because until then even the company did not know the identity of the land to which the restriction was to apply. Indeed he placed the beginning of the scheme as at September 1977, that is to say, after the reserved matters had been approved. There was, he said, no suggestion or project of acquisition until then. What mattered, he said, was the scheme as it existed at the valuation date. He referred to internal memoranda of H the county council and documents passing between the county council and the district council. The first mention of the possibility of making a compulsory purchase order in respect of the 1.23 ha had been in a county council internal memorandum dated November 27 1981.

Mr Burke-Gaffney, too, referred to internal memoranda of the county council and, more particularly, to a memorandum from the chief education officer to the county architect dated May 23 1973 saying: "As far as the size of site is concerned I should like you to reserve sufficient land for an annual intake of 80 primary pupils a year plus a nursery unit amounting to 3.25 acres. In addition a further area of 3 acres will be required for a playing field . . .". The J memorandum also referred to timing and commented: "The proposed sites, taking into account the position of the existing school, will be in a good position to serve the whole of Towcester as at present planned . . .". Reference was made to the same subject in a memorandum from the county education officer to the county planning officer dated February 6 1976, to which was attached a copy of the memorandum dated May 23 1973.

Mr Burke-Gaffney said that the covenant and the scheme were one and the same thing. The covenant was there because the scheme was there. The covenant was part, and an essential part, of the scheme because without the reservation the land would not have been used as a school. Further, the covenant was part of the scheme by express K reference: see the statement of reasons accompanying the compulsory purchase order. There was no doubt, he said, that the scheme including the covenant diminished the value of the land.

Mr Burke-Gaffney asked why the company should not have the residential value of its land. The reservation of the school site did not mean that the company was giving away the right to compensation for what it had lost. The company accepted the reservation but not the diminution in value of the 1.23 ha.

Some authorities

In *Re Lucas and Chesterfield Gas & Water Board* [1909] 1 KB 16 at p 29, Fletcher Moulton LJ said:

The principles upon which compensation is assessed when land is taken under L compulsory powers are well settled. The owner receives for the lands he gives up their equivalent, ie, that which they were worth to him in money. His property is therefore not diminished in amount, but to that extent it is compulsorily changed in form. But the equivalent is estimated on the value to him, and not on the value to the purchaser, and hence it has from the first been recognized as an absolute rule that this value is to be estimated as it stood before the grant of the compulsory powers. The owner is only to receive compensation based upon the market value of his lands as they stood before the scheme was authorized by which they are put to public uses. Subject to that he is entitled to be paid the full price for his lands, and any and every element of value which they possess must be taken into consideration in so far as they increase the value to him.

In *Fraser* v *City of Fraserville* [1917] AC 187 at p 194, Lord M Buckmaster said:

The principles which regulate the fixing of compensation of lands compulsorily acquired have been the subject of many decisions, and among the most recent are those of *In Re Lucas and Chesterfield Gas and Water Board, Cedars Rapids Manufacturing & Power Co* v *Lacoste* [1914] AC 569, PC and *Sidney* v *North Eastern Railway Co* [1914] 3 KB 629, DC. The principles of those cases are carefully and correctly considered in the judgments the subject of appeal, and the substance of them is this: that the value to be ascertained is the value to the seller of the property in its actual condition and at the time of expropriation with all its existing advantages and with all its possibilities, excluding any advantage due to the carrying out of the scheme for which the property is compulsorily acquired, the question of what is the scheme being a question of fact for the arbitrator in each case.

Lord Denning MR in *Myers* v *Milton Keynes Development Corporation* (1974) 230 EG 1275 at p 1281 said:

It was apparent, therefore, that the valuation in the present case had to be done in an imaginary state of affairs in which there was no scheme. The valuer must cast aside his knowledge of what had in fact happened in the last eight years due to the scheme. He must ignore the developments which would in all probability take place in the future 10 years owing to the scheme. Instead, he must let his imagination take flight to the clouds. He must conjure up a land of make-believe, where there had not been, nor would be, a brave new town: but where there was to be supposed the old order of things continuing — a county planning authority which would grant planning permission of various kinds, at such times and in such parcels as it thought best, but with an assurance that in March, 1980, planning permission would be available for the residential development of the Walton Manor Estate. In this imaginary state of affairs, the valuer had then to ask himself what was the appropriate way of valuing the land with this assumed planning permission.

In *Birmingham City District Council* v *Morris & Jacombs Ltd* (1976) 33 P&CR 27 at p 33*, Sir John Pennycuick reviewed the authorities and said:

There has been some discussion as to what is meant by "a scheme" in this connection. A scheme means, I think, no more than a project on the part of the authority concerned to acquire land, and of course to acquire it for some purpose for which it was authorised so to acquire it.

He then referred to *Wilson* v *Liverpool Corporation* [1971] 1 WLR 302† and *Salop County Council* v *Craddock* (1969) 213 EG 633.

Ormrod LJ said at p 39:

There is clearly a danger that the word "scheme" is acquiring or will acquire a mystique of its own and be used as a starting point for highly elaborate — I almost said "sophisticated" — argument, whereas if one looks at the authorities the scheme referred to is the scheme for which the property is being compulsorily acquired.

Third, Stamp LJ, after referring to the authorities, added:

These principles, in my judgment, run through the authorities to which we have been referred as far as the authorities are relevant to the issue in the instant case. The word "scheme" in the context in which it is used in the cases, in those circumstances, refers, in my judgment, to a scheme pursuant to which the property is being compulsorily acquired, and not, as was submitted by counsel for the respondents, to a policy or, if you will, a "scheme" whereunder permission to develop the property as building land would if applied for have been refused.

Some excerpts from the judgments of the Court of Appeal in the present case are:

(1) Lawton LJ (1986) 53 P&CR 1 at p 4‡:

On July 7, 1983 the county council made a compulsory purchase order . . . stating it was for a primary school for 245 children. On June 6, 1984 it gave development permission for the erection of a primary school and unit for 315 children. This number was probably greater than was necessary to accommodate the children on the developers' estate. It was intended to accommodate children from an adjacent housing estate as well as from the developers' estate. In my judgment this fact does not affect the planning purpose which paragraph 5 was intended to achieve.

(2) *Ibid*, at p 5:

As I have already stated paragraph 5 was put into the first schedule with the object of securing the proper planning of the area. As long as there were 343 houses on the estate there would be need for a primary school somewhere on it. The need would be a continuing one. Proper planning for the area would not have been achieved had the covenant allowed the county council, or a private person, to start a school on the 1.30 hectares, and then, after a short time, close it down and use the buildings for some other purpose.

(3) Parker LJ at p 7:

The respondents' contention that the covenant if it existed was of limited duration is, as I understood it, based on the notion that its purpose was merely to enable the local education authority to acquire the land for a school and accordingly that once that authority had acquired it the covenant was spent. If this was indeed the contention it is in my judgment misconceived. It was for the district council to determine, as a matter of planning, where they wished a school to be. It was and is the duty of the local education authority to secure that there are available for its area sufficient primary and secondary schools. It is however not obliged itself to provide all the schools and it is specifically empowered by section 6(1) of the Education Act 1953 to make arrangements for the provision of primary and secondary schools not maintained by it. There is therefore no reason why the local education authority should not have made arrangements with, for example, some charitable body that it should purchase the land in question and provide a school. If such a body did purchase the land but the covenant then lapsed that body would be possessed of land upon which there was no fetter. Furthermore even if the purchaser is the local education authority it by no means follows that the covenant has no further use, for such authority might, after a time, decide that in accordance with its duties it could close down one school and wished to close down this particular school. In such circumstances the covenant could remain of much importance to the district council.

If then, as I think, the covenant binds the land, what the respondents had, as soon as the land was identified, was land subject to a permanent covenant which would clearly affect its value in the open market. They contend however that it must be valued on the basis that it could be sold on the open market free of the covenant.

The only bases upon which, in my view, such a contention could succeed are: (i) if the covenant had been discharged before the acquisition of the land under the compulsory purchase order, which it had not, or (ii) if the circumstances were such as to justify discharge at that time.

(4) Nourse LJ at p 9:

The next question is whether the obligation imposed by condition 5, which admittedly is in form positive, is in substance positive or negative. If, as I have earlier identified it, the effect of the condition is that the development must be so restricted and regulated that the specified area is left available for school purposes, the substance of it is that the development must not be carried out so as to interfere with that requirement. That is in substance a negative and not a positive obligation.

(5) As to the declarations granted by the Court of Appeal in the present case, see p 120 of this decision and (1986) 53 P&CR 1 at p 9.

Mann LJ in *Hertfordshire County Council* v *Ozanne* [1989] 2 EGLR 19; [1989] 43 EG 182 said at p 184:

The *Pointe Gourde* principle must operate to prevent a scheme enhancing value and equally to prevent the scheme diminishing value. It is to be remembered that a scheme is a creature of evolution (see Lord Denning MR [1971] 1 WLR 302* at p 309D). It is also to be remembered that one authority can inherit or adopt the scheme of another (see *Bird* v *Wakefield Metropolitan District Council* (1978) 37 P&CR 478 at p 487). The change from Bishop's Stortford UDC to East Hertfordshire DC thus has no materiality.

Finding as to the identity of the scheme underlying the acquisition

The scheme underlying the acquisition in the present case consists of or includes physical development in the form of the erection of a primary school on the 1.23 ha. That is shown by any one or more of the following facts:

(1) that a primary school, providing 315 places and known as Nicholas Hawksmoor School, has been erected on the 1.23 ha and is in use as such; it was built in 1985;
(2) that the compulsory purchase order was made under the Education Act 1944 and other enabling powers;
(3) that the statement of reasons which accompanied the compulsory purchase order stated its purpose.

A scheme of which the imposition of a restrictive covenant is part does not readily fit the definition of "scheme" as found in the authorities, but that may be because the facts of this case have never before arisen. If the question "What is the scheme?" is one of fact for the arbitrator, the answer must vary from case to case.

There was a seven-year interval between the making of the section 52 agreement and the compulsory purchase order and there seems to have been no express mention of the possibility that a compulsory purchase order would be necessary in aid of the reservation in the agreement. However, the section 52 agreement was expressly mentioned in the statement of reasons for the making of the compulsory purchase order, notwithstanding that an authority different from the one which made the order had the benefit of it. The purpose of the covenant, that the specified area be left available for school purposes (see *per* Nourse LJ), did not require the company to build the school; a primary school to supply the needs of the education authority was in the circumstances contemplated; such a school could only be a school provided by the education authority, that is to say, a state school or, to use Parker LJ's example, a school with the same function as that of a state primary school but provided by a charity. If the company was unwilling to sell its land to the education authority for such a purpose, a compulsory purchase order would have been necessary.

Accordingly, para 5 of the first schedule to the section 52 agreement must have contemplated that a compulsory purchase order would or might be required. For that reason, I hold that the imposition of the restrictive covenant was part of the scheme

---

*Editor's note: In report at [1986] 1 EGLR 24 see respectively pp 25L, 26D, 26L-27B and 27H for quotations in paras (1)-(4).
†Editor's note: Also reported at [1976] EGD 513 at p 520; (1976) 240 EG 539 at p 541.
‡Editor's note: Also reported at [1971] EGD 144; (1970) 217 EG 987.

*Editor's note: *Wilson* v *Liverpool Corporation (supra).*

underlying the acquisition. Similar observations as those which I have made with regard to the section 52 agreement are applicable to the draft section 52 agreement delivered by the rural district council, and it seems to me that the scheme originated with the letters dated July 27 1973 and August 3 1973 and with the delivery of the draft. The scheme originated with the rural district council and was taken over by the district council in 1974 and by the county council on the making of the compulsory purchase order: see *Bird* v *Wakefield Metropolitan District Council* (1978) 37 P&CR 478, at p 487*.

In summary, the scheme consisted of the erection of a primary school of at least 245 places on land which was the subject of the requirement in para 5 of Schedule 1 to the section 52 agreement, the imposition of which requirement was itself part of the scheme.

Section 9 of the Land Compensation Act 1961

No account shall be taken of any depreciation of the value of the relevant interest which is attributable to the fact that (whether by way of designation, allocation or other particulars contained in the current development plan, or by any other means) an indication has been given that the relevant land is, or is likely, to be acquired by an authority possessing compulsory purchase powers.

The expression "the relevant land" in the present case means the 1.23 ha. In *Jelson Ltd* v *Blaby District Council* (1977) 34 P&CR 77 at p 81†, Stephenson LJ said:

It seems to me that what flowed from the indication included a reasonable response by the owners to the indication.

According to Mr Burke-Gaffney, the letters dated July 27 1973 and August 3 1973 together with the draft which contained the same covenant as that ultimately included in para 5 of Schedule 1 to the section 52 agreement constituted the indication. He said that the company responded and reasonably responded to that indication by entering into the section 52 agreement itself. Accordingly, the depreciation in value suffered by the company's interest in the 1.23 ha was attributable to the indication as he had defined it. Mr Whittaker said that nothing in those letters or the draft indicated that the 1.23 ha were going to be or were likely to be acquired by an authority possessing compulsory purchase powers. Moreover, he said, if there was such an indication, there was no evidence that the company entered into the section 52 agreement in response to the indication.

The rural district council, no doubt, had no compulsory purchase powers enabling them to acquire land for a school site. If, as was likely, the company was unwilling to transfer its interest in the 1.23 ha to the education authority, compulsory purchase powers would be necessary; they could be exercised only by the education authority. Accordingly, it seems to me that the documents relied upon by Mr Burke-Gaffney did constitute an indication that the land in question was likely to be acquired by an authority possessing compulsory purchase powers. I regard the execution by the company of the section 52 agreement as a response, and reasonable response, within the meaning given to the word "indication" by Stephenson LJ in *Jelson*. The depreciation in value of the company's interest in the 1.23 ha was attributable to the restrictive covenant contained in that agreement and, in the alternative, it seems to me that that agreement itself constituted an indication within the meaning of section 9 or, in the further alternative, that the proffering by the district council to the company of the draft of the section 52 agreement, or the engrossment of it, ready for execution by the company constituted the indication and the execution constituted the response. For these reasons I hold that section 9 of the Act of 1961 applies.

Award

I have given careful attention to the judgments of the Court of Appeal in the present case and to the declaration (that the amount of compensation to be paid by the county council for the compulsory acquisition of the 1.23 ha be determined on the basis that such land is affected by the restrictive covenant) which I have mentioned on p 120 of this decision and I have given careful thought to the question whether the Lands Tribunal is inhibited by the judgments and the declaration in the assessment of compensation. Mr Whittaker has not taken that point and it appears to me that Mr Burke-Gaffney was right when he said that, although the interest to be valued is the freehold subject to the restrictive covenant, and although the Court of Appeal was dealing with the real world, the Lands Tribunal in assessing compensation is not concerned with the real world and is unfettered by the decision and declaration of the court.

I can, therefore, proceed to assess compensation in the normal way. Having regard to my assessment of the full open market residential value of the company's interest acquired, to my finding as to the identity of the scheme and to my decision with respect to section 9 of the Land Compensation Act 1961, I award to the company by way of compensation for that interest the sum of £300,000 together with surveyor's fees on Ryde's Scale (1984).

Mr Westoby's alternative valuation

I consider this valuation in case I should be wrong in my reasoning leading to the above award. I consider that valuation in the light of the company's third basis of claim, alternatively its fourth basis, without distinguishing between them. That was Mr Westoby's approach. He put the matter in this way:

(1) The interest for sale in the open market comprises the freehold of the 1.23 ha with vacant possession having a planning permission for use as a primary school and the assumption of a planning permission for residential development. The land is affected by a restrictive covenant "reserving the land for school purposes".
(2) For the purposes of calculating the compensation, any decrease in value entirely due to the scheme underlying the acquisition is to be ignored; so, too, is any decrease attributable to the giving of any indication by any means that the land is or is likely to be compulsorily acquired.
(3) In the open market the prospective purchasers or bidders would include private individuals, housebuilding companies, housing associations and others interested in purchasing land for residential development; school site purchasers, including possibly the county council and possibly also certain church bodies which operate primary schools; the district council; the vendor company; and land investors and speculators.
(4) It was to be assumed that the county council was in the market for a school site but not the 1.23 ha. If it did bid for that land it would have to bid alongside other bidders and be prepared to pay, like them, the full market price less a small percentage.
(5) The county council had an urgent need for a school site in Towcester and it would consider other suitable sites, of which there were six in addition to the 1.23 ha. Two of those sites were particularly suitable: "Site No 1", namely a piece of agricultural land of more than 3 acres adjoining and immediately to the south-west of the 1.23 ha and zoned for residential use in the Towcester District Plan for 1981 and "Site No 6", an area of agricultural land to the north-east of the 13.88 ha also zoned for residential use. His preferred site was site no 1. If the county council were to bid for the 1.23 ha, that would encourage other bidders to bid more, since the county council were not going to exercise compulsory purchase powers.
(6) In the no-scheme world he would expect the county council to purchase one of the six alternative sites,if necessary, using compulsory purchase powers.
(7) A purchaser of the 1.23 ha in the open market would expect that, so soon as the county council had bought one of the other sites, the district council, acting reasonably, would be willing to lift the covenant. Alternatively, he would assume that he would have to go to the Lands Tribunal and that, if so, he would be successful there.
(8) Such a purchaser would be willing to pay the full open market residential value for the 1.23 ha less 5%, that is to say, £350,550.
(9) He would be patient just as he would be if there were some problem relating to access: he would be prepared to "bank" the land and wait.

The need for a school site in Towcester

The provision of a new primary school of 315 places in Towcester was planned in the 1970s. The number of places was calculated by applying to the estimated number of houses, namely 900, the formula:

$$\frac{3 \times 7}{60}$$
$$900 \times 0.35 = 315$$

Since 1984 a different formula has been used. Prior to the opening of the Nicholas Hawksmoor School there was a shortage of school

*Editor's note: Also reported at [1978] EGD 533; (1978) 248 EG 499.
†Editor's note: Also reported at [1977] EGD 309 at p 313; (1977) 243 EG 47 at p 49.

A places in Towcester and temporary measures, such as "mobile" classrooms had to be used. The intake of pupils at the bottom end of the school range of ages was 80 pa. The intake of pupils generated by the company's development on the 13.88 ha was 16 pa, so that the intake for Towcester as a whole was five times that for the 13.88 ha. Thus, while a new school was necessary in order to accommodate the primary school children coming from the company's development, it was by no means entirely necessary for that reason.

Alternative award

Mr Whittaker said that even in the no-scheme world it was permissible to assume that the county council might make a compulsory purchase order for the 1.23 ha. Mr Burke-Gaffney said
B that that was not permissible. I agree with him: see the passage from the judgment of Lord Denning MR in *Myers* which I have quoted. It is to be assumed that the county council has no scheme of acquisition for the 1.23 ha. Subject to quantum, I accept Mr Westoby's thesis. I agree with Mr West [for the claimants] that if there were no proposal to put a school on the 1.23 ha then the purchaser in the open market would assume that the original purpose of the covenant had gone. Accordingly, the purchaser would believe that the district council, acting reasonably, would agree to the discharge of the covenant. If the matter had to go to the Lands Tribunal, then the purchaser would assume that the covenant would be discharged under para (*a*) of section 84(1) of the Law of Property Act 1925. It is not necessary to consider the other grounds available in that subsection, but I accept
C Mr West's argument in relation to them.

I think that Mr Westoby is rather optimistic in assuming that the prospective purchaser would be willing to pay as much as 95% of the full open market residential value of the land. As a matter of caution, I would substitute 85%. Therefore, if it were necessary for me to do so I would make an alternative award to the company of £300,000 less 15%, namely £255,000 together with fees as before.

The acquiring authority will pay two-thirds of the claimants' costs of the proceedings, such costs if not agreed to be taxed by the Registrar of the Lands Tribunal on the High Court standard scale.

D

WROTHAM PARK SETTLED ESTATES v HERTSMERE BOROUGH COUNCIL

(REF/2/1987)

March 26 1991

(His Honour Judge MARDER QC)

Estates Gazette June 1 1991 and June 8 1991

[1991] 21 EG 123 and [1991] 22 EG 135

**Claim for injurious affection under s 10 of Compulsory Purchase Act 1965 — Land at Elstree, Hertfordshire, subject
E to burden of restrictive covenant — Land compulsorily acquired and developed under statutory powers without compliance with restrictive covenant — Preliminary issue of measure of compensation — Whether claimants entitled to diminution in value of benefited land or amount representing price for relaxation of covenant**

**Covenant affecting restricted land provided that the land could not be developed for building purposes except in strict accordance with a layout plan approved by the claimants and that no building could be erected except in accordance with plans and elevations previously approved — Council acquired restricted land in 1976 under Housing Act 1957 powers and in
F 1984 commenced development of part without submitting a layout plan — Claimants' application for interlocutory injunction dismissed in High Court in 1985 — Claim made for £307,500, calculated as proportion of difference in value between site as developed by council and site if developed in accordance with maximum permitted density if layout plan submitted to claimants**

Detailed examination of authorities carried out by tribunal — Those reported cases directly arising under s 68 of Lands Clauses Consolidation Act 1845 (now re-enacted as s 10 of 1965 Act) showed it had consistently been held that (a) a claim in respect of injurious affection is sustainable only where the claimant would have a right of action for damages but for the G **exercise of statutory powers and (b) where the claim is sustainable in respect of an interference with the claimant's rights in his land, the right to compensation has been determined by reference to the loss in value of the claimant's property — Review of modern cases in which damages had been awarded by reference to "loss of bargain" (none of them concerned with s 68 claims) led tribunal to conclude there was no rule, and no authority, requiring compensation payable under s 10 of 1965 Act for injurious affection in present circumstances to be calculated by reference to the amount likely to be receivable by the claimant in free negotiation for relaxation of the restrictive covenant — True rule, shown by** H **s 68 cases, is that claimant is entitled to be compensated either for actual damage caused to his land or, in an appropriate case, for loss of value of his land attributable to works executed on servient land — Parliament had conferred on the council statutory powers in relation to housing and the council were entitled in exercise of those powers to carry out works and to disregard or override restrictive covenants where necessary to do so for the purposes of carrying out those works — The so-called "breach" of covenant was accordingly lawful and not wrongful — *SJC Construction Ltd* v *Sutton Borough Council* and *Wrotham Park Estate* v *Parkside Homes Ltd* distinguished — In those and other "licence"** J **cases the awards were in recognition of the defendants' wrongdoing or their proposed breach of covenant and were an endeavour to do justice by avoiding unjust enrichment of infringers**

Held: Where compensation is payable under s 10 in respect of a breach of covenant comprising execution of works in exercise of statutory powers authorising those works, that compensation is to be assessed by reference to the diminution in value of the benefited land consequent on the carrying out of the authorised works on the restricted land K

The following cases are referred to in this report.

Argyle Motors (Birkenhead) Ltd v *Birkenhead Corporation* [1975] AC 99; [1974] 2 WLR 71; [1974] 1 All ER 201; (1973) 27 P&CR 122; 72 LGR 147, HL

Bracewell v *Appleby* [1975] Ch 408; [1975] 2 WLR 282; [1975] 1 All ER 993; [1976] EGD 190; (1974) 237 EG 731

Caledonian Railway Co v *Walker's Trustees* (1882) 7 App Cas 259

Carr-Saunders v *Dick McNeil Associates Ltd* [1986] 1 WLR 922; [1986] 2 All ER 888; (1986) 53 P&CR 14; [1986] 2 EGLR 181; 279 EG 1359

Horn v *Sunderland Corporation* [1941] 2 KB 26; [1941] 1 All ER 480, CA

Kirby v *Harrogate School Board* [1896] 1 Ch 437

Long Eaton Recreation Grounds Co v *Midland Railway Co* [1902] 2 KB 574

Metropolitan Board of Works v *McCarthy* (1874) LR 7 HL 243; 31 LT 182, HL L

Penny and South Eastern Railway Co, Re (1857) 7 EL&BL 660; 26 LJ QB 225; 5 WR 612

Ricket v *Metropolitan Railway Co* (1867) LR 2 HL 175

SJC Construction Co Ltd v *Sutton London Borough Council* (1975) 29 P&CR 322; [1975] EGD 725; 234 EG 363, CA

Stockport Metropolitan Borough Council v *Alwiyah Developments* (1986) 52 P&CR 278

Stoke-on-Trent City Council v *W&J Wass Ltd* [1988] 1 WLR 1406; [1988] 3 All ER 394; (1988) 87 LGR 129, CA

Wadham v *North Eastern Railway Co* (1884) 14 QBD 747

Wrotham Park Settled Estates v *Maclean Homes (North London) Ltd* unreported, February 1985

Wrotham Park Estate Co Ltd v *Parkside Homes Ltd* [1974] 1 WLR 798; [1974] 2 All ER 321; (1973) 27 P&CR 296

M

George Bartlett QC and Michael Druce (instructed by Farrer & Co) appeared for the claimants; John Howell (instructed by Carter Lemon) represented the compensating authority.

Giving his decision, **Judge Marder** said: This is a reference to determine the compensation payable by the compensating authority to the claimants by virtue of section 10 of the Compulsory Purchase Act 1965 for injurious affection to the claimants' land [forming part of the Studio Estate, Elstree, Hertfordshire] caused by the execution of works on adjoining land, consisting of housing development carried out by the compensating authority in breach of covenants imposed for the benefit of the claimants' adjoining estate.

At the instance of the parties, the tribunal agreed to determine as a

A preliminary point of law the appropriate basis for assessment of the compensation. The parties have formulated a question for the tribunal which may be summarised in its simplest form as whether the claimants are to be compensated (a) by reference to any diminution in value of their estate consequent on the carrying out of development in breach of covenant, as the compensating authority contends, or (b) by reference to the amount which could reasonably have been expected to be paid to the claimants as the price for permitting that development to take place, as the claimants contend.

The parties put forward a statement of facts which were agreed only for the purpose of determining the point at issue. Those facts may be summarised as follows:

(1) The claimant is a corporate body and owner of the Wrotham
B Park Estate. The estate extends to about 2,400 acres and comprises a number of blocks of land. Those parts of the estate closest to the restricted land are shown edged black on the plan attached to this decision [not reproduced here]. The estate includes a principal mansion house and a number of smaller houses and numerous cottages, together with substantial areas of agricultural land.

(2) By a conveyance dated July 1944 the predecessor in title to the estate sold an area of some 51 acres ("the restricted land") being formerly part of the estate near Elstree, Hertfordshire. The restricted land is shown hatched on the plan. The conveyance contains a covenant by the purchaser for the
C benefit of the owner or owners for the time being of the Wrotham Park Estate:

> Not to develop the said land for building purposes except in strict accordance with the layout plan to be first submitted to and approved in writing by the vendor or his agent (who shall not unreasonably withhold his consent to any such development) nor shall any building be erected except in accordance with the plans and elevations to be previously approved by the vendor or his agent . . .

(3) The Hertsmere Borough Council ("the council") acquired the restricted land by virtue of powers conferred upon it by Part V of the Housing Act 1957. The acquisition was completed in July 1976.

(4) In October 1984, in the exercise of statutory powers to provide
D housing accommodation, the council started to develop part of the restricted land, an area of some 5.5 acres known as H3 (and so marked on the plan). No layout plan for that area had been submitted to the claimant or its agent. The development thus started constituted a breach of covenant, but was not actionable by reason only of the fact that it was carried out by the council in pursuance of its statutory powers.

(5) In *Wrotham Park Settled Estates* v *Maclean Homes (North London) Ltd and Hertsmere Borough Council* (unreported) the claimants sought an injunction restraining the builders from developing site H3 in breach of the covenant. An application for an interlocutory injunction was dismissed by Mervyn Davies J in the Chancery Division of the High Court
E in February 1985. It is clear from a transcript of the judgment in that case that although the development was being carried out by the builders under an agreement with the local authority, partly for the purpose of enabling the restrictive covenant to be disregarded, the learned judge refused an interlocutory injunction on the sole ground that the council were in occupation of the land and were carrying out the development in the exercise of their statutory powers under the Housing Act.

(6) On November 13 1986 the claimants claimed compensation in respect of the injurious affection to the Wrotham Park Estate through breach by the council of the restrictive covenant in
F developing H3 for building purposes without first having submitted or having approved a layout plan. The compensation claimed was £307,500, calculated as a proportion of the difference in value between site H3 as developed by the council and its value if developed in accordance with the maximum density the claimants would have permitted if a layout had been submitted to them for approval.

The question now formulated by the parties for the tribunal's determination is as follows:

> Whether the basis upon which the claim for compensation in respect of injurious affection of the interest should be assessed is:
>
> (a) The amount of the difference (if any) between
>
> (i) the value of the interest immediately before the commencement of the development with the benefit of the restrictive covenant (but disregarding the power of an authority to override the restrictive covenant); and G
>
> (ii) the value of the interest immediately thereafter given the carrying out of the development; or
>
> (b) The amount (if any and if different from (a)) which could reasonably have been expected to be paid at the time when the development had begun to the owner of the interest for allowing such amount of the development (if any) as was in excess of that which under the restrictive covenant it could not reasonably have withheld its consent to when considering any layout plan submitted to it for approval.

Claimants' submissions H

Mr George Bartlett QC, on behalf of the claimants, submitted that as owners of the dominant tenement the claimants would have a right of action for breach of covenant were it not for the fact that the council were executing works under statutory powers. Accordingly, the claimants were entitled to compensation under section 10 of the Compulsory Purchase Act 1965 for injurious affection caused by the execution of the works. The principles established in respect of such claims under section 68 of the Lands Clauses Consolidation Act 1845 (now re-enacted as section 10 of the 1965 Act) entitle the claimants to damages assessed in the same way as a plaintiff's claim for damages if the works had not been carried out under statutory powers. He submitted that there was no reason in fairness or in law why the J claimants should be any worse off than if they had been in a position to sue for damages in respect of a breach of covenant.

Mr Bartlett referred me to a passage in the judgment of Scott LJ in the case of *Horn* v *Sunderland Corporation* [1941] 2 KB 26 at p42. Scott LJ, after dealing with payment of compensation for the acquisition of land, went on to say:

> There is a third kind given by the Act, namely by section 68, but that has nothing to do with compulsory acquisition. It is a remedy for injuries caused by the works authorised by the Act to the lands of an owner who has none of his land taken in that locality. The remedy is given because Parliament by authorising the works has prevented damage caused by them from being actionable, and the compensation is given as a substitute for the damages at law. K

Mr Bartlett also referred me to *Kirby* v *Harrogate School Board* [1896] 1 Ch 437 and to *Metropolitan Board of Works* v *McCarthy* (1874) LR 7 HL 243. He cited a passage from the speech of Lord Penzance in the latter case at p262:

> It may reasonably be inferred that the legislature, in authorising the works and thus taking away any rights of action which the owner of land would have had if the works had been constructed by his neighbour, intended to confer on such owner a right to compensation co-extensive with the rights of action of which the statutes had deprived him.

He referred also to *Long Eaton Recreation Grounds Co* v *Midland Railway Co* [1902] 2 KB 574 and *Argyle Motors (Birkenhead) Ltd* v *Birkenhead Corporation* [1975] AC 99. L

Mr Bartlett further submitted that the appropriate basis for the assessment of damages was what the parties would have agreed as the price for relaxation of the covenant. In support of that proposition he relied on the following cases: *Wrotham Park Estate Co Ltd* v *Parkside Homes Ltd* [1974] 1 WLR 798, *Bracewell* v *Appleby* [1975] Ch 408, *Carr-Saunders* v *Dick McNeil Associates Ltd* [1986] 2 All ER 888 and *Stoke-on-Trent City Council* v *W & J Wass Ltd* [1988] 1 WLR 1406.

Submissions of respondent compensating authority

Mr John Howell, who represented the council, submitted that where statutory powers had been exercised so as to override the provisions of a restrictive covenant, the only remedy of a person entitled to the benefit of the covenant was the statutory entitlement to M compensation under section 68 of the 1845 Act, now in section 10 of the 1965 Act. Both statutes provided only for "compensation in respect of any lands, or of any interest therein, which shall have been . . . injuriously affected by the execution of the works". Mr Howell submitted there was no power to award compensation other than for the diminution in value of the land for the benefit of which the covenant was taken. He said that it was clear in the light of well-established principles that such compensation is payable only in respect of either (a) physical damage to the benefited land or (b) the loss of value to the benefited land. On the authorities, compensation was not payable in respect of all losses caused by the execution of the works. Hence the loss of business profits was not recoverable as such:

A see the *Argyle Motors* case. Nor was compensation payable for "mere personal obstruction or inconvenience or damage occasioned to a man's trade or the goodwill of his business": (see the *McCarthy* case, *op cit* p 256), nor if the damage in respect of which compensation is sought "is only incident to the particular purpose for which the owner is occupying the land": see the *Long Eaton* case *per* Collins MR at p 581.

Mr Howell asserted that, since there was no claim made in this case in respect of physical damage, the only claim open to the claimants was for the loss of value caused by execution of the works.

Mr Howell further submitted that the loss of the power to "extract a ransom" was not in any event the result of the execution of the works but rather the result of the grant of statutory powers to do the
B works. It is the grant of statutory powers which authorises the local authority to disregard or override the restrictive covenant. If the claimants have suffered any loss by being deprived of the right to bargain for the release of the covenant, that is not in consequence of the building works carried out on the servient land but because the servient land is owned by a local authority possessing statutory powers. Execution of the works may be the occasion for the right to compensation to arise, but those works do not themselves cause any loss of the claimants' bargaining powers. Under section 10, compensation may be awarded only in respect of the injurious effect on the land of the execution of the authorised works.

It was also contended by Mr Howell that, in any event,
C compensation in respect of any diminution in value is not payable to any greater extent than damages would be payable if the statutory authority did not exist, and the compensation may indeed be less than such damages. In support of that proposition Mr Howell cited the case of *Re Penny and South Eastern Railway Co* (1857) 7 EL & BL 660, a case approved by Lord Chelmsford LC in *Ricket* v *Metropolitan Railway Co* (1867) LR 2 HL 175.

As to the measure of damages, Mr Howell examined in detail the cases cited by Mr Bartlett in support of the proposition that the appropriate method of calculating damages was by reference to the amount the plaintiff might receive in return for a relaxation of the covenant. Mr Howell submitted that the established and normal rule was to place the plaintiff in the same position as if the covenant had
D not been broken. He said that the basic approach at common law is to compensate for the breach and not for its release. Accordingly, it was contrary to that basic principle to assess damages by "loss of ransom", and where no diminution of value of the benefited land is proved, nominal damages only are recoverable. Where, on the other hand, a loss of value of the land is shown, then it is in accord with the basic principle that damages be measured by reference to that loss of value.

Mr Howell submitted that on a proper analysis of the cases it was clear that in no case was damages awarded on the basis of a loss of bargain or "ransom", save in what are referred to as "user cases", eg *Bracewell* v *Appleby*, or in cases akin to the award of exemplary damages where the defendant was in effect to be deprived of unjust
E enrichment, eg the *Parkside Homes* case. Mr Howell relied in particular on the analysis of the cases in the judgment of Nourse LJ in the *Stoke-on-Trent* case.

Finally, he submitted that the present case of the exercise of statutory powers was not to be equated with the "wrong-doing" in *Parkside Homes*. The local authority were not unlawfully in breach of the restrictive covenant but were entitled by virtue of their statutory powers not to be bound by it.

Decision

The relevant statutory provision in this case in section 10 of the Compulsory Purchase Act 1965, which so far as material provides:

> (1) If any person claims compensation in respect of any land, or any interest in
> F land, which has been taken for or injuriously affected by the execution of the works, and for which the acquiring authority have not made satisfaction under the provisions of this Act . . . any dispute arising in relation to the compensation shall be referred to and determined by the Lands Tribunal.
> (2) This section shall be construed as affording in all cases a right to compensation for injurious affection to land which is the same as the right which section 68 of the Lands Clauses Consolidation Act 1845 has been construed as affording . . .

I accept the basic proposition put forward by the compensating authority that the only right of the claimants is derived from the statute and is a right to be compensated to the extent that their interest in the benefited land has been injuriously affected by the execution of the works, that is to say, by the building operations carried out on the servient land by the compensating authority.

That right to compensation is expressed in section 10(2) to be co- G extensive to the right "which section 68 of the 1845 Act has been construed as affording". The first question, therefore, is whether, as the council contend, section 68 has been construed as affording a right to claim damages assessed by reference to the diminution of value in the benefited land consequent upon the carrying out of the works or whether, as the claimants contend, the cases in which section 68 has been construed authorise the assessment of damages based on what the claimants would, but for the exercise of statutory powers, have been entitled to receive for relaxation or release of the covenant.

It is relevant in this context to note that none of the modern cases cited by Mr Bartlett in which damages have been awarded by reference to "loss of bargain" was concerned with claims for H compensation for injurious affection under section 68. In *Wrotham Park Estate* v *Parkside Homes* the damages were awarded in lieu of a mandatory injunction requiring the demolition of 14 houses erected in breach of covenant. In *Bracewell* v *Appleby*, likewise damages were awarded in lieu of a mandatory injunction for the loss of amenity caused by the unauthorised increased user of a right of way. In *Carr-Saunders* v *Dick McNeil Associates*, the plaintiff sued in respect of the obstruction of his right of light and again was awarded damages in lieu of an injunction. Finally, in *S J C Construction Co Ltd* v *Sutton London Borough Council* (1975) 29 P&CR 322, compensation was awarded upon the discharge of a restrictive covenant under the provisions of section 84 of the Law of Property Act 1925. J

However, it is the principal submission of the claimants that section 68 is to be construed as affording the same right to damages as they would have in the absence of statutory powers and to be assessed on the same basis. It is necessary to examine the authorities to determine whether or not that is correct.

Examination of authorities — section 68 cases

In *Re Penny and South Eastern Railway* (1857) 7 EL & BL 660, in a claim brought by virtue of section 68, damages had been awarded by a jury as compensation for a house being overlooked from a newly constructed railway. The Queen's Bench quashed the jury's verdict. Lord Campbell CJ at p 668 stated:

> . . . I am clearly of opinion that he is not entitled to any compensation for the K overlooking of his premises . . . unless the particular injury would have been actionable before the company had acquired their statutory powers it is not an injury for which compensation can be claimed.

The claim for compensation under section 68 is thus co-extensive with the right of action for damages which would exist but for the exercise of statutory powers.

The judgment of Lord Campbell CJ in *Re Penny* was expressly approved and applied by Lord Chelmsford MR in *Ricket* v *Metropolitan Railway*. In that case the owner of a public house sustained a loss of business profits when access to his premises was obstructed for a period during the course of public works carried out under statutory powers. Lord Chelmsford at p 196 stated: L

> Upon a review of all the authorities and upon a consideration of the sections of the statutes relating to this subject, I have satisfied myself that the temporary obstruction of the highway which prevented the free passage of persons along it, and so incidentally interrupted the resort to the plaintiff's public house, would not have been the subject of an action at common law as an individual injury sustained by the plaintiff distinguishing his case from that of the rest of the public. Therefore he altogether fails to bring himself within the general principle upon which a claim to compensation under the Acts in question has been determined to depend.

In the same case, Lord Cranworth at p 198 stated:

> Both principle and authority seem to me to show that no case comes within the purview of the statute, unless where some damage has been occasioned to the land itself, in respect of which, but for the statute, the complaining party M might have maintained an action. The injury must be actual injury to the land itself, as by loosening the foundation of buildings on it, obstructing its light or its drains, making it inaccessible by lowering or raising the ground immediately in front of it, or by some such physical deterioration. Any other construction of the clause would open the door to claims of so wide and indefinite a character as could not have been in contemplation of the legislature.

The House of Lords again considered the application of section 68 in the case of *Metropolitan Board of Works* v *McCarthy* (1874) LR 7 HL 243. The claim was that of the owner of premises close to a draw-dock on the Thames. The value of the premises was diminished when the Thames Embankment was constructed under statutory powers, thus destroying the dock and the claimant's access to the river which

A he had enjoyed in common with the public. On those facts the claimant was held entitled to compensation under section 68.

In his speech, Lord Cairns LC adopted the test which had been formulated by Mr A H Thesiger, counsel for the claimants. At p 253 he said:

> . . . in his very able argument, Mr Thesiger stated what he would rely upon as a definition of the right to compensation, and having considered this case very fully I myself should not be disposed to find fault with any part of that definition, although definitions are always matters of very considerable difficulty. Mr Thesiger stated that the test which he would submit as one which he thought would explain and reconcile the various cases upon this subject was this, that where by the construction of works there is a physical interference with any right public or private which the owners or occupiers of
> B property are by law entitled to make use of in connection with such property, and which right gives an additional market value to such property, apart from the uses to which any particular owner or occupier might put it, there is a title to compensation if, by reason of such interference, the property as a property is lessened in value.

Lord Chelmsford likewise adopted that definition subject to a qualification at p 256:

> I think the rule as thus stated may be accepted with this necessary qualification, that where the right which the owner of the house is entitled to exercise is one which he possesses in common with the public, there must be something peculiar to the right in its connection with the house to distinguish it from that which is enjoyed by the rest of the world.

Lord Penzance at pp 261-2 said:

> C A rule has been established in these cases from which it would I think be disastrous to depart. It is this, that whether damage can be recovered under the words "injuriously affected" depends upon whether it might have been the subject of an action if the works which caused it had been done without the authority of Parliament . . . it may reasonably be inferred that the legislature in authorising the works and thus taking away any rights of action which the owner would have had if the works had been constructed by his neighbour intended to confer on such owner a right to compensation co-extensive with the rights of action of which the statute had deprived him. But on no reasonable ground, as it seems to me, can it be inferred that the legislature intended to do more and actually improve the position of the person injured by the passing of the Act. There is another rule, which is, I conceive, well settled in these cases, namely that the damage or injury which is to be the subject of compensation must not be of a personal character but must be a
> D damage or injury to the "land" of the claimant considered independently of any particular trade that the claimant may have carried on upon it.

Finally, Lord O'Hagan agreed also with the definition put forward by counsel but added at p 267:

> When as here, the injury is particular, consists in the diminution of the value of a holding, is perfectly appreciable, and in the particular case has actually been appreciated to a considerable amount, I am strongly of opinion that it gives a clear title to compensation under the statute.

In the later case of *Caledonian Railway Co* v *Walker's Trustees* (1882) 7 App Cas 259 the case of *McCarthy* was held indistinguishable and followed. Lord O'Hagan at p 287 said in reference to *McCarthy*'s case:

> E I see no reason to alter in any respect the view which I then adopted or the reasons which I urged in support of it. In that case, as in this, there was a public way and in that as in this access to it was obstructed by the works complained of. There, as here, the obstruction produced inconvenience to the general public who had the use of the way, but there, as here, there was in addition a particular and appreciable damage to the premises of an individual, rendering them less available for his purposes and therefore diminishing their value.

Mr Howell, on behalf of the council, relied on these cases as demonstrating that section 68 provided a substitute to a right of action for damages but that the statutory right was not necessarily co-extensive with the right to civil damages. The claimant cannot recover more in compensation than he might recover in damages, but he may not recover as much, since the compensation claim is limited
F to the depreciation in value of the property.

It is sufficient to say that the passages which I have just quoted may suggest that that is the position but do not, in my view, establish that as a principle of law. In none of the speeches in these cases was the *method* of assessing damages under discussion.

The case of *Wadham* v *North Eastern Railway Co* (1884) 14 QBD 747 was not cited to me by counsel, but it appears to me to be relevant to the issue. This was a special case stated by an arbitrator who had determined that the stopping-up of a street under statutory powers had diminished the value of a hotel and public house for using, selling or letting as such. In approving the award based on the depreciation of value of the property, Mathew J said, after referring to the earlier cases (p 752):

> . . . I do not understand the learned judges to have intended to lay down more G
> than this, viz. that you are not in calculating the damage for injuriously affecting the premises, to take into account any special and exceptional value which the premises may have in the possession of the then proprietor, but you are to see whether or not the value of the property as a marketable article to be employed for any purpose to which it may legitimately and reasonably be put, has been interfered with or not.

I was referred to *Kirby* v *Harrogate School Board* [1896] 1 Ch 437, which is primary authority for the proposition that where property is acquired under statutory powers, the covenantees with the benefit of a restriction over the land cannot maintain an action to restrain the breach of covenant but are restricted to the claim for compensation in respect of injurious affection. In the course of his judgment, Lindley LJ said at p 449: H

> It follows that the School Board are perfectly right in contending that an action for an injunction or damages is out of the question. If the plaintiffs can make out that they are injuriously affected, they will be entitled to compensation under section 68 of the Lands Clauses Act. They are not in the position of a person who is in the enjoyment of some easement, but they have a right conferred upon them by that covenant. What is its value in money is another matter. It is quite possible that in asking for compensation and in putting into force the machinery for obtaining it their damages may be assessed at nil; but if they can make out that they are damnified they are entitled to compensation under section 68.

It must be said, however, that in so far as this passage appears to indicate that compensation is to be limited to the loss of value of the benefited land it is, in my view, plainly *obiter*. The issue in the case J
was restricted to whether or not the owners of the land could maintain an action for an injunction to restrain the breach of covenant by the statutory authority.

The next case is *Long Eaton Recreation Grounds Co* v *Midland Railway Co* [1902] 2 KB 574, strongly relied upon by the council in this case. The plaintiffs owned land with the benefit of a restrictive covenant over adjoining land "not to erect any erection or building of any kind". A railway company erected an embankment on the subject land and this was held by the Court of Appeal to be a breach of covenant in respect of which a claim was maintainable under section 68.

Sir Richard Henn Collins MR at p 581 said:

> Then we come to the real point in the case, whether a breach of a restrictive K
> covenant is the subject of compensation under section 68. In point of principle I cannot see why it should not be . . . if the damage in respect of which compensation is sought is only incident to the particular purpose for which the owner is occupying the land, that is held to be excluded from compensation; but if the market value of the land is affected then the land is affected. It is a matter of common sense that when by covenant, by an easement being attached, or in any other way the value of the land is enhanced and the Railway Company subtract a portion of the value by making their railway, they are injuriously affecting the land.

Later in the judgment, the Master of the Rolls said at p 583:

> It seems to me that it can make no difference whether the right was an affirmative one . . . or a negative one . . . in either case the value of the land to which the benefit was attached would be reduced. L

Mathew LJ in the same case at p 586 stated:

> Then the last point is that, granting that the covenant had been entered into and been violated still that is not the subject of compensation under section 68. No case was cited to us to justify such a conclusion . . . over and over again there is an intimation from the highest authority that a restrictive covenant of this sort is the subject of compensation under the Act. It is extremely difficult to see why it should not be so. There is in this case a negative easement adding to the monetary value of the estate. Why that should not be the subject of compensation it is difficult to conceive . . . Statutory powers were never intended to affect a covenant of this sort, which is part of the value of the estate.

The latest case referred to in which a claim under section 68 was directly in issue is *Argyle Motors (Birkenhead) Ltd* v *Birkenhead* M
Corporation [1975] AC 99. In that case access to a car dealer's premises was obstructed by works carried out by the local authority under statutory powers, and the question raised on a preliminary issue was whether the local authority was liable under section 68 to pay compensation in respect of the consequent loss of business profits. Both the Court of Appeal and the House of Lords held that compensation under section 68 could be obtained only in respect of some loss of value of land or some damage to land and not for any loss personal to the owner or related to a particular use of the land. Accordingly, compensation was not payable under section 68 for loss of profits as such, although, in so far as loss of profitability was an element in valuing the owner's interest, the owner would be entitled

A to be compensated for any resultant loss of value.

In the judgment of the Court of Appeal delivered by Buckley LJ it is stated at p 114G:

In our judgment, no claim to compensation can be brought within section 68 relating to injury to a business. To avoid confusion, however, we add that this does not mean that, if injury to a business can be shown to have occasioned a diminution in the value of the land where the business is carried on, compensation cannot be recovered for that injurious affection of the land.

In the House of Lords the leading speech was delivered by Lord Wilberforce, who at p 129F-G said:

. . . by a series of judicial observations of high authority it is well established that the only compensation which can be obtained under this section is "in respect of . . . lands", ie in respect of some loss of value of land, or (what is a
B branch under this same heading) in respect of some damage to lands, and that compensation cannot be obtained for any loss which is personal to the owner or which is related to some particular user of the land.

Later in his speech, Lord Wilberforce stated at p 130F:

. . . the pronouncements by the eminent members of this House are in such explicit terms that a clear conviction of their error or, possibly, the most powerful considerations of policy would need to be present before so strong a current of authority could be turned back. As a matter of language, though language is an uncertain guide in this statute, it must be said that the words I have quoted "in respect of any lands . . ." support the exclusion of claims for personal loss. And, though it might be said that a generous policy of compensation would favour compensation for losses caused to individuals through works of social benefit, a policy to this effect, however just it might
C appear in a particular case, involves too great a shift in financial burden and too many adjustments or qualifications (if it were to be workable) to be suitable for introduction by judicial decision.

I must, therefore, agree with both courts below that if the right to compensation in the present case depends upon section 68 the appellants cannot succeed in obtaining compensation for business losses as such. I make it clear, as did the Court of Appeal, that if they can prove that a loss of profitability affects the value of their interest in the land they can recover compensation for this loss of value.

This examination of the authorities shows that in reported cases directly arising from claims under section 68 in the past 130 years it has consistently been held (a) that a claim in respect of injurious affection is sustainable only where the claimant would have a right of
D action for damages but for the exercise of statutory powers and (b) that where the claim is sustainable in respect of an interference with the claimant's rights in his land, the right to compensation has been determined by reference to the loss in value of the claimant's property. Two of the cases (namely *Kirby* and *Long Eaton*) were claims made by owners entitled to the benefit of restrictive covenants over the land on which the works were executed. In neither case was the question of how compensation was to be assessed or calculated directly in issue, but it is to be inferred that the Court of Appeal in the *Long Eaton* case assumed that it was directly related to the loss of value of the benefited land. In any event there was no suggestion in either case, or indeed in any of these cases, that compensation might be assessed in the manner contended for by the claimants in the
E present case.

It appears, therefore, that there is no warrant in any of the reported cases under section 68 of the 1845 Act for an assessment of damages otherwise than by reference to the loss of value of the claimants' land consequent upon execution of works by the compensating authority.

Examination of authorities — "loss of bargain" cases

It remains to be seen whether the later cases relied on by the claimants justify the contention that the appropriate measure of compensation under section 68 may be by reference to the price obtainable by the claimants for relaxation or release of the covenant.

The first such case was *Wrotham Park Estate Co Ltd* v *Parkside Homes Ltd* [1974] 1 WLR 798. In that case the covenant in question
F was the same as in the present case, ie a covenant not to build except in accordance with a layout plan previously approved by the covenantee. In breach of that covenant the defendants proceeded to build 14 houses and took deposits from prospective purchasers. Brightman J refused to grant a mandatory injunction to pull down the offending houses but awarded damages to the plaintiffs instead under Lord Cairns' Act.

In the course of his judgment, Brightman J said at p 812:

I turn to the consideration of the quantum of damages . . . the basic rule in contract is to measure damages by that sum of money which will put the plaintiff in the same position as he would have been in if the contract had not been broken. From that basis the defendants argue that the damages are nil or purely nominal, because the value of the Wrotham Park Estate as the plaintiff concedes is not diminished by one farthing in consequence of the construction G
of a road and the erection of 14 houses on the allotment site. If therefore, the defendants submit, I refuse an injunction I ought to award no damages in lieu. That would seem on the face of it a result of questionable fairness on the facts of this case . . . if for social and economic reasons, the court does not see fit in the exercise of its discretion to order demolition of the 14 houses, is it just that the plaintiff should receive no compensation and that the defendants should be left in undisturbed possession of the fruits of their wrong-doing? Common sense would seem to demand a negative answer to this question.

Later, after referring to the so-called "user cases", Brightman J continued at p 814H:

The facts of the cases I have mentioned are a long way from the facts of the case before me. Should I as invited by the plaintiffs apply a like principle to a case where the defendant in defiance of protest and writ has invaded the H
plaintiffs' rights in order to reap a financial profit for itself? . . . in the present case I am faced with the problem what damages ought to be awarded to the plaintiffs in the place of mandatory injunctions which would have restored the plaintiffs' rights. If the plaintiffs are merely given a nominal sum or no sum, in substitution for injunctions it seems to me that justice will manifestly not have been done . . . In my judgment a just substitute for a mandatory injunction would be such a sum of money as might reasonably have been demanded by the plaintiffs from Parkside as a quid pro quo for relaxing the covenant.

Brightman J went on to award £2,500 by way of damages, calculated as 5% of the defendant's anticipated profit, as a fair measure of compensation in a case which, on its particular facts, required the court "to act with great moderation". J

The case of *Bracewell* v *Appleby* [1975] 1 Ch 408 is a decision of Graham J at first instance. The defendant had by the time of the trial built a house on land to which he was held to have no right of way and to which he could obtain access only by means of the plaintiffs' right of way along a private road. Although no question arose in this case of a mandatory injunction to pull down the house, the effect of an injunction against trespassing over the plaintiffs' right of way would have rendered the house uninhabitable. The learned judge refused to grant such an injunction and awarded damages in lieu.

In the course of his judgment, Graham J said at p 419D:

After consideration I propose to approach the question of damages and assess the amount, which I was requested to do by both parties, along the same lines as those followed by Brightman J in *Wrotham Park Estates* v *Parkside* K
Homes. It seems to me that the defendant must be liable to pay an amount of damages which in so far as it can be estimated is equivalent to a proper and fair price which would be payable for the acquisition of the right of way in question.

Later, at p 420, Graham J went on:

The circumstances here are very different from those in the *Wrotham Park* case and I think the proper approach is to endeavour to arrive at a fair figure which, on the assumption made, the parties would have arrived at as one which the plaintiffs would accept as compensating them for loss of amenity and increased user, and which at the same time whilst making the blue land a viable building plot, would not be so high as to deter the defendant from building at all.

In the event the learned judge awarded the plaintiffs £400 each by L
way of damages, being one-fifth of the amount which he estimated the defendant would have been prepared to pay to the five owners of the right of way in order to get his right of access.

The next case is that of *SJC Construction Co Ltd* v *Sutton London Borough Council* (1975) 29 P&CR 322, a decision of the Court of Appeal on an appeal from this tribunal. The appellant company proposed to build a block of flats in breach of a restrictive covenant benefiting the respondent council's adjoining land. When an application to relax the restriction was made under section 84 of the Law of Property Act 1925, the tribunal held the proposed development to be a reasonable user of the restricted land; that the continued restriction would be contrary to the public interest; and that the respondent council could be adequately compensated in M
money for the discharge of the covenant. Compensation was then assessed by reference to notional friendly negotiation between the parties for the removal of the restriction, "the most likely outcome" of which would have been an agreement to split the development value equally between the company and the council. Accordingly, the Lands Tribunal awarded substantial compensation in the sum of £9,500.

That method of assessing compensation was unsuccessfully challenged in the Court of Appeal. In his judgment, Lord Denning MR said at p 325:

Now Mr Newsom, on behalf of the company has submitted that the President was wrong in principle in awarding compensation on that basis. He said . . .

A that it was inconsistent with the case decided by the Lands Tribunal for Scotland *Re McVey's Application*. . . . It seems to me as it did to the President that it is distinguishable . . . the Glasgow Corporation who were entitled to the benefit of the restriction did not suffer any loss or disadvantage from its removal except that they were deprived of the right or power to obtain money. The Lands Tribunal for Scotland held that a right or power to obtain money was not the subject of compensation under the Scottish Act when no other loss or disadvantage had been incurred. That seems to me entirely right. I think however that the position would have been very different if there had been any loss or disadvantage such as injury to the amenities or obstruction of view or increase in noise or anything of that kind . . . the only thing that the Glasgow Corporation lost was the opportunity of extracting money as the price of removing the restriction. But such was not the subject of compensation. So we come back to the question: what is the basis or proper basis of compensation?
B It is simply to make up "for the loss or disadvantage suffered" by the borough council. There is no method prescribed by the Act by which it is to be assessed; it is essentially a question of quantum. It is however to be assessed for loss of amenities, loss of view and so forth which are things which it is hard to assess in terms of money . . . so here the loss was the disadvantage to the borough council the occupiers of Whitehall and their tenants . . . They are entitled to be compensated for the loss of amenities, the loss of view across to the park and so forth. That has to be translated into money terms. Some calculation has to be made. The President did it by finding the realisable development value, and splitting it equally. I see nothing wrong in his approach.

In a concurring judgment, Stephenson LJ said at p 327:

The modification of this restrictive covenant clearly resulted in the respondents suffering the loss of a practical benefit of substantial value or
C advantage. For that substantial loss or disadvantage they were entitled to substantial compensation. That they got from the President, by a method which I would not say was the only method but could not say was unlawful or contrary to the Act or to any authority which has been cited to us.

The decision in the *SJC Construction* case was considered by the Court of Appeal in *Stockport Metropolitan Borough Council* v *Alwiyah Developments* (1986) 52 P&CR 278. In that case the appellant local authority owned a large housing estate with the benefit of a covenant restricting the use of an adjoining area of open space for any purpose other than agriculture. The respondent company had applied to the Lands Tribunal under section 84 to modify the restriction to enable them to build 42 houses on the restricted land, planning permission having been obtained for that
D development. The modification was granted and the tribunal awarded the council compensation "to make up for the loss or disadvantage" suffered in consequence of the removal of the restriction. The award of the tribunal was calculated as the diminution in the potential sale value of those few houses on the council estate directly affected by the respondent's development. The council appealed, contending (*inter alia*) that the power to bargain for a price as consideration for the discharge of the covenant was a benefit of substantial value or advantage, which therefore justified refusal to discharge the covenant. The Court of Appeal rejected that submission.

Eveleigh LJ, in the course of his judgment, at p 281 stated:

E The benefit envisaged must be a practical one as opposed to a pecuniary one, that is the practical benefit which is afforded by observation of the covenant. Thus the subsection exempts from discharge or modification those covenants whose preservation will secure a practical benefit. Bargaining power is only a benefit when it results in the receipt of the price upon the covenant being discharged. Such a benefit cannot be of the kind contemplated by the subsection for it results from the discharge and not the continuance of the covenant.

Eveleigh LJ went on to say at p 282:

We are therefore concerned to discover the loss or disadvantage suffered in consequence of the discharge or modification. If it is the loss of the ability to have houses with a view as part of a housing stock, then from what I have already said it would not appear that such a loss is of any real value. If it is the diminished price of the houses then I would not disturb the figure arrived at by
F the Tribunal for I can see no error in the Tribunal's approach to the assessment of this sum.

Dillon LJ expressed a similar view and then referring directly to the *SJC Construction* case said at p 284:

I turn to the question of compensation. Compensation is to be awarded for the loss or damage which the person entitled to the benefit of the covenant will suffer from a release or modification of the restriction. In *SJC Construction Ltd* v *Sutton Borough Council* Lord Denning MR commented that a right or power to obtain money — the opportunity of extracting money as the price of removing the restriction — was not the subject of compensation when no other loss or disadvantage had been incurred. He went on to say that the compensation was to be assessed for the loss of amenities, loss of view and so forth. In the same judgment Lord Denning did indeed also say that he saw no error of law in the decision of the President of the Lands Tribunal in the actual *SJC Construction* case. . . . The President had said that "it is a disadvantage G
resulting in a loss to be deprived of the right to bargain for the price to be paid for a modification" but that was the case where there had been a very serious loss of amenity, and what the Court of Appeal was approving was the President's valuation of the loss of amenity, and not the valuation of the loss of a right to bargain or blackmail position. The loss of the right to bargain is not itself a subject for compensation, although the development value may be a relevant factor for consideration in some circumstances in valuing a loss of amenity.

Mr Bartlett also referred to the decision of Millett J in *Carr-Saunders* v *Dick McNeil Associates* [1986] 2 All ER 888, a case concerned with an infringement of the right of light. I confess that I have found this decision, in so far as it relates to the assessment of damages, not easy to understand. Millett J at p 896 expressed himself H
as adopting the same approach as in *Wrotham Park Estate* v *Parkside Homes* and *Bracewell* v *Appleby* for the purpose of assessing damages in lieu of an injunction. The learned judge went on to say that he had no evidence as to the defendant's profits from the development but that he did have evidence of a general loss of amenity resulting from the infringement of the right to light. It would appear that he then proceeded, with "little material to guide him", to adopt a robust round-figure approach resulting in the award of £8,000.

Finally, I was referred by Mr Howell to the Court of Appeal decision in *Stoke-on-Trent City Council* v *W & J Wass Ltd* [1988] 1 WLR 1406, which concerned the infringement of the plaintiff council's market rights. The trial judge had found that the council as J
owners of the market franchise had suffered no loss from the defendants' rival market, but he nevertheless awarded damages calculated by reference to the licence fee which the council could reasonably have obtained. On appeal by the defendant company, the Court of Appeal held that the council were entitled to nominal damages only and thus refused to extend the "licence" or "ransom" principle to the case of infringement of a market right.

Nourse LJ stated the problem thus at p 1408D:

The levying of an unlawful same day market within six and two-third miles of a franchise market is actionable at the suit of the market owner without proof of loss. At trial he will usually be awarded a permanent injunction and nominal damages. Where an interlocutory injunction has been running before trial, no further question of damages will usually arise. In respect of K
any period before or after trial where an injunction is in force substantial damages will be recoverable if loss can be proved. But suppose a case where there has been no loss. Is the market owner kept to his nominal damages or can he recover substantial damages on the footing that if his leave and licence had been sought he could have required a fee to be paid to him? That is the novel and somewhat surprising question with which we are confronted on this appeal.

Nourse LJ proceeded to state the general rule as to damages in tort at p 1410 as follows:

The levying of an unlawful rival market is a tort . . . the general rule is that a successful plaintiff in an action in tort recovers damages equivalent to the loss which he has suffered, no more and no less. If he has suffered no loss the most he can recover are nominal damages. A second general rule is that where the L
plaintiff has suffered loss to his property or some proprietary right, he recovers damages equivalent to the diminution in value of the property or right.

Nourse LJ thereafter pointed out that there were exceptions to these general rules in tort, including the decision of Brightman J in *Wrotham Park Estate* v *Parkside Homes* and the decisions of Graham J in *Bracewell* v *Appleby* and of Millett J in *Carr-Saunders* v *Dick McNeil*. Having recited the effect of those decisions, Nourse LJ went on at p 1413H:

As I understand these authorities their broad effect is this. In cases of trespass to land and patent infringement and in some cases of detinue and nuisance the court would award damages in accordance with what Nicholls LJ has aptly termed "the user principle". On an analogous principle, in a case where there M
was a breach of a restrictive covenant the court has in lieu of a permanent mandatory injunction to restrain the breach, awarded damages equivalent to the sum which the plaintiffs might reasonably have demanded for a relaxation of the covenant. But it is only in the last-mentioned case and in the trespass cases that damages have been awarded in accordance with either principle without proof of loss to the plaintiff. In all the other cases, the plaintiff having established his loss, the real question has not been whether substantial damages should be awarded at all but whether they should be assessed in accordance with the user principle or by reference to the diminution in value of the property or right. In other words, those other cases are exceptions to the second but not to the first of the general rules stated above.

Nourse LJ went on to consider further the *Wrotham Park Estate* case, which he described as standing "very much on its own". At p 1414 he went on to say:

> The practical result of the *Wrotham Park* decision was something akin to an award of exemplary damages for breach of contract, albeit that their amount bore no relation to the loss which would have been suffered by the defendant purchasers if they had to demolish their houses. In saying this I do not wish to suggest that the case was wrongly decided. Indeed I regard the result as having been entirely appropriate and I see no reason why it should not serve as a precedent for other cases of the same kind. I merely wish to emphasise that it stands a long way away from the present problem and does not assist in its solution.

Nourse LJ concluded his judgment at p 1415 by stating:

> It is possible that the English Law of Tort, more specially of the so called "proprietary tort" will in due course make a more deliberate move towards recovery based not on loss suffered by the plaintiff but on the unjust enrichment of the defendant. But I do not think that that process can begin in this case and I doubt whether it can begin at all at this level of decision.

In an assenting judgment, Nicholls LJ expanded upon what he described as the "user principle" as a method of calculating damages for the invasion and abstraction by an infringer of a property right, such as a patent. Nicholls LJ concluded his judgment by referring to the *Wrotham Park Estate* case in the following terms:

> I do not think that decision takes the matter any further so far as the present case is concerned. That case concerned an award of damages for breach of a restrictive covenant in lieu of a mandatory injunction. It is far removed from the present case. Wherever precisely the boundary, separating causes of action in which the user principle will be applied from those in which it will not, is to be drawn, for the reasons I have sought to state above that principle in my view has no application to a cause of action for disturbance of a market where during the pre-injunction period, the unauthorised market did not in fact disturb the lawful market.

This examination of the cases leads me to conclude that there is no rule, and no authority, requiring me to hold that the compensation payable under section 10 of the Compulsory Purchase Act 1965 for injurious affection to the claimants' land in the circumstances of this case is to be calculated by reference to the amount likely to be receivable by the claimant in free negotiation for relaxation of the covenant. The true rule, in my judgment, is shown by the cases arising under section 68 to be that the claimant is entitled to be compensated either for the actual damage caused to his land or, in an appropriate case, for the loss of value of his land attributable to the works executed on the servient land.

All the later cases in which a share of development profit or "ransom" has been used as a method of calculating either compensation or damages are shown to be, as explained by Nourse LJ in *Stoke-on-Trent City Council* v *Wass*, special cases which are either akin to the award of exemplary damages or where the calculation by reference to a share of development profit has been held to be a convenient means of measuring the actual loss of amenity suffered by the dominant owner in the absence of evidence of actual pecuniary loss.

I appreciate the force of the claimants' argument that they should be no worse off by reason of the exercise of statutory powers than they would have been if they had been able to license or permit the release of the covenant. However, I consider that there are two decisive answers to that particular proposition. First, that the statutory provisions as interpreted by the courts entitle them only to compensation limited to the loss of value of the interest which has been injuriously affected. In the passages which I have already cited from the judgments of Eveleigh LJ and Dillon LJ in the *Stockport* case (*op cit*) it was made clear that the loss of right to bargain was not "a loss or disadvantage suffered" by the covenantee in the context of section 84 of the Law of Property Act 1925. In my judgment, the principle expressed by Dillon LJ and Eveleigh LJ in that case is equally applicable to the award of compensation under section 10 of the 1965 Act "in respect of any land or interest in land which has been . . . injuriously affected by the execution of the works".

The second point is closely related and arises from the submission made by Mr Howell on behalf of the local authority that, in any event, the overriding of a restrictive covenant by the local authority acting in accordance with its statutory powers is not to be equated with the wrong-doing of the defendants in, for example, the *SJC Construction* case and *Wrotham Park Estate* v *Parkside Homes*. I agree with that submission. Parliament has conferred on the council statutory powers in relation to housing, and the council are entitled in exercise of those powers to carry out works and to disregard or override restrictive covenants where necessary to do so for the purposes of carrying out those works. The so-called "breach" of covenant is accordingly lawful and not wrongful. Parliament has also provided a statutory remedy for the owners of land injuriously affected by that lawful exercise of statutory powers. That, in my view, is wholly different from the situation which arose in the *SJC Construction* case or in the *Parkside Homes* case. In those and the other "licence" cases the awards made were in recognition either of the defendants' wrong-doing or of their proposed breach of covenant, and were an endeavour to do justice by avoiding unjust enrichment of the infringers.

I have already referred to the concluding passage from the judgment of Nourse LJ in the *Stoke-on-Trent* case in which, in effect, his lordship stated that if the "user/licence" principle were to be extended, that was a matter for Parliament and not for the courts. I acknowledge the different context in which that view was expressed, but it seems to me that it is an apt observation in the circumstances of this case and that I ought respectfully to adopt it in relation to a claim for compensation under section 10 of the 1965 Act.

I therefore hold that where compensation is payable under section 10 in respect of a breach of covenant comprising the execution of works in the exercise of statutory powers authorising those works, that compensation is to be assessed by reference to the diminution in value of the benefited land consequent upon the carrying out of the authorised works on the restricted land.

Accordingly, I determine the preliminary issue in favour of the compensating authority and answer the question posed in the application as follows:

The basis upon which the claim for compensation in respect of injurious affection of the claimants' interest should be assessed is the amount of the difference (if any) between

(i) the value of the interest immediately before the commencement of the development with the benefit of the restrictive covenant (but disregarding the power of an authority to override the restrictive covenant) and

(ii) the value of the interest immediately thereafter given the carrying out of the development.

The claimants will pay the costs of the compensating authority such costs if not agreed to be taxed by the Registrar of the Lands Tribunal on the High Court Scale of Costs.

LEASEHOLD VALUATION TRIBUNALS

LANDLORD AND TENANT ACT 1987

A GREGORY AND ANOTHER v SADDIQ

(Y/LVT/0032)

February 26 1991

(MRS S R HOWDLE (chairman) and J A G LEES FRICS)

Estates Gazette April 27 1991 and May 4 1991

[1991] 16 EG 141 and [1991] 17 EG 97

Price payable for landlord's leasehold interest in three-storey, semi-detached Huddersfield house converted into two self-contained flats — Each flat held on a regulated tenancy — B **Original disposal a sale at £45,000 by vendors who had purchased at auction some 18 months earlier for £39,500 — New landlord bought under misapprehension that he could obtain vacant possession — Valid purchase notice served by tenants under s 12(3)(*b*) — Tenants contended for price of £25,000 — Decision in *Cousins* v *Metropolitan Guarantee* adopted**

Tenants sought to distinguish *Cousins*, submitting that the tribunal had jurisdiction to investigate the surrounding circumstances of the original disposal in an appropriate case and that the present case was an appropriate one — Unlike *Cousins* the price had not been set at auction and tenants' C **expert advice was that property worth only £25,000 with protected tenants in possession — Landlord relied strongly on *Cousins* for general principle that there was no power to investigate the circumstances surrounding the sale, whether at auction or otherwise**

As on other occasions, tribunal comment, the legislation proved difficult to interpret — Crux of the matter was the extent, if at all, to which the tribunal have the power to determine the price payable when the purchase notice served pursuant to s 12(3)(*b*) provides for the "terms" on which disposal is to be made to be determined by tribunal — There D **were two possible constructions of s 12 provisions: (1) s 12(3)(*b*) read in isolation or in conjunction with s 13(1)(*b*) would not imply any limitation on a tribunal's power to substitute their own *de novo* valuation as decided in *Cousins*; (2) s 12(1) provides for the tenants' rights to buy and lays down the limits within which this right can be exercised whether the purchase notice is served under s 12(1) itself or under s 12(3)(*b*) — On balance, tribunal decided to adopt latter construction, ie the *Cousins* view — Tribunal consider present case by no means the typical case envisaged by those enacting the legislation**

Five different situations which had arisen for argument or E **decision identified by tribunal as possible instances where they might be called upon to investigate the surrounding circumstances — Four of these not applicable here — Even if the tribunal were convinced that the fifth situation, the present case, provided such an instance (which they were not), the evidence as to market value did not lead them to conclude that the price of £45,000 should be disturbed**

Parties' experts gave valuations on an investment basis of around £25,000-£30,000 — But tribunal adopted wording of tribunal in *30 Upperton Gardens Management* case: "Whatever might be said, the best evidence of the value of F **anything is what it will fetch in the open market" — As in that case, there was evidence in present case of an auction price and to view expert evidence without taking that into account was not appropriate — It was clear in auction conditions that the sale was subject to existing protected tenancies — "The market" in the particular area could reasonably be viewed as paying more than traditional investment value for properties such as subject property — Even if tribunal entitled to look at surrounding circumstances because of landlord's mistake, the price of £45,000 was not out of line with open market value**

Held: normally the tribunal has no jurisdiction in G **determining the terms of a disposal under a purchase notice pursuant to s 12(3)(*b*) to do other than ascertain the consideration payable under the original disposal and determine that to be the price payable — Determination, accordingly, of £45,000**

The following cases are referred to in this report.

Cousins v *Metropolitan Guarantee Ltd* [1989] 2 EGLR 223; [1989] 31 EG 56, LVT

30 Upperton Gardens Management Ltd v *Akano* [1990] 2 EGLR 232; [1990] 45 EG 121 & [1990] 46 EG 131

Wilkins v *Horrowitz* [1990] 2 EGLR 217; [1990] 29 EG 57 H

Andrew Beck (of Walker Morris Scott Turnbull, of Leeds) appeared for the applicants; James Behrens (instructed by Fowler & Crossley, of Huddersfield) appeared for the respondent.

Giving their decision, THE TRIBUNAL said: This is a decision upon an application to a leasehold valuation tribunal ("the tribunal") under section 13 of the Landlord and Tenant Act 1987 ("the Act") by the applicants, being the nominated persons within the meaning of section 13, for the determination of the price to be paid for, and the other terms of the conveyance of, the estate or interest of the respondent landlord [Mahommed Saddiq] in the land and premises known as 83 New North Road, Huddersfield (the "subject premises"), consequent upon the service upon the respondent of a J purchase notice under section 12 of the Act.

The background

The subject premises consist of a three-storey Victorian semi-detached house with basement, the whole being converted into two self-contained flats. The tenants of these flats are respectively Mr M Gregory and Mr F Rahimi, each of them holding under a regulated tenancy. The landlord's title is leasehold, deriving from a 999-year lease taking effect from March 25 1919 at a ground rent of £8 pa, and a good leasehold title was first registered at HM Land Registry, title no WYK430305, on December 2 1988, consequent upon the purchase at auction of the premises on October 18 1988 by Mrs Jasbir Kaur for £39,500. On November 27 1989 the respondent entered into a written K contract to purchase the premises for £45,000, and the transfer to him was executed by Mrs Kaur on March 30 1990, the respondent being registered as proprietor on May 11 1990.

On March 21 1990 Messrs Gregory and Rahimi served a notice upon the respondent's then solicitors under section 11 of the Act requiring information about this transaction. By letters dated April 5 and 25 1990, they were informed that the property was held under a 999-year lease from March 25 1919, that formal completion took place on March 30 1990 and that the price was £45,000.

On or about June 11 1990 Messrs Gregory and Rahimi served a

purchase notice under section 12 of the Act, requiring that the estate or interest and the terms of the disposal be determined by a rent assessment committee under section 13 of the Act (that is, constituted as a leasehold valuation tribunal) and nominating themselves as the persons to whom disposal should be made. The application was made to the tribunal on July 13 1990. The sum stated to be paid by the landlord was £45,000. To the question: "What price do you consider payable under the Act?" the applicants answered: "£25,000."

At the hearing the following points were agreed by the parties:

(i) the subject premises were premises to which Part I of the Act applied;

(ii) the tenants of the subject premises were qualifying tenants under section 3 of the Act, and for all purposes formed the "requisite majority" of such tenants, and they properly nominated themselves as the nominated person(s) under sections 12 and 13;

(iii) the transfer of the leasehold interest on March 30 1990 to the respondent was a relevant disposal under section 4 by a landlord as defined by section 2;

(iv) that disposal was made in breach of section 1, in that no prior notice complying with the terms of section 5 was given to the qualifying tenants (although certain informal negotiations took place, as outlined below);

(v) the applicants served a valid purchase notice under section 12 in good time and in due form upon the respondent.

The tribunal conducted an inspection of the exterior and interior of the premises prior to the hearing. At the hearing the applicants were represented by their solicitor Mr Beck and the respondent by Mr Behrens of counsel.

The hearing

Mr Beck indicated that the crucial issue was the basis upon which the tribunal could deal with the question of the price payable. He put forward three principal contentions:

First, the tribunal had jurisdiction to look at the circumstances of the original disposal in an appropriate case. Second, the tribunal had jurisdiction to consider the genuineness of the transaction which formed the original disposal. Third, the tribunal had jurisdiction under section 12(4)(*b*) to look at the effect of an incumbrance upon the price payable, and that should be done here.

He recognised that the decision of another such tribunal in London in *Cousins* v *Metropolitan Guarantee Ltd* [1989] 2 EGLR 223* did not assist him on the first two points, but he sought to distinguish the present case. In *Cousins* four questions were put to counsel, but consequent upon the ruling of the tribunal upon the first of them, that it had no power to determine a value for the premises *de novo*, the parties reached agreement and further argument was not heard.

His argument that this was an appropriate case to look at the circumstances was this: he could accept that the policy of the legislation, as reflected in *Cousins*, was that the tribunal should not have discretion to substitute its own valuation and compel the new landlord who had paid a genuine price to sell for less. This, he contended, was because the practical effect of the whole scheme of the legislation was that a landlord wishing to sell would normally receive the open market investment value of the property, either from the tenants or from a purchaser, provided that sections 5 to 10 had been complied with. But the tenants had more than simply a "right of first refusal": they had the right to compel a sale when their right of first refusal had been infringed. In this latter situation they should not be in any worse situation than if the proper procedure had been followed, involving offer, rejection, counter offer etc; such a procedure would normally result in a sale at the true open market value. Where the tenants had not had such an opportunity, they should not simply be presented with a *fait accompli* where the new landlord had made a bad bargain, paying twice as much as the property was worth; the burden of that bad bargain should fall rather upon him — he was more likely to have other avenues to recoup his loss. He would distinguish *Cousins,* where the price had been set at auction; here the tenants had not had the protection of an auction and their expert advice was that the property was worth only £25,000 with tenants in possession. He also pointed to the form of the purchase notice and application, which clearly envisaged the question of price being open to argument, and questioned the need for the tribunal to be set up to include an expert in valuation if no such function existed.

With regard to genuineness, he submitted that this was within the tribunal's power to investigate, as counsel had suggested in *Cousins*. Here the tenants were in difficulty: the only evidence as to price which they had seen was an open contract in the following terms:

> I Mahommed Saddiq of 176 Trinity Street, Huddersfield agree to buy 83 New North Road, Huddersfield from Mr Kuldip Singh for the sum of £45,000. I agree to pay a £5,000 non-returnable deposit in case I fail to complete the deal within 3 months the deposit is non-refundable. The deal is subject to me receiving a satisfactory mortgage at current martgage (*sic*) rates on 176 Trinity Street, Huddersfield. The house is sold with sitting tenants and in its present condition of which I am fully aware.
>
> I Kulbir Singh of 23 Greyfriars Avenue, Deighton agree to sell 83 New North Road to Mr Mohammed Saddiq for the sum of £45,000 and agree to accept a £5,000 deposit and agree that if I fail to complete the deal then I shall return the deposit in full as well as pay the purchaser £5,000 compensation. A period of 3 months is to be allowed for the completion.

This contract was dated November 27 1989 and was signed by Mr Singh and Mr Saddiq with K S Uppal of 122 Lockwood Road, Huddersfield, named as witness. The registered proprietor at that time was Mrs Kaur. The respondent had not produced a copy of the transfer to him or any other evidence.

With regard to section 12(4)(*b*), he produced a repair notice, served under section 190 of the Housing Act 1985 upon Mrs Kaur in June 1990, requiring certain repair works listed in the schedule to be carried out by a specified date. He submitted that this was an incumbrance within the meaning of the section and that the burden of the repairs would run with the land and fall upon the tenants upon purchase, so that this should be taken into account in determining a price.

Mr Behrens responded to Mr Beck's first argument by rejecting the general proposition that the purpose and effect of the Act was to allow the tenants to purchase at market value. The Act gave no more than a right of first refusal and the whole core was to be found in section 6(3), which allowed the landlord, where the tenants were not prepared to agree to his price, to dispose to such person as he thought fit provided that the consideration required was not less than that specified in the offer notice. This principle should be borne in mind even where there was a breach, since the Act was not intended to have penal consequences: the burden of the bad bargain did not fall upon the tenants, as they were not forced to purchase at all, whereas it would be unfair on the new landlord to make him suffer a loss. The limit of the tenants' rights was quite clear in section 12(1): they could require the new landlord "to dispose of the estate or interest that was the subject-matter of the original disposal, on the terms on which it was made (including those relating to the consideration payable)". Clearly there were circumstances where adjustments could be made, eg under section 12(3)(*a*), and this accounted for the wording on the form and the use of valuation expertise on the tribunal. In any case, the wording on the form should not be used as an aid to the construction of the words of the statute itself. He relied strongly upon *Cousins*, which he cited at length, for the general principle that there was no power to investigate the circumstances surrounding the sale, whether at auction or otherwise. In any case, he pointed out that the subject premises had been sold at auction as recently as October 1988 for £39,500.

On the question of genuineness, he produced copy correspondence passing between solicitors acting for Mrs Kaur and the respondent, in support of the contention that the respondent had indeed paid £45,000 for the property. After a short adjournment to allow Mr Beck to consider the effect of these letters, Mr Beck agreed that he was in no position to contradict the inference that the transaction had been a genuine arm's-length sale for £45,000. (It should be mentioned here that although the name of Mr Singh appeared at various points in the correspondence and he was clearly involved with Mrs Kaur in both the purchase and sale of the property, only Mrs Kaur was registered as proprietor and she alone has been treated in these proceedings as the landlord making the original disposal.)

On section 12(4)(*b*), Mr Behrens pointed out that the repair notice produced was dated before the disposal, whereas section 12(4) referred to the property becoming subject to an incumbrance "at any time since the original disposal". Mr Beck responded that the local authority had subsequently served notice on the respondent seeking details of ownership of the land, under section 16 of the Local Government (Miscellaneous Provisions) Act 1976, but he conceded that no new repair notice had been served since the disposal.

The tribunal at this point drew attention to another reported decision upon the Act, *30 Upperton Gardens Management Ltd* v

*Editor's note: Also reported at [1989] 31 EG 56.

A *Akano* [1990] 45 EG 121*, and invited comment. Mr Beck again emphasised the different approach to be adopted where the disposal had been by auction, but adopted it as authority for the view that such circumstances could be investigated. Mr Behrens did not accept such a distinction in the present case, where there was ample evidence of what the market would pay.

In view of the complexity of the legislation, the tribunal wished to give further consideration to the question of the extent of its powers to investigate the circumstances of the disposal in determining the price to be paid. However, it was agreed by the parties that it would be preferable to take evidence as to valuation at this stage, *de bene esse*, rather than adjourning and reconvening at a later date.

For the applicants, Mr M J Sellers FSVA, with 20 years' experience B in selling properties in the area, produced his written report and gave oral evidence as to his knowledge of the property. He had first visited it in June 1988 to carry out a building society valuation. Although the property had been in a slightly better condition then, it was sadly neglected in a number of major areas which he outlined. He had advised then that any mortgage offer should be conditional upon a structural engineer's report, plus investigation and treatment of the woodworm, dry rot and wet rot, and his valuation was then £22,000. His only explanation for the figure of £39,500 reached at auction shortly afterwards was on the basis that the purchaser thought that vacant possession was available. When he revisited for the present valuation in July 1990 he found few, if any, repairs had been carried C out, and his valuation, taking into account the need for those major repairs, the fact that it was a listed building in a conservation area, and that it was occupied by two protected tenants on registered rents, was £25,000. That the respondent had paid £45,000 was beyond his comprehension, in terms of return on capital investment and the amount of repairs needed. In his view, the presence of the sitting tenants was the major factor in keeping the valuation down, in this case, to about 50% of the vacant possession value. The price paid by the respondent would be a fair market price with vacant possession, and the market had not really moved significantly since November 1989. He disagreed with Mr Newton, the respondent's expert, who valued it with vacant possession at £60,000, in view of the scale of repairs required on a listed building; however, he would not disagree D with Mr Newton's investment valuation of £27,150.

Cross-examined, he said that, as an investment, one would take the rent (plus some element to allow for imminent re-registration) less repairs, insurance and management — say £3,000 pa — and a multiplier of somewhere between five and eight years' purchase: recent sales of investment property at auction had showed such a variation. He agreed that the presence of sitting tenants had a variable effect, depending upon age and circumstances, but here there were two fairly young tenants who had not indicated any intention to move. He agreed that it was possible that an auction price of £39,500 implied the existence of another underbidder prepared to go quite high (or a reserve set at that level). He accepted that E Bramleys, an established and qualified local firm, might, as alleged, have valued the property at £50,000 in May 1989, but he questioned the basis upon which the valuation was requested. He also suggested that the tenants, if they offered £44,000 to Mrs Kaur, might have been using the vacant possession approach. Questioned by the tribunal, he rejected the possibility that the price obtained reflected a possibility of conversion to offices, because in his view there was not the car-parking space required by the planners for use as either office or bed-sit accommodation.

At this point the tribunal raised the question of the appropriate valuation date: it was agreed that the relevant time should be that of the original disposal, not the hearing date, in accordance with the general wording of section 12, but what was the disposal date? Was it F the date of the contract, the execution of the transfer or the registration of the new proprietor? It was agreed, in view of the earlier decision, upon the status of a contract in *Wilkins* v *Horrowitz* [1990] 29 EG 57†, and the general law as to the position between completion and registration, that the appropriate date to take, if valuation were to be permissible, was the date of execution of the transfer, ie March 30 1990.

Mr Gregory, the tenant of the first-floor flat, then gave evidence as to the circumstances in which the tenants had made offers to purchase. He agreed that in May 1989 they had offered £44,000 as a result of a meeting with Mr Uppal, the estate agent who was seeking G to achieve a resolution of the disputes which had arisen between them and Mrs Kaur and Mr Singh. That offer was made without professional advice or a mortgage application, and was made because they were prepared to buy at whatever price to end the harassment they were suffering. He agreed also that on June 5 1989 an offer of £40,000 was made by their solicitors, again without professional valuation advice, which offer was rejected in a letter from Mrs Kaur and Mr Singh's solicitors on June 8 stating that their clients "will not accept less than £50,000". He had also been told in May 1989 that Bramleys had valued it at £50,000, but that must have been without internal inspection, since only he and Mr Rahimi had the keys to allow access. He denied receiving any letter from Mr Uppal in September 1989 stating that the asking price was £50,000 H and inviting their highest offer, and said that at no stage was he aware that the property was actually being marketed.

The valuation evidence for the respondent consisted of the written report and accompanying letter of Mr Robert Newton FRICS FAAV, written in October 1990, and the oral evidence below. On Friday January 11 1991 the respondent had sought a postponement of the hearing because Mr Newton was unable to attend on the 15th; this was opposed by the applicants. The tribunal did not accede to this request, having regard to the grounds laid down in regulation 8(1) of the Rent Assessment Committee (England and Wales) (Leasehold Valuation Tribunal) Regulations 1981. However, Mr Beck agreed that Mr Newton's report and letter should form part of the valuation J evidence, albeit untested by cross-examination. Mr Newton's conclusions were that as an investment property its value was in the region of £27,150 (based upon an annual return of £2,715), that the vacant possession value in its present condition was about £60,000, and that a negotiated figure for a purchase by sitting tenants would have been somewhere midway beween these figures, say £43,575.

Mr Uppal, owner of the Parade Estate Agency with offices in Huddersfield and Leeds, gave evidence that he had been involved when Mr Singh and Mrs Kaur were having difficulty with the tenants and had arranged a meeting in May 1989. Mr Gregory seemed willing to make an offer at that stage but for substantially less than £50,000 and it was not acceptable. He then received instructions to market the property in September 1989, and he produced the particulars. The K property was described as containing two flats "let on an unfurnished registered basis", details being given of the rents and registration dates. The price sought was offers around £50,000 "Subject to tenancy". He had thought that this was the best price obtainable. Although he understood the logic of the investment calculation, people were in fact willing to pay somewhat more for the possibility that the tenants might vacate. At least two other parties had been interested at that price but had not been able to proceed because the tenants would not allow access for inspection. He himself had got in only by pretending to take a builder round. The respondent had first indicated an interest about November 15 and the contract was signed for sale at £45,000 on November 27. If the property were L offered with vacant possession he would look for at least £60,000. The adjoining property had been offered recently at £130,000, and even though that was in a good state of repair and had been converted to bed-sits on assured lettings the gap between the two could not be so enormous. He insisted that he had contacted the tenants in September about buying the property, as that would always be the easiest way to achieve a sale.

In cross-examination, he said that he had five years' experience of the Huddersfield market and that he had sold commercial property all over Yorkshire. If he had simply marketed it at the investment valuation of £25,000 there would have been a stampede in his office. He agreed that he had no evidence that the tenants might move out, but over the last five years he had sold about half a dozen properties M subject to tenancies and there had always been the "hope element" over and above the investment formula.

The respondent, whose command of English is somewhat limited, then gave evidence that he had been looking for a house to move into with his wife and two children, as he was living in rented property. He went to various estate agencies including Mr Uppal's, and went to look round this property with Mr Uppal. The tenants showed him round the inside of both flats, Mr Uppal having introduced him as a builder. He and his wife decided to offer £45,000. They felt there was no need to seek valuation advice, as it was in quite a good state inside and was quite big; they were surprised the price was not higher. He was aware that the tenants were there, but he thought that he would

---

*Editor's note: Also reported at [1990] 2 EGLR 232.

†Editor's note: Also reported at [1990] 2 EGLR 217.

be able to take them to court through solicitors and get them out. He first saw a solicitor after he had signed the contract at Mr Uppal's office, and the solicitor told him that he could not remove the tenants. However, he went ahead with the purchase because under the contract he would have lost his £5,000 non-refundable deposit. He stated that the solicitor did not tell him that the tenants had a right to buy from him.

Mr Behrens' submissions as to the valuation basis were these: first, one must take into account what actually happened. He accepted that two professionals said that valuation should be upon an investment basis, say £25,000 to £30,000. But it was known here that it was sold at auction in October 1988 for £39,500, which was good evidence of market value. It was possible that Mrs Kaur and Mr Singh had not realised about the sitting tenants, but there was no evidence about that and the contract on that occasion, which he produced, clearly stated that the property was sold subject to the tenancies. Bramleys had valued it at £50,000, and there was no evidence that that was on the wrong basis. There was interest expressed by others besides the respondent, and the tenants themselves had offered more. Finally, it actually sold for £45,000. All this surely amounted to good evidence as to the actual market value, and, therefore, he contended that the appropriate market value was what was actually paid.

Second, there was nothing in the Act to say that, if the tribunal embarked upon the determination of the price to be paid, this must be at the market value. In view of the fact that the tenants were themselves prepared to pay £40,000 to £44,000, it would be fair to both sides, taking all the circumstances into account, to insert such a figure so as to ensure that they did not receive a windfall profit at the expense of the landlord.

Mr Beck argued that the tribunal should proceed upon the basis of the market value and the best evidence of that was the valuation evidence of the professional valuers, Mr Sellers and Mr Newton, in preference to Mr Uppal's views. The only person as to whom there was clear evidence of being interested in the property was the respondent, who had bought under a misapprehension about the status of the tenants, and the fact that he was willing to pay £45,000 was, therefore, irrelevant as to value.

The tribunal then considered how to deal with the question of the other terms of the conveyance which it had been requested to settle. Mr Beck said that, although one would normally expect the terms to be the same as those of the original disposal, he had not been in a position to propose that in the purchase notice, since he had not had sight of anything other than the very brief and unhelpful open contract and the office copy of the entry on the register. During the course of the hearing, various documents had been produced, including the transfer, the original lease, and the formal contract entered into pursuant to the auction sale in October 1988. He did not feel in a position to agree the terms immediately, but he was hopeful that, if the tribunal gave its determination as to price, the parties would be able to agree the other terms without the need to return to the tribunal.

The tribunal agreed to this general course of action and ruled that it would give its interim determination as to price in due course to the parties, which determination could be the subject of an appeal to the Lands Tribunal; if the parties were not able to agree the other terms within a period of 14 days after the expiry of the time allowed for lodging notice of appeal, no such notice having been lodged, either party could apply for the tribunal to reconvene to make its determination thereon. If an appeal upon the interim determination was made, the tribunal would, subject to any directions to the parties by the Lands Tribunal, reconvene on the application of either party if terms could not be agreed after disposal of the appeal.

The decision

This tribunal has considered very carefully the effect of the legislation with reference to the price to be paid by the applicants in purchasing the subject premises, such right to purchase clearly being exercisable in this case. As on other occasions, the legislation has proved difficult to interpret and we are grateful for the assistance of those who appeared for the parties.

The crux of the matter is the extent, if at all, to which the tribunal has the power to determine the price payable when the purchase notice served pursuant to section 12(3)(b) of the Act provides for the "terms" upon which the disposal by the new landlord is to be made to be determined by the tribunal. Mr Beck's first and principal submission, strongly opposed by Mr Behrens, was that the tribunal has wide power in this regard. He also made two more limited submissions which can here be disposed of at the outset quite briefly:

First, the question of the genuineness of the price paid upon the original disposal: in view of the evidence produced at the hearing, Mr Beck did not feel in a position to pursue this point. Nor do we, save to comment that, had a live issue emerged, we would have considered very carefully whether this was something into which we should inquire, involving as it might have done allegations of bad faith, or whether the proper course was for the matter to be adjourned for determination of this question under the more formal process of the county court. There being no such issue here, we are satisfied that £45,000 was the price actually paid upon the original disposal.

Second, the effect of the repair notice under section 12(4)(b): we would certainly be of the opinion that such a notice which has been served and become operative under section 190 of the Housing Act 1985 could constitute an incumbrance, particularly in view of the effect of section 190(5), added under section 130 of the Housing Act 1988, which provides for it to be a local land charge. However, the only repair notice here was served and became operative before the original disposal and therefore section 12(4) is not applicable thereto.

Returning to the principal argument, Mr Beck's contention was not that in all cases the tribunal has power to put its own valuation upon the property and order disposal at that price, but that in what for present purposes we might label an appropriate case the tribunal should go into the circumstances surrounding a disposal when the price paid was completely out of line with what an expert valuation of the open market price would indicate and substitute that expert valuation. However, we have given some consideration to the broader approach, because we have felt it right to form our own view as to the correct basis for the exercise of our jurisdiction and because arguments about this and the narrower approach are inextricably linked.

Much argument centred on *Cousins* and its status in relation to the present case. Before considering the substantive arguments it might be helpful to say this: although the decision of another leasehold valuation tribunal is not as such binding upon us (and therefore it is not perhaps of such crucial importance to identify what could strictly be regarded as the *ratio decidendi* and *obiter dicta*), where we have a decision reached after careful argument and discussion, and not dissented from, so far as we are aware, by any tribunal before whom this question has arisen, we would not readily adopt a different approach. *Cousins* clearly decided that the tribunal does not have the power to substitute its own valuation but must proceed upon the basis of the price actually paid (with possible exceptions to be dealt with further below). This conclusion was arrived at partly on a close reading of the various statutory provisions and partly on the perceived policy of the Act. We would make the following observations.

A reading of section 12(3)(b) in isolation, or indeed in conjunction with section 13(1)(b), would not imply that there is any such limitation upon the tribunal's powers in determining the terms of the disposal. However, the alternative reasoning is that section 12(1) does not simply lay down one possible form of purchase notice (ie where the terms are known, with section 12(3)(b) providing for the other form as used here) but is the fundamental provision which provides for the tenants' right to buy and lays down the limits within which this right can be exercised, whether the purchase notice invoking the tribunal's assistance is one served under section 12(1) itself or under section 12(3)(b). If this view (the *Cousins* view) be accepted, then clearly the price is to be "the consideration payable" under the original disposal, and the tribunal has no general jurisdiction to modify that particular term by substituting its own *de novo* valuation. The provisions are by no means easy to construe, but on balance we had decided to adopt the latter construction.

Consideration of the policy of the legislation does not yield clear guidance in the present case. We accept that the Act should not readily be construed to have a penal effect upon a landlord who has sold after genuine negotiation in the open market by making him sell at a lower valuation. (It may be of interest in this regard to note that section 12(6) prevents the tenants from taking a "windfall" profit where the property has increased in monetary value owing to a change in circumstances, but there is no corresponding provision for reduction, other than for an actual charge or incumbrance under subsection (4).)

On the other hand, where the property has been sold in clear contravention of section 1 of the Act and at a price which is higher

A than what could have been expected to be obtained on the open market, should the tenants be presented with a "take it or leave it" choice? We recognise, however, in that instance that had the landlord followed the correct statutory procedures and the tenants not proceeded upon his final offer he would be able to sell within 12 months, under section 8(3), at a higher price (but not lower) if he could find a willing purchaser.

We might comment here that although we have not regarded it as a consideration which should properly influence our conclusions, we think the present case is by no means the typical case envisaged by those enacting this legislation. Here we have a property containing solely these two tenanted flats. If we are to accept the experts' view, the price which a seller could expect to receive on the open market B would be governed by the existence of the protected tenants, and therefore the "investment value" would dictate the price around which those tenants can negotiate with the selling landlord if he serves a notice under section 5; that is different from the "sitting tenant value", which normally results in a higher price. The investment value does not reflect the possibility that, with only two tenants involved, they may well be able to resell shortly afterwards with vacant possession. Nevertheless, these tenants clearly have the right to avail themselves of the provisions of the Act, and we have attempted to construe that wording in the light of what we infer to be the general policy of the Act and carefully considering the arguments put forward by the representatives of both parties.

C In the end, we have concluded that we should proceed upon the basis that the Act does not give a tribunal, wherever there is a purchase notice served under section 12(3)(b), the power to embark upon a *de novo* valuation exercise to determine the price which is to be paid.

We have gone on to consider the occasions where the tribunal may, or must, be involved in the determination of a question relating to the price payable, to see whether there is scope for such an exercise of our jurisdiction in the present case. Mr Beck put this in terms of the "investigation of the surrounding circumstances". We have made a tentative attempt to clarify that rather vague phrase by identifying different situations which have already arisen for argument or decision.

D 1. It may be that where the consideration payable on the original disposal was not actually expressed in monetary terms, the tribunal might be called upon to determine a dispute as to the monetary price now payable, for instance where the original disposal involved a land exchange deal or transfer of shares in an unquoted company. That point does not arise for decision here.

2. Where the question arises as to whether there was a genuine arm's-length transaction involving a real bargain at the price alleged, this must clearly be a matter which requires investigation if the Act is not to be circumvented by fraud. However, the question whether this is properly a matter for the tribunal or for the county court did not arise in this case.

E 3. There are the circumstances expressly laid down in the Act where the consideration payable is modified: section 12(3)(a), where the original disposal included property additional to the subject premises; section 12(4)(b), where the property has subsequently become subject to an incumbrance reducing its value; section 12(6), where the property has increased in value owing to a change in circumstances ("other than a change in the value of money"). In these circumstances, it seems to be implicit, and has so been accepted by tribunals, that it is for the tribunal to determine the adjustment of price, in accordance with the varying formulae laid down by the respective subsections. These are themselves by no means easy to interpret, but none of them has been found to be applicable here.

4. There are occasions where it may be found that the property was F the subject of misdescription on the original disposal, so that the estate or interest now to be disposed of is not precisely the same as that which was then offered for sale. This was the situation which arose in *Cousins*; in the event the parties reached agreement as to the abatement of price, but the tribunal was of the opinion that its jurisdiction would have extended as far as determining the appropriate abatement in relation to the actual price paid. Such a view was based in part upon an analogy with section 12(3)(a). This situation has not, so far as we are aware, arisen in the present case. We add that qualification because it will be recalled that the application to this tribunal provided for the estate and interest to be disposed of and the terms of the disposal to be determined by the tribunal, as the applicants were not in a position at that stage to ascertain the full details of the original disposal. The applicants have G now been provided with the appropriate details, and have indicated that such determination will in all probability not be necessary. Indeed it is very difficult to envisage circumstances in a situation such as this, involving a registered transfer preceded by the terms of such sparse contractual documentation, whereby the purchaser would be able to seek reduction of price on the basis of misdescription. However, we would indicate at this point that the price which we have proceeded to determine is founded upon the premise that there has been no misdescription which would have entitled a purchaser to abatement. If a later hearing upon the other matters becomes necessary, and the highly unlikely fact of a misdescription is proved, then an adjustment of price to reflect the difference in estate or interest to be disposed of might need to be further considered. H

5. We have considered whether this case may present another instance where the tribunal may modify the price payable, in accordance with Mr Beck's argument. We have phrased this exception in the narrowest terms possible, and have felt justified in regarding it as within our jurisdiction at least to consider its applicability. This instance would be where two requirements were satisfied: first, the price paid was totally at variance with the open market value and, second, it was paid by a purchaser who was suffering under a misapprehension (as in this case where the purchaser thought he could obtain vacant possession while in fact there were protected tenants). We stress that we are not dealing with a case of misdescription as under 4 above; nor was there any evidence J before us of a misrepresentation of the position by Mrs Kaur or her agents or of an acquiescence by her or her agents in the purchaser's unilateral mistake (whether it be labelled a mistake of fact or law) such as to have entitled him to relief against her.

This would clearly go further than the previous instances. In the "misdescription" case, it could be argued that from the outset the "consideration payable", under section 12(1), was actually the price paid less the amount of the abatement to which the purchaser was legally entitled. In the present case it is difficult to see how, legally, the purchaser would have been able to look to the vendor for a reduction in price. On the other hand, is it right to adopt such a rigorous interpretation of section 12(1) that the tenants who have already been denied their right to be offered the property initially are K then told that they may exercise their right only on the terms which a mistaken purchaser accepted after failing to seek professional valuation or legal advice?

In the event, we do not find it necessary to decide that point in isolation, since in this case the other requirement comes into play. Mr Beck was not suggesting, and we certainly would not accept, that the mistake alone was sufficient. He contended that the other relevant "surrounding circumstance" was the fact that the price was completely out of line with the open market valuation made by experts. We recognise that for the tribunal to inquire into this aspect might perhaps be regarded as simply another attempt to import a *de novo* valuation by the back door. We do not necessarily accept this L view. The purpose on this occasion would simply be part of the background investigation to establish that this was one of the very limited number of cases where consideration of surrounding circumstances was appropriate.

If such an investigation is permissible, we would put the question to be asked by the tribunal in these terms: "Was the price paid in this case totally out of line with the open market value of the property, sold as an investment property with existing protected tenants?" This would accord with the approach laid down in the provision which is perhaps the most closely analogous, section 12(6): "the amount that might reasonably have been obtained on a corresponding disposal made on the open market at the time of the original disposal". The professional experts gave a theoretical investment valuation, in the M region of £25,000 to £30,000, based upon the income/years of purchase approach. That would obviously carry much weight against the unsupported opinion of a local estate agent and would accord with the valuation experience of the tribunal. However, we adopt the wording of the tribunal in *30 Upperton Gardens Management Ltd* v *Akano*, "Whatever might be said, the best evidence of the value of anything is what it will fetch in the open market". Here, as in that case, there is evidence of what the property did fetch, at auction, in October 1988, ie £39,500, and to view the expert evidence without taking that into account would not be appropriate. The conditions of sale at the auction made it clear that the sale was subject to the existing protected tenancies. There is no evidence that the market in

A investment property slumped in the period between then and the relevant disposal for £45,000. Even without taking into account other evidence as to persons offering prices around that region, we are led to the view that "the market" in this particular area can reasonably be viewed as paying more than the traditional investment property valuation for properties such as the subject property, and it is not particularly helpful to speculate further as to the reasons for this.

Therefore, even if we accept Mr Beck's argument, that this would be one of the instances where we are entitled to look at the surrounding circumstances because of the mistake of the respondent and the level of price, we are of the opinion that he has not succeeded in showing that the price paid, £45,000, was out of line with the open B market value.

As we stated at the outset, we have not found the matter easy to determine, and look forward to a time when authoritative guidance will have been provided on various points by Lands Tribunal C decisions. However, the conclusion that we have arrived at is that normally the tribunal has no jurisdiction, in determining the terms of a disposal under a purchase notice served pursuant to section 12(3)(b), to do other than ascertain the consideration payable under the original disposal and determine that to be the price payable. We have suggested some possible instances (nos 1 to 4) where the tribunal may be called upon to do more than that, but none of those is applicable here. We are not convinced that instance no 5 provides another such occasion, but even if we were, the evidence here as to market value would not lead us to the conclusion that we should disturb the price of £45,000.

We have therefore determined that the price payable by the D applicants for the disposal by the respondent of his interest in the subject premises is £45,000.

SUPPLEMENT

CASES NOT REPORTED IN "ESTATES GAZETTE"

AUCTIONS

Chancery Division

February 7 1990

(Before Mr John LINDSAY QC, sitting as a deputy judge of the division)

SWINGLER v KHOSLA

Vendor and purchaser — Sale by auction — Failure of purchaser to complete — Alleged misrepresentations in catalogue inducing sale — Purported rescission — Planning permissions — Exclusion clauses — Whether representations relied on — Forfeiture of deposit — Decision in vendor's favour — Section 49(2) of Law of Property Act 1925 considered

The present proceedings arose from the abortive sale by auction of a freehold property, a listed building — Property was knocked down at the auction to the defendant for £275,000 — The deposit of £27,500 was paid after the defendant's first cheque had bounced — Sale never completed — Vendor sought a declaration that defendant had wrongfully failed to complete and claimed the forfeiture of the deposit, interest and damages — Defendant alleged that the contract had been rescinded on the ground of misrepresentation and non-disclosure and claimed the return of the deposit and other relief — The auction took place on April 23 1987 — After a second abortive auction sale to another purchaser whose cheque for the deposit was not honoured, the vendor sold the property by private treaty for £360,000

The defendant's case, which was in large part rejected by the judge, was that he had relied on misrepresentations of various kinds inducing him to enter into the contract of sale — Some of these allegations fell to the ground during the hearing; those that remained concerned mainly alleged misrepresentations as to the existence of planning permissions — It became evident that the defendant wished to get out of his bargain, which he had entered into with little forethought or experience, and was looking round for various grounds to cover his withdrawal — These included alleged failures to bring matters to his attention, including an enforcement notice, the fact that the hotel started in the subject premises was not a going concern, and certain special conditions; and in the end he complained particularly about alleged misrepresentations concerning the number of bedrooms likely to be available — The judge concluded that, even if any of the alleged misrepresentations could be established, the defendant had clearly failed to demonstrate that he had relied on them

The result was that there had been no effective rescission of the contract of sale by the defendant — It followed that the plaintiff was entitled to forfeit the deposit unless there were any grounds for exercising discretion in his favour under section 49(2) of the Law of Property Act 1925 — It was suggested on behalf of the defendant that as the plaintiff had made a "profit" of £85,000 on the eventual sale of the property by private treaty for £360,000, the deposit should be returned to the defendant with only deductions for the plaintiff's expenses on resale — The judge rejected this suggestion — He also rejected a claim for damages under section 2(1) of the Misrepresentation Act 1967 — Finally, although this was not necessary to decide in view of the grounds of his decision, he rejected a submission by the defendant that the plaintiff was not entitled to rely on certain exclusion clauses incorporated in the auction contract on the ground that they were unreasonable — The judge would have held, if it had been necessary, that the plaintiff had demonstrated their reasonableness

The deposit was declared forfeited to the plaintiff and the agreement for sale held to be terminated — The plaintiff was entitled to interest and an inquiry as to damages — The defendant's counterclaim was dismissed

The following cases are referred to in this report.

Dimsdale Developments (South East) Ltd v *De Haan* (1983) 47 P&CR 1
South Western General Property Co Ltd v *Marton* [1982] EGD 113; (1982) 263 EG 1090
Walker v *Boyle* [1982] 1 WLR 495; [1982] 1 All ER 634; (1981) 44 P&CR 20; [1982] EGD 1149; 261 EG 1090

In this action the plaintiff, Raymond John Peter Swingler, sought against the defendant, Hukam Chand Khosla, a declaration that a deposit of £27,500 paid by the defendant under the contract for the sale to him by auction of Wicken Hall, Wicken, Cambridgeshire, was forfeit to the plaintiff and that he was entitled also to other relief. The defendant claimed the return of the deposit and damages on the basis that he had been entitled to rescind the contract for misrepresentation.

Nicholas Le Poidevin (instructed by Rustons & Lloyd, of Newmarket) appeared on behalf of the plaintiff; Wayne Clark (instructed by Hillman Trivedi & Co) represented the defendant.

Giving judgment, **MR JOHN LINDSAY QC** said: This case concerns the sale by auction on April 23 1987 by the plaintiff, Mr Swingler, of his freehold property known as Wicken Hall, Wicken, Cambridgeshire. The property was knocked down to the defendant, Mr Khosla, for £275,000. The 10 per cent deposit was eventually paid. The sale did not proceed. Mr Swingler says the deposit is forfeit to him and claims a declaration that that is so and also some other relief. Mr Khosla claims that he was entitled to rescind the contract for misrepresentation and that the deposit should be repaid to him and that damages for that misrepresentation should also be paid to him.

I shall first describe the parties and their respective witnesses in a little more detail. Mr Raymond John Peter Swingler is the plaintiff. He is the assistant director of the Press Council and was in the employ of the Press Council at the time of the auction. Before that he had been a journalist. His planning consultant and professional designer at all material times was Richard Thomas Ambrose, a sole practitioner who provides an architectural, planning and design service under the name Richard Ambrose Associates. Mr Swingler's solicitor in connection with the sale of Wicken Hall was Michael John Drake, a partner in the firm of Rustons & Lloyd, of Newmarket. The auctioneers he used in the sale were Barnard Marcus and the auction was conducted from the rostrum by their Mr Robin Douglas Cripp, an experienced auctioneer, with the assistance, of course, of other members and staff of that firm.

All I have so far named gave evidence before me, as also did John Weatherall, Nathan Anthony Smith and Derek Michael Bishop, all of Barnard Marcus. Mr Weatherall apart, their evidence was largely directed to points which ceased to be contentious and which for that reason I shall not need to refer to in any detail.

On the defendant's side there were three witnesses only: the defendant himself, Mr Hukam Chand Khosla gave evidence. He has been a factory worker at Ford's factory for over 20 years. While his spoken English is not good it has not been suggested that he does not understand either written or spoken English. His son, Sunil Khosla, also gave evidence. He is a self-employed site manager, now 27 years of age. The defendant's wife is the sister of the third witness on the defendant's side, Mr Arjundev Ghai, a successful property developer who came to this country in the 1950s with next to nothing and who now has companies with gross assets in the order of £4 to 5m.

The property with which the case is concerned, Wicken Hall, is a listed building set in its own four acres of land in the village of Wicken, Cambridgeshire. It is some 10 miles from Cambridge and six miles north-west of Newmarket. It is said to date back in part to the 14th and 15th centuries, although it was largely remoulded in the 1760s and is, accordingly, Georgian in style. It had, when the relevant events begin, a driveway stretching in a straight line from close to the house to the nearby road. The driveway ran towards the eastern side of the Hall's own land, meeting the public highway close to the church land immediately to the east.

Mr Swingler acquired the property in 1966. It was for a time during his ownership used partly as his, his wife's and their family's home and partly for temporary accommodation for up to 26 US Air Force personnel to whom, I think, rooms were let, room by room. Mrs Swingler died in 1980 and it was thereabouts that thoughts of selling the property came to Mr Swingler's mind. He hoped to keep the property, and indeed his children wished that he should. He thus turned his mind to devising ways in which he could either keep the house or, if he had to sell it, sell it to best advantage. He thought the grant of planning permission for its use as an hotel would be likely to increase its sale value.

In May 1980 there was a planning application for change of use from multi-occupancy accommodation, which referred to the US Air Force short-term lettings, to use as an hotel. Mr Swingler had, in March 1980, consulted Mr Ambrose as planning consultant and he had prepared the application. The local authority granted such planning permission on August 1 1980. The grant referred to the plans which had then been submitted. It was a condition of the grant that the old straight driveway's access to the highway should be closed "permanently and effectively" before any development work was commenced. The plan referred to in the grant showed a new crooked driveway. It also showed a projected house for the manager of the hotel. It showed the site of the existing building, describing it as follows: "Existing three-storey building, The Hall, contains sleeping accommodation to 10 people."

The plan also showed what might be described as three prospective extensions, to the east, to the west and to the south of the existing building. As to the eastern development, this was shown on the plan and was described as "New two-storey wing with five double bedrooms over public rooms." The western development was shown and described as "New two-storey wing with five double bedrooms over reception rooms, bar etc." A dining-room and kitchen were also indicated as part of the western development. The southern development, an L-shaped development joining the eastern development at the southern end of that development and then running first north/south and then dog-legging westwards, was described as to the north/south stretch as "New two-storey wing with staff accommodation over garaging", and as to the dog-leg as "New single-storey stabling for six horses."

Although the 1980 planning permission was in terms as to use only "Change of use from multi-occupancy accommodation to hotel", a number of the conditions related to the developments shown on the plans. Condition 1 required that no work should begin until detailed plans and elevations had been approved. Condition 7 specified the facing materials to be used in the development. Another condition described the application as having been made for outline permission only and stated that it gave, in some respects, perhaps as to parking, insufficient detail of the proposed developments.

On a fair reading of the 1980 planning permission, the reader would take it that subject to the provision of further detailed drawings and to the making of suitable arrangements for joining the highway and some other minor details, the local authority had no objection to substantial development of the Wicken Hall site along the lines broadly shown as the eastern, western and southern developments, nor to the use of the existing building, nor as and when developed, the extensions, as an hotel. Mr Ambrose's evidence was that the 1980 planning consent would expire on August 1 1983.

Notwithstanding the grant of the 1980 planning permission, Wicken Hall continued to be described as being used for multi-occupancy occupation. At any rate it was so described in the next relevant planning application, which was made, again as prepared by Mr Ambrose, in December 1983. On February 13 1984 the local authority gave permission for "Renewal of application for change of use to hotel incorporating alterations and extensions." The grant referred to a plan dated December 19 1983. As had been the case in the 1980 grant, the local authority were concerned about access to the highway and required the then still existing straight driveway's junction with the highway to be permanently closed and the new one to be opened prior to the commencement of the proposed works of development.

This 1983 grant was not for a change of use only but, more plainly than had been the 1980 one, gave consent for the extensions shown on the accompanying plan. Moreover, it was not expressed in any way as being merely an outline consent. Mr Ambrose saw this permission as running to February 13 1989. It was thus current at the date of the auction. There were in fact four plans describing the approved proposals and they, as it were, amended and fleshed out the proposals for the eastern development as proposed in 1980. The plans did not deal with the proposed southern development at all and only barely with the development to the west of the existing house.

Plan 10, E1, showed detailed proposals for the ground-floor level. Plan 11, E2, showed the proposals for the first-floor level. These included five bedrooms within the existing building, one bedroom as being a new first storey over an existing ground-floor building, and another bedroom, and a large hall-like area with three double French windows leading on to a terrace. This area was marked "Billiards". There was also provision for male and female lavatories and the staircase. It thus showed seven bedrooms as such and the large hall at first-floor level. The second floor plan, E3, plan 12, showed, *inter alia*, a further three bedrooms, all within the existing building. Accordingly, detailed planning permission was in 1983 obtained for, *inter alia*, a large area marked "Billiards" and for 10 bedrooms, of which two were in new buildings and eight in the existing building. The billiards area would, however, be capable of amendment within its exterior size and shape to contain another five bedrooms and associated corridor space. Mr Ambrose said it had been designed so it could be converted. New windows would be necessary if bedrooms were to be installed in that area, and an amendment to the approved plans would be needed for that. Mr Ambrose said a new planning application would not be warranted.

The five bedrooms would be double rooms, each surpassing in area the then requirements of the East Anglian Tourist Board. The layout would need fire approval, but in view of the proposals containing a reference to a terrace at first-floor level and a staircase to the ground floor, Mr Ambrose thought the conversion from proposed billiards area to proposed five bedrooms had the potential to comply with the fire prevention officer's requirements. Mr Ambrose said that the local authority as planning authority were in general not concerned with the interior layout of the proposed buildings and he saw the amendment for five bedrooms in the area marked "Billards" as a minor amendment which the planners would indicate would be an acceptable one.

Accordingly, as far as bedrooms and detailed planning permission are concerned, the position in February 1984 was that there was planning permission for the use of 10 bedrooms as hotel bedrooms, two of which were unbuilt, eight of which were in the existing building, and there was detailed planning permission for the erection of a building originally described for billiards which could be expected, by simple amendment, readily to achieve permission for use as a further five bedrooms, making 15 in all — eight in the existing building, and seven in areas as yet unbuilt.

In July 1985 Mr Ambrose, on Mr Swingler's instructions, made another planning application. It was for a renewal of the 1980 permission for change of use from multi-occupancy to hotel. The permission, which referred to a plan, was granted subject to conditions on October 10 1985. It was still current at the date of the

A auction. No works were to be done until detailed plans were approved. A new driveway access was to be provided and the old one to be closed within 28 days after the bringing into use of the new one. As in 1980, although nominally relating only to use, the conditions included ones relating to building works. The grant made reference in one of its conditions to the application having been for outline planning permission only. Mr Ambrose regarded it as a grant of outline planning permission.

The plan referred to in this grant was the same as that in the 1980 grant which had by then lapsed. It thus contemplated, *inter alia*, the western and southern developments which I have earlier explained. In the western development five bedrooms could be provided as the plan indicated. Indeed, Mr Ambrose was of the view that five was not B the maximum there, but eight to nine bedrooms to Tourist Board standards could be contained in the exterior shell indicated on the plan and which I described as the western development.

As for the southern development, Mr Ambrose was of the view that notwithstanding the description "Staff accommodation over garaging", the permission did not restrict the building to that purpose but the consent was for a building for the broad planning purpose of an hotel and that at least a further 20 bedrooms to the Tourist Board standard could be contained in the exterior shell of the L-shape which I have called the southern development. He was of the view that more than 20 double bedrooms to Tourist Board standards, and with their own en suite bathrooms and wcs, could be C accommodated within the outline, even if the dog-leg stretch originally described as stabling was kept as a single- rather than a double-storey building.

Mr Ambrose's view was that the 1985 outline planning consent was for further accommodation "On the lines of what was shown on the plan". He would not expect an outline consent to specify the precise number of bedrooms consented to unless a number had been specified in the application, which had not here been the case. The 1985 consent was thus, in his view, a consent which, although expressed only as to use, was properly to be regarded in practical terms as an outline approval for further hotel accommodation of the order of five double rooms, and perhaps as many as eight or nine in the western development, and at least 20 in the southern.

D I have had no evidence other than from Mr Ambrose on planning matters. In particular, I have heard nothing from the local authority. I accept Mr Ambrose's evidence. To collect the effect of his evidence and of the planning permissions so far as concerns the number of hotel bedrooms as they stood at the time of the auction, and dividing the effect between existing buildings, the eastern and western developments together, and the southern development, the position is as follows:

(a) *Detailed permission*

(i) eight in the existing building;

(ii) two in unbuilt areas designated as bedrooms within the eastern development;

E (iii) the prospect of five in the area designated as the billiards area, but for which bedroom conversion would be obtainable upon a simple amendment of the eastern development.

(b) *Outline permission*

For development on the lines of:

(i) five at least, and possibly eight to nine in the western development;

(ii) at least 20 in the southern development.

On September 19 1985 Mr Swingler had been served with Notice Requiring Provision of Sanitary Appliances served under the Local Government (Miscellaneous Provisions) Act 1976. It required the installation of two water closets and three wash-hand basins and efficient hand-drying facilities. On November 7 1985 Mr Swingler F was served with an enforcement notice specifying breaches of planning control and requiring the permanent closure of the straight access route and the construction of the proposed crooked one. This enforcement was the next day entered upon the register of planning charges at the local Land Charges Registry. There was later some correspondence with the local authority on the subject of these notices and a threatened prosecution.

The position was finally dealt with by Mr Swingler's giving a personal undertaking on March 18 1987 that he would not open the premises as an hotel unless "An apposite lavatory block is erected and the driveway completed". An alteration was made to the course of the driveway. Instead of running as it had in a straight line, two bends were put in it so that moving from the Hall to the road one would first travel down the old straight driveway and then bend first G left and then right and meet the road not at the eastern end of the Hall's land, as had previously been the case, but very close to its western boundary. The work had been completed by the summer of 1986.

The plan on p 71 of the auction particulars showed this new crooked driveway. That plan did not show the superseded northernmost part of the erstwhile straight driveway. That northernmost part had not been permanently closed. There were iron gates at the point at which the old driveway (to call it that) met the road. The siting of these gates occasioned some difficulty, with both the local authority and those responsible for the adjacent church taking an interest in their position. Eventually Mr Swingler gave an undertaking to remove the gates. He had to close off the old driveway H in order to comply with his undertaking.

At the time of the auction the new driveway had long since been constructed but the old one had not been finally blocked off. Wicken Hall had operated as an hotel only briefly while in Mr Swingler's hands. It was closed as an hotel several months before the auction.

At some stage Mr Swingler decided to sell Wicken Hall. He chose to sell it by auction and he chose Barnard Marcus of Shepherd's Bush, London W6, as the auctioneers. They sent a representative to the property. Mr Swingler showed him the planning permissions which Mr Swingler had obtained, and also a brochure which had been prepared in 1984 when there had been a proposal for the property to be developed by some third parties with Mr Swingler participating. J The text of the page which later became the page in the auction particulars referring to Wicken Hall and headed "Planning" was prepared by Mr Swingler in association with Barnard Marcus's representative. Mr Swingler believed the southern development could contain a further 20 bedrooms. He sanctioned the descriptions used.

In this way the auction catalogue references to Wicken Hall were prepared. Wicken Hall was one only of over 180 lots being offered for sale in the same catalogue. On p 70 of the catalogue there were four colour photographs, one of the exterior and three of the interior, of Wicken Hall. The lot, which was lot 109, was headed "Substantial Freehold Period Hotel, Freehouse and Restaurant set in approximately four acres with planning permission/vacant K possession". There then followed a general description which, so far as concerns bedrooms, wrote of two on the second floor, described as "owner's private accommodation", and five on the first floor. On p 71 there was a coloured axonometric drawing, a colour key to that drawing, and a small plan. The drawing, key and plan had been prepared by Mr Ambrose. The small plan showed the existing building in blue, the eastern and western developments in pink and the southern in yellow. The prospective manager's house was shown in pink. The only driveway shown was the new crooked one.

The axonometric drawing used the same colour scheme. It had lines pointing to parts of the building with words purporting to describe them; hence, of the blue of the existing building it said L "Hotel bedrooms" and "Bar". Of the eastern development it said "Functions Room". Of the western it said "Hotel reception and entrance, dining room and kitchen, bedrooms over" and "Conservatory bar". The southern development which, as to its dog-leg, was now shown as a two-storey building, was described as "Motel bedrooms". Alongside the drawing and the plan was a colour key with three boxes. Blue indicated "Existing house"; pink showed "Detailed planning permission, 10 further bedrooms"; and yellow "Outline planning permission".

Square footages were given for the existing house and for each of the eastern and western developments. None was given for the southern one. This key was inaccurate. As I have explained above, there was no detailed planning permission in the pink areas for 10 M further bedrooms but only for two with the prospect of an added five in the so-called billiards area, and outline planning permission for at least five but possibly up to eight or nine in all in the western development.

The pink was further inaccurate in showing the western development as being the subject of detailed planning consent in respect of its whole area. In fact, only its conservatory had received detailed planning consent. Inaccuracy was unfortunately also found in the text. It read as follows: "Planning permission has been granted by the East Cambridgeshire District Council for the erection of an extension comprising 14 bedrooms, bar, restaurant, functions room and kitchen, and a separate manager's/owner's home. Outline

A planning permission has been granted for a further 20 bedrooms. Copies of the plans and planning permission are available for inspection at the auctioneer's office."

I shall later need to refer to this subject in some detail, but when I refer to the 14-bedroom point and to the 20-bedroom point, it is to argument relating to these references and to those numbers of bedrooms that I shall be referring.

There was no detailed planning permission for the manager's house. The extensions for which detailed consent had been obtained did not comprise 14 bedrooms and the southern development's outline planning permission did not in terms refer to "a further 20 bedrooms". Mr Ambrose said the 14 figure was incorrect, as was, he
B also accepted, the reference to 10 further bedrooms in the key. He accepted the pink colouring of the western development was wrong in its signification of detailed planning permission, whereas only outline had, for the large part, been obtained. There was also, of course, a conflict between the key's reference to 10 further bedrooms in the east and west developments, and the text's reference to "An extension comprising 14 bedrooms". Plainly there were a number of inaccuracies in the descriptions given in the auction particulars.

Next I turn to the terms on which and the manner in which Barnard Marcus offered properties generally, and Wicken Hall in particular, for sale. The auction was one of a series of regular auctions conducted by Barnard Marcus. Their auction office is in the Shepherd's Bush Road. The sale in question was a two-day sale. The
C range of properties sold was wide, from relatively inexpensive single freeholds with vacant possession to blocks of flats, freehold ground rents issuing out of blocks of properties, commercial properties and development plots. The public which the auction aimed to attract was no doubt correspondingly wide.

The auction particulars contained a notice at its p 4 headed in red "Important notice to be read by all bidders". It draws to the reader's attention that general and special conditions were to be found at the back of the catalogue. It stated that prospective purchasers were deemed to have made all usual and necessary searches and enquiries with all relevant authorities. It pointed out that as the gavel fell there would be a binding contract to buy in accordance with the provisions of the catalogue. It stated that a memorandum of contract would
D need to be signed and a 10 per cent deposit paid.

Prospective purchasers were strongly advised to check all matters to which the properties were expressed to be subject and of which they were to have the benefit. At the back of the catalogue, on its p 96, was a further notice in red in large print headed "Notice to all bidders". It read: "Please ensure that you have your copy of the general conditions and special conditions as these are now printed separately from the catalogue". The final page of the catalogue was the form of memorandum which, upon its being suitably completed, a successful bidder would be required to sign.

The general and special conditions so referred to were comprised in a separate document some 56 pages long. The first two pages thereof
E were described as "General conditions of sale applicable to all lots" and the rest dealing with different lots respectively were headed "Special conditions of sale". Of those general conditions I need refer to general condition 1(i), which incorporated National Conditions of Sale, 20th ed with certain provisions as to amendments and inconsistency; general condition 1(iii), by which the purchaser was deemed to have full knowledge of the conditions subject to which the property was sold; general condition 6, the 10 per cent deposit to be held by Barnard Marcus Estate Holders; and general condition 10(i), completion four weeks from the date of the auction.

General condition 11 at subpara (i) stated that the purchaser must himself take all steps which may be necessary to verify the accuracy of the statements in the particulars and conditions of sale. It warned
F that any incorrect statement would not annul the sale or entitle the purchaser to be discharged from his purchase. It stated at subpara (ii) that the purchaser would be deemed to have made local land charge searches and enquiries and to have knowledge of the matters that would thereby be disclosed. At subparas (iii) and (vi), and para 12 there were further provisions to which I refer but need not summarise.

At general condition 16 there was a provision relating to the Misrepresentation Act. As for the National Conditions of Sale, 20th ed, thus incorporated, condition 17 is entitled "Errors, Misstatements and Omissions". I need not refer to it in any more detail at this stage.

It will have been seen that, leaving aside how far they took effect, the broad tenor of the auction general conditions was to cast upon the G
prospective purchasers the requirement that they should "look before they leap" and to deny them remedy should they fail to do so.

The special conditions relating to lot 109, Wicken Hall, were, in the initial print, quite short. However, I find on the evidence that it is far from unusual for there to be modifications in special conditions and that when that occurs it is dealt with in a number of ways. First of all, there are addendum sheets put on each seat in the auction room and laid out in piles at the back and front of that room. This was done in this case. The addendum sheet in the present case is two pages long and states as to lot 109 that "There are revised special conditions of sale containing details of outstanding notices on the property". It advises prospective purchasers to obtain a copy prior to the auction and warns that a copy of them will be attached to the contract to be H
signed by the purchaser.

Second, an announcement is made orally by the auctioneer at the beginning of the session of the auction. That, too, was done in this case, at a little after one o'clock in the afternoon. Mr Cripp referred to the addendum sheets for that day's sale. The text of what he said is available because a tape recording is habitually made and one still exists. The text is not disputed. It makes specific reference to there being revised conditions of sale as to Wicken Hall and that they were available at the auctioneer's right hand.

Third, the auctioneer, if there has been an amendment relating to a particular lot, makes an announcement to that effect when he comes to that lot. In the present case when Mr Cripp came to lot 109 he made J
such an announcement. Its text is not in dispute, nor is it disputed that it was adequately broadcast throughout the auction room. It draws attention, *inter alia*, to the fact that there are revised special conditions for lot 109. It mentioned a prosecution in respect of the driveway and sanitary provisions as having been withdrawn. The auctioneer said he presumed the revised special conditions had been read by proposed purchasers.

The revised special conditions so referred to, the text of which Mr Cripp had received from the vendor's solicitor about two weeks before the sale, stated that an enforcement notice had been issued on December 7 1985 under section 87 of the Town and Country Planning Act 1971 and went on "in respect of which the vendor has given and has had accepted an undertaking that the property will not be used as K
a hotel until such time as the enforcement notice is complied with". Copies of these revised special conditions were available at the auction itself and had been made available to those making enquiries of the auctioneers ahead of the auction.

It is difficult to see what more could, in practical terms, be expected of auctioneers in connection with drawing to the attention of those likely to bid that there were revised special conditions of sale applicable to lot 109.

Mr Khosla had never been to any property auction before this one. A couple of days before the auction his son, Sunil Khosla, brought him a copy of the Barnard Marcus catalogue. He telephoned his brother-in-law, Mr Ghai, the evening before the auction. He had not spoken to him before that on the subject of Wicken Hall. L
Throughout the rest of the events as they unfold it is to be remembered that Mr Khosla and Mr Ghai were closely associated in the venture which I shall describe. Both were emphatic that it was Mr Khosla who bid for Wicken Hall and that he did so in his own right and not as nominee for or partner of Mr Ghai. I accept that, but even so Mr Ghai's role is hereafter that of close associate as well, of course, as brother-in-law, of Mr Khosla.

Mr Khosla saw the "Notice to all bidders" in the catalogue. He saw the general conditions of sale but did not read them. He saw the original print of the special conditions relating to lot 109. He arranged to meet Mr Ghai at the auction itself. When Mr Khosla arrived the auction, which was held at the Kensington Hilton Hotel, M
was already in progress. He would, therefore, have missed the announcement made by Mr Cripp shortly after 1 pm. Mr Ghai was already there and as Mr Khosla arrived was at the refreshment counter. The auctioneer was operating from a rostrum with a desk or table to his left and right at which auction staff were present. The table to the right of the auctioneer had papers on it. The one on the left had ladies of the auctioneers' staff sitting at it. They keep files on the properties coming up for sale and keep them with them throughout the sale. Addendum sheets were available in piles at the back of the room by the entry and at a desk or table at the front.

The auction was very well attended. Many, if not all, of the seats were occupied. There was no problem in hearing the public address system. After standing for some 10 to 15 minutes at the back, Mr Khosla and Mr Ghai took seats. By then any addendum sheets placed on or near to their chairs had been picked up by earlier occupants and had not been replaced with fresh ones. Mr Cripp told me that auctions of this kind ran at about 20 lots an hour. The day's sale began with lot 81 at or about 1 pm, so I can take it that lot 109 came up somewhere roughly between 2.15 and 2.30 pm. I hold that the existence of special conditions of sale referable to lot 109 was adequately brought to the notice of those in the auction room by the time lot 109 was bid for and that those persons included Mr Khosla and Mr Ghai.

Bidding for lot 109 was reasonably brisk. Mr Khosla had not discussed borrowing with any banker before the auction. He had not visited Wicken Hall or even Wicken before the auction, nor had he instructed any solicitors or others to make searches or enquiries on his behalf. He tells me that he was looking for a business, one to be run by him and his family. However, he had no experience of running an hotel, nor was it suggested there was any such expertise among his immediate family. His business experience was limited to the shared running of a shop in Nuneaton. He had not looked at any other hotel businesses with a view to buying. He would, he said, have hoped to develop the business gradually, implementing the indicated planning permissions for extensions over a period. He would borrow to pay for the extensions.

He did not know the difference between outline and full planning permission at the time. He did not know whether the bedrooms mentioned in the catalogue as having attracted planning permission were large or small, or with or without their own bathrooms or showers. While he regarded the on licence which the premises still had as valuable, he made no enquiries as to whether it could be transferred.

He had, at the time of the auction, no idea what any of the extensions would cost. He had no idea what profits the business would generate. He did not know what rate of interest he would have to pay on his borrowings. He had no idea whether he could attract a lot of customers. He thought he probably could because of the proximity of Cambridge University but he did not know what other hotels there were in the area. He did not know what staff might be available in the area. He expected to repay borrowings by instalments paid out of the business, both the sums borrowed to buy and those to extend. He recognised that, limited to the existing building and occupying part of it as his own and his family's home, he could not at first let more than six bedrooms.

He had never run a restaurant, nor was I told he had ever run a bar. He did not know whether the premises would need repairs. He said at first, and incredibly, that he did not expect to make a living from the letting of only seven bedrooms, but this of course led him into difficulties as how in that case he would ever have become able to expand and develop and turn to the implementation of the planning permissions.

Later, his evidence was that he did expect to be able to make a living out of a business consisting of a maximum of six bedrooms with the possibility of a restaurant and possibly a bar. Mr Ghai's view was that if Mr Khosla could let six bedrooms then he was positive that Mr Khosla could make a living. Bearing in mind the large number of factors unknown to both of them, I cannot accept these unsubstantiated assertions, but even if I had, merely making a living could not lead to the generation of such profits as would enable the building extensions to be embarked upon.

When they met at the auction Mr Ghai was, for the first time, made aware that Mr Khosla was interested in a particular lot, lot 109. Mr Khosla said it looked very nice to him. Mr Ghai tells me he told Mr Khosla that lot 109 seemed nice because it had outline planning permission for 20 bedrooms, and also that he mentioned to Mr Khosla that there was planning consent for a further 14 bedrooms. Mr Ghai had good contacts with a Mr Manning, the manager of a branch of Barclays Bank at Stevenage, and Mr Ghai was confident that the bank would lend money to Mr Khosla. Mr Ghai said he told Mr Khosla that the bank would lend because of the planning permissions. Mr Ghai tells me that he could get a loan to the extent of 70 to 75 per cent of the surveyor's valuation from Barclays *provided* the particulars given in the auction catalogue were true. I do not accept that Mr Ghai believed that. As an experienced property man he would have known that the bank would be likely to require and would be guided by a professional valuation. That being so, I see no reason why the bank should add that proviso. I shall return to this point later. If, as I hold, Mr Ghai did not believe that, he would have had no reason to say it to Mr Khosla.

Mr Ghai told Mr Khosla that he would reckon to go to about £280,000 to £285,000 on the lot. They were together in the auction room for at least 10 to 15 minutes before the lot came up. Mr Ghai also told Mr Khosla after the sale that he would get him the shortfall between what he could borrow from the bank and the purchase price. Mr Khosla bid for lot 109. In doing so he relied on what Mr Ghai had told him was his, Mr Ghai's, idea of what the value of the property was. Mr Ghai did not bid. Lot 109 was knocked down to Mr Khosla for £275,000. A deposit of £27,500 was accordingly payable forthwith. Mr Khosla gave his cheque in that sum to the auctioneers' staff. He knew that he did not have the funds in his account sufficient to meet it.

At the time, Mr Khosla was earning about £190 a week net after all deductions. He had less than £1,000 in the bank, although he had more, but not enough to enable him to meet the deposit, at a building society. Mr Ghai told me that while at the auction, but after lot 109 had been knocked down to Mr Khosla, he, Mr Ghai, noticed the discrepancy between the reference in the catalogue to planning permission for 14 bedrooms and the pink box referring to only 10 further bedrooms. He claimed he did not tell Mr Khosla of this discrepancy because in his, Mr Ghai's, view there was no point in grumbling at that stage. Mr Khosla had by then left his side to sign and had signed the memorandum, and the property was, thought Mr Ghai, worth the price paid, even with detailed permission for only 10 further bedrooms.

I do not need to find whether Mr Ghai did or did not notice the discrepancy, but I do not believe his evidence that if he had, as he claimed, he would not have mentioned it to Mr Khosla. If, as he told me he had, he had specifically told Mr Khosla that there was planning permission for a further 14 bedrooms, I have no doubt that he would have drawn the discrepancy to Mr Khosla's attention, either the moment he noticed it or, at any rate, when, as will later appear, Mr Khosla began, to Mr Ghai's knowledge, to make complaints as to the contract. Yet he plainly did neither.

Mr Michael Drake, as vendor's solicitor, had taken the trouble to attend the auction with an assistant of his. After the auction Mr Drake indicated that his office would send a wad of details to Mr Khosla's solicitors. I have no evidence that shortly after the auction of lot 109 any request was then made by Mr Khosla or by any of the auctioneers' staff to Mr Drake for any copy of the relevant planning consents, or indeed any other information relating to Wicken Hall.

This is a case in which witness statements were exchanged before the hearing and form part of the material put before me. Mr Khosla's witness statement makes no mention of his seeking but failing to obtain a copy of the planning permissions at the auction. His solicitors made no such point in their correspondence. His pleadings, even in the form as amended after the plaintiff's case was concluded, make no such allegation, nor does the affidavit he swore herein on March 11 1988. The allegation that he had so tried and failed was not put to any of the plaintiff's witnesses. However, a number of the auctioneers' staff, having very credibly told me during the plaintiff's case that they had no specific recollections of dealings with lot 109, there was then made for the first time the allegation by Mr Khosla that he had tried and failed to see the planning permissions at the auction.

Mr Ghai claimed to have heard Mr Khosla's request and also to have himself made a request for the outline planning permission to Mr Weatherall before the lot had come up. I do not accept Mr Khosla's or Mr Ghai's evidence on this point. The auctioneers' staff at one of the tables by the rostrum generally held all particulars relating to the various properties for sale, kept in order for ease of reference. Mr Weatherall, an auctioneer's assistant, would have been called over to the table had any query arisen which had not been there and then satisfied, and he had no recollection of being called over in relation to lot 109, nor, it would seem, was any difficulty mentioned to Mr Drake, as I have mentioned.

Moreover, Mr Weatherall, whom I permitted to be recalled, said that if Mr Ghai, who was well known to him, *had* said he had needed to see the planning details and they had not been available at the auction room itself, he would have remembered this event as he would then have sent someone round from the auction room to the office to get them.

Mr Khosla signed a copy of the revised conditions of sale relating to lot 109, specifically referring to the enforcement notice of December

A 7 1985. He signed a page referring to the existence of revised conditions of sale as to lot 109 and he signed a printed memorandum as to his purchase of lot 109 at £275,000. These three signed papers are comprised in exhibit P1 marked "Auction Contract" which also repeats the "Important notice to be read by all bidders", the photographs and text in the auction catalogue referring to lot 109, the page headed "Planning" which I have already described, and the general conditions of sale all bound together in the form in which such of them as were signed were presented for his signature.

The day after the auction Mr Khosla asked Sunil to go to the auctioneers' office to pick up the planning permissions and all other papers relating to Wicken Hall. Mr Khosla did not indicate to his son that he was worried about planning permission. He did not leave B Sunil with the view that it was a topic of importance to him. There is no evidence that he indicated that he had already tried to see, but had failed to obtain a sight of, the planning permissions, which would have been a natural thing to mention had it been so.

Sunil went to the office and asked a female member of the staff. She made further inquiries within the office but returned to say the papers had been sent back to the vendor's solicitor. That, indeed, would have been the usual course and it was followed here. There was no evidence from Sunil that his father was concerned at this news, nor were any letters then written on the subject.

The next relevant event occurred on Sunday, April 26 1987 when, by arrangement, a party visited Wicken Hall. It consisted of Mr and C Mrs Khosla, Sunil Khosla and Mr and Mrs Ghai. While it is likely, and I accept, that from the purchaser's side request was made (be it in response to Mr Swingler's offer or not does not, I think, matter) that they be supplied with such documents containing information as to the property generally as Mr Swingler might have, I accept Mr Swingler's evidence, first, that no particular interest was then shown in planning as an immediate matter and, second, that as at that visit the premises quite plainly did not contain a trading business, a going concern. Further, I accept Mr Swingler's evidence that by the date of that visit the new roadway had long since been completed, that both routes were then open and that barring off the old route involved only a minor effort or expense.

During this visit there was no complaint that there was no hotel in D operation, yet it must have been quite obvious that that was so. The subject of the enforcement notice and Mr Swingler's personal undertaking came up. There was no complaint about either's existence, or that its existence was then being learned of for the first time. Mr Swingler explained that the undertaking was easy to resolve and that no great financial outlay was involved. Mr Swingler had made up bundles of papers for the information of whoever transpired to be the purchaser. On April 27, in response to the request made at the visit, he sent some documents to 161 Sharrard Road, E7, the Khoslas' address. It was addressed to S Khosla because Mr Swingler mistakenly thought S was the initial letter of the defendant's forename.

E While Mr Swingler was in the witness box being cross-examined, Mr Sunil Khosla produced to his father's advisers what he said was the bundle which he had received from Mr Swingler. It includes an incomplete version of the planning permission of October 1985 without plans and a copy of Mr Swingler's undertaking of March 18 1987. Mr Khosla had not, when it was received, looked through it to see if any planning permission was in it. Sunil had not had the impression that his father was worried about planning permission, and Sunil had not checked to see what, as to planning permission, was in the bundle he had received.

Mr Swingler could not recollect whether one or more of the planning permissions had been included in the papers which he had sent. He sent everything that seemed to him relevant but he could not F be certain on that particular point. At all events, his letter of April 27 attracted no response. There was no complaint that planning permissions were not included or were incomplete, or that they had transpired not to be what they had been described as being.

On April 28 Mr Swingler's solicitor sent a number of documents, including copy planning permissions and the enforcement notice, to Mr Khosla's solicitors. The associated plans were not sent. This letter and its enclosures was acknowledged on April 29 by a letter which said: "We are taking our client's instructions and will write to you as soon as possible". I have no evidence of any delay or difficulty experienced by Mr Khosla's solicitors in their obtaining those instructions. There was, however, no early complaint as to the planning permissions or the enforcement notice, which one might reasonably have expected had the terms of the former or the existence G of the latter furnished any grounds for complaint or doubt. Nor was the point taken that the plans had not been included — a reaction one might perhaps expect in a recipient interested in planning.

The £27,500 cheque given for the deposit by Mr Khosla was dishonoured on presentation. He told me it bounced because of his confusion about planning permission. He would have obtained assistance from Mr Ghai towards the deposit cheque earlier, he said, had there not been this confusion. I find that it bounced simply because there were not funds to meet it and without that being related to any planning difficulties or confusions. Mr Swingler's solicitors had sent the copy planning permissions to Mr Khosla's solicitors on April 28 1987. Had Mr Khosla been then confused on the subject, I have no doubt they would have raised the subject on his behalf. H However, they had made no request for clarification and no inquiries at that stage, yet the deposit cheque was not met until second or subsequent presentation on May 7. I do not accept Mr Khosla's reasons for the dishonouring of the cheque. The deposit was met after Mr Ghai gave Mr Khosla £7,500 to help him meet it. Mr Ghai has not been repaid.

Mr Ghai telephoned and visited his bankers in Stevenage after the auction and, on the second visit, in the second week after the auction, Mr Khosla went with him. There was a discussion as to the bank lending to assist or enable Mr Khosla's purchase. The bank manager telephoned surveyors for valuation purposes while Mr Ghai was there. I hold that the bank said that it would lend 70 to 75% of a J valuation of the premises. I do not believe that the bank indicated that it would, without more, lend 70 to 75% of the purchase price nor, as I have previously mentioned, do I accept evidence that the bank indicated that it would lend, or lend any particular proportion (either of the price or of a valuation), if and only if the descriptions given in the catalogue were correct, as Mr Khosla told me it had. If the bank manager had determined, as apparently he had, to rely on the valuation, I can see no reason for his further stipulating as to the accuracy of the catalogue conditions.

Mr Khosla sought to persuade me that the bank's view was that if the particulars in the catalogue were correct then the bank would lend one proportion but only a lower one if they were not correct. I do not accept that evidence either. It makes no commercial sense in the K context of a bank which was insisting and intending to rely upon a valuation. I add that Mr Khosla did not form any impression of what the lower proportion would be.

Mr Khosla told me that he did not think that the bank would be greatly interested in the building on the site as it was, but rather in the potential of the site. I cannot accept that evidence as being the bank's position, nor do I believe that Mr Khosla had any reason to believe it to be so. The bank may well have been interested in both what was the actual and what was the potential in the premises and would have been concerned, I would expect, to see a valuation which dealt with both, but I cannot see why the bank should pay little or no regard to the premises as they stood when considering whether to lend and L what to lend. The premises as they stood, after all, would be their only initial security. Mr Khosla told me that the bank sent a surveyor to the premises but it could, I think, have been arranged through Mr Ghai. The valuation was to be for the bank, whoever arranged or paid for it.

Mr Khosla's solicitors by letter required that Mr Swingler should be registered as proprietor of a sliver of the land sold as to which his registered title was formally incomplete. They were thus indicating an insistence on full and literal performance by Mr Swingler of his side of the contract. I shall need to refer later to this letter, which I shall call the May 12 letter. The sliver of land was duly registered, but this need to register it delayed the point at which the plaintiff was able to describe himself as ready, willing and able to complete. M

On May 13 there was a hint of trouble. Mr Khosla's solicitors wrote on that day to Barnard Marcus. The complaint was three-fold. First, that the revised special conditions had not been drawn to Mr Khosla's attention prior to his bidding; second, that they were not drawn to his attention when he signed the contract; and third, that the contents of the enforcement notice were not drawn to his attention. No point was taken as to the nature of the planning permission, neither as described in the auction particulars nor as had transpired to be the case.

Barnard Marcus answered that letter on May 22. They sent a transcript of the opening oral announcement which had been made at the auction drawing attention to the revised conditions of sale. They could have added that Mr Khosla had signed a copy of the revised conditions of sale as part of the auction contract which is now exhibit P1. Mr Ghai told me that in the second week of May his architect had advised him that there was, as Mr Ghai put it, no planning, by which I think he meant not that there was no planning permission at all but that the ones literally as described in the catalogue did not exist. He said he passed the information to Mr Khosla shortly thereafter and that both of them visited Mr Khosla's solicitor towards the end of May.

If all this was so it would put difficulties in the way of accepting that Mr Khosla was, and was to Mr Ghai's knowledge, relying on the catalogue description on the 20- and 14-bedroom points because no complaint as to those points was raised until July 30. On May 26 Mr Khosla's solicitor, still corresponding with the auctioneers, stated that Mr Khosla was interested in the property "as a going concern" which he expected to be able to "run almost immediately". It must have been glaringly apparent from what all the purchaser's party had seen on their visit on April 26 that there was no going concern, yet no comment had been made until this letter. The complaint was made that the premises could not be used as an hotel because a roadway had to be constructed first and that Mr Khosla was entitled to withdraw on the basis of the vendor's misrepresentations, which were not spelled out.

The letter made no complaint as to the terms of the planning permissions and proceeded on the basis, which was false, that a roadway had to be constructed. It had not. It had already been substantially constructed. All that remained were the very minor works involved in closing off the old roadway. That, too, must have been apparent at the visit on April 26.

On the same day, May 26, Mr Khosla's solicitors wrote to Mr Swingler's enclosing the Barnard Marcus correspondence. Their letter includes the sentence "It seems to us that the *whole point*" (my emphasis) "of our client's disappointment is the matter of the roadway which will prevent anyone from opening an hotel until it is built." No complaint was made as to the number of bedrooms for which planning permission had, in full or in outline, been obtained.

On May 29 Rustons & Lloyd, on Mr Swingler's behalf, served a condition 22 notice to complete. By a letter of June 11 Mr Khosla's solicitors argued that the notice to complete was invalid because Mr Khosla had been induced to enter into the contract "by a number of misrepresentations". They requested the return of the deposit although, of course, it was not the plaintiff's solicitors but Barnard Marcus who held it. There was no suggestion that the misrepresentations included matters not already raised in the correspondence.

Mr Khosla told me at one stage that his solicitors had told him that there was no planning permission for 20 bedrooms before they asked for the deposit back, in other words before June 11. However, the point was not taken in the letter at this stage. If Mr Khosla's evidence was correct and he had been so told by his solicitors, then it would have been remarkable, he having relied on the 20-bedroom representation, that they should not have complained on that score at this time. Later, Mr Khosla told me that he found out about the inaccuracy of the 20-bedroom point sometime around July 28 when he saw or heard of a letter I shall refer to later. I find Mr Khosla's evidence unreliable on these points.

On June 18 1987 Mr Swingler's solicitors indicated that the deposit was forfeit. Despite this, a little later in June Mr Swingler was approached to permit an inspection of the premises with a view to their valuation. This was agreed on a "without prejudice" basis. It seems a Mr Jeremy Whitton-Spriggs [FRICS] was instructed to prepare a report and valuation for Barclays Bank, Stevenage Branch, and that he did visit the premises. Before preparing the report he took up the question of the planning permission with the local authority. What he discovered is set out in a letter of July 28 1987 from Smith-Woolley [chartered surveyors] to Mr Ghai, a copy of which was sent on July 30 to Mr Swingler's solicitors by Mr Khosla's, who say: "You will see that there is no record of any planning permission for a further 20 bedrooms". They go on to say that that had been a representation "upon which our client would be entitled to rescind the contract". They asked for the return of the deposit. The only complaint in relation tq planning was the absence of outline planning permission for a further 20 bedrooms.

It may be that Mr Whitton-Spriggs did not complete any written report or valuation. It looks as if he did not. At all events, I have seen none. Mr Khosla's evidence was that he knew from the letter of July 28 1987 that the 14-bedroom representation was wrong. He had no explanation of how, that being so and he having, as he claimed, relied upon that representation, his later affidavit sworn for Ord 14 purposes had not mentioned the point. There would also be the difficulty of why, if that had been so, the point was not pleaded until after the plaintiff's case was concluded — a point to which I shall later come. I do not accept Mr Khosla's evidence on this point.

Mr Ghai told me that he knew of the information in Smith-Woolley's letter of July 28 before its receipt, perhaps three weeks before.

The sale to Mr Khosla did not, of course, proceed. Mr Swingler put up the property for auction again on September 23 1987. It was knocked down to a purchaser for £290,000, but the cheque for the deposit was not honoured and the sale did not proceed. The purchaser under the contract was not pursued. There was, it seems, some difficulty as to his mental capacity. Finally, perhaps disillusioned with auctions, Mr Swingler sold the property by private treaty for £360,000 on June 9 1988. It might be thought at first blush, as he eventually sold the property for substantially more than Mr Khosla would have paid, that Mr Swingler had lost nothing and indeed had gained from Mr Khosla's not having proceeded. That does not follow. Given the market conditions of the period, which Mr Khosla described as "booming", it does not follow, especially in terms of the ability to buy other property out of the proceeds, that the £360,000 received in the summer of 1988 was better than £275,000 which should have been received in May or so of 1987.

Mr Swingler launched these proceedings in August 1987. The defence with a counterclaim was served on October 7 1987. Little has been said in the argument about the availability of innocent misrepresentation as being a ground for rescission. It has, I think, been tacitly accepted that this is so. Further, it was I think common ground that if Mr Khosla was entitled to rescind and had rescinded the contract which he had made at the auction room then he would have been entitled to require the repayment of his deposit. The defence in its earliest form relied as grounds for rescission on, *inter alia*, the following:

(i) the falsity of a representation in the auction particulars as to outline planning permission for a further 20 bedrooms;

(ii) non-disclosure of the enforcement notice.

Rescission of the contract was said to have occurred upon the defendant's solicitor's letter of June 11 1987. The counterclaim was for return of the deposit and other ancillary relief. There have been Ord 14 and other interlocutory proceedings, both as to the claim and the counterclaim. I have not seen them in any detail, but in each case I understand unconditional leave to defend was granted. In the course of the Ord 14 proceedings Mr Khosla swore evidence, *inter alia*, that:

(i) he was interested in purchasing the property as a going concern;

(ii) he was particularly interested in the fact that outline planning permission had been granted for a further 20 bedrooms;

(iii) he had relied on the representation as to 20 bedrooms;

(iv) there had been a failure to disclose the enforcement notice.

Very shortly before the hearing before me, the defence was amended upon leave being given in that behalf by the master. The defence was amended substantially, but I need draw attention to only four points arising out of the amendments, namely:

(i) there was no allegation of any reliance upon any representation as to 14 bedrooms;

(ii) there was an allegation that although the revised special conditions were attached to the written acknowledgement of contract signed by Mr Khosla, they had not been brought to his attention before he signed and he did not know what they contained;

(iii) there was a new complaint of non-disclosure as to the property not being able to be used as an hotel until the enforcement notice was complied with;

(iv) there was a new plea of fundamental mistake based on the inability to use the premises as an hotel.

The plaintiff had not amended his reply in defence to counterclaim before the hearing began before me, but on the first day Mr Le Poidevin, for the plaintiff, applied to amend, and the defendant did not — indeed, in view of his own late amendments, could hardly — object. I gave leave to the plaintiff. The amended reply took the point which I shall call the waiver point, that by requiring the plaintiff fully to perform his side of the bargain after the plaintiff's undertaking as to not using the premises as an hotel had been brought to the defendant's notice, the defendant had waived any right to rescind

A based on its earlier non-disclosure. It also asserted what I might call the exclusion clauses and the reasonableness of their terms.

After the plaintiff had completed his case, Mr Clark, for the defendant, applied to amend the defence and counterclaim by adding a reliance upon the falsity of a representation that planning permission had been granted for the erection of an extension comprising, *inter alia*, 14 bedrooms. He said he wished to add this complaint because of Mr Ambrose's evidence. Despite Mr Le Poidevin's understandable objections to amendment at so late a stage, I allow the amendment.

Happily, a number of matters canvassed in the pleadings and in the evidence ceased to be materially in issue. For example, it is now common ground that on the fall of the hammer an oral contract was
B made. Its terms can be collected from the various documents circulated before the sale and to which a reference was made by the auctioneers. Mr Le Poidevin argued in his opening that the revised special conditions were incorporated into the contract as made on the basis, *inter alia*, that reasonable steps had been taken to draw them to Mr Khosla's attention. Mr Clark accepts this, so this, too, is now common ground.

Moreover, the enforcement notice and its non-disclosure fell away as complaints when, after the plaintiff's case was concluded and during the examination in chief of Mr Khosla, Mr Clark conceded that as the new access way had been constructed before the auction and the further steps needed to comply with the notice were so
C insubstantial, the matter was not such as to amount to a complaint of a kind such as could lead to rescission. At the same time, the defendant's allegations as to misrepresentation and mistake as to the premises being an hotel were abandoned.

The remaining significance of the enforcement notice and hotel points is that on the facts both must have been seen by Mr Khosla to have been bad points as early as the group visit to the site on April 26. It must have been then apparent that the enforcement notice needed only insubstantial things to be done for it to be complied with and that the premises were plainly not in business as an hotel, yet both complaints were initiated and persisted in until the late stage that I have mentioned.

The defendant's arguments from this point on were those and only
D those relating to a misrepresentation as to planning in relation to 14 bedrooms, a misrepresentation as to planning in relation to 20 further bedrooms and, third, as to the exercise of the statutory discretion conferred by section 49(2) of the Law of Property Act 1925. As for one of the plaintiff's arguments, the waiver point, Mr Clark accepts that if the May 12 letter represents a true waiver then Mr Khosla could not have rescinded on June 11. He refutes, however, that the May 12 letter could represent a true waiver. One cannot waive, he said, what one does not know, and he goes on that as the falsity of the 14 and 20 bedroom representations was not known to Mr Khosla on May 12 it could not then have been waived by him or on his behalf. He argues that Mr Khosla's own evidence shows him not
E to have been aware of the planning discrepancies until the Smith-Woolley letter of July 28 1987. He goes on to argue that in any event the terms of the May 12 letter were neutral in character and do not amount to so clear an affirmation of the contract as is necessary if it is to be taken to have waived earlier grounds for complaint or avoidance.

I need make no further findings on these points, as in his final speech Mr Le Poidevin indicated that, the complaint as to the enforcement notice having been abandoned, the waiver point is no longer maintained.

I have, in the course of my judgment so far, made a number of observations on the evidence of witnesses on specific points. My views of the oral evidence I have heard have not been based solely on
F any such analyses of what has been said because I have also, of course, had the advantage of seeing and hearing the witnesses in the course of their evidence. Even making every allowance for the possibility, or even the likelihood, that some of Mr Khosla's hesitancy arose from a weak command or understanding of spoken English, I have formed a poor view of both Mr Khosla and Mr Ghai as witnesses. I did not see either as much concerned with assisting the court in its task of arriving at the truth. I regard neither as a reliable witness.

Mr Swingler I find to be a man doing his best truthfully to recall events and conversations to which there was no particular reason to attach much importance at the time and which are already over two and a half years old.

A number of matters having fallen away, as I have mentioned, the G
remaining issues may be broadly summarised as follows:

1 Were there any, and if so what, representations made as to 20 and 14 bedrooms?

2 How far, if at all, were any such representations as were made materially false?

3 Had the defendant relied on the truth of what transpired to be false?

4 If he had, was he thus entitled, exclusion clauses apart, to have rescinded the contract before the supposed forfeiture of the deposit to the plaintiff, and, if the defendant was so entitled, did he duly exercise that right?

5 Is the defendant, exclusion clauses apart, entitled to damages for misrepresentation? H

6 How far, if at all, would the exclusion clauses operate on the above matters?

7 How ought the discretion under section 49(2) of the Law of Property Act to be exercised?

Mr Clark properly accepts that the burden of the case as to misrepresentation, be it to ground rescission or for damages, falls upon the defendant.

First, then, there is the 20-bedroom point. Page 71 of the auction catalogue contains a clear statement that outline planning permission has been granted for a further 20 bedrooms. That I hold to have been a representation of fact. Mr Swingler believed it to be true. Was it a false representation? Mr Clark helpfully drew my attention to J
Halsbury's Laws of England, 4th ed, vol 31, paras 1044-1047. From those paragraphs I conclude that for a representation to be false for present purposes it must be false in substance. Put another way, if the picture the representation would convey to a reasonable representee is accurate in substance, in the sense that he would not consider the discrepancy between that picture and the actual position as it is to be material, then no relevant falsity is shown.

I hold that the 20-bedroom representation was not literally true. There was no outline planning permission which in terms specifies 20 further bedrooms as permissible. However, accepting as I do Mr Ambrose's evidence as to how the 1988 grant was in practical terms to be regarded as permitting at least 20 further bedrooms in the southern development, I do not find the representation to have been false in K
substance. It must be noted that I have heard no evidence from the local authority or any other planning expert to disturb my reliance on Mr Ambrose's views on this point. Mr Khosla made no approach to the local authority to find out just what they would permit within the permissions which they had granted and it is certainly not established that 20 bedrooms would not be permitted to be erected in the southern development upon an implementation of the present outline planning permission by anyone who chose to present drawings and to ask for detailed permission for that number.

Thus, while there is a representation as to the 20 bedrooms, it is not, for present purposes, a misrepresentation.

Next there is the 14-bedroom point. Page 71 of the auction L
catalogue states: "Planning permission has been granted . . . for the erection of an extension comprising 14 bedrooms". That is clear enough, but the colour key adds a degree of confusion by referring to only "10 further bedrooms" in the pink area, which was plainly the extension being referred to. I think I should in the circumstances assume against the plaintiff that the key's reference to 10 did not qualify the earlier clear reference to 14. I thus find there to have been a representation of fact that there had been a grant of planning permission for an extension comprising 14 bedrooms.

As the 14-bedroom point was not raised until after the plaintiff's case had been concluded, there was no evidence from Mr Swingler on the subject of whether he believed that the catalogue was correct. There was no application that he should be recalled to deal with this M
point and in the circumstances I think I ought to assume that he believed the catalogue to be correct. I have certainly no reason to think otherwise.

Was there a false representation? I hold it was not literally true. Leaving aside the southern development separately referred to in the catalogue, and again referring to the effect of Mr Ambrose's evidence which I have accepted, there had been detailed planning permission for two bedrooms in the eastern development with the prospect of a further five therein upon a simple and acceptable amendment being sought, and outline planning permission on the lines of five, or indeed up to eight to nine, bedrooms, in the western development. Thus, although it might well have proved easy to

A obtain full planning permission for the erection of the extension comprising 14 bedrooms, no such permissions, be they detailed or outline, had in fact been granted by the date of the auction.

Was the representation false in substance? Would a reasonable representee consider the discrepancy between a grant of full planning permission for an extension of 14 bedrooms on the one hand, and detailed planning permission for two with full planning permission to be readily obtainable for a further five, and existing outline for between five and nine, as material? The answer may depend on what characteristics one is to give the reasonable representee. It would seem right to attribute to him some of Mr Khosla's features. I thus have in mind a reasonable representee looking for a business to run, needing to borrow and hoping gradually to develop his premises out
B of future profits. Such a man would recognise that whether he would be financially able to develop the premises and, if he was, when he would become so able, and what shape any developments he would then wish to make or would need to take would all be matters lying in the unknowable future. He would, I think, recognise that the planning permissions the property actually had permitted substantial development, quite as much as he could hope to afford or would wish for for quite a while, and that they showed that the planning authority was in principle agreeable to substantial development. This subject is inevitably related to the discussion I shall come to later as to whether Mr Khosla in fact relied on certain misrepresentations. Reverting to the reasonable representee, I am unconvinced that such
C a representee would consider any such planning discrepancy material. However, I shall proceed with the matter on the basis that the 14-bedroom point was materially false, or false in substance; in other words, that it was a misrepresentation.

Accordingly, I turn to the next question — reliance. Although I have held neither the 14-bedroom nor the 20-bedroom representations to be false for the present purposes, I shall assume for the moment and go ahead on the basis that they were both misrepresentations.

More fully spelled out, the issue of reliance is this: Was the truth of the 20- and/or 14-bedroom representations one of the factors which induced Mr Khosla to make the contract which he did? Mr Clark urged upon me that Mr Khosla's evidence and other evidence clearly
D showed that Mr Khosla had relied on the truth of the 20-bedroom and 14-bedroom representations as conferring benefits upon him in three ways:

1 because the planning permissions they describe conferred upon Mr Khosla the opportunity to expand;

2 because they conferred upon him, if they existed, an ability to obtain finance, because the obtaining of finance would be easier if there was a development potential in the premises;

3 because, should he fail as an hotelier, their existence would make the premises more attractive than would otherwise be the case and so would attract a better price from, and a readier sale to, the next owner.

E Given my observations about the quality of Mr Khosla's and Mr Ghai's evidence, I am not able to accept that there was reliance simply because it has been asserted that there has been; nor did that simple assertion become more acceptable upon its frequent repetition. Moreover, the reliance claimed is not that without thought of why he should do so Mr Khosla simply believed that he would benefit in these three ways. It is not said, for example, that an examination of the three ways of benefit is irrelevant because they were relied on without regard to whether they were possible. Rather, the claim is that these three ways of benefit were believed in and relied on because they were true and because their truth was apparent to Mr Khosla. In other words, I am asked to examine not only Mr Khosla's state of mind but also the alleged facts said to lie behind and to bring about that state of
F mind.

It is a consequence of the way the matter has been put to me that if the facts do not support the claimed state of mind I will have some reason to doubt the state of mind. I therefore turn to the three forms of benefit.

As for an opportunity to expand, the planning permissions which in fact existed gave ample opportunity to expand. While I would accept that Mr Khosla may have been in part attracted or induced to buy by the opportunity to expand, it seems to me to be in the highest degree improbable that he should have been at all influenced by the truth or falsity of the excess of the represented planning situation over the factual planning situation. Indeed, given the long catalogue of relevant matters as to which he had no information and had made no inquiries, I see no good reason to treat planning as such a marked G
exception to his general uninformed and unreflective approach to the purchase. It will be remembered that he did not know of the difference between full and outline planning permission.

In any event, given that any expansion was to be paid for by further borrowings which, as well as the borrowings for purchase, were to be repaid out of profits generated by the business, I am not satisfied that such planning permission as there was would have inhibited the carrying out of any such expansion as would have been at all likely to have been attempted over the whole of the then foreseeable future.

As for the existence of the represented planning permissions enabling or facilitating the financing of the purchase or of any extension, I have had no independent or credible evidence that the discrepancy between the represented and actual position would have H
had any effect at all on any banker or other lender. Mr Khosla told me that if the premises had not had planning permission for a further 20 bedrooms then the bank would not have lent him enough to enable him first to buy and then to develop. I do not accept that evidence. I have heard nothing from the bank on any subject at all, let alone to suggest either that they would have lent Mr Khosla enough, however much that might be, if the represented planning permissions had existed or that they would not if it did not. A bank may well have been influenced by the existence of an opportunity to expand, but that opportunity did exist. I would need clear and credible evidence before me if I was to be able to hold, given the non-existence of a going concern at Wicken Hall and Mr Khosla's lack of research and J
information relative to any future business at the premises, that the bank or its surveyors would have been at all affected by the opportunity to expand being limited to the forms of expansion described in the actual rather than in the represented planning permissions. There was no such clear evidence.

As to resale should the business have failed, I have had no informed evidence that, either in the booming conditions of the time or later, the discrepancy in the planning representations would have had any effect on either the ability to resell or the speed or price at which a resale could be effected. I can quite see that a prospective purchaser might be affected by the ability to expand, but, as on other aspects, I would have needed clear and credible evidence, which I have not received, before I could accept that the discrepancy would K
have had any effect.

Nor is there anything in the way in which Mr Khosla has dealt with the planning point which leads me to believe that there was any material reliance. The planning permissions were sent to Mr Khosla's solicitors on April 28, even if they had not formed part of the bundle sent to Sunil on April 27. The defendant did not trouble even to look at the documents sent to Sunil. They excited no questions or complaints of any kind that one might expect if the 20- or 14-bedroom representations were indeed being relied upon.

Mr Khosla never saw the plans referred to in the permissions. I have had no evidence of any approach to the local authority to find whether, whatever the present planning permission, permission for L
extension for 14 and 20 bedrooms could be obtained or would be likely to be obtainable, if not immediately at least by the time at which a purchaser, having re-established an hotel business at Wicken Hall, might become minded to expand. I have had no independent evidence of the likelihood of any extension of the premises as an hotel being or becoming commercially viable in Mr Khosla's circumstances.

When complaint was first made by the defendant, as it was on May 13, it was not as to these representations but as to the enforcement notice. When the complaint was amplified, as it was on May 26, still these representations were not mentioned. For a period it could be argued that naturally enough there was no investigation of the planning documents as they came to light and there was no complaint M
as to their contents because the catalogue representations were being relied on as true. Indeed, the absence of complaint, it could be argued, was an indication of the reliance that had been placed on the catalogue. It may be said that it is only he who does not rely on the catalogue who scrutinises the planning documents and so finds he has grounds for complaint.

However, leaving aside the uncertain evidence about when first the catalogue descriptions were known to be inaccurate, even in very late July 1987 when the defendant had certainly been advised as to planning consent, it was only the 20-bedroom representation that was referred to, rather than the 14-bedroom one. The 14-bedroom point was unmentioned in Mr Khosla's Ord 14 affidavit or in his witness statement exchanged with the other side. It was added only after the plaintiff's side had completed its evidence and the amendment was applied for because the inaccuracy of the 14-bedroom representation had emerged from Mr Ambrose's evidence. While I would not wish to attach any blame to Mr Clark in his seeking to make the amendment, I cannot see the supposed reliance on the 14-bedroom point as being other than opportunistic. My overriding impression upon the hearing and sight of the witnesses, in particular of Mr Khosla, is that Mr Khosla, having determined, for whatever quite different reason of his own, that he either could not or ought not to proceed, he then looked about him for whatever reasons *ex post facto* might clothe his failure to proceed with respectability. He lit upon the enforcement notice, a supposed failure to have constructed the new driveway, the 20-bedroom point, the personal undertaking, the fact that the hotel was not a going concern, the alleged failure to bring the revised special conditions to his notice and finally, as I have explained, upon the 14-bedroom point.

The fact that the other assertions came to nothing does not deny his reliance on the representations as to bedroom numbers and, had reliable evidence been given, then whatever doubts the failure of his other assertions might have created, could have been dispelled. However, I heard no evidence from Mr Khosla's bankers, or from any surveyor or any other expert, or from Mr Ghai's architects. I add that, detecting signs of what I thought might prove to be unreliability, at the conclusion of Mr Khosla's evidence-in-chief — and he was the defence's first witness — I mentioned that consideration would need to be given by the defence to whether it would be necessary or desirable for the defence to call someone from the bank or surveyors.

I am not satisfied that Mr Khosla relied on either the 20-bedroom or the 14-bedroom representations. Indeed, at one point in his evidence Mr Khosla told me that if he could get 20 bedrooms into the yellow, the southern, development, then he would have no complaint. I hold on the evidence of Mr Ambrose that he could have got 20 bedrooms into the yellow.

Accordingly, innocent misrepresentation fails as a ground either for rescission or for damages because, even if one assumes misrepresentation, I am not satisfied that the reliance by Mr Khosla that needs to be shown has been demonstrated.

As for the exercise of any right to rescind, it is common ground that if there was no effective rescission by the defendant, then the plaintiff's forfeiture takes effect subject only to section 49(2) of the Law of Property Act 1925. I do hold there was no effective rescission. Had there been due grounds for rescission I would have held that the defendant's letter of June 11 1987 requesting the return of the deposit was a sufficient election for or notice of rescission. Its terms may be precatory, but as between solicitors the clear request for the return of the deposit should, in my judgment, have been fairly taken to be an election to rescind. Rescission thus fails not because of want of notice of it but for want of due grounds for it.

Accordingly, unless section 49(2) of the 1925 Act otherwise provides or requires, the deposit is forfeit to the plaintiff.

As for damages for misrepresentation under section 2(1) of the Misrepresentation Act 1967, in the light of my findings on other aspects of the case I make no award in the defendant's favour. Even assuming or accepting that Mr Khosla entered into a contract after a misrepresentation had been made to him by another party to it, I have no material on which I could hold that as a result thereof he suffered loss, nor could I conclude in the light of my other findings that had the representations only been made fraudulently the other party would have been liable to him in damages.

As for the exclusion clauses, having regard to my conclusions on other points, I believe I may deal with the exclusion clauses far more shortly than I would otherwise have had to. Mr Le Poidevin bowed to Mr Clark's argument that National Condition 17 is inconsistent with general condition 11(i), that the latter takes precedence and hence that National Condition 17 is not available to the plaintiff. That leaves only, in effect, general conditions 11, 12(3) and 16 (which I shall call "the remaining exclusion clauses") as required to be considered.

Put simply, the defendant's case is that the plaintiff cannot rely on the remaining exclusion clauses as they are not reasonable within the Misrepresentation Act 1967 and the Unfair Contract Terms Act 1977. The burden is put by statute on the plaintiff to demonstrate their reasonableness and, says Mr Clark, that the plaintiff has failed to do. In the course of his thoughtful argument he has drawn my attention in detail to *South Western General Property Co Ltd* v *Marton* (1982) 263 EG 1090 and to *Walker* v *Boyle* [1982] 1 WLR 495. I shall not refer to the statutory provisions, which are fully set out in the cases to which I shall now turn.

In the *South Western General* case Croom-Johnson J was dealing with a defendant who claimed to have been induced to enter into a property contract at auction by false though innocent representations made in the auctioneer's catalogue. The land there sold had been described as "Long leasehold building land". The exclusion clauses the learned judge was there dealing with, though not identical to those in the present case, were to substantially similar effect. Croom-Johnson J said: "I have to deal with *this* contract, with the parties to this contract in the circumstances in which *this* contract was made" (my emphasis). He was not, in other words, purporting to describe or apply some rule of general application. He was also plainly affected by the consideration that the exclusion clauses he was considering, if they were permitted to take effect, would in that case exclude liability for a failure adequate to inform the purchaser on a subject — the ability to build on the land — which was plainly central to the whole contract.

While the case is, of course, instructive, I do not regard it as binding me to arrive at a like conclusion in the different circumstances of the present case in which, even if I were to be wrong about reliance, it would on the facts be impossible to elevate the 14- and 20-bedroom representations and the degree to which they were inaccurate to the central position with which Croom-Johnson J was dealing in his case.

As for *Walker* v *Boyle*, it dealt with National Condition 17, 19th ed, in relation to an innocent misrepresentation that the vendor was not aware of any boundary dispute. Dillon J (as he then was) held that to the representor's knowledge there was such a dispute. He held that in the circumstances of the case he did not regard that form of condition 17 as satisfying the statutory requirements of reasonableness: see p 507. Again, no general test was said to exist or purported to be applied, and indeed it may be that, given the nature of the type of case and the statutory language, an *ad hoc* case-by-case approach is unavoidable. I would thus have regarded myself as materially unbound by authority and, thus free, I would, had it been necessary, have held that in the circumstances of this case the remaining exclusion clauses did satisfy the statutory test of reasonableness and hence could be relied on by the plaintiff.

It is important to note that here the representor is to be taken to have believed in the truth of the representations. Moreover, the purchaser, even had he — contrary to my earlier findings — relied on the 20-bedroom representations, would not have relied upon them to the extent demonstrated in *Walker* v *Boyle*, namely that had the truth been disclosed there would have been a refusal to complete a contract which otherwise would have been completed. I should add that in reaching this conclusion I have considered the remaining exclusion clauses not generally or in the abstract as to what remedies or liabilities they might have excluded but specifically in relation to their exclusion of the remedy of rescission on the specific ground of such falsity as there was in the 20- and 14-bedroom representations. I believe the statute enables or requires such a specific approach, but if that were to be wrong and a general or abstract approach was necessary, I would have found the test of reasonableness not satisfied.

Viewing the remaining exclusion clauses as I have done, I would, had it been necessary, have then gone on to hold that the remaining exclusion clauses, having survived the statutory test, would have been effective to deny any claim for rescission or damages based on such falsity as there was in the 20- or 14-bedroom points. I am conscious that I have not done full justice to all the arguments I heard on the exclusion clauses, but as, in the light of other findings, they have been relegated to the periphery of the case, I hope I shall be forgiven in not further lengthening an already long judgment.

Finally, as to issues, I turn to section 49(2) of the Law of Property Act 1925*. Mr Clark urges upon me that as Mr Swingler made a "profit" upon his later resale of some £85,000, it would be right that the deposit should be returned to Mr Khosla, albeit with deductions to meet expenses incurred by Mr Swingler incidental to there having been a resale. He drew my attention to the decision of Mr Gerald

*Editor's note: The subsection reads: (2) Where the court refuses to grant specific performances of a contract, or in any action for the return of a deposit, the court may, if it thinks fit, order the repayment of any deposit.

A Godfrey QC, sitting as a deputy High Court judge, in *Dimsdale Developments (South East) Ltd* v *De Haan* (1983) 47 P&CR 1. In that case the learned deputy judge took the view that the vendor had made a profit on the resale in the market in 1980-81. However, I am unconvinced that in the market circumstances at the time I am dealing with, Mr Swingler did indeed make a "profit" from Mr Khosla's failure to complete and from the consequential later resale.

It seems to me simplistic to point to the increase in the price received and to say ergo there was a benefit to Mr Swingler. I would have needed to be told more about what Mr Swingler would have done with the net proceeds had he received them punctually from Mr Khosla. Would he have reinvested them in another home or, for example, in the Stock Exchange? By the time he did eventually B receive the proceeds of sale, could he still have bought the property he might have bought earlier or the investments he might have chosen earlier if only he had been paid on time?

If Mr Khosla wishes to assert that the statutory discretion should be exercised in his favour because of the profit Mr Swingler had made, then he must accept, too, that it was incumbent upon him to explore these and other similar questions. There was no such exploration and I am unconvinced that any great or sufficient weight should be attached to this possibility.

Doing the best I can to consider all relevant factors in exercising this admittedly broad discretion, I do not feel able to exercise it by ordering the return of the whole or any part of the deposit to Mr Khosla.

Turning, then, to the relief claimed by the parties, and dealing first with the counterclaim, I do not make the declaration sought at para 1. Para 2 thus falls away and I do not make any award of damages under para 3. Paras 4 and 5 fall away. I thus dismiss the counterclaim. C

Turning to the claim, I am prepared in principle to declare under para 1 of the prayer that the deposit has been forfeited and the agreement for sale terminated. I order under para 2 that the defendant do join in instructing Barnard Marcus to release the deposit to the plaintiff. Moreover, the parties are, I think, agreed that an award should be made in the plaintiff's favour as to interest. Mr Clark conceded the principle on this subject, which arises under para 3 of the prayer, but I was not told of any figures which had been agreed. I would expect that such figures could be agreed. D

As for the other relief claimed in the prayer to the statement of claim, while there was some mention of a number of other ways in which the plaintiff has allegedly suffered from the defendant's wrongful failure to complete, there was nothing further explored before me. As to such damages, I direct an enquiry under para 4 of the prayer, while expressing the hope that, having come this far, the parties would see their way to agreeing figures without the expensive pursuit of all claims to their utterance. I would think that the most convenient course would be for me to direct a minute to be drawn by Mr Le Poidevin and circulated to Mr Clark for his agreement. If the parties are agreed, the minute can be initialled by me and the matter need be restored for hearing only if agreement cannot be reached.

For a further case on this subject see p 27

COMPULSORY PURCHASE AND COMPENSATION

Chancery Division

February 22 1990

(Before Mr Justice ALDOUS)

WESTMINSTER CITY COUNCIL v QUERESHI AND ANOTHER

Compulsory purchase — Compulsory Purchase Act 1965, section 4 — Acquisition of Land Act 1981, section 26(1) — Compulsory Purchase (Vesting Declarations) Act 1981, sections 3 and 4(1) — Whether compulsory purchase powers were exercised within the time-limit of three years from the date when the compulsory purchase order became operative — Effect of preliminary notices served under section 3 of the Compulsory Purchase (Vesting Declarations) Act 1981 — Analogy between such notices and a notice to treat discussed — Authorities reviewed — Decision in favour of plaintiff acquiring authority's contention that powers were exercised in time

This case raised questions of law arising out of compulsory purchase orders made by the acquiring authority, Westminster City Council — The owner of properties affected by the orders had contended that vesting declarations made by the authority were of no effect because the orders in question had lapsed — The ground for this contention was that the authority's powers had not been exercised before the expiration of three years from the date when the relevant compulsory purchase orders became operative — This date was July 19 1985 when notice of the confirmation of the orders was published in the local press — It was therefore clear that the three-year period started to run on that date — The acquiring authority had proceeded in due course to implement the provisions of the Compulsory Purchase (Vesting Declarations) Act 1981 — The preliminary notices required by section 3 of that Act were served eventually on February 1 1988 — The authority were then in a position to execute a vesting declaration under section 4 of that Act — The exact date of vesting is not entirely clear from the judgment, but it is clear that it was later than three years from the operative date of the compulsory purchase order

The question for decision by the judge was whether the acquiring authority, by serving the section 3 preliminary notices on February 1 1988, had exercised the power for the compulsory purchase of the properties — If so, the time-limit laid down by section 4 of the Compulsory Purchase Act 1965 was satisfied — If not, the act of making the vesting declarations was too late and the compulsory purchase orders lapsed — The plaintiff council submitted that there was a close analogy between the preliminary notice required by section 3 of Compulsory Purchase (Vesting Declarations) Act 1981 and a notice to treat — Relying on the cases which held that the notice to treat was an exercise of the power of acquisition, it was argued that the preliminary notices served in February 1988 were also an exercise of the power of acquisition — After considering a number of authorities at length, Aldous J concluded that the analogy, although attractive, was not close enough to resolve the question in the plaintiffs' favour — However, although rejecting the analogy, he did conclude that the section 3 notices were the key to the solution — Once these notices were served the formal transfer of title by the execution of a vesting declaration followed — Section 4 of the 1965 Act stated that the powers for the compulsory purchase of land shall not be exercised after three years, not that all the formalities for the vesting of title must be completed within the three-year period — Thus the service of the preliminary notices activated the powers of compulsory purchase — The plaintiff council had therefore complied with the section 4 time-limit and were entitled to the relief they sought — Declarations accordingly

The following cases are referred to in this report.

Advance Ground Rents Ltd v *Middlesbrough Borough Council* [1986] 2 EGLR 221; (1986) 280 EG 1015, LT

Grice v *Dudley Corporation* [1958] Ch 329; [1957] 3 WLR 314; [1957] 2 All ER 673; (1957) 55 LGR 493; 9 P&CR 58; [1957] JPL 723

Salisbury (Marquis of) v *Great Northern Railway Co* (1852) 17 QB 840; 21 LJQB 185; 16 Jur 740

Simpsons Motor Sales (London) Ltd v *Hendon Corporation* [1964] AC 1088; [1963] 2 WLR 1187; [1963] 2 All ER 484; (1963) 62 LGR 1; 14 P&CR 386; [1963] EGD 207; 187 EG 581; [1963] RVR 522, HL; [1963] Ch 57; [1962] 3 WLR 666; [1962] 3 All ER 75; (1962) 60 LGR 393, CA

This was an originating summons by which Westminster City Council sought declarations that their powers of compulsory purchase in respect of certain properties in London W9 had been exercised within the time-limit set by section 4 of the Compulsory Purchase Act 1965, and other associated relief. The first defendant to the summons was Mr Quereshi, the owner of the properties; the second defendants, Barclays Bank plc, were the mortgagees.

Andrew Collins QC and Robin Campbell (instructed by the solicitor to Westminster City Council) appeared on behalf of the plaintiffs; the first defendant appeared in person; Ms Margaret McCabe (instructed by Harry I Alkin & Co) represented the second defendants.

Giving judgment, **ALDOUS J** said: An application has been made by Mr Quereshi for an adjournment. The case in which I am to give judgment is a case brought by the plaintiffs seeking declarations by way of an originating summons. Although the originating summons was properly served, the first defendant failed to file the appropriate acknowledgement of service. He did, however, attend with a solicitor at a hearing before the master when directions were given for the summons to be set down for hearing in the non-witness list. Because he failed to file an acknowledgement of service the court did not have available to it a document with his recorded address upon it. Further, there is no solicitor on the record who could be notified of the date.

On the date for the hearing being appointed, the plaintiffs made all reasonable attempts to notify him of the date of the hearing, including sending to a variety of addresses notification of the hearing. Further, the court wrote a letter telling him of the date of the hearing. That letter was addressed to him at 77 Elgin Avenue. Mr Quereshi has told me that he is at the moment residing at 71 Elgin Avenue and that he did not hear of the appointment of the hearing in time for him to attend. He did in fact come on the last day of the hearing but by that time the hearing had been concluded.

He now attends in person on the date when the case is listed for judgment and has applied to me for an adjournment to enable him to be represented and the case in effect to be reheard. The fault that he was not notified of the hearing date, as he says he was not, is entirely his. He did not give to the court an address which could be used to notify him of the hearing. Further, he has no solicitors upon the record. In those circumstances I would only adjourn this case upon terms that he did file an acknowledgement of service and upon terms that he should pay all the costs that have been thrown away by the adjournment. It would seem that those costs will be substantial, and the case would need to be re-argued, probably in front of another judge.

The dispute raised is a matter of law, not fact, and if Mr Quereshi had appeared in person he could not have helped me. His English is poor and he has no legal training. Because the dispute is one of law I have come to the conclusion that the best course is to deliver the judgment which I have prepared. If Mr Quereshi does not accept my judgment then he should appeal and instruct solicitors and counsel on his behalf. In my view, that may well be the cheapest and best course for him to adopt if he wishes to contest the matter further. Such a course would prevent the costs which have already been incurred being thrown away and would also keep open to him the opportunity of having his case properly and fully argued before a court, namely the Court of Appeal.

In those circumstances I refuse Mr Quereshi his request for an adjournment. In my view, the plaintiffs were entitled to proceed with the hearing for the declaration even though Mr Quereshi did not attend because he failed to file the appropriate acknowledgement of service. I, therefore, will deliver the judgment which I have prepared.

By an originating summons dated March 30 1989 the plaintiffs, Westminster City Council, seek the following declarations:

1 That the powers of the City of Westminster, as acquiring authority for the compulsory purchase of land at 210 Ashmore Road, London W9, were exercised prior to the expiration of three years from the date on which the City of Westminster, 210 Ashmore Road W9, Compulsory Purchase Order 1984 became operative.

2 That the powers of the City of Westminster as acquiring authority for the compulsory purchase of land at 75 and 77 Elgin Avenue, London W9, were exercised prior to the expiration of three years from the date on which the City of Westminster, 75 and 77 Elgin Avenue W9 Compulsory Purchase Order 1984 became operative.

3 That the City of Westminster have vested in themselves the land known as 210 Ashmore Road, London W9 and also the land known as 75 and 77 Elgin Avenue, London W9 respectively, by virtue of Part III of the Compulsory Purchase (Vesting Declarations) Act 1981.

4 That the City of Westminster are entitled to enter upon and take possession of respectively the land known as 210 Ashmore Road, London W9 and also the land known as 75 and 77 Elgin Avenue, London W9.

The first defendant is the owner of the properties at 210 Ashmore Road, London W9, and 75 and 77 Elgin Avenue, London W9, mentioned in those declarations. He did not attend the hearing and was not represented, but it is clear that he was properly served with the summons. I understand from the evidence that his ability to read and write English is limited.

The second defendants are mortgagees. They were represented at the hearing but took no part. As the first defendant did not appear, I was taken in detail through the documents, the relevant statutes and such cases as there are which might have a bearing on the dispute between the plaintiffs and the first defendant, and I am indebted to counsel for their assistance.

The facts are not in dispute. On May 16 1984 the plaintiffs served upon the first defendant a compulsory purchase order in respect of 210 Ashmore Road, and on July 5 1984 served an appropriate preliminary statement of reasons for making the order. The compulsory purchase order and preliminary statement of reasons for the properties in Elgin Avenue were served on July 5 1984 and November 29 1984 respectively.

The plaintiffs applied, as is required by the statute, for confirmation of those orders by the Secretary of State for the Environment. The order in respect of 210 Ashmore Road was confirmed on September 24 1984 and the order in respect of Elgin Avenue on January 25 1985. On July 18 the plaintiffs sent to the first defendant, by recorded delivery, notices of confirmation of both orders and informed him that press notices of confirmation of the orders would be advertised in the *Kilburn Times* and in the *Paddington Mercury* on July 19 1985. This was done.

On August 29 1985 the first defendant issued a notice of motion in the divisional court, seeking an order that the compulsory purchase orders be quashed on grounds which I need not go into. Roch J dismissed the motion with costs on May 18 1987. The first defendant appealed. That appeal was dismissed for failure by the first defendant to comply with Ord 59, r 9 of the Rules of the Supreme Court, but before the order had been issued solicitors acting for the first defendant applied to have the appeal reinstated. The Court of Appeal held up the issue of the order and relisted the case for hearing on October 28 1987. On that date the Court of Appeal heard the matter and I believe read the evidence filed. They refused to alter the decision reached and therefore the first defendant's appeal was finally dismissed.

During the period when the first defendant's application to quash the order was before the court, the plaintiffs took no action to enforce the compulsory purchase orders. On February 1 1988 they served a statement concerning general vesting declarations and advised the first defendant that a press notice of the statement would be advertised on February 11 and 18 1988. These advertisements appeared on those dates. On September 30 1988 the plaintiffs executed vesting declarations vesting the properties in the plaintiffs as from the end of the period of 28 days from the date on which the notices were served. On October 28 1988 the plaintiffs served upon the defendant notices of the vesting declarations and a notice stating the effect of such declarations and informing the first defendant that the properties would vest in the city council on December 5 1988.

On November 30, solicitors acting for the first defendant wrote to the plaintiffs contending that the vesting declarations were of no effect as the compulsory purchase orders in question had lapsed. They said:

> We have taken counsel's advice and it appears these vesting declarations are of no effect as the compulsory purchase orders in question have lapsed. Section 4 of the Compulsory Purchase Act 1965 requires that the powers of the acquiring authority for the compulsory purchase of land shall not be carried out after the expiration of three years from the date on which the compulsory purchase order becomes operative. By section 26 of the Acquisition of Land Act 1981 a compulsory purchase order becomes operative on the date on which the notice of the confirmation or making of the order is first published in accordance with the Act.
>
> We understand that so far as 75/77 Elgin Avenue W9 is concerned, the compulsory purchase order became operative when publication took place in the Kilburn Times and Paddington Mercury on July 19 1985. We are not sure when publication took place in respect of 210 Ashmore Road but we understand that the order was confirmed by the Secretary of State in September 1984, and we imagine therefore that it would have been published in a local newspaper within a reasonable time thereafter. Please let us know the date of publication as regards this property. Certainly as far as 75/77 Elgin Avenue W9 is concerned, the vesting declaration made on September 30 1988 cannot be valid as more than three years have elapsed since publication in July 1985.
>
> We draw your attention to the decision in *Advance Ground Rents Ltd* v *Middlesbrough Borough Council* [1986] 2 EGLR 221, in which it was held that a notice to treat, served on a mortgagee more than three years after the compulsory purchase order became operative was invalid, and the Lands Tribunal had no jurisdiction to determine compensation under section 1 of the Act of 1961. In the circumstances would you please confirm that your council accepts that the compulsory purchase orders made in respect of the above properties no longer have any effect. Otherwise we must ask you to let us have your detailed grounds for any assertions to the contrary.

Having considered the evidence in this case, I have come to the conclusion that the only possible argument that could be raised against the plaintiffs' case is that set out in the letter which I have read. Thus the question for decision is whether the compulsory purchase orders had lapsed.

The statutory provisions for compulsory acquisition of land are contained in the Acquisition of Land Act 1981. Section 10 requires that a compulsory purchase order shall be in the prescribed form and shall describe the property by reference to a map. The relevant compulsory purchase orders complied with this requirement. Section 11 provides for advertisement of the orders so that objections can be raised. Such advertisements were placed by the plaintiffs and no objection was raised. Further, the requirements of section 12 were met in that the appropriate notices were served on the first defendant. Pursuant to section 13 the Secretary of State confirmed the orders. On July 19 1985 advertisements of the confirmation of the orders were published as provided for in section 15. Section 3 of the Compulsory Purchase (Vesting Declarations) Act 1981 precluded the plaintiffs from executing the vesting declarations without having included certain particulars either in the statutory notice of confirmation of the orders, or in another notice given afterwards. The notice of confirmation which had been served did not contain the relevant particulars and therefore these were included in the

A statement served on February 1 1988, to which I have referred.

That being done, section 4, which is in these terms, enabled the acquiring authority to execute a vesting declaration:

> 4.—(1) The acquiring authority may execute in respect of any of the land which they are authorised to acquire by the compulsory purchase order a declaration in the prescribed form vesting the land in themselves from the end of such period as may be specified in the declaration (not being less than 28 days from the date on which the service of notices required by section 6 below is completed).

I need not read subsections (2) and (3).

Section 4 of the Compulsory Purchase Act 1965 is in these terms:

> The powers of the acquiring authority for the compulsory purchase of the land shall not be exercised after the expiration of three years from the date on
> B which the compulsory purchase order becomes operative.

Section 26(1) of the Acquisition of Land Act 1981 is as follows:

> Subject to section 24 above, a compulsory purchase order, other than one to which the Statutory Orders (Special Procedure) Act 1945 applies, shall become operative on the date on which notice of the confirmation or making of the order is first published in accordance with this Act.

Having regard to those sections, there can be no dispute that the three-year period mentioned in section 4 of the 1965 Act started running from July 19 1985. The dispute turns on whether the plaintiffs had prior to July 1989 exercised the power of compulsory purchase of the properties or not. The plaintiffs submit that the powers were exercised by service of the notices on February 1 1988 as
C required by section 3 of the Compulsory Purchase (Vesting Declarations) Act 1981 and the first defendant contends that the powers were not purported to be exercised until the vesting declarations were made under section 4 of that Act in July 1989. Thus the question for decision is whether the plaintiffs, by serving the section 3 notice under the 1981 Act in February 1988, had exercised the power for the compulsory purchase of the properties. If they did not, then the act of making the vesting declarations was too late and the compulsory purchase orders lapsed as the solicitors for the first defendant contend.

I did not have any case drawn to my attention which directly bears upon the question I have to decide and I believe there is no such case. *Advance Ground Rents Ltd* v *Middlesbrough Borough Council*,
D which was referred to in the first defendant's solicitor's letter, does not help. In that case the confirmation of the order was advertised on June 24 1977 and the council elected to proceed by way of notice to treat which was served on February 26 1985. Thus, on any basis, the three-year period had elapsed. That case has no bearing upon whether a notice under section 3 of the 1981 Act is an exercise of the compulsory purchase power.

Mr Collins, who appeared for the plaintiffs, submitted that some guidance could be obtained from the cases dealing with notices to treat. He drew to my attention the procedure for notices to treat which is laid down in the 1965 Act. Section 5 of that Act lays down that when the acquiring authority require to purchase any land
E subject to compulsory purchase, they shall give a notice to treat setting out certain particulars and the intention of the authority. Section 6 provides for the amount of compensation, if not agreed, to be settled by the Lands Tribunal. Section 9 deals with the case where the owner refuses or fails to give good title to the property. In such cases the authority may pay the compensation into court and, having done so, execute a deed poll. Upon execution of the deed poll all estate and interest in the property vests in the authority.

Mr Collins submitted that a notice to treat was equivalent to a section 3 notice and the payment into court and deed poll were equivalent to the vesting declaration. If so, relying on authorities which held that the notice to treat was an exercise of the power of acquisition, he submitted that the section 3 notice served in February
F 1988 was also an exercise of the power of acquisition. Thus, the plaintiffs had exercised the powers of compulsory purchase prior to a lapse of the three-year period.

In *Salisbury (Marquis of)* v *Great Northern Railway Co* (1852) 17 QB 840; 117 English Reports 1503, the defendants on May 21 1851 gave to the plaintiff a notice to treat. No agreement was reached as to the compensation to be paid and a surveyor appointed by two justices of the peace fixed that at £158. That sum was deposited with the Bank of England and on June 23 1851 the defendants delivered to the plaintiff a bond as prescribed by the relevant statute. In July 1851 the defendants threatened to enter upon the land and thereafter the plaintiff sought an injunction to prevent this happening.

The relevant statute provided that the powers of the defendants of compulsory purchase should not be exercised after the expiration of G five years from the passing of the Act, and therefore this period ended on June 26 1851. Thus the plaintiff contended that the defendants had not exercised the compulsory purchase powers in time, as they had only served a notice to treat and the five-year period had elapsed.

Lord Campbell CJ at p 1507 of English Reports, (p 853 of the QB report) said:

> We have, therefore, at present, to consider what would be the effect of a company not entering within the prescribed period. This depends much upon whether the entry is an exercise of one of the powers of compulsory purchase. In my opinion it is not. I think the power of entry is a power necessary for the completion of the purchase, but is not itself one of the powers of compulsory purchase. Those powers had been, I think, exercised within the five years. Strictly speaking, there is no purchase, and no contract, created by the notice H under section 18; but the company and the landowner are placed by the notice in the same position as if a contract of purchase had actually been entered into by them.

After considering certain authorities, he concluded:

> But I consider that I am justified, both by the language of the statute and by the existing decisions upon it, in holding that the contract of purchase must be looked upon as completed upon the notice being given to the company, the amount of purchase money alone remaining undetermined. I am therefore of the opinion that we should certify that the defendants had a right to take the lands in question.

Patteson J at p 1508 said:

> The question therefore is, whether, not having entered within the prescribed period, they are precluded by section 27 of their special Act, from doing so J now. That depends upon whether the entry upon the lands is or is not an exercise of one of their compulsory powers. I am of the opinion that it is not, and that the company have, within the prescribed period, taken all the steps necessary for the compulsory purchase of the lands. I do not at all dissent from the decisions in which it is laid down that the mere notice by the company is of itself sufficient to create a contract of purchase between them and the landowner. But here much more has been done; all the steps prescribed by section 85 have been taken by the company. The principle, however, established by the cases is that the notice is all that is necessary as an exercise of the compulsory powers; the steps which follow are not of a compulsory character, but are merely for the purpose of completing the compulsory purchase to which the company had acquired a right by the notice.

Coleridge J was of the same view. He said at p 1509: K

> The question turns on the construction of section 27 of the company's special Act. What are "The powers of the company for the compulsory purchase of lands" there mentioned? A distinction must be drawn between the powers for compulsory purchase, and powers for completing a compulsory purchase. In a large sense both may be called powers for compulsory purchase and that may account for the word "powers", in the plural, being used in section 27 of the special Act. But a power for compulsory purchase, strictly speaking, is a power which enables one to purchase land from an owner who is unwilling to part with it. I consider the giving the notice to be a sufficient exercise of that power: the notice does enable the company to purchase, whether the landowner chooses or not; and all that is done by them afterwards is done only for the purpose of completing the title, and is done as much for the benefit of the landowner as of themselves. L

Wightman J's judgment was to the same effect. He said at p 1509:

> The only question is, whether, after the expiration of the period prescribed for the exercise of the compulsory powers, the company could enter upon the lands and hold them as against the plaintiff. I think this case is governed by Doe dem *Armstead* v *North Stafford Railway Company* (16 QB 526). It is admitted by the plaintiff that all steps prescribed by section 85 have been taken by the company, and that entry by them, within the five years, would have made the exercise of their compulsory powers complete. I do not see how it can make any difference whether the taking possession of the land occurs before or after the expiration of that period. If the company, after what they had done, had a right to enter at once, the entry is not an exercise of their compulsory powers, but an act which the previous exercise of those powers entitles them to do. Here they had acquired the right of entry; and it is immaterial whether they exercised that right within the five years or not. M

The ratio of that case is that the power of entry was not one of the powers of compulsory purchase. The judges drew the distinction between powers for compulsory purchase and powers for completing the purchase. In that case, as was pointed out by the Chief Justice and Wightman J, the defendants had a right to enter after service of the notice; thus entry was not an exercise of the compulsory powers but an act which the previous exercise of the powers entitled them to do.

In the present case the section 3 notice gave to the plaintiffs the right to make a vesting order and in that sense is equivalent to a notice to treat. However, the procedure laid down in the Compulsory Purchase (Vesting Declarations) Act 1981 is different. Section 7 of that Act provides that on the vesting date certain statutory provisions

A shall apply as if the vesting declaration was a notice to treat. Further, section 8 provides that on the vesting date the land and the right to enter shall vest in the authority. Thus until a vesting declaration has been made an authority has no right to enter, and that is different to the position considered in the *Marquis of Salisbury* case.

In *Grice* v *Dudley Corporation* [1958] Ch 329 Upjohn J considered a case where the defendants served a compulsory purchase order in 1937. That was approved by the minister on December 2 1938 and in July 1939 the compensation was agreed at £6,000. The war intervened and nothing happened until August 1954, when the defendants raised the question of purchasing the property for the sum that had been agreed in 1939. The plaintiffs objected. The judge held that the notice to treat served in 1938 was for the purposes of the B relevant Act an exercise of compulsory purchase powers and therefore the defendants had exercised their powers within the statutory time-limit. However, he held that the defendants had, by reason of the delay, abandoned their rights.

Section 123 of the Lands Clauses Consolidation Act 1845 was in these terms:

> And be it enacted, that the powers of the promoters of the undertaking for the compulsory purchase or taking of lands for the purposes of the special Act shall not be exercised after the expiration of the prescribed period, and if no period be prescribed not after the expiration of three years from the passing of the special Act.

At p 336 the judge set out the position that is arrived at on service C of a notice to treat. He said:

> Upon the first point, provision is made for the service of the notice to treat by section 8 of the Lands Clauses (Consolidation) Act, but the operation of a notice given under that section is now so well settled that I need not read it. It may have any one of the following effects. First, the landowner may be a willing vendor and the parties may agree upon the price or compensation and enter into a binding agreement of sale and purchase whence the usual results flow as in the case of any other contract of sale and purchase.
>
> Secondly, the landowner may be a willing seller but the parties may fail to agree upon the price, in which case either party can proceed to have the compensation payable assessed by the appropriate authority; originally it was by arbitration or by summoning a jury; then by an official arbitrator under the Acquisition of Land Act 1919, and, finally, now by the Lands Tribunal. In this case there is a limited right in the acquiring authority to withdraw the D notice to treat on seeing the claim for compensation delivered by the landowner: see the Acquisition of Land Act 1919, section 5(2).
>
> The third position is this: the landowner may be entirely hostile, in which case the promoter can proceed to have compensation assessed, and if the landowner refuses or neglects to make title, or to convey the land, the promoter can pay the compensation money into court and execute a deed poll vesting the property in himself. (That is by sections 76 and 77 of the Lands Clauses (Consolidation) Act.) The relationship between the parties thus created by the service of the notice to treat has sometimes been described as quasi-contract. It is, further, clear that where the parties have agreed upon the compensation, or it has been assessed, but not before, the remedy of specific performance is open to either side: see *Fry on Specific Performance*, 6th ed, p 62. That state of affairs has sometime been referred to as a parliamentary contract.
>
> That then is the nature of the legal relationship created by the parties, and E the question is whether the giving of such a notice is the exercise of compulsory powers of acquisition of land for the purposes of section 123 of the Act.

Having set out that section, he continued:

> Apart altogether from authority, I should be of the opinion that the service of a notice to treat is an exercise of a compulsory power because it enables the promoter to proceed to acquire the land, whether the owner likes it or not.

Mr Collins relied on that passage. He said that the section 3 notice enabled the plaintiffs to proceed to acquire the land, whether the owner liked it or not. That is true, but it is also true of the other steps in the procedure for the compulsory purchase of land. The actual purchase is dependent upon the steps required by the statute being carried out, and if carried out enable the authority to purchase the F land. Further, as I pointed out, the actual right of entry is dependent upon a vesting declaration being made and that was not executed until after the three-year period had elapsed. With a notice to treat, the authority has the right of entry pursuant to section 11 of the 1965 Act, but to obtain title to the land it has to carry out further acts, namely paying money into court and executing a deed poll. Only in that respect is the procedure the same as the compulsory purchase route using a section 3 notice.

Upjohn J continued on p 339:

> So far therefore the result seems to be to be this: the promoters, ie the corporation, served a valid effectual notice exercising compulsory powers which led to the creation of a legal relationship between the parties which I have already described. Once that notice was served, section 123 was out of the picture and no further period is laid down by the statute within which the G next step to acquire the property must be taken.

The legal relationship that the judge referred to must be one of the relationships referred to in the passage which I have read. I do not understand him to be saying that there was something in the nature of a contract for sale but only that the relationship was one in which legal rights were given. That was, I believe, made clear by the Court of Appeal in *Simpsons Motor Sales (London) Ltd* v *Hendon Corporation* [1963] Ch 57. That case went to the House of Lords [1964] AC 1088, and the judgment of Upjohn LJ was accepted as correct. In that case a compulsory purchase order was made on March 25 1952 and on August 13 1952 a notice to treat was served. Provisional agreement on compensation was agreed at £1,400. The acquiring authority changed their mind about their plans but in 1958 H sought to acquire the land for the sum previously agreed. The court held that the notice to treat was valid. Upjohn LJ, giving the judgment of the court, said at p 82:

> Upjohn J stated the principles applicable to this type of case in the recent case of *Grice* v *Dudley Corporation*. As the relevant principles do not seem to have been considered in this court for many years restatement thereof seems not out of place. It is not in doubt that service of the notice to treat confers legal rights on both parties: on the acquiring authority the right to acquire the land on payment of the proper compensation to be assessed if necessary by the Lands Tribunal; on the owner, the right to have the compensation assessed and paid. Mere service of a notice does not give the acquiring authority any estate or interest in the land, nor of itself constitutes a contract. The various ways in which the compulsory purchase procedure may be completed after J service of the notice were discussed fully in the judgment in *Grice's* case, and it is not necessary to restate them here, because no binding agreement as to the amount of compensation was reached in this case, and it falls to be decided upon the footing that there is no consensual or parliamentary contract in evidence.

There is similarity between a section 3 notice and a notice to treat in that they basically contain the same essential particulars. Further procedure thereafter is similar, but in both cases execution of a document has to take place before title passes. There is, however, a basic difference. After a notice to treat a legal relationship exists which enables the authority to enter upon the land. That right is not given by service of a section 3 notice. Further, with a notice to treat both parties have rights in that, save where the authority elects to K withdraw which is given in limited cases, the landowner has a right to compensation; whereas after service of a section 3 notice the landowner has no right to compensation. His right is dependent upon a vesting declaration being executed.

I conclude that the analogy which the plaintiffs seek to draw between a notice to treat and a section 3 notice is not so close as to resolve the dispute in the plaintiffs' favour. I must therefore go back to the statutes to try to ascertain what was the intention of Parliament.

Section 26 of the Acquisition of Land Act 1981, which I have read, provides that a compulsory purchase order shall become operative on the date that the confirmation of the order or the making of the order L is first published. Section 5 of the Compulsory Purchase (Vesting Declarations) Act 1981 prevents the authority, save with consent, executing a vesting declaration before two months have elapsed from the date of publication of the section 3 notice, and in any case before the compulsory purchase order has become operative. Section 3 of that Act lays down that certain preliminary notices have to be given before a vesting declaration can be made. Thus the compulsory purchase order became operative in 1985 and enforceable subject to a valid section 3 notice being given. Those notices were given within the three-year limit laid down in section 4 of the 1965 Act. Thus the compulsory purchase orders were operative and effective within the three-year limit. All that required to be done was formally to transfer the title by executing the vesting declarations. Section 4 of the 1965 M Act states that the powers for the compulsory purchase of land shall not be exercised after three years, and not that all acts to vest title must be completed within the three-year period. Thus, what is covered by the section is the powers of compulsory purchase.

I believe that once the compulsory purchase orders had become operative and all conditions had been met which made them effective so that the authority could vest the land in themselves, then the powers of compulsory purchase had been exercised. Having exercised such powers, section 4 of the Compulsory Purchase (Vesting Declarations) Act 1981 entitled them to execute the vesting declarations. That entitlement was dependent upon the authority validly exercising the powers given to them for compulsory purchase.

A The word "declaration" also suggests to me that the powers of compulsory purchase have been exercised and that the vesting declarations are documents declaring that the compulsory purchase has taken place and having the effect of formally transferring the title to the land.

I have, therefore, come to the conclusion the plaintiffs had, prior B to the lapse of the three-year period, exercised the power of compulsory purchase for the properties and that they are entitled to the declarations sought.

For a further case on this subject see p 31

LANDLORD AND TENANT

GENERAL

A **Queen's Bench Division**

May 16 1990

(Before Mr Recorder Thayne FORBES QC, sitting as a deputy judge of the division)

IRVINE v MORAN

Landlord and tenant — Section 32 of the Housing Act 1961 (and provisions replacing it in recent legislation) — Landlord's implied obligations in short leases of dwelling-
B **houses — A case of practical importance on a subject with little authority on the issues raised — Meaning of "structure" of dwelling-house — Impact of landlord's implied obligations on tenant's express covenants in lease — Analysis of section 32(1) and judge's rulings on a variety of its applications**

This report is concerned with the determination of two preliminary issues in a landlord's action for damages in respect of the tenant's alleged failure to comply with repairing and decorating covenants in a lease — The issues concerned the construction and application of section 32 of the Housing Act 1961, the lease in question being for less than seven years
C **from September 1 1978 — Section 32 of the 1961 Act was replaced by section 11 of the Landlord and Tenant Act 1985, which was in turn in some respects amended by section 116 of the Housing Act 1988, but the points dealt with in the present judgment remain applicable to the re-enacted legislation — The questions raised were of a detailed nature**

The express covenants in the tenant's lease could be summarised generally as imposing on him extensive repairing and decorating obligations — The broad effect of section 32(1) was that, in order to ascertain what the tenant's obligations were, the obligations imposed by the section on
D **the landlord had to be subtracted from the tenant's obligations under the lease — That would show what residual obligations the tenant had for repair and decoration after the landlord had fulfilled his statutory obligations**

The judge began with an examination of the word "structure" in the phrase "the structure and exterior of the dwelling-house" — It did not mean the entire dwelling-house or the entire constructed building — The judge concluded that the structure of the dwelling-house consisted of those elements of the overall dwelling-house which gave it its essential appearance, stability and shape — The expression
E **did not extend to the many and various ways in which the dwelling-house would be fitted out, equipped, decorated and generally made to be habitable — Having discussed the meaning of structure in section 32(1), the judge proceeded to give his rulings on the particular questions raised in the two issues — The judgment needs to be studied for the details of these answers, which are of considerable practical interest**

The following cases are referred to in this report.

Campden Hill Towers Ltd v *Gardner* [1977] QB 823; [1977] 2 WLR 159; [1977] 1 All ER 739; (1976) 34 P&CR 175; [1977] EGD 151; 242 EG 375; 13 HLR 64, CA

Pearlman v *Keepers and Governors of Harrow School* [1979] QB 56; [1978] 3 WLR 736; [1979] 1 All ER 365; (1978) 38 P&CR 136; [1978] EGD 721; 247 EG 1173; [1978] JPL 829, CA F

In this case Mr Recorder Thayne Forbes QC, sitting as a deputy judge of the Queen's Bench Division, was asked to determine two preliminary issues of law in advance of the trial in an action by the plaintiff, Cecile Godman Irvine, against the defendant, John Denton Moran, for alleged breach of covenants relating to the property known as Little Oak Hall, Wivelsfield, Sussex.

Jonathan Brock (instructed by Fynmores, of Bexhill-on-Sea) appeared on behalf of the plaintiff; Jonathan Ferris (instructed by Keene Marsland, of Croydon) represented the defendant.

Giving judgment, **MR RECORDER THAYNE FORBES QC** said: I have G
been asked in this case to determine two preliminary issues which it is hoped will shorten the time and expense of the trial as a whole. The action is one brought for the recovery of damages on the part of the landlord (I shall simply use the expression "the landlord" at this stage) in respect of a tenant's failure to comply with repair and decoration covenants in a lease made on October 18 1978 whereby the property known as Little Oak Hall, Wivelsfield, Sussex, was demised to the defendant for a term of less than seven years from September 1 1978. It is common ground between the parties, and conceded in argument before me, that section 32 of the Housing Act 1962, an Act of Parliament which has since been repealed, did apply to that lease. What that means is that the landlord's implied covenants, which are imposed upon a lease by section 32 of the Housing Act 1961, applied H
to this lease.

The preliminary issues of law which I have been asked to determine, and which I have ordered in the course of the hearing should be determined, have been agreed between the parties and by counsel. They are to be found in the plaintiff's statement of issues, which has been helpfully typed and handed to me at the outset of this trial, as issues 3 and 5. For the avoidance of doubt I shall read those preliminary issues out so that it is quite clear to what I have addressed my mind. The first of the preliminary issues, preliminary issue no 1, which is issue no 3 of the plaintiff's statement of issues, reads as follows:

> Which items fall within the landlord's implied covenants under section 32 of J
> the Housing Act 1961? In particular, but without limitation to these matters, to what extent do the garage, gates, driveway, gardens and grounds, windows including sashes, cords, frames, glazing and furniture, internal wall plaster and door furniture, fall within the section?

Preliminary issue no 2, which is plaintiff's statement of issues no 5, reads as follows:

> Do the tenant's covenants to paint and decorate the exterior of the dwelling-house and other relevant installations, such as radiators, remain wholly or partially in effect on a true construction of section 32(1)?

Again that is a reference to section 32 of the Housing Act 1961.

Those are the two preliminary issues which fall to be determined. As far as I feel I can, I shall now rule upon them and I turn first of all K
to preliminary issue no 1. In order to determine that issue it is necessary to remind myself of the relevant section, section 32, of the Housing Act 1961, which reads as follows:

> (1) In any lease of a dwelling-house, being a lease to which this section applies,

— and in parenthesis I repeat that the parties accept that this section does apply —

> there shall be implied a covenant by the lessor —
> (*a*) to keep in repair the structure and exterior of the dwelling-house (including drains, gutters and external pipes); and
> (*b*) to keep in repair and proper working order the installations in the dwelling-house —

(i) for the supply of water, gas and electricity, and for sanitation (including basins, sinks, baths and sanitary conveniences but not, except as aforesaid, fixtures, fittings and appliances for making use of the supply of water, gas and electricity), and

(ii) for space heating or heating water, and any covenant by the lessee for the repair of the premises (including any covenant to put in repair or deliver up in repair, to paint, point or render or to pay money in lieu of repairs by the lessee or on account of repairs by the lessor) shall be of no effect so far as it relates to the matters mentioned in paragraphs (*a*) and (*b*) of this subsection.

It is not necessary for me to read from that section any further.

It is also necessary to remind myself, in particular for the purposes of preliminary issue no 2, what the tenant's covenants in the lease were. They are fully set out in para 2 of the statement of claim in this case and are to be found in the lease as clauses 2(iii), (iv), (v), (vi) and (vii). I do not propose to read those covenants out in full. I think they can be summarised generally as imposing upon the defendant extensive repairing and decorating obligations.

The question to which both preliminary issues are addressed is to what extent the repairing and decorating obligations by the defendant (the tenant) as set out in those various clauses of the lease to which I have referred have been cut down by the implied covenant to repair imposed upon the landlord by section 32 of the Housing Act 1961. It will be recalled that part of subsection (1) of that section expressly provides that, to the extent that there is a covenant by the lessee to carry out various works to the premises of repair, decoration and the like, those covenants shall be of no effect so far as those covenants on the part of the tenant relate to the matters mentioned in paras (*a*) and (*b*) of the subsection. The practical effect is that, to the extent that section 32 has imposed upon the landlord certain implied obligations to repair, those obligations must be subtracted from the tenant's express obligations under the lease to see what residual obligation the tenant has for repair, redecoration and the like after the landlord has fulfilled his statutory repairing obligations.

Preliminary issue no 1 has directed my attention to determining to what extent certain aspects of the demised premises can be said to be either part of the structure of the dwelling-house or part of the exterior of the dwelling-house, because it is in relation to the structure and exterior of the dwelling-house that the landlord's implied repairing covenants arise. Also, it is to certain specified installations in the dwelling-house that his obligations are addressed by the section. I am asked to consider in particular the garage, the gates, the driveway, the gardens, the grounds, the windows and so forth, the internal plaster and door furniture. I am asked to give a ruling in relation to those as a matter of law. The preliminary issue also asks me to rule upon the landlord's implied covenants without limiting my ruling to those items, although those are the ones of prime concern to the parties. It is of interest, and somewhat surprising, that there is very little authority on this matter at all and so I have to decide this preliminary issue without significant assistance from the authorities.

The first thing that I have to address my mind to is what is meant by section 32(1)(*a*) as keeping in repair the structure and exterior of the dwelling-house. I think that Mr Brock is correct in his submission, and I so hold, that the word "structure" in subsection (1)(*a*) should not be construed in isolation but should be construed as part and parcel of the overall expression "the structure of the dwelling-house" as distinct from "the exterior of the dwelling-house". In my judgment, it is clear that the legislature did not intend by the use of the word "structure" in that subsection to mean the entire dwelling-house or the entire constructed building. Had the legislature intended to give the very wide meaning to "structure" that Mr Ferris argued for, then I agree with Mr Brock that the words that we find in para (*a*) of subsection (1) in parenthesis "including drains, gutters and external pipes" would be mere surplusage and totally unnecessary. I also agree with Mr Brock that by using the words "the structure of the dwelling-house" and "the exterior of the dwelling-house" the legislature in both cases was referring to something less than the dwelling-house as a whole. The structure of the dwelling-house is something less than the overall dwelling-house itself and the exterior of the dwelling-house is also something less than the overall dwelling-house itself. Of course, the difficulty that is posed is deciding to what more limited aspects of the overall dwelling-house the word "structure" is addressed. I have come to the view that the structure of the dwelling-house consists of those elements of the overall dwelling-house which give it its essential appearance, stability and shape. The expression does not extend to the many and various ways in which the dwelling-house will be fitted out, equipped, decorated and generally made to be habitable.

I am not persuaded by Mr Brock that one should limit the expression "the structure of the dwelling-house" to those aspects of the dwelling-house which are load-bearing in the sense that that sort of expression is used by professional consulting engineers and the like; but what I do feel is, as regards the words "structure of the dwelling-house", that in order to be part of the structure of the dwelling-house a particular element must be a material or significant element in the overall construction. To some extent, in every case there will be a degree of fact to be gone into to decide whether something is or is not part of the structure of the dwelling-house. It is not easy to think of an overall explanation of the meaning of those words which will be applicable in every case and I deliberately decline to attempt such a definition. I am content for the purposes of this case to say that I accept Mr Brock's submission that "structure of the dwelling-house" has a more limited meaning than the overall building itself and that it is addressed to those essential elements of the dwelling-house which are material to its overall construction.

That I think is as far as I am able to go. In coming to that conclusion I have been guided to some extent and have sought to follow the approach suggested by Megaw LJ in *Campden Hill Towers Ltd* v *Gardner* [1977] QB 823, in particular the passage that appears at p 834F. I did not find that I was greatly assisted by the various authorities cited to me by Mr Ferris, and in particular the case of *Pearlman* v *Keepers and Governors of Harrow School* [1979] QB 56 at p 79. I agree with Mr Brock that it would be very dangerous to try to extrapolate from definitions referable to the word "structural" any meaning which would enable me to be satisfied that I was properly interpreting a different concept, namely "the structure of the dwelling-house". It seems to me that "structural" as an adjective has acquired a very much more limited meaning than the one for which Mr Ferris has been arguing in this case and I did not find the citation of *Pearlman* particularly helpful to me in my deliberations.

Approaching the matter in the light of what I consider to be the meaning of subsection (1), para (*a*) of section 32, I now address the question posed in preliminary issue no 1. I have been helped to some extent by concessions made by Mr Ferris at the outset of his submissions, concessions which have been made throughout by the defendant, namely that the gardens and grounds and driveway are not within section 32 of the 1961 Act. Mr Brock made the suggestion, which I gratefully adopt, that the further concession made by the defendant in relation to glazing can be most easily reconciled with the legislation by accepting, as I do, from Mr Ferris that he conceded liability in relation to glazing on instructions in circumstances which would probably be a proper application of section 32(2)(*a*). The glazing concessions that have been made relate to damage to windows caused by the tenants during their occupation. Thus the concession was not a concession that glazing could not fall within section 32(1) in other circumstances.

Having dealt with those concessions and dealt with my approach to the general question of what is the structure of the dwelling-house, I turn to the listed items in the preliminary issue no 1. The driveway, gardens and grounds have been excluded by concession. The glazing has been excluded by concession. I see no reason at this stage not to follow the natural approach and say that a separate garage and separate gates do not ordinarily, in my judgment, form part of the dwelling-house. Since no factual material has been placed before me which would enable me to deal with the argument that there may be factual material which would rebut the natural approach, I am driven to the conclusion that they do not fall within section 32 because they do not form any part of the dwelling-house, no factual material having been put to me to suggest otherwise.

So far as windows are concerned, I find this a difficult matter to decide and will come to that last in relation to preliminary issue no 1. I therefore turn to internal wall plaster and door furniture. As I have said, section 32(1)(*a*) and the words "structure of the dwelling-house" mean something less than the dwelling-house overall and are limited to the essential material elements that go to make up the structure of the dwelling-house. It seems to me that internal wall plaster is more in the nature of a decorative finish and is not part of the essential material elements which go to make up the structure of the dwelling-house. I therefore hold that internal wall plaster and, for the same reason, the door furniture do not form part of the structure of the dwelling-house, bearing in mind I have held that those words mean something less than the overall construction.

Windows pose a slightly different problem. I have some hesitation about this, but bearing in mind that one is talking about a dwelling-house, and rejecting as I do the suggestion that one should use "load-bearing" as the only touchstone to determining what is the structure of the dwelling-house in its essential material elements, I have come to the conclusion that windows do form part of the structure of the dwelling-house. My conclusion might be different if one were talking about windows in, let us say, an agricultural building. The essential material elements may change, depending on the nature and use of the building in question. In the case of a dwelling-house, it seems to me that an essential and material element in a dwelling-house, using ordinary common sense and an application of the words "structure of the dwelling-house" without limiting them to a concept such as "load-bearing", must include the external windows and doors. Therefore, I hold that windows themselves, the window frames and the sashes do form part of the structure. It follows that, since these are sash windows, it would be invidious to separate the cords from the sashes and the essential furniture from the frames. So, in my judgment, the windows including the sashes, the cords, the frames and the furniture are part of the structure of the dwelling-house.

Again, there is room for factual differences. For example, though it has not been suggested that such is the case here, there could be furniture which is added to a window which is not essential for the operation of that particular window. In such circumstances I think that it would be doubtful if such non-essential furniture were to be regarded as part of the structure. However, I do not have to consider such nice questions as that and so I do not make any specific ruling on it. In this case, it not having been suggested that there is any circumstance which does not justify the particular furniture being placed on the particular windows, I am satisfied that the windows together with their various accoutrements are part of the structure.

The external windows, in my view, do form part of the exterior of the building, at least on their outer face. If I am wrong about regarding the windows as part of the structure, I am satisfied that at least on their outer face the windows are part of the exterior. If I am wrong about the windows being part of the structure, then in that more limited sense the windows still fall within section 32(1)(*a*). If I am wrong about the windows being part of the structure and I am only right that the window frames and so forth form part of the exterior of the dwelling-house it would follow, on that more limited basis, that the cords and furniture, all of which would be internal, would not be part of the exterior. It does happen (and I speak from personal experience) that some part of the window furniture can be on the exterior. If that happens to be the case here, it is part of the exterior of the building as well. I do not know whether there actually are any parts of the window furniture on the outside of the building.

That enables me to say that I have dealt with preliminary issue no 1. I turn to preliminary issue no 2. The issue that has to be determined is to what extent the tenant's covenants to paint and decorate the exterior of the dwelling-house and other relevant installations, such as radiators, remain wholly or partially in effect on a true construction of section 32(1). This preliminary issue is directed at elements in the claim where the plaintiff has conceded that the element is within section 32(1) and is thus subject to the landlord's covenants to repair. The plaintiff submits that notwithstanding the landlord's covenants to repair and notwithstanding the "non-effect" of the tenant's covenants to put in repair, deliver up in repair, to paint, to point or render, and so forth, there is still a liability on the part of the tenant to paint and decorate elements of the building which are the subject-matter of the landlord's repairing covenants, because of clauses 2(iv) and (v) of the lease. The submission is that the painting and decoration of the exterior of the dwelling-house and other relevant installations is purely decorative and does not involve any "keeping in repair". In my judgment, so far as the exterior of the dwelling-house is concerned, the tenant's obligation to paint and decorate has been eliminated by section 32(1) of the Act. It seems to me that painting and decorating the exterior of the dwelling-house must inevitably and invariably involve a degree of protection against the elements and against the processes of rot and the like. Inevitably it is part and parcel of the process of keeping the exterior of the dwelling-house in repair. There are aesthetic benefits and pleasure to be gained from a high state of decorative order, but in my judgment, once I am satisfied that the obligation that would otherwise be imposed on the tenant by the express terms of the tenancy involves a degree of keeping in repair the exterior of the dwelling-house, then the tenant's obligation is subsumed into the landlord's implied obligations under the Act.

So far as what are described in the preliminary issues as "other relevant installations" such as radiators and, on my findings, the internal faces of windows are concerned, I find myself in some difficulty in giving a definitive ruling. If all that is to be achieved by the painting is to bring the item in question, whether it be a window frame or a radiator, up to a high state of decorative order so as to make it attractive to an incoming tenant, that does not involve any keeping in repair. It is requiring the tenant to carry out something that is purely decorative. If, on the facts relating to any particular item, that were held to be the case, then the obligation of the tenant would continue because the express covenant of the lease required the tenant to do something other than that which is the subject of the landlord's implied obligations, the landlord's obligations being to keep the relevant elements in repair. In other words, to the extent that what is required of the tenant is purely decorative it seems to me that the landlord will be entitled to performance of the covenant. If in relation to any relevant item, such as the radiators or internal faces of the windows, the painting is required for the purposes of repair or keeping in repair, then this is the obligation of the landlord under section 32(1) of the Act. This may be a somewhat unrealistic distinction in this case because probably most if not all of the internal decorative works are referable to elements which fall within the ambit of the landlord's repairing covenants. In painting such elements, there will be, as in the case of the exterior, always an element of repair, protection, call it what you like. The decorative benefits that come from keeping an element in repair are subsumed into the repairing covenant of the landlord. Thus it may be that the obligation that may still exist, so far as the tenant is concerned, under the painting and decorating covenant in the lease may have no practical consequences at all. That obligation would exist only where all that was required of the tenant was something which was purely decorative. In my view, such a residual obligation on the part of the tenant could still exist, although having little practical significance.

Court of Appeal

October 12 1990

(Before Lord Justice SLADE, Lord Justice NICHOLLS and Lord Justice BINGHAM)

HAMMERSMITH AND FULHAM LONDON BOROUGH COUNCIL v MONK
BARNET LONDON BOROUGH COUNCIL v SMITH AND ANOTHER

Landlord and tenant — Periodic tenancies — Joint tenants — Notice to quit — Two appeals raising the same question — Whether under the general law, in the absence of a contrary provision in a tenancy agreement, a notice to quit given by one only of two joint tenants, without the authority of the other, will bring the tenancy to an end — The appeal in one case was by landlords against the dismissal of a claim for possession, in the other case by a tenant against the dismissal of proceedings to set aside an order for possession — The court considered a number of authorities from *Doe d Aslin* v *Summersett* onwards — In the end, despite criticisms, *Greenwich London Borough Council* v *McGrady* was followed — After a detailed examination, *Howson* v *Buxton* held not to be an authority supporting a contrary view — Limits as to a *per incuriam* challenge considered — Tenant's notice to quit in each of the present cases upheld

In the *Monk* case the notice purporting to terminate the tenancy of the subject flat was given by the joint tenant who had been the respondent's cohabitee — On the respondent's refusal to vacate the flat the council landlords sought possession, but Judge Roger Cooke, at West London County Court, refused the order and the council appealed — In the *Smith* case, Mrs Smith, who was a joint tenant with her husband of a council house, gave the council notice to terminate the tenancy — On Mr Smith's refusal to vacate, the

A **council sought and obtained from the registrar, at Barnet County Court, an order for possession — An appeal by Mr Smith was dismissed by the assistant recorder and Mr Smith appealed to the Court of Appeal — In each case the tenancy was a weekly tenancy and in neither case was the notice to terminate given with the authority of the other joint tenant**

The appeals raised the same two issues — The first was whether under the general law a notice to quit given by one only of two joint tenants, without the authority of the other, was effective to bring the joint tenancy to an end — The second issue was whether there was anything in the provisions of the particular tenancy agreement which affected the
B **position under the general law — On the general issue it had been thought that the decision of the Court of Appeal in 1982 in *Greenwich London Borough Council* v *McGrady* had decided, approving the old case of *Doe d Aslin* v *Summersett* and the more recent decision in *Leek & Moorlands Building Society* v *Clark*, that a notice to quit given by one of several joint tenants could bring the tenancy to an end — However, in the present appeals the decision in *McGrady* was attacked as being inconsistent with the earlier Court of Appeal decision in *Howson* v *Buxton* and also as having been given *per incuriam***

As regards *Howson* v *Buxton*, an agricultural holdings case
C **with special facts which keeps on raising its head, Slade LJ's analysis in the present appeals made it clear that the validity of the joint tenant's notice was not in issue there and that Scrutton LJ's statement that one joint tenant cannot effectively give notice to quit without the authority of the other was a "mere *obiter dictum*" — Having thus disposed of the criticism based on *Howson* v *Buxton*, Slade LJ examined the suggestion that the *McGrady* decision had been given *per incuriam* and considered an article to that effect in *The Conveyancer* — After a discussion of the true nature of a periodic tenancy as continuing indefinitely unless terminated**
D **by a notice to quit, not as re-created by a notional reletting at the commencement of each new period, and an examination of the *per incuriam* doctrine, Slade LJ rejected the submission that *McGrady* can be regarded as a decision given *per incuriam* — *McGrady* must be regarded (unless or until the House of Lords decides otherwise) as a binding authority that under the general law, in the absence of a contrary provision in the tenancy agreement, a notice to quit given by one of the joint tenants under a periodic tenancy, without the consent of the other or others, will have the effect of bringing the tenancy to an end**

The court held that there was nothing in the provisions of
E **either tenancy agreement in the present appeals to displace the application of the general principle — There was a somewhat stronger argument on this point in the *Smith* case, but it was not strong enough to be effective — The result was to allow the appeal in the *Monk* case and to dismiss the appeal in the *Smith* case — Suggestion by Nicholls LJ in regard to the forms of tenancy agreement used by local authorities should be noted**

The following cases are referred to in this report.

Alford v *Vickery* (1842) Car & M 280
Annen v *Rattee* [1985] 1 EGLR 136; (1984) 273 EG 503; 17 HLR 323, CA
Doe d Aslin v *Summersett* (1830) 1B&Ad 135
F *Doe d Kindersley* v *Hughes* (1840) 7 M & W 139
Doe d Whayman v *Chaplin* (1810) 3 Taunt 120
Duke v *Reliance Systems Ltd* [1988] QB 108; [1987] 2 WLR 1225; [1987] 2 All ER 858, CA
Edell v *Dulieu* [1923] 1 KB 533
Gandy v *Jubber* (1870) 9 B&S 15; (1864) 5 B&S 78; 33 LJQB 151; 9 LT 800; 10 Jur NS 652; 12 WR 526
Gray v *Spyer* [1922] 2 Ch 22
Greenwich London Borough Council v *McGrady* (1982) 81 LGR 288; 46 P&CR 223; [1983] EGD 519; 267 EG 515, CA
Howson v *Buxton* (1928) 97 LJKB 749
Leek & Moorlands Building Society v *Clark* [1952] 2 QB 788; [1952] 2 All ER 492; [1952] 2 TLR 401, CA
Opoku-Forfieh v *Haringey London Borough Council* unreported October 17 1988.
Parsons v *Parsons* [1983] 1 WLR 1390; (1983) 47 P&CR 494; [1984] EGD 281; 269 EG 634 G
Rakhit v *Carty* [1990] 2 WLR 1107; [1990] 2 All ER 202; [1990] 2 EGLR 95; [1990] 31 EG 40, CA
Rickards v *Rickards* [1989] 3 WLR 748; [1989] 3 All ER 193, CA
Viola's Indenture of Lease, Re [1909] 1 Ch 244

The first appeal was by the London Borough of Hammersmith and Fulham from an order of Judge Roger Cooke, at West London County Court, dismissing their claim for possession of the ground-floor flat at 35 Niton Street, Fulham, London SW6, occupied by the respondent, Kenneth John Monk. The second appeal was by James Martin Smith against the order of Mr A A Goymer, sitting as an assistant recorder at Barnet County Court, dismissing an appeal H
from an order of the registrar refusing to set aside an order for possession of the council house occupied by the appellant at 119 Alexandra Road, Muswell Hill, London N10. The respondents to this appeal were the London Borough of Barnet.

In the first appeal Mark Strachan QC and Lincoln Crawford (instructed by the director of legal services, London Borough of Hammersmith and Fulham) appeared on behalf of the appellants; Paul Staddon (instructed by Oliver O Fisher & Co) represented the respondent, Kenneth John Monk.

In the second appeal Anthony Radevsky (instructed by Rhodes Barlow, of New Barnet) appeared on behalf of the appellant, James Martin Smith; Charles Salter (instructed by the controller of legal services, London Borough of Barnet) represented the respondent J
council.

Giving judgment, **SLADE LJ** said: There are before the court two appeals which have been heard consecutively and both raise a common question: under the general law, in the absence of provision to the contrary in a tenancy agreement, does a notice to quit given by one only of two joint lessees under a periodic tenancy, without the authority of the other, have the effect of bringing the tenancy to an end? In one case this question was answered by the court below in the affirmative. In the other it was answered in the negative. In each case the further question arises whether the provisions of the particular tenancy agreement alter the position which would otherwise arise under the general law. K

The appellants and respondents to the two appeals have all appeared by different counsel, who have assisted us by full and careful arguments. However, the submissions for the respective lessors and lessees in the two cases have all followed much the same lines and I do not propose to differentiate between them in referring to them in this judgment.

HAMMERSMITH and FULHAM LONDON BOROUGH COUNCIL v MONK

In the first of the two appeals the London Borough of Hammersmith and Fulham ("the council") appeal from an order of His Honour Judge Roger Cooke made in the West London County Court on February 15 1990 whereby he dismissed their claim for possession of certain premises known as ground-floor flat, 35 Niton L
Street, London SW6 ("the flat").

The facts of this case can be shortly stated. The respondent to the appeal, Mr Kenneth John Monk, cohabited with a Mrs Powell. In the summer of 1987 Mrs Powell and Mr Monk signed a written tenancy agreement, which was also signed by a representative of the council, under which they agreed to take a tenancy of the flat from the council with August 24 1987 as its commencement date.

The agreement was in a standard printed form of tenancy agreement, with the names of the tenants and of the premises and the tenancy commencement date inserted in manuscript. The front page of the agreement, headed "Hammersmith & Fulham Tenancy Agreement", began with the words:

> The Conditions of Tenancy and rights and duties of both parties are set out below. If you are joint tenants the term "tenant" refers to both tenants or all of you . . . M

Save that clause 3 is headed "Tenants' obligations" (with the apostrophe after the "s" in "Tenants'") the agreement used the expression "tenant" in the singular throughout. That positioning of the apostrophe is, in my opinion, clearly due to a clerical error.

Clause 3(a) provided:

> Rent. (a) To pay the weekly rent and other charges regularly and promptly. Rent is due in advance every Monday.

Clause 3(t) provided:

> Termination of tenancy. (t) To give four weeks' notice, (or such lesser period

A as the Council has agreed to accept), to expire on a Monday, when giving up the tenancy . . .

It is common ground that the effect of this agreement was to create a weekly tenancy to which the provisions of the Housing Act 1985 applied.

Subsequently there were differences between the two tenants. In 1988 Mrs Powell left the flat. She went to see the council and made certain allegations against Mr Monk. As the judge found, whether or not these allegations were justified, the council accepted them and agreed to rehouse her, provided that she terminated the current tenancy. For the purpose of doing so she came to see Miss Smith, an employee of the council. Miss Smith drew up, in the form of a letter, a notice to quit. Mrs Powell signed this letter and gave it to Miss Smith
B who, as the judge found, had authority to receive it on behalf of the council and did so receive it. The notice to quit took the following form:

> I hereby give you four weeks notice in writing of my intention to terminate my tenancy of the ground floor flat, 35 Niton Street, London, SW6. My tenancy will end on Monday the 9th of January 1989.

This notice was unquestionably effective to terminate the tenancy on the last-mentioned date if Mrs Powell had the legal right to terminate it without Mr Monk's authority or consent. However, he had not given her any authority or consent to serve the notice. He refused to vacate the flat. On February 14 1989, the council instituted proceedings against him.

C The judge regarded two issues as falling for his decision. The first of them was the question of general law set out at the beginning of this judgment. The second was: did this particular tenancy agreement on its true construction require that a notice terminating the tenancy had to be signed by both tenants if it was to be effective?

As to the first of these two questions, the judge referred to the decision of this court in *Greenwich London Borough Council* v *McGrady* (1982) 267 EG 515 (''*McGrady*''), as to which he said:

> the facts . . . are similar in all material respects to the present case and . . . the Court of Appeal held in firm terms that one of two joint tenants can give notice to quit terminating the letting.

As the judge observed: ''It might have been thought that such
D authority would have brought the case to a premature end''. However, he was persuaded that the earlier decision of the Court of Appeal in *Howson* v *Buxton* (1928) 97 LJKB 749 was ''clear authority for the proposition that one joint lessee alone cannot give the requisite notice''. Furthermore, the judge regarded the decision in *McGrady* as inconsistent with the modern concept of a periodic tenancy as expounded by the Court of Appeal in *Gray* v *Spyer* [1922] 2 Ch 22.

In these circumstances, the judge said:

> I am therefore faced with the choice, either the *Gray* v *Spyer* and *Howson* v *Buxton* line of authority ought to be followed or *McGrady* ought to be followed. The earlier line of authority appears to me to be consistent with principle and it was not cited to the Court of Appeal in *McGrady*. The stark
> E choice therefore arises. Which do I [choose]? In my judgment I ought to choose the *Gray* v *Spyer* and *Howson* v *Buxton* line and I do.

His answer to the first of the two principal issues which fell to be considered by him was therefore that under the general law Mrs Powell did not have power to give a valid notice to determine the letting and did not determine it.

However, in case this conclusion was wrong, the judge proceeded to consider the second issue. As to this, he concluded that, on the true construction of the tenancy agreement, a notice of termination of tenancy by virtue of clause 3(t) had to be given by both tenants if it was to be effective.

On both these grounds the judge decided that no effective notice to quit had been given, that Mr Monk remained a tenant of the premises
F and that the council's claim to possession should be dismissed. The council now appeal from that decision.

A periodic tenancy such as a yearly or weekly tenancy is a peculiar form of tenancy. It differs from a tenancy for a fixed number of years in that, unless determined by a proper notice to quit, it may last indefinitely. Section 5(1) of the Protection from Eviction Act 1977 provides (*inter alia*) that no notice to quit premises let as a dwelling shall be valid unless it is given not less than four weeks before the date on which it is to take effect. At common law, while it is open to the parties to make such arrangements as they please, the length of notice must, in the absence of agreement to the contrary, in the case of a weekly tenancy be not less than the length of the tenancy and the notice must purport to terminate the tenancy at the end of the current period: see *Cheshire & Burn's Modern Law of Real Property* (14th ed) at p 443. G

The Court of Exchequer Chamber in a judgment in *Gandy* v *Jubber*, somewhat unusually reported at (1870) 9 B&S 15 as ''agreed to'' by Erle CJ, Martin B, Willes J, Channell B, Keating J, Pigott B and Smith J but ''not delivered'', in describing the nature of a tenancy from year to year, stated at p 18:

> There is not in contemplation of law a recommencing or reletting at the beginning of each year.

In *Gray* v *Spyer* [1922] 2 Ch 22 at p 38 Scrutton LJ, likewise referring to the nature of a tenancy from year to year as known to lawyers, described it as

> defined by Parke B in *Oxley* v *James* as ''a lease for a year certain, with a H growing interest during every year thereafter, springing out of the original contract, and parcel of it''. Such a tenancy is for a year certain, terminable by either side by six months' notice expiring at the end of the year, involving, if such notice is not given, another year's tenancy terminable in the same way.

In *Edell* v *Dulieu* [1923] 1 KB 533 at p 539, McCardie J similarly referred to the true principle of law as being ''not that there was a reletting each year (in the case of a weekly tenancy), but a springing interest in the tenancy which could only be determined by a notice to quit''. Since periodic tenancies will continue indefinitely unless and until determined by a valid notice to quit served by landlord or tenant, notices to quit will assume particular significance in their case.

In *Doe d Aslin* v *Summersett* (1830) 1 B&Ad 135 (''*Doe d Aslin*'') it J was held that a notice to quit served by one of two joint lessors without adoption by the other joint lessor put an end to a tenancy from year to year. Lord Tenterden CJ, delivering the judgment of the court, said at p 140:

> When joint-tenants join in a lease, each demises his own share, Co Litt 186 a, and each may put an end to that demise as far as it operates upon his own share, whether his companions will join with him in putting an end to the whole lease or not . . .

A little later (at p 140) Lord Tenterden CJ said:

> Upon a joint demise by joint-tenants

— that is the lessors in that case —

> upon a tenancy from year to year, the true character of the tenancy is this, not K that the tenant holds of each the share of each so long as he and each shall please, but that he holds the whole of all so long as he and all shall please; and as soon as any one of the joint-tenants

— that is the lessors in that case —

> gives a notice to quit, he effectually puts an end to that tenancy . . .

In *Leek & Moorlands Building Society* v *Clark* [1952] 2 QB 788 (''*Leek & Moorlands*'') the question for decision was whether a contract by one of two joint lessees to purchase the reversion terminated a subsisting tenancy before the full period of the tenancy had run. This court held that it did not. Somervell LJ, in the course of delivering the reserved judgment of this court, drew a distinction between a right to determine within the period of a lease and a right to L terminate a periodic tenancy. He described (at p 793) the ratio of *Doe d Aslin* as to be found in the second of the passages from Lord Tenterden CJ's judgment which I have quoted. He continued:

> It is to be noted that Lord Tenterden was dealing with a notice to quit in respect of a periodic tenancy. He was not dealing with a right to determine a lease for say 21 years at the end, say, of the seventh or fourteenth year. Nor was he dealing with surrender.
>
> There is, we think, force in the submission made on behalf of the plaintiffs, that in the case of a periodic tenancy Lord Tenterden's principle would apply when there were joint lessees. A periodic tenancy continues from period to period unless the notice agreed or implied by law is given. But if one of two joint lessees who ''hold the whole'' wishes it not to continue beyond the end of a period, it might well be held that it did not continue into a new period. That would happen only if all, that is, the joint lessees, shall please. M
>
> If one considers a lease to joint lessees for a term certain with a right of renewal, it would be obvious, we think, that both must join in requiring a renewal. A periodic tenancy renews itself unless either side brings it to an end. But if one of two or more joint lessees does not desire it to continue, we would have thought that it was in accordance with Lord Tenterden's principle, and with common sense, that he should be able to make that effective.

Somervell LJ, however, concluded that that principle had no application to a surrender. Such a right, in his view, would be exercisable only if all desired to exercise it. In his words (at pp 794-795):

> They all have the right to the full term, and all must concur if this right is to be abandoned . . . If property or rights are held jointly, *prima facie* a transfer

A must be by or under the authority of all interested.

Finally, Somervell LJ said (at p 795) that *Doe d Aslin* was not an exception to the rule just stated but that

> It is an illustration, in a highly technical field, of the general principle that if a joint enterprise is due to terminate on a particular day, all concerned must agree if it is to be renewed or continued beyond that day. To use Lord Tenterden's phrase, it will only be continued if "all shall please."

In *McGrady* the facts were similar to those of the present case. In 1976 the local authority had let premises to Mr and Mrs McGrady on a joint weekly tenancy. In 1980 the parties separated. On May 11 1981 Mrs McGrady gave the authority a notice to quit taking effect on June 15, one of the periodic tenancy days. Mr McGrady refused to
B leave and proceedings were brought in the county court. So far as the report shows, the tenancy agreement contained no relevant provisions relating to notices to quit. Sir John Donaldson MR, in a short judgment with which May LJ agreed, described the issue for decision as being

> whether where there is a letting to joint tenants the act of one of the joint tenants in giving notice to quit both brings the tenancy to an end and deprives the other joint tenant of the protection of the Housing Act 1980.

He said that the law could "for practical purposes" be taken from the decision of this court in *Leek & Moorlands*. He said that it was "quite clear" that in that case this court was approving the decision in *Doe d Aslin*. Having referred to Somervell LJ's description of the ratio of the latter decision and quoted the passage from his judgment in which
C he had again adverted to *Doe d Aslin* (at p 795 of the report), the Master of the Rolls continued:

> In my judgment it is clear law that if there is to be a surrender of a joint tenancy, that is a surrender before its natural termination, then all must agree to the surrender.
>
> If there is to be a renewal, which is the position at the end of each period of a periodic tenancy, then again all must concur. In this case Mrs McGrady made it quite clear by her notice to quit that she was not content to renew the joint tenancy on and after June 15 1981. That left Mr McGrady without any tenancy at all.

Thus, in *McGrady* the principle that a notice to quit given by one of several joint lessees under a periodic tenancy brings the tenancy to an end, which had evolved from *Doe d Aslin* and *Leek & Moorlands*,
D emerged as the ratio of the decision of this court.

This principle has been followed and applied in at least three subsequent cases, namely by Mr Donald Rattee QC, sitting as a deputy judge of the Chancery Division, in *Parsons* v *Parsons* [1983] 1 WLR 1390; by this court in *Annen* v *Rattee* (1984) 273 EG 503 (a licence case) and by this court in *Opoku-Forfieh* v *Haringey London Borough Council*, October 17 1988 (unreported). In the latter case, which concerned a notice to quit given by one of two joint tenants under a periodic tenancy, Ralph Gibson LJ, giving the leading judgment, said in terms that *McGrady* constituted authority binding the Court of Appeal and that if the law were to be changed this could require legislative intervention.

E On the face of it, the decision in *McGrady* is quite clear authority binding this court to hold that under the general law, in the absence of provision to the contrary in a tenancy agreement, a notice to quit of proper length, given by one only of two joint lessees without the authority of the other, has the effect of bringing a periodic tenancy to an end.

Counsel appearing for the respective occupiers in the two appeals now before the court have, however, submitted that *McGrady* should not be followed in effect on two grounds, namely:

(a) because it is inconsistent with *Howson* v *Buxton* (*supra*), a decision which, it is said, is binding on this court and should be preferred to *McGrady*; and/or

(b) because it was a decision given *per incuriam*, both in ignorance of
F or failure to advert to both the provisions of the Law of Property Act 1925 relating to joint tenancies and the true nature of a periodic tenancy as illustrated by cases such as *Gray* v *Spyer*.

I will begin by considering the first of these two grounds, which was favoured by the learned judge. This will necessitate a detailed and rather tedious examination of the decision of this court in *Howson* v *Buxton*.

The somewhat complicated nature of the facts of *Howson* v *Buxton* were as follows. By an agreement of September 29 1917 a farm was let to Mr Walker and his nephew, Mr Buxton, for a term of one year from March 25 1918 and so on from year to year determinable at the first or any other year by either party giving to the other six months' written notice to quit. Mr Walker had been sole tenant of the farm for 27 years prior to March 25 1918. As part of the G
arrangements under which the joint tenancy was granted to uncle and nephew on the termination of Mr Walker's sole tenancy, Mr Buxton paid him a sum of money in respect of tenant-right compensation and provided the whole of the finance for carrying on the farm. Mr Walker was made a joint tenant with Mr Buxton at the request of and for the protection of the landlord. Mr Buxton resided at the farmhouse on the farm and Mr Walker resided in a cottage belonging to the landlord not on the demised premises. In the words of Scrutton LJ (at p 752):

> Obviously therefore, the person interested in the tenancy was Buxton, and Walker was there as a surety for the protection of the landlord.

Some years later, on February 21 1924, Mr Walker alone gave H
notice to quit and deliver up possession of the farm on March 25 1925. The statement of facts in the report of the case states (at p 750):

> This notice *which was admitted to be bad as a notice to quit* was given on the alleged ground that the farm was being badly cultivated.

(Emphasis added.)

By a notice of March 21 1924, the landlord's agent served on both tenants a notice to quit the farm on March 25 1925. It was stated that this notice was given because at its date the tenants were not cultivating the farm according to the rules of good husbandry. The two tenants duly quitted the farm in pursuance of the notice.

Section 12(1) of the Agricultural Holdings Act 1923, so far as material, provided: J

> Where the tenancy of a holding terminates by reason of a notice to quit given by the landlord, and in consequence of such notice the tenant quits the holding, then, unless the tenant —
>
> (a) was not at the date of the notice cultivating the holding according to the rules of good husbandry . . . and unless the notice to quit states that it is given for one or more of the reasons aforesaid, compensation for the disturbance shall be payable by the landlord to the tenant in accordance with the provisions of this section.

Section 12(7), so far as material, provided:

> Compensation shall not be payable under this section . . .
>
> (b) unless the tenant has, not less than one month before the termination of the tenancy, given notice in writing to the landlord of his intention to make a claim for compensation under this section. K

An arbitrator having found that the farm was *not* cultivated badly, on January 3 1925 a notice claiming compensation for disturbance under section 12 of the 1923 Act was served by Mr Buxton on behalf of himself and Mr Walker upon the landlord's agent. In due course the arbitrator stated a special case for the opinion of the county court on two questions, which were in substance (1) whether the notice of January 3 1925 claiming compensation for disturbance was valid: and (2) whether the notice to quit of February 21 1924 estopped Mr Buxton from making that claim.

The Court of Appeal upheld the decision of the county court judge, who had answered the first question in the affirmative and the second in the negative. In the course of summarising the facts, L
Scrutton LJ, who delivered the leading judgment, referred to Mr Walker's notice to quit of February 21 1924 and made this observation in passing (at p 752):

> I personally take the view that one joint tenant cannot give a notice to terminate the tenancy unless he does so with the authority of the other joint tenant, and Walker had not the authority of Buxton to terminate the tenancy.

At the end of his summary of the facts, Scrutton LJ said (at p 752):

> Thereupon Buxton alone gave a notice of his intention to claim compensation, and the question which has been argued by the landlord throughout, and has now been argued before us, is whether one joint tenant, who is the only person who has in fact suffered damage, can give a notice claiming arbitration and compensation under the Agricultural Holdings Act 1923, without the concurrence of the other joint tenant, the other joint tenant M
> having suffered no damage.
>
> That turns on the construction of the Agricultural Holdings Act 1923.

On this first issue, Scrutton LJ concluded (at p 754) that having regard to the mischief which the 1923 Act intended to remedy, the notice of January 3 1925 was on the true construction of section 12(7) a valid notice. He then turned to the second issue, which (at p 754) he described as raising the point that "when Mr Walker gave to the landlord notice to quit because the farm was being cultivated badly, the other joint tenant was in some way estopped from proving that the farm was not cultivated badly". In rejecting this argument, he said (at p 754):

> . . . it appears to me that the statement made by one joint tenant, though it

A may possibly be evidence — in this case apparently wrong evidence — certainly is not an estoppel of the other joint tenant in proving, as he did before the arbitrator, that the farm had been properly cultivated.

Greer LJ decided the first issue in favour of Mr Buxton, but on a different ground, namely that Mr Buxton, in serving the notice of January 3 1925, "was acting, and was authorised to act, and did act on behalf of the tenants, who were himself and Walker" (see p 754). Having dealt with this first issue, he dealt with the second issue by simply saying (at p 755): "On the other point, which seems to me to be conclusive, I agree".

Sankey LJ began his judgment by referring to the two questions which fell to be decided by the court. Having accurately identified the first question, he said (at p 755):

B The other question is a question whether the notice to quit, dated February 21 1924, given by Mr Walker, was a valid notice, and so estopped Mr Buxton from claiming compensation for unreasonable disturbance. On that question I desire to add nothing; I agree with what has been said with regard to it. My remarks are in respect of the notice given by Mr Buxton only.

Sankey LJ ultimately decided (at p 757) that for the reasons given by Scrutton LJ the proper construction of the section was the one contended for by Mr Buxton and that the notice given by him alone was a valid one.

I think it clear that Sankey LJ's description of "the other question" referred to by him was not quite accurate. If the notice to quit given by Mr Walker had been a valid notice, no questions of C estoppel would have arisen. On this footing, no possible right to compensation under section 12 of the 1923 Act could have arisen because the tenancy would not have terminated on March 25 1925 by reason of a notice to quit *given by the landlord* within the meaning of section 12(1); it would have terminated by reason of Mr Walker's notice. Questions of estoppel arose only because, as was stated in the statement of facts (at p 750), the notice to quit given by Mr Walker on February 21 1924 "was admitted to be a bad notice", but the landlord was contending that the notice nevertheless estopped Mr Buxton from proving that the farm was not cultivated badly.

The admission that that notice was a bad notice must have been an admission by the appellant landlord. No doubt, in the light of that admission, the summary of the argument of counsel for the landlord, D at p 751 of the report, shows that he was not seeking to argue that the notice to quit given by Mr Walker on February 21 1924 had been sufficient to determine the tenancy or that his claim against the landlord failed for that reason. The validity of that notice to quit was not an issue in the case because its invalidity was conceded.

In these circumstances it is, in my judgment, plain that the crucial sentence in Scrutton LJ's judgment, on which Judge Cooke placed so much reliance, introduced by the words "I personally take the view . . .", was a mere *obiter dictum*. And indeed the very form of that introduction to my mind constitutes an implicit recognition by Scrutton LJ that this was so. Even if Sankey LJ, in the passage from his judgment which I have already quoted, must be read as E concurring with this sentence from Scrutton LJ's judgment, Sankey LJ's opinion on this point was likewise expressed *obiter*.

In dealing with the decision in *Howson* v *Buxton*, Judge Cooke said:

There were two joint lessees, one of whom gave a tenant's notice to quit and the other did not. If that notice had been effective to determine the letting no compensation would have been payable and the claim would have failed in limine. It is therefore central to the decision of the court to decide whether the tenant's notice to quit was validly given.

On this approach to the issues in the case, he regarded the sentence in Scrutton LJ's judgment at p 752 beginning with the words "I personally take the view . . ." and the two sentences in Sankey LJ's judgment at p 755 beginning with the words "The other question F . . ." (incorrectly transcribed in Judge Cooke's judgment as "the only question") as together constituting binding authority for the proposition that one of two joint tenants under a periodic tenancy cannot without the authority of the other give a notice determining a joint tenancy.

For the reasons given, I respectfully disagree with this conclusion. They do not, in my view, constitute any such binding authority. *Howson* v *Buxton* is not a decision which entitled the learned judge, or entitles this court, to decline to follow *McGrady*.

I now turn to consider whether the *McGrady* decision was given *per incuriam*. We have had the benefit of very full and detailed arguments to this effect addressed to us by counsel for the tenants in both appeals. These arguments were in part based on the submission that *Howson* v *Buxton* was not cited in argument to the Court of G Appeal in *Leek & Moorlands* or in *McGrady* and was binding on the court in both those cases. We cannot tell from the report of *McGrady* whether *Howson* v *Buxton* was cited in argument. However, since the relevant dicta of the court in the latter case were *obiter*, a failure to cite it in argument could not itself have rendered the *McGrady* decision *per incuriam*.

The rest of the submissions of counsel for the tenants in this context have been largely based on some further cogent criticisms of the *McGrady* decision contained in a long and learned article by Mr Frank Webb in [1983] 47 Conv 194-210. I hope that I will do no injustice to the points made if I summarise their principal features fairly briefly.

H Sir John Donaldson MR in *McGrady* did not think it necessary to examine the case law in detail because he thought that the law could for practical purposes be taken from the decision of this court in *Leek & Moorlands*. However, in that case the nature of a periodic tenancy was not in issue and the relevant dicta of Somervell LJ were no more than *obiter dicta*. Furthermore, they were, it was said, based on two fundamental misconceptions, first as to what was decided in the 19th-century cases, and second as to the nature of a periodic tenancy. The 19th-century cases and in particular *Doe d Aslin*, it was submitted, are based on reasoning which, even if correct in law at the time, is not applicable following the enactment of the Law of Property Act 1925 which, by sections 34 and 36, now makes it J impossible for undivided shares to exist in a legal estate in land, such as a lease, and requires that such shares shall subsist merely as equitable interests behind a trust for sale.

Close analysis of *Doe d Aslin*, it was said, shows that its ratio was that as a notice to quit by one joint landlord would be effective to end the tenancy of his severable share, either the tenant or the other landlords could, on the service of such notice, treat the tenancy as totally at an end. It was submitted that the reasoning of that decision was wrong, or at least suspect, even under the pre-1926 law, but certainly could not apply since 1925, when a lease by joint landlords is unquestionably one joint grant and not a set of separate grants. Thus, it was said, the statement in *Doe d Aslin* that "the tenant holds the whole of all so long as he and all shall please", to which Somervell LJ K and subsequently Sir John Donaldson MR attached such importance, is based on reasoning which, if it was ever correct, cannot apply since 1925.

Somervell LJ in *Leek & Moorlands*, in holding that rights held jointly can only be surrendered jointly, distinguished *Doe d Aslin* not on these grounds but on the grounds that that decision had established a principle applicable to notices to quit under a periodic tenancy given by either one joint landlord or one joint tenant, but not to surrenders or break notices. The principle was that service of a notice to quit under a periodic tenancy was equivalent to the failure to exercise an option to renew. As Somervell LJ accepted, the exercise of an option to renew undoubtedly has to be a joint act. A positive act L is required to cause the new tenancy to arise. Similarly, with a surrender or a break notice, a joint positive act is needed to cause the tenancy to terminate. It was submitted, however, that Somervell LJ (whose reasoning was subsequently adopted by the court in *McGrady*) erred in drawing the analogy between a notice to quit a periodic tenancy and an option to renew. The true analogy, it was submitted, is between a notice to quit under a periodic tenancy and a notice taking advantage of a "break" clause. So far as appears from the reports, the line of cases exemplified by *Gray* v *Spyer* was not drawn to the attention of the court either in *Leek & Moorlands* or in *McGrady*. This, it was submitted, led both courts to misinterpret the nature of a periodic tenancy and to overlook the fact that a periodic tenancy continues indefinitely from the time of the original grant M unless and until it is determined by a notice to quit and that in the eyes of the law there is not a notional reletting at the commencement of each new period.

I was impressed by at least some of these submissions, in particular the submission that the true analogy is between a notice to quit and a notice taking advantage of a break clause. It is common ground that under the general law a notice taking advantage of a "break" clause during the currency of the fixed term of a lease has to be given by both tenants if it is to be effective. In these circumstances it is not perhaps at first sight entirely easy to rationalise a rule which in contrast provides that a notice to quit, given during the currency of a periodic tenancy by one lessee only, will suffice to determine the tenancy (which, in default of the notice, would continue indefinitely until

A properly determined by a subsequent notice). Such a rule will enable one of the two joint lessees, who by virtue of the Law of Property Act 1925 notionally hold the lease as trustees, effectively to destroy this trust asset without the consent of the other. No one doubts that an ordinary *disposition* of this asset (eg by sale) would require the consent of both trustees.

Nevertheless, there are strict limits to the circumstances in which a decision of this court may be held to have been given *per incuriam*. In *Rickards* v *Rickards* [1989] 3 WLR 748 at p 755 Lord Donaldson of Lymington MR (in a statement cited more recently by Russell LJ in *Rakhit* v *Carty* [1990] 2 WLR 1107* at p 1114) said:

> The importance of the rule of stare decisis in relation to the Court of Appeal's
> B own decisions can hardly be overstated. We now sometimes sit in eight divisions and, in the absence of such a rule, the law would quickly become wholly uncertain. However the rule is not without exceptions, albeit very limited. These exceptions were considered in *Young* v *Bristol Aeroplane Co Ltd* [1944] KB 718; *Morelle Ltd* v *Wakeling* [1955] 2 QB 379 and, more recently, in *Williams* v *Fawcett* [1986] QB 604, relevant extracts from the two earlier decisions being set out at pp 615-616 of the report. These decisions show that this court is justified in refusing to follow one of its own previous decisions not only where that decision is given in ignorance or forgetfulness of some inconsistent statutory provision or some authority binding upon it, but also, in rare and exceptional cases, if it is satisfied that the decision involved a manifest slip or error. In previous cases the judges of this court have always refrained from defining this exceptional category and I have no intention of departing from that approach save to echo the words of Lord Greene MR in
> C *Young's* case, p 729, and Sir Raymond Evershed MR in *Morelle's* case, p 406, and to say that they will be of the rarest occurrence.

In *Duke* v *Reliance Systems Ltd* [1988] QB 108 the Master of the Rolls, with whose judgment Ralph Gibson LJ and Bingham LJ agreed, said this (at p 113):

> I have always understood that the doctrine of *per incuriam* only applies where another division of this court has reached a decision in the absence of knowledge of a decision binding upon it or a statute, and that in either case it has to be shown that, had the court had this material, it *must* have reached a contrary decision. That is *per incuriam*. I do not understand the doctrine to extend to a case where, if different arguments had been placed before it or if different material had been placed before it, it *might* have reached a different conclusion.

D As I have already indicated, we do not know what authorities were cited to this court in *McGrady* or what arguments were addressed to it. However, whatever course those arguments took, I am certainly not prepared to assume that the members of the court did not have in mind either the relevant effect of the Law of Property Act 1925 or that the relevant dicta of Somervell LJ in *Leek & Moorlands* were *obiter*. The court in *McGrady* was, in my judgment, faced with a general question of law as to which there was no binding authority, so that it was free to decide this question either way. That question concerned the relationship of lessor on the one side and joint lessees on the other under a periodic tenancy. Though I have indicated certain considerations which might point to a different conclusion, it was, in my opinion, open to the court to take the view that if, by way
E of a joint enterprise, two or more persons enter into a tenancy agreement as joint lessees, it is inherent in the nature of the arrangement that, unless the contrary is expressly agreed, the arrangement will continue only so long as both lessees wish it to do so; and that correspondingly *vis-à-vis* the landlord either lessee will have the right to determine the tenancy on proper notice, even though his co-lessee might conceivably have a claim against him for breach of trust in the event of such unilateral notice being given — see *Parsons* v *Parsons* [1983] 1 WLR 1390 at p 1400B-D; and that the landlord will correspondingly have the right to accept such notice. It should be observed that acceptance of the contrary view would mean that one joint lessee under a periodic tenancy, even as short as a weekly
F tenancy, could find himself or herself exposed to obligations to pay rent and to perform the other tenant's obligations for an indefinite period of time unless he or she could persuade the other lessee to join in serving a notice to quit.

There is much to be said on both sides, but I am not persuaded that the members of the court in *McGrady* necessarily would have come to a different decision, even if all the authorities cited and all the arguments addressed to this court had been drawn to their attention, though I accept that this is a possibility.

Accordingly, in my judgment, *McGrady* cannot be regarded as a decision given *per incuriam*. As from its delivery on November 22 1982 it constituted binding authority for the proposition that under G
the general law, in the absence of provision to the contrary in a tenancy agreement, a notice to quit given by one only of joint lessees under a periodic tenancy without the authority of the other will have the effect of bringing the tenancy to an end, whatever personal rights or remedies such other lessee may possess against the lessee serving the notice.

In the present case, in my judgment, the parties to the tenancy agreement entered into in the summer of 1987 must be deemed to have done so in the knowledge that this was the state of the general law as established by Court of Appeal authority. The only provision of this agreement which might arguably be said to have the effect of negativing the general law is clause 3(t) quoted above. It is submitted H
on behalf of Mr Monk that this subclause, coupled with the statement on the front page of the agreement that "if you are joint tenants the term 'tenant' refers to both tenants or all of you" negatived the right of either tenant acting on his or her own to serve a notice to quit and correspondingly the right of the lessor council to accept such a notice.

The point of construction is a short one but I cannot, for my part, accept this submission. In my judgment, clause 3(t), like the other provisions of clause 3, was simply directed to specifying *obligations* of the two tenants *vis-à-vis* the lessor council. It was not intended to confer rights on the tenants *vis-à-vis* the council — they had their ordinary rights to determine the tenancy under the general law — or to regulate their rights *inter se*. It was merely intended to make it clear that the council was entitled to expect that any notice served on them J
purporting to terminate the tenancy should be of at least four weeks' duration and should expire on a Monday. In my judgment, clause 3(t) cannot in its context properly be read as making it obligatory that both tenants should serve any such notice — it was not directed to the question who should serve the notice — or as negativing the position under the general law which had been clearly established by *McGrady*. The decision in *Re Viola's Indenture of Lease* [1909] 1 Ch 244, which concerned the construction of a "break" clause in a lease which entitled the "lessees" to determine it by notice, though relied on by the learned judge, does not, in my opinion, assist Mr Monk's argument on the construction of the particular tenancy agreement now under consideration.

For the reasons stated, I would allow this appeal. I would set aside K
the judge's order and make an order granting the council possession of the flat.

BARNET LONDON BOROUGH COUNCIL v SMITH

In the second of the two appeals before the court Mr James Martin Smith appeals from an order made on June 5 1990 in Barnet County Court by Mr A A Goymer, sitting as an assistant recorder, whereby he dismissed Mr Smith's appeal against an order of the registrar dated January 31 1990 refusing to set aside an order for possession of a council house known as 119 Alexandra Road, Muswell Hill, London N10 ("the house").

The facts of the case were as follows. On January 21 1985, Mr L
Smith and his wife, Mrs Tina Glenys Smith, entered into a tenancy agreement with the Barnet London Borough Council (the respondents to this appeal and in this judgment called "Barnet Council") under which they agreed to take a weekly tenancy of the house from Barnet Council with January 28 1985 as its commencement date. The agreement was in a standard printed form of tenancy agreement, with the names of the tenants and of the premises and the tenancy commencement date inserted in manuscript. After naming the parties and identifying the demised premises, it said:

> If two or more people are included in the expression "the Tenant" it refers to both or all of them. M

There then followed section A headed "Letting of the Property", section B headed "Obligations of the Council" and section C headed "Obligations of the Tenant". The agreement used the expression "tenant" in the singular throughout. Clause 5 of section C provided for the tenant to pay the rent weekly in advance. Clause 18 of section C provided:

> The Tenant may end the tenancy on a Monday in any week by giving the Council four weeks' previous written notice.

On September 11 1989 Mrs Smith gave Barnet Council a notice in the following form:

> I hereby give notice to terminate the tenancy of 119 Alexandra Road, Muswell Hill, London N10 giving 28 days notice as from September 11 1989.

---

*Editor's note: Also reported at [1990] 2 EGLR 95.

A Mr Smith, who had not authorised the service of this notice and had no knowledge of it, remained in the house.

On November 14 1989 Barnet Council issued proceedings seeking possession of the house as against Mr and Mrs Smith, relying on Mrs Smith's notice to quit. On January 31 1990 the registrar ordered that possession should be given on February 28 1990. Mr Smith appealed from that order. By his order of June 5 1990 the assistant recorder dismissed that appeal, but further suspended the order for possession (which had already been suspended) pending the hearing of the appeal by this court.

In his succinct judgment, the assistant recorder accepted Barnet Council's submission that he was bound by *McGrady* to hold that the notice to quit was valid. He rejected an argument advanced on behalf
B of Mr Smith by reference to Mr Webb's article in *The Conveyancer*, referred to above, to the effect that *McGrady* had been decided *per incuriam*. It appears that it was not argued that the particular terms of the tenancy agreement on their true construction required that a notice terminating the tenancy had to be signed by both tenants if it was to be effective. Though this point was not taken in the court below and not taken explicitly in Mr Smith's notice of appeal, we have allowed it to be taken in this court.

For the reasons which I have stated in regard to the other appeal, the submission that the decision in *McGrady* was given *per incuriam* is, in my judgment, ill-founded. We are therefore left with the question of construction.

C I think that the arguments available to Mr Smith on this question are a little stronger than those which were available to Mr Monk. In the instant case, the relevant clause (clause 18) does appear in form to be drafted so as to confer a right on "the tenant" to end the tenancy on a Monday in any week by giving the council four weeks' previous notice; and the opening provisions of the agreement state that "if two or more people are included in the expression 'the tenant' it refers to both or all of them." On the face of it the case is thus rather closer to *Re Viola's Indenture of Lease*. Nevertheless, once again the parties must be deemed to have entered into the tenancy agreement in the knowledge that under the general law, in view of *McGrady*, either tenant would have the right to serve a notice to quit without the consent of the other. In my judgment, clause 18 cannot be read as
D evincing a sufficiently clear intention of the parties to depart from this position. Though, as I have said, the clause is in form drafted so as to confer a right on "the tenant", the very nature of the tenancy as a weekly tenancy meant that it could be determined by either side on proper notice. Clause 18 is included in a section of the agreement headed "Obligations of Tenant" and forms part of a series of clauses (5 to 20) which impose duty obligations on "the tenant". Though I see the force of the argument to the contrary, in my judgment clause 18 in its context should be read in the same sense as clause 3(t) of the tenancy agreement in the other appeal, namely as conferring no rights on the tenants beyond those conferred by the general law and doing no more than making it clear that the lessor was entitled to
E expect that any notice served on it purporting to terminate the tenancy should be of at least four weeks' duration and should expire on a Monday. Clause 18 cannot, I think, be properly read in its context as making it obligatory that both tenants should serve any such notice if the tenancy was to be effectively determined; it was not directed to the question who should serve the notice or to negativing the position under the general law which had been established more than two years before by the decision of this court in *McGrady*.

For these reasons, I think that the assistant recorder reached the right conclusion and I would dismiss this appeal.

Agreeing, **Nicholls LJ** said: The framework within which these appeals fall to be decided comprises two legal principles, neither of
F which is in dispute. The first principle is that, at common law, when a lease expires or is determined any renewal requires the concurrence of all concerned. If, as frequently occurs, the lessor comprises two or more persons, both or all of the joint lessors must concur. Similarly if the lessee comprises two or more persons. If one of the joint lessors, or one of the joint lessees, refuses to agree, that is, in principle, fatal to the renewal of the lease at common law.

The second relevant principle is this. Leases commonly contain a break clause, giving one party or the other an option, or right, to end the lease before it has run its full term. When a break option is granted to joint lessors or to joint lessees, *prima facie* the exercise of the option requires the concurrence of all the joint lessors or joint lessees as the case may be. The principle is that, unless otherwise agreed expressly or by necessary implication, the exercise of a right G
conferred by a lease on persons jointly requires the positive concurrence of all those to whom the right was jointly given.

The primary question raised by these appeals is how these two principles fall to be applied to a notice to quit given by joint lessors or joint lessees of a periodic tenancy. At first sight the comparison between such a notice to quit and a break clause is compelling. A periodic tenancy is not a succession of relettings, period by period. A periodic tenancy is a single letting which continues until it is terminated (*Gandy* v *Jubber* (1864) 5 B&S 78). A notice to quit determines a tenancy which otherwise would have continued, in the same way as a notice given pursuant to a break clause brings to an end a lease for a term of years.

There is much to be said for this approach. But for over 150 years H
the courts have adopted a different approach, not without its own rational basis. The alternative approach is founded on a particular view of the effect of a periodic tenancy when granted by persons holding their estates as joint tenants. Although a periodic tenancy is one entire letting of indefinite duration, it is a letting which continues, from period to period, only so long as both parties shall please. The view which has been adopted by the courts is that this means that, where there are joint lessors or joint lessees holding as joint tenants *all* must concur in the continuation of a periodic tenancy. Such concurrence is assumed unless due notice of dissent is given by any person whose concurrence would have been required for a renewal. If due notice of dissent (namely a notice to quit) is given by J
any one such person, the tenancy does not continue into the next period. Thus if there are joint lessors, notice of dissent by any of them precludes the continuation of the tenancy. Likewise if there are joint lessees. Only if there is no dissent will the tenancy go on. Thus service of a notice to quit in such cases has been equated to refusal to agree to a renewal.

Before turning to the authorities I should make two introductory points. First, to understand the old authorities it is necessary to keep in mind that before 1926 undivided shares in land existed at law as well as in equity. Joint lessors could hold either as tenants in common or as joint tenants. It is important to have this in mind because a distinction was drawn between the legal effect of a lease granted by tenants in common, each owning an individed share in the land, on K
the one hand and a lease granted by joint lessors holding under a joint tenancy on the other hand. Today, of course, joint lessors can hold the legal estate only as joint tenants. Thus it is with the law relating to joint lessors and joint lessees who hold their estates as joint tenants that these appeals are concerned.

The second preliminary observation is that, to avoid confusion, throughout this judgment I shall use the expressions "joint lessors" and "joint lessees". I shall use the phrase "joint tenants" only when referring to the estate held by co-owners and it is necessary to distinguish between a joint tenancy and a tenancy in common.

Doe d Aslin v *Summersett*

The relevant authorities are few in number. I can start in 1830. In L
Doe d Aslin v *Summersett* (1830) 1 B & Ad 135 the court of King's Bench reviewed the earlier authorities. The question before the court concerned the effectiveness of a notice to quit signed by one of two joint lessors. Lord Tenterden CJ delivered the reserved judgment of the court. Central to his exposition of why a notice to quit given by one of two joint lessors holding as joint tenants did not operate merely to determine the tenancy so far as his share was concerned was Lord Tenterden's appraisal of "the true character" of a tenancy from year to year in such cases. He stated (at p 140) that "the true character" of a tenancy from year to year granted by joint lessors is that "the tenant holds the whole of all so long as he and all shall please, and as soon as any one of the joint-tenants gives a notice to M
quit he effectually puts an end to that tenancy". The material passage reads:

> . . . But, though upon a joint lease by joint-tenants each demises his own share, this is not the only operation of such a lease. Joint-tenants are seised not only of their respective shares, per my, but also of the entirety, per tout; Litt s 288 . . . Upon a joint demise by joint-tenants upon a tenancy from year to year, the true character of the tenancy is this, not that the tenant holds of each the share of each so long as he and each shall please, but that he holds the *whole* of *all* so long as he *and all* shall please; and as soon as any one of the joint-tenants gives a notice to quit, he effectually puts an end to *that* tenancy . . .

Evaluation of this decision is not easy for modern lawyers because the reasoning is based on some very ancient technicalities of the law

A of joint tenants. But the *Summersett* case is a clear and considered decision that, unless otherwise agreed, service of a notice to quit by one of several joint lessors whose legal estate is that of joint tenants operates to determine the whole lease. The case was so understood at the time. In *Doe d Kindersley* v *Hughes* (1840) 7 M & W 139 one of the issues concerned the validity of a notice to quit given by an agent authorised by one of three joint lessors. Parke B said (at p 141):

> Here the party was previously authorized by one of the trustees: That is the same as if that trustee had given the notice: and *Doe d Aslin v Summersett* (1 B & Ad 135), is an express authority that a notice to quit by one of several joint-tenants, purporting to be given on behalf of them all, is good for all, because the tenant holds the premises only so long as he and they all shall agree.

B In that case the joint lessors were not beneficial co-owners: the property was held by them as trustees for the benefit of a Miss Shipley. In 1842, in *Alford* v *Vickery* (1842) Car & M 280, Coleridge J also applied the principle expounded by Lord Tenterden in *Doe* v *Summersett* in the case of a notice to quit given by three of four joint lessors who held land as joint tenants as trustees of a local charity.

In *Platt on Leases*, published in 1847, the law regarding joint lessors was summarised thus: tenants in common are considered as holding separate interests, the only unity between them being that of possession. If all of several tenants in common concur in one lease, the lease will operate as "the distinct demise of each tenant of his part, and not as the joint demise of all" (p 131). By way of contrast is the anomalous position of joint lessors whose estate is that of joint C tenants (p 124):

> Joint-tenants, holding (according to the technical phrase) per my et per tout, enjoy a singular and anomalous species of tenure. During their joint ownership they constitute but one tenant of the whole land, and are then said to be seised per tout; but, for the purposes of alienation, each has an exclusive right to and dominion over a moiety; and in this sense, adopting Lord Coke's exposition, which is more simple and intelligible than Littleton's, or Blackstone's, must be understood the expression that joint-tenants are seised or possessed per my.

The *Summersett* case and the decision of Mansfield CJ in *Doe d Whayman* v *Chaplin* (1810) 3 Taunt 120 are then cited as authority for the following propositions (at p 126):

> Though the title as well as the estate of joint-tenants is undivided, yet each has D his own particular portion. Their joint demise operates to a certain extent as a demise by each of his share; and, therefore, one of several joint-tenants who had joined in a demise from year to year, on giving due notice to quit, may recover his several share in ejectment on his several demise, without the concurrence of his co-tenants. But the true character of a tenancy from year to year under a joint demise by joint-tenants is not that the tenant holds of each the share of each so long as he and each shall please, but that he holds the whole of all so long as he and all shall please. It is competent, therefore, to either of the lessors to put an end to the tenancy as to the whole; and an ejectment may be maintained on the joint demise of all the lessors, though the notice to quit be given by one of them only.

So far as the reported decisions go, that view of the law seems never thereafter to have been challenged in any case, apart from one dictum E by Scrutton LJ in *Howson* v *Buxton* (1928) 97 LJKB 749. I agree with Slade LJ's observations on that case. Throughout the 19th century and subsequently the *Summersett* principle, if I can so describe it, was set forth as representing the law, even if on occasion this was accompanied by a note of caution. For instance, in an early edition of *Woodfall on Landlord and Tenant* (2nd ed 1834), the editor (at p 236) stated the *Summersett* principle, adding "but a notice to quit by joint tenants had better be signed by all the joint tenants". Likewise in the first edition (1911) of *Halsbury's Laws of England*, vol 18, p 452, the proposition in the text was that "where the demise is by joint tenants, one may give notice on behalf of all, but this doctrine is confined to a common law notice to quit", but a footnote was appended: "It is not, perhaps, finally settled that it is unnecessary to F prove the authority of the other joint tenants."

The law having stood thus for 160 years, in a field where much importance is to be attached to the existence of a clear rule, for my part I think this court should be very slow indeed to venture upon a reappraisal of the *Summersett* principle.

On these two appeals the defendant tenants, being the tenants who did not concur in giving notice to quit, submitted that, whatever may have been the true view of the law before 1926, the *Summersett* principle did not survive the 1925 property legislation. Tenancies in common can no longer exist as legal estates in land, and a joint tenancy of a legal estate cannot be severed so as to create a legal tenancy in common: see sections 34 and 36 of the Law of Property Act 1925. Thus, so the argument runs, the mischief which caused Lord Tenterden to hold that a notice to quit served by one joint lessor G ended the whole lease no longer exists. Unlike in 1830, a notice served by one joint lessor would not operate now to terminate the lease as to the share of the joint lessor who gave the notice.

I am not greatly impressed with this argument. By 1925 the *Summersett* principle had been in place and, no doubt, in operation unchallenged for nearly a century. The underlying rationale of the principle, based on the true character of periodic tenancy granted by joint tenants, was unaffected by the 1925 property legislation. I recognise that after 1925, and leaving aside settled land, joint lessors and joint lessees will invariably hold the reversion or lease in question as trustees for sale, and that, in general, action by trustees requires unanimity. But, as already noted, the *Summersett* principle was applied by the court before 1926 to joint lessors who were trustees as H well as to joint lessors who were beneficial co-owners. Trust law was not regarded as impinging on the validity, as between landlord and tenant, of a notice to quit not given by all the joint lessors. The explanation for this may be that, as all must concur if a periodic tenancy is to continue, one of several trustees can withhold his consent from a continuation just as much as he can withhold his consent from any other proposed trust transaction. Service of a notice to quit is the means by which a trustee communicates his refusal to agree. Furthermore, so far as I am aware, no suggestion was ever made by anyone that the 1925 property legislation might have had the effect now suggested before Mr Frank Webb's valuable article appeared in [1983] 47 Conv 194. J

Per incuriam

I have considered these criticisms of the *Summersett* principle because of the argument advanced that decisions of this court, on the very point which arises on these appeals, are not binding because they were decided *per incuriam*. I am unable to accept this argument. In *Greenwich London Borough Council* v *McGrady* (1982) 81 LGR 288, Sir John Donaldson MR applied the reasoning of Somervell LJ, delivering the judgment of this court, in *Leek & Moorlands Building Society* v *Clark* [1952] 2 QB 788. I do not think there can be any question of characterising Somervell LJ's approval of the *Summersett* principle, or the Master of the Rolls' subsequent adoption of Somervell LJ's observations, as a "manifest slip or K error", so that in consequence the *McGrady* case is to be regarded as having been decided *per incuriam* and therefore not binding on this court: see *Rickards* v *Rickards* [1989] 3 WLR 748, at p 755. I do not accept the criticism that the court in the *Leek* and *McGrady* cases misunderstood the continuing nature of a periodic tenancy. It is clear that in the *Leek* case the court accepted Lord Tenterden's analysis of the "true character" of a periodic tenancy granted by joint tenants. In doing so they adopted one of two different possible analyses of the effect of such a tenancy. That was their view. The fact that, as with many legal points, there is something to be said for another, alternative, view does not mean that their view was a manifest error. Having adopted Lord Tenterden's approach, the court in the *Leek* L case was surely right to consider that the *Summersett* principle was equally applicable to a case of joint lessees. The underlying rationale, that a periodic tenancy granted by joint tenants continues only so long as all shall please, dictates that the same answer shall be given in respect of a notice to quit served by one of several joint lessees as in respect of a notice to quit served by one of several joint lessors. Somervell LJ said (at p 793):

> If one considers a lease to joint lessees for a term certain with a right of renewal, it would be obvious, we think, that both must join in requiring a renewal. A periodic tenancy renews itself unless either side brings it to an end. But if one of two or more joint lessees does not desire it to continue, we would have thought that it was in accordance with Lord Tenterden's principle, and with common sense, that he should be able to make that effective.

Subsequently, in the *McGrady* case, this court adopted that M approach. That is a binding decision on this point.

The two tenancy agreements

Once it is accepted that the *Summersett* principle is applicable today to a notice to quit given by one of two joint lessees unless displaced by contrary agreement, the instant appeals present no difficulty. In both cases there is an express provision in the written tenancy agreements, concerning termination of the tenancies. In each case, as one would expect, the two joint lessees fall within the expression "the tenant". In the *Barnet* case, clause 20 provides that "the tenant may end the tenancy on a Monday in any week by giving the Council four weeks' previous written notice." In the

A Hammersmith and Fulham case, clause 3(t) provides for the tenant "to give four weeks' notice (or such lesser period as the Council has agreed to accept), to expire on a Monday, when giving up the tenancy." I think that in each case the termination clause is directed exclusively at the length of notice required, and the need for the notice to expire on a Monday and to be in writing. I do not think that, in either case, the clause was aimed at regulating, one way or the other, what was to be the position regarding who could give a notice to quit in those cases where "the tenant" comprised two or more persons. On that question the tenancy agreements are silent. The ordinary rule applies.

Some argument was advanced to us on the consequences for couples living together in local authority housing if the contentions B of these two local authorities were upheld. We were told that when one partner, frequently the wife and often with young children, chooses or is compelled to leave a house or flat provided by the local authority, it is the practice of at least some local authorities to require the partner who has left the accommodation to give notice terminating the jointly held tenancy as a prelude to rehousing her. Thus, it was said, the man sometimes finds himself, without any fault on his part, losing his home as well as his wife and his children. He may well not have a priority need for rehousing. Against this it was argued that, with limited housing facilities available to local authorities, it is undesirable that the man should continue indefinitely to occupy accommodation in excess of his reasonable C requirements. Arguments such as these can have no place in the present case. The point at issue applies to all periodic tenancies, to joint lessors as well as joint lessees, and to all types of property: agricultural and business property as much as residential property. Considerations peculiar to tenants of local authority housing really cannot carry weight in the circumstances.

But the *Summersett* principle does have one unsatisfactory consequence of general application. The present cases are good illustrations of this. If one read the two tenancy agreements and applied ordinary canons of construction but without having the *Summersett* principle in mind, one would not understand from the terms of the agreements that an effective notice to quit could be given by one of two joint lessees. It is only when the reader has the D *Summersett* principle in mind that the agreements would be read in that way. Plainly this is not satisfactory. In the present cases the local authorities have sought to set out the parties' respective rights and obligations in a simple written form. But unless he had by his elbow a lawyer skilled in the law of landlord and tenant, or at least a comprehensive textbook on that subject, a tenant who read the form and then signed it would not appreciate that an effective notice to quit could be given by one of two joint lessees acting contrary to the wishes of the other. In the case of local authorities, this is a matter which can be resolved easily by their making suitable amendments to their forms of tenancy agreement or the accompanying literature. Whether the somewhat technical *Summersett* principle itself calls for E reform is not a matter on which this court has sufficient information to express a view. I, too, would dismiss the defendant's appeal in the *Barnet* case and allow the local authority's appeal in the *Hammersmith and Fulham* case.

Bingham LJ also agreed and did not add anything.

The first appeal was allowed and the order below set aside, possession being granted to the council of the ground-floor flat at 35 Niton Street, London SW6. The second appeal was dismissed. Applications for leave to appeal to the House of Lords were refused but provision was made for staying possession orders in the event of petitions to the House. No order was made for costs except for legal aid taxation for the respondent in the first case and the appellant in F *the second case.*

Court of Appeal G

October 25 1990

(Before Lord Justice SLADE, Lord Justice NICHOLLS and Lord Justice FARQUHARSON)

WILLIAM HILL (SOUTHERN) LTD v WALLER AND OTHERS

Landlord and tenant — Original tenants called upon to pay rent after assignment of lease — Action against guarantors to recover such payments — Summary judgment in favour of H tenants — Interlocutory appeal — Whether triable issue — Notices purporting to exercise break clause — Whether notices defective — Whether landlords impliedly represented that they accepted validity of notices — Question as to estoppel by convention — Held, allowing appeal, that there were arguable issues to be tried

The plaintiffs were the original tenants of certain business premises, holding a 25 years' term subject to a break clause on six months' notice — The present litigation arose from the financial difficulties of a company to which the plaintiffs had assigned the lease — That company eventually ceased to trade J — Two of its directors, the first and second defendants in the present proceedings, had entered into covenants with the landlords guaranteeing payment of the rent and performance of the lessees' other obligations under the lease — For a time the guarantors paid the rent due but later ceased to pay — The landlords (who became the third defendants) then resorted to the plaintiffs, as the original tenants, for satisfaction and the plaintiffs paid the rent due for five quarters plus the amount due on an insurance premium — In this action the plaintiffs claimed against the first two defendants, the guarantors, the recovery of these payments and obtained summary judgment from the master under Ord 14 in respect of the claim — This K judgment was upheld by Judge Butler, sitting as a High Court judge — The first two defendants brought the present appeal to the Court of Appeal, asserting that the lease had been duly determined at Michaelmas 1986 by a notice under the break clause in the lease — In view of the way in which the litigation was developing, the plaintiffs added the landlords as the third defendants

There had in fact been two notices purporting to exercise the break clause, but both, as it turned out, were defective for various reasons, one being that they had not been given by the L tenant under the lease but by the guarantors — In reply to this difficulty the first two defendants pleaded that, if the notices were defective, the landlords were estopped from so asserting — The factual basis for this estoppel contention was, as Nicholls LJ said, "slim" — It was founded on disputed evidence of telephone conversations between the second defendant and a representative of the landlords in the course of which, it was alleged, the first two defendants were given grounds for believing that the landlords accepted the notices as valid and effective — This allegation was denied by the third defendants and was clearly one which could only be satisfactorily resolved at a trial — The judge below, however, M had felt able to take a strong line and to dismiss the estoppel point at the interlocutory stage on the ground that the foundation was so weak and vague that a trial judge would feel compelled to reject it — The Court of Appeal differed from the judge in this respect — Despite the sparse factual basis, and some justified criticism of the pleadings, the court did not feel able to hold that there was no arguable defence on the merits — The matter must therefore be allowed to go to trial — The court also rejected a too literal application of the precondition of the break clause, namely that the tenant should have paid the rent and observed and performed all the

A **relevant covenants in the lease before being entitled to invoke the clause — Appeal of first and second defendants allowed**

The following case is referred to in this report.

Lloyds Bank Ltd v *Ellis-Fewster* [1983] 1 WLR 559; [1983] 2 All ER 424, CA

This was an interlocutory appeal by the first and second defendants, Bertram George Waller and Keith Richard William Russell, in the action brought by the plaintiffs, William Hill (Southern) Ltd, against these defendants, as guarantors, and Guardcliffe Properties Ltd, the third defendants, the lessors of the business premises concerned in the action at 189–191 New Kent
B Road, London SE1. The appeal was from a decision of Judge Butler QC, sitting as a judge of the Queen's Bench Division, dismissing an appeal from a judgment of Master Creightmore granting summary relief under Ord 14 to the plaintiffs.

Roger Henderson QC and Simon Buckhaven (instructed by Thomas Boyd White, of Bexleyheath) appeared on behalf of the appellants; Charles Sparrow QC and Stephen Jones (instructed by Titmuss, Sainer & Webb) represented the respondent plaintiffs; Ellis Meyer (instructed by Farmar Miller) represented the third defendants.

Giving the first judgment at the invitation of Slade LJ, **Nicholls LJ** said: This case concerns a claim by the original tenant of a lease to
C recover from subsequent guarantors rent which the original tenant has been called upon by the landlord to pay and has paid. The plaintiff, William Hill (Southern) Ltd, was the original tenant of a lease dated July 22 1977 of business premises, 189-191 New Kent Road in the London Borough of Southwark. The lease was for a term of 25 years from Michaelmas 1976. On February 18 1983 the plaintiff assigned the lease to Pan Litho Sales Ltd, a printing company. At the time of the assignment the first and second defendants, Mr Bertram George Waller and Mr Keith Richard William Russell, who were the directors and majority shareholders of that company, entered into covenants with the then landlord, Elphington Ltd, guaranteeing payment of the rent and the due performance and observance of the tenant's covenants and obligations. On May 1 1984, after an
D intermediate assignment to another landlord, the reversion to the lease was assigned to the third defendant, Guardcliffe Properties Ltd.

In 1984 Pan Litho got into financial difficulty. In March the company stopped trading. Thereafter from time to time the first and second defendants as guarantors paid the rent, which was due quarterly in advance. They paid the rent due up to and including the rent for the quarter March to June 1986. They did not pay the rent which fell due at Midsummer 1986 or make any payments thereafter. The third defendant, the landlord, turned to the plaintiff as the original tenant and called upon it to pay, which it did. The plaintiff paid the rent for five quarters, starting with the payment due in June
E 1986, plus an insurance premium, making a total altogether of £14,021 odd. In this action the plaintiff claims payment of that sum, with interest, from the first two defendants. On January 18 1989, Master Creightmore gave the plaintiff summary judgment on that claim. On July 6 1989 Judge Butler QC, sitting as a judge of the Queen's Bench Division of the High Court, dismissed an appeal by the first two defendants. They have now appealed to this court.

The defence of the first two defendants is that the lease was determined at Michaelmas 1986 by the due exercise of a break clause contained in clause 6 of the lease. They say further that if the notice given was defective, the landlord Guardcliffe Properties Ltd is estopped from so asserting. I pause to observe that this line of defence provides no answer in respect of the rent paid by the plaintiff for the
F quarter from June to September 1986. The first and second defendants accept liability to the plaintiff in respect of that sum plus interest.

At an early stage in this action the first two defendants took the point that the lease had been determined by notice under clause 6. This prompted the plaintiff to add the third defendant as an additional party. Having paid sums as rent pursuant to demands by the third defendant in respect of the four quarters from and after Michaelmas 1986, the plaintiff advanced in the action an alternative claim against the third defendant for repayment of those sums if the contentions of the first two defendants regarding termination of the lease at Michaelmas 1986 were well founded. For its part, the third defendant counterclaimed against the plaintiff for payment of further instalments of unpaid rent, namely the rent falling due on and G
since Michaelmas 1987.

The master gave judgment for the third defendant on the counterclaim in the sum of £8,250 and the plaintiff's appeal to the judge against that order was dismissed.

The break clause is in a common form. It reads:

Provided further and it is hereby further agreed and declared that if the Landlord or the Tenant shall be desirous of determining the term hereby granted at the expiration of the tenth year of the said term and of such desire shall give to the other party not less than six calendar months prior written notice and (in case of the Tenant giving the notice) the Tenant shall up to the date of such determination pay the rents and observe and perform the covenants on its part herein-before reserved and contained then immediately on the expiration of such tenth year the present demise and everything herein H
contained shall cease and be void but without prejudice to the rights and remedies of either party against the other in respect of any antecedent claim or breach of covenant.

Two notices were given in relation to that clause. The first one was dated July 12 1984. It was sent and signed by the second defendant and addressed to the third defendant. It was sent by recorded delivery post. The notice was in the form of a letter written from the second defendant's home address and contained no heading in respect of the Pan Litho Company. The letter read — addressed, as I have said, to the third defendant —

Dear Sir

— and there was then a heading stating the address of the property, J
189-191 New Kent Road SE1 —

I am writing to serve notice that as of 29th September 1986 Mr B G Waller and I will terminate the lease.
This formal notice is in accordance with the terms and conditions of the lease which requires us [as] Guarantors to give six months' prior written notice.
Yours faithfully.

I agree with the judge that this was not an effectual notice exercising the break clause. Clearly the notice was intended by the first two defendants to be an exercise of that clause. But to be valid what was required was a notice in writing given by the tenant. This written notice was not given by Pan Litho. Nor was it given on behalf of Pan Litho. It was given by the first two defendants personally in their capacity as guarantors. That is clear from the second sentence of K
the letter. There is no ambiguity. But the guarantors themselves had no standing in their own right to give a notice under clause 6.

Mr Henderson advanced an argument based on subrogation. Having paid rent as guarantors, the first two defendants became entitled to repayment from Pan Litho. They became entitled to take over the lease, and to exercise all the tenant's rights thereunder, including the right conferred on the tenant by clause 6.

I am unable to accept this. Subrogation in order to achieve reimbursement is one thing. Taking preventive action to stop a potential liability of the debtor, and hence of the guarantors, arising or crystallising in the future is quite another. The argument is that the guarantors themselves were entitled to intervene in the landlord- L
tenant relationship and determine their liability as guarantors by ending the tenant's continuing rights and obligations as tenant, and that they, the guarantors, were entitled to do this without any authority from the tenant. No decision has been cited to us which assists Mr Henderson's submissions on this point. Suffice it to say, I do not think that this subrogation argument affords the first two defendants an arguable defence in this case.

The second notice was dated January 22 1986. That notice also was ineffectual. It suffered from more than one defect. It suffered from the same defect as the earlier notice. But I need not read its terms, because in addition this notice, whatever its terms, could not have been effectual as a notice by "the tenant", ie Pan Litho, because Pan Litho was, by then, in the course of compulsory winding up and the M
first two defendants had no authority in January 1986 to give any such notice on behalf of the company.

Before the master the first two defendants rested their defence solely on the validity of the two notices. Further evidence was produced to the judge. The evidence of the second defendant is to the effect that after the hearing before the master the first two defendants changed their solicitors and took advice from different counsel. The upshot of this was that before the judge the first two defendants set up estoppel as a further ground of defence. What is said is that after the collapse of Pan Litho, and from time to time when the first two defendants were personally paying the rent, the second defendant, Mr Russell, and also the secretary of the company, a Mrs Carol

A Hutley, had telephone conversations with Mr Ferster, a director of the third defendant. The second defendant cannot now recall the name of the man to whom he spoke, but on the evidence this must have been Mr Ferster. The thrust of the second defendant's evidence is that he spoke quite often to this executive. In paras 4 and 5 of his affidavit he says:

> Throughout my entire contacts with this man, a period which spanned 1984–1986, far from being given any intimation that Guardcliffe regarded the Notice which I had served as a nullity, I was given every impression that my notice was valid and acceptable and that our responsibilities as sureties under the licence would cease at the end of September, 1986 in accordance with the option which we had exercised.
>
> Our entire conversations were conducted on that basis and indeed I remember
> B on one occasion the remark being made by this man that soon I would finally be finishing with my responsibilities for the rent.

Then, a little lower down, the second defendant continued:

> Had we any intimation at any stage that our Notice of July 22 1984 was regarded by Guardcliffe as ineffective or a nullity, we would certainly have immediately contacted the Official Receiver and pressed him to exercise the option in good time. Bearing in mind that we had a continuing liability there would have been no sense in not doing so. Because the time was fast approaching our release as sureties, as I thought, I wrote the second Notice dated January 22 1986 as a reminder and confirmation of our earlier Notice. Even after this Notice was served, I continued to have contact with the executive dealing with our case and he in turn continued to allow me to believe that there was absolutely nothing wrong with either of these two notices.

C Mrs Hutley said in para 6, and continuing into para 7 of her affidavit:

> I remember equally distinctly talking to him

— that is Mr Ferster —

> after the service of the July 1984 notice upon Guardcliffe and discussing with him the fact that with the determination of The Lease at the end of September 1986, it was necessary for us to find a new tenant as quickly as possible, because as each month passed, the terms which we were going to be offering would be getting shorter and shorter. I remember telling him that we were using several different estate agents in order to achieve a quick result. I also remember his telling me what a good idea it was. Indeed on one occasion he even asked me if we would put up a "For Sale" sign as Guardcliffe were interested in selling the premises. I told him that we were not interested in a
> D sale as our only concern was in relieving the two sureties from their liabilities for the two-year period. Indeed after some time had elapsed and no tenant had been found, I remember telling him in one of our telephone conversations, that the worry would be his when the two years were up. Subsequently I remember Mr Ferster telephoning me to tell me that he had managed to sell the property. I congratulated him and then went on to discuss how lucky he had been in finding a purchaser in view of the fact that the property would be reverting to him in September 1986.
>
> At no time during any of these conversations did Mr Ferster ever say anything to me or indicate to me or even suggest that he or Guardcliffe did not accept the service of the July notice as valid. Indeed quite to the contrary all our conversations assumed the effectiveness of the notice and the determination of the liabilities of the two sureties at the end of September 1986.

E In his affidavit evidence Mr Ferster firmly disputes this version of events. He did speak to the second defendant on the telephone to remind him quarterly that another payment of rent was due, but he is sure that no conversations as alleged in the relevant parts of the second defendant's evidence took place. Mr Ferster's evidence is that he is not aware of giving the first two defendants any grounds for believing that the notices were valid or were going to be accepted by the third defendant. He also denies the thrust of the assertions in Mrs Hutley's affidavit. He has no recollection of ever discussing the notices with Mrs Hutley.

That dispute, of course, is not one which can be resolved without a trial. The judge nevertheless felt able to reject this estoppel defence, considering that it had all the hallmarks of a number of vague
F assertions having been hastily put together, which a trial judge would feel compelled to reject. He was unable to find in the evidence any clear, unequivocal promise or assurance or statement of fact of the kind necessary to found an estoppel. There was a startling lack of particularity in the draft amended defence.

I recognise that the factual basis on which this defence of estoppel is sought belatedly to be raised is slim. In particular, evidence of what Mr Ferster is alleged to have said is sparse. Despite this, and despite the inadequacies in the draft amended defence to which Mr Sparrow has drawn our attention, I do not feel able to conclude that there is no arguable defence on the merits on this estoppel point. This is a case where I feel bound to reach a different assessment of the affidavit evidence from the judge. Mr Ferster has given no evidence of what was his state of mind regarding the validity or invalidity of the first G
notice when he received it. Mr Meyer, for the third defendant, submitted that the defect in the first notice was obvious, and Mr Meyer was content that for the purposes of this appeal against summary judgment, it should be taken that Mr Ferster appreciated at the time that the notice was bad and that he knew that the first two defendants mistakenly believed that the notice was good. But if that be so, and if that were Mr Ferster's state of mind from 1984 to 1986, then the case comes perilously close to being a case of trickery.

As a matter of strict law Mr Ferster was under no obligation to tell the first two defendants that their notice was bad. Mere silence could not give rise to an estoppel in this case. But this is not a case of mere silence. Mr Ferster telephoned the second defendant quarter after quarter and asked for a further payment of rent, knowing that the H
first two defendants mistakenly believed that, come June 1986, the second defendant and his co-guarantor would be rid of this burden. In such circumstances it would not be surprising if Mr Ferster gave the impression of concurring in and agreeing with the second defendant's mistaken belief and thereby impliedly represented that he accepted that the notice was valid. Depending upon what the second defendant said, what else could Mr Ferster do if he was to avoid alerting the second defendant to the need to serve a fresh notice, this time on behalf of Pan Litho, as to which there would have been no difficulty even after Pan Litho went into liquidation? I am very far from saying or suggesting that Mr Ferster did give any such impression at all, but in my view there is here a matter fit for trial. A J
trial with oral evidence is needed to resolve this type of dispute satisfactorily.

Let me assume, however, that Mr Ferster's state of mind was such that there was no question whatever of any devious conduct on his part. Let me assume that he had not spotted the defect in the notice and that up to the middle of 1986 or so he, too, thought that the notice was valid. Even on this footing it seems to me that there is a triable issue. On this footing there is nothing inherently improbable in the evidence of the second defendant and Mrs Hutley. The payments made to the third defendant by the first two defendants, even though they were already obliged to make those payments, and their abstention from taking steps to serve a fresh notice, seem to me to form an arguable basis for estoppel by convention, as it is now called. K
Here also I am very far from saying or suggesting that the first two defendants will succeed. The case of the first two defendants on this point is a difficult one. But on this Ord 14 application the defendants need show only an arguable defence which needs to be resolved at a trial.

On the estoppel point Mr Sparrow also raised, but later abandoned, a point to the effect that his clients were strangers to the estoppel. So I need say no more about this point.

I turn now to a different point, taken both by Mr Sparrow and by Mr Meyer. Indeed, Mr Meyer put this point at the forefront of his argument. It was his primary, indeed his principal, point. He submitted that if the third defendant failed on every other point, the L
third defendant was still bound to succeed at the trial, because a precondition to the termination of the lease at Michaelmas 1986 was, as provided in clause 6, that "the Tenant shall up to the date of such determination pay the rents and observe and perform the covenants on its part herein-before reserved and contained". But, Mr Meyer pointed out, in September 1986 the tenant was Pan Litho and Pan Litho had not paid the rent. The rent had been paid for the last quarter by the original tenant, namely the plaintiff and, for what it might matter, the plaintiff had not been reimbursed either by Pan Litho or by the guarantors. Mr Sparrow, for his part, submitted that to comply with the precondition Pan Litho should have tendered the rent to the third defendant before September 29, even though the rent had been paid earlier in that month by the plaintiff. M

It is established law in this field that there must be complete compliance with the stipulated precondition before the right, whether to determine the lease or renew the lease, arises. But for my part I am unable to accept that the precondition in clause 6, in the common-form words which I have read, does require any such literal degree of performance. The substance of the condition is that up to the date of termination the rents shall have been paid and the tenant's covenants observed and performed. I cannot regard the particular form of words used in clause 6 as importing a need for the rent to have been paid by the current tenant itself rather than, for example, a guarantor or, still less, that after the rent has been paid by a guarantor or the original tenant, the current tenant should go through the idle,

A and indeed absurd, ceremony of tendering the same rent yet again to the landlord. What matters to the landlord is that there are no subsisting breaches at the material date. I am unable to accept a construction of clause 6 which would have the consequence that if, at any time prior to the date of termination, the payee of the rent was someone other than the tenant for the time being, that would forever preclude the exercise of the right of termination unless the current tenant went through the suggested formality of tendering a sum which is no longer payable. That cannot have been the intended effect of this form of words. That would be formalism and literalism run amok.

I should mention one last point. In July 1984 a receiver of Pan Litho was in office under a debenture granted to a bank. I do not
B think that this furnishes an insuperable obstacle for the first two defendants. There is some evidence, sketchy though it is and unsupported though it is by a statement of the source of Mrs Hutley's belief, that before the July 1984 notice was served the lease had been released from the charge. Some support for this can, perhaps, be derived from the further evidence that steps were taken to put the residue of the lease on the market. Clearly, if there had been such a release, the existence of the receiver in July 1984 in respect of all the other assets of Pan Litho would not have negatived the ability of the board of Pan Litho to give a notice under clause 6 in respect of the lease. For these reasons, for my part, I would allow the appeal of the first and second defendants.

C **FARQUHARSON LJ** agreed and did not add anything.

Also agreeing, **SLADE LJ** said: I only wish to add a few words of my own.

I cannot, for my part, accept Mr Sparrow's submission that the decision of this court in *Lloyds Bank Ltd* v *Ellis-Fewster* [1983] 1 WLR 559 effectively precludes us from taking a view of the evidence, as opposed to the law, different from that taken by the judge. Sir John Donaldson MR said in that case (at p 562): "If one judge thinks there is a triable issue, it would be surprising if two or three judges think there is not." I respectfully agree, but I do not think the converse is true. If, on an Ord 14 application, one judge thought there was not a triable issue and his view depended upon the evidence, it
D would seem to me not in the least inherently surprising if two or three judges, perhaps on fuller argument, subsequently took a different view. If sitting in this court, they are, in my opinion, entitled to do so for good and sufficient reason; they are in as good a position to assess the weight of the affidavit evidence as the judge in the court below, though his views are of course entitled to respect.

In my judgment, in the present case the criticism made by the judge and by the respondent's counsel of the vagueness and lack of particularity of the first and second defendants' draft amended pleading and their affidavit evidence is largely justified. The judge was, I think, quite right to approach their projected defence with caution.

E Nevertheless, the ultimate question for him, as it is for us, was whether, with respect to the claim made on the plaintiff's application for summary judgment against the first and second defendants, there is an issue in dispute which ought to be tried or whether there ought, for some other reason, to be a trial of their claim, within the meaning of Ord 14, r3. With respect to the judge and despite the cogent submissions made to this court by Mr Sparrow and by Mr Meyer, I take a different view of the evidence from that taken by the judge. After some initial doubts I am satisfied, for the reasons given by Nicholls LJ, that the issues raised by the first and second defendants in their evidence do show sufficient cause against the application for summary judgment to entitle them to leave to defend. In particular, it appears to me that the issue of estoppel is at least arguable on the
F evidence now before the court. It seems to me at least arguable on such evidence that the conduct of Guardcliffe by Mr Ferster has estopped it from denying that the notice of July 12 1984, whatever its defects, was effective to determine the lease with effect from September 29 1986.

It is quite true that silence on its own will not give rise to an estoppel in a case where there is no duty to speak, and there was no duty to speak here. But on the evidence before us it seems to me arguable that Mr Ferster's conduct in his conversations with Mr Russell and Mrs Carol Hutley — and he does not deny that he did have conversations with them during the material period —was actively misleading. Furthermore, Mr Russell has deposed, rightly or wrongly, that on one occasion an executive of Guardcliffe, whose name he did not remember — I think it must have been Mr Ferster to whom he G
referred — made the remark that he, Mr Russell, would soon be finally finishing with his responsibilities for the rent.

Mr Ferster, for his part, strongly denies that he gave either the first or the second defendant any grounds for believing that their notices were valid or that they were accepted as valid by Guardcliffe. It may be that at the trial Mr Ferster's evidence will be preferred to that of Mr Russell and Mrs Hutley. But that will be a matter for determination by the trial judge, who will have the advantage of hearing all the oral evidence. As things are, the point is, in my judgment, one that ought to go to trial, as ought the other points raised by the defendants which Nicholls LJ has indicated are arguable.

On this interlocutory appeal I do not think it would be helpful for me to add further to his judgment, with which I am in full agreement. H
The first and second defendants' appeal will accordingly be allowed and we will hear the parties as to the form of order which it will now be appropriate for us to make.

The appeal was allowed in part. The master's costs order was to remain undisturbed but the order for costs made by the judge below was to be set aside. The costs of all three defendants below were to be costs in the cause. The first and second defendants' costs of the appeal were to be paid by the plaintiff. No order was made as to the costs of the third defendant or of the plaintiff in respect of the appeal.

J

Chancery Division

November 8 1990

(Before Mr Justice HOFFMANN)

WROTHAM PARK SETTLED ESTATES v NAYLOR

Landlord and tenant — 999-year lease of farmhouse — Alleged breach of covenant by lessee — History of differences between the parties — Conversion of cowshed into cottage and store — Cottage occupied by couple performing services for lessee and his wife — Breach alleged was of covenant that the premises demised should be used only for the sole occupation and use of the tenant and the family of the tenant — Meaning of "family" — Whether, if there was a breach of covenant it had been waived or acquiesced in — Whether, if not, relief against forfeiture should be granted without a condition requiring cessation of breach — Action dismissed K

The demised premises in this case comprised a substantial 17th-century Hertfordshire farmhouse called Knightsland Farm, with outbuildings, gardens and some 15 acres of agricultural land — The lease was for 999 years — The relationship between the landlords and the tenant had been difficult — The tenant had wished to convert part of a dilapidated cowshed into a cottage, his intention being to offer the accommodation to a couple who would do some work in the house and grounds and whose presence would provide some security against burglars or prowlers, the farmhouse being relatively isolated — The tenant succeeded in obtaining planning permission for the conversion but failed to obtain the landlords' consent — However, considering the refusal of consent unreasonable, the tenant went ahead with his plan — The cowshed was duly converted and the tenant engaged service occupiers to work for himself and his wife — Eventually the landlords served a notice under section 146 of the Law of Property Act 1925 on the tenant alleging that the occupation of the cottage by service occupants was a breach of the covenant to use the premises for the sole use and occupation of the tenant and his family — An action claiming forfeiture followed L M

Was there a breach of covenant? Could the service occupants be regarded as part of the tenant's family? The judge was referred to the definition of "family" in the *Shorter Oxford English Dictionary*, which gave as one of its primary

A **meanings "the body of persons who live in one house or under one head, including parents, children, servants etc" — It was objected on behalf of the landlords that the service occupants in the cottage did not qualify as "family" — They constituted a household of their own and their work was mainly outside, in the gardens and grounds — It was suggested that the dictionary definition contemplated indoor servants in the employer's house — Hoffmann J rejected this distinction as material and decided that the service occupants were part of the tenant's family within the meaning of the covenant — Hence there was no breach of covenant — This was sufficient**
B **to dispose of the matter, but in case it went further the judge dealt with two other points which had been argued — He would have held that, if there had been a breach of covenant, there was not evidence that it had been waived or acquiesced in by the landlords — He also expressed the view that if he had been called upon to grant relief against forfeiture he would have done so only on condition that the breach of covenant should cease**

The action was dismissed

No cases are referred to in this report.

C This was an action by the plaintiffs, Wrotham Park Settled Estates, seeking the forfeiture of the lease of Knightsland Farm, Barnet, Hertfordshire, held by the defendant tenant, Christopher Naylor, on the ground of breach of covenant.

Kirk Reynolds (instructed by Boodle Hatfield) appeared on behalf of the plaintiffs; Simon Berry QC (instructed by Denton Hall Burgin & Warrens) represented the defendant.

Giving judgment, **Hoffmann J** said: This is an action for forfeiture of a 999-year lease for breach of covenant. The lease was granted by the plaintiff, which then had a slightly different name, to the defendant on September 6 1968.

The demised premises are a substantial 17th-century Hertfordshire farmhouse called Knightsland Farm, with its
D outbuildings, gardens and some agricultural land amounting in all to about 15 acres. The reversion forms part of a large family estate called the Wrotham Park Estate in which the beneficial interest is held by a company controlled by Mr Julian Byng.

The background to the grant of the lease is that Mr Julian Byng and the defendant, Mr Naylor, are cousins and were both beneficiaries under the will of their late grandfather, the Earl of Strafford. By an arrangement under the Variation of Trusts Act 1958, which was approved some months before the grant of the lease, Mr Naylor and his mother and brother were each given options to acquire 999-year leases of houses on the estate. The terms of the leases, which were in very similar form, had been negotiated as part of the arrangement.
E All three beneficiaries exercised their options and the leases were duly granted.

For the purposes of this action I need consider only two of the tenants' covenants in the lease. First, by clause 3(13)(i) the tenant covenanted not without the licence in writing of the lessor first obtained, which consent shall not be unreasonably or vexatiously withheld, to erect or suffer to be erected any new buildings on the premises or without such licence as aforesaid make any alteration or addition whatsoever in or to the exterior of any building now on the premises. The other was by clause 3(22) a covenant to observe certain regulations specified in the fifth schedule to the lease, of which I need to refer only to the first two.

1. The premises shall be occupied and used only as a private dwellinghouse
F with garage and gardens for the sole occupation and use of the tenant and the family of the tenant.

2. That neither the premises nor any part thereof shall be used for business or trade purposes other than agriculture and horticulture.

In 1970 Mr and Mrs Naylor decided to convert part of a dilapidated cowshed into a cottage. The advantage for the Naylors was that they would be able to offer the cottage as accommodation to a couple who would in return act as caretakers and do some work for them in the house and grounds. The gardens and hedges were at that time very much overgrown and the ditches needed to be cleaned out. The house itself required to be maintained. Mr Naylor, who had then recently left the army and gone into industry, did not have much time to do the work himself. Mrs Naylor had a new baby, and it was also
particularly important for the Naylors that they should have G
someone living near the house. Knightsland Farmhouse is a relatively isolated building and a resident caretaker and his wife would be a protection against burglars and prowlers.

There were, however, some potential problems about planning permission. The farmhouse was in the green belt and the planning authority were very unlikely to give consent for the creation of a new residence, other than for agricultural workers. But Mr Naylor appears to have been advised that someone who did some work of an agricultural nature would for this purpose count as an agricultural worker even though most of his work had nothing to do with agriculture at all. There was some work of an agricultural nature to be done on the ditches, hedges, fields and the farm track which gave
access to the house, and Mr Naylor thought that this would be enough H
to satisfy the planning condition. I do not have to decide whether this was a correct interpretation of the law. It was a view which Mr Naylor genuinely and honestly held and in the event he has had no trouble from the planning authority over the 20 years in which he has acted on that view.

This was not in fact the first time that the possibility of having resident employees at Knightsland Farmhouse had been considered. Back in 1959, shortly before the Naylors actually moved into the house, there had been discussion between them and Mr Naylor's aunt and Mr Julian Byng's mother, Lady Elizabeth Byng, who was then tenant for life under the settlement and farming Knightsland Farm
and other holdings, about the possibility of having farm employees J
residing near the farmhouse. With that in mind, Lady Elizabeth applied in May 1959 for planning permission for the erection of a pair of semi-detached agricultural cottages at the farmhouse. That permission was granted subject to a number of conditions, including a standard condition that the "buildings be used only as dwellings to be occupied by persons engaged or employed in the use of Knightsland Farm for agriculture as defined in the Town and Country Planning Act, and during such occupation by their spouses and issue".

Mr Naylor's first thought in 1970 had been to implement that permission. A change in the law in 1968 had introduced a time-limit for old planning permissions and Mr Naylor was aware that unless
implemented within the near future it would expire, and so on June 28 K
1970 he wrote to Mr J B Farnsworth, who was a partner in the firm of Berry Bros & Legge, who were acting as land agents to the estate at the time, saying that he was thinking about erecting cottages under that permission. He said that he thought that the effect would be to improve the value of the estate, that it would make use of a permission which was otherwise about to expire and "It would mean that the farmhouse is not so cut off and would have some resident farm labour".

Mr Farnsworth referred this letter to Mr Byng and Mr Byng declared himself puzzled about what Mr Naylor could have in mind. The effect of the grant of the lease was that Knightsland Farmhouse
had been severed from the agricultural holdings of the estate. Mr L
Naylor was not a farmer. Even the 15 acres which were comprised in the lease were being farmed by Lady Elizabeth, together with the rest of her holdings. The farmhouse had therefore no need for resident farm labour.

It seems to me, however, that provided one does no more than credit Mr Naylor with an intention to make a meaningful statement, the meaning is plain enough. The purpose of building the cottages was to have employees resident at the farmhouse so that it would not be so cut off, and they would also be able to work as agricultural workers for the estate. No doubt, in addition, Mr Naylor expected that they would be able to render some services for him as well.

Mr Byng's reaction to this proposal was to instruct Mr Farnsworth
to write back and suggest instead that the estate would be willing itself M
to construct the cottages in return for Mr Naylor's agreeing to the lease being terminable on its 21st anniversary and every subsequent 21 years thereafter after the death of Lady Elizabeth. This proposal did not appeal to Mr Naylor. He felt that he had taken some trouble to negotiate a 999-year lease and was not going to see it cut down to 21. He also took the view that refusal of consent would be unreasonable. By October, however, he had changed his mind about implementing the planning permission and decided instead to convert one half of the cowshed into a cottage. The other half was to serve as a garden store. On October 8 1970 he wrote to Mr Farnsworth enclosing the outline plans which he was submitting for planning permission. It was part of the planning application that Mr Naylor

A was willing to surrender the benefit of the permission which had been obtained by Lady Elizabeth. Mr Byng objected to this course and made representations against it to the planning authority. He felt that the permission had been obtained for the benefit of the estate and that if Mr Naylor was not going to use it at Knightsland Farm it should be transferred to somewhere else on the estate. Mr Naylor, on the other hand, not only took the legal point that the permission enured for the benefit of the land of which he was tenant but also said that the permission had been originally obtained by Lady Elizabeth specifically so that she could have employees living at the farmhouse.

In the event, the planning authority granted permission in accordance with Mr Naylor's application, once again subject to their standard condition for agricultural occupation. That permission was B granted on August 11 1971.

On May 15 1972 Mr Naylor wrote to Mr Farnsworth formally applying for consent under the lease for the conversion of the cowshed, and on May 24 Mr Farnsworth wrote back refusing permission. No further reasons were given and Mr Farnsworth merely referred back to his earlier letter in which he said that the estate would refuse permission for the construction of the two cottages originally proposed, and one must therefore assume that the ground for refusal was that the planning permissions ought to have been transferred to somewhere else on the estate. Mr Naylor, having been advised that the refusal of permission was unreasonable, went ahead and carried out the conversion.

C On November 1 1972 the plaintiff served a notice under section 146 of the Law of Property Act 1925 alleging that the conversion was a breach of the covenant against alterations. No action was, however, taken upon that notice.

Mr and Mrs Naylor engaged a couple to come and live in the farmhouse. They were Mr and Mrs Steed. They stayed a year or two and have since been succeeded by others. Meanwhile the dispute between the Naylors and Mr Byng rumbled on but did not ever come to a head until in 1979 new agents were appointed, Bidwells, and the senior partner, Sir Francis Pemberton, advised Mr Byng that something ought to be done to regularise the relationship between the plaintiff and Mr Naylor. The ground rent of £15 a year had not been accepted since about 1970 and Sir Francis thought that a licence D should be negotiated and the matter once more put upon a proper legal footing.

Negotiations for the licence were carried on over the next few years but came to nothing, and eventually, on November 12 1985, Mr Naylor issued a summons in the Barnet County Court in which he claimed declarations, first, that the right to forfeit the lease pursuant to the 1972 notice was statute barred and, second, that consent to the alterations had been unreasonably withheld. There was no answer to the first claim for a declaration that forfeiture was statute barred and so on January 9 1986 Mr Byng was constrained to write a letter confirming that he was not entitled to forfeit under that notice. Within a few days, however, Mr Byng struck back. He asked his E agents to arrange for an inspection of the premises in order to discover whether there were any other breaches of covenant, and that inspection took place on April 11 1986.

Mr William Shearer [FRICS] of Bidwells carried out the inspection. He looked over the cottage and asked who was occupying it. Mr Naylor told him that they were Mr and Mrs Larman, who were there as service occupants performing work for the Naylors. The result of that inspection was a further section 146 notice dated December 29 1986 alleging that the occupation by the Larmans was a breach of the regulation to use the premises for the sole occupation and use of the tenant and the family of the tenant. The Larmans subsequently moved out and another section 146 notice was served alleging a breach in terms sufficiently general to cover any couple who might be F living as service occupants in the cottage. The present action is based upon that notice.

The first question which I have to decide is whether on those facts there has been a breach of covenant. It was at first alleged that there had been a parting of possession or occupation, but the terms upon which the Larmans and their successors, who are at present a couple called Mr Sheffield and Miss Flint, occupy the cottage show that the possession remains in the Naylors and that the Naylors equally occupy the cottage by their employees. The dispute therefore comes down to the question of whether the cottage occupant can be described within the meaning of regulation 1 in the fifth schedule as being a part of the family of the tenant.

What in this context is the meaning of the word "family"? I was referred to the entry in the *Shorter Oxford English Dictionary* which G tells me that the English word "family" comes from the Latin word *familia*, meaning a household, and it gives as one of its primary meanings "the body of persons who live in one house or under one head, including parents, children, servants, etc."

Mr Reynolds, who appeared for the plaintiff, and Mr Byng in his own evidence, accepted that family in this context had to have a wider meaning than blood relations and that it was a meaning more in accordance with the passage from the dictionary which I have quoted. Mr Byng said that the farmhouse had originally accommodation for staff on the second floor and that he could not contend that indoor servants living in that accommodation would not fall within the description of family in the lease. But Mr Reynolds submitted that the couple in the cottage fell outside this description H for two reasons: first, because they constituted a separate household. They had their own family life in the cottage. Their relation to the Naylors was entirely one of employer and employee, and the nature of their work had meant that they had little social contact with the Naylors. The second reason was that apart from coming into the house occasionally to do odd repairing or decorating jobs, all their work was outside in the gardens and grounds. They were, in the old-fashioned distinction, outdoor servants rather than indoor servants.

I have to look at what this word "family" may have been intended to mean in a lease granted in 1968 having regard to the social conditions of those times. As part of the factual background I think that I am also entitled to have regard not only to what was actually J going on at the time at Knightsland Farmhouse but also to what was happening in the other two houses which were the subject of leases in this respect in identical terms. One of the other leases was for a house called Dancers Hill, which Mr Byng said was a somewhat grander house occupied by Mr Naylor's mother, Lady Mary Naylor. She, it appears, had at the time of the grant of the lease a couple named Mr and Mrs Paine living in a cottage in the grounds. I do not know what Mr Paine did but Mrs Paine had apparently for many years been Lady Mary's cook. It is quite clear that nobody thought that the continued occupation of the cottage by the Paines would be a breach of covenant on the part of Lady Mary, and that, in my view, provides some guidance to what the parties intended it to mean. It does not affect the distinction which Mr Byng drew between indoor and K outdoor servants, but it does show that the fact that the servants did not live within the house and take their meals in the same kitchen was not thought to affect the question of whether they were in this sense a part of the family. Mr and Mrs Paine no doubt had their own little household in the cottage at Dancers Hill in much the same way as the couples living in the cottage at Knightsland Farmhouse.

That seems to me to be in accordance with the passage from the dictionary which speaks of the body of persons who live in one house *or* under one head, including parents, children, servants, etc. It suggests that one can have a house with other cottages or dependent accommodation in which servants live and that they will constitute a family provided that they are all subject to one head, which in this L case would be Mr and Mrs Naylor. That, therefore, leaves the distinction drawn by Mr Byng between indoor and outdoor servants. In my judgment, no such distinction was intended in this lease in 1968. I very much doubt whether it was contemplated in 1968 that any of the occupants of this house during its 999-year term would necessarily have indoor servants at all. The kind of services commonly performed by couples who are given accommodation in largish houses in the country in this way, namely a bit of help in the house and work in the grounds and in the garden, was a perfectly common form of employment even in 1968.

The evidence shows that almost invariably the occupants of the cottage have not been solely employed by Mr and Mrs Naylor. That would have been quite impracticable because they were not paid any M wage. They were simply given the right to reside in the cottage in return for a certain amount of work. So the present incumbent, Mr Sheffield, in fact has a full-time job working for London Buses and also, in return for his accommodation, does a certain number of hours a week for Mr and Mrs Naylor. None the less, the occupant couple are, in my judgment, a part of the Naylor household or family living in dependent accommodation and working very much subject to the directions of Mr, and in particular Mrs, Naylor as to the things which they do from day to day.

For those reasons I do not consider that the position of the occupants of the cottage can really be distinguished from what would have been the position of employees living within the house and, consequently, I have come to the conclusion that they are a part of the family of the tenant within the meaning of the regulation, and that their occupation does not give rise to any breach of covenant. That conclusion is really sufficient to dispose of the case but, in case it should go any further, I should say something about the other two points which were argued, namely that if there was a breach of covenant it had been waived or acquiesced in and, second, that even if it had not, relief against forfeiture should be granted without any condition requiring the breach of covenant to cease in the future.

Let me deal first with the question of waiver and acquiescence. Mr Byng's evidence was that until he received the report from Bidwells in 1986 telling him that the cottage was occupied by the Larmans he had no knowledge that there was an occupant there such as to constitute a breach of the regulation. The defendants on the other hand say that Mr Byng must have known quite enough to constitute knowledge that someone doing the kind of work which the Larmans were doing and having the kind of relationship with the Naylors which they had was occupying the cottage. I think that in the end the dispute is not so much over what Mr Byng actually knew but about the amount of detail and certainty which is necessary to constitute knowledge for this purpose. For example, while Mr Byng very likely had no first-hand knowledge of who was occupying the cottage, or indeed that anybody was occupying the cottage, it seems to me that if at any moment he had addressed his mind to the question he must have acknowledged that it was highly improbable that the cottage was unoccupied. If it was occupied, who would Mr Byng have thought it was occupied by?

In an internal memorandum dated April 7 1975 Mr Byng, in the course of saying various uncomplimentary things about Mr Naylor, said that he had made alterations without consent so as to convert the "hideous existing cowshed into a bungalow for a so-called agricultural worker". That is a reference to the planning consent, but it also suggests that Mr Byng was scornful about the occupant of the flat being capable of being described as an agricultural worker in the true sense. If that is so, who did Mr Byng think he was? Mrs Naylor said in her evidence that in the spring or early summer of 1971 she had had a conversation with Mr Farnsworth, the agent, and had on that occasion explained to him exactly why they wanted to convert the cowshed. She had told him about the need to have someone to work in the grounds and in particular about the need to have someone to caretake the premises. It seems to me unlikely that in some form or another that information was not conveyed to Mr Byng. Equally, when Sir Francis Pemberton went to see the Naylors in 1981 and had a general discussion about the matters in dispute, it seems to me unlikely that the question of the nature of the occupation of the cottage did not surface or that it was not touched upon in Sir Francis' report to Mr Byng. In so far as it was not specifically touched upon, the only reason why mention of it would have been omitted, it seems to me, was that there was nothing which either party thought needed to be discussed. The nature of such occupation was perfectly clear to everyone.

Accordingly, at all material times, in my judgment, Mr Byng knew enough to enable him to allege that the occupation was a breach of covenant if, upon the true construction of the regulation and contrary to the finding which I have made, a couple living there and doing the work which the occupants were actually doing at the time did not constitute a part of the family. Knowledge, however, is not sufficient in itself to give rise to an inference of acquiescence. The authorities to which I have been referred show that the question is whether in the end it is permissible to draw the inference that the breach has in fact been acquiesced in. Inaction for a very long time in the face of knowledge of the breach is certainly a fact from which such an inference can be drawn.

In this case, however, I find it very difficult to draw the inference that Mr Byng ever acquiesced in anything. On the contrary, it seems to me that throughout the relevant period Mr Byng was making it clear that he objected, and in so far as he did not object specifically to the occupation of the cottage, it was only because he was objecting to the very existence of the cottage and the nature of its occupation was therefore not material.

There came the moment in 1986 when Mr Byng had to acknowledge that he could no longer object to the existence of the cottage because his right to complain of a breach of covenant, if any, was statute barred. Accordingly, he acknowledged that in his letter of January 1986, to which I have referred, and it seems to me that the only way in which a case on acquiescence can be put is that in all the circumstances that acknowledgment also necessarily involved a concession that he could no longer object to the way in which the cottage had been occupied over the previous 16 years. Now if that withdrawal of objection had taken place before the building of the cottage and if in reliance upon such withdrawal the Naylors had built the cottage then I think there would be a strong case for saying that Mr Byng must be taken also to have consented to the way in which it was clear that the Naylors would use it. But that was not the case here. This was a case in which Mr Byng was forced to admit that by reason of the Limitation Act he had lost the right to object to a once-and-for-all breach, if such it was, which occurred in 1972. That admission, not in any way thereafter separately acted upon by the Naylors, does not, in my judgment, necessarily involve an admission that he was unable to enforce any other breach of covenant which might exist or which might occur in the future.

Consequently, although it is not necessary for me to decide this point, I think that if I had been required to decide the question of acquiescence I would have been constrained to say that the breach of covenant had not been acquiesced in.

Finally, there is the question of relief from forfeiture. Mr Reynolds realistically conceded that the defendants would be entitled to relief. The only issue is whether it would be a condition of such relief that the hypothetical breach of covenant should not take place in the future. Mr Berry submitted that since the court has a complete discretion in the matter it would be proper in this case not to impose such a condition. I accept that the discretion of the court is perfectly general, but I find it impossible to see how it could be a proper exercise of the discretion not to impose such a condition in this case. It would involve my saying that although I found that there had been a breach and although I found that there had been no acquiescence in the continuation of that breach, nevertheless the tenant could have the lease reinstated without any assurance that no breaches could take place in the future. In those circumstances it seems to me that the inevitable consequence would be that the landlord would be entitled to serve another notice under section 146 and the parties would find themselves back in court again. Mr Berry suggested that if I had on this occasion granted relief without imposing such a condition a subsequent court would be bound also to do so and it would therefore be frivolous and vexatious for the plaintiff to start proceedings which would inevitably end in the same way. I find that a very unappealing prospect because it involves the proposition that it is frivolous and vexatious to attempt to enforce a covenant which the court has specifically held to be still valid and enforceable. That seems to me an impossible position for the court to put itself in, and, therefore, if I had been called upon to exercise my discretion and to grant relief I would have done so only upon the condition that the breach of covenant should cease. That too, however, is not a matter on which I am, in the event, called upon to decide.

For those reasons the action will be dismissed.

Court of Appeal

November 22 1990

(Before Lord Justice MUSTILL and Lord Justice STAUGHTON)

WESTMINSTER CITY COUNCIL v BASSON

Landlord and tenant — Possession claim — Whether occupier a tenant — Claim to council tenancy — Conflicting indications from council — Assessments of rent rebate — Issue of rent book — Payments for "use and occupation" — Suggestion that the three indicia recognised in *Street* v *Mountford* were present — Reference by council to "Your unlawful occupation" and to forthcoming legal proceedings for recovery of possession — Appeal against county court judge's possession order dismissed

The dwelling-house in question had been the home for a number of years of a couple to whom the council had granted a tenancy — Eventually the wife ceased to live with the husband and the appellant (defendant below) moved into the house, the idea being that the husband and the appellant

A **should live there as man and wife — This relationship, however, did not last long; the husband moved out after some months and did not come back — The appellant remained in occupation — After a break-in she had occasion to visit the council offices and was told by an official that she had no lawful right to reside in the premises, that an action would be taken against her if she stayed and that any money she paid would be regarded as for use and occupation — Later the husband's tenancy was terminated and the appellant received a letter from the council referring to her "unlawful occupation" and confirming that legal proceedings had been**
B **instituted — It went on to say that she would be responsible for damages by way of use and occupation charges, stated what the charge would be per week and mentioned the times when the council offices would be open to receive payment — The letter ended with the sentence: "In making the payments as Use and Occupation Charges this arrangement is not intended as the creation of a tenancy or a Licence akin to a tenancy in any way whatsoever" — After that letter, however, the appellant applied for a rent rebate and was assessed and reassessed under the council's rent rebate scheme — Later she was issued with a rent book; this was after proceedings against her for recovery of possession had begun**

C **The county court judge made a possession order — On appeal it was submitted on behalf of the appellant that among the conflicting signals emanating from the council there were three indicia recognised in *Street* v *Mountford* as the hallmarks of a tenancy, the grant of permission to occupy, exclusive occupation and regular payments referable to that occupation — The court did not accept this interpretation — It was negated by the last sentence of the council letter referred to above — Although there were signs that the council's left hand did not know what their right hand was doing, the sense of the letter was: "We desire you to vacate the premises. We**
D **trust that you will do so voluntarily. If not we shall take steps to remove you. Meanwhile we are not going to let you remain there free of charge and to make sure that we are paid we shall expect to receive payment at the stated rate in the stated manner until we succeed in regaining possession" — Such things as the issue of a rent book, which taken by themselves might point the other way, did not prevail against the clear evidence of the council's attitude — The judge below had come to the right conclusion — Appeal dismissed**

The following case is referred to in this report.

E *Street* v *Mountford* [1985] AC 809; [1985] 2 WLR 877; [1985] 2 All ER 289; [1985] 1 EGLR 128; (1985) 274 EG 821, HL

This was an appeal by the defendant, Beatrice Basson, against the possession order made by Judge Finestein QC, at Bloomsbury County Court, in respect of 12 Aldsworth Close, Amberley Estate, London W9, in favour of the plaintiffs, Westminster City Council.

Angus J Macpherson (instructed by Gillhams, of Willesden), appeared on behalf of the appellant; Miss Erica Foggin (instructed by the solicitor to Westminster City Council) represented the respondent council.

Giving judgment, **Mustill LJ** said: This is an appeal against a possession order made by His Honour Judge Finestein QC made in
F the Bloomsbury County Court on October 1 1989. The order was in favour of Westminster City Council. The subject-matter was council accommodation at 12 Aldsworth Close, Amberley Estate, London W9. The defendant was Miss Beatrice Basson. The issue before the judge was whether, as Westminster Council maintained, Miss Basson was a trespasser or whether, as she maintained, she had since September 1985 been a tenant of the premises or alternatively became a tenant in April 1987.

The history of the matter is as follows. During December 1977 the council granted to a Mr and Mrs Simpson the tenancy of the premises in question. In 1984 Mrs Simpson ceased to live with her husband at the premises and in July 1984 the appellant moved into those premises with Mr Simpson so they should live together as man and wife. That relationship appears to have come to an end some months G
later, for in February 1985 Mr Simpson moved out and never moved back. In September 1985 there was a break-in at the premises and various items, including items of Mr Simpson's furniture, were taken away. This prompted the appellant to visit the council offices and have a conversation with Mr Merriman, a council official. The judge at the hearing of the application for possession believed Mr Merriman's account of what was then said and according to the judge's notes of evidence Mr Merriman told the appellant that she had no right to reside in the premises, that action would be taken if she stayed and that any money paid would be for use and occupation. On September 23 Mr Simpson's tenancy was terminated and on September 27 a letter was written by the district housing officer of H
Westminster Council to the appellant in the following terms:

> Dear Sir/Madam
> Address: 12 Aldsworth Close
> I refer to your unlawful occupation of the above accommodation and confirm that legal proceedings have been instigated for the recovery of vacant possession of the dwelling.
> As the previous tenancy was terminated on September 23 1985 and you have been in occupation since this date I hold you responsible for damages by way of Use & Occupation Charges until such time as you leave the accommodation.
> The charge per week is £30.13 and I enclose a payment card for your use in paying these charges. The payments should be made to the office either by post or in person at the following times:

Monday	9.00 am–4.00 pm	J
Tuesday	9.00 am–2.30 pm	
Wednesday	9.00 am–2.30 pm	
Friday	9.00 am–12.00 noon	

> In making the payments as Use and Occupation Charges this arrangement is not intended as the creation of a tenancy or a Licence akin to a tenancy in any way whatsoever.

I shall return to this letter at a later stage, but I make two points upon it. In the first place, although the letter stated that legal proceedings had been instigated, in fact proceedings were not taken until more than a year later. Second, the payment card which was enclosed with the letter was not put in evidence at the hearing before the county court judge and we have not ourselves seen it.

After that letter the appellant made an application to the K
respondents for a rebate under their rent rebate scheme and for this purpose she completed two documents. One was a request to her employers for them to furnish to the local authority particulars of her earnings and the second was a verification form intended to be completed by those employers. The appellant also completed an application form in relation to the rent rebate. The only materiality of this, so far as I can see, is that in the hand of some unknown person, no doubt an employee of the council, there was written at the top of the form — "1st App [short for "application"] commencement of tenancy."

In due course a rent rebate was assessed on a form designed for the internal use of the council, which made more than one reference to L
the rent and also stated that the tenancy commenced on September 23 1985. There was more than one round of applications by the appellant for a reassessment of the rent rebate and reassessments were indeed made. On October 3 1986, in the middle of the second round, the council issued a possession summons and shortly after the reassessment was made there was a conversation between an official of the plaintiff council and the appellant in which he told her that she must vacate the premises. Finally, in the middle of November 1986, the possession proceedings were served. This does not seem to have made any difference to the practice of the appellant to apply for, and of the council to grant, a rent reassessment; indeed such a rebate reassessment was in progress when on February 9 1987 the defence was served in the possession proceedings. Finally, in the spring of M
1987 the respondents issued the appellant with a rent book. There was then a long interval until the matter came before the county court judge with the result which I have already described.

I now return to the letter of September 27 1985, which has been the focal point of this appeal. As has been recorded, it begins with the words, "I refer to your unlawful occupation of the above accommodation and confirm that legal proceedings have been instigated for the recovery of vacant possession of the dwelling". These words are plainly inconsistent with any notion that the appellant was in occupation of the premises with the consent of the council. It is, however, argued by Mr Macpherson on her behalf that the effect of the first paragraph is nullified by what followed. I need not read it again, but Mr Macpherson submits that we here find the office making administrative arrangements in advance for the payment of monetary consideration labelled "Use and Occupation Charges". That expression is an accurate description of sums paid in respect of a tenancy. It is not apt terminology, so the submission goes, for amounts payable by a trespasser. Thus, it is submitted, the letter must have been intended to communicate the consent of Westminster Council to the appellant's continued occupation of the premises. Accordingly, we have present all three indicia of a tenancy, recognised in *Street* v *Mountford**, namely the grant of permission to occupy, exclusive occupation and regular payments referable to that occupation.

That argument would have had at least something to commend it if the letter had not begun in the way which I have described and if it had not ended with the words "In making the payments as Use and Occupation Charges this arrangement is not intended as the creation of a tenancy or a Licence akin to a tenancy in any way whatsoever". That seems to me to show plainly that the draftsman of this letter was not using the expression "Use and Occupation Charges" in a technically correct manner. The sense of the letter is perfectly clear, particularly when it is read against the background of the earlier conversation with Mr Merriman. The council were saying "We desire you to vacate the premises. We trust that you will do so voluntarily. If not we shall take steps to remove you. Meanwhile we are not going to let you remain there free of charge and to make sure that we are paid we shall expect to receive payment at the stated rate in the stated manner until we succeed in regaining possession". Speaking for myself, I feel no doubt that this letter is entirely inconsistent with the proposition that the appellant was remaining in the council premises with the consent of the council.

Reliance was placed on subsequent events to displace the plain inference to be drawn from the conversation with Mr Merriman and the letter. There are, it is true, isolated references to rent and a tenancy, but, in my view, these merely show that one department of the council was not fully aware of what was happening in the other departments. That this is so is clearly demonstrated by the fact that although proceedings for possession had been taken in October 1986, the same grounds of rent rebate applications and reassessments employing the same terminology proceeded as if the action had never started and it seems to me impossible to hold that at the time when the possession action was well under way the council can possibly have intended, by the use of these isolated turns of phrase, to convey that the character in which the appellant was in occupation of the premises was radically different from the character which the council were asserting in the pending proceedings. In agreement with the learned judge, I am satisfied that nothing which happened after September 27 1985 served to demonstrate that the letter did not mean what it said or, alternatively, that there had in some way been a change in the character of Miss Basson's occupation.

There is a separate point to which I should briefly draw attention. I have mentioned that in April 1987 the council issued the appellant with a rent book. In the cover of the rent book there are certain instructions for payment and conditions for tenancy which, if taken in isolation, would have given strong support to a contention that the appellant was indeed a tenant of the council. There is an unfortunate degree of obscurity as to what use was made of the issue of the rent book at the hearing before the county court judge. It was argued before us that even if the appellant was wrong in maintaining that she had been a tenant from the outset, the issue of the rent book marked a change in her status. I think that is an impossible argument to maintain once it is recognised that the rent book was issued not only after the possession action had begun but after it had reached the stage of defence. It is quite inconceivable that the council should have intended, or the appellant understood, that by the issue of this rent book the ground should be cut entirely from under the feet of the council as regards their pending proceedings for possession.

Since I have formed this clear opinion I do not think it necessary to enter into any details as to the manner in which this particular feature of the history was deployed in the county court. It is not referred to in the judgment or in the judge's note. We have been told by counsel that, in spite of the fact that the point was not pleaded, reliance was placed on it in the county court and we are perfectly prepared to proceed on the assumption that that is so. We would not wish the appellant to think that she has lost these proceedings because of an omission in the pleadings or in the judge's notes of evidence. Even on the assumption that the point was made, in my opinion it is quite unsustainable.

Accordingly, for these various reasons I would hold that the appeal must be dismissed.

STAUGHTON LJ agreed and did not add anything.

The appeal was dismissed with costs, not to be enforced without the leave of the court. Legal aid taxation was ordered.

*Editor's note: Reported at [1985] 1 EGLR 128.

Court of Appeal

January 23 1991

(Before Lord DONALDSON OF LYMINGTON MR, Lord Justice RUSSELL and Lord Justice NOLAN)

TAGRO v CAFANE AND ANOTHER

Landlord and tenant — Housing Act 1988 — Wrongful eviction — "A cautionary tale" — Background of harassment — Tenant of bed-sitting-room locked out and locks changed — Offer of a key which was not real reinstatement — Room wrecked in tenant's absence — *Ex parte* injunction not complied with — Quantum of damages under sections 27 and 28 of 1988 Act — Whether £31,000 too high — No ground for holding that judge below was at fault — Landlord's appeal dismissed

This appeal was by the landlord against a decision in the county court awarding the tenant of a bed-sitting-room £31,000 damages for unlawful eviction — The room was in a building leased on a monthly business tenancy to the landlord, who carried on a second-hand furniture business on the ground floor — The upper floors were let in rooms — There had been an unpleasant history of harassment of the tenant, culminating one day in her finding herself locked out of her room and the locks changed — Next day she obtained an *ex parte* mandatory injunction, which was renewed when the landlord failed to comply with it — Some days later the landlord offered the tenant a key, but when she returned to the building she found the front door lock broken and her own room in chaos, with some of her belongings stolen — Despite the authority of her mandatory injunction she decided not to risk a return to the premises; she was afraid of the landlord and his agent — The landlord's suggestion that the chaos and wreckage were due to a burglary was not accepted by the judge, who took the view that there was a determination to ensure that the tenant did not return — The county court pleadings were amended to include a claim by the tenant under sections 27 and 28 of the Housing Act 1988 — The judge awarded her damages of £31,000, from which judgment the landlord appealed — There was no appeal from an additional sum of £15,588.28 damages for trespass to the tenant's personal belongings, for which judgment was given against the landlord and his agent

So far as the question of liability was concerned, the Court of Appeal was satisfied that the landlord was liable under section 27 of the 1988 Act — The court rejected a defence that the tenant had been reinstated within the meaning of section 27(6)(a) — It was not reinstatement to give the tenant a key to a lock which did not work, inviting her to resume occupation of a room which had been totally wrecked — The court also rejected a submission that the tenant had no right to choose whether to accept an offer of reinstatement — The court also found no effective ground of defence under section 27(7)

The judge's decision on the quantum of damages, which he put at £31,000, was also criticised on behalf of the landlord — The amount had been based on a valuation by the tenant's surveyor, who had subtracted the value of the landlord's

A **leasehold interest subject to the tenant's occupation (£44,000) from the value of the interest with vacant possession (£75,000), the difference being the £31,000 awarded — No evidence had been called to dispute these figures — The landlord had not been represented below — The criticism in the Court of Appeal was therefore confined to suggestions that no reasonable judge could have accepted the tenant's figures and that the judge had failed to appreciate that the landlord's interest was only a monthly tenancy — The first of these suggestions did not stand up and the second overlooked the possible durability of the landlord's tenancy under the** B **Landlord and Tenant Act 1954 — The damages of £31,000 might seem high but in the absence of evidence there were no grounds for saying that the judge was at fault — The court did not think that the case should be sent back to give the landlord a second chance to call valuation evidence which he did not choose to present the first time — Landlord's appeal dismissed**

No cases are referred to in this report.

This was an appeal by the landlord, John Cafane, from the decision of Judge Simpson, at Lambeth County Court, awarding £31,000 damages to the respondent (the plaintiff below), Miss C Aurelie Tagro, for unlawful eviction from her first-floor front bed-sitting-room in premises at 116 Landor Road, Stockwell, London SW9. Mr Cafane's agent, Mr Patel, who was the second defendant below, did not take part in the appeal.

Robert Carnwath QC and William Geldart (instructed by John Gittens & Co) appeared on behalf of the appellant; David Neuberger QC and Hugh Jackson (instructed by Anthony Gold Lerman & Muirhead) represented the respondent.

Giving judgment, **Lord Donaldson of Lymington MR** said: This appeal is a cautionary tale for landlords who are minded unlawfully to evict their tenants by harassment or other means.

The facts are these. Miss Tagro was the residential tenant of a first D floor front-room in premises at 116 Landor Road, Stockwell, London SW9. That was a bed-sitting-room and she had the use of the kitchen, bathroom and toilet jointly with the occupier of another room on that floor. Mr Cafane was Miss Tagro's landlord. He was the tenant of the whole building, the ground floor of which was used for his second-hand furniture business. The first floor, to which I have already referred, was used to provide these two bed-sitting-rooms, and there was a second floor which was similar to the first floor. The freehold owner of the premises was Lambeth London Borough Council and they had let the whole premises to Mr Cafane on a tenancy, which was determinable on one month's notice, subject always, of course, to the effect of the Landlord and Tenant Act 1954.

E Mr Cafane appeals against a judgment of His Honour Judge Simpson, given in the Lambeth County Court on November 24 1989, whereby he awarded £31,000 damages to Miss Tagro for unlawful eviction, those damages being claimed by her under sections 27 and 28 of the relatively new Housing Act of 1988. In addition, he awarded her £15,588.28 damages for trespass to her personal belongings which were on the premises. This latter judgment was also given against the second defendant, a Mr Patel, who was the agent of Mr Cafane. Mr Patel does not appeal in respect of that part of the judgment, nor does Mr Cafane. The appeal is limited to the judgment for £31,000, which was solely against Mr Cafane because he alone was the landlord.

The defendants were not represented below, but we have been F greatly assisted by the fact that Mr Cafane has been represented in this court by Mr Robert Carnwath QC and Mr William Geldart, who have put in an admirable skeleton argument which, of course, Mr Carnwath has supplemented orally.

The further facts are these. It was in August 1989 that war really broke out, but some mention should be made of the previous history because it obviously affected the reaction of the tenant. For example, on one occasion she had been woken up at two o'clock in the morning with a demand for rent. There was some dispute as to whether the rent was due, but it really does not matter because two o'clock in the morning is not a time at which any landlord is entitled to demand rent from his tenant or even remind the tenant that rent is due. On another occasion there was even more outrageous conduct. Miss Tagro was visiting the lavatory, which was situated in the bathroom, and the G landlord, or somebody on his behalf, came and kicked the door down. She complained previously of having been referred to as a black bastard and of loud music being used to disturb her enjoyment of the flat and so on. That was the background.

On August 3 she returned to the room to find herself locked out and the locks changed. There has been no challenge in this court, and there could not be any serious challenge, to the fact that the locks were changed with the intention of evicting her. There was some suggestion at one stage that a number of keys had been floating around and got into the possession of various other people and that the lock was changed for security purposes. The judge rejected that on his assessment of the evidence in view of the fact that it was some days later before Miss Tagro was ever offered a key which actually H fitted the lock.

Faced with that situation, on August 4, the next day, she obtained an *ex parte* injunction to readmit her to the premises and allow her access to her belongings. That injunction was very properly for a limited period, being given *ex parte*, and it came up for renewal on August 7, when it was continued, Mr Cafane having taken no step whatever to comply with it in the meantime. The application to continue the injunction was coupled with an application to commit for contempt of court.

On August 8 or 9 — it matters not which — Mr Cafane relented to the extent that he offered Miss Tagro a key. Her evidence about this I can take from her affidavit: J

I collected the key late morning on Wednesday August 9 1989 in the office of Messrs Mohabir & Co and went to the building. The lock to the front door was broken and, in fact, I did not need the key to enter. I went upstairs to the premises. Everything inside was in chaos. Many things were broken. A large amount of my belongings were stolen. On the advice of my solicitor I called the police, who arrived, then had to leave on an emergency call. They said they would be back to take fingerprints and that I should not touch anything shiny or any door or wardrobe or handle until they were able to return which at the time of swearing this affidavit they have not done although I have been reminding them. In the late afternoon my solicitor's surveyor arrived to take photographs and draw up a schedule of broken items.

Later in her affidavit, she stated:

I did not sleep at the premises on the night of Wednesday August 9 1989. This K was partly because the police told me not to touch any of the objects in the premises including everything which was shiny and partly because the lock to the front of the building was damaged and would not "lock", leaving me with greatly reduced security. The main reason however was that I had by then become too frightened of the defendants. I have not returned to the premises to sleep and do not intend to do so. After the recent events outlined above I cannot conceive of returning to the premises.

I would add that the state of the premises to which Miss Tagro swore in her affidavit was attributed by the landlord to the fact that there had been a burglary. The learned judge wholly rejected the suggestion that there had been a burglary. It is perhaps inherently unlikely that there was one, in that the damage done to the flat was not the sort of damage that one would expect from a burglar who was L intent upon removing goods. Such an intruder does not smash the goods and the flat itself. Any judge would be entitled to take judicial notice that that was an unusual way of proceeding. At all events, the judge clearly formed the view that this so-called burglary never took place and that what had happened was that, having locked Miss Tagro out, Mr Cafane or those acting on his behalf were determined to make certain that under no circumstance would she be prepared to return even if they were eventually obliged to let her in again as a result of the court order; and of course, they were brilliantly successful.

One other date needs to be mentioned because it is relevant to Mr Carnwath's arguments and that is August 25 1989. The proceedings originally were in support of the *ex parte* injunction and claimed an M injunction ordering the defendants to readmit the plaintiff to the building and premises forthwith and to deliver keys to the building to the plaintiff and further restraining the defendants, their servants or agents or otherwise from further interfering with the plaintiff's quiet enjoyment of the premises and from further interfering with the plaintiff and from further trespassing on the premises or the plaintiff's goods and for damages. There was no claim under, or for damages under, section 27 of the Housing Act 1988. That claim was introduced by fairly extensive amendment on August 25 1989.

As I have already made clear, this is a claim under the Housing Act and the defences raised also arise under that Act. The principal defences arise under section 27(6)(b) and section 27(7)(b) of the Act.

A So let me turn to the wording of the Act itself. Section 27 is headed "Damages for unlawful eviction" and provides:

> (1) This section applies if, at any time after 9th June 1988, a landlord (in this section referred to as "the landlord in default") or any person acting on behalf of the landlord in default unlawfully deprives the residential occupier of any premises of his occupation of the whole or part of the premises.

I do not think I need refer to subsection (2) save to say that it provides for similar treatment where there is what might be described as constructive eviction of the residential occupier as contrasted with an actual eviction, namely conduct which is such that the occupier cannot reasonably be expected to remain, although physically he or she could remain, and the occupier accordingly gives up his B occupation of the premises. The subsections continue:

> (3) Subject to the following provisions of this section, where this section applies, the landlord in default shall, by virtue of this section, be liable to pay to the former residential occupier, in respect of his loss of the right to occupy the premises in question as his residence, damages assessed on the basis set out in section 28 below.
>
> (4) Any liability arising by virtue of subsection (3) above —
>
> (a) shall be in the nature of a liability in tort; and
>
> (b) subject to subsection (5) below, shall be in addition to any liability arising apart from this section (whether in tort, contract or otherwise).
>
> (5) Nothing in this section affects the right of a residential occupier to enforce any liability which arises apart from this section in respect of his loss of the right to occupy premises as his residence; but damages shall not be awarded both in respect of such a liability and in respect of a liability arising by C virtue of this section on account of the same loss.
>
> (6) No liability shall arise by virtue of subsection (3) above if —
>
> (a) before the date on which proceedings to enforce the liability are finally disposed of, the former residential occupier is reinstated in the premises in question in such circumstances that he becomes again the residential occupier of them; or
>
> (b) at the request of the former residential occupier, a court makes an order (whether in the nature of an injunction or otherwise) as a result of which he is reinstated as mentioned in paragraph (a) above;
>
> and, for the purposes of paragraph (a) above, proceedings to enforce a liability are finally disposed of on the earliest date by which the proceedings (including any proceedings on or in consequence of an appeal) have been determined and any time for appealing or further appealing has expired, except that if any appeal is abandoned, the proceedings shall be taken to be D disposed of on the date of the abandonment.

I will come to subsection (7) in a moment. I would like first of all to deal with the argument on subsection (6).

Before doing so I ought to mention that in subsection (9) "'residential occupier', in relation to any premises, has the same meaning as in section 1 of the 1977 Act", that is to say the Protection from Eviction Act 1977, which in subsection 1(1) provides:

> In this section "residential occupier", in relation to any premises, means a person occupying the premises as a residence, whether under a contract or by virtue of any enactment or rule of law giving him the right to remain in occupation or restricting the right of any other person to recover possession of the premises.

E There is no dispute, of course, that Miss Tagro was a residential occupier, and immediately after August 3 she was a former residential occupier.

Mr Carnwath says in relation to subsection (6) that, in the case of Miss Tagro, the court made an order as a result of which she was reinstated as mentioned in para (a), that is to say she was reinstated in the premises in question in such circumstances that she again became the residential occupier of them. He says it cannot seriously be suggested that a tenant who has been unlawfully evicted and who the landlord is able and willing to reinstate simply has an option whether to accept reinstatement or not.

There are two quite separate questions there. As to the first — was Miss Tagro actually reinstated? — for my part I have no doubt or F hesitation in saying that she was not and could not be at that stage. Reinstatement does not consist in merely handing the tenant a key to a lock which does not work and inviting her to resume occupation of a room which has been totally wrecked. Therefore, on the facts of this case, that is an argument which simply does not run.

On the question of whether the tenant has a right to choose whether to accept the offer, subject to a point which arises under a later subsection, I think, for my part, that the tenant has. Mr Carnwath says that cannot be right because it means to say that a tenant who is unlawfully evicted, perhaps for a relatively short period, is able to achieve some enormous financial reward which is only available to him if he does not accept the offer of reinstatement. There is, of course, a good deal to be said for Mr Carnwath's argument, as there always is for Mr Carnwath's arguments, but there is no indication in G the statute that the tenant does not have that choice. It is difficult to see how you can reinstate a tenant who does not wish to be reinstated. The apparent intention of Parliament is not as unreasonable as it might appear at first when you take into account the fact that, under section 28, to which I will come in a moment, the damages are designed to be equal in amount to the benefit which the landlord gets by having had the tenant removed from the premises, namely the increase in the value to him of the premises. It is therefore quite intelligible that Parliament should have said: "Well, if the tenant does not choose to go back, at least the landlord shall not benefit by it", and to have proceeded on that basis. So, for my part, I think that the argument that there is any defence for Mr Cafane under subsection (6) is quite untenable. H

I turn, therefore, to subsection (7), which provides:

> If, in proceedings to enforce a liability arising by virtue of subsection (3) above, it appears to the court —
>
> (a) that, prior to the event which gave rise to the liability, the conduct of the former residential occupier or any person living with him in the premises concerned was such that it is reasonable to mitigate the damages for which the landlord in default would otherwise be liable, or
>
> (b) that, before the proceedings were begun, the landlord in default offered to reinstate the former residential occupier in the premises in question and either it was unreasonable of the former residential occupier to refuse that offer or, if he had obtained alternative accommodation before the offer was made, it would have been unreasonable of him to refuse that offer if he had not obtained that accommodation, J
>
> the court may reduce the amount of damages which would otherwise be payable to such amount as it thinks appropriate.

Let me say straightaway that para (a) has no application whatever to this case. No suggestion has ever been made that Miss Tagro's conduct was such that it could be criticised and would make it reasonable to mitigate the damage. The two paragraphs seem to me to involve two quite different concepts.

Para (a) is dealing with what one might describe as a contributory negligence concept, although contributory negligence, in the context of an unlawful eviction, is clearly not a happy phrase. But it is the idea that in some measure, although not in such measure as in any way to excuse the conduct of the landlord, the tenant has brought the problem on his own head and the paragraph enables the court to K mitigate the consequences for the landlord.

Para (b) seems to me to borrow something from the concept of the defence of tender in relation to a debt, a defence which is only available if tender takes place before the proceedings are begun. It is, therefore, necessary to consider what is meant by the phrase "before the proceedings were begun" in para (b). It will be remembered that subsection (7) begins with the words, "If, in proceedings to enforce a liability by virtue of subsection (3) above . . ."

Mr Neuberger would have argued, if he had been given the opportunity of doing so, that that meant the totality of the proceedings which were begun at the beginning of August with the claim for an injunction and that, therefore, there is no question here L of any offer of reinstatement before the proceedings were begun. Mr Carnwath argues strongly that these were not proceedings to enforce a liability until an amendment took place later in August and that para (b), with its reference to "before the proceedings were begun", cannot be referring to proceedings which had nothing whatever to do with the enforcement of a liability under subsection (3).

I am bound to say that, for my part, although I have to accept that it is an *obiter* expression of opinion on the facts of this case, I think Mr Carnwath is wholly correct in his argument. I say it is *obiter* because the question would arise as a matter of decision only if Mr Carnwath could make good his argument that there was an offer to reinstate the former residential occupier in the premises in question at any time and, on the facts as I see them, there never was any such M offer. All that was on offer was to give her the key and to readmit her. Certainly, there was no suggestion that the room was put into proper order, the locks repaired or any offer made to Miss Tagro to allow her to resume occupation in any realistic sense of the word. But, had it arisen, as I say, and it may arise in other cases, my view would have been that the subsection is referring to a time before there is a claim in the proceedings under sections 27 and 28 of the Act.

So much for the appeal in respect of liability. But Mr Carnwath, on behalf of Mr Cafane, has also addressed arguments to us on the subject of quantum. That is governed by section 28 and I think I need only read subsections (1) and (2):

> (1) The basis for the assessment of damages referred to in section 27(3)

A above is the difference in value, determined as at the time immediately before the residential occupier ceased to occupy the premises in question as his residence, between —

(a) the value of the interest of the landlord in default determined on the assumption that the residential occupier continues to have the same right to occupy the premises as before that time; and

(b) the value of that interest determined on the assumption that the residential occupier has ceased to have that right.

(2) In relation to any premises, any reference in this section to the interest of the landlord in default is a reference to his interest in the building in which the premises in question are comprised (whether or not that building contains any other premises) together with its curtilage.

On the facts of this case, the learned judge was concerned to consider not the value of the room which had been let to Miss Tagro but the
B value of the whole building, including four bed-sitting-rooms and the business premises below.

Miss Tagro, or those acting on her behalf, called a surveyor, who was an associate member of the Royal Institution of Chartered Surveyors, to give evidence. He had earlier given them a valuation before the judge and he had revised his valuation because he thought there had been a change in market conditions by the time of the hearing. In this he was wrong because the relevant time is immediately before the residential occupier ceased to occupy the premises (subsection (1)). He then gave oral evidence, and his evidence was, first of all, as to a freehold value, which was really immaterial because Mr Cafane was not a freehold owner, and he
C went on to consider the rental value of the whole building, including the shop. He capitalised a profit rental and he considered the extent to which the value of Mr Cafane's interests would be reduced by the presence of Miss Tagro. He said:

I start with freehold with occupancy ie £89,000 then made deduction of £45,000 ie amount of rent × 10 being capital value of Mr Cafane's interest. This produces £44,000 which represents value of leasehold interest with Miss Tagro as sitting tenant. £75,000 [which was the figure that he had valued Mr Cafane's interest with vacant possession] minus £44,000 equals £31,000. This represents the difference in value of the leasehold interest with or without Miss Tagro. £31,000 is my final figure for damages.

Mr Cafane, who, I imagine, knows something about properties, was invited to cross-examine and asked whether the valuation took
D account of disrepair because, in the indirect way in which judges normally take notes of evidence, the following sentence appears:

Cross-examined — this takes into account disrepair as in my report.

Then there was a re-examination on the same point.

Mr Cafane called no evidence to dispute this figure. Mr Carnwath is therefore reduced to saying that the learned judge can be faulted on the footing that no reasonable judge could have accepted that evidence. That submission he made simply and with force and with slightly more difficulty when Russell LJ said, "Well, what should he have done?" It is true, I suppose, that he could have adjourned the hearing and urged the landlord to call expert evidence. But is difficult to see how he could have said, "I am not satisfied that any damages
E are due on this evidence" and, if he is not to say that, then he would have in some way to reduce the surveyor's figure on the basis, I suppose, of what he (the judge) thought were proper values. Had he done so, this court would probably have quashed his decision on the grounds that he was acting not upon evidence and not upon something of which a judge should take judicial knowledge but upon some extraneous view as to a matter of fact.

But, essentially, what Mr Carnwath concentrates upon is his submission that there is no evidence that the judge fully appreciated that this was only a monthly tenancy as far as Mr Cafane was concerned and that, had he appreciated that, he could not have F
accepted this evidence. The difficulty about that argument — apart from any other difficulties and there are many — is that the surveyor was entitled to have regard to the realities of the situation and to have regard to the extent to which Lambeth Council were likely to be minded to serve any notice to quit and would be successful in any event, bearing in mind the provisions of the Landlord and Tenant Act 1954. If the surveyor, applying his local knowledge and professional expertise, came to the conclusion that the lease, other things being equal, notwithstanding that it could be theoretically determinable on a month's notice, would go on virtually for ever, there could be no possible faulting of that aspect of his valuation.

He also says that the surveyor should have approached the matter G
on the basis that the value of the property without Miss Tagro would have been virtually nil because the lease contains prohibitions against assignments and subletting and, says Mr Carnwath, if we were deciding what Mr Cafane could have got for premises without Miss Tagro, no one would have bought it because, as soon as they took an assignment, they would be faced with an application for forfeiture by Lambeth London Borough Council.

There is the scintilla of a point of law there in that clearly the court would have to have regard to what was the interest of the landlord whose value had to be determined under section 28(1)(b), but they would have had some guidance from section 28(3), which provides:

For the purposes of the valuations referred to in subsection (1) above, it shall be assumed — H

(a) that the landlord in default is selling his interest on the open market to a willing buyer;

(b) that neither the residential occupier nor any member of his family wishes to buy; and

(c) that it is unlawful to carry out any substantial development of any of the land in which the landlord's interest subsists or to demolish the whole or part of any building on that land.

I do not understand that section to contemplate, as Mr Carnwath's argument contemplates, that the premises will be treated as virtually inalienable and having no value in consequence. Subsection (3) clearly contemplates that there shall be no increase in the damages because the effect of the tenant being dispossessed is that it enables some very valuable development to take place. But the whole concept J
of the landlord in default selling his interest on the open market to a willing buyer assumes that he can sell it on the open market to a willing buyer, and that involves the subsidiary proposition on the facts of this case that the willing buyer would take a lease from Lambeth Council on a monthly basis subject to the Landlord and Tenant Act with a covenant against subletting or assignment in exactly the same way as Mr Cafane had done. In my judgment, there is nothing in that point.

I accept that the damages do seem to be high, but I have to warn myself against using any knowledge that I may have gained in other ways to support that view, and I am quite unable to say that the judge was at fault. If we were to interfere on this ground, it could only be on K
the basis of sending it back for a rehearing designed to enable Mr Cafane to call valuation evidence. It is not clear to me why he should have a second opportunity to call valuation evidence when he had the opportunity originally and did not choose to avail himself of it.

Accordingly, and for those reasons, I would dismiss the appeal.

Russell and **Nolan LJJ** agreed and did not add anything.

The appeal was dismissed with costs, such order not to be enforced without further order. Legal aid taxation was ordered.

For further cases on this subject see p 57

TOWN AND COUNTRY PLANNING

A **Queen's Bench Division**

December 11 1987

(Before His Honour Judge MARDER QC, sitting as a judge of the High Court)

CRANFORD HALL PARKING LTD v SECRETARY OF STATE FOR THE ENVIRONMENT AND ANOTHER

Town and Country Planning Act 1971, sections 36 and 245 — Motion to quash inspector's decision on an appeal under section 36 against refusal of planning permission — Inspector
B **misdirected himself — Wrong approach — Matter seen by inspector in terms of onus of proof as if he were considering perhaps a piece of civil litigation or possibly even a criminal prosecution — The fact that the proposed development is in green belt land does not in itself constitute a sound reason for refusing planning permission — Correct approach set out in Department of the Environment Circular 14/85 — Inspector's decision quashed**

The land in question in this case consisted of about two acres off the Bath Road, near Heathrow Airport, long used for off-airport parking — There had been a series of
C **temporary planning permissions for that use but there had then been a refusal, which gave rise to the present appeal — The land was within the metropolitan green belt — In rejecting the appeal against the planning refusal, the inspector, in paras 6 and 7 of his decision letter, gave himself directions as to the methods of approach which were criticised by the appellants as erroneous — He stated that there was a general presumption against development in green belts**

The judge agreed with the criticisms which had been made of the inspector's approach as expressed in his decision letter — In the judge's view, the true issues which the inspector was called upon to decide could be expressed in two propositions:

D **First of all, was the use which was proposed for car parking an appropriate and acceptable use of land in this part of the green belt; or was that use such in this particular case as to cause demonstrable harm to the green belt's function and purpose?**
Second, if the use that was proposed was inappropriate and damaging to the green belt, were there exceptional reasons, such as an overriding need for the facility, for permitting it in the face of the presumption against such inappropriate development within a green belt area?

The inspector appeared in his letter to see the matter in terms of onus on proof as if he were considering perhaps a piece of civil litigation or possibly even a criminal prosecution — In *J A Pye (Oxford) Estates Ltd* v *West Oxfordshire*
E ***District Council* the judge had said that the term "burden of proof", as it is used in civil litigation between parties, is not appropriate in the context of planning appeals — In the present case the inspector had appeared to cast the whole burden on the appellants and had in effect refused planning permission because the appellants had failed to discharge the onus on them**

The correct approach was set out in the Department of the Environment Circular 14/85 — This was, of course, ministerial policy guidance, but in this respect it correctly reflected the statutory position — It pointed out that the planning system fails in its function whenever it prevents, F
inhibits or delays development which could reasonably have been permitted

The principles on which the court is prepared to intervene were set out in the well-known decision of Forbes J in *Seddon Properties Ltd* v *Secretary of State for the Environment* — Notwithstanding the inspector's error of approach in the present case, Judge Marder would not necessarily have interfered with the decision if the letter had demonstrated clear reasoning and sound grounds for dismissing the appeal — Unfortunately, he found the issues discussed in a confusing manner and the reasoning obscure — It was difficult to tell from the letter why the inspector decided as he did — There G
was no alternative but to quash the decision

The following cases are referred to in this judgment.

Pye (JA) (Oxford) Estates Ltd v *West Oxfordshire District Council* (1982) 47 P&CR 125; 264 EG 533; [1982] JPL 577
Seddon Properties Ltd v *Secretary of State for the Environment* (1978) 42 P&CR 26; 248 EG 950; [1978] JPL 835

This was a motion by Cranford Hall Parking Ltd under section 245 of the Town and Country Planning Act 1971 (now section 288 of the Town and Country Planning Act 1990) to quash the decision of an inspector appointed by the Secretary of State for the Environment to H
determine an appeal by the applicants under section 36 (now section 78 of the 1990 Act) of the Town and Country Planning Act 1971. The appeal related to an area of about two acres of land near Heathrow Airport, which had been used under temporary planning permissions for car parking until the refusal of permission which gave rise to the present litigation.

Anthony Dinkin (instructed by Reginald Johnson & Co) appeared on behalf of the applicants, Cranford Hall Parking Ltd; Guy Sankey (instructed by the Treasury Solicitor) represented the respondent Secretary of State; the second respondents, Hounslow London Borough Council, were not represented and took no part in the proceedings. J

Giving judgment, JUDGE MARDER QC said: This is a motion brought under the provisions of section 245 of the Town and Country Planning Act 1971 to quash the decision of an inspector, appointed by the Secretary of State for the Environment, to determine an appeal under section 36 of the Act against the refusal of planning permission. The land in question is about two acres just off the Bath Road, close by Heathrow Airport, and an area of land long used for off-airport parking. There had been a series of temporary planning permissions for that use, but the renewal of permission had been refused, and hence the appeal to the Secretary of State from that refusal.

The site, which is shown on a plan that was before me, is part of a narrow corridor of open land, which has the status of approved K
"metropolitan green belt", the corridor providing separation between Cranford to the east and the huge bulk of developed land comprising Heathrow Airport to the west.

As it seems to me, the true issues which the inspector was called upon to decide in the circumstances of this case could be expressed in two propositions: first of all, was the use which was proposed for car parking an appropriate and acceptable use of land in this part of the green belt; or was that use such in this particular case as to cause demonstrable harm to the green belt's function and purpose? Second, if the use that was proposed was inappropriate and damaging to the green belt, were there exceptional reasons, such as overriding need for the facility, for permitting it in the face of the presumption against such inappropriate development within a green belt area?

The inspector gave his decision on the appeal by letter dated November 18 1986, and it is apparent to me beyond doubt that the appellants' complaints about that letter are well founded. Paras 6 and 7 of the letter constitute, in effect, directions which the inspector gave himself as to the method of approach to the issues in this appeal and to determining the appeal.

In para 6 he said:

> It is a general principle of planning that there is a presumption in favour of development unless there are sound and clear cut reasons to the contrary. The onus of proof rests with the council. This argument was promoted in support of

— the appellants' —

> scheme.
> However, the approach is different within Green Belts. Here the principle which has been laid down in successive circulars since the 1950s, and recently reaffirmed in Circular 14/84 [*Green Belts*], is that there is a general presumption against development in Green Belts. Development is only allowed in exceptional circumstances. The onus of proof is thus with the prospective developer.

In para 7, he went on:

> Accordingly, from my inspection of the site and surroundings, the on-airport parking provision and other off-airport carparks, together with the consideration of all the evidence and representations, I deem that the main issue in this appeal is whether or not the need for airport parking facilities is sufficient to justify an exception to the strong presumption against development in the Green Belt.

It will be noticed from that formulation in para 7 that he does not appear to consider it to be an issue for him to determine, or at any rate, not a main issue, whether this development would bring about demonstrable harm to the green belt, and para 6, which I have just read, is clearly a misdirection, first, because the inspector appears to see the matter in terms of onus of proof as if he were considering perhaps a piece of civil litigation or possibly even a criminal prosecution.

Mr David Widdicombe QC, sitting as a deputy judge of the High Court, in the case of *Pye*, which is reported at (1982) 47 P&CR 125, the full name being *J A Pye (Oxford) Estates Ltd* v *West Oxfordshire District Council*, had this to say on this very question of onus of proof, at p 130:

> I agree with Mr Laws

— who was counsel for the Secretary of State in that case —

> that the term "burden of proof" as it is used in civil litigation between parties is not appropriate in the context of planning appeals.

On p 131, he went on to say:

> In my judgment, the task of the inspector on an appeal is to consider the facts and contentions put before him by the parties at the inquiry (including any third parties) and in the light of what he ascertains at the inquiry and his view of the site to advise the Secretary of State as to whether there are any sound and clear-cut reasons for refusal of planning permission. In doing this, there is nothing objectionable in the inspector saying in his report "I do not accept the contention" of one party or another, or "I am not satisfied that such and such a point has been made out", provided that it is clear that it is in the overall context referred to above. It would often, certainly, be helpful if inspectors made a clear finding as to whether there is or is not a sound and clear-cut objection to the grant of permission, thus showing that they have followed the Secretary of State's policy, but a decision is not invalidated because it does not use that precise language.

That is the end of the quotation from Mr Widdicombe's judgment. I am happy to adopt every word he said on that matter and apply it to the circumstances of this case, and it follows that the inspector placing the onus of proof on the prospective developer in this instance was clearly in error.

If that were a mere matter of words, then the court would not interfere, but it is clear from the conclusions expressed by the inspector and, in para 13 in particular, by the use of words like "the need for parking should be definite and overwhelming" and the phrase "I am not convinced that such an argument has been presented in this case", by those expressions and by para 13 in general, that this incorrect approach has coloured the inspector's assessment of the evidence. In essence, he has cast the whole of the burden on the appellants and he has, in effect, refused planning permission because the appellants have failed to discharge the onus on them or, at any rate, that is what it seems to me that the inspector has done.Second, and it may be perhaps that this is merely another way of saying the same thing, the approach of the inspector in para 6, the passage I have read, indicates also that the inspector has either failed to have regard to current ministerial guidance in Circular 14/85 or has misunderstood or misinterpreted that guidance.

The Circular 14/85 is headed *Development and employment* and was issued in July 1985 to demonstrate the Government's policy of seeking to promote enterprise and job creation. Para 3 appears to me to be quite specific. It says:

> Development proposals are not always acceptable. There are other important objectives to which the Government is firmly committed: the need to preserve our heritage, to improve the quality of the environment, to protect the Green Belts and conserve good agricultural land.

It goes on:

> There is therefore always a presumption in favour of allowing applications for development, having regard to all material considerations, unless that development would cause demonstrable harm to interests of acknowledged importance.

In parenthesis, I can add that "interests of acknowledged importance" clearly refers back to such policies to which the Government is committed as the protection of the green belts.

The matter was expressed, both in the appellants' motion and, indeed, by Mr Widdicombe in the decision in the case of *Pye* to which I have just referred, as one of failure to construe properly ministerial policy guidance, but I remind myself that in this instance, at any rate, that ministerial policy guidance is no more than a reflection of the statutory position or the legal position, for at the very heart of the control of development in the 1971 Act, and of course in its predecessors, there is the discretionary power to refuse permission for development of an owner's land and that power is exercisable, as a matter of law, only on good, sufficient and lawful reasons being shown for the refusal.

As I have said, the passage in the Circular 14/85, to which I have just referred, sets out a correct approach as a reflection of that statutory position and it is precisely because the correct approach is currently set out there in the circular that I ventured to formulate the real issues before this inspector in the way I did at the beginning of this judgment.

Of course there is a presumption, derived from earlier circulars that are still current, that inappropriate development will not be permitted within the green belt and there must, no doubt, frequently be cases where the fact that the proposals are inappropriate and damaging to the interests of green belt policy will be self-evident, but the circular points out, and I think correctly points out, quite clearly that the fact that it is green belt land does not in itself constitute a clear-cut and sound reason for refusing permission. The right approach, as I said earlier, is a presumption that planning permission will always be allowed, which will be overridden where it is shown that the development would cause demonstrable harm to the green belt policy. It is clear that the inspector, in the passages in paras 6 and 7 of his decision letter, has adopted a quite different and incorrect approach.

I am conscious of the restricted jurisdiction that the court exercises in matters of this kind. The principles on which the court is prepared to intervene are clearly set out in the oft-cited decision of the late Forbes J in the case of *Seddon Properties Ltd* v *Secretary of State for the Environment** and, notwithstanding the inspector's error of approach in paras 6 and 7 of his letter, I would still not necessarily have interfered with his decision if the remainder of the decision letter had indicated clear reasoning and sound reasons for dismissing the appeal, or if the decision of the appeal must inevitably be the same if a correct approach had been adopted, but I regret that I cannot say that in this case.

Looking in particular at the passages in paras 11 to 13 of the letter, I find the issues discussed confusing and the reasoning obscure. The appellants justly complain that they cannot tell from that letter why the planning permission has been refused. Mr Guy Sankey, on behalf of the Secretary of State, has endeavoured with considerable skill and, if I may say so, with considerable patience, to assist me on this matter, but I am bound to say without a great deal of success. I still am unable to extract precisely why the inspector decided as he did.

I do not think that detailed analysis by me of the passage in question at this stage will be of assistance since, in my judgment, the

*Editor's note: Reported at (1978) 248 EG 950.

whole matter requires to be looked at again.

I see no alternative but to quash the decision. The motion, therefore, succeeds and an order will be made accordingly.

The inspector's decision was quashed with costs against the Secretary of State.

Court of Appeal

(CRIMINAL DIVISION)

February 20 1990

(Before Lord Justice RUSSELL, Mr Justice TURNER and Mr Justice POTTS)

R v ALATH CONSTRUCTION LTD AND ANOTHER

Town and Country Planning Act 1971, sections 60(1) and (6) and 102(1) — Tree preservation order — Prosecution for cutting down tree without consent of local planning authority — Gale on October 16 1987 — Question as to burden of proof — Whether burden of proof as to condition of tree lay with prosecution or whether, as a true exception or exemption, it lay with defendants — Speeches in House of Lords in *R* v *Hunt* considered — Ruling by recorder in present case that burden of proof as to condition of tree lay with defendants held to be correct — Appeal against convictions dismissed

The appellants in this case, a company involved in property development and the company's chairman or managing director, were convicted at St Albans Crown Court of cutting down a beech tree, which was the subject of a tree preservation order, without the consent of the planning authority — The tree was cut down on the day after the violent gale of October 16 1987 — At the trial the recorder ruled, on a preliminary point, that the burden of proof that the defendants' actions came within the exemptions set out in section 60(6) of the 1971 Act (trees which were dead or dying or had become dangerous or where action was necessary for prevention or abatement of a nuisance) rested on the defendants — It was not, in the recorder's view, for the prosecution to establish, as an ingredient in the offence, the absence of the conditions mentioned in section 60(6) — The recorder's ruling was reflected in the summing-up and in due course the jury convicted the defendants — It was contended on appeal that the recorder's ruling was wrong and that he had misdirected the jury

On appeal the relevant provisions in section 60(1), section 60(6) and section 102(1) were considered by the court — The power to make a tree preservation order was conferred on a local planning authority by section 60(1) of the 1971 Act — The section creating the offence of contravening a tree preservation order was section 102 — The ingredients of the offence were (1) the existence of a tree preservation order, (2) proof that the defendant had cut down the tree or carried out one or more of the other activities mentioned in subsection (1), and (3) proof that the defendants' action was done without the permission of the planning authority — Provided that these three ingredients are present, section 102 created an offence of absolute liability — Section 60(6), which refers to exemptions from the tree preservation order in the cases of cutting down, uprooting, topping or lopping of dying, dead or dangerous trees, *inter alia,* does not create "negative ingredients of the offence" to be proved by the prosecution — It is a "free-standing exemption clause", quite apart from the statutory definition of the offence and proof of its application was a matter for the defendants

This was the view taken by the recorder, "in a concise and reasoned, albeit short, ruling, distinguished by its clarity", with which the Court of Appeal entirely agreed — The recorder had referred to the case of *R* v *Hunt,* in which, although the actual decision as to what the prosecution had to prove was different, the speeches in the House of Lords were helpful in distinguishing between essential elements in the proof of the offence and true exceptions and exemptions — The burden of proof of the latter lay on the person asserting them — It was pointed out in *R* v *Hunt* that when the cases are analysed those in which the courts have held that the burden lies on the defendant are those in which that burden can be more easily discharged by him — In the present type of case it would be far easier for the owner of the land, or the person responsible for cutting down a tree, to prove its condition at the time of the action than it would be for the local planning authority

The recorder here came to the correct conclusion — Appeal dismissed

The following case is referred to in this report.

R v *Hunt* [1987] AC 352; [1986] 3 WLR 1115; [1987] 1 All ER 1; (1986) 84 Cr App R 163; [1987] Crim LR 263, HL

This was an appeal by Alath Construction Ltd and Grenville Geoffrey Brightman, described as the company's chairman or managing director, from a decision of the St Albans Crown Court, presided over by Mr Recorder Zucker QC, whereby the appellants were held to have cut down a beech tree subject to a tree preservation order without the permission of the local planning authority. The appellants claimed that a preliminary ruling given by the recorder as to the burden of proof was wrong in law and led to a misdirection to the jury.

David Anthony Pearl (instructed by Parrott Coales, of Aylesbury) appeared on behalf of the appellants; Philip Waller (instructed by Pickworths, of St Albans) represented the Crown.

Giving judgment, **RUSSELL LJ** said: These appeals raise a point of statutory construction arising out of provisions relating to tree preservation orders contained in the Town and Country Planning Act 1971 (as amended).

The first appellant, Alath Construction Ltd, is, as its name implies, a company involved in property development. The second appellant, Mr Grenville Geoffrey Brightman, is the managing director or chairman of that limited company. At all material times the company owned a plot of land in Berkhamsted whereon there were a number of trees. The company developed the site, constructing a number of dwellinghouses upon it.

In May 1984 the local planning authority, the Dacorum District Council, made a tree preservation order entitled the "Dacorum District Council (Halcyon, Shootersway Lane, Berkhamsted) Tree Preservation Order 1984", which embraced the land to which we have referred. Among the trees standing on the land was a beech tree which photographs disclose was close to one of the dwellinghouses erected by the company.

After the making of the order the appellants made a number of applications for permission to fell the beech tree, but all those applications were refused.

On October 16 1987 there was a great gale which ravaged large areas of the United Kingdom. On October 17 1987 it is common ground that the beech tree on the appellants' land was cut down upon the instructions of the appellants, that being without the permission of the planning authority.

In due course, therefore, a prosecution ensued at the instance of the planning authority. The appellants elected trial and the case came before the St Albans Crown Court presided over by Mr Recorder Zucker QC.

After arraignment a preliminary point was raised with the learned recorder as to the burden of proof, it being contended on behalf of the appellants that the burden of proving that at the material time the beech tree was not dying, or dead, or dangerous, or creating a nuisance, lay throughout with the prosecution. The learned recorder ruled that the prosecution did not have to prove that at the material time the beech tree was in any of the conditions to which we have just referred. He ruled, conversely, that there was a burden upon the defendants to establish one or other, or all, of those conditions, and the trial proceeded in the light of that ruling.

We have not got a copy of the summing-up but it is common

A ground that the ruling was reflected in the summing-up and in due course the jury convicted. It is against those convictions that this appeal is brought on a point of law only, it being contended that the preliminary ruling made by the learned recorder was wrong and that consequently he misdirected the jury upon the task that confronted them.

The indictment alleged a contravention of the tree preservation order to which we have referred, the section of the statute breached being section 102(1)*. That section reads:

> If any person, in contravention of a tree preservation order, cuts down, uproots or wilfully destroys a tree or wilfully damages, tops or lops a tree in such a manner as to be likely to destroy it, he shall be guilty of an offence . . .

B As it seems to us, that section creating the offence has within it the following ingredients. First, there has to be a tree preservation order. Second, it has to be proved that the appellants cut down the tree or carried out the other activities referred to in the section. And, third, it has to be shown that whatever was done by the appellants in the way of cutting down or engaging in other activities relative to the tree was done without the permission of the planning authority. No more than that is apparent from the section.

It is necessary now to go to the section which is the enabling provision entitling the planning authority in appropriate circumstances to make a tree preservation order. That is an entirely different section of the Act. Section 60(1)† provides that a local authority may make a tree preservation order if it appears "expedient C in the interests of amenity". The section goes on to say that they may prohibit:

> (subject to any exemptions for which provision may be made by the order) the cutting down, topping, lopping, uprooting, wilful damage or wilful destruction of trees except with the consent of the local planning authority.

That, as it seems to us, is the section which empowers the local authority to make a tree preservation order, the breach of which gives rise to an offence under an entirely different section, namely section 102(1), the offence under that section being, in our judgment, an absolute offence, provided always that the prosecution prove the three ingredients to which we have referred.

We move on to consider section 60(6), to which our attention has D been directed. That provides:

> Without prejudice to any other exemptions for which provision may be made by a tree preservation order, no such order shall apply to the cutting down, uprooting, topping or lopping of trees which are dying or dead or have become dangerous, or the cutting down, uprooting, topping or lopping of any trees in compliance with any obligations imposed by or under an Act of Parliament or so far as may be necessary for the prevention or abatement of a nuisance.

That subsection, in our judgment, provides an exception to the absolute liability created by the offence rehearsed in section 102. Section 60, it is to be observed, does not create the offence. Section 60 deals with the circumstances in which the local authority may make the order and section 60(6), in particular, is, in our judgment, a free- E standing provision quite independent of section 102 and defines in terms the exceptions to the normal consequences of cutting down a tree which is the subject-matter of an order. That being so, as it seems to us, section 60(6) is truly an exception to the definition of criminal liability to be found in section 102 and it does not create negative ingredients of the offence in respect of which a burden of proof rests upon the prosecution.

It is to be observed that in the tree preservation order itself, para (2) reads:

> Subject to the provisions of this Order and to the exemptions specified in the Second Schedule hereto, no person shall, except with the consent of the authority . . .

The paragraph goes on to deal with the prohibited activities. F Turning to Schedule 2, that provides:

> This Order shall not apply so as to require the consent of the authority to . . .

In subpara (3) of Schedule 2 there are set out those activities referred to in section 60(6). They are only part of the exemptions created by Schedule 2 and from the section of the Act itself we observe that the words of subsection (6) are prefaced by:

> Without prejudice to any other exemptions for which provision may be made by a tree preservation order . . .

The wording of section 60(6) itself reinforces our view that it is indeed a free-standing exemption clause quite apart from the G statutory definition of the offence to be found in section 102.

In a concise and reasoned, albeit short, ruling, distinguished by its clarity, the learned recorder set out his analysis of the law and his conclusions. We shall refer to them in a moment. We entirely agree with them.

During the course of submissions made to the learned recorder he was referred to the case of *R* v *Hunt* (1986) 84 Cr App R 163, a decision of the House of Lords. It is unnecessary in this judgment to rehearse the facts of *Hunt*; it concerned drugs. It had nothing whatever to do with tree preservation orders, but some observations of their lordships in that case were of general application. At p 174 of the opinion of Lord Griffiths this passage appears:

> I would summarise the position thus far by saying that *Woolmington* v *DPP* H did not lay down a rule that the burden of proving a statutory defence only lay upon the defendant if the statute specifically so provided: that a statute can, on its true construction, place a burden of proof on the defendant although it does not do so expressly: that if a burden of proof is placed on the defendant it is the same burden whether the case be tried summarily or on indictment, namely a burden that has to be discharged on the balance of probabilities. The real difficulty in these cases lies in determining upon whom Parliament intended to place the burden of proof when the statute has not expressly so provided. It presents particularly difficult problems of construction when what might be regarded as a matter of defence appears in a clause creating the offence rather than in some subsequent proviso from which it may more readily be inferred that it was intended to provide for a separate defence which a defendant must set up and prove if he wishes to avail himself of it.

Later, Lord Griffiths said: J

> However, their Lordships were in agreement that if the linguistic construction of the statute did not clearly indicate upon whom the burden should lie the court should look to other considerations to determine the intention of Parliament, such as the mischief at which the Act was aimed and practical considerations affecting the burden of proof and, in particular, the ease or difficulty that the respective parties would encounter in discharging the burden. I regard this last consideration as one of great importance for surely Parliament can never lightly be taken to have intended to impose an onerous duty on a defendant to prove his innocence in a criminal case, and a court should be very slow to draw any such inference from the language of a statute. When all the cases are analysed, those in which the courts have held that the burden lies on the defendant are cases in which the burden can be easily discharged. K

Finally, at p 176, Lord Griffiths said:

> In *Edwards* (1974) 59 Cr App R 213, 221; the Court of Appeal expressed their conclusion in the form of an exception to what they said was the fundamental rule of our criminal law that the prosecution must prove every element of the charge. They said that the exception:
>
> is limited to offences arising under enactments which prohibit the doing of an act save in specified circumstances or by persons of specified classes or with specified qualifications or with the licence or permission of specified authorities.

Having considered the case of *Hunt,* the learned recorder in the instant case towards the end of his ruling, at p 23 of the transcript, said: L

> Applying the decisions as outlined in those speeches to the facts in this case, in my judgment on a proper construction of section 102 and section 60 of the Town and Country Planning Act 1971, section 60(6) provides for what are indeed true exceptions or exemptions. The subsection says in so many words: "Without prejudice to any other exemptions for which provision may be made by a tree preservation order". I am quite satisfied that this is a true exception or exemption, and unlike the case of *Hunt,* section 60(6) does not provide for one of the essential ingredients of the offence. This being an exemption or exception, the burden of proof, in my judgment, lies on the person who asserts it, who in this case is the defendant. I am reinforced in that view because from a practical viewpoint I have no doubt that it is far easier for the owner of the land, or the person responsible for the cutting down of a tree, to prove its condition at the time when such action was taken than the local authority. It is to be noted in particular that section 60(6) provides for a tree becoming dangerous after an order has been made, and it has been correctly, M in my judgment, submitted to me by Mr Waller that it would place an impossible burden upon a local authority to check every tree which is to be the subject of a tree preservation order to determine what its condition is and to continue to check it after an order has been made to determine whether it has become dangerous. I accept Mr Waller's submissions that the defendant responsible for the cutting down of a tree or otherwise dealing with it, contrary to the terms of the tree preservation order, is in a unique position to know the true condition of the tree at the time of his act. So that practical considerations reinforce the view that on the true construction of this Act the burden of proving the matters which come within section 60(6) lies upon the defendant, lies upon him to prove them on the balance of probabilities, and that is my ruling on the submissions that have been made to me.

With respect, we simply cannot improve upon those observations,

---

*Editor's note: Now section 210 of the Town and Country Planning Act 1990.
†Editor's note: Now section 198 of the Town and Country Planning Act 1990.

A which are clear and to the point and, in our judgment, accurately analyse the legal position in this appeal. Despite the persuasive submissions made to the learned recorder (and recorded in the transcript) by Mr Pearl, and repeated before this court, we are abundantly satisfied that the learned recorder came to a correct conclusion. The appeals are dismissed.

The appeals were dismissed with costs.

Court of Appeal

B May 9 1990

(Before Lord Justice LLOYD and Lord Justice BELDAM)

NEWPORT BOROUGH COUNCIL v KHAN AND OTHERS

Town and Country Planning Act 1971 — Tree preservation order — Grant of injunction in support of order — Jurisdiction to grant such an order — *Kent County Council* v *Batchelor* and *Stoke-on-Trent City Council* v *B&Q (Retail) Ltd* — Section 22 of County Courts Act 1984 — County court
C judge had the necessary jurisdiction but failed to direct himself as to the principles which should guide the exercise of his discretion — Grant of injunction not justified in present case

The events leading up to this appeal began when the appellant cut the branches of a yew tree in his garden when he considered that the tree was in a dangerous condition — The planning authority made a tree preservation order covering some 15 trees in the garden — There followed an unfortunate history of differences, of delay on the part of the authority in replying to applications and notices by the appellant, ending
D with proceedings in the county court — The judge accepted undertakings from the appellant and his wife but granted an interlocutory injunction against the person engaged to carry out the cutting operations — Subsequently the appellant applied to be released from the undertakings and to have the injunction discharged — He claimed that the court had no jurisdiction to grant the injunction — The judge held that he had jurisdiction under section 22 of the County Courts Act 1984 — He refused to release the appellant and his wife from the undertakings and continued the injunction until trial — Hence the present appeal

As the appellant was appearing in person at the appeal, and
E it seemed to the court that the question of jurisdiction was of some importance, arrangements were made for an *amicus curiae* to attend — Section 22 of the County Courts Act 1984 clearly gave the county court the same jurisdiction as the High Court "to grant an injunction or declaration in respect of, or relating to, any land, or the possession, occupation, use or enjoyment of any land" — It appeared, however, that there was no reported case in which the jurisdiction of the High Court to grant an injunction at the behest of a local authority in support of a tree preservation order had been directly in issue — The nearest case was *Kent* v *Batchelor* in which, in the
F second proceedings, Talbot J felt bound (in the light of the earlier proceedings) to hold that there was such a jurisdiction — Lloyd LJ in the present case held that Talbot J had been right and that any remaining doubt had been removed by the approval given to Talbot J's decision by Lord Templeman in his speech in *Stoke-on-Trent City Council* v *B&Q (Retail) Ltd* — Lloyd LJ also accepted that the county court had a similar jurisdiction in a tree preservation order case

The more difficult question, so far as the present case was concerned, was whether the judge had exercised this undoubted jurisdiction on the correct principles — In fact the judge had given himself no directions as to the principles on which he should exercise his jurisdiction — He had not G
referred to the cases in which guidance had been given and it thus fell to the Court of Appeal to exercise the discretion themselves — The cases had laid down the following guidance, set out in *City of London Corporation* v *Bovis Construction Ltd*: (1) the jurisdiction, where Parliament has prescribed a maximum penalty in the criminal law, must be exercised sparingly and with great caution; (2) there must be something more than a mere infringement of the criminal law, though obviously a deliberate and flagrant flouting would suffice; (3) an important ground is the inference that the offender will be likely to continue his objectionable activities H
unless he is restrained by injunction

In the present case there was no question of a deliberate or flagrant flouting of the law by the appellant, nor was there any reason to infer that he would commit further breaches of the tree preservation order — There was a suggestion that if the appellant failed to comply with the tree preservation order the harm done would be irreparable — In a sense this was true — A tree once felled cannot be restored, it can only be replaced by another tree — One must, however, look also at the comparative gravity of the harm that would be done — The facts in the present case fell far short of the gravity and persistence of the conduct of the farmer in the case of *Kent* J
***County Council* v *Batchelor* — Appeal allowed**

The following cases are referred to in this report.

Attorney-General v *Chaudry* [1971] 1 WLR 1614; [1971] 3 All ER 938; (1971) 70 LGR 22, CA
Barking & Dagenham London Borough Council v *Home Charm Retail Ltd* [1984] Ch 1; [1983] 3 WLR 78; [1983] 2 All ER 787; (1983) 82 LGR 10, CA
City of London Corporation v *Bovis Construction Ltd* (1988) 86 LGR 660
Gouriet v *Union of Post Office Workers* [1978] AC 435; [1977] 3 WLR 300; [1977] 3 All ER 70, HL
Kent County Council v *Batchelor* (1976) 33 P&CR 185; 75 LGR 151; 242 EG 287, CA
Kent County Council v *Batchelor (No 2)* [1979] 1 WLR 213; [1978] 3 All ER K
980; (1977) 38 P&CR 185; 76 LGR 714; [1978] JPL 179
Runnymede Borough Council v *Ball* [1986] 1 WLR 353; [1986] 1 All ER 629; (1986) 84 LGR 481, CA
Smith v *Oliver* [1989] 2 PLR 1
Stoke-on-Trent City Council v *B&Q (Retail) Ltd* [1984] AC 754; [1984] 2 WLR 929; [1984] 2 All ER 332, HL

This was an appeal by the first defendant, Sabz Ali Khan, from interlocutory orders made by Judge Tom Crowther QC, at Newport (Gwent) County Court, in regard to matters concerning a tree preservation order made by Newport Borough Council in respect of trees on the appellant's land at 2 Oakfield Road, Newport, Gwent. The second and third defendants were Mrs Khan and Mr R T
Roderick. L

The appellant, Sabz Ali Khan, appeared in person; Keith Bush (instructed by the solicitor, Newport Borough Council) represented the respondent council; Guy Sankey appeared as *amicus curiae.*

Giving judgment, LLOYD LJ said: In this case we have been concerned with methods of enforcing a tree preservation order. In November 1988 the first defendant, Mr Sabz Ali Khan, purchased a property at 2 Oakfield Road, Newport, Gwent. There are a number of trees of different species on the property which are said to add greatly to the amenity of the area. There was no tree preservation order in force when Mr Khan bought the property, but the house is situated in a conservation area. Immediately opposite the house stands the new crown court, and the municipal offices are also to be M
found in the neighbourhood.

Soon after purchasing the property Mr Khan decided to cut the branches of a yew tree standing in the front garden which he regarded as dangerous. There was a complaint. Mr Hasan, the senior development control officer employed by Newport Borough Council, visited the property on April 18 1989. He found workmen engaged in cutting branches. He told them to stop. He made a tree preservation order the same day with the authority of the chairman of the planning committee. It covered 15 trees in the garden to the front and side of the house. The yew tree was T15. By clause 13 of the tree preservation order, it was provided that it should take immediate effect as a provisional order under section 61 of the Town and

A Country Planning Act 1971 [section 201 of the Town and Country Planning Act 1990]. The date stated in clause 13 is April 18 1989, so the order was still in force the following September, when the events occurred out of which these proceedings arise.

On April 19 the tree preservation order was served on Mr and Mrs Khan. On April 20 Mr Khan applied to the council for leave to fell two of the 15 trees which he regarded as dangerous. In addition to the yew, there was a redwood, T6. That application did not come before the council apparently until August 11, when consideration of the application was deferred pending a site inspection. On September 5 Mr Khan, having heard nothing from the council, wrote, saying that he intended to fell the yew tree on September 10, thus giving the council a further five days' notice. He employed a contractor for that
B purpose. It cost him £400.

Meanwhile, on September 8 the council at long last refused consent to Mr Khan's application and indeed resolved to prosecute Mr Khan for the work which he had already carried out without authority, in breach of the tree preservation order. That must be a reference to a mountain ash, T11, which Mr Khan says was little more than a bush.

In the very early morning of Sunday September 10, Councillor Graham telephoned Mr Tapp, the chief executive, to tell him that men were at work cutting down the yew tree. Mr Tapp arrived on the scene at 8.10 am. Mr Khan was not there, but Mr Tapp introduced himself to the third defendant, Mr R T Roderick, who was the man in charge of the operation. He told the third defendant to stop. But the
C third defendant refused, whereupon the plaintiffs obtained an interim injunction from Judge Morgan the same day. That injunction was served on the defendants at 12 noon. But by that time the yew tree was no more than a stump.

On Monday September 11 the case came before His Honour Judge Crowther. The defendant gave an interim undertaking to the court, pending a further hearing on September 14. On September 12 the plaintiffs served particulars of claim, in which they say they are entitled to damages in the sum of £5,000 for loss of or damage to the amenity of the area. It is now conceded that such a claim for damages cannot possibly be sustained. On September 14 Judge Crowther accepted a further undertaking from Mr and Mrs Khan and granted an interlocutory injunction against the third defendant.

D On October 13 Mr Khan gave notice of an application to be released from his undertaking and to discharge the injunction. His grounds were, first, that the statement of claim disclosed no cause of action known to the law and, second, that the county court had no jurisdiction to grant an injunction, since the claim for an injunction was not ancillary to a viable claim for damages or other relief within the jurisdiction of the county court: see section 38 of the County Courts Act 1984.

The hearing of that application came before Judge Crowther on November 22. He held that he had jurisdiction to grant an injunction, on the ground that the injunction was in respect of or relating to any land or the use or enjoyment of any land within the meaning of
E section 22 of the County Courts Act 1984. Accordingly, he refused to release Mr and Mrs Khan from their undertaking and continued the injunction until trial. There is now an appeal to this court.

When the case came on for hearing before us, Mr Khan appeared in person to support his appeal. It seemed to us that the question of the jurisdiction of the county court to grant an injunction in support of a tree preservation order might be one of some importance, as also might be the scope of the local authority's powers under section 222 of the Local Government Act 1972 and the proper exercise of the judge's discretion in a case such as the present. So we adjourned the hearing in order that we might seek the help of an *amicus.* At the resumed hearing Mr Guy Sankey has appeared in that capacity and has given us every possible assistance.

F Section 22(1) of the County Courts Act 1984 provides:

> Subject to the provisions of this section, a county court shall have the same jurisdiction as the High Court to grant an injunction or declaration in respect of, or relating to, any land, or the possession, occupation, use or enjoyment of any land.

I need not refer to subsections (2) or (3). Section 222 of the Local Government Act 1972 provides:

> Where a local authority consider it expedient for the promotion or protection of the interests of the inhabitants of their area—
> (*a*) they may prosecute or defend or appear in any legal proceedings and, in the case of civil proceedings, may institute them in their own name.

I need not read the rest of that section.

Logically, the first question for our consideration is whether the High Court has jurisdiction to grant an injunction in a case such as G
this, seeing that the breach of a tree preservation order is a criminal offence for which Parliament has provided specific penalties. For if the High Court has no such jurisdiction, neither has the county court.

It appears that there is no reported case in which the jurisdiction of the High Court to grant an injunction at the behest of the local authority in support of a tree preservation order has been directly in issue. The nearest case is *Kent County Council* v *Batchelor* (1976) 75 LGR 151. In that case, Cusack J or Chapman J — it is not clear from the reports which — granted the plaintiffs an injunction to restrain a farmer from cutting down trees in breach of a tree preservation order. There was then an application to commit the farmer for contempt on the ground that he had disobeyed the injunction. H
Caulfield J committed the farmer to prison for a month. His appeal was allowed by the Court of Appeal, but not, be it noted, on the ground that the injunction should never have been granted, but on the ground that there was insufficient evidence to establish a breach of the injunction. There was then a further application to commit. This time it was argued that there was no jurisdiction to grant the injunction. But Talbot J held, in *Kent County Council* v *Batchelor (No 2)* [1979] 1 WLR 213, that in view of the course that the previous proceedings had taken in the Court of Appeal, where it was assumed on all sides that there is jurisdiction to grant an injunction in a tree preservation order case, he was bound so to hold in the case before him.

In my judgment Talbot J was right so to hold. If there remained J
any doubt, it was dispelled by Lord Templeman's speech in *Stoke-on-Trent City Council* v *B&Q (Retail) Ltd* [1984] AC 754, where he referred at p 774 to the decision of Talbot J with evident approval as a typical case in which the local authority are in the best possible position to judge whether it is expedient, in the interests of a particular locality, to apply for an injunction under section 222 of the Local Government Act.

So I would hold that the High Court does indeed have jurisdiction to grant an injunction at the behest of a local authority in support of a tree preservation order in an appropriate case. I will return to Lord Templeman's speech later, when I come to consider the principles which ought to govern the exercise of the jurisdiction.

The next question is whether, if the High Court has jurisdiction to K
grant an injunction in a tree preservation order case, as I have just held, the county court has a similar jurisdiction by virtue of section 22 of the County Courts Act 1984. This depends on whether such an injunction would, if granted, be an injunction in respect of or relating to any land or the use of any land, having a rateable value of less than £1,000. Once again there appears to be no reported case on this point, in other words, no case in which section 22 has been applied in the case of a tree preservation order. But the explanation for that dearth of authority may be, as Mr Bush pointed out for the council, that the section has been in force only since 1977.

It seems to me that in principle section 22 ought to apply to such a case. A tree preservation order comes within the ordinary meaning of L
the words of the section. Such an order almost always identifies the trees which are subject to the order by reference to the land on which the trees are situated. Thus, in the present case the trees are identified by reference to land at 2 Oakfield Road. This is true even when the order applies only to a single tree. The tree preservation order thus clearly affects the land on which the tree or trees are situated. As a matter of ordinary language it is thus an order in respect of or relating to that land.

This is borne out by section 60(1A) of the Town and Country Planning Act 1971, which refers to a tree preservation order being made in relation to land in the three cases covered by subparas (*b*) to (*d*). It is also borne out by section 60(8), which refers to a tree preservation order being made in respect of land. It may be no more M
than a coincidence that the same language is used in section 22 of the County Courts Act as is used in section 60 of the Town and Country Planning Act. But even if so, it at least shows that it is a perfectly ordinary and natural use of words in this context. If I am right that a tree preservation order is an order in respect of or relating to land, it must follow that an injunction to enforce a tree preservation order is also an injunction in respect of or relating to land within the meaning of section 22.

The matter can be tested in this way. Suppose my neighbour threatens to come on my land and cut down my tree. There could be no doubt that the court would have jurisdiction to grant an injunction in such a case. Otherwise, one would be depriving section

22 of much, or at any rate some, of its beneficial effect. But if, in such a case, the application for an injunction would be an application in respect of land, how could the case differ if it is not a neighbour but the owner himself who threatens to cut down his own tree and is then sought to be restrained by the local authority? So I would hold, contrary to Mr Khan's argument and without much hesitation, that Judge Crowther did indeed have jurisdiction to make the order under section 22 of the Act.

But that is not the end of the case. There then comes the more difficult question whether the judge exercised the jurisdiction on the right principles. A glance at the judgment shows that the judge did not direct himself at all as to the proper principles which should have guided the exercise of his discretion. He regarded it as sufficient to hold that he had jurisdiction to make the order. Nowhere does he refer to the decision of the House of Lords in *Stoke-on-Trent City Council* v *B&Q (Retail) Ltd*. Nowhere does he refer to the cases which have followed that decision, namely, among others, *Runnymede Borough Council* v *Ball* [1986] 1 WLR 353 and *City of London Corporation* v *Bovis Construction Ltd* (1988) 86 LGR 660.

This may not have been altogether the judge's fault, since as Mr Bush pointed out, the judge was never asked by Mr Khan to exercise his discretion in his, Mr Khan's, favour. But equally it may not have been Mr Khan's fault, since until the judge granted leave to amend the statement of claim in the course of the hearing on November 22 Mr Khan had a complete, if technical, answer on the pleading as it stood. But wherever the fault may have lain, the fact remains, as indeed Mr Bush concedes and avers, that the judge never exercised any discretion in this case at all. It follows that we have no alternative but to exercise the discretion ourselves. Mr Bush submits that we do not have sufficient material to exercise our discretion. I do not agree. The existing material, including as it does three affidavits sworn on behalf of the council as to the council's reasons for applying for an injunction, is amply sufficient to enable us to form a fair view of the matter.

How then should we exercise our discretion? The theme which underlies all the speeches in the *Stoke-on-Trent* case is that the jurisdiction to grant an injunction to prevent an infringement of the criminal law, where Parliament has itself enacted, as here, a maximum pecuniary penalty, must be exercised sparingly and with great caution. The reasons were explained by Lord Wilberforce in *Gouriet* v *Union of Post Officer Workers* [1978] AC 435, in a passage immediately following the passage quoted by Lord Templeman from Lord Wilberforce's speech in the *Stoke-on-Trent* case at p 776. Something more than an infringement of the criminal law is required. But what more? A deliberate and flagrant flouting of the law will clearly suffice. But on no view could Mr Khan be said to have flouted the law in the present case. On the contrary, he had been seeking permission to fell the two trees, which he regarded as dangerous, since April 20. He had received advice from an independent tree surgeon that the yew and the redwood both contained much decaying and dead wood and were an obvious danger to passers-by. He had been told that he would hear from the council within six weeks of making his application on April 20. It was only when he did not hear for nearly six months that he wrote his letter on September 5, saying that unless he heard within five days he would have no alternative but to cut the yew for the sake of public safety. This conduct seems to me to be the very reverse of a deliberate flouting of the law.

But the subsequent cases show that the flouting of the law is not an exhaustive test. Indeed the use of the phrase itself, though it has the support of much very high authority, has been criticised by Sir Roger Ormrod in the *Runnymede* case. What else, then, was there here? The ground on which the case was put forward by the plaintiffs in all three affidavits sworn on the council's behalf is that Mr Khan was likely to commit further breaches of the tree preservation order in the future, and that nothing less than an injunction would stop him: see the *City of London Corporation* case *per* Bingham LJ at p 682.

But I see no reason to infer that Mr Khan will commit further breaches of the tree preservation order. Apart from the mountain ash T11, as to which the evidence was inconclusive, there is no evidence that Mr Khan ever intended to cut down any trees other than the yew and the redwood. The only other tree of which we have heard tell is a pine tree, T12. It is true that that tree may have to come down, as the council accept; it is now leaning badly. But that is all. Mr Khan tells us that he does not regard any of the other trees covered by the tree preservation order as dangerous. Contrast the failure of the defendants to answer a similar question asked by Lord Diplock at the commencement of the hearing of the *Stoke-on-Trent* case in the House of Lords.

If Mr Khan is right that the yew and the redwood are or were dangerous, he would have a defence to criminal proceedings under section 60(6) of the 1971 Act. He would be guilty of no infringement: see *Smith* v *Oliver* [1989] 2 PLR 1. If, on the other hand, he is wrong, then, on the council's own case, the value of the yew tree is £5,000 and therefore the fine in summary proceedings in relation to that tree alone could be as high as £10,000. So this is obviously not a case where the financial penalty enacted by Parliament is likely to prove inadequate. So the plaintiffs have not, on the facts of this case, made good the grounds which they put forward in their evidence in support of their claim for an injunction.

Finally, there is the argument which Mr Bush, on behalf of the council, put at the forefront of his case, although it is not reflected in the evidence, that the harm done by Mr Khan if he fails to comply with the order would be irreparable. In one sense this is of course true. A tree, once felled, cannot be restored. It can only be replaced by another tree. But irreparability on its own does not necessarily justify an injunction. One must also have regard to the gravity of the harm likely to be done and the scale of the operations as a whole: see *per* Ackner LJ in the *Stoke-on-Trent* case in the Court of Appeal [1984] Ch 1. In *Attorney-General* v *Chaudry* [1971] 1 WLR 1614, the granting of an injunction was upheld by the Court of Appeal because the risk of the hotel in that case catching fire was serious and any consequential loss of life resulting from the fire would have been very grave indeed. The present case is not in that class. Nor does it have the features to which Purchas LJ drew attention in the *Runnymede* case. I am not saying that there may not be a case where, exceptionally, an injunction may not be justified in support of a tree preservation order. *Kent County Council* v *Batchelor* was just such a case. But the facts of the present case fall far short of the gravity and persistence of the conduct of the farmer in that case. Accordingly, I would reject Mr Bush's argument based on the irreversible nature of a tree-cutting operation.

In the end, the question for us is quite simple. Is there anything sufficiently exceptional in the facts of the present case to justify resort to the civil law in support of the criminal law? I have no doubt that if the judge had had his attention directed to the relevant authorities to which I have referred, instead of being asked to concentrate on the question of his jurisdiction, he would have declined to exercise his discretion in favour of the plaintiffs in the present case. In the event, he never exercised his discretion one way or the other, so the discretion devolves upon us. The judge was right to hold that he had jurisdiction under section 22 of the County Courts Act 1984 to grant an interlocutory injunction, but, on the evidence before him and before us, an injunction is not justified.

In those circumstances it is not necessary for me to deal with the other ground relied upon by Mr Khan, that the action of the chief executive in applying for an interlocutory injunction was never properly considered or authorised by the council themselves, or any relevant committee in the light of the council's discretion under section 222 of the Act: see in that connection *Barking & Dagenham London Borough Council* v *Home Charm Retail Ltd* [1984] Ch 1. Since I am in favour of Mr Khan on the broader point I need not consider that narrow argument.

I would allow this appeal for the reasons I have given and discharge the injunction.

Agreeing, **BELDAM LJ** said: I add my own observation only on the proper approach to the use by a local planning authority of an injunction to enforce a tree preservation order. It is an example of seeking the assistance of the civil courts to forestall the commission of an offence under the criminal law. I consider that the correct principles are those set out by Bingham LJ in the case of *City of London Corporation* v *Bovis Construction Ltd* (1988) 86 LGR 660. Bingham LJ at p 682, said:

> The guiding principles must, I think, be (1) that the jurisdiction is to be invoked and exercised exceptionally and with great caution: see the authority already cited; (2) that there must certainly be something more than mere infringement of the criminal law before the assistance of civil proceedings can be invoked and accorded for the protection or promotion of the interests of the inhabitants of the area: see *Stoke-on-Trent City Council* v *B&Q (Retail) Ltd (supra)* at pp 767 and 415, 776 and 485, and *Wychavon District Council* v *Midland Enterprises (Special Events) Ltd* (1987) 86 LGR 83, at p 87; (3) that the essential foundation for the exercise of the court's discretion to grant an injunction is not that the offender is deliberately and flagrantly flouting the

A law but the need to draw the inference that the defendant's unlawful operations will continue unless and until effectively restrained by the law and that nothing short of an injunction will be effective to restrain them.

In the case of a local planning authority who is seeking the aid of a civil court to enforce a tree preservation order, it is important to bear in mind the nature of the sanctions and remedies which are available to the local planning authority under the Town and Country Planning Act 1971 before deciding whether it is reasonable in all the circumstances that they should seek the aid of the civil courts.

Section 102(1) of the Act as substituted by section 10(3) of the Town and Country Amenities Act 1974 provides that the offence of contravening the tree preservation order by cutting down or wilfully destroying a tree or damaging it is triable either way, that is to say,
B summarily or on indictment, so that in serious cases the offender may be committed for trial to the crown court and be liable to an unlimited fine. In the case of a summary conviction he is liable to a very substantial fine or twice the sum which appears to the court to be the value of the tree. In addition to the penalties available, if a person in fact damages or cuts down a tree, even in those cases in which he would be justified in doing so because the tree was, for example,
C dangerous, he may still be liable to an order that he should replace the tree under section 62 of the Act, and section 103 provides the local planning authority with extensive powers for enforcing the duty to replace such trees.

The local planning authority have available more varied and potent sanctions and remedies than, for example, are available to a local authority in the enforcement of laws prohibiting Sunday trading, and thus, before considering whether it is appropriate in a particular case to apply for a civil injunction, it seems to me that it would be essential for the local planning authority at least to consider whether their existing remedies offered them an adequate solution to the difficulties which they faced.

D For the reasons which Lloyd LJ has given, in the present case the evidence before the learned judge did not, in my view, justify the granting of a civil injunction and I agree that this appeal should be allowed.

The appeal was allowed with costs in the Court of Appeal and below; application for leave to appeal to the House of Lords was refused.

For further cases on this subject see p 175